T5-CVP-855

ORGANOTIN

ORGANOTIN

Environmental fate and effects

Edited by

M.A. Champ

and

P.F. Seligman

CHAPMAN & HALL

London · Weinheim · New York · Tokyo · Melbourne · Madras

Published by Chapman & Hall, 2–6 Boundary Row, London SE1 8HN, UK

Chapman & Hall, 2–6 Boundary Row, London SE1 8HN, UK

Chapman & Hall GmbH, Pappelallee 3, 69469 Weinheim, Germany

Chapman & Hall USA, 115 Fifth Avenue, New York, NY 10003, USA

Chapman & Hall Japan, ITP–Japan, Kyowa Building, 3F, 2–2–1 Hirakawacho, Chiyoda-ku, Tokyo 102, Japan

Chapman & Hall Australia, 102 Dodds Street, South Melbourne, Victoria 3205, Australia

Chapman & Hall India, R. Seshadri, 32 Second Main Road, CIT East, Madras 600 035, India

---

First edition 1996

Typeset in 10/12 Palatino by Photoprint, Torquay, Devon
Printed in Great Britain by the University Press, Cambridge

ISBN 0 412 58240 6

A catalogue record for this book is available from the British Library

Library of Congress Catalog Card Number: 95–78838

∞ Printed on acid-free text paper, manufactured in accordance with ANSI/NISO Z39.48–1992 (Permanence of Paper).

CONTENTS

CONTRIBUTORS

R. ABEL
Toxic Substances Division,
Department of the Environment,
Marsham Street,
London SW1P 3PY, UK

C.M. ADEMA
Naval Surface Warfare Center,
Carderrock Division,
Annapolis Detachment Code 2834,
Maryland 21402, USA

M.E. BENDER
Virginia Institute of Marine Science,
School of Marine Science,
College of William and Mary,
Gloucester Point,
Virginia 23062, USA

S. BERMAN
Institute for Environmental Chemistry,
National Research Council of Canada,
Montreal Road,
Ottawa,
Ontario K1A 0R9, Canada

G.W. BRYAN
Plymouth Marine Laboratory,
Citadel Hill,
Plymouth,
Devon PL1 2PB, UK

STEVEN J. BUSHONG
Holme, Roberts & Owen LLC,
401 Pearl St, Suite 400,
Boulder,
Colorado 80302, USA

MICHAEL A. CHAMP
Texas Engineering Experiment Station,
Washington, DC Office,
Texas A&M University System,
4601 North Fairfax Drive, Suite 1130,
Arlington,
Virginia 22203, USA

CESAR CLAVELL
Naval Command Control and Ocean Surveillance Center,
RDT&E Division,
Environmental Sciences (Code 52),
San Diego,
California 92152-5000, USA

J.J. CLEARY
Plymouth Marine Laboratory,
Prospect Place,
West Hoe,
Plymouth, PL1 3DH, UK

S.Y. COLA
Computer Sciences Corporation,
4045 Hancock Street,
San Diego,
California 92110-5164, USA

BRADLEY M. DAVIDSON
Computer Sciences Corporation,
4045 Hancock Street,
San Diego,
California 92110-5164, USA

P.F. DE LISLE
Coastal Bioanalysts, Inc.,
PO Box 626,
Gloucester Point,
Virginia 23062, USA

C.A. DOOLEY
Naval Command Control and Ocean Surveillance Center,
RDT&E Division,
Environmental Sciences (Code 52),
San Diego,
California 92152-5000, USA

F.A. ESPOURTEILLE
Rutgers University,
New Brunswick,
New Jersey 07162, USA

D.A. EVANS
Virginia Institute of Marine Science,
School of Marine Science,
College of William and Mary,
Gloucester Point,
Virginia 23062, USA

KARL FENT
Swiss Federal Institute for Environmental Science and Technology (EAWAG) and Swiss Federal Institute of Technology Zurich (ETH),
CH-8600 Dübendorf, Switzerland

ROY L. FRANSHAM
Computer Sciences Corporation,
4045 Hancock Street,
San Diego,
California 92110-5164, USA

P.E. GIBBS
Plymouth Marine Laboratory,
Citadel Hill,
Plymouth,
Devon PL1 2PB, UK

J. GREAVES
Department of Chemistry,
University of California,
Irvine,
California 92717, USA

JOSEPH G. GROVHOUG
Naval Command,
Control and Ocean Surveillance Center,
RDT&E Division, Environmental Sciences (Code 52),
San Diego,
California 92152-5000, USA

LENWOOD W. HALL, JR
The University of Maryland,
Agricultural Experiment Station,
Wye Research and Education Center,
Box 169, Queenstown,
Maryland 21658, USA

J.R.W. HARRIS
Plymouth Marine Laboratory,
Prospect Place,
West Hoe,
Plymouth, PL1 3DH, UK

R. SCOTT HENDERSON
Building 216,
Marine Corps Base Hawaii,
Kaneohe Bay,
Hawaii 96863-3002, USA

K.R. HINGA
Graduate School of Oceanography,
University of Rhode Island,
Narragansett,
Rhode Island 02882, USA

E. HIS
Institut Français de Recherche pour l'Exploration de la Mer,
Quai du Commandant Silhouette,
33120 Arcachon, France

R.J. HUGGETT
Virginia Institute of Marine Science,
School of Marine Science,
College of William and Mary,
Gloucester Point,
Virginia 23062, USA

ROY B. LAUGHLIN, JR
Azurea, Inc.,
PO Box 561178,
Rockledge,
Florida 32955, USA

RICHARD F. LEE
Skidaway Institute of Oceanography,
10 Ocean Science Circle,
Savannah,
Georgia 31411, USA

W.G. MACINTYRE
Virginia Institute of Marine Science,
School of Marine Science,
College of William and Mary,
Gloucester Point,
Virginia 23062, USA

R. JAMES MAGUIRE
National Water Research Institute,
Department of Environment,
Canada Centre for Inland Waters,
Burlington,
Ontario L7R 4A6, Canada

CHERYL L. MATTHIAS
Mine Safety Appliances Co.,
38 Loveton Circle,
Sparks,
Maryland 21152, USA

FREDERICK C. NEWTON, III
Kinnetic Laboratories, Inc.,
5225 Avenida Encinas,
Suite H,
Carlsbad,
California 92008, USA

C.D. RICE
Department of Anatomy,
Medical College of Virginia,
Richmond,
Virginia 23298, USA

M.H. ROBERTS, JR
Virginia Institute of Marine Science,
School of Marine Science,
College of William and Mary,
Gloucester Point,
Virginia 23062, USA

MICHAEL H. SALAZAR
Applied Biomonitoring,
11648 72nd Place NE,
Kirkland,
Washington 98034, USA

SANDRA M. SALAZAR
EVS Consultants,
200 West Mercer Street,
Seattle,
Washington 98102, USA

PAUL SCHATZBERG
Naval Surface Warfare Center,
Annapolis,
Maryland 21402-5067, USA

PETER F. SELIGMAN
Naval Command, Control, and Ocean Surveillance Center,
RDT&E Division, Environmental Sciences,
US Navy,
San Diego,
California 92152-6335, USA

K.W.M. SIU
Institute for Environmental Chemistry,
National Research Council of Canada,
Montreal Road,
Ottawa,
Ontario K1A 0R9, Canada

H. SLONE
Virginia Institute of Marine Science,
School of Marine Science,
College of William and Mary,
Gloucester Point,
Virginia 23062, USA

MARTHA O. STALLARD
Computer Sciences Corporation,
4045 Hancock Street,
San Diego,
California 92110-5164, USA

P.M. STANG
PRC Environmental,
4065 Hancock Street,
San Diego,
California 92110, USA

A.R.D. STEBBING
Plymouth Marine Laboratory,
Prospect Place,
The Hoe,
Plymouth PL1 3DH, UK

JOHN E. THAIN
Ministry of Agriculture,
Fisheries and Food,
Fisheries Laboratory,
Remembrance Avenue,
Burnham-on-Crouch,
Essex CM0 8HA, UK

M.A. UNGER
Virginia Institute of Marine Science,
School of Marine Science,
College of William and Mary,
Gloucester Point,
Virginia 23062, USA

ALDIS O. VALKIRS
Computer Sciences Corporation,
4045 Hancock Street,
San Diego,
California 92110-5164, USA

M.E. WAITE
National Rivers Authority,
South Western Region,
Manley House,
Kestrel Way,
Exter,
Devon EX2 7LQ, UK

M.J. WALDOCK
Ministry of Agriculture, Fisheries and Food,
Fisheries Laboratory,
Remembrance Avenue,
Burnham-on-Crouch,
Essex CM0 8HA, UK

JAMES H. WEBER
University of New Hampshire,
Chemistry Department,
Parsons Hall,
Durham,
New Hampshire 03824-3598, USA

FOREWORD: ORGANOTINS – WHAT HELP FROM HINDSIGHT?

A.R.D. Stebbing

Plymouth Marine Laboratory, Prospect Place, Plymouth PL1 3DH, UK

INTRODUCTION

In any discipline there are occasional issues on which the advancement of a subject turns. TBT (tributyltin) has provided problems that have been important for the development of marine environmental toxicology far beyond the issue of one specific microcontaminant. TBT was first introduced as a biocide in antifouling paints about 30 years ago; it is said to be the most toxic material ever deliberately introduced into the marine environment (Goldberg, personal communication). Now the story of TBT and its related compounds provides an exemplar, just as DDT was for terrestrial systems, that will long be instructive in science policy and regulatory toxicology.

The scientific account of the identification of the curious abnormalities induced in molluscs by TBT, and the development of chemical techniques sufficiently specific and sensitive to detect the minuscule concentrations that can cause such effects, was in itself fascinating. As instructive in another sense was the way in which the growing case for legislation was strenuously countered by the paint industry. But the weight for control measures of evidence became overwhelming and eventually led to the introduction of legislation, which is now resulting in a gradual recovery of affected populations in coastal ecosystems and shell fisheries. TBT provides a rare instance, for those involved in the management of contamination, of a complete cycle through the phases of identifying a problem, establishing its cause-and-effect relationships, introducing legislation, and confirming the effectiveness of legislation through monitoring. Now, with the help of hindsight, it can be recognized that TBT was probably as great a hazard to water quality in estuarine and coastal waters as any single toxicant before it, yet one that has now been brought under effective control – at least in estuaries and the nearshore environment.

The problem with TBT and its cause was first recognized in France, then in the United Kingdom and the United States of America; and in these and other countries legislation is now in place (see Abel, Chapter 2; Champ and Wade, Chapter 3), but in many countries the hazard is only now being identified. This volume has the important function of making available to all a summary of the results of work on TBT and the main conclusions. It will help to minimize the duplication of research and speed the introduction of legislation around the world to control organotin pollution. It is the more valuable because research on TBT has often been published in less accessible journals and symposium proceedings. This volume brings together accounts of these findings by the major contributors to the TBT story, providing the most comprehensive account to date.

The TBT problem has proved to be instructive in a number of different ways beyond the bounds of the specific issue (Stebbing, 1985). Most important is that TBT can be seen as a challenge to monitoring systems for nearshore waters, by which it can be judged how effective monitoring has been in fulfilling its purpose, and what improvements should be made. Most instructive was the time it took to bring TBT under control. It was first used in the marine environment about 30 y ago, although it did not really begin to have significant biological effects until the mid-1970s, and those effects were not linked to TBT until the early 1980s. The French acted first, adopting what is now called a 'precautionary' approach (Stebbing, 1992), and introduced legislation on rather meager scientific evidence. Their mariculture industry was thereby protected while elsewhere, awaiting stronger evidence of causality, estuarine and coastal water quality continued to decline. An incontrovertible causal relationship between TBT and its toxic effects on a broad, taxonomic spectrum of marine species took until 1985 to establish, and legislation to ban the use of TBT on small vessels was not introduced in the United Kingdom until 1987. It is only now that affected populations and shell fisheries are recovering.

As TBT has often been the only contaminant present in otherwise unpolluted estuaries, the problem would appear to have been straightforward, but it took over 10 y to establish the case for legislation in this relatively simple case; and the last 5 y were after the French had established grounds sufficient for their 'precautionary' TBT legislation. This begs the question: how much more difficult might it have been to establish causality and grounds for legislation in more typical cases, where a highly toxic microcontaminant is obscured within a complex mixture of compounds in more polluted waters?

Why was the threat posed by TBT not recognized sooner? The reasons are largely historical and are mainly because pollution legislation has been and is directed primarily to the control of chemical contaminants, with its implicit assumption that those that are likely to be biologically significant are known and monitored chemically. The weakness in this assumption is only now being accepted by the recognition that biological monitoring is the way to detect toxic contaminants. Obviously it is not their mere presence in the marine environment that is important, but their potential biological activity that should determine the best approach. From the environmental point of view, the use of biota is indispensable in monitoring the marine environment, and chemical effort is most efficiently directed to where there are demonstrable problems. As with so many environmental contaminants in the past, it was the occurrence of biological effects that first brought TBT to the notice of the scientific community. This clearly indicates the part that biological techniques should play in monitoring, in a reconnaissance role to direct chemical effort more effectively, rather than as an optional extra. In this case, and perhaps others in the future, the biota were the more important, because, when the effects of TBT were first noticed, chemical techniques were not sufficiently sensitive to detect the concentrations that were biologically active. With thousands of new compounds synthesized each year, it is inevitable that other contaminants, like TBT, have or will escape present controls and consent procedures, and find their way into the marine environment at concentrations that are biologically active. It must be asked whether the ability to detect, control, and monitor new and previously unsuspected pollutants is improved as a result of the experience with TBT.

With hindsight it is probable that the recurrent failure of recruitment of the East Coast oyster fisheries through the 1970s was because the planktonic larvae did not develop to metamorphosis. There is little doubt that TBT played a part in causing recruitment failure, and it is ironic that the oyster embryo development is now proving to be such an effective method to bioassay the effects of contaminants on water quality (Thain, 1992), when for over 10 y their failure to survive the planktonic phase in the same waters was not thought to be related to toxic contamination. Had the same technique been deployed then as now, in conjunction with water and tissue chemistry, there would not have been such a long delay between the detection of biological effects and the introduction of legislation.

The purpose of the discussion that follows is to compare the respective roles of contaminant ('factor') monitoring with biological ('target') monitoring, and to ask how they should be viewed in the light of what has been learned from TBT.

'FACTOR' VS 'TARGET' MONITORING

The purpose of monitoring is to provide information at regular intervals to environmental managers regarding the health of nearshore waters, and the compliance of environmental concentrations to agreed standards and legislation (the 'Action Cycle,' see Abel, Chapter 2). Monitoring can relate to 'factors,' which are usually the specific contaminants, or to 'targets,' which are typically the biota that the monitoring strategy is designed to protect (Holdgate, 1979). Often the link between them is implicit in that the significance of chemical data is interpreted in relation to an EQS (Environmental Quality Standard), which is designed to protect the biota by being set at a lower level than toxicological threshold concentrations. The environmental significance of monitoring factors lies in the detection of actual or potential effect on targets, and, without this link, factor monitoring is wasted effort.

A target is defined as anything that may be liable to show change in distribution, abundance or performance (e.g. biota or ecosystems). A factor is defined as anything liable to cause changes in the environment or to living targets (e.g. chemical contaminants). Hence come the terms target monitoring of biota or factor monitoring of contaminants.

Target monitoring has the fundamental weakness that it cannot stand alone. Where the performance of target organisms leads one to believe there is a problem, the cause of the problem remains to be demonstrated by positively identifying the factor(s) responsible. Only then can appropriate control action be taken. Nevertheless, target monitoring provides the best means of measuring the totality of environmental quality in a biological sense, simply because only the status of the target indicates whether any factor is impinging on it. For such reasons, target monitoring is essential (Stebbing and Harris, 1988), yet biological monitoring remains a minor part of the total effort, and resources are wasted on monitoring chemical contamination at biologically inconsequential concentrations.

Chemical monitoring predominates for the historical reason that pollution legislation until recently has been formulated in chemical terms, and the principal role of monitoring is to serve the legislation, demonstrating the extent to which policy and protocols for managing contaminating inputs are effective, or should be improved.

But for the purposes of monitoring environmental health, a monitoring strategy that relied on target monitoring to provide overall geographic coverage would be preferable, identifying sites where there are actual or incipient problems. Factor monitoring could then be directed to where there are demonstrable problems and used to establish causal relationships and evidence for control measures. Factor monitoring of innumerable contaminants is much too expensive to be used in the unfocused manner that is often typical at present. The high cost of chemical monitoring has often meant that the measurements have been too infrequent to be really effective.

The use of organisms for factor monitoring of chemical contaminants by analysis of their tissue burdens, as this approach is exemplified in the Global Mussel Watch Programme, circumvents the bioavailability question, since whatever is taken up and induces effects or responses is inevitably bioavailable. Despite repeated efforts to develop suitable extraction techniques, they cannot be expected to mimic reliably the bioavailability of contaminants in water and sediments, due to the confounding effects of variations in uptake, metabolization, sequestration, and excretion.

But the use of contaminant burdens in organisms to estimate environmental concentrations misses the point to some extent, since it is the biological effects of contaminants that are of most concern. The more instructive relationship is not between tissue burdens and environmental concentrations, but between tissue burdens and their physiological effects (see below).

TARGET MONITORING

It is proposed that target, or biological, monitoring be used as a precursor to chemical monitoring effort, enabling it to be directed more effectively to where there are demonstrable biological problems. The criteria by which environmental quality is ultimately judged are primarily biological, so it is logical to use biological monitoring techniques. With hindsight, it is quite clear that if pollution monitoring had incorporated the use of sensitive and appropriate biological methods to monitor water quality, whether as water quality bioassays (such as the oyster embryo bioassay) or by the use of 'sentinel organisms' (e.g. the mussel *Mytilus edulis*), the decline of water quality due to TBT contamination could have been detected as much as a decade sooner. Damage caused to coastal and marine ecosystems could then have been reversed at a much earlier stage, preventing significant loss of estuarine and coastal resources.

It is not surprising that TBT, as a previously unsuspected toxic contaminant in the marine environment, should be detected first by mariculturalists, who depend on consistently high-quality seawater for their livelihood. In effect, to culture organisms is to monitor biological water quality. Oyster growers in France were the first to notice the effects of TBT on the recruitment and shell growth of the Japanese oyster (*Crassostrea gigas*), although the early evidence linking cause and effect was far from conclusive (Alzieu and Heral, 1984; His, Chapter 12). Similar deformities were noticed in the United Kingdom (Key *et al.*, 1976), and the multi-layered shell thickening and 'balling' of the shells was sufficiently characteristic to suggest a common cause (Waldock and Thain, 1983; Waldock *et al.*, Chapter 11). For some time the implication was that the oysters alone were affected, and the mistaken belief was held that a case for control of TBT should be weighed only against the value of the oyster industry and attempts to introduce the Japanese species. But it is highly improbable that one contaminant could affect a single species to the exclusion of all others; besides which TBT was selected for use in antifouling paints because of its properties as a broad-spectrum biocide. TBT was initially added as a supplement to copper-based antifouling paints to enhance their algicidal properties. It is curious therefore that work on the environmental effects of TBT did not reveal effects on macroalgae (but see Lee, Chapter 17).

It soon became apparent that other molluscs were affected, sometimes in equally unexpected ways. Imposex in gastropods linked to boating activity and inducible by TBT was demonstrated first in the United States of America (Smith, 1981), and this phenomenon became widely used in the United Kingdom first to show the impact of TBT (Bryan *et al.*, 1986; Gibbs and Bryan, 1986, and Chapter 13), causing the decline of many populations close to centers of boating activity by sterilizing females. Now imposex is recognized and used as a specific index of TBT pollution. While some other organotin compounds may induce imposex (Bryan *et al.*, 1988), no others are apparently effective at environmental concentrations. The severity and frequency of imposex are being used to determine the effectiveness of existing legislation aimed at smaller vessels, and the possible effects of TBT originating from ocean-going ships (Bailey and Davies, 1988, 1991). The specific effects of organotins on gastropod molluscs have been crucial in detecting the toxicity of TBT, establishing its geographic scale and now the effectiveness of legislation. Imposex is used as a bioassay *sensu stricto* as an alternative to chemical analysis, circumventing the high cost and complexity of chemical analysis at nanogram-per-liter concentrations.

The following points form a rationale for target monitoring using biological techniques:

1. Criteria for environmental quality are primarily biological.
2. Biological techniques are therefore the most appropriate to measure and monitor environmental quality.
3. This utilizes the capacity of the biota to integrate the effects of all toxic contaminants, whether they are known and chemically monitored or not.
4. It provides a means of integrating fluctuations in concentrations over time to provide a realistic environmental measure of contamination.
5. It also provides a means of detecting new or unsuspected contaminants by their effects.
6. Such an approach directs chemical monitoring effort more efficiently and effectively to where there are demonstrable pollution problems.

While imposex has been established as a specific index of TBT pollution, there are important advantages to using generalized biological stress indices. Organisms respond to the totality of their environment, acting as both 'filters' and 'integrators,' filtering out those environmental contaminants that are biologically inconsequential, while integrating the combined effects of those that are not. In addition, there are the variations in concentration with time, over a tidal cycle for example, that organisms can integrate. Seawater contains numerous contaminants, many of which are known, and some of which are included in chemical monitoring programs. Other compounds are not monitored chemically, but may contribute to the depression of water quality. A sensitive and 'calibrated' indicator organism, used as a 'sentinel,' can be expected to detect the effects of any contaminant of biological significance, whether its presence as a specific contaminant is known or not. Such an approach is essential to detect new, unsuspected contaminants, just as TBT was when first introduced as a biocide in antifouling paints.

DEFICIENCIES OF 'FACTOR' MONITORING ALONE

As a means of effecting protection of the marine environment, the use of factor or chemical monitoring alone is flawed, because implicit in this approach are a number of implicit assumptions that are often not justified:

1. All those contaminants (or homologous groups) likely to be toxic are known and can be measured at concentrations likely to be biologically active.
2. Chemical analyses are of the bioavailable phase(s) and species of contaminants, or can be expressed in terms representing a bioavailable fraction.
3. Interactions among contaminants, and between contaminants and natural environmental factors, do not affect bioavailability and toxicity.
4. Degradation products of contaminants that are potentially hazardous are either monitored or are non-toxic.
5. Analyses are carried out on samples that are representative with respect to both spatial and temporal variations in contaminant concentrations.
6. Concentrations of contaminants that are potentially toxic are analytically detectable.
7. Samples for analysis are representative (spatially or temporally) of water or sediment to which organisms are likely to be exposed.

'Factor monitoring' therefore has inherent weaknesses related to assumptions about the extent to which it can be expected to monitor the number of contaminants present. Factor monitoring presupposes all the significant factors are known, an impossible assumption given only the thousands of microcontaminants and their degradation products known to be present in the effluent of major estuaries. TBT was certainly not a 'known' contaminant until its biological effects were identified; pollution regulatory strategies patently failed, and accepted practices have been modified in most countries to prevent a recurrence. The other assumptions related to factor monitoring also make assumptions related to bioavailability, interactions among contaminants, coverage of degradation products, analytical sensitivity, and representativeness of sampling; difficulties that are largely circumvented by the use of target monitoring closely coupled to analytical chemistry to establish causality.

It is not my intention to devalue the key role of chemical analysis, but to put its purpose into perspective. It is clear that, with respect to TBT and many of the organotin compounds, each of the assumptions listed above has at some time proved invalid. Just as chemical data have their limitations, so too do biological data in isolation. The focus of environmental toxicology should be the study of the interface between environmental contamination by toxic materials and the organism, and to this end, data relating to each are of limited utility without the other.

TBT illustrates this point well. Once the French had linked shell deformities and recruitment failure in *C. gigas* to TBT, it was important to establish what concentrations of TBT occurred in estuarine and coastal waters where biological effects were suspected. Finding these TBT concentrations posed particular problems because initially the only technique suitable was for determining total organotins by atomic absorption spectrophotometry and had a detection limit of 0.11 $\mu g\ l^{-1}$ (Waldock and Miller, 1983; Cleary and Stebbing, 1985), a concentration in excess of those that subsequently proved to be biologically active. Data using this technique were related to toxicity data from experiments with TBT, although the environmental data inevitably included other organotin compounds. It was clear that the chemical data needed to be more specific and so chromatography techniques developed by Matthias and colleagues (1986; Matthias, Chapter 5) were adopted, providing for the first time specific analyses for TBT and its less toxic breakdown products, dibutyltin (DBT) and monobutyltin (MBT). It also became necessary to improve sensitivity to detect concentrations of TBT of a few nanograms per liter, because toxicological thresholds for the induction of imposex by TBT fell to a few nanograms per liter (Gibbs *et al.*, 1988; Gibbs and Bryan, Chapter 13). It is important to recognize the achievement of these levels of analytical sensitivity, which are three orders of magnitude below typical water quality chemistry.

One approach used at an early stage in the United Kingdom to circumvent some of the limitations of chemical analysis of water samples was to analyze tissue burdens in oysters. The analytical advantage was that since oysters and other lamellibranchs accumulate TBT to concentrations thousands of times those in seawater (Waldock *et al.*, 1983; Laughlin *et al.*, 1986), analysis was easier and the data more precise. TBT contamination could also be studied over wider areas where it was undetectable from analyses of water samples alone. More importantly, the use of oysters or mussels to monitor environmental concentrations of TBT avoids the limitations of individual analyses of water samples, representative of one point at one moment in time. Studies of temporal and spatial variability in contaminant concentrations of many contaminants have shown the dangers in extrapolating from single samples. Clavell and colleagues (1986) showed how important it is to have a time-integrating water sampler, as they found water concentrations could vary 20-fold during the course of a single tidal cycle. While such samplers are technically feasible (Schatzburg *et al.*, 1986), they are costly; and in principle the use of organisms such as oysters, which accumulate contaminants as time-integrating samplers, offers a similar solution. Perhaps most important of all is that the use of organisms in this way circumvents the problems of extracting from an environmental matrix the proportion of a contaminant that is bioavailable, since that which occurs in tissues is by definition biologically available.

There remain potential uses of data on tissue burdens of considerable potential that have not yet been exploited. As it is possible to establish experimentally the relationship between tissue burdens of toxic contaminants and their effects, it follows that it must be possible to predict the toxic effects of the same contaminant from tissue concentrations in organisms from the field (Stebbing *et al.*, 1980). In recent times this approach has been developed by Widdows and co-workers, who have established relationships experimentally between scope for growth in the mussel *Mytilus edulis* and tissue burdens of petroleum hydrocarbons (Widdows and Donkin, 1988). Using such relationships it should be possible to use data from experiments, such as those to study the accumulation and depuration of TBT in *Crassostrea* (Waldock *et al.*, 1983; Laughlin *et al.*, 1986) where its sublethal effects are also determined, to predict the same effects from tissue burden data for organisms from the field. The approach also allows synergistic effects to be dealt with effectively.

THE PROBLEM OF ESTABLISHING CAUSALITY

Tissue burden data and ranges also offer important evidence of causality, because to induce effects upon organisms, contaminants must be present in their tissues, even if only transiently until they are metabolized. Simple relationships are sometimes complicated by adaptive mechanisms for sequestering or storing toxic materials, but Widdows and co-workers have now developed a methodology for establishing causality using tissue burdens and a quantitative structure–activity relationship (QSAR) approach in *Mytilus* and other lamellibranchs (Widdows and Donkin, 1992).

If tissue concentrations of a toxic contaminant exceed a threshold established experimentally that is capable of inducing an observed effect, that in itself can be taken as sufficient evidence of causality. In experiments with dogwhelks, the coincident increase in tissue organotin concentrations and induction of imposex in those organisms transplanted to a contaminated site provided evidence for a causal relationship (Bryan *et al.*, 1986), but what would have removed the need to invoke any other causal agent would be the knowledge that tissue concentrations of TBT exceeded those necessary to elicit the observed effect.

The question of establishing causality with TBT has depended primarily on the relationship between environmental data for water concentrations and toxicological threshold concentrations determined in the laboratory (Cleary and Stebbing, 1985). The validity of doing so depends on the assumption that each type of data relates to the same compound, a difficult assumption when analyses of AAS were of total organotins and when TBT in any static toxicological test may soon begin to degrade to the less toxic DBT and MBT, or become adsorbed to the walls of experimental vessels. However, as these difficulties were dealt with (by improved specificity of analysis and the design of experiments), and the weight of evidence implicating TBT accumulated, there was no room for doubt that environmental concentrations were at many times and places exceeding toxicological threshold concentrations. What must be regretted is the long time taken to reach this point of certainty, where the occurrence of TBT in relatively unpolluted waters should have assisted in its rapid identification as a significant pollutant. There is an urgent need for much more rapid techniques to establish causality unequivocally. 'Specific' indices such as imposex are always likely to be a rarity, but many biomarkers go some way to identifying their cause, while chemical manipulations of water samples in conjunction with bioassays (Stebbing, 1980; Bening *et al.*, 1992), and toxicological interpretation of tissue burdens (Widdows and Donkin, 1992) show promise.

Absolute proof of causality for contaminant(s) and their effect(s) in the environment is an unattainable goal, and a burden of evidence approach must remain the norm. In the case of TBT, despite the fact that it often occurred in otherwise uncontaminated waters, it took a decade to assemble sufficient evidence for the case for a ban on its use on small vessels to become compelling. For TBT there were 'specific' biological effects, although any assertion of absolute specificity is practically impossible to establish. That it took so much time to assemble sufficient evidence of causality is clearly unsatisfactory. There is therefore a requirement for an agreed set of operational criteria for establishing causality that can be applied more rapidly with existing methods (Stebbing, 1992). A list of candidate criteria might include the usual spatial and temporal correlation of evidence, while environmental contaminant concentrations in relation to laboratory toxicity thresholds will remain important. More important is the adoption of methods to provide a toxicological interpretation of tissue burden data and the development of biomarkers where the effect or response helps identify its cause(s).

TOWARD BETTER PREDICTION AND ANTICIPATION

Ecotoxicological research with TBT has been in part retrospective, attempting to establish whether or not the TBT in estuaries and coastal waters was responsible for the decline in biological water quality indicated by the work on oysters, dogwhelks, and the like. I have tried to point out how both the current approach to monitoring water quality, and the techniques themselves, are inadequate, as it took over a decade to detect the problem and identify its cause beyond doubt. In any similar case in the future, we can at best expect only a reduction in the lag time taken to feed information back to those with the responsibility for managing the marine environment. While it is obviously an advance to minimize environmental damage in this way, the ideal situation must be to be able to avoid damaging ecosystems altogether, and it should be the objective of ecotoxicologists to provide environmental managers with the means of predicting, anticipating, and thereby preventing pollution. The ability to detect new environmental contaminants and limit their unintended effects on natural ecosystems needs to be as effective as the capacity to synthesize new compounds like TBT.

In the case of marine applications of TBT it should have been possible to predict the consequences of using a broad-spectrum biocide that had already become widely used in agriculture, without making invalid extrapolations about its toxicity and degradation. It is important to provide those who have to address requests to use or release toxic materials into the marine environments with the right tools, so that accurate predictions can be made in advance of polluting discharges to the marine environment.

Central to this requirement is the need to identify and extrapolate the effects on toxicants from one taxon of organisms to another. This is more feasible at lower levels of biological organization, closer to the site of action where mechanistic interpretation of effects is possible, but indices at the cellular and biochemical levels continue to attract criticism for having little relevance to populations and ecosystems. The call for simple 'litmus test' indices continues to hold back the development of techniques with an interpretive capability.

Even more important is the need to predict the effects of new toxicants based upon the known effects of others, as the number of organic compounds in polluted waters can be so large. Important advances have been made in the QSAR field in recent years, which are making it possible to make predictions of toxicity within families of compounds. The rapid growth of interest in this approach in marine ecotoxicology has been due in part to early studies of the family of organotin compounds by Laughlin and his co-workers (Laughlin *et al.*, 1984a,b), which now make it possible to predict toxicities with confidence within homologous series of compounds. Now the potential use of QSARs in environmental toxicology is widely recognized

Marine environmental and ecotoxicological problems are invariably interdisciplinary problems, with elements of hydrography, sedimentology, chemistry, and biology needed to provide a realistic prediction of the consequences of releasing unknown toxic substances into coastal waters. The environmental resource or pollution manager is confronted with the problem of bringing together such information to make realistic predictions of the consequences of new effluents and contaminants. In the past, consents and provisions have depended heavily on experience, limited databases, and large safety margins.

It is now widely accepted that computer simulation models provide the best vehicle for the complex set of processes that determine the way contaminants behave in estuarine and coastal waters. Models provide the most satisfactory expression of current and evolving understanding for scientists. As important is the fact that such models, suitably adapted, provide the best means of transferring that understanding to those who need to apply it – the regulators and managers with responsibility for environmental decision making. There are numerous applications: developing appropriate legislation for pollution control, consent setting, planning monitoring programs, exploring worst-case scenarios, identifying the best strategy for dealing with pollution incidents, communicating with those involved in decision making, and many others.

On both sides of the Atlantic, modelling has been used to assist with predicting the behavior of TBT in inshore waters (Harris and Cleary, 1987; Seligman *et al.*, 1987). For example, the TBT Simulator (now available as the Estuarine Contaminant Simulator – ECoS), developed for the UK Department of the Environment by the Plymouth Marine Laboratory, provided an accessible means for the non-expert to set up a one-dimensional model in a personal computer for any estuary to assess the fate and dispersal of any contaminant. One application has been to run the model to assess the effectiveness of TBT legislation and improved dockyard practices (Harris *et al.*, 1991), and the extent to which predicted declines related to actuality.

RESIDUAL ASPECTS OF THE TBT PROBLEM

Levels of TBT in the water column remain higher than might have been expected in United Kingdom waters, although there is recovery from the effects of TBT pollution in areas where legislation has controlled the use of antifouling paints containing TBT on small vessels. While inputs from the illicit use of TBT are probably slight, there remain problems in enclosed waters frequented by large vessels to which existing legislation does not apply. In addition concentrations of TBT in the sea surface microlayer 50 km offshore in the German Bight (North Sea) have been shown to exceed the United Kingdom EQS of 2 ng l^{-1} These concentrations of TBT and its degradation products (Hardy and Cleary, 1992) are most likely to have originated from oceangoing vessels to which current legislation does not apply. Accumulation of TBT, metals (Cu, Pb, Zn), and other contaminants in the sea-surface microlayer have been shown to be toxic when microlayer samples were exposed to bioassay organisms such as echinoderm larvae (Hardy and Cleary, 1992), oyster and clam larvae (McFadzen, 1992; Thain, 1992), and copepod larvae (Williams, 1992). It remains to establish how toxic TBT and its degradation products, together with other contaminants that accumulate at the sea surface, may be to the indigenous biota. The neuston community that inhabits the surface waters includes the embryos and larvae of many pelagic and benthic species, including the larvae of many commercially exploited fish species. It is noteworthy that elevated frequencies of developmental abnormalities were found in dab (*Limanda limanda*) embryos in the surface plankton during the same period (Cameron and Berg, 1992). While further evidence is needed, such observations raise again the question whether the use of TBT in antifouling paints should cease altogether.

Overall concentrations in the water column suggest that the remobilization of sediments contaminated by TBT provides a continuing input, as it has become clear from field data that the degradation of organotins in sediments is $\sim$ 5 y, while laboratory experiments indicate a half-life of at least 7–8 y, whether or not the sediments are anaerobic (Langston, personal communication). Tidal pumping of suspended sediments in estuaries can result in higher concentrations of TBT attached to sediments at the turbidity maximum. This does not result in an accumulation in the benthic sediments, probably because of the mobility of these sediments and the low stability constant for TBT (Langston and Harris, personal communication).

It is clear therefore that aspects of the TBT problem will remain for some time due to TBT locked in sediments that is degrading so slowly that inputs to the water column will continue for some years to come. Rapid light-mediated degradation was promoted by the paint industry in favor of its environmental use, with no understanding of the stability of TBT in marine sediments. Recovery from this extremely conservative microcontaminant is likely to take decades, particularly where large vessels enter enclosed waters, shipyards and ports, and given that a significant proportion of the TBT in estuarine sediments may be in the form of solid paint particles (Harris *et al.*, 1991); however, improvements in working practices for drydocks will help to reduce inputs.

CONCLUSIONS

The TBT problem has established more emphatically than any before it that biological techniques have a primary role in monitoring programs for the detection and measurement of pollution. Monitoring strategies are directed to serve legislation formulated largely in chemical terms, but if the primary objective of environmental monitoring is to help sustain the health of marine ecosystems, then the adoption of biological effects techniques from the wide range now available is an important priority (Stebbing *et al.*, 1992).

If the criteria for environmental quality are primarily that the seas support life, then the relative ease with which biological monitoring can be conducted makes it the most appropriate option. Biological monitoring, compared with monitoring all the chemical contaminants likely to be significant, makes it highly cost-effective, allowing chemical effort to be focused more closely on the problems of establishing causality where demonstrable pollution problems occur.

Within this volume are chapters by analytical chemists, biologists, ecologists, mathematical modellers, hydrographers, and toxicologists. It seems a truism, but an ability to solve environmental problems, like those provided by TBT, depends not merely on the application of these separate disciplines, but on their integration. Then this interdisciplinary approach must be directed in a coordinated way to understanding key processes: that TBT undergoes leaching from painted surfaces, dispersion, partitioning to different phases, degradation, bioavailability, bioaccumulation, and toxicity. The example of TBT shows clearly how much more readily ecotoxicological problems can be solved when research is carried out in a context where academic disciplinary boundaries can be bridged.

ACKNOWLEDGEMENTS

I am grateful to my colleagues in the Ecotoxicology Group at Plymouth Marine Laboratory, who contributed so much to the scientific underpinning necessary for the UK TBT legislation. In particular I acknowledge the help of Drs Peter Gibbs, Bill Langston, John Harris, and John Widdows, who have provided recent information that has been incorporated in this prologue. Recollections of Dr Dick Dally helped to provide a balanced account, but the views expressed are my own. This prologue is dedicated to the memory of Dr Geoff Bryan, whose death has been a sad loss to all who knew him; the volume is a fitting memorial to his contribution to TBT research as his contribution will be clear to all who read it.

REFERENCES

Alzieu, C. and M. Heral. 1984. Ecotoxicological effects of organotin compounds on oyster culture. In: *Ecotoxicological Testing for the Marine Environment*, Vol. I. G. Persoone, E. Jaspers and C. Claus (Eds.). State University of Ghent, Ghent, and Institute for Marine Scientific Research, Bredene, Belgium, pp. 187–196.

Bailey, S. and I.M. Davies. 1988. Tributyl tin contamination around an oil terminal in Sullom Voe (Shetland). *Environ. Pollut.*, **55**, 161–172.

Bailey, S. and I.M. Davies. 1991. Continuing impact of TBT, previously used in mariculture, on dog whelk (*Nucella lapillus* L.) populations in a Scottish sea loch. *Mar. Environ. Res.*, **32**, 187–199.

Bening, J.-C., L. Karbe and G. Schupfner. 1992. Liquid/solid phase extraction of water samples used for toxicity testing in the German Bight. *Mar. Ecol. Prog. Ser.*, **91**, 233–236.

Bryan, G.W., P.E. Gibbs, L.G. Hummerstone and G.R. Burt. 1986. The decline of the gastropod *Nucella lapillus* around southwest England: evidence for the effect of tributyl tin from antifouling paints. *J. Mar. Biol. Ass. UK*, **66**, 611–640.

Bryan, G.W., P.E. Gibbs and G.R. Burt. 1988. A comparison of the effectiveness of tri-n-butyltin chloride and five other organotin compounds in promoting the development of imposex in the dogwhelk, *Nucella lapillus*. *J. Mar. Biol. Ass. UK*, **68**, 733–744.

Clavell, C., P.F. Seligman and P.M. Stang. 1986. Automated method of organotin compounds: a method for monitoring butyltins in the marine environment. In: Proceedings of the Oceans '86 Organotin Symposium, Vol. 4, M.A. Champ (Ed.). Institute of Electrical and Electronics Engineers, New York, pp. 1152–1154.

Cleary, J.J. and A.R.D. Stebbing. 1985. Organotin and total tin in coastal waters of southwest England. *Mar. Pollut. Bull.*, **16**, 350–355.

Gibbs, P.E. and G.W. Byran. 1986. Reproductive failure in populations of the dogwhelk, *Nucella lapillus*, caused by imposex induced by tributyltin from antifouling paints. *J. Mar. Biol. Ass. UK*, **66**, 767–777.

Gibbs, P.E. and G.W. Bryan. 1987. TBT paints and the demise of the dogwhelk, *Nucella lapillus* (Gastropoda). In: Proceedings of the Oceans '87 Organotin Symposium, Vol. 4, M.A. Champ (Ed.). Institute of Electrical and Electronics Engineers, New York, pp. 1482–1487.

Gibbs, P.E., P.L. Pascoe and G.R. Burt. 1988. Sex change in the female dog whelk, *Nucella lapillus*, induced by tributyltin from antifouling paints. *J. Mar. Biol. Assoc. UK*, **68**, 715–731.

Hardy, J.T. and J. Cleary. 1992. Surface microlayer contamination and toxicity in the German Bight. *Mar. Ecol. Prog. Ser.*, **91**, 203–210.

Harris, J.R.W. and J.J. Cleary. 1987. Particle–water partitioning and organotin dispersal in an estuary. In: Proceedings of the Oceans '87 Organotin Symposium, Vol. 4, M.A. Champ (Ed.). Institute of Electrical and Electronics Engineers, New York, pp. 1370–1374.

Harris, J.R.W., C.C. Hamlin and A.R.D. Stebbing. 1991. A simulation study of the effectiveness of legislation and improved dockyard practice in reducing TBT concentrations in the Tamar Estuary. *Mar. Environ. Res.*, **32**, 279–292.

Holdgate, M.W. 1979. *A Perspective of Environmental Pollution*. Cambridge University Press, Cambridge.

Key, D., R.S. Nunny, P.E. Davidson and M.A. Leonard. 1976. Abnormal shell growth in the Pacific oyster (*Crassostrea gigas*). Some preliminary results from experiments undertaken in 1975. International Council for the Exploration of the Sea, Copenhagen, C.M. Papers and Reports, K:11, 1976 (mimeo).

Laughlin, R.B., R.B. Johannesen, W. French, H. Guard and F.E. Brinckman. 1984a. Structure–activity relationships for organotin compounds. *Environ. Toxicol. Chem.*, **4**, 343–351.

Laughlin, R.B., R.B. Johannesen, W. French, H. Guard and F.E. Brinckman. 1984b. Predicting toxicity using computed molecular topologies: the example of triorganotin compounds. *Chemosphere*, **13**, 575–584.

Laughlin, R.B., W. French and H.E. Guard. 1986. Accumulation of bis tributyltin oxide by the marine mussel, *Mytilus edulis*. *Environ. Sci. Technol.*, **20**, 884–890.

Matthias, C.L., G.J. Olsen, J.M. Bellama and F.E. Brinkman. 1986. Comprehensive method for determination of aquatic butyltin and butylmethyltin species at ultratrace levels using simultaneous hydridization/extraction with GC/FPD detection. *Environ. Sci. Technol.*, **20**, 609–615.

McFadzen, I.R.B. 1992. Growth and survival of cryopreserved oyster and clam larvae along a pollution gradient in the German Bight. *Mar. Ecol. Prog. Ser.*, **91**, 215–220.

Schatzburg, P., C.M. Adema, W.M. Thomas and S.R. Mangum. 1986. A time-integrating, remotely moored, automated sampling and concentrating system for aquatic butyltin monitoring. In:

Proceedings of the Oceans' 86 Organotin Symposium, M.A. Champ (Ed.), Institute of Electrical and Electronics Engineers, New York, pp. 1152–1154.

Seligman, P.F., C.M. Adema, P.M. Stang, A.O. Valkirs and J.G. Grovhoug. 1987. Monitoring and prediction of tributyltin in the Elizabeth River and Hampton Roads, Virginia. In: Proceedings of the Ocean' 87 Organotin Symposium, Vol. 4, M.A. Champ (Ed.). Institute of Electrical and Electronics Engineers, New York, pp. 1357–1363.

Smith, B.S. 1981. Tributyltin compounds induce male characteristics in female mud snails *Nassarius obsoletus = Ilyanassa obsoleta*. *J. Appl. Toxicol.*, **1**, 141–144.

Stebbing, A.R.D. 1980. The biological measurement of water quality. *Rapp. P. - v. Réun. Cons. Int. Explor. Mer*, **179**, 310–314.

Stebbing, A.R.D. 1985. Organotins and water quality – some lessons to be learned. *Mar. Pollut. Bull.*, **16**, 383–390.

Stebbing, A.R.D. 1992. Environmental capacity and the precautionary principle. *Mar. Pollut. Bull.*, **24**, 287–295.

Stebbing, A.R.D. and J.R.W. Harris. 1988. The role of biological monitoring. In: *Pollution of the North Sea: An Assessment*. W. Salomens, B.L. Bayne, E.K. Duursma and U. Forstner (Eds). Springer-Verlag, Berlin, 687 pp.

Stebbing, A.R.D., B. Akesson, A. Calabrese, J.H. Gentile, A. Jensen, and R. Lloyd. 1980. The role of bioassays in marine pollution monitoring: bioassay panel report. *Rapp. P.-v. Réun. Cons. Int. Explor. Mer*, **179**, 322–332.

Stebbing, A.R.D., V. Dethlefsen and M. Carr. (1992). *Biological Effects of Contaminants in the North Sea. Results of the ICES/IOC Bremerhaven Workshop*. Special Volume of Marine Ecology Progress Series, Vol. 91. Inter-Research, Germany, 361 pp.

Thain, J.E. 1992. Use of the oyster *Crassostrea gigas* embryo bioassay on water and sediment elutriate samples from the German Bight. *Mar. Ecol. Prog. Ser.*, **91**, 211–213.

Waldock, M.J. and D. Miller. 1983. The determination of tin and tributyl tin in seawater and oysters in areas of high pleasure craft activity. International Council for the Exploration of the Sea, Copenhagen, C.M. Papers and Reports, E:12, 1983 (mimeo).

Waldock, M.J. and J. Thain. 1983. Shell thickening in *Crassostrea gigas*: organotin antifouling or sediment induced? *Mar. Pollut. Bull.*, **14**, 411–415.

Waldock, M.J., J. Thain and D. Miller. 1983. The accumulation and depuration of bis (tributyltin) oxide in oysters: a comparison between the Pacific oyster (*Crassostrea gigas*) and European flat oyster (*Ostrea edulis*). International Council for the Exploration of the Sea, Copenhagen, C.M. Papers and Reports, E:52, 1983 (mimeo).

Widdows, J. and P. Donkin. 1988. Interpretation of the relationship between growth and concentration of aromatic hydrocarbon in the tissue of *Mytilus edulis*: mechanisms of toxicity and ecological consequences. *Mar. Environ. Res.*, **24**, 254.

Widdows, J. and P. Donkin. 1992. Mussels and environmental contaminants: bioaccumulation and physiological aspects. In: *The Mussel Mytilus: Ecology, Physiology, Genetics and Culture*, E. Gosling (Ed.). Elsevier, Amsterdam.

Williams, T.D. 1992. Survival and development of copepod larvae *Tisbe battaglia* in surface microlayer, water and sediment elutriates from the German Bight. *Mar. Ecol. Prog. Ser.*, **91**, 221–228.

Dedication: Geoffrey William Bryan, B.Sc., Ph.D., D.Sc.

The sudden death of Geoff Bryan on 17 September 1993 whilst playing badminton has left a deep void in the life of the Plymouth Laboratory, both scientifically and socially. His quiet, dedicated approach to his research was widely admired and respected; his laconic humor enriched many a discourse. On both counts he will be fondly remembered by staff and visitors alike, as well as by colleagues worldwide.

Geoff came to Plymouth in 1958, after completing his Ph.D. at the University of Bristol. Initially, he was employed on a Special Appointment funded by the UK Atomic Energy Authority, subsequently joining the staff of the Marine Biological Association of the UK to continue his radioisotope studies of ionic regulation in decapod crustaceans. His interests broadened to encompass diverse aspects of the effects of metallic contamination on estuarine organisms: few estuaries in the United Kingdom escaped his attention. His recall of field trips, of metal sources from long-abandoned mines, and, above all, of data relating to metal concentrations in animals and sediments was truly remarkable. He was awarded his D.Sc. degree in 1980.

Geoff was drawn to the tributyltin question in 1984 when he discovered that the *Nucella lapillus* population below the laboratory had virtually disappeared and that imposex, the early signs of which were first noticed in this same population by Blaber in 1969, was developed to an advanced stage in the few surviving females. Some five years earlier, Smith had uncovered evidence of the link between TBT pollution and imposex in *Ilyanassa obsoleta*, but the nature and the extent of the problem had yet to be unravelled. Its elucidation was to become a focal point of Geoff's research until his death.

Arguably, his best paper was his first on the subject; this appeared in 1986 and presented the evidence linking widespread extinctions of populations of *Nucella lapillus* to TBT leachates from antifouling paints (*J. Mar. Biol. Ass. UK*, **66**, 611–640). This clear exposition demonstrated Geoff Bryan's rare versatility in combining detailed ecological observation, field and laboratory experimentation, and analytical chemistry. In this case he caused us all to revise our concept of toxicity down to the 1 ng l^{-1} level and precipitated the introduction of restrictions on TBT paint usage as well as the resetting of the UK Environmental Quality Standard for TBT in seawater to just 2 ng l^{-1}. Globally, subsequent studies have demonstrated consistent results for imposex in many neogastropods.

Geoff was a retiring character; despite his considerable achievements, he never sought the limelight and positively shunned acclaim. Nevertheless, he was always approachable and many staff, students, and visitors have reason to be grateful for his sound advice and help. Throughout his 35 years at the laboratory he was always productive, and his name appears on over 90 papers. Inevitably, a large amount of the data he gathered had to be side-lined in favor of more urgent projects. With his retirement at 60 in March 1994 just six months away, he was beginning to plan the publication of this material. Some will still appear, but there can be no doubt the papers will lack the depth and clarity of analysis Geoff would have imparted to their writing.

Our deepest sympathy and condolences are extended to his wife Janet and daughter Jane.

Peter Gibbs

GEOFF BRYAN: PUBLICATIONS RELATING TO TRIBUTYLTIN POLLUTION

Bryan, G.W., P.E. Gibbs, L.G. Hummerstone and G.R. Burt. 1986. The decline of the gastropod *Nucella lapillus* around south-west England: evidence for the effect of tributyltin from antifouling paints. *Journal of the Marine Biological Association of the United Kingdom*, **66**, 611–640.

Gibbs, P.E. and G.W. Bryan. 1986. Reproductive failure in populations of the dog-whelk, *Nucella lapillus*, caused by imposex induced by tributyltin from antifouling paints. *Journal of the Marine Biological Association of the United Kingdom*, **66**, 767–777.

Bryan, G.W., P.E. Gibbs, G.R. Burt and L.G. Hummerstone. 1987. The effects of tributyltin (TBT) accumulation on adult dog-whelks, *Nucella lapillus*: long-term field and laboratory experiments. *Journal of the Marine Biological Association of the United Kingdom*, **67**, 525–544.

Bryan, G.W., P.E. Gibbs, L.G. Hummerstone and G.R. Burt. 1987. Copper, zinc, and organotin as long-term factors governing the distribution of organisms in the Fal Estuary in southwest England. *Estuaries*, **10**, 208–219.

Gibbs, P.E. and G.W. Bryan. 1987. TBT paints and the demise of the dog-whelk, *Nucella lapillus* (Gastropoda). In: Proceedings of the Oceans '87 International Organotin Symposium, Vol. 4, Institute of Electrical and Electronics Engineers, New York, pp. 1482–1487.

Gibbs, P.E., G.W. Bryan, P.L. Pascoe and G.R. Burt. 1987. The use of the dog-whelk, *Nucella lapillus*, as an indicator of tributyltin (TBT) contamination. *Journal of the Marine Biological Association of the United Kingdom*, **67**, 507–523.

Bryan, G.W., P.E. Gibbs and G.R. Burt. 1988. A comparison of the effectiveness of tri-n-butyltin chloride and five other organotin compounds in promoting the development of imposex in the dog-whelk, *Nucella lapillus*. *Journal of the Marine Biological Association of the United Kingdom*, **68**, 733–744.

Bryan, G.W. and P.E. Gibbs. 1989. Water pollution. In: *McGraw-Hill Yearbook of Science and Technology 1990*. McGraw-Hill, New York, pp. 357–359.

Bryan, G.W., P.E. Gibbs, R.J. Huggett, L.A. Curtis, D.S. Bailey and D.M. Dauer. (1989). Effects of tributyltin pollution on the mud snail, *Ilyanassa obsoleta*, from the York River and Sarah Creek, Chesapeake Bay. *Marine Pollution Bulletin*, **20**, 458–462.

Bryan, G.W., P.E. Gibbs, L.G. Hummerstone and G.R. Burt. (1989). Uptake and transformation of ^{14}C-labelled tributyltin chloride by the dog-whelk, *Nucella lapillus*: importance of absorption from the diet. *Marine Environmental Research*, **28**, 241–245.

Gibbs, P.E., G.W. Bryan, P.L. Pascoe and G.R. Burt. 1990. Reproductive abnormalities in female *Ocenebra erinacea* (Gastropoda) resulting from tributyltin-induced imposex. *Journal of the Marine Biological Association of the United Kingdom*, **70**, 639–656.

Langston, W.J., G.W. Bryan, G.R. Burt and P.E. Gibbs. 1990. Assessing the impact of tin and TBT in estuaries and coastal regions. *Functional Ecology*, **4**, 433–443.

Spence, S.K., G.W. Bryan, P.E. Gibbs, D. Masters, L. Morris and S.J. Hawkins. 1990. Effects of TBT contamination on *Nucella* populations. *Functional Ecology*, **4**, 425–432.

Bryan, G.W. and P.E. Gibbs. 1991. Impact of low concentrations of tributyltin (TBT) on marine organisms: a review. In: *Metal Ecotoxicology: Concepts and Applications*, M.C. Newman and A.W. McIntosh (Eds). Lewis Publishers, Ann Arbor, pp. 323–361.

Gibbs, P.E., G.W. Bryan and P.L. Pascoe. 1991. TBT-induced imposex in the dogwhelk, *Nucella lapillus*: geographical uniformity of the response and effects. *Marine Environmental Research*, **32**, 79–87.

Gibbs, P., G. Bryan and S. Spence. 1991. The impact of tributyltin (TBT) pollution on the *Nucella lapillus* (Gastropoda) populations around the coast of south-east England. *Oceanologica Acta, Special Volume*, **11**, 257–261.

Gibbs, P.E., P.L. Pascoe and G.W. Bryan. 1991. Tributyltin-induced imposex in stenoglossan gastropods: pathological effects on the female reproductive system. *Comparative Biochemistry and Physiology*, **100C**, 231–235.

Spooner, N., P.E. Gibbs, G.W. Bryan and L.J. Goad. 1991. The effect of tributyltin upon steroid titres in the female dogwhelk, *Nucella lapillus*, and the development of imposex. *Marine Environmental Research*, **32**, 37–49.

Bryan, G.W., D.A. Bright, L.G. Hummerstone and G.R. Burt. 1993. Uptake, tissue distribution and metabolism of ^{14}C-labelled tributyltin in the dog-whelk, *Nucella lapillus*. *Journal of the Marine Biological Association of the United Kingdom*, **73**, 889–912.

Bryan, G.W., G.R. Burt, P.E. Gibbs and P.L. Pascoe. 1993. *Nassarius reticulatus* (Nassariidae: Gastropoda) as an indicator of tributyltin pollution before and after TBT restrictions. *Journal of the Marine Biological Association of the United Kingdom*, **73**, 913–929.

Gibbs, P.E. and G.W. Bryan. 1994. Biomonitoring of tributyltin (TBT) pollution using the imposex responses of neogastropod molluscs. In: *Biological Monitoring of Coastal Waters and Estuaries*, K.J.M. Kramer (Ed.). CRC Press, Boca Raton (in press).

Gibbs, P.E. and G.W. Bryan. 1995. Reproductive failure in the gastropod *Nucella lapillus* associated with imposex caused by tributyltin pollution: a review. (Chapter 13, this volume).

PREFACE

Organotin compounds comprise a large group of organometallic moieties characterized by a tin atom covalently bonded to various organic groups (e.g. methyl, ethyl, butyl, propyl, phenyl, etc.). The use of organotin compounds has increased dramatically over the last 30 years. The biocidal use of organotin, representing about 20% of the total production, has included development of insecticides, fungicides, bactericides, wood preservatives and antifouling agents in the form of triorganotins, primarily triphenyl- and tributyltin. The largest use of organotins today is in the stabilizing of polyvinylchloride (PVC) plastics, primarily from low toxicity di- and monoorganotin derivatives.

In the 1960s, tributyltin (TBT) compounds were first used as molluscicides to kill several species of freshwater snails that are the intermediate host of the parasitic worm *Schistosoma*, which transmits the disease schistosomiasis to humans. Shortly hereafter, triorganotin compounds were used as paint additives in antifouling coatings and proved to be highly effective biocides in preventing the attachment and growth of fouling organisms such as barnacles and tube worms on the hulls of vessels. The growth of fouling organisms produces increased roughness and concomitant increases in turbulent flow and drag across the hull, increasing fuel consumption and reducing speed. Antifouling paints prevent or slow this growth, thus providing substantial economic benefit through reduced fuel expenditure. Tributyltin compounds have been one of the most effective biocides used in antifouling paints. Prior to regulation, about 2.6 million liters of TBT-containing paints were used in the US.

Early paints contained simple mixtures of tributyltin oxide or tributyltin fluoride in a soluble matrix and were referred to as 'free association' paints. More advanced coating systems were later developed which chemically integrated the tributyltin within a copolymer, such as tributyltin methacrylate, which allows for self polishing or ablation to occur and for lower release rates, thus renewing the surface and maintaining the antifouling capacity for a number of years. These antifouling paints are effective because hydrolysis occurs which releases tributyltin (TBT) cations into the water creating a toxic envelope near to the surface of the hull. The release of TBT from hulls, sediments, drydocks or from sewage treatment plants constitutes the loading sources of the compound into receiving waters. The resultant distribution, fate and effects of the compound in the aquatic environment is the subject of this book. The genesis of this volume was from the research and resultant papers presented at the International Organotin Symposia at the Oceans 1986 and 1987 Conferences in Washington DC and Halifax, Canada, respectively, and from subsequent work in the field. These conferences were the first forums where the chemistry, toxicity and fate and behavior of organotin compounds, and specifically TBT, in the environment were extensively addressed as a single subject.

The compound tributyltin (TBT) is emphasized in this book as an example of a biocide whose application preceded both the regulatory framework and the scientific capacity to fully evaluate the ecological risks from its use as an antifouling agent in hull paints. Tributyltin was initially believed to be toxic only to fouling organisms on the painted surface and thus not an environmental risk. However, there was little scientific data available to confirm the environmental fate and effects and behavior of this compound, particularly in the marine environment. Until the early 1980s there was virtually no exposure assessment, largely because no ultra-trace (ng l^{-1}) speciation methods were available and dose–response assessment was limited primarily to acute testing with target fouling organisms and mammalian toxicity studies. In addition, very little was known about the loading, fate, persistence, partitioning and bioavailability of the compound, particularly in the marine environment.

Initial regulatory and legislative action in several countries was based primarily on a retrospective assessment of the apparent effects of TBT at low concentrations (100–500 ng l^{-1}) on non-target molluscan organisms. Specifically, shell thickening and reduced productivity in the cultured non-native Pacific oyster culture (*Crassostrea gigas*) in France (late 1970s) and the UK (early 1980s) were attributed to relatively low concentrations of TBT resulting in a shell thickening response and in some cases reduced productivity. In addition, it was shown that much of the problem relative to TBT loading was from the recreational boating sector resulting in high concentrations in yacht harbors, at times exceeding acute thresholds for sensitive species.

The developing case for TBT-caused imposex (imposition of male sexual characteristics) in female dogwelks (marine snails) in the UK at low parts per trillion (ng l^{-1}) concentrations added to the concern about the risk of the compound to the aquatic environment. These concerns led to regulatory and legislative actions in a number of countries without formal or well defined risk assessments. These regulatory and legislative actions all took an intermediate and similar course of action: the reduction of TBT environmental loading by allowing use only on larger (>25 m) vessels because of the substantial economic benefit for ocean going vessels, and the elimination of higher release rate paints and those with high concentrations of TBT, rather than a total ban. As documented by a number of monitoring studies in the UK and US, this approach has been successful in substantially lowering environmental levels of TBT in the water column and in bivalve tissues.

This book summarizes over 30 years of research and development and presents the current state of knowledge about the toxicity, fate, transformation, loading and distribution of tributyltin in the aquatic environment. The book is organized into six sections: 1. Introduction – Historical and Regulatory Perspectives, 2. Analytical Methodology, 3. Biological Effects (including acute and chronic effects, metabolism and bioaccumulation), 4. Distribution and Fate (including release rates, environmental loading, degradation, partitioning and monitoring studies), 5. Regional Monitoring Studies, and 6. Future Research Needs: Science and Policy. The book's 29 chapters, with broad international authorship, are intended to be read either as a comprehensive compendium of the environmental implications of the use of organotins or as individual references for a particular topic area. Accordingly, there is slight redundancy in some chapters, particularly in introductory sections, to maintain their status as independent chapters. However, this has been minimized to the degree possible. It is hoped that this book will be broadly useful to regulators, environmental policy makers and environmentalists, as well as scientists, academicians, students and specialists in aquatic environmental sciences. The combined work in this volume represents a broad approach to ecological risk assessment of a particular man-made xenobiotic compound (TBT) and it is hoped that it can act as a guide to future evaluations of new compounds that may impact aquatic ecosystems.

The TBT Conference held in December 1995 in Malta was the first major effort to assess and document the cost benefits from the use of organotins in antifouling paints. Estimates of these benefits to vessels greater than 25 m in length range as high as 2.7 billion US dollars annually on a global basis (when consideration is given to fuel avoidance savings, boat bottom cleaning, dry dock costs, sanding and repainting, and disposal). This economic benefit was created by advanced technologies. However, there are additional technological advances that should be explored in developing an economically and environmentally sound regulatory strategy. For a national regulatory strategy to be supportive of the creation of high technology chemicals and products, the strategy must include the promotion of continued research and development to push these compounds to additional refinements that enhance environmental attributes and improve competitiveness in the global market place. The difficult global decisions are those that balance environmental stewardship and excessive regulation, which can inhibit economic development of products and technologies.

The current regulatory strategies for organotins have several major shortcomings. The regulatory strategies of developed nations unfortunately focus on short-term national self interests and may not represent a 'think globally, act locally' philosophy. The principal regulatory approach is to reduce organotin concentration in the local environment by reducing the concentration in the paint (or in the release rate) and in the concentrations discharged to the environment from shipyards. Regulators in developed nations, in setting regulatory environmental concentrations (water quality standards) to protect local coastal waters are, in effect, encouraging shipping companies to take their antifouling repainting business abroad at the economic loss of domestic shipyards.

US and European shipyards cannot effectively compete in the lesser developed non-environmentally regulated marketplace if, in addition to high labor and operational costs, they must also shoulder the expense of waste treatment and disposal of antifouling residues from removal of spent antifouling paints to achieve a regulated discharge (environmental water quality standard) level to protect local waters. Consequently, large vessel owners can enjoy the double cost benefit of being able to have their vessel painted by cheap labor without having to be responsible for environmental degradation and human health hazards (externalities) in lesser developed countries.

We must use the same advanced technologies that produced organotin compounds to create antifouling boat bottom coatings that are, in a commonly used phrase, sustainable, economically viable, environmentally sound, and socially acceptable.

ACKNOWLEDGEMENTS

We would like to thank Joyce G. Nuttall for her tremendous assistance as technical editor in the preparation of this book. Others to be thanked include Dr Iver W. Duedall of the Center for Academic Publications at Florida Tech (Florida Institute of Technology) for his generous support and assistance in this project. We also thank the US Navy, NOSC, NOAA Office of the Chief Scientist, the National Ocean Pollution Monitoring Program, the NOAA National Sea Grant Program, and the University of Florida Sea Grant Program for providing partial support for the editing of this volume.

A large group of unnamed peer reviewers spent countless hours on the chapters of this book, which has been one of our main stays over the past six years in producing this book. Another was the authors of the individual chapters; not one pulled their paper as we encountered publication delays and rewrites. During this period, we have had book contracts with two publishers, three publishing editors, worked with three different publisher copy-editors and in our own careers have changed organizations in name or position at least twice. As the book grew, our publishers became concerned about the costs for producing such a large book. However, we must take a moment to acknowledge that Chapman & Hall always believed in the quality of the content of this book and Dr R.C.J. Carling (Senior Editor, Life Sciences) should be commended for supporting such a large reference work in an age where such quality can be threatened by the ever increasing costs of production.

The Editors

ACRONYMS

AAS	Atomic Absorption Spectrophotometry
AAS–GF	Atomic Absorption Spectrophotometry With Graphite Furnace
ACP	Advisory Committee on Pesticides (UK)
ACS	American Chemical Society
ADI	Acceptable Daily Intake
ANOVA	Analysis of Variance
ASTM	American Society for Testing and Materials
BAF	Bioaccumulation Factors
BATNEEC	Best Available Technology Not Entailing Excessive Cost
BCFs	Bioconcentration Factors
CBI	Chesapeake Bay Institute
CEC	Commission of European Communities
CEFIC	European Chemical Industry Council
CEQ	Council on Environmental Quality (US)
C_{eq}	Equilibrium Water Concentration
CFR	Code of Federal Regulations (US)
CI	Condition Indices
COE	Army Corps of Engineers (US)
COPA	Control of Pollution Act 1974 (UK)
CRMs	Certified Reference Materials
CV	Coefficient of Variation
CWA	Clean Water Act of 1972 (US)
DBT	Dibutyltin
DCI	Data Call-In Notice
D Larvae	Straight-Hinge Oyster Larvae
DOM	Dissolved Organic Matter
DTRC	David Taylor Research Center
DWT	Dry Weight Tonnage
EA	Environmental Assessment
EAAS	Electrothermal Atomic Absorption Spectrophotometry
EAWAC	Swiss Federal Institute for Water Resources and Water Pollution Control
EC	European Community
EC_{50}	Toxic Concentrations Which Cause 50% Abnormal Larval Development
ECoS	Estuarine Contaminant Simulator
EDL	Electrodeless Discharge Lamp

EIMS	Electron Impact Mass Spectrometry
EIS	Environmental Impact Statement
EMAP	Environmental Monitoring and Assessment Program (USEPA)
EMAP–E	Environmental Monitoring and Assessment Program – Estuarine
EPA	Environmental Protection Agency (US)
EQOs	Environmental Quality Objectives
EQS	Environmental Quality Standard
EST	Environmental Quality Target
FAO	Food and Agriculture Organization (UN)
FEPA	Food and Environment Protection Act of 1985 (UK)
FIFRA	Federal Insecticide, Fungicide, and Rodenticide Act of 1972 (US)
FONSI	Interim Finding of Non Significant Impact
FPD	Flame Photometric Detection (method) or Detector (instrument)
FR	Federal Register (US)
GC	Gas Chromatography (separation science) or Chromatograph (instrument)
GC–AAS	Gas Chromatography–Atomic Absorption Spectrophotometry or Spectrophotomer
GC–ECD	Gas Chromatography–Electron Capture Detection (method) or Detector (instrument)
GC–FID	Gas Chromatography–Flame Ionization Detection [or Detector]
GC–FPD	Gas Chromatography–Flame Photometric Detection [or Detector]
GC–MS	Gas Chromatography–Mass Spectrometry [or Spectrometer]
GC–TCD	Gas Chromatography–Thermal Conductivity Detection [or Detector]
GERG	Geochemical and Environmental Research Group (TAMU)
GESAMP	Joint Group of Experts on the Scientific Aspects of Marine Pollution
HDAA	Hydride Derivatization and Atomic Absorption Detection [or Detector]
HMIP	Her Majesty's Inspectorate of Pollution
HMSO	Her Majesty's Stationary Office
HP	Hewlett-Packard
HPLC	High-Pressure Liquid Chromatography
IAEA	International Atomic Energy Agency (UN)
ICP	Inductively Coupled Plasma
IC_{50}	Toxic Concentration Inhibiting Primary Productivity by 50%
ICES	International Council for the Exploration of the Sea
ICP–AES	Inductively Coupled Plasma–Atomic Emission Spectrometry [or Spectrometer]
ICP–MS	Inductively Coupled Plasma–Mass Spectrometry [or Spectrometer]
IFREMER	Institut Français de Recherche pour l'Exploration de la Mer
IMO	International Maritime Organization
IP–ICP–MS	Ion Pairing–Inductively Coupled Plasma–Mass Spectrometry [or Spectrometer]
ISMS–MS	Ionspray Mass Spectrometry–Mass Spectrometry [or Spectrometer]
ISO	International Standards Organization
K_{app}	Apparent Sorption Coefficient
K_{oc}	Organic Carbon–Water Partition Coefficient

K_{ow}	Octanol–Water Partitioning Coefficient
K_p	Equilibrium Partition Coefficients
LC	Liquid Chromatography
LC_{50}	Median Tolerance Limit (Lethal Concentration for 50% of Test Animals)
LC_{100}	Lethal Concentration for 100% of Test Animals
LOD	Limit of Detection
LOQ	Limit of Quantitation
MAFF	Ministry of Agriculture, Fisheries, and Food
MATC	Maximum Acceptable Toxicant Concentration
MBT	Monobutyltin
MEPC	Marine Environmental Pollution Committee
MERL	Marine Ecosystem Research Laboratory, University of Rhode Island
MFO	Mixed-Function (Mono) Oxygenase System
MIBK	Methyl-Isobutyl Ketone
MLLW	Mean Lower Low Water
MMA	Methylmethacrylate
MPRSA	Marine Protection, Research, and Sanctuaries Act of 1972 (US)
MS	Mass Spectrometry [or Spectrometer]
NAVSEA	Naval Sea Systems Command (US)
NBS	National Bureau of Standards (US) renamed NIST
NC	Nominal Concentrations
ND	Not Detectable or Detected
NEPA	National Environmental Policy Act of 1969 (US)
NIST	National Institute of Technology and Standards (US)
NMR	Nuclear Magnetic Resonance
NOAA	National Oceanic and Atmospheric Administration (US)
NOEC	No Observed Effects Concentration
NOEL	No Observable Effects Level
NOPPA	National Ocean Pollution Planning Act of 1978 (US)
NOSC	Naval Ocean Systems Center (US)
NRC	National Research Council (US, Canada)
NS&T	National Status & Trends Program (US NOAA)
OAD	Ocean Assessments Division (US NOAA)
OAPCA	Organotin Antifouling Paint Control Act of 1988 (US)
OECD	Organization for Economic Cooperation and Development
ORTEP	Organotin Environmental Programme Association
OSHA	Occupational Safety and Health Act of 1976 (US)
PACS-1	Marine Harbor Sediment Certified Butyltin Reference Material (NRC Canada)
PAH	Polycyclic Aromatic Hydrocarbons
PCB	Polychlorinated Biphenyls
PCI	Positive Chemical Ionization
PD	Position Document (US EPA)

P.L.	Public Law
PLYMSOLVE	Contaminant Numerical Simulation Model
PML	Plymouth Marine Laboratory (UK)
ppm	Part-Per-Million
ppt	Part-Per-Trillion
PVC	Polyvinyl Chloride
QSAR	Quantitative Structure–Activity Relationship
RPAR	Rebuttable Presumptions Against Registration
RPS	Relative Penis Size
SAL	Salinity
SAS	Statistical Analysis System
SCX–ICP–MS	Strong Cation Exchange–Inductively Coupled Plasma–Mass Spectrometry [or Spectrometer]
SD	Standard Deviation
SIM	Selected Ion Monitoring
SNK	Student–Newman-Keuls Multiple Range Test
TAMU	Texas A&M University
TBT	Tributyltin
TBTA	Tributyltin Acetate
TBTBr	Tributyltin Bromide
TBTCl	Tributyltin Chloride
TBTF	Tributyltin Fluoride
TBTH	Tributyltin Hydride
TBTM	Tributyltin Methacrylate
TBTO	Bis(Tributyltin)Oxide
TBTS	Tributyltin Sulfide
TI	Thickness Index
TPT	Triphenyltin
TPrT	Tri-n-propyltin Chloride
TSCA	Toxic Substances Control Act of 1974 (US)
TTBT	Tetrabutyltin
UKDoE	Department of Environment (UK)
UNEP	United Nations Environmental Programme (UN)
USC	US Code
UV	Ultraviolet Irradiation
VDS	Vas Deferens Sequence
VIMS	Virginia Institute of Marine Science (US)
WHO	World Health Organization (UN)

AN INTRODUCTION TO ORGANOTIN COMPOUNDS AND THEIR USE IN ANTIFOULING COATINGS

1

Michael A. Champ[1] *and Peter F. Seligman*[2]

[1]*Texas Engineering Experiment Station, Washington, DC Office, Texas A&M University System, 4601 North Fairfax Drive, Suite 1130, Arlington, Virginia 22203, USA*

[2]*Naval Command, Control, and Ocean Surveillance Center, RDT&E Division, Environmental Sciences, US Navy, San Diego, California 92152-6335, USA*

Organotin. Edited by M.A. Champ and P.F. Seligman. Published in 1996 by Chapman & Hall, London. ISBN 0 412 58240 6

ABSTRACT

This chapter has been prepared as an historical overview of organotin compounds and a discussion on their use as biocides in antifouling coatings (boat-bottom paints). It is also a summary of the environmental effects of organotin compounds on nontarget organisms, ranging from oyster shell thickening and growth anomalies to imposex in the common dogwhelk *Nucella lapillus*. A comparative summary of the toxicity of environmental concentrations of TBT to different groups of marine organisms is also presented. Organotin compounds are today one of the most studied groups of organometallic chemicals. Organotin compounds have demonstrated significant economic benefits, as well as environmental costs, as a result of their varied industrial and agricultural uses and applications. The emphasis of this chapter and this book is on tributyltin (TBT), the principal biocidal ingredient of most organotin antifouling paints. The widespread use of this modern, high-technology group of chemicals in industrial, agricultural, and public health applications poses a potential global dilemma, which requires a better understanding of how to control and utilize the unique properties of organotin compounds.

1.1. INTRODUCTION

Since the first days of sail, mariners have been battling 'fouling' – the growth of barnacles, seaweeds, tubeworms, and other organisms on boat bottoms. The Phoenicians, realizing that smoother bottoms translated into easier rowing and faster sailing, were the first to nail copper strips on the hulls of their ships to inhibit fouling. In naval actions, the cleaner, faster vessel often escaped stronger forces or caught up with slower ones. Although copper strips have been replaced by advances in high-technology chemicals in which antifouling coatings incorporate biocides in the paint matrix, fouling remains more important today than ever to both naval and commercial fleets. Fouling affects all vessels, small and large, from supertankers to fishing and sailing boats. Fouling produces roughness on the surface of the bottom of the vessel, thereby increasing turbulent flow and drag. Organotin compounds are one of the most effective classes of biocides utilized against fouling; tributyltin (TBT), the major biocidal ingredient of most organotin antifouling paints, is the major subject of this book.

Organotins have three primary uses: as catalysts to stabilize polyvinyl chloride (PVC) polymers; as preservatives to protect wood, ceramics, plastics, and fabrics from fungal damage; and as biocides to protect plants from insects and boat hulls from fouling organisms. As biocides, the triorganotins were found to be toxic at very low levels and to target specific groups of organisms. In addition to preventing the growth of unwanted organisms, tributyltin was found to induce gross physiological and anatomical deformations in molluscs. The two most pronounced effects studied were oyster shell thickening and growth anomalies and imposex in the common dogwhelk *Nucella lapillus*, a marine snail.

In the early 1970s, organotin compounds became the modern high-technology wonder chemicals and were marketed as the perfect biocides. They appeared to be very toxic and specific to molluscs (Duncan, 1974, 1980), yet an effective biocide against a wide range of marine fouling organisms (Evans and Smith, 1975; van der Kerk, 1975). Organotin compounds were first used by mixing them into a can of paint and then applying the paint (and they were used with a philosophy similar to fertilizers: 'the more the better'). Boaters take very seriously and personally the intrusion of fouling on their boat bottoms and hold the belief that fouling is a personal invasion; most are totally devoid of any environmental appreciation for fouling organisms.

TBT was found to be an excellent antifouling agent. It was effective at very low concentrations, which later turned out to be at the part-per-trillion (nanogram-per-liter) levels, and required the development of greater sensitivity and precision in analytical protocols and methodologies. TBT was toxic to certain species at levels below analytical detection, for most of the analytical laboratories in the world, during most of the 1970s and middle 1980s.

In 1984, the US Navy issued an 'Environmental Assessment: Fleetwide Use of Organotin Antifouling Paint' (NAVSEA, 1984). The Navy assessment addressed the work by Thain (1983), Waldock and Miller (1983), Waldock *et al.* (1983), Alzieu and Portman (1984), and the French studies on the impact of TBT on oysters in Europe. These studies on abnormal oyster growth, however, appeared to be linked to excessive use of free-association TBT-based paints (section 1.6.1) on small recreational boats, which were primarily used in shallow coastal and estuarine waters where oysters were grown. The US Navy perspective on the utilization of TBT was based on three critical determinants: (1) Navy ships were mostly seagoing vessels and spent only minimal periods of time in harbors or shallow coastal waters; (2) the Navy was proposing to use TBT-based copolymer paints (section 1.6.2) with low release rates (Schatzberg, Chapter 19), so the impact on nontarget organisms would be very limited; and (3) the cost benefits from the use of these coatings was estimated to be from \$100 to \$130 million annually in fuel avoidance (savings) costs and \$5 million in annual maintenance costs (NAVSEA, 1984, 1986; Bailey, 1986; Eastin, 1987; Ricketts, 1987).

The Navy's proposed action, combined with similar proposals in Europe, focused extensive US and international scientific and political attention for the next decade on the potential effects of the use of organotin compounds as biocides in antifouling paints. The results of this international scientific and regulatory interest are presented in this book. The book focuses on the biological activity, environmental partitioning, bio-geochemical fate, and ecological implications of organotins, principally in the marine and estuarine environments, with some freshwater data from Canada and Europe. Other chapters discuss analytical protocols, environmental concentrations, distribution and sources of tributyltins, and regulatory and legislative actions. The development of regulatory strategies for organotin compounds in antifouling coatings are discussed in Abel (Chapter 2) for Europe and Champ and Wade (Chapter 3).

Until equally effective coatings are developed, the economic benefits from the use of TBT-based antifouling paints for ocean-going vessels are considerable and will not diminish in the future. A large international market has sprung up in developing countries for applying organotin-based coatings to ships. This shift of the application from regulated to non-regulated countries (due to environmental constraints and human health risks and the attractiveness of lower labor costs in developing countries) will keep the use of organotin-based antifouling paints as a future global environmental and economic issue for many years to come.

1.2. INTRODUCTION TO ORGANOTIN CHEMISTRY

Organometallic chemistry encompasses compounds in which a metallic (or metalloid) element is covalently bonded directly to one or more carbon atoms. By 1830 both main group metals and certain transition metals were known to bond to carbon. In the succeeding 150 y many thousands of compounds have been prepared and studied; some of these compounds have extensive industrial and agricultural applications.

Organotin compounds are substances in which at least one direct tin–carbon bond is present. The great majority of organotin compounds have tin in the +4 oxidation state. Tin–carbon bonds are, in general, weaker and more polar than those formed in organic compounds of carbon, silicon, or germanium; organic groups attached to tin are more readily removed. This higher reactivity does not, however, imply an instability of organotin compounds under ordinary conditions.

The first organotin compound was described in 1852 by Lowig, and the first comprehensive study of organotin compounds was conducted by Sir Edward Frankland, who prepared diethyltin diiodide in 1853 and tetraethyltin in 1859 (van der Kerk and Luijten, 1954, and papers cited therein; Champ and Bleil, 1988). Despite this early discovery, organotins remained mere curiosities for nearly a century. The biocidal uses of organotins stem from a systematic study of these compounds at the Institute of Organic Chemistry T.N.O., Utrecht, sponsored by the International Tin Research Council (M. Harriott, Queens University of Belfast, School of Chemistry, Belfast, Northern Ireland, personal communication).

1.2.1. TRIBUTYLTIN ANTIFOULING BIOCIDES

Tributyltin compounds used in antifouling paints consist of a tin (Sn) atom covalently bonded to three butyl ($C_4H_9^-$) moieties and an associated anion (X; Fig. 1.1). A number of organotin compounds have been used as

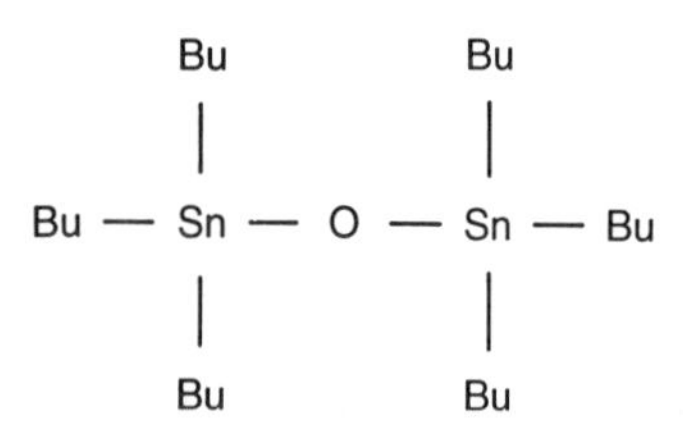

Figure 1.1 Chemical structure of bis(tributyltin) oxide, the active ingredient of some antifouling paints (BU = $C_4H_9^-$).

ingredients in antifouling paints (Kuch, 1986):

Trialkyltins
- bis(tributyltin) oxide
- bis(tributyltin) sulfide
- tributyltin acetate
- tributyltin acrylate
- tributyltin fluoride
- tributyltin naphthenate
- tributyltin resinate
- tributyltin methacrylate/methylmethacrylate copolymer
- bis(tributyltin) adipate
- tricyclohexyltin hydroxide

Triaryltins
- triphenyltin hydroxide

Dialkyltins
- dibutyltin dilaurate
- dibutyltin isooctyl mercaptoacetate
- dibutyltin maleate
- dioctyltin isooctyl mercaptoacetate
- dioctyltin maleate

Monooctyltins
- monooctyltin tris isooctyl mercaptoacetate

1.2.2. AQUATIC CHEMISTRY OF TRIBUTYLTIN

The behavior of tri-n-butyltin biocides (Bu_3SnX), where X is the anion (e.g. oxide, sulfide, acetate, fluoride), may be described by the following equation for the undisassociated (neutral) compound and for the dissociated (positively charged) cation upon reaction with water:

$$Bu_3SnX + H_2O \rightarrow Bu_3Sn(H_2O)^+ + X^-$$

It is widely accepted that tributyltin toxicity is ascribed to the cation (TBT^{++}) and not to which anion is associated with the biocide in the neutral compound.

Copolymer paints allow manufacturers to formulate paints whose biocide release rates in seawater can be controlled. The most com-

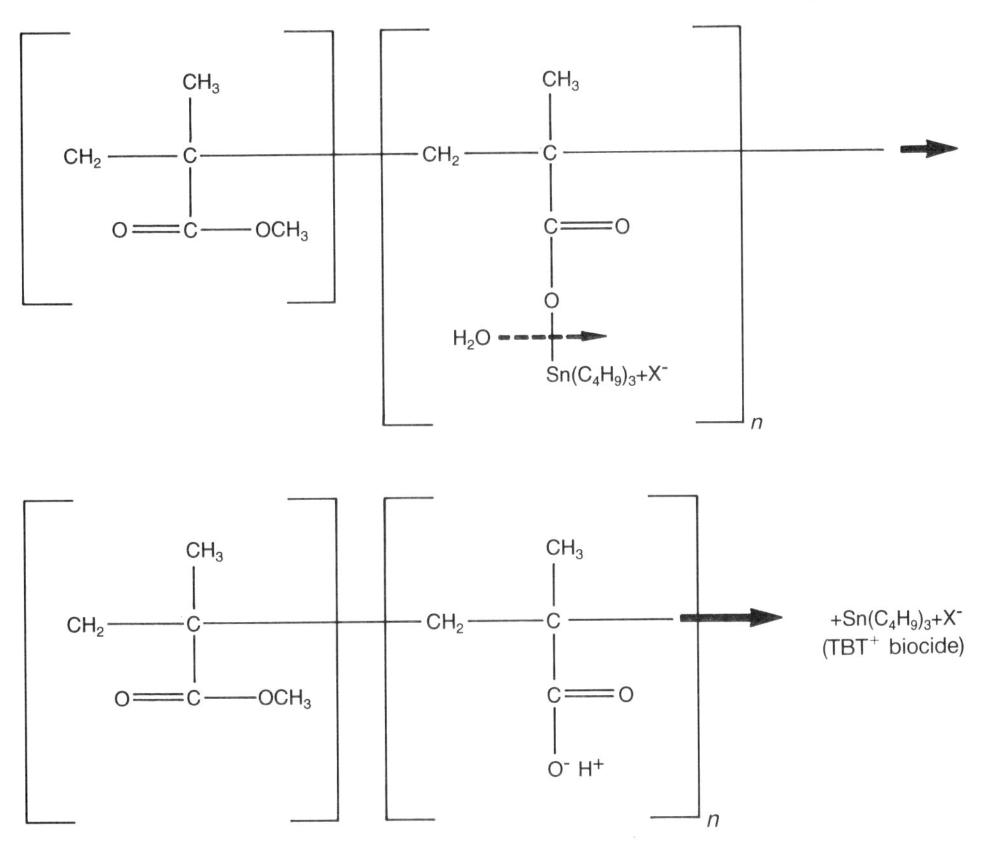

Figure 1.2 The hydrolysis reaction that takes place between the tributyltin methacrylate/methylmethacrylate copolymer and seawater releases the TBT^+ cation into the seawater. Upon immersion in seawater, the chemical bond holding the tributyltin cation onto the copolymer is broken and TBT is released, to prevent fouling (from Anderson and Dalley, 1986).

mon copolymer is tributyltin methacrylate (TBTM)/methylmethacrylate (MMA). Because the antifouling paint coating is hydrophobic, seawater interacts with the copolymer at the surface only, initiating a hydrolysis reaction that cleaves TBT from the copolymer backbone, releasing it in seawater (Fig. 1.2).

In aquatic systems, the distribution of TBT species is dependent on pH and salinity. In seawater, the hydrated TBT cation, tributyltin chloride, bis(tributyltin carbonate), and tributyltin hydroxide (Guard *et al.*, 1982) are in equilibrium. At pH 8, the normal pH of seawater, the major species are tributyltin hydroxide and tributyltin carbonate.

1.2.3. REPORTING CONVENTIONS FOR ORGANOTIN CONCENTRATIONS AND UNITS

Researchers have been inconsistent and at times ambiguous in reporting concentrations of organotins and in their use of units in the literature. In some instances, authors have favored reporting measurements as tin (Sn). Others have reported TBT concentrations as the compound TBTX (where X is for example chloride or acetate) or as the cation (TBT^+). The expression of concentration has, in part, depended on how and for what purpose the measurement was made. In the case of

Table 1.1 Comparison of units

parts per thousand	[PPT]	grams/liter	$g\ l^{-1}$
parts per million	[PPM]	milligrams/liter	$mg\ l^{-1}$
parts per billion	[PPB]	micrograms/liter	$\mu g\ l^{-1}$
parts per trillion	[PPTr]	nanograms/liter	$ng\ l^{-1}$

toxicity experiments, the specific compound is known and may be reported.

Where environmental samples (water, tissue, or sediment) are measured, the butyltin species may be determined chromatographically (i.e. mono-, di-, or tributyltin). The anion is not known, however, so it is more appropriate to express concentration in terms of the cation, as Sn, or as the standard used. The most accurate, and, unfortunately, least used, means of expression is as molar quantities (M). The use of moles would clarify the issue, allow better intercomparison among studies and different biocides, and allow mass balance determination experiments.

In any case, authors and readers should use and be aware of appropriate conversion factors based on molecular weights. In this book, the authors have individually defined their reporting of concentration and units. A comparison of units is used in the organotin literature and in this book is shown in Table 1.1.

1.3. INDUSTRIAL AND AGRICULTURAL APPLICATIONS OF ORGANOTINS

There are three major uses of organotin compounds. Dibutyltins are used as a catalyst for stabilizing PVC polymers; as wood preservatives against wood-reducing fungi (such as *Poria monticola*, *Coniophora olivacea*, and *Formes lividus*), in bathrooms (shower stalls and flooring), and on decks and exterior siding and fences; as fungicides in the textile industry to control odor-producing fungi in socks, undergarments, and children's pajamas; and as biocides for agricultural applications and in antifoulant paints.

The first industrial application for organotin compounds was as a mothproofing agent in 1925. In 1932, organotin compounds were used for stabilizing chlorinated benzenes and the biphenyls used in transformers and capacitors (Champ and Bleil, 1988). This was followed by the use of dibutyltin dilaurate and other dibutyltin salts in 1936 as PVC stabilizers. The biocidal properties of diverse organotin molecules were first discovered in the 1950s by a group headed by G.J.M. van der Kerk at the Institute for Organic Chemistry, T.N.O., Utrecht, Holland, under the sponsorship of the International Tin Research Institute in Greenford, Middlesex, England (van der Kerk and Luijten, 1954; van der Kerk, 1975).

1.3.1. STABILIZING PVC POLYMERS

The largest use of organotin compounds today is in stabilizing PVC polymers with diorganotin compounds. PVC is used extensively in the construction industry (flooring, fencing, and piping) and in the food-packaging industry (bottles and films) to allow customers visual inspection of one side of food products such as poultry, pork, or beef. Exposure of PVC to heat or ultraviolet light for prolonged periods causes diminished optical clarity and undesirable coloration. Stabilizers are required if the end products containing PVC must be colorless or transparent, as in the case of bottles, films, or sheets used for food packaging. In the 1970s, about 10% of all PVC being produced was stabilized with organotin compounds (Champ and Bleil, 1988).

1.3.2. BIOCIDAL APPLICATIONS

Since the early studies of Buckton on the toxicity of alkyltin compounds in 1858, bio-organotin chemistry has attracted many researchers. The first application of an organotin compound was in 1929 by an American company. In 1954, the first organotin used in medicine for the treatment of staphylococcal infections led to the death of 102 people (Saxena, 1987). Systematic studies of the toxicity of organotin compounds did not occur until van der Kerk and Luijten's (1954) discovery of the fungicidal properties of organotin compounds and that these properties were dependent upon the number and kind of organic groups bonded to tin. Since then bactericidal and fungicidal uses in agriculture and industry have been developed. Both triphenyltin hydroxide and triphenyltin acetate are used to control fungi that cause potato blight (leaf spot) on sugar beets, celery, carrots, onions, and rice. Organotin compounds are also used as fungicides to prevent tropical plant diseases in peanuts, pecans, coffee, and cocoa. As insecticides, triorganotin compounds have been used against houseflies, cockroaches, mosquito larvae, cotton bollworms, and tobacco budworms. Triorganotin compounds also act as chemosterilants and have antifeeding effects on insects (Thayer, 1984).

1.3.3. MOLLUSCICIDES

In the early 1960s, two organotin compounds (tributyltin oxide and tributyltin fluoride) were first used as molluscicides to kill several species of freshwater snails that were the intermediate hosts of the worms of the genus *Schistosoma*, which transmit the disease schistosomiasis to humans. This immediately led to the use of TBT compounds as paint additives in 1961 for their biocidal properties in antifoulant boat-bottom paints.

The toxicity of organotin compounds to aquatic organisms is thought to increase with the number of butyl substitutions from one to three, and then to decrease with the addition of a fourth butyl group. Elemental or inorganic forms of tin (as in mineral deposits or tin cans) appear to cause negligible toxicological effects in humans or wildlife. In contrast, organotins display an increased fat solubility and, consequently, an enhanced ability to penetrate biological membranes, thereby posing a greater toxicity potential.

In general, the toxicity of TBT or other organotins is considered to be independent of the anion (e.g. OH, F^-, Cl^-, OH) and solely due to the TBT cation. Triorganotins are the most biologically active, with substantial reduction in toxicity as alkyl groups are removed to di- and monoorganotins. In order to assess the fate of a particular tributyltin to degradation in water, one must consider the dissociated active form, the TBT cation (Bu_3Sn^+), and its major degradation products presumably formed by progressive debutylation to inorganic tin (Brinckman, 1981).

In general, the level of toxicity for an aquatic organism to organotins is as follows (Laughlin *et al.*, 1985; Walsh, 1986):

$$R_4Sn < R_3Sn^+ > R_2Sn^{2+} > RSn^{3+} > Sn^{4+}$$

where R represents an organic group (i.e. butyl, propyl).

1.4. FOULING ORGANISMS

The growth of marine organisms on boat bottoms has been the mariner's curse ever since humans first set sail. Typical examples of fouling organisms are shown in Figs 1.3 and 1.4; the latter also shows the life cycle of the acorn barnacle. The growth of barnacles, seaweeds, or tubeworms on a boat's bottom is referred to as 'fouling.' This fouling produces roughness that increases turbulent flow, acoustic noise and drag, and can increase fuel consumption. A 10-μm (one-thousandth of a centimeter) increase in average hull roughness on a power vessel can result in a 0.3–1.0% increase in fuel consumption. For large vessels (bulk carriers),

Figure 1.3 Marine fouling. Photo from a painting by Lisa Haderlie Baker, Lawrence Hall of Science, University of California at Berkeley. Copyright 1980 by the Regents of the University of California, courtesy National Science Teachers Association, the Carolina Biological Supply Company, and Ms Lisa Haderlie Baker.

fuel costs can amount to as much as 50% of the total operating costs (Bailey, 1986; NAVSEA, 1986; Champ and Lowenstein, 1987). For example, the 1985–86 fuel bill of the *Queen Elizabeth II* (One of the world's largest ships) was $17 million (US dollars), thus a 1% increase (10 μm = 0.01 mm) in average hull roughness could cause an increase in fuel consumption that would amount to $170 000. In 1990, the estimated annual worldwide fuel

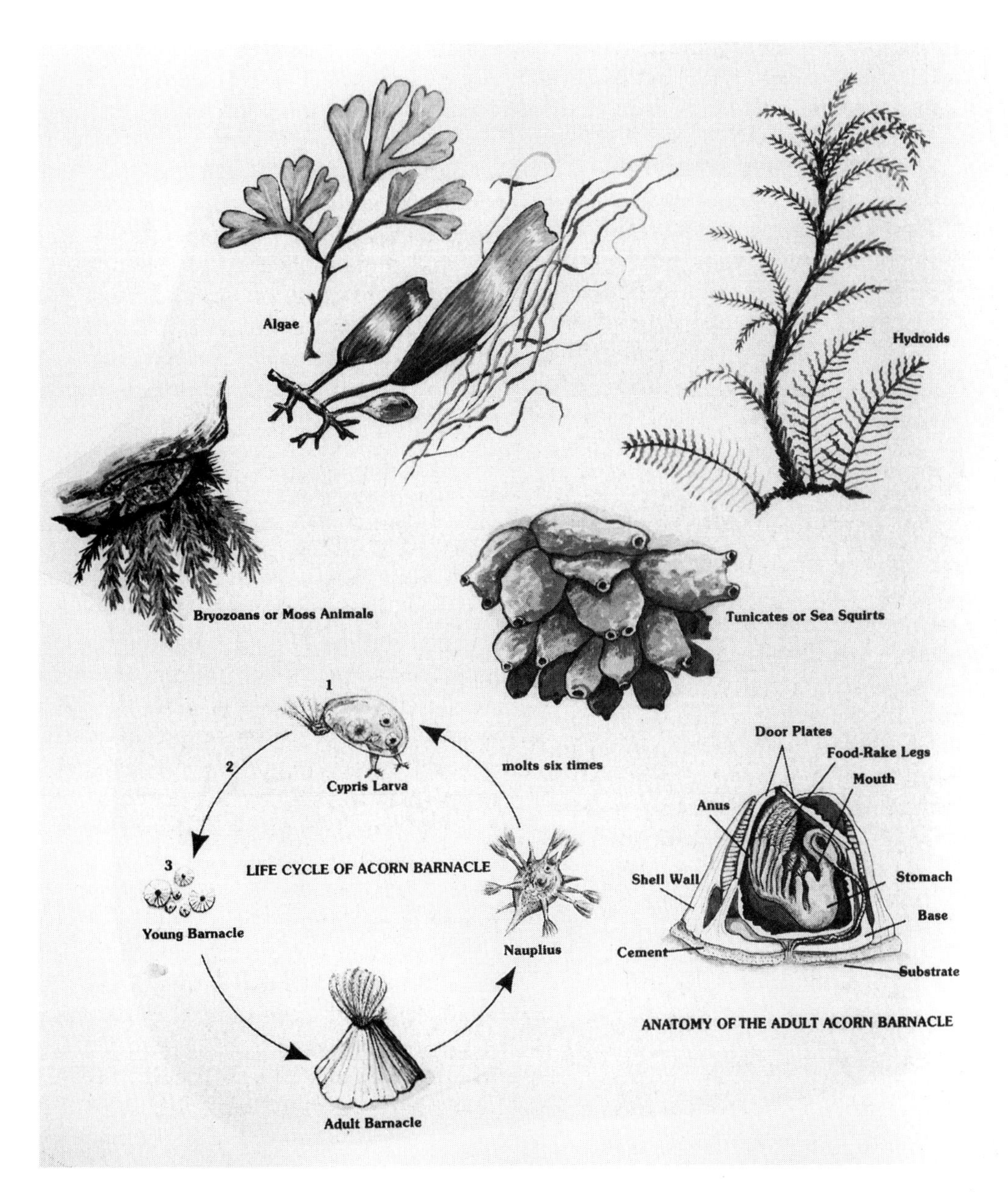

Figure 1.4 The marine hitchhikers: examples of the animals and plants that grow on underwater surfaces such as ship hulls – resulting in marine fouling. On the bottom left-hand side the life cycle of the acorn barnacle is depicted: (1) free-swimming larva must attach to a substrate in order to develop and feed (if it does not attach, it will die in a matter of days); (2) the larva produces a strong, long-lasting adhesive for attaching itself to substrates; (3) the larva attaches head first, forms a conelike shell around itself, and uses its feet to propel food into its mouth. The anatomy of the resulting adult is shown on the lower right. From a drawing by Lisa Haderlie Baker, Lawrence Hall of Science, University of California at Berkeley. Copyright 1980 by the Regents of the University of California, courtesy National Science Teachers Association, the Carolina Biological Supply Company, and Ms Lisa Haderlie Baker.

savings (in 1990) for the world's shipping industries would be about 2.5×10^9 US dollars by the use of tributyltin antifouling compounds, principally through reductions in fuel consumption of some 7.2×10^6 metric tons per year (Milne, 1990).

The rate and extent of fouling on hulls depends upon the marine exposure experienced during vessel operations, water temperature, and particularly the geographic areas of the vessel. If vessels are stationary much of the time (e.g. yachts and some classes of naval ships) or are very active (such as container ships), fouling type and extent will differ greatly. The paint industry has developed a range of antifouling paints to meet these different user requirements. A key factor in the release of biocides from antifouling paints is ship speed. When ships are at rest in harbors, copolymers exhibit the slowest release rates of all antifouling paints. This property represents a major improvement in environmental safety for ship hull coatings.

For the US Navy, which has ships having bottom (wetted hull) areas of almost 14 000 m^2, maintaining a fouling-free bottom becomes a major task. In tropical oceans, the Navy has found that ships may begin to experience significant bottom fouling in<1 y if painted with copper-based antifoulant paints as compared with the 5–7 y if painted with TBT-based antifoulant paints.

1.5. ANTIFOULING MATERIALS/COATINGS/PAINT FILMS

In the 17th and 18th centuries, the use of copper on boat bottoms was found to be effective in preventing growth of fouling organisms. Cuprous oxide paints were introduced and widely used by the turn of this century. Organo-mercury compounds and stearoarsenicals were used until the 1970s to increase the biocidal properties of cuprous oxide paints. These compounds are no longer used in antifoulant paints because of their toxicity and environmental contamination problems.

Most antifouling paints contain cuprous (copper) oxide as the main active ingredient. These paints are effective against barnacle fouling. Many such copper-based paints also contain tributyltin oxide or fluoride (both being called TBT antifouling paints) to inhibit grass and slime fouling, which often leads to extensive barnacle fouling. The most commonly used and most effective TBT-containing paints today are known as copolymers and work by the controlled release of cuprous oxide and TBT.

1.5.1. REGULATIONS

Widespread use of TBT-based antifouling boat-bottom paints began in the early 1970s. It wasn't until the early 1980s that researchers in France and the United Kingdom began to suggest that the use of TBT in antifouling paints was impacting a number of marine species other than fouling organisms, including economically important species such as oysters (Alzieu, 1981, 1986, 1991; Thain, 1986; Abel, Chapter 2; Waldock *et al.*, Chapter 11; His, Chapter 12). The effects on untargeted species attracted increasing international concern. A number of countries have adopted policies regulating or restricting the use of TBT antifouling paints. Many states in the United States have also taken action, and a Federal regulation of TBT by the US Congress has been signed into law (P.L. 100-333, The Organotin Antifouling Paint Control Act of 1987; see Champ and Wade, Chapter 3). This law has restricted organotin paint use to the low-release-rate coatings and to non-aluminum vessels >25 m in length (US Senate, 1988). In the United States, all antifouling paints, whether they contain TBT or not, are required by law to be registered with the US Environmental Protection Agency and all pertinent state agencies (see Champ and Wade, Chapter 3).

1.5.2. ANTIFOULING COATING BENEFITS

Over the past 200 y, naval fleets with superior hull antifouling strategies have often proved more effective in combat. Some examples include:

- Nelson's victory over the French fleet at Trafalgar in 1805. The British fleet was 'copper bottomed' and foul free; the French fleet was heavily fouled and hence less maneuverable.
- In World War II, US naval antifouling technology was more effective at controlling fouling than that used by the Japanese. This advantage gave America's Pacific fleet longer range deployment than the Japanese fleet.
- During the Falklands War of 1982, the cruise liner *Queen Elizabeth II* was converted to troopship status in days and required virtually no hull coating work prior to dispatch to the Falklands, and arrived there ahead of schedule because she was coated with an organotin copolymer antifouling paint.

The economic benefits to navies that use TBT copolymers have not only increased combat performance but include the following:

- Extended in-service employment periods of 5–7 y between drydockings versus 18–30 months at present. Improved ship operating readiness, which is a critical factor in a time of national emergency, allowing ships to be available for deployment with clean, foul-free hulls at short notice without requiring drydocking to remove fouling and the need to repaint hulls.
- Increased operating range, which is important in distant waters such as the Indian Ocean, Persian Gulf, and South Pacific.
- Maintenance of top speed and lower fuel consumption during extended high-speed operations.
- Elimination of underwater hull cleaning, which is a time-consuming activity, to remove fouling during deployment. Copolymers self 'polish' and smooth, providing the possibility of reducing underwater hull noise.
- Application to special underwater advanced sonar and electronic communication and defense systems.

The US Navy calculated in 1985 that if the entire fleet (600 ships) were painted with TBT antifoulant paints, the new fuel avoidance costs would exceed $120 million annually (calculated with fuel costing ~$18 per barrel). Because of improved copper-based antifouling coatings, more recent estimates have reduced this cost avoidance estimate (NAVSEA, 1986).

The use of TBT antifouling paints on commercial ships, fishing vessels, and private boats in the United States could add another $300 to $400 million in fuel savings annually. Worldwide annual fuel savings are estimated to be almost 2 billion gallons of fuel. These figures do not include the savings from the decreased wear on propulsion machinery and the down time for hull scraping, cleaning, and painting that would result from the use of TBT paints.

1.5.3. TYPES OF ANTIFOULING PAINTS

An antifoulant paint consists of a film-forming material (matrix/binder/resin/medium) and a pigment. The film-forming material and pigment can affect the following paint properties: strength, flexibility, water absorption, and color. An antifouling paint is similar to any other paint (matrix plus pigment); however, the paint film is biocidal due to properties of the biocide added to the matrix. The antifoulant paint works by releasing small amounts of biocide at the paint surface that kill or repel the settling stages of fouling organisms (Lewis and Baran, 1993; Kuballa *et al.*, 1995).

There are three types of antifoulant paints: (1) conventional or referred to as 'free association,' in which the biocides are simply mixed in the paint matrix and are released by contact leaching; (2) soluble matrix and ablative, which peels off; and (3) self-polishing, in which the biocides are added in free association or chemically integrated within a matrix as in the organotin copolymer paints. Types 1 and 2 are referred to as conventional paints; both use the biocide in the free-association form. Type 3, a copolymer paint, is the most commonly used type of antifouling paint.

1.5.3a. Type 1 – free association

Type 1 (free-association) antifouling paint uses contact leaching to release the biocide. In free-association paints TBT is physically mixed with the paint and is released, or leached, into the aquatic environment by diffusion through the paint matrix. Biocides leach exponentially from the paint with time. In this process, seawater percolates slowly through a tough, insoluble paint matrix over time (Fig. 1.5). This category of TBT antifoulant coatings has traditionally posed a problem of high early release rate with a subsequently shortened time period of protection from attachment and growth of fouling organisms. After a period of <2 y, the paint film ages, calcium carbonate ($CaCO_3$) clogs the microchannels in the paint surface and inhibits the release of biocide, then the surface becomes biofouled. This leaves a quantity of biocide that remains unused on the vessel hull, which must be removed prior to the next painting.

The removed paint film must be properly disposed of or it may then be a source of environmental contamination. The usual response of the recreational boater and small commercial operator to calcareous fouling of the paint film is to abrade the surface or remove the paint film and reapply. There may be a significant amount of TBT remaining in the film when the fouled film is removed. There have been few studies directed to the extent to which the old paint film is a source of TBT to the environment. The British Royal Yachting Association, in close consultation with other concerned fisheries associations and the UK Department of the Environment (UKDoE), produced a public education document directed at controlling the introduction of old paint into coastal waters.

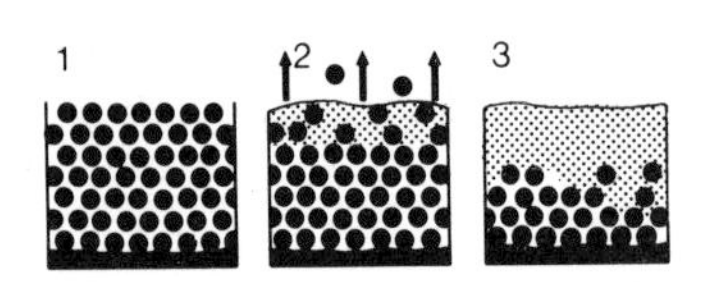

Figure 1.5 In free-association TBT paints, the TBT molecules are leached from a permeable matrix by seawater percolating through the paint (from Anderson and Dalley, 1986).

In the late 1980s, about one-half of the registered paint formulations were free association with TBT mixed with copper compounds. An example of this is a vinyl copper formulation that uses a very low concentration (1%) of TBT in the film as an antibacterial–antislime agent, which increases the effectiveness of the copper oxide, the primary antifouling material. Other free-association formulations varied in TBT concentrations from 2.5% to 15.7% with varying concentrations of copper. Following regulation by many nations (see Champ and Wade, Chapter 3) the free-association TBT-type antifouling paints were banned.

1.5.3b. Type 2 – ablative

Type 2 antifouling paint is commonly referred to as an 'ablative' (or shedding) paint. It is a slightly seawater-soluble matrix paint that sheds during use – as the paint surface roughens, paint particles (very thin microlayers) peel off, exposing a fresh supply of biocide (Fig. 1.6). The biocides are added in the free-association form, leaching exponentially over time. The release of biocide also is inhibited by the formation of surface-insoluble calcium carbonate. The lifetime of this paint is ~2 y.

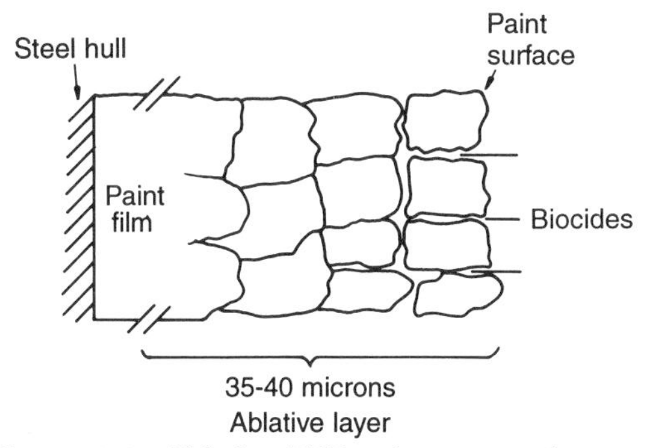

Figure 1.6 Ablative TBT paints use a less permeable matrix that gradually flakes off, exposing new leaching surfaces (from Anderson and Dalley, 1986).

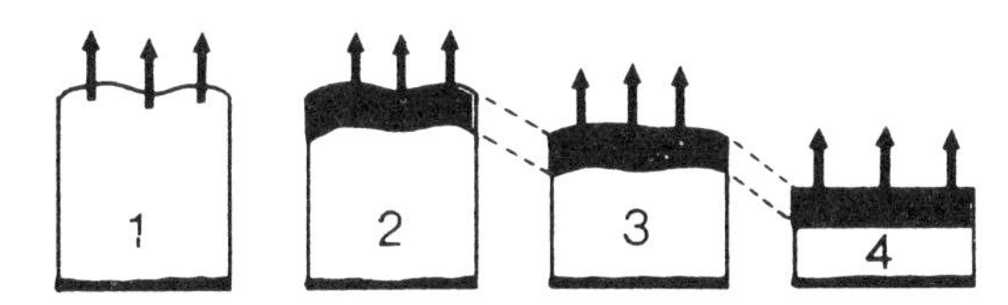

Figure 1.7 In copolymer TBT paints, the TBT is part of an impermeable matrix and is released through a chemical reaction with seawater at the paint surface over time (1–4). New TBT is exposed by gradual erosion of the paint (from Anderson and Dalley, 1986).

1.5.3c. Type 3 – self-polishing copolymer

Type 3 antifouling paint, commonly referred to as 'self-polishing' copolymer paint, is the most commonly used in the 1990s and the most effective TBT-containing paint. In copolymer paints, TBT is chemically bonded with the polymer. This chemical bond tends to retard the release of TBT into the aquatic environment. Copolymer paints were introduced in 1974. The paint is hydrophobic (i.e. seawater does not enter into the paint matrix). The seawater/paint reaction layer occurs at the surface of the paint; the paint has an unstable release layer that gradually erodes (Fig. 1.7). Unlike conventional antifouling paints that leach large amounts of toxic ingredients very rapidly after application, the slow, controlled release of active ingredients by copolymer paints makes them more effective. The copolymer paint formulation is therefore able to utilize low levels of biocide because the biocide is only released at the paint surface as the paint erodes back to the hull. As a result of this unique property, copolymer paints have antifouling lifetimes of 5–7 y.

The advantages of copolymers are that they allow the formulation of antifouling paints that (1) have a true, low, and controlled release rate for the TBT cation over time (Fig. 1.8); (2) have a demonstrated life span well in excess of conventional and ablative coatings; (3) have extremely low levels of free biocide; and (4) have their 'activity' at the surface layer with a coating chemistry promoting constant renewal of the surface layer (M&T Chemical Co. – International Paints Company, 1986).

The coating chemistry promotes constant renewal of the surface layer. The TBT moiety is chemically bonded through an ester linkage to a polymer backbone (e.g. TBT methacrylate copolymer). This bond is designed to be hydrolytically unstable under slightly alkaline conditions. This controlled release process has two major advantages: (1) release is governed by hydrolysis of the TBT group rather than dissolution of paint particles, and (2) the release rate is more effectively controlled (slowed down) by altering the polymer's water absorption characteristics.

Compared with Type 1 or 2 paints, these polymeric, film-forming resin coatings are also characterized by an initial higher release rate (higher in free-association paints than in copolymer paints) during the 'conditioning' period (approximately the first month after the freshly painted hull is placed in the water), followed by a constant low release rate of antifouling biocide (Fig. 1.8). A portion of this high initial release rate is due to unbound (free) TBT in the paint. Paint manufacturers have indicated that these levels should drop for the copolymer paints as better quality control practices are implemented.

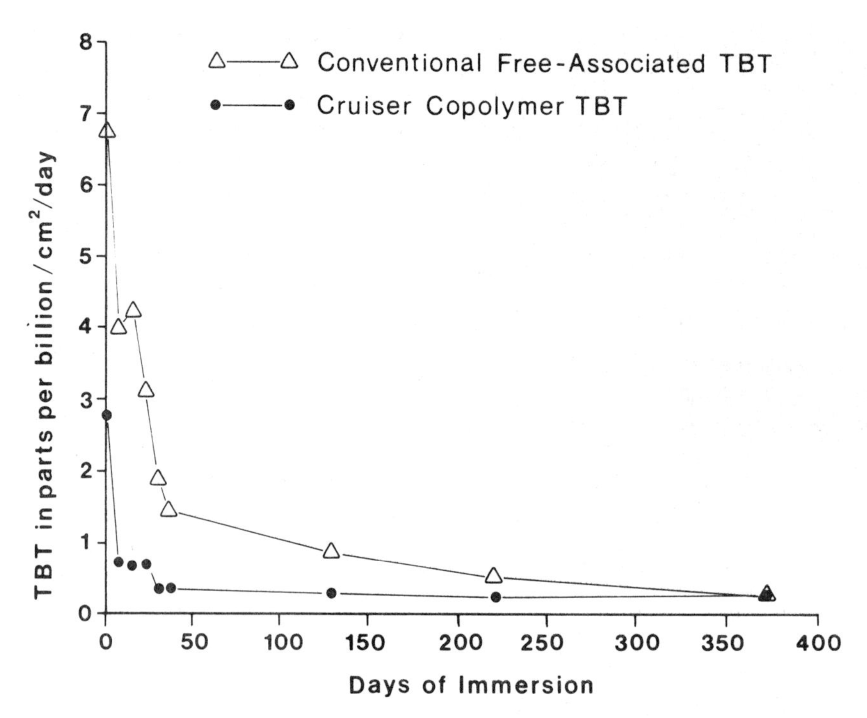

Figure 1.8 Comparison of measured release rates for free-association and copolymer TBT antifouling paints over 375 d. Except for an initial spike of rapid release of TBT from fresh paint, copolymer paints show a lower release rate than free-association paints (from Anderson and Dalley, 1986).

The controlled biocide release also gives the antifoulant paint a controlled life span (the thickness of the initial application of the paint determines the life span, see Fig. 1.9). Current data suggest that 5–7 y is the average life span of an application of copolymer TBT paint. Also, copolymer paints can be applied directly to the ship's hull surface without having to sandblast previous copolymer layers away. This reduces the shipyard costs of removing old paint, and, subsequently, less TBT is released into the environment as spent paint waste.

1.6. SUMMARY OF ENVIRONMENTAL EFFECTS

Tributyltin was found to induce gross physiological and anatomical deformations in non-target (non-fouling organisms) molluscs (for a summary of environmental effects see UK DoE, [1986]). The two most pronounced effects were (1) oyster shell thickening and growth anomalies, and (2) imposex in the common dogwhelk, a marine mollusc. For specific details and discussions on environmental effects, see Waldock *et al.* (Chapter 11), His (Chapter 12), Gibbs and Bryan (Chapter 13), Stewart *et al.*, Alzieu (1993), Bailey *et al.* (1995), Guolan and Yong (1995), Matthiessen *et al.* (1995), Ruiz *et al.* (1992) and Strben *et al.* (1995). An abbreviated summary of biological effects and measured concentrations in marine systems, before and after legislation, is presented in Fig. 1.10.

1.6.1. OYSTER SHELL THICKENING AND GROWTH ANOMALIES

In 1974, oyster planters from the east coast of England were finding abnormal shell growth in *Crassostrea gigas*, the Pacific oyster. Initial

Figure 1.9 A photograph showing the effectiveness of growth inhibition of a TBT copolymer antifouling paint on underwater colonization by marine fouling organisms on a paint test panel in which a plate containing a narrow section (bracketed) painted with a copolymer paint (Poly-Flo-4024) was submerged for 21 months in the ocean. Photograph courtesy of M&T Chemicals Inc., Woodbridge, New Jersey (Panel 8-83-7a, Notebook 4733-35).

field surveys were conducted and reported by the Ministry of Agriculture, Fisheries and Food Fisheries Laboratory at Burnham-on-Crouch in Essex (Key, 1975). When the shells were sectioned, this 'abnormal thickening' of the shell was seen to be caused by the formation of a series of shell cavities (Fig. 1.11). In 1976, Key and co-workers conducted field experiments at five growth sites around the coasts of England and Wales. These studies suggested that the abnormal thickening in the valves of *C. gigas* was related to growth rate, and that both shell thickening and growth rate were related to the concentration of fine inorganic particles in suspension in the water.

Concentrations of accumulated metals and pesticides in the oysters and associated with sediments did not correlate with the degree of thickening (Key *et al.*, 1976). Farming of the Pacific oyster was attempted unsuccessfully in many areas of Britain, particularly in the estuaries of the rivers Crouch and Roach on the east coast of England. These areas later were found to be highly contaminated with TBT (Waldock and Miller, 1983).

In 1976, French scientists observed this abnormal thickening of the shells of juvenile Pacific oysters in Arcachon Bay, located in Southern France (Alzieu *et al.*, 1980; Alzieu, 1981). The Pacific oyster had been introduced in 1968 to replace declining populations of the European flat oyster (Alzieu *et al.*, 1981–1982). Arcachon Bay is an elongated body of water with restricted circulation and has large numbers of marinas and docks with between 10 000 and 15 000 boats during the summer boating season (Alzieu, 1991).

In addition to anomalies in shell growth, the shells also took on a ball shape (Fig. 1.12, top photograph) owing to the formation of numerous 'chambers' (Fig. 1.12, bottom photograph). Early researchers noticed an association between marinas and boat moorings and shell malformations. The degree of shell thickness in different areas of the bay correlated well with the numbers of boats in an area, suggesting a relationship between the use of TBT antifouling paints and growth deformations. The early French and British studies found that the shell anomalies occurred when chambers formed between the calcified layers of the shell as the shell growth occurred (see Fig. 1.11 for impact on different-aged oysters). The shell thickening process is described by Heral *et al.* (1981) and Krampitz *et al.* (1983).

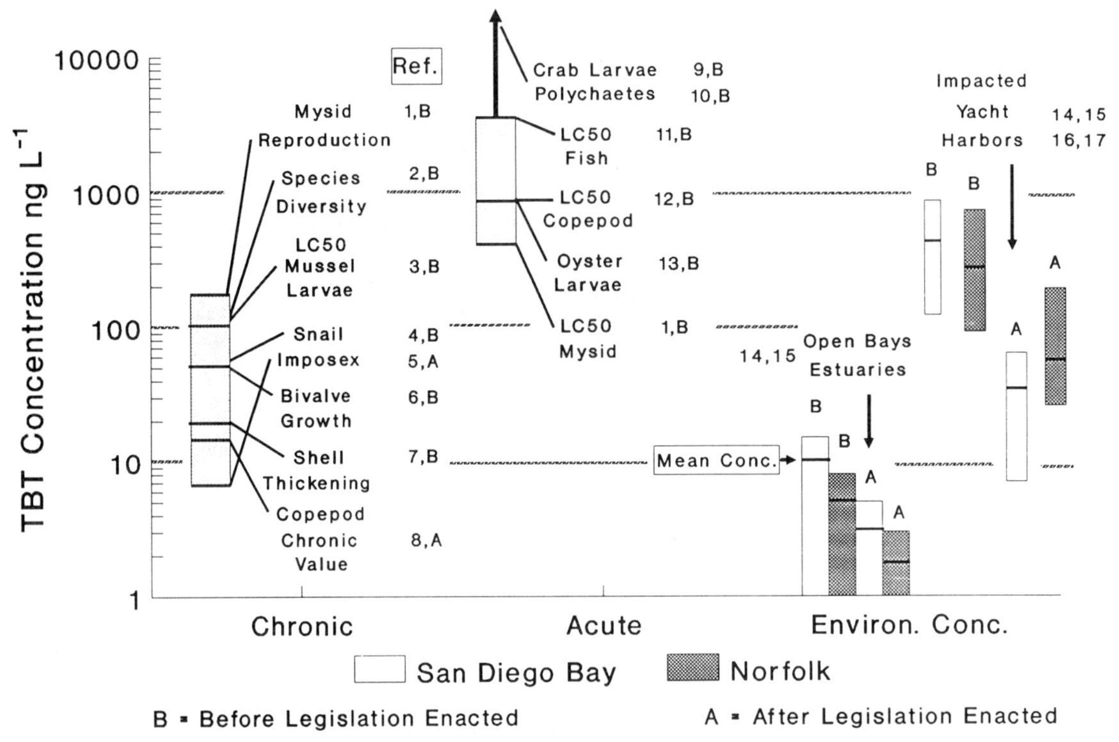

Figure 1.10 Summary of selected TBT toxicity data compared to measured TBT concentration in receiving water (updated and modified from Huggett *et al.*, 1992). Key to references cited: (1,B) Davidson *et al.*, 1986; (2,B) Henderson, 1986; (3,B) Beaumont and Budd, 1984; (4,B) Bryan *et al.*, 1986; (5,A) Gibbs *et al.*, 1988; (6,B) Laughlin *et al.*, 1987; (7,B) Thain *et al.*, 1987; (8,A) Bushong *et al.*, 1990; (9,B) Laughlin and French, 1980; (10,B) NAVSEA, 1984; (11,B) Ward *et al.*, 1981; (12,B) U'Ren, 1983; (13,B) Roberts, 1987; (14) Seligman *et al.*, 1990; (15) Valkirs *et al.*, 1991; (16) Valkirs *et al.*, 1986; (17) Federal Register, 1986. Reprinted with permission from *Environ. Sci. Technol.*

In the most acute malformations in Pacific oysters, the anomalous thickening of the oyster shell was more rapid than its lengthwise growth, and the oyster took on a ball shape. These malformations were not observed in the European flat oyster. Subsequent studies by Alzieu *et al.* (1989) found that there was a good agreement between TBT levels in seawater and the frequency of oyster shell anomalies monitored at the same time.

The influence of TBT on calcification processes in shells of *C. gigas* was confirmed by Thain and Waldock (1983); however, the mechanism of TBT action on calcification remains unknown today (Alzieu, 1991). Studies by Chagot *et al.* (1990) found that TBT would induce shell abnormalities by inhibiting calcification of *C. gigas* (i.e. the formation of gelatinous pockets or chambers between layers of the shell) at concentrations down to 2.0 ng l^{-1}, the lowest concentration they tested. Okoshi *et al.* (1987) have reported that sensitivity to chambering was not identical for two different genetic stocks of *C. gigas* bred in Japan.

A year later, in 1976, these anomalies were detected throughout Arcachon Bay. From 1976 to 1981, oyster spatfall was very low, and in some years failed completely, whereas it remained satisfactory in neighboring areas (Alzieu, 1986, 1991; His, Chapter 12). Some

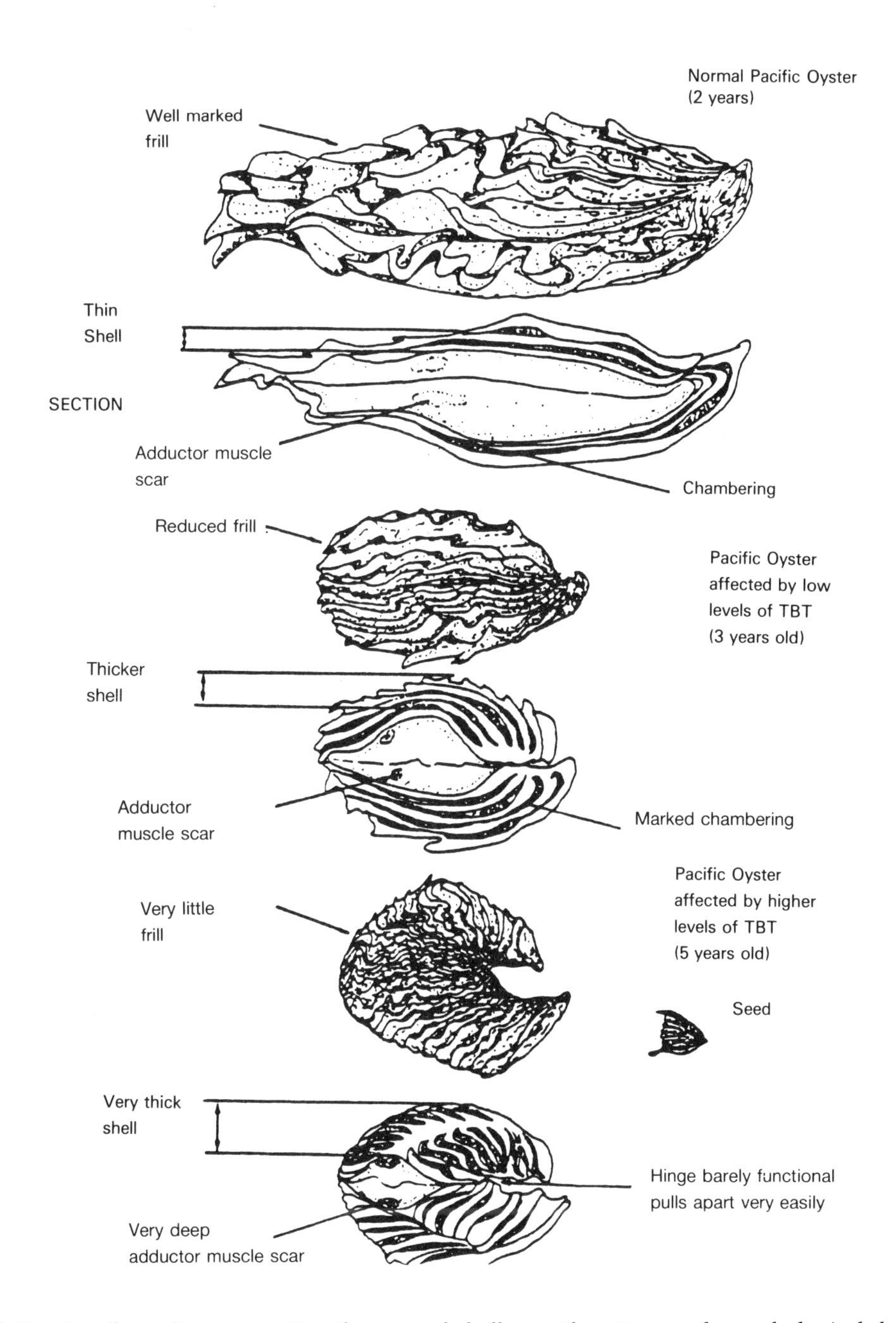

Figure 1.11 An illustration presenting the normal shell growth pattern and morphological shape and a cross section of shell layers of an adult 2-y-old *Crassostrea gigas*, which has not been exposed to TBT, followed by examples of older Pacific oysters with 'abnormal thickening' subsequent to exposure to increasing concentrations of TBT. Illustration courtesy of John E. Thain and Michael J. Waldock of the UK Ministry of Agriculture and Food, Burnham-on-Crouch Fisheries Laboratory.

Figure 1.12 Top, photograph of a Pacific oyster (*Crassostrea gigas*) collected from Coos Bay, Oregon, following exposure to TBT in the bay, illustrating the characteristic ball shape that inhibits shell closure. Bottom, cross sections of Coos Bay Pacific oyster (*Crassostrea gigas*) shells: the first shell has been collected from a control area that has had no exposure to TBT and does not exhibit chambering, whereas the second shell (lower one in the bottom photograph) has been collected in the same area as the top photograph and was exposed to high levels of TBT as evidenced by the degree of chambering. Photographs courtesy of Krystyna Wolniakowski, Department of Environmental Quality, State of Oregon.

areas of the bay had little or no natural oyster larvae settling on hard substrates, suggesting toxic effects in the early life stages. Over time, such high spat mortality would adversely affect oyster populations in the bay.

French researchers found that oyster larvae were surviving only a few days in the waters of the bay while growth from similar representative samples under laboratory conditions remained normal in clean water (Alzieu *et al.*, 1980). Subsequently, Alzieu *et al.* (1981–1982, 1986, 1989) produced abnormalities in oysters experimentally exposed to TBT compounds similar to the abnormalities found in field-reared oysters. The determination of the role of TBT in the anomalies was further established experimentally in the bay of Marennes-Oleron from the study of batches of anomaly-free oysters placed in a marina and in experimental tanks supplied with clean tidal seawater in which plates coated with TBT-based paint were placed (Heral *et al.*, 1981).

His *et al.* (1983, 1984) and His and Robert (1985) found that these anomalies were not tied to the fecundity of the breeding stocks, since parent oysters having reached sexual maturity, whether within the bay or outside, generated viable larvae with normal growth under the same laboratory conditions. The findings suggested that some type of contaminant was being introduced into the bay that was the responsible pollutant. This hypothesis was strengthened when French researchers found that the tendency for shell thickening could be reversed by moving the oysters to an area far removed from boating activity. His and Robert (1980, 1985) using the *C. gigas* embryo–larval bioassay demonstrated that concentrations ($<1.0\ \mu g\ l^{-1}$) of tributyltin acetate were also blocking the development of fertilized eggs (see His and Robert 1987a,b; His, Chapter 12).

Waldock *et al.* (1983) and Waldock and Thain (1983), in testing comparative accumulation and depuration of TBTO by *C. gigas* and *O. edulis* under the same environmental conditions, found 2- to 4-fold differences in tissue concentrations of TBTO. This suggested that the 3- to 9-fold differences found in the field earlier by Waldock and Miller (1983) were valid even though sample sizes varied and were limited. The data suggest that *C. gigas* accumulates TBT more readily than *O. edulis*, the European flat oyster. In a subsequent study by Thain and Waldock (1985) with a variety of commercially important bivalve spat that were exposed for 7 weeks to concentrations of organotin leachates, they found that, at $0.24\ \mu g\ l^{-1}$ TBT, growth of *C. gigas*, *Mytilus edulis*, and *Venerupis decussata* was severely reduced, but that, in *Ostrea edulis* and *V. semidecussata*, weight increase was similar to control values. Thain (1986) further delineated effects on reproduction, growth, and survival.

In the United States, examples of Pacific oysters with malformations similar to those in Britain and France were found in 1987 in the South Slough National Estuarine Research Reserve near Charleston, Oregon, an arm of Coos Bay, Oregon (Wolniakowski *et al.* 1987), and in San Diego Bay (Stephenson *et al.*, 1986). South Slough is adjacent to a boatyard where improper boatyard practices with TBT- based antifouling paints were believed to be a major contributor of TBT to South Slough at that time. Seawater concentrations of TBT were up to $14\ ng\ l^{-1}$, and oyster tissue concentrations ranged from 50 to $102\ \mu g\ kg^{-1}$ (wet) within South Slough in Coos Bay. The correlation of TBT concentrations and shell thickening, however, is not clear since the TBT levels were measured well after the shell thickening had occurred.

1.6.2. IMPOSEX IN THE COMMON DOGWHELK

More extreme deformations occur in the common dogwhelk (*Nucella lapillus*), a species of thick-shelled snails found around the southwest peninsula of England. These deformations are referred to as imposex, which is the development of male character-

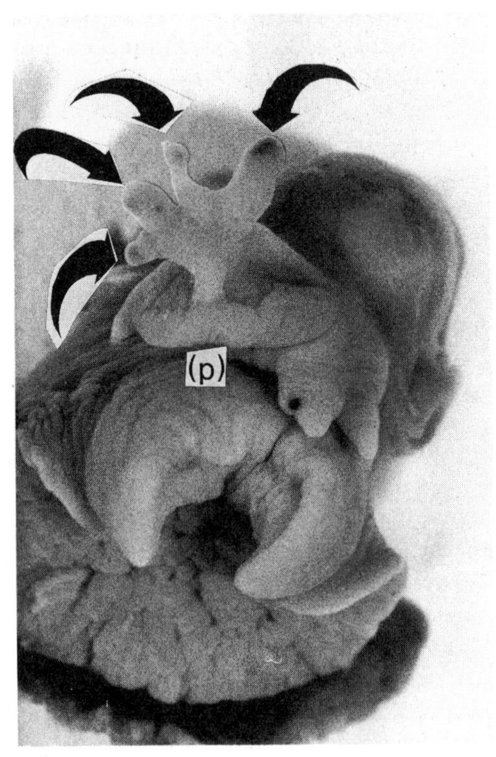

Figure 1.13 Photograph of a female dogwhelk (*Nucella lapillus*) with induced imposex showing development of large penis (p) behind the right tentacle. Further malformation is shown by the growth of four accessory tentacles (arrowed). This example was produced in 77 d after the shell spire had been coated with TBT antifouling paint. Photograph courtesy of Dr P.E. Gibbs and Dr G.W. Bryan of the UK Plymouth Marine Laboratory and the *Marine Pollution Bulletin*.

istics, notably a penis and a vas deferens (sperm duct) by females (Fig. 1.13). This phenomenon is also referred to as pseudohermaphroditism. In this condition, the female opening (vulva) is occluded by overgrowth of vas deferens tissue. This tissue growth promotes blockage of the pallial oviduct and prevents the release of egg capsules, rendering the female sterile.

Common dogwhelk snail populations are found on rocky shores on both sides of the North Atlantic (northern Russia to Portugal and southern Newfoundland to New York). Imposex in populations off the southwest peninsula of England has been studied by Bryan *et al.* (1986, 1987a,b), Gibbs and Bryan (1986, 1987 and Chapter 13), Proud (1994), Bailey *et al.* (1995) and Axiak *et al.* (1995). These researchers found a high occurrence of imposex close to centers of boating and shipping activity. The occurrence of imposex correlated significantly with tin concentrations in dogwhelks (up to 2.0 $\mu g\ gm^{-1}$ of dry tissue). Laboratory studies have confirmed that exposure to 0.02 $\mu g\ l^{-1}$ of tin leached from a TBT antifouling paint will readily induce imposex (Gibbs and Bryan, Chapter 13). Their studies suggest that seawater concentrations of Sn as low as 2 $ng\ l^{-1}$ or 5 ng^{-1} as TBT can initiate imposex in *N. lapillus* populations (Bryan *et al.*, 1986, 1988; Gibbs and Bryan, 1986; Gibbs *et al.*, 1987, 1988, 1991).

This sensitivity is in part due to critical exposure pathways to *N. lapillus* because of their carnivorous feeding habits; therefore dogwhelks accumulate concentrations of TBT higher than that of the water surrounding them. TBT bioaccumulation levels in dogwhelks can be 1000 times more than the concentration in the surrounding water (Gibbs and Bryan, 1986).

In the case of the dogwhelk, imposex has a potential population significance: those populations in which individuals frequently exhibit imposex show signs of decline. In these populations, fewer females occur than would be expected; and juveniles and deposited egg capsules are scarce or absent, indicating a low reproductive capacity. Female dogwhelks exhibiting imposex had oviducts clogged with decomposing eggs that could not be released because the newly formed male reproductive tissues blocked the oviducts (Gibbs and Bryan, Chapter 13).

In addition to the low-level effects of TBT on dogwhelks and oysters discussed above, numerous laboratory studies have identified acute and chronic effects from exposure to TBT concentrations of 50 $ng\ l^{-1}$ to microgram-per-liter levels (Hall and Bushong, Chapter 9; Laughlin *et al.*, Chapter 10).

1.7. CONCLUSIONS

Antifouling coatings and paints are needed, and at present alternatives to them that rely on non-toxic compounds are in their infancy. Efforts to use TBT-based antifouling paints safely require accurate and comprehensive environmental and toxicological data. TBT-based organotin-based coatings are toxic to sensitive nontarget organisms at levels that approach the analytical detection limits; therefore, researchers are limited in their ability to study the toxic effects. The extensive research (>20 y) from many nations on the environmental fate and effects of organotins summarized in this book has expanded the knowledge about organotin compounds in the marine environment and has provided regulatory agencies with much of the data necessary to develop environmental control strategies.

The economic benefits from the use of TBT antifouling paints present policy and decision makers with the dilemma of needing to apply today's advanced technologies to produce modern chemicals that are needed for a wide range of industrial, agricultural, and human health applications. Organotin compounds are among the first chemicals to have so many applications in different areas with so many economic and human health benefits. Today's scientists with multidisciplinary approaches must focus on ways to utilize the combined knowledge about physical, chemical, and biological processes with the environmental fate and effects of these modern chemicals to develop ways to maximize their benefits and minimize their external (environmental and social) costs. For the TBT-based antifouling coatings, the development of the copolymer antifouling paints was a long step in the right direction.

It must not be forgotten that organotin compounds were extensively used for several decades and were 'registered pesticides' by government authorities all over the world for use in antifouling paints for over a decade before environmental scientists began to realize that they were affecting oysters. Until more is known about the organotin compounds, evaluation of the effectiveness of innovations, such as new paint formulations or matrices that have lower release rates, will be tentative and inconclusive. Organotins are a 'lost leader' in the realm of chemistry today because they are a future view of new high-technology chemicals yet to be developed (Frederick E. Brinckman, National Institute of Standards and Technologies, personal communication). As such, their study now offers tremendous opportunities to gain knowledge and understanding for future advances in chemistry. Without a full understanding of the environmental mechanisms and processes that control their distribution, critical pathways of toxicity, and biological uptake mechanisms, the high technology that created TBT antifouling paints cannot be unleashed to make them more effective and safer.

ACKNOWLEDGMENTS

The authors thank Joyce G. Nuttall, Nuttall Editing, for her contributions to the editing of this manuscript and the entire book, and Carol Dooley and Aldis Valkirs, NCCOSC, RDT & E Division, for their contributions to the preparation of this chapter. Ms Lisa Haderlie Baker, Lawrence Hall of Science, University of California at Berkeley, is to be thanked for permission to use her illustrations on fouling organisms. We also thank Colin D. Anderson and Richard Dalley, International Paints, for permission to reproduce their illustrations on release mechanisms for antifouling paints. We also thank Michael Harriott for his assistance in developing the introductory materials on the chemistry of organotin compounds. We would have liked to have had a special chapter dedicated just to the chemistry of organotin compounds; however, since the book focuses on the environmental fate and effects of organotins with many chapters, we do not have the space. We also think several extensive books could be written on the chemistry of organotins.

REFERENCES

Alzieu, Cl. 1981. Evaluation des risques dus à l'emploi des peintures anti-salissures dans les zones conchylicoles. Anomalies des calcification. Thème surveillance continue mécanismes d'actions des polluants. Addendum du rapport du 15 Juin 1981. Institut Scientifique et Technique des Peches Maritimes, Nantes, le 15 December 1981.

Alzieu, Cl. 1986. TBT detrimental effects on oyster culture in France – evolution since antifouling paint regulation. Proceedings of the Oceans '86 Organotin Symposium, Vol. 4, Marine Technology Society, Washington, DC, pp. 1130–1134.

Alzieu, Cl. 1991. Environmental problems caused by TBT in France: assessment, regulations, prospects. *Mar. Environ. Res.*, **32**, 7–17.

Alzieu, Cl. 1993. Antifouling systems and environment. *Oebalia*, **19** (Supplement), 423–429.

Alzieu, Cl. and J.E. Portman. 1984. Effect of tributyltin on the culture of *Crassostrea gigas* and other species. Proceedings of the 50th Annual Shellfish Conference, pp. 87–104.

Alzieu, Cl., Y. Thibaud, M. Heral and B. Boutier. 1980. Evaluation des risques dus à l'emploi de peintures anti-salissures dan les zones conchylicoles. *Rev. Trav. Inst. Pêches Marit.* **44**(4), 301–349.

Alzieu, Cl., M. Heral, Y. Thibaud, M.J. Dardignac and M. Feuillet. 1981–1982. Influence des peintures anti-salissures à base d'organostanniques sur la calcification de la coquille de l'huître *Crassostrea gigas*. *Rev. Trav. Inst. Peches Marit.*, **45**(2), 101–116.

Alzieu, Cl., J. Sanjuan, J.P. Deltreil and M. Borel. 1986. Tin contamination in Arcachon Bay: effects on oyster shell anomalies. *Mar. Pollut. Bull.*, **17**, 494–498.

Alzieu, Cl., J. Sanjuan, P. Michell, M. Borel and J.P. Dreno. 1989. Monitoring and assessment of butyltins in Atlantic coastal waters. *Mar. Pollut. Bull.*, **20**(1), 22–26.

Anderson, C.D. and R. Dalley. 1986. Use of organotins in antifouling paints. Proceedings of the Oceans '86 Organotin Symposium, Vol. 4, Marine Technology Society, Washington, DC, pp. 1108–1113.

Axiak, V., A.J. Vella, D. Micallef, P. Chircop and B. Mintoff. 1995. Imposex in *Hexaples trunculus* (Gastropoda: Muricidae): First results from biomonitoring of tributyltin contamination in the Mediterranean. *Mar. Biol.*, **121**(4), 423–429.

Bailey, W.A. 1986. Assessing impacts of organotin paint use. Proceedings of the Oceans '86 Organotin Symposium, Vol. 4, Marine Technology Society, Washington, DC, 1101–1107.

Bailey, S.K., I.M. Davies and M.J.C. Harding. 1995. Tributyltin contamination and its impact on *Nucella lapillus* populations. *Proc. Roy. Soc. Edinburgh B*, **103**, 113–126.

Beaumont, A.R. and M.D. Budd. 1984. High mortality of the larvae of the common mussel at low concentrations of tributyltin. pp. *Mar. Pollut. Bull.*, **15**, 402–405.

Brinckman, F.E. 1981. Environmental organotin chemistry today: experiences in the field and laboratory. *J. Organometal. Chem.*, **12**, 343–376.

Bryan, G.W., P.E. Gibbs, G.L. Hummerstone and G.R. Burt. 1986. The decline of the gastropod *Nucella lapillus* around southwest England: evidence for the effect of tributyltin from antifouling paints. *J. Mar. Biol. Ass. UK*, **66**, 611–640.

Bryan, G.W., P.E. Gibbs, G.R. Burt and L.G. Hummerstone. 1987a. The effects of tributyltin (TBT) accumulation on adult dog-whelks, *Nucella lapillus*: long-term field and laboratory experiments. *J. Mar. Biol. Ass. UK*, **66**, 525–544.

Bryan, G.W., P.E. Gibbs, L.G. Hummerstone and G.R. Burt. 1987b. Copper, zinc, and organotin as long-term factors governing the distribution of organisms in the Fal Estuary in southwest England. *Estuaries*, **10**(3), 208–219.

Bryan, G.W., P.E. Gibbs and G.R. Burt. 1988. A comparison of the effectiveness of tri-n-butyltin chloride and five other organotin compounds in promoting the development of imposex in the dog-whelk, *Nucella lapillus*. *J. Mar. Biol. Ass. UK*, **68**, 733–744.

Bushong, S.J., M.C. Ziegenfuss, M.A. Unger and L.W. Hall, Jr. Chronic tributyltin toxicity experiments with the Chesapeake Bay copepod, *Aractia tonsa*. *Environ. Toxicol. Chem.* **9**, 359–366.

Chagot, D., Cl. Alzieu, J. Sanjuan and H. Grizel. 1990. Sublethal and histopathological effects of trace levels of tributyltin fluoride on adult oysters *Crassostrea gigas*. *Aquat. Living Resour.*, **3**, 121–130.

Champ, M.A. and F.L. Lowenstein. 1987. TBT: the dilemma of hi-technology antifouling paints. *Oceanus*, **30**(3), 69–77.

Champ, M.A. and D.F. Bleil. 1988. Research needs concerning organotin compounds used in antifouling paints in coastal environments. NOAA Technical Report published by the Office of the Chief Scientist, National Ocean Pollution Office. 5 parts plus appendices.

Davidson, B.M., A.O. Valkirs and P.F. Seligman. 1986. Acute and chronic effects of tributyltin on mysid, *Acanthomysis sculpta* (Crustacea, Mysidacea). Proceedings of the Oceans '86 Organotin Symposium, Vol. 4, Marine Technology Society, Washington, DC, pp. 1219–1225.

Duncan, J. 1974. The ideal molluscicide. In: *Molluscicides in shistosomiasis control*, T.C. Cheng (Ed.). Academic Press, New York, pp. 249–258.

Duncan, J. 1980. The toxicology of molluscicides: the organotins. *Pharmacol. Therapeut.*, **10**, 407–429.

Eastin, K.E. 1987. Tributyltin paint – the navy perspective. *Sea Technology*, **28**(3), 69.

Evans, C.J. and P.J. Smith. 1975. Organotin-based antifouling systems. *J. Oil Col. Chem. Assoc.*, **58**, 160–168.

Gibbs, P.E. and G.W. Bryan. 1986. Reproductive failure in populations of the dog-whelk, *Nucella lapillus*, caused by imposex induced by tributyltin from antifouling paints. *J. Mar. Biol. Ass. UK*, **66**, 767–777.

Gibbs, P.E. and G.W. Bryan. 1987. Tributyltin paints and the demise of the dog-whelk *Nucella lapillus*. Proceedings of the Oceans '87 International, Organotin Symposium, Vol. 4, Marine Technology Society, Washington, DC, pp. 1482–1487.

Gibbs, P.E., G.W. Bryan, P.L. Pascoe and G.R. Burt. 1987. The use of the dog-whelk, *Nucella lapillus*, as an indicator of tributyltin (TBT) contamination. *J. Mar. Biol. Ass. UK*, **67**, 507–523.

Gibbs, P.E., P.L. Pascoe and G.R. Burt. 1988. Sex change in the female dog-whelk, *Nucella lapillus*, induced by tributyltin from antifouling paints. *J. Mar. Biol. Ass. UK*, **68**(4), 715–731.

Gibbs, P.E., G.W. Bryan and P.L. Pascoe. 1991. TBT-induced imposex in the dogwhelk, *Nucella lapillus*: geographical uniformity of the response and effects. *Mar. Environ. Res.*, **32**, 1–5.

Guard, H.E., W.M. Coleman and A.B. Colbert. 1982. Speciation of tributyltin compounds in seawater and estuarine sediments. Abstract, 185th National Meeting, Division of Environmental Chemists. American Chemical Society, Las Vegas, Nevada.

Guolan, H. and W. Yong. 1995. Effects of tributyltin chloride on marine bivalve mussels. *Water Res.*, **29**(8), 1877–1884.

Henderson, R.S. 1986. Effects of organotin antifouling leachate on Pearl Harbor organisms: a site specific flow through bioassay system. Proceedings of the Oceans '86 Organotin Symposium, Vol. 4, Marine Technology Society, Washington, DC, pp. 1226–1233.

Heral, M., J.P. Berthome, E. Polanco-Torres, C. Alzieu, J.M. Deslou-Paoli, D. Razet and J. de Garnier. 1981. Anomalies de croissance de la coquille de *Crassostrea gigas* dans le bassin de Marennes-Oleron. Bilan de trois années d'observation. ICES C.M. Paper (K), 31 pp.

His, E. and R. Robert. 1980. Action d'un sel organo-metallique, l'acetate de tributyle-etain sur les oeufs et les larvae de *Crassostrea gigas* (Thurnberg). ICES C.M. Paper (F27), 10 pp.

His, E. and R. Robert. 1985. Developpement des veligeres de *Crassostrea gigas* dans le Bassin d'Arcachon. Etudes sur les mortalites larvaires. *Revue Trav. Inst. Peches Marit.*, **47**, 63–88.

His, E. and R. Robert. 1987a. Comparative effects of two antifouling paints on the oyster *Crassostrea gigas*. *Mar. Biol.*, **95**, 83–86.

His, E. and R. Robert, 1987b. Impact des facteurs anthropiques sur le recrutement de l'huitre: example du Bassin d'Arcachon. *Oceanis*, **13**, 317–335.

His, E., D. Maurer and R. Robert. 1983. Estimation de la teneur en acetate de tributyltin dans l'eau de mer, par une methode biologique. *J. Mollusc. Stud. Suppl.*, **12A**, 60–68.

His, E., D. Maurer and R. Robert. 1984, 1986. Observations complementaires sur les causes possibles des anomalies de la Bassin D'Arcachon. *Revue Trav. Inst. Pêches Marit.*, **48**, 45–54.

Huggett, R.J., M.A. Unger, P.F. Seligman and A.O. Valkirs. 1992. The marine biocide tributyltin. *Environ. Sci. Technol.*, **26**(2), 232–237.

Key, D. 1975. Investigations into the growth characteristics of the shell of *Crassostrea gigas*. ICES C.M. Paper (K28).

Key, D., R.S. Nunny, P.E. Davidson and M.A. Leonard. 1976. Abnormal shell growth in the Pacific oyster (*Crassostrea gigas*): some preliminary results from experiments undertaken in 1975. ICES Paper (K11), 12 pp.

Krampitz, G., H. Drolshagen and J.P. Deltreil. 1983. Soluble matrix components in malformed oyster shells. *Experientia*, **39**, 1105–1106.

Kuballa, J., R.-D. Wilken, E. Jantzen, K.K. Kwan and Y.K. Chau. 1995. Speciation and genotoxicity of butyltin compounds. *Analyst*, **120**(3), 667–673.

Kuch, P.J. 1986. Survey of antifouling paint use at boatyards and shipyards. Proceedings of the Oceans '86 Organotin Symposium, Vol. 4, Marine Technology Society, Washington, DC, pp. 1114–1116.

Laughlin, R.B. Jr and W.J. French. 1980. Comparative study of the acute toxicity of a homologous series of trialkyltins to larval shore crabs, *Hemigrapsus nudus*, and the lobster, *Homarus americanus*. *B. Environ. Contam. Tox.*, **25**, 802–809.

Laughlin, R.B. Jr, R.B. Johannesen, W. French, H.E. Guard and F.E. Brinckman. 1985. Structure–activity relationships for organotin compounds. *Environ. Tox. Chem.*, **4**, 343–351.

Laughlin, R.B. Jr, P. Pendoley and R.G. Gustafson. 1987. Sublethal effects of tributyltin on the hard shell clam, *Mercenaria mercenaria*. Proceedings of the Oceans '87 International Organotin Symposium, Vol. 4, Marine Technology Society, Washington, DC, pp. 1494–1498.

Lewis, J.A. and I.J. Baran. 1993. Biocide release from antifouling coating. *Oeballa*, **19** (Supplement), 449–456.

Matthiessen, P., R. Waldock, J.E. Thain, M.E. Waite and S. Scropte-Howe. 1995. Changes in periwinkle (*Littorina littorea*) populations following the ban on TBT-based antifoulings on small boats in the United Kingdom. *Ecotoxicol. Environ. Safety*, **30**(2), 180–194.

M&T Chemical Co. (International Paints). 1986. US EPA Briefing Mechanics of Antifouling. EPA Office of Pesticide Programs, Washington, DC, 26 pp.

Milne, A. 1990. Roughness and drag from the marine chemist's viewpoint. Proceedings of the International Workshop on Marine Roughness and Drag. Royal Institution of Naval Architects, London.

NAVSEA (US Naval Sea Systems Command). 1984. Environmental assessment of fleetwide use of organotin antifouling paint. NAVSEA, Washington, DC, 128 pp.

NAVSEA (US Naval Sea Systems Command). 1986. Organotin antifouling paint: US Navy's needs, benefits, and ecological research. A Report to Congress. NAVSEA, Washington, DC, 38 pp. + appendix.

Okoshi, K., K. Mori and T. Nomura. 1987. Characteristics of shell chamber formation between the two local races in the Japanese oyster, *Crassostrea gigas*. *Aquaculture*, **67**, 313–320.

Proud, S.V. 1994. Tributyltin pollution and the bioindicator *Nucella lapillus*: population recovery and community level responses. Thesis, University of Liverpool, Port Erin, Isle of Man, xviii, 244 pp.

Ricketts, Rear Admiral M.V. 1987. The effects of the chemical tributyltin (TBT) on the marine environment. Hearing Record. Senate Subcommittee on Environmental Protection of the Committee on Environment and Public Works. April 29. S.HGR 100-89. US Government Printing Office, 73-832, Washington, DC, pp. 28–30 (oral) and pp. 76–84 (written).

Roberts, M.H. Jr. 1987. Acute toxicity of tributyltin chloride to embryos and larvae of two bivalve mollusks, *Crassostrea virginica* and *Mercenaria mercenaria*. *B. Environ. Contam. Tox.*, **39**, 1012–1019.

Ruiz, J.M., G.W. Bryan and P.E. Gibbs. 1995. Effects of tributyltin (TBT) exposure on the veliger larvae development of the bivalve *Scrobicularia plana* (da Costa). *J. Experimental Mar. Biol. Ecol.*, **186**(1), 53–63.

Saxena, A.K. 1987. Organotin compounds: toxicity and biomedicinal applications. *Appl. Organometal. Chem.*, **1**, 39–56.

Seligman, P.F., J.G. Grovhoug, R.L. Fransham, B. Davidson and A.O. Valkirs. 1990. US Navy Statutory Monitoring of Tributyltin in Selected US Harbors, Annual Report 1989. Naval Ocean Systems Center Technical Report No. 1346. Naval Ocean Systems Center, San Diego, California, 32 pp.

Stephenson, M.D., D.R. Smith, J. Goetzl, G. Ichikawa and M. Martin. 1986. Growth abnormalities in mussels and oysters from areas with high levels of tributyltin in San Diego Bay. Proceedings of the Oceans '86 Organotin Symposium, Vol. 4, Marine Technology Society, Washington, DC, pp. 1245–1251.

Stewart, C., S.J. de Mora, M.R.L. Jones and M.C. Miller. 1992. Imposex in New Zealand neogastropods. *Mar. Pollut. Bull.*, **24**(4), 204–209.

Strben, E., U. Schulte-Oehlmann, P. Fioroni and J. Oehlmann. 1995. A comparative method for easy assessment of coastal TBT pollution by the degree of imposex in prosobranch species. *Haliotis*, **24**, 1–12.

Thain, J.E. 1983. The acute toxicity of bis (tributyl tin) oxide to the adults and larvae of some marine organisms. ICES C.M. Paper (E13), 5 pp.

Thain, J.E. 1986. Toxicity of tributyltin to bivalves: effects on reproduction, growth and survival. Proceedings of the Oceans '86 Organotin Symposium, Vol. 4, Marine Technology Society, Washington, DC, pp. 1306–1313.

Thain, J.E. and M.J. Waldock. 1983. The effect of suspended sediment and bis (tributyltin) oxide on the growth of *Crassostrea gigas* spat. ICES C.M. Paper (E10), 10 pp.

Thain, J.E. and M.J. Waldock. 1985. The growth of bivalve spats exposed to organotin leachates from antifouling paints. ICES C.M. Paper (E28), 10 pp.

Thain, J.E., M.J. Waldock and M.E. Waite. 1987. Toxicity and degradation studies of tributyltin (TBT) and dibutyltin (DBT) in the aquatic environment. Proceedings of the Oceans '87 International Organotin Symposium, Vol. 4, Halifax, Nova Scotia, 28 September – 1 October 1987, Marine Technology Society, Washington, DC, pp. 1398–1404.

Thayer, J.S. 1984. *Organometallic Compounds and Living Organisms*. Academic Press, New York.

UK Department of the Environment. 1986. Organotin in antifouling paints, environmental considerations. Her Majesty's Stationery Office, London, England. Pollution Paper No. 25. 82 pp.

U'Ren, S.C. 1983. Acute toxicity of tributyltin oxide to a marine copepod. *Mar. Pollut. Bull.*, **14**(8), 303–306.

US Federal Register. 1986. **51**(5), 778–779.

US Senate. 1988. Organotin Antifouling Paint Control Act of 1987. Senate Amendments. Congressional Record – Senate. US Government Printing Office, Washington, DC, April 18, S. 4204–4211.

Valkirs, A.O., P.F. Seligman, P.M. Stang, V. Homer, S.H. Lieberman, G. Vafa and C.A. Dooley. 1986. Measurement of butyltin compounds in San Diego Bay. *Mar. Pollut. Bull.*, **17**, 319–324.

Valkirs, A.O., B.M. Davidson, L.L. Kear, R.L. Fransham, J.G. Grovhoug and P.F. Seligman. 1991. Long-term monitoring of tributyltin in San Diego Bay, California. *Mar. Environ. Res.*, **32**, 151–167.

van der Kerk, G.J.M. 1975. Present status of the use of organotin compounds (in German). *Chem. Ztg*, **99**, 26–32.

van der Kerk, G.J.M. and J.G.A. Luijten. 1954. Investigations on organo-tin compounds III. The biocidal properties of organo-tin compounds. *J. Appl. Chem.* (4 June), 314–319.

Waldock, M.J. and D. Miller. 1983. Determination of total and tributyl tin in seawater and oysters in areas of high pleasure craft activity. ICES C.M. Paper (E12), 17 pp.

Waldock, M.J. and J.E. Thain. 1983. Shell thickening in *Crassostrea gigas*: organotin antifouling or sediment induced? *Mar. Pollut. Bull.*, **14**(11), 411–415.

Waldock, M.J., J.E. Thain and D. Miller. 1983. The accumulation and depuration of bis (tributyltin) oxide in oysters: a comparison between the Pacific oyster (*Crassostrea gigas*) and European flat oyster (*Ostrea edulis*). ICES C.M. Paper (E52), 8 pp.

Walsh, G.E. 1986. Organotin toxicity studies conducted with selected organisms at EPA's Gulf Breeze Lab. Proceedings of the Oceans '86 Organotin Symposium, Vol. 4, Marine Technology Society, Washington, DC, pp. 1210–1212.

Ward, G.S., G.C. Cramm, P.R. Parrish, H. Trachman and A. Slesinger. 1981. Bioaccumulation and chronic toxicity of tributyltin oxide: tests with salt water fish. In: *Aquatic Toxicology and Hazard Assessment*, D.R. Branson and K.L. Dickson (Eds). American Society for Testing and Materials. ASTM STP737. Philadelphia, PA, pp. 183–200.

Wolniakowski, K., M.D. Stephenson and G. Ichikawa, G. 1987. Tributyltin concentrations and Pacific oyster deformations in Coos Bay, Oregon. Proceedings of the Oceans '87 International Organotin Symposium, Vol. 4, Marine Technology Society, Washington, DC, pp. 1438–1442.

EUROPEAN POLICY AND REGULATORY ACTION FOR ORGANOTIN-BASED ANTIFOULING PAINTS 2

R. Abel

Toxic Substances Division, Department of the Environment, Marsham Street, London SW1P 3PY, UK

Organotin. Edited by M.A. Champ and P.F. Seligman. Published in 1996 by Chapman & Hall, London. ISBN 0 412 58240 6

ABSTRACT

Antifouling paints incorporating tributyltin (TBT) compounds were introduced in Europe in the 1960s and within a decade were in widespread use. The French shellfish industry experienced problems with the cultivation of oysters in the 1970s, and scientific investigations suggested these were caused by TBT. Many of the problems were resolved when the use of TBT paints on small boats was banned in 1982. Similar effects were noted in the United Kingdom, and controls were first introduced to limit the TBT content of paints. These controls proved insufficient to protect shell fisheries and the general environment, and new problem areas were seen in fresh waters and where TBT was used on nets in salmon farms. The United Kingdom therefore banned the use of TBT, except on vessels >25 m, in 1987. A number of European countries have taken some action, or are considering doing so, and a European Community Directive will harmonize the regulations of member states by 1991. Scientists and policy makers are still undecided on whether TBT antifoulants used on large vessels constitute a risk to the environment during normal operations; current research is addressing this issue. Unregulated drydock practices involving cleaning and repainting can, however, lead to discharges containing unacceptably high concentrations of TBT; steps are therefore being taken to control these discharges. The need for further action is being considered within the international conventions (Paris, Helsinki, and Barcelona) and in the International Maritime Organization.

2.1. INTRODUCTION

Underwater surfaces provide good substrata for the attachment and development of a community of 'fouling' organisms that are of significance for a number of reasons. The shipworm (*Teredo* spp.), for instance, can cause structural damage by boring into untreated wooden boats and harbor piers. Barnacles and weed growth, as well as being unsightly, will roughen a hull surface, increase drag, and affect performance by reducing speed and increasing fuel costs. Weed growth on net enclosures of fish farms can reduce water circulation and lead to deoxygenated conditions. Removal of such encrusting growths on boats and nets can be expensive and time consuming.

Boat builders used lead and copper sheathing in previous centuries for control of boring and fouling organisms, but really effective remedies have been available only for ~100 y since the introduction of paints containing biocides such as cuprous oxide (Stebbing, 1985). Many different toxic substances have been tried, but by far the most successful have been the tributyltin (TBT) compounds. First added as boosters to cuprous oxide paints in the 1960s, they became increasingly popular during the next decade as the main or sole biocide. They achieved a very substantial penetration of the pleasure boat market in the United Kingdom, and at the present time many commercial and naval vessels are treated with TBT. The advantages in terms of effectiveness, operational efficiency, military preparedness, and maintenance costs are considerable. Ludgate (1987) has estimated for the United States that a commercial vessel that extends its time between drydockings from 24 to 30 months makes annualized savings of $50 000 (US) per year, while savings in fuel can be worth $200 000–400 000 (US) per year.

Three main types of paint were available by the 1980s (Anderson and Dalley, 1986; Champ and Seligman, Chapter 1). In conventional 'free-association' paints, the TBT is physically dispersed in a hard matrix and released by diffusion. Ablative paints of this type have also been developed, in which the spent surface layer dissolves away to expose a new active layer from which further leaching can take place. In both types, however, it is difficult to control the release rate precisely over the life of the paint – generally 1–2 y. A real breakthrough was the development of self–polishing 'copolymer' formulations. In these, the TBT is chemically bound within a methacrylate matrix and released by a chemical reaction with the water at the surface of the paint. Removal of the TBT cation changes the properties of the surface film, admitting water and allowing the exposed backbone of the copolymer chain to dissolve. The polishing action gives a much lower and more constant release rate over the life of the paint, which can be extended up to 5 y by applying a sufficiently thick layer.

Much emphasis is placed nowadays on prior assessment of chemicals before they reach the marketplace and, increasingly, on applying precautionary principles to environmental protection matters (Champ and Wade, Chapter 3). However, at the time that TBT-based paints were introduced in the United Kingdom, there was no *formal* screening procedure for new chemicals. The toxicity of TBT to fouling organisms was appreciated, but its persistence, the rapid growth in its use, and the possibility that concentrations would be maintained in the marine environment that could affect non-target species were not foreseen. It was not until the late 1970s that French and British scientists noted problems with the growth of Pacific oysters (*Crassostrea gigas*), an important cultivated species (His, Chapter 12). The investigation into the cause of these problems led ultimately to the banning of TBT use on small boats in France, the United Kingdom and Ireland, and to the European Community Directive that will harmonize European legislation.

Holdgate (1979) has drawn attention to the central role of surveillance and monitoring in the pattern of activities leading from the recognition of an environmental problem to its solution. His diagramatic representation of the process, like that of Stebbing and Harris (1988), recognizes feedback mechanisms among policy, legislation, and monitoring. The concept of the 'Action Cycle' (Department of the Environment, 1982), developed by the UK's Steering Committee on Environmental Monitoring and Assessment, captures the idea in idealized and simplified form (Fig. 2.1).

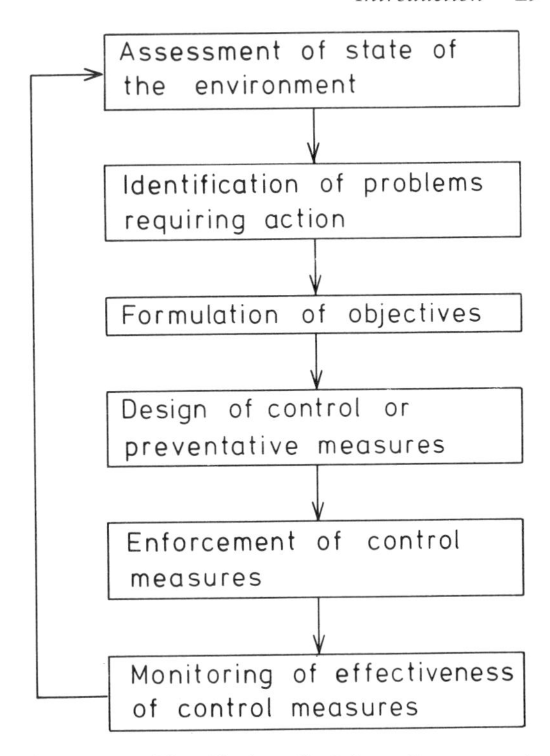

Figure 2.1 The 'Action Cycle', a diagrammatic representation of feedback mechanisms among policy, legislation, and monitoring (after Department of the Environment, 1982).

The first stage is an assessment of the state of the environment, which might be based on the results of a general surveillance program or on specific chemical or biological monitoring. There might be a general concern that things were not as they should be, or more concrete evidence that environmental conditions had deteriorated. Identification of a problem should lead to a search for its causes and to decisions on a policy level about the extent to which improvements are desirable and can be achieved. This stage can be fraught with difficulties since it may be necessary to reconcile the views and interests of lobby groups who create pressures in different directions. The objectives set must be translated into action, by designing appropriate control or preventative measures. Although the enforcement of control or preventative measures is often achieved through legislation, there is also a place in some circumstances for voluntary agreements. Monitoring is the final step, allowing a judgement to be made of the effectiveness of the measures taken. The cycle is closed by the results of the monitoring program leading to a reassessment of the state of the environment and a decision either to do nothing further or to initiate a further cycle of activity.

The framework outlined above will be used in this chapter to describe the action taken on TBT in France and the United Kingdom. Inevitably, it will be imperfect since events in the real world do not proceed in such an orderly linear or cyclical pattern. For this reason there is some condensation of the headings in Fig. 2.1. However, it is probably fair to say that the French government has completed one cycle of activity, which included their ban on the use of TBT on small boats; and the British have completed two cycles to reach essentially the same point. The contribution made by large vessels to environmental contamination by TBT is still being assessed, and this could lead to further cycles of action.

2.2. ACTION TAKEN IN FRANCE

2.2.1. ASSESSMENT OF THE STATE OF THE ENVIRONMENT AND IDENTIFICATION OF THE CAUSE OF THE PROBLEM

Alzieu (1986) gives a concise summary of the scientific evidence that led to the ban on TBT use in France. Further detail is provided in Alzieu *et al.* (1980, 1982), Alzieu (1981, 1991), and Alzieu and Héral (1984), and by His (Chapter 12). A significant role was played by the shellfish industry, plagued since the 1950s by climatic and disease problems that had severely affected the cultivation of traditional species. The successful introduction of Pacific oysters in the early 1970s promised a new start. When this species in turn showed growth anomalies and reproductive failures in the mid- to late 1970s, the industry demanded that the cause be found.

The most severe effects were seen in Arcachon Bay, on the Atlantic coast of France. Naturally produced larvae (spat) of the Pacific oyster suffered high mortality in three successive years, and there was consequently very little spat settlement and regeneration (His, Chapter 12). At the same time, the shells of adult oysters frequently grew abnormally (see Laughlin *et al.*, Chapter 10; Waldock *et al.*, Chapter 11) with the calcified layers being separated by cavities containing a gelatinous material. The thickening could be so extreme that the oysters became ball shaped with a substantially reduced meat yield. The effects were noted to be more severe close to boat moorings; this directed the attention of the French scientists toward evaluating the possible risks from the substances used in antifouling paint.

The investigation of possible risk from antifouling paint started in 1979 and concentrated on Arcachon Bay (His, Chapter 12), comparing the results with other areas that differed both in boating activity and in the extent of problems with shellfish culture. The more seriously malformed oysters tended to have high concentrations of tin, but there were no consistent relationships with lead, cadmium, zinc, or copper levels. In Arcachon Bay, there were local variations related to the proximity of moorings. In La Vigne marina, for example, tin levels were 14 times higher than at La Hillon in the center of the bay. The source of the tin was assumed to be the tributyltin antifouling paints, although the French scientists were not equipped to carry out more specific analyses at this stage.

Samples of Pacific oysters collected in the Marennes–Oleron area showed progressive changes in shell shape with distance from a marina. Those taken within 1 km of the marina were particularly thickened, but samples collected 2 km away were still abnormal. Similar variation in other areas could be consistently explained in terms of the presence of concentrations of boats and the pattern of water currents. Further, malformed oysters from Arcachon relaid in unaffected areas returned to a normal growth pattern, and reciprocal transplants became stunted and thickened.

None of these observations, however, was sufficient to confirm TBT as the cause of the growth problems or to exclude other possible pollutants. A critical experiment was therefore carried out between June and December 1981 in the Marennes–Oleron area (Alzieu and Héral, 1984). Normally growing oysters were exposed under laboratory conditions to marina water or to water in contact with surfaces painted with TBT. Control animals survived the experiment, contained <1 mg kg^{-1} of tin in their tissues, and showed no growth anomalies. Those maintained in water dosed with TBT or in marina water, on the other hand, showed significant mortality, accumulated high concentrations of tin, and developed gel-filled chambers. These effects were seen at TBT concentrations estimated at 200 ng l^{-1} (Alzieu and Portmann, 1984; His, Chapter 12).

By this stage, literature surveys had also been carried out (Alzieu, 1981) that indicated that TBT was more persistent than originally thought, with a half-life in water of perhaps 8–9 months. On the other hand, little was known about the risk of bioaccumulation. Work published on acute toxicity had led some authors to propose maximum acceptable concentrations for the aquatic environment in the range 100–3000 ng l^{-1} TBT. Indeed, concentrations of 50 ng l^{-1} had been shown (see His, Chapter 12) to interfere with larval feeding and development of Pacific oysters from Arcachon. Alzieu (1981) also concluded that concentrations of copper and zinc in Arcachon Bay were not directly responsible for the mortalities in oyster larvae seen in the 1970s.

2.2.2. OBJECTIVES, CONTROLS, AND ENFORCEMENT

The studies carried out up to the end of 1981 pointed to antifouling paints, specifically those containing TBT, being the cause of the problems in oyster growth and reproduction. The evidence, however, could best be described as circumstantial (Alzieu and Portmann, 1984) since no specific measurements of TBT had been made in water or oyster tissues, and other possible causes had not been completely excluded. The French government was nevertheless concerned about the state of the oyster industry and ordered a selective ban on marine antifouling paints from 19 January 1982 (Ministry of the Environment, France, 1982a). The use of paints containing >3% by wt of organotin on the hulls of boats <25 t was prohibited. The ban applied for a 3-month period to the Atlantic coast, later extended to 26 July 1982 (Ministry of the Environment, France, 1982b).

The Evaluation Committee on the Ecotoxicity of Chemical Substances considered the evidence on organotin antifouling paints and delivered its opinion on 18 June 1982. Subsequently, the French government revised the ban (Ministry of the Environment, France, 1982c) so that it related to the application of all antifouling paints containing organotin compounds to the hulls of ships and boats with an overall length <25 m. It was effective for 2 y from 14 September 1982, but paints containing <3% organotin were exempted until 1 October 1982 to allow clearance of stocks. There were no geographical restrictions, but a special dispensation was allowed for continued use of organotin paints on ships and boats with light alloy hulls. This was necessary because alternative paints based on copper could not be used safely on aluminium since they would induce electro-chemical corrosion.

The 14 September decree also provided that the Minister for the Environment should lay down the labeling and packaging conditions for paints containing organotins. These applied from 1 October 1982 and required that containers for organotin-based antifouling paints of <20 l should carry a warning on the label that it was forbidden to use the product on boats <25 m overall length except where the hull was made of light alloy (Ministry of the Environment, France, 1982d).

On the recommendation of the Evaluation Committee the ban was again extended for a further 2 y, on 12 February 1985 (Ministry of the Environment, France, 1985a) and the labeling requirement from 15 June 1985 (Ministry of the Environment, France, 1985b). Finally, the ban was made permanent on 10 March 1987 (Ministry of the Environment, France, 1987). The ban is enforced by the police, who rely on local shell fishers informing them when they think boat owners are not obeying the warning on the label.

A new decree was awaiting Ministerial approval in 1990 which would incorporate the requirements of Directive 89/677/EEC (see section 2.4.2, European Community). The decree extended existing French controls on organotin-based paints to use on mariculture equipment and went beyond the provisions of the directive in prohibiting the sale and use of paints containing aldrin, dieldrin, endrin, pentachlorophenol and its derivatives, chlordane, heptachlor, hexachlorobenzene, camphechlor, and DDT. Local authorities can also prohibit navigation in shellfish areas by all ships coated with organotin-based paints and can prohibit boatyards in shellfish culture areas from applying or removing such paints (Alzieu, 1991).

2.2.3. MONITORING

The effectiveness of the ban has been monitored from 1982 to the present time (Alzieu *et al.*, 1986, 1989) by measuring concentrations of organotins in water and oyster tissues and by following changes in shell thickness. During this period, the analytical techniques were improved so that from 1986 onward it has been possible to report values for tri-, di-, and monobutyltins separately, and the limit of detection has been improved from 150 ng l^{-1} to 1–2 ng l^{-1}.

Between 1982 and 1985, the concentrations of total tin in the water of La Vigne and Arcachon marinas dropped from 4000–5000 to <1000 ng l^{-1}. By November 1985 the levels in La Vigne were similar to those at the La Hillon reference site in the center of the basin, which is well away from any moorings. Organotin values at both marina sites showed a similar trend. The total tin content of oyster tissues decreased at all nine sites sampled in Arcachon Bay from 2–7 mg kg^{-1} in August 1982 to <0.3 mg kg^{-1} throughout 1985, with one exception. The drop was most obvious in the samples collected after 1984, which indicated that the ban only became really effective after ~2 y. Parallel changes also occurred in the proportion of oysters in a sample exhibiting gross shell deformities (chambering in both valves). By 1985 the severe malformations had disappeared, but 40% still showed at least some evidence of chambering. Perhaps most encouraging was the observation that spatfall was again satisfactory in Arcachon in 1982 and abundant in subsequent years. Between 1982 and 1985, monitoring studies in Arcachon Bay found that concentrations were from five to ten times lower than those recorded in 1982; in addition, a decrease in shell malformations, both in terms of importance and spread, was observed (Alzieu, 1991).

The survey was expanded in 1986–87 to include other marina and oyster cultivation areas and sites where commercial shipping might have been expected to contribute to TBT levels. The water concentrations of TBT found in marinas were generally <100 ng l^{-1}, although values up to 1500 ng l^{-1} have been found. In well flushed areas and those not directly affected by inputs, values were generally <5 ng l^{-1}. The incidence of severe malformations of oyster shells in Arcachon Bay remained low, but the proportion of individuals with minor deformities increased again from 40% in 1985 to ~70% in 1986 and 1987. Results for 1988 and 1989 (Alzieu, 1991; C. Alzieu, IFREMER, personal communication) are more encouraging, showing some further recovery, though not yet to the 1985 values.

2.2.4. INTERIM REASSESSMENT

The French ban has resulted in a significant reduction in the contamination of Arcachon Bay. Recovery is obviously not complete, but the local oyster industry has returned almost to normal in this area (Alzieu *et al.*, 1986), justifying the legislative action taken. However, recent work has shown that gel formation in Pacific oysters may be initiated at TBT concentrations as low as 2 ng l^{-1} (Alzieu *et al.*, 1989). The levels found in Arcachon Bay in 1986–87 (mostly <2–5 ng l^{-1}) are thus high enough to account for the less severe malformations that are still seen. High concentrations of TBT in water samples in marinas along the Atlantic and Mediterranean coasts and the changes in the frequencies of malformations suggest that there are continuing abnormal inputs of TBT despite the ban. They also imply a need to improve the enforcement of the regulations.

2.3. ACTION TAKEN IN THE UNITED KINGDOM

2.3.1. FIRST TURN OF THE ACTION CYCLE

The early work on TBT in the United Kingdom is described in Waldock and Thain (1983), Alzieu and Portmann (1984), Thain (1986), Thain and Waldock (1986), Waldock (1986), and Waite *et al.* (1991). Summaries have been published by the UK Department of the Environment (1986a) and Abel *et al.* (1986). Figure 2.2 shows the action taken up to the end of 1986.

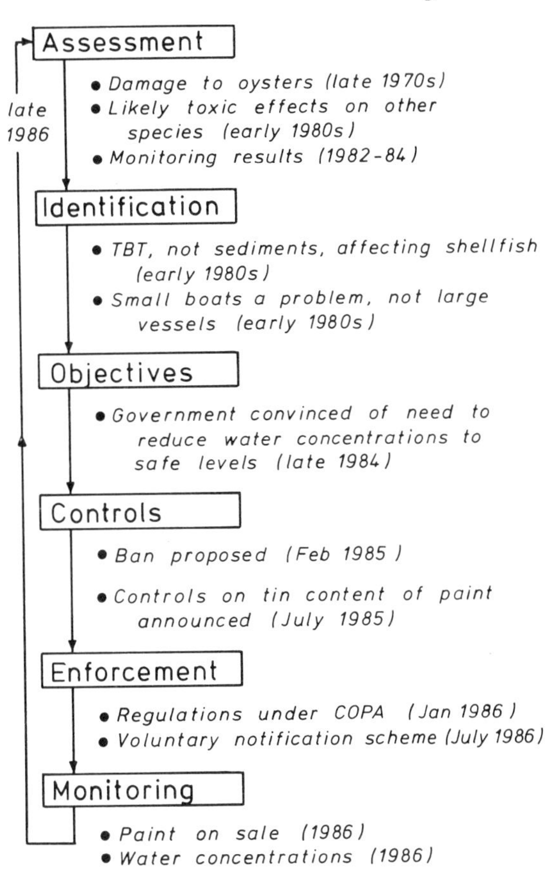

Figure 2.2 Action taken to control TBT in the UK: first turn of the Action Cycle. (COPA = Control of Pollution Act 1974).

2.3.1a. Assessment of the state of the environment and identification of the cause of the problem

As in France, Pacific oysters were introduced into the United Kingdom in the 1970s, but did not thrive in some estuaries, particularly on the east coast. Reduced growth and shell thickening were at first thought to be related to high concentrations of fine silt particles in the water (Alzieu and Portmann, 1984). It was not until scientists in the United Kingdom saw the results of the French work on shell thickening in 1980 that the influence of TBT was suspected. Calculations suggested that concentrations of TBT that would exceed those known from the literature to be toxic (Waldock, 1986; Waite *et al.*, 1991) could indeed build up in estuaries.

During 1981, methods were developed for the analysis of TBT and total tin, and preliminary measurements were made in east coast estuaries in 1982. Water samples from the River Crouch contained 80–430 ng l^{-1} as tributyltin oxide (TBTO). Higher concentrations were found elsewhere at sites where oyster growth trials had failed, including values up to 2250 ng l^{-1} in marinas. Samples of Pacific oysters and the European flat oyster (*Ostrea edulis*) were exposed at several sites and were found to accumulate tin to similar levels, but the Pacific oyster had significantly more tin present as TBT. Shell thickening in this species was observed at most sites and was associated with the presence of TBT; normally growing oysters had no detectable TBT.

Laboratory and field experiments were conducted in 1982 and 1983 (Waldock and Thain, 1983; Alzieu and Portmann, 1984) that clearly demonstrated that TBT, rather than sediment, caused the shell thickening and reduced growth of Pacific oysters. In the field experiments, high densities of recreational craft were consistently associated with poorer growth of three oyster species. Although several environmental variables were measured and the oysters were analyzed for metals, organochlorines, and hydrocarbons, as well as for total tin and TBT, only TBT concentrations were capable of accounting adequately for the variation in growth among the sites (Thain and Waldock, 1986).

A survey of oyster growers carried out in 1984 (Waldock, 1986) demonstrated that there did indeed appear to be a conflict of interest between shellfish cultivation and yachting. Many traditional shellfish areas had shown shell thickening after the late 1970s, and the scale of the problem was probably underestimated since some cultivators were known to have relocated their stocks or gone out of business when they could no longer produce marketable oysters.

Much of the emphasis had been placed on deformities in Pacific oysters up to this time, partly because of the dependence of the industry on this species, but also because the signs of damage were so obvious. Laboratory work, however, had demonstrated that other species were also affected by TBT at low concentrations. Thain (1983), for instance, quotes acute LC_{50} values for larvae of mussels, shrimp, crab, and sole in the microgram per liter range. Beaumont and Budd (1984) showed high mortality of mussel larvae after 15 d at concentrations as low as 100 ng l^{-1}; and Beaumont and Newman (1986) reported earlier work of theirs showing growth reductions in three species of microalgae at 100 ng l^{-1}. Other results from the world literature indicated the particular sensitivity of molluscs to TBT (UK Department of the Environment, 1986a). It was therefore of some interest to determine the susceptibility of other shellfish species that might be exploited by mariculturists. Laboratory tests (described in Laughlin *et al.*, Chapter 10) showed substantial variation in growth and survival among five species. None grew well at 2600 ng l^{-1}, and only two survived and grew well at 240 ng l^{-1}.

Thus, by the end of 1984, growth failures had been observed in Pacific oysters cultivated in France and the United Kingdom, and these growth failures had been associated with the presence of boats bearing TBT-based antifouling paint. TBT had also been shown in laboratory experiments to have toxic effects on a range of species at low concentrations, and the UK surveys of the River Crouch and other sites had demonstrated that these concentrations could be exceeded in the field. The French ban on TBT-based paints, resulting in a recovery of the Pacific oyster, could be interpreted as a massive field experiment strengthening the causal link between TBT and the effects previously seen. It seemed likely that not just commercial species but other components of the marine ecosystem as well might be affected wherever large numbers of small boats congregated in shallow enclosed waters with poor exchange. Large boats were considered unlikely to present the same hazard since they use deeper waters with better opportunities for dilution and dispersal of the toxin.

2.3.1b. Objectives and controls

An attempt was made in 1984 to persuade paintmakers in the United Kingdom to withdraw TBT-based paints voluntarily. Their view, however, was that use of the new copolymer paints, which were replacing free-association paints at that time, would result in a 3- to 5-fold reduction in environmental concentrations of TBT (Waldock, 1986). Government scientists, on the other hand, on the basis of their own tests of the leach rate of paints, suggested that the amount of TBT released might be 5–10 times higher than claimed by the manufacturers under the flow conditions experienced in an estuary.

This difference in perception of the seriousness of the problem led directly to the government's resolve to legislate to control the use of TBT paints in order to reduce environmental concentrations to safe levels. A consultation paper (UK Department of the Environment, 1985a) was issued, and announced in Parliament (Waldegrave, 1985a), in February 1985. It proposed that regulations should be made under the Control of Pollution Act 1974 (COPA). These regulations would limit the tin content of antifouling paints containing organotins for retail sale or for application to boats under 12 m to 0.4 g per 100 ml of paint. A derogation was to be allowed for boats with aluminum and light alloy hulls and transmission housing systems, with the intention that the derogation would be withdrawn when suitable non-TBT alternatives were available.

The proposals, however, provoked a strong and coordinated response from the paint industry and yachting interests (Side, 1987). The limit of 0.4% tin would effectively have prevented the boat owner from buying any of the copolymer paints or the TBT-boosted copper-based paints then on the market. A persuasive case was made for allowing time for the new generation of copolymer paints to be properly evaluated and for the paint industry to develop and test paint systems incorporating other biocides. There was also a strong lobby group in Parliament against the proposals. The available evidence of environmental damage was just not sufficient to outweigh the arguments put by yacht owners and the paint industry, and a compromise was necessary.

The Environment Minister made a statement in Parliament on 24 July 1985 (Waldegrave, 1985b; Abel *et al.*, 1986) in which he announced the government's intention to control the sale of organotin-based antifouling paints, to introduce a notification scheme for new antifouling agents, and to establish an ambient water quality target for TBT. Guidelines would be prepared for cleaning and repainting small boats, and the effectiveness of the action taken would be monitored through an enhanced research program.

2.3.1c. Enforcement of controls and other actions

The first UK regulations were made under COPA and came into force on 13 January 1986 (Anon, 1985). They prohibited the retail sale, and supply for retail sale, of antifouling paints containing organotin compounds if the amount of tin (measured in the dried film) exceeded 7.5% in copolymers or 2.5% in other types of paint. These were the lowest concentrations that the three major UK paint manufacturers could achieve at that time without unduly compromising the efficacy of the paints.

Local authority Trading Standards Officers enforced the regulations by checking stocks on the shelves of chandlers' shops and other retail outlets. A list of paints that complied with the regulations was published to help them in this task (UK Department of the Environment/Welsh Office, 1986) and updated at intervals. In addition, samples were analyzed by the Laboratory of the Government Chemist for compliance. The government placed particular weight on having an effective means of enforcement in place and on ensuring that nobody should be able to claim ignorance of the Regulations (Rumbold, 1986).

The government and the paint companies agreed to introduce a voluntary notification scheme for new antifoulants from July 1986. The intention was to make sure that antifouling agents were assessed for environmental safety before being marketed in order to prevent, as far as possible, the sorts of problem to which TBT compounds had given rise. The detailed data requirements are set out in Abel *et al.* (1986). These arrangements ran for a year and were eventually replaced by a statutory scheme (see section 2.3.2c Enforcement and other actions).

Cleaning the hulls of yachts with high-pressure hoses appeared likely to lead to large inputs of TBT into rivers and estuaries (Waldock *et al.*, 1987a). A guidance leaflet was therefore produced entitled 'Don't Foul Things Up' (Royal Yachting Association, 1986). This included advice on minimizing losses to the environment during cleaning and repainting and on the proper means of disposing of debris. About 175 000 copies were distributed to boat owners through a variety of outlets in time for the 1986 repainting season.

An expert meeting was convened on 20 September 1985 (UK Department of the Environment, 1985b) to discuss the scientific issues and to ensure that all those carrying out research in the area were aware of each other's work. On the basis of the toxicity data available at that time, the experts agreed that it would be reasonable to adopt an Environmental Quality Target (EQT) for TBT in coastal and estuarine waters of 20 ng l^{-1}. This was a value roughly 3–5 times lower than the concentration at which harmful effects had been recorded. The EQT was to be regarded as a target against which to monitor, and which should be aimed at, rather than a standard to be complied with. The intention, however, was to replace it with a statutory Environmental Quality Standard (EQS) when the supporting research had been completed.

At the same meeting experts also reviewed the scientific studies being carried out in the United Kingdom and endorsed the government's plans for chemical and biological monitoring to determine the effectiveness of the controls. A number of research requirements were identified for the meeting (Abel *et al.*, 1986). These included improving the detection limits for organotins and checking the levels of TBT in fresh water, around Scottish fish farms, and in the vicinity of large vessels, both during cleaning and when laid up. These research requirements and others were incorporated into a coordinated government program.

Finally, because of the controversy that had surrounded the whole issue of controlling the use of TBT, the government issued its own report in September 1986 (UK Department of the Environment, 1986a). This set out the background to the concern shown by mariculturists, conservationists, and scientists; reviewed the evidence; and explained the government's response.

2.3.1d. Monitoring

At the beginning of 1986, reports from shellfish cultivators and conservation bodies suggested that paints that did not conform to the regulations were still available in chandlers' shops. There may well have been some illegal stocking and sale of the prohibited paints in the period immediately after the regulations came into force. However, by April, Trading Standards Officers were reporting that chandlers were aware of the regulations and abiding by them.

The Ministry of Agriculture, Fisheries and Food (MAFF) monitored water quality at 14 sites (41 stations) around the UK coast in 1986 (Waldock *et al.*, 1987a; Waite *et al.*, 1991; and see Waite *et al.*, Chapter 27). Most of these sites were along the south coast of the United Kingdom with two on the east coast and two in Wales. These are, by and large, the areas of the United Kingdom with the highest densities of yacht moorings and thus are under the greatest pressure from small boats. Over 250 samples were taken altogether, and the development of the analytical method over the winter of 1985–86 allowed unambiguous identification of mono-, di-, and tributyltin with a detection limit of ~1 ng l^{-1}. Some of the samples were taken from the tops of estuaries and had TBT concentrations below the limit of detection; at the other extreme, marina samples could be as high as 1500 ng l^{-1}. Nearly half the samples exceeded the EQT of 20 ng l^{-1} TBT.

Detailed sampling in the River Crouch showed a gradual decrease in environmental concentrations over the period 1982–85, probably reflecting the replacement of free-association paints by copolymers. Against this trend, there was no marked decrease following the introduction of the regulations in January 1986. Organotin concentrations were also measured by the Plymouth Marine Laboratory at a number of sites around the south coast of England from 1984 (Cleary and Stebbing, 1985, 1987a; Langston *et al.*, 1987). Values fell in the same range as the MAFF results, and again there was little hard evidence of a decline in concentration between 1984 and 1986.

Preliminary studies were carried out in 1986 on freshwater areas and around large vessels in the River Fal estuary to determine whether more detailed assessments were needed. High TBT levels were found in the freshwater Norfolk Broads (Waldock *et al.*, 1987b; Waite *et al.*, 1989; and see Waite *et al.*, Chapter 27). Values up to 898 ng l^{-1} were measured in open water, while one marina sample contained as much as 3260 ng l^{-1}. Of the 30 samples taken, 50% had >80 ng l^{-1}. On the other hand, a limited number of samples taken from the estuary near anchored commercial vessels contained only 6–17 ng l^{-1} of TBT (Waldock *et al.*, 1988). Despite difficulties of interpretation in the presence of yacht moorings, the results suggested that large vessels would have only a limited effect on inshore TBT levels.

2.3.2. SECOND TURN OF THE ACTION CYCLE

The new scientific information that became available during 1986 had been commissioned to allow a reappraisal of the state of the environment so that decisions could be made about the need for further controls. The Environment Minister had promised (Waldegrave, 1985b) that the effectiveness of the action taken and of the progress made toward achieving the environmental quality target would be reviewed before the end of 1987. The evaluation and subsequent controls that constitute the second turn of the cycle are shown in Fig. 2.3. The main elements of the argument that led to the 1987 ban are summarized in Abel *et al.* (1987) and Waldock *et al.* (1987a,b).

2.3.2a. Assessment of the state of the UK environment and of the scientific evidence and identification of the problem

By the autumn of 1986, much more toxicity data had been collected from the scientific literature. A very marked overlap was demonstrated between concentrations of TBT known to be toxic to a range of organisms

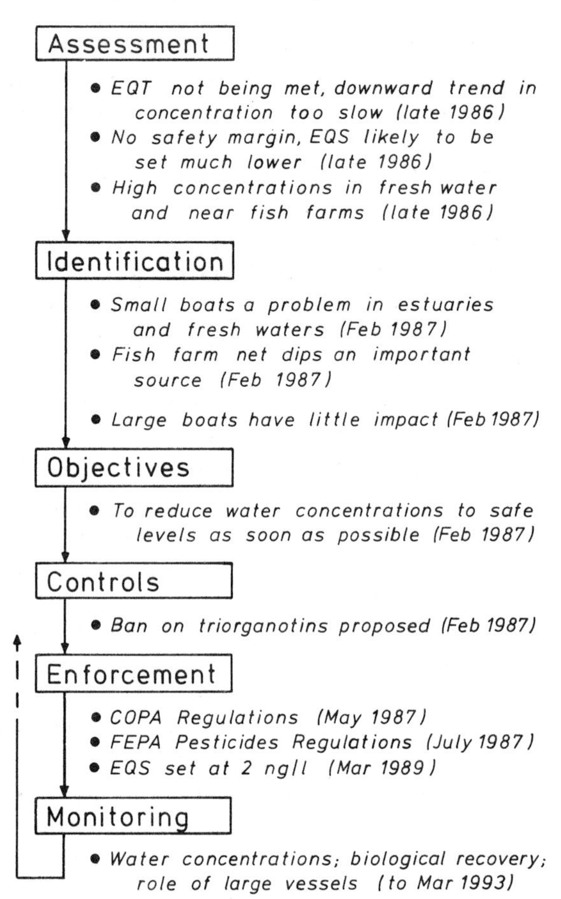

Figure 2.3 Action taken to control TBT in the UK: second turn of the Action Cycle (EQT = Environmental Quality Target; EQS = Environmental Quality Standard; COPA = Control of Pollution Act 1974; FEPA = Food and Environment Protection Act 1985).

and those commonly found in the field, not just in marinas, but in open estuarine sites (Waldock *et al.*, 1987a; Cleary and Stebbing, 1987a). The conclusion had to be that a variety of marine organisms were likely to have been damaged. There was also evidence that TBT concentrations were enhanced in the surface microlayer, potentially increasing the exposure of many species (Cleary and Stebbing, 1987a,b; and see Waite *et al.*, Chapter 27).

Molluscs and algae appeared to be the most sensitive groups and one species in particular, the dogwhelk (*Nucella lapillus*), received much attention at this time. Populations of dogwhelk had been declining around the south and southwest coasts of England (Bryan *et al.*, 1986) for a number of years. The cause proved to be the induction of male characteristics in females ('imposex') at extremely low (nanogram per liter) concentrations of TBT (Gibbs and Bryan, 1986, 1987). In many places this ultimately produced a failure of juvenile recruitment. The phenomenon is described more fully by Gibbs and Bryan in Chapter 13, but the main points of interest here are that the effect was shown to be specific to TBT (Bryan *et al.*, 1987) and that a graded response could be seen with increasing concentration. This made the dogwhelk a sensitive indicator of low levels of TBT contamination (Gibbs *et al.*, 1987).

In Scotland, attention was directed mainly at the activities of salmon farms where antifoulants were used on the netting panels of sea cages. Most of these farms are located in sheltered inlets or sea lochs on the west coast. Refurbishing nets by dipping them in TBT-based antifouling treatments could result in extremely high concentrations of TBT in the water within the cages. One day after treatment, levels were ~6000 ng l^{-1}, falling to 150 ng l^{-1} over the course of the next 6 weeks. Cultivated salmon kept in treated cages accumulated 0.5–1.0 mg kg^{-1} TBT in their muscle tissue, compared with 0.10–0.16 mg kg^{-1} TBT in fish from untreated net enclosures (Balls, 1987; Davies and McKie, 1987). Treated nets were also shown to be capable of affecting the local environment to a considerable extent. Pacific oysters deployed at various distances away from fish farms in 1985 and 1986 showed significant accumulation of TBT within 200 m and growth reduction and shell thickening up to 1 km away (Davies *et al.*, 1987a,b).

Toward the end of 1986, therefore, several conclusions could be drawn. The EQT was clearly being exceeded in areas used by large numbers of pleasure craft, and the regulations introduced in January had not resulted in a significant reduction in environmental concentrations of TBT. Furthermore, the new toxicity data available, particularly those for the dogwhelk, suggested that the Environmental Quality Standard would need to be set substantially below 20 ng l^{-1} to protect some important non-commercial marine species. It was unlikely, on the basis of the trends shown by the monitoring data, that progress toward meeting the EQT, or a lower figure, would be rapid enough to satisfy Ministers' expectations. Indeed, there now appeared to be an unacceptable absence of a safety margin between actual concentrations in the marine environment and those known to be toxic. New problems had been recognized with the use of TBT on net enclosures of fish farms and on pleasure boats used in fresh water, but the available evidence still suggested that large commercial vessels in coastal and estuarine waters had relatively small impact.

2.3.2b Objectives and controls

Revised regulations were introduced on 30 January 1987 in which the limit for the tin content of copolymer paints was reduced from 7.5% to 5.5% (Anon, 1986a) in line with advances in paint technology. The associated local authority circular (UK Department of the Environment/Welsh Office, 1987a) again specified the paints that would conform. By the time these regulations were laid, however, it was already appreciated that they were unlikely to achieve the environmental improvement required. The paint industry and others with an interest were informed of this conclusion in November 1986 (UK Department of the Environment, 1986b).

The unexpectedly high concentrations of TBT in fresh water like the Norfolk Broads led the Environment Minister to write to the paint makers (Waldegrave, 1987a) asking them to consider immediate restrictions on the marketing of TBT-based paints in this area. An informational note was produced for boat owners in January 1987 (UK Department of the Environment, 1987a) drawing attention to the problem that existed in the Norfolk Broads and asking them to use antifouling paint only when strictly necessary and to avoid the use of TBT-based paints.

The Scottish National Farmers Union and the Clyde River Purification Board had become so concerned about the implications for the environment of the application of TBT-based net dips that they wrote to fish farmers in 1986 advising strongly against their continued use. Although many farmers took this advice, it was clear at the beginning of 1987 that TBT treatments were still available and being used by some. There seemed to be a need for these products to be included within the scope of the existing controls, and indeed the government made it clear in February 1987 that all antifoulants, whatever their use and composition, would be brought within a statutory pesticides approval scheme from 1 July 1987 (Gummer, 1987).

Ministers were keeping all these issues under review, and an announcement in Parliament in January 1987 (Waldegrave, 1987a) signaled that a ban on the use of organotin compounds on small boats was likely. By the end of February 1987 the Ministers had decided that the existing controls had not been effective in reducing contamination of the aquatic environment. In addition, there had been some public concern expressed about the potential hazard to human health from the ingestion of TBT in shellfish and some farmed salmon (e.g. Duff, 1987). The government therefore proposed (Waldegrave, 1987b) that the best long-term solution was to ban, as soon as possible, the retail sale of antifouling paints containing TBT, thus preventing their use on small boats. TBT net treatments would be banned at the same time.

2.3.2c. Enforcement and other actions

The Control of Pollution (Anti-Fouling Paints and Treatments) Regulations 1987 came into force on 28 May 1987 (Anon, 1987). These regulations prohibit the retail sale, and supply for retail sale, of antifouling paints containing triorganotin compounds. Also banned are the wholesale and retail sale of antifouling treatments containing triorganotins intended for use on nets, cages, floats, or other apparatus in connection with fish or shellfish propagation or cultivation. Previous legislation relating to the tin content of antifouling paints was revoked. As before, enforcement was ensured by Trading Standards Officers referring to a list of approved paints (UK Department of the Environment/ Welsh Office, 1987b). No provision was made in the regulations for continued use of TBT-based paints on aluminum hulls and engine parts. The government decided against a derogation because two UK paint firms were marketing non-copper, tin-free antifouling paints.

The Control of Pollution Act was initially the only primary legislation available under which TBT could be effectively controlled, but was not flexible enough to allow restrictions to be placed on the *use* of antifouling paints by individuals. The Food and Environment Protection Act 1985 (FEPA), however, contained provisions in Part III that would empower Ministers to control the marketing and use of pesticides. A consultative document on the implementation of Part III was issued in November 1985 (MAFF, 1985) in which it was proposed that the marketing of new antifouling paints and compounds should be brought under the same degree of control as pesticides generally. This would ensure greater environmental safety by subjecting antifoulants to the procedures and committee examination being proposed for pesticides in general. Those involved in the manufacture, sale, and use of antifouling paints were consulted more directly in April 1986 (MAFF, 1986).

The Control of Pesticides Regulations 1986 were made under FEPA on 29 August 1986 (Anon, 1986b), with the antifouling paint provisions coming into operation on 1 July 1987. These regulations prohibit the advertisement, sale, supply, storage, and use of any pesticide – including antifoulants – unless Ministers have given prior approval to the product and a consent to the activity. In reaching decisions about approvals, Ministers are able to call on advice from the Advisory Committee on Pesticides (ACP), which was established as a statutory body under FEPA in October 1986. Antifoulants already on the market at the time the Pesticides Regulations came into force could be given a provisional approval on the basis of information supplied under the voluntary notification scheme (see section 2.3.1c, Enforcement of controls and other actions).

Following a review carried out by their Scientific Subcommittee, the ACP recommended on 16 April 1987 that products containing TBT should not be granted approval for use on small vessels or fish farming apparatus when antifouling paints were brought within the scope of FEPA (Thompson, 1987). The ACP also recommended that provisional approval could be given to the use of TBT on large vessels. Ministers acted on this advice and, on 26 June 1987, published a list of 305 antifoulants that had been given provisional approval (Health and Safety Executive, 1987). Triorganotins have only been approved for sale by wholesale and in drums containing 20 l or more. Products containing triorganotins must be labeled 'Not for retail sale; not for use on vessels less than 25m length or fish/shellfish farming equipment'. Such products have only been approved for deepsea use, including use on oil rigs.

The provisional approvals were granted originally for 2 y, to July 1989, but have since been extended for a further 3 y on the recommendation of the ACP. It is intended that a fuller data set will be required from registrants in due course, and this would certainly be needed for new antifoulants that fell outside the compositional or use limits of existing formulations. These latter cases would need to be considered in detail by the ACP under the 'Committee procedure.' As far as yacht paints are concerned, 21 new approvals have been added to the original list of 152; but, since no new active ingredients have been involved, these approvals have all been dealt with by a Technical Secretariat under a less stringent 'Secretariat procedure.' Of course, all approvals can be reviewed at any time and withdrawn if scientific evidence is produced of unforeseen risks to human health or the environment. Fuller information about all pesticides that have been approved, including their active ingredients, is published annually (e.g. Ministry of Agriculture, Fisheries and Food/Health and Safety Executive, 1989).

Although the ban on the use of TBT paints on small boats came into effect in 1987, most boats will have had TBT on their hulls throughout that year, either on the surface or below new paint layers. It was therefore particularly important to reinforce the message about cleaning off old paint layers before repainting began in 1988. The 'Don't Foul Things Up' leaflet was therefore revised and reissued in January 1988 (UK Department of the Environment, 1988).

Guidance has also been published (UK Department of the Environment, 1987b) on the cleaning and painting of large vessels in shipyards. Unregulated drydock practices have been shown to release large quantities of TBT to the environment (Waldock *et al.*, 1988), but some yards are giving serious attention to reducing their emissions. Devonport Management Ltd., for instance, have demonstrated (C. Hamlin, personal communication) that the principles contained in the guidelines can be applied successfully and can result in very significant improvements. The simple containment measures introduced have won the company an international pollution abatement award (Anon, 1989). The Ministry of Defence is monitoring the effects of its own activities (Allison and Sawyer, 1987), and those of non-Ministry of Defence shipyards undertaking work on naval vessels, and is developing its own code of practice.

An EQS of 2 ng l^{-1} TBT was set in March 1989 for the protection of saltwater life (UK Department of the Environment/Welsh Office, 1989). This level was based on recommendations in a report by the Water Research Centre (Zabel *et al.*, 1988). An EQS of 20 ng l^{-1} TBT was also set at the same time for the protection of freshwater life. The standards apply for the purpose of controlling discharges to the aquatic environment under a 1976 European Community Directive (76/464/EEC). Organotins are included in List II of the directive, which implies that all discharges likely to contain them must receive prior authorization and must conform to emission standards that will be based on the EQS. The implication, in relation to dry docks, is that no discharge should raise the concentration in the receiving water above 2 ng l^{-1} outside a mixing zone.

Under the Water Act 1989, the control of pollution will be the responsibility of the newly established National Rivers Authority. The National Rivers Authority will have to submit information demonstrating the implementation of a national program to the UK Department of the Environment by January 1991 for onward transmission to the Commission of the European Communities. The National Rivers Authority will need to indicate those surface waters affected by discharges of organotins and where the standards have been applied. In some instances it may recognize that the EQS is unattainable in the immediate future, but in these cases it must ensure a 'standstill' – there must be no further deterioration in water quality.

A new Environment Protection Bill is scheduled to be introduced into the UK Parliament during 1990. Proposals were published in July 1988 (UK Department of the Environment/Welsh Office, 1988) for tighter controls over the most dangerous substances entering the aquatic environment. In April 1989 the government announced the inclusion of tributyl- and triphenyltin compounds in the list of 23 substances (the UK 'Red List') thought to represent the best candidates for priority action (Caithness, 1989). It is intended that discharges of Red List substances from 'prescribed processes' should be controlled by Her Majesty's Inspectorate of Pollution (HMIP) under proposed arrangements for integrated pollution control. There would be a requirement that the process be operated according to the 'best available techniques not entailing excessive cost' ('BATNEEC'). Whether HMIP would consider 'scheduling' operations within dry docks and calling for the use of best available techniques has, however, not yet been decided.

2.3.2d. Monitoring

The main thrust of the government's research program since 1987 has been directed toward assessing the impact of the ban. Results of these studies are reported by Waldock *et al.* and Waite *et al.* (Chapters 11 and 27). They indicate that there has been a steady decrease in water concentrations in the River Crouch, with most summer values now below 20 ng l^{-1}. In other sites, TBT values in the summer of 1988 were around a half the 1986 concentrations, and greater reductions (to a third or a quarter) have been seen in some marinas. Similar trends are seen in concentrations of TBT in oyster tissue.

The expected recovery of partially sterilized populations of dogwhelks (*N. lapillus*) and the potential recolonization of areas denuded of dogwhelks in relation to ambient concentrations of TBT are also being followed. Although it has now been demonstrated that imposex develops more readily in juveniles than adults (Gibbs *et al.*, 1988) and that two other organotins also have a limited capacity to induce imposex (Bryan *et al.*, 1988), the dogwhelk remains a useful and sensitive indicator of TBT contamination. It will continue to be used in the United Kingdom to monitor recovery (see Gibbs and Bryan, Chapter 13).

Scientists in Scotland have, in fact, relied heavily on dogwhelk monitoring to identify and discriminate between the main sources of TBT (small boats, large vessels, and fish farming activities) in seven study areas (Bailey and Davies, 1988a,b, 1989). Similar effects have been seen in dogwhelk populations whatever the source, although the duration of exposure is important and may affect the detail of the response. Thus female dogwhelks from fish farming areas show less development of male tissue for a given body concentration of tin than those from boating areas. TBT was used as a net treatment mainly between 1983 and 1987, while boating and shipping areas have been exposed to TBT probably since 1969–70. The degree of imposex does not appear to decrease in individual adults as the concentrations of TBT in water (and their tissues) decline. Hence, at a population level, only the recruitment of less-affected juveniles to the adult population will reduce the overall values of the imposex index. Monitoring the performance of juveniles should therefore give a better indication of recovery.

The extent to which populations are self sustaining has been investigated by volunteers at ~130 sites around the UK coast between 1986 and 1988 (Spence *et al.*, 1987; Hayter, 1988). Many populations seem to be sparse with little or no recruitment, especially in sheltered locations, although conditions appear more normal in exposed areas, particularly on the west coast. It is still too early with such a long-lived species to judge whether particular populations are declining or recovering since the ban. However, surveys conducted in a Scottish sea loch for the following consecutive years where TBT was used in mariculture on nets until banned in 1987 found that low residual concentrations of TBT were still present in the loch during 1988 and 1989. It was found that the reproductive condition of adult female dogwhelks did not improve, but the rate of imposex development in juveniles was decreasing (Bailey and Davies, 1991).

Studies aimed at determining the inputs of TBT from large vessels have concentrated on harbors, anchorages, and shipyards and are described by Waite *et al.*, Chapter 27. Waldock *et al.* (1988) found that TBT concentrations close to the hulls of commercial vessels within harbors and at anchorages were generally low (<20 ng l^{-1}) and that there was rapid attenuation with distance from the ship. Where concentrations around vessels were higher, this seemed to be associated with recent painting. Because many waters used by large vessels are also frequented by high densities of small boats, it is not possible to be certain of the relative contributions until the 1987 ban is fully effective. However, surveys of imposex in dogwhelk populations around oil terminals in Sullom Voe (Shetland) and Scapa Flow (Orkney) have confirmed that the effects of TBT released from large vessels in coastal waters can be recognized (Davies and Bailey, 1991). Cleaning off old antifouling paint in shipyards was also shown to be capable of giving rise to extremely high concentrations of TBT in the wash water and in the plume leaving the yard (Waldock *et al.* 1988). Simulations of TBT dispersal in the Tamar Estuary in southwest England have been carried out by Harris *et al.* (1991). They suggest that levels probably halved within a year because of the ban on TBT use on small craft, with inputs of dissolved TBT from dockyards being insignificant. Release of paint chips, prior to the introduction of containment measures, could have contributed four times as much TBT to the estuarine loading as that leached from small boats.

2.3.2e. Interim reassessment

The time series of water quality measurements for the River Crouch and other sites and the concentrations of TBT in oyster tissue all indicate improvements since the ban. Recent water quality studies have found that both the surface microlayer and the subsurface water organotin concentrations began to decline in 1988 and continued to do so in 1989. The greatest changes occurred in marinas with a maximum 10-fold decline in sub-surface waters and a 20-fold decline in the microlayer. Nevertheless, these TBT concentrations are still above the EQS and toxicity threshold values for a number of marine species (Cleary, 1991). The general trend in water concentrations, however, suggests that it may yet be some time before concentrations fall to a level that will protect the most sensitive species (the dogwhelk). Other species have already fared better. Experimentally laid Pacific oysters showed more normal growth in the River Crouch in 1988 than in 1987, and commercially grown oysters also seem to be recovering in areas previously abandoned because of the effects of TBT (Dixon, 1989).

However, there is no evidence yet of a decrease in concentrations of TBT in sediments (see Waite *et al.*, Chapter 27). Its slow degradation, especially under anaerobic conditions, could mean that benthic infauna and filter feeders will be exposed to substantial amounts of TBT for some time, but there is very little information on its bioavailability when bound to sediment particles. It is possible that sediments around boatyards contain flakes of active paint and that many boats still bear old TBT layers that have not yet been fully expended. Such sources may be preventing TBT levels in water from falling as rapidly as anticipated (see Harris *et al.*, 1991). More worrying are the results quoted by Waite *et al.* (1991; Chapter 27) for Hythe marina, which suggest that there may still be some illegal application of TBT-based paints. Although TBT-based paints are not now available through retail outlets in the United Kingdom, some hoarding may have occurred, and they can be bought elsewhere in Europe. Strict enforcement of the 'use' provisions in the legislation would be difficult, and it may be that eliminating TBT from waters used for yachting will not be completely possible until Europe-wide controls are in place (see section 2.4.2, European Community).

A Parliamentary select committee report on antifouling paints (House of Lords, 1989) suggested that the government should be pressing for international action to ban the use of TBT in all marine antifouling paints, including those applied to vessels over 25 m. The committee was particularly influenced by results from Sullom Voe, Shetland (Bailey and Davies, 1988b), where virtually all the TBT input comes from some 750 oil and gas tanker movements each year at the oil terminal.

The government's monitoring program should provide sufficient information within the next 3 y to decide whether TBT use on large vessels poses a real threat to the environment. The current view is that, during normal operation, the latest generation of slow-release TBT paints designed for large vessels is likely to give rise to concentrations less than the EQS of 2 ng l^{-1} except in the immediate vicinity of the vessel. This assessment will need to be kept under review, and reports will be made to the international bodies that are currently considering the need for further action on TBT (see section 2.4.3, International conventions).

2.4. OTHER EUROPEAN LEGISLATION

2.4.1. NATIONAL LEGISLATION

Apart from France and the United Kingdom, several European countries have taken some action on TBT or intend to do so in the near future. In Italy organotins are prohibited for use as biocides in water used for shellfish farming (Official Gazette No. 123 of 4 May 1985, quoted by Bressa and Cima, 1986). Discharges of industrial cooling water containing organotins are regarded for all purposes as sources of chemical pollution. In the Federal Republic of Germany, the paint industry made a voluntary commitment to the Paris Commission in the autumn of 1986 in which industry members renounced the use of organotins in free-association antifouling paints. The industry also undertook to limit the organotin content of copolymers to 3.8% of the ready-to-use paint. Switzerland is reported (Champ and Pugh, 1987) to have banned use of TBT in antifouling paints in freshwater environments. Germany has also indicated that it believes controls should be coordinated internationally and has therefore postponed national legislation while awaiting agreement at the EC level (see section 2.4.2, European Community). The Netherlands has similarly supported action in the EC, but will probably introduce its own legislation well before the implementation date of the directive (Ritsema *et al.*, 1991).

The Republic of Ireland introduced a by-law on 9 April 1987 (Department of the Marine, Ireland, 1987). This prohibited the application of organotin antifouling compounds to boats, piers, jetties, buildings, or other structures that are in or on the sea, or a river, lake, or canal, or are likely to be covered at any time by such waters. The ban also applies to fishing engines, machines, or any apparatus or other objects intended for use in the sea or in fresh water. Boats to be used entirely on the sea and either over 25 m or made of aluminum alloy are exempted.

Interestingly, a defendant prosecuted under this by-law will be presumed to have applied the antifouling paint, and to have done so after 9 April 1987, unless the contrary can be proved. Manufacturers must also submit data on antifoulants to the pesticides section of the Department of Agriculture for review and approval.

In Scandinavia, Sweden restricted the organotin content of antifouling paints (Linden, 1987) according to the active ingredient. Tributyltin oxide was limited to 2% by weight (as tin in the wet paint), the fluoride to 4%, while triphenyltin hydroxide, chloride, and fluoride were not to exceed 3%. For copolymers based on TBT and triphenyltin, the maximum values were 4% and 6%. The Swedish National Chemical Inspectorate issued a draft ordinance in 1988 that proposed restricting TBT use to boats greater than 25 m and prohibiting retail trade and use on nets. Restrictions on use of TBT came into effect on 1 January 1989 and from 1 January 1992 all antifouling products had to be approved by the Chemicals Inspectorate before they can be imported, offered for sale, transferred, or used (Organisation for Economic Co-operation and Development, 1988). The Norwegian Ministry of the Environment has drafted similar regulations and proposed that, in special cases, the State Pollution Control Authority should be able to extend the prohibition to large vessels. These regulations came into force in 1989. The Danish Environmental Protection Agency is also reported to be likely to move toward a ban (Zuolian and Jensen, 1989) and as a member of the European Community will have to implement the provisions of the recently adopted directive (see section 2.4.2, European Community).

2.4.2. EUROPEAN COMMUNITY

Before introducing national legislation, member states of the European Community are obliged under the terms of Directive 83/189/EEC to inform the Paris Commission of the EC of their intentions. The Commission must then decide within 3 months whether the national measures will create a barrier to trade and whether they therefore wish to bring forward proposals for a directive that will harmonize controls throughout the EC. The original French ban predated this procedure, but the UK regulations of January 1986 and January 1987 were only introduced after the Commission had been informed and had decided not to act.

The United Kingdom, however, felt that EC action was necessary to deal with the environmental problems caused by organotin compounds since other states were probably affected in a similar way. During the United Kingdom's tenure of the presidency of the EC in the second half of 1986, the Government wrote to the Commission inviting it to take action (UK Department of the Environment, 1986b). At the November 1986 Council of Environment Ministers meeting, the Commission responded by agreeing to draft a proposal for a directive. This proposal was submitted to the Council on 4 February 1988 (Commission of the European Communities, 1988) as part of the proposed Eighth Amendment to the 'Marketing and Use' Directive (76/769/EEC).

After consideration and amendment in the Council and Parliament, the directive was adopted on 21 December 1989 (Commission of the European Communities, 1990). Organotin compounds may not be used to prevent fouling on the hulls of boats of an overall length less than 25 m, nor on cages, floats, nets, or equipment used for fish or shellfish farming, nor on any totally or partly submerged appliances or equipment. Organotin antifoulants may not be sold to the general public and may only be placed on the market in packages of 20 l or greater capacity. These must be marked 'Not to be used on boats of an overall length of less than 25 metres or on any appliances or equipment used in fish or shellfish farming; restricted to professional users.' Use of organotins in the treatment of industrial waters, irrespective of their use, is also prohibited. Arsenic and mercury compounds are similarly proscribed in all antifouling paints (including those for use on large vessels) to prevent any possibility of their reintroduction. These provisions are thus very close to the UK model, and there was indeed some measure of convergent thinking in their development. At one stage, an exception was proposed for use of organotins on aluminum boat hulls and other components (such as engines and propulsion gear), but the Commission would not accept this as substitutes were available (Commission of the European Communities, 1989). Member states were obliged to implement the directive by June 1991.

2.4.3. INTERNATIONAL CONVENTIONS

Several international treaties have been agreed since 1972 on the pollution of water by dangerous substances (Vrijhof, 1985). The Oslo and London Dumping Conventions established the principle of listing those substances that should be prohibited in dumped wastes (the black list) and those that should not be disposed of in significant quantities without a permit (the gray list). Later Conventions have copied the two-tier black and gray list concept, adapting it to their particular requirements. Three of these Conventions in particular are relevant to the development of controls on organotins used in antifouling paints.

2.4.3a. The Paris Convention

The Paris Convention for the Prevention of Marine Pollution from Land-based Sources came into effect on 6 May 1978 and applies to the northeastern parts of the Atlantic and Arctic oceans. Organotin compounds (as a class) were included in the gray list of the Convention on the basis of their persistence, toxicity, and tendency to bioaccumulate. Biologically harmless organotin compounds, or those readily degraded to harmless substances, were to be regarded as excluded; but the amount of information on individual compounds was relatively limited at that time.

At the beginning of 1987, the United Kingdom drew attention, in the Convention's Technical Working Group, to the existing French ban on organotin-based antifouling paints and the United Kingdom's imminent ban on triorganotin-based paints and net treatments. It was suggested that any country with concentrations of pleasure craft or a mariculture industry using netting cages was likely to suffer the same problems as France and the United Kingdom. Within the Paris Convention area it was also likely that a significant number of pleasure craft would visit countries other than their country of registration. This indicated that controls would have to be applied internationally to be fully effective.

As a result, the Commission Recommendation 87/1 of 3 June 1987 (Paris Commission, 1988) recognized that the use of TBT compounds in antifouling paints on boats, ships, underwater structures, and fish net cages was causing serious pollution in the inshore areas of the Convention waters. The Contracting Parties agreed to take steps without delay to eliminate such pollution by banning retail sale or use of organotin paints for pleasure craft and fish net cages. They also agreed to consider restrictions on the use of organotins on sea-going vessels and underwater structures and the adoption of a quality standard for coastal and marine waters.

Action taken by the Contracting Parties on Recommendation 87/1 was reported through the Technical Working Group in Spring 1988. The Netherlands, in particular, suggested that organotins should be treated as black list substances. The Commission agreed (Paris

Commission, 1989) that organotin compounds should be placed on the 'waiting list of hazardous substances' to be reviewed by the Technical Working Group with a view to their being included in Part I of Annex A to the Convention. The Commission also decided to inform the International Maritime Organisation (IMO) of the controls already applied to small boats and to ask them to consider the appropriateness of restricting use of organotins on seagoing vessels. The Marine Environment Protection Committee of IMO took note at its September 1988 meeting and is expected to return to the issue at future meetings.

Paris Commission Recommendation 88/1 of 17 June 1988 recognized that it was probably impossible to apply a ban to large vessels (as implied in Recommendation 87/1) immediately, for economic reasons. The influence of activities within dry docks, including sand blasting, on the amount of organotins reaching the aquatic environment was, however, recognized as very important. The Contracting Parties agreed to develop procedures and technology for reducing emissions from dry docks, to stimulate their implementation, and to report back to the Commission in 1990.

2.4.3b. The Helsinki and Barcelona Conventions

The Helsinki Convention on the Protection of the Marine Environment of the Baltic Sea Area, adopted in 1974, places organotins in Annex II (the gray list). The Commission's Recommendation 9/10 of 17 February 1988 noted that organotin compounds, particularly tributyltin compounds, were causing pollution in some inshore areas of the Baltic Sea area when used in antifouling paints. They recommended that Contracting Parties take action as soon as possible, and no later than 1991, to eliminate such pollution. A first step should be a ban on the retail sale or use of organotin-based antifoulants for pleasure boats or fish net cages. Contracting Parties were also invited to consider the need for restrictions on other uses of organotin antifouling paint, for example on seagoing vessels and underwater structures.

The Barcelona Convention for the Protection of the Mediterranean Sea against Pollution was agreed in 1976. Its third protocol, for the Protection of the Mediterranean Sea against Pollution from Land-based Sources, came into effect in June 1983 (United Nations Environment Programme, 1988). Organotin compounds (apart from those that are biologically harmless or rapidly converted to harmless substances) are included in Annex I to the Protocol (the black list). The Contracting Parties have undertaken to eliminate pollution by Annex I substances and have agreed that they should base their action on an assessment document produced for each group of substances by the Secretariat.

The organotin document (United Nations Environment Programme, 1988) was considered by the Scientific and Technical Committee in May 1988 along with results from a pilot survey of the Mediterranean Sea. As a result, the Contracting Parties to the Convention agreed in October 1989 that antifouling preparations containing organotins should not be used on the hulls of boats less than 25 m or on mariculture equipment after 1 July 1991. The measure would not apply to ships owned or operated by states who were party to the Protocol where used only on government, non-commercial, service. A further 2-y grace could be allowed for states without access to substitute products by July 1991. It has also been agreed that a code of practice should be developed for the removal of old antifouling paint and subsequent repainting in order to minimize contamination around boatyards and drydocks. In the meantime, studies conducted by Alzieu *et al.* (1991) in the Mediterranean have found that, for the great majority of the marina and harbor waters sampled, the 'no observable effect level' for oyster larvae of 20 ng l^{-1} was exceeded. This probably results from the extensive movement of vessels between France and countries which do not legislate against TBT, and reinforces the need for further legislation on a Mediterranean regional basis.

2.5. CONCLUSIONS

In banning specific uses of organotin-based antifoulants, France, Ireland, and the United Kingdom have each taken action against those sources that were thought to be the greatest contributors to TBT concentrations in the aquatic environment. The evidence of damage at the time of the French ban was thin, but sufficient to justify action to protect important shellfish cultivation interests. For the United Kingdom, the argument was centered to a greater extent on the environmental rather than the commercial impact of organotins, and more evidence had to be adduced to support a ban. The interest of both scientists and legislators in the problems, their causes, and their solutions has increased dramatically during the 1980s. It is now generally accepted that the use of TBT-based antifoulants on small boats and fish farming equipment can lead to high concentrations of TBT in sheltered inshore waters, which can exceed toxicity thresholds for a variety of species. This view is reflected in the recently adopted EC Directive and in the Recommendations of the Paris, Helsinki, and Barcelona Conventions. That it is correct is evidenced by the recovery that has taken place to date in France and the United Kingdom following national legislation.

Attention has now shifted to exploring the implications of allowing continued use of organotins on large vessels. There is little doubt that cleaning off or refurbishing old antifouling paint on ships in drydocks, shipyards, on the foreshore, or while afloat can create wash waters with very high concentrations of TBT. These have the potential to cause severe damage locally, and a number of initiatives have been taken to bring this source of pollution under control.

The extent to which the normal operation of large vessels raises concentrations of TBT in inshore waters is still unclear. If any action is to be taken on large vessels, it will have to be practicable in relation to operational requirements. For instance, a large proportion of vessels using British (and EC) waters and terminals such as Sullom Voe are foreign owned and likely to be built or refitted elsewhere. Preventing just British (or EC) vessels from using TBT would not provide a significant environmental benefit. It would, on the other hand, have the effect of driving refit business away from EC shipyards. Clearly, further developments should take place in an international context. The international community will require justification for restrictions on the use of TBT on large vessels, and the environmental benefits will need to be weighed carefully against the large cost penalties claimed by the paint and shipping industries for giving up TBT paints. Research is under way in the United Kingdom and elsewhere to study the impact of large vessels in coastal waters and in harbors. This work should contribute to an eventual assessment, which will either complete the Action Cycle or drive it through one more turn to the universal abandonment of organotins in antifouling paints.

ACKNOWLEDGMENTS

The structure of this chapter as developed in 1990 was suggested by the idea of the 'Action Cycle' developed by Dr D.L. Simms in the late 1970s as a framework for UK monitoring activities. I am also grateful to Dr N.J. King and Dr I.M. Davies for their comments on the manuscript. The opinions expressed in this chapter are those of the author and do not necessarily reflect UK Government policy.

REFERENCES

Abel, R., N.J. King, J.L. Vosser and T.G. Wilkinson. 1986. The control of organotin use in antifouling paint – the UK's basis for action. In: Proceedings of the Oceans '86 Organotin Symposium, Vol. 4, Washington, DC, 23–25 September 1986, Institute of Electrical and Electronics Engineers, New Jersey, pp. 1314–1323.

Abel, R., R.A. Hathaway, N.J. King, J.L. Vosser and T.G. Wilkinson. 1987. Assessment and regulatory actions for TBT in the UK. In: Proceedings of the Oceans '87 International Organotin Symposium, Vol. 4, Halifax, Nova Scotia, 29 September–1 October 1987. Institute of Electrical and Electronics Engineers, New Jersey, pp. 1314–1319.

Allison, D.M. and L.J. Sawyer. 1987. The UK Ministry of Defence's experiences, practices, and monitoring programmes for the application, maintenance and removal of erodable organotin antifouling paints. In: Proceedings of the Oceans '87 International Organotin Symposium, Vol. 4, Halifax, Nova Scotia, 29 September–1 October 1987. Institute of Electrical and Electronics Engineers, New Jersey, pp. 1392–1397.

Alzieu, C. 1981. Evaluation des risques dus à l'emploi des peintures anti-salissures dans les zones conchylicoles. Thème surveillance continue et mécanismes d'action des pollutants. Institute Scientifique et Technique des Peches Maritimes, Nantes, le 15 Decembre 1981, 84 pp.

Alzieu, C. 1986. The detrimental effects on oyster culture in France – evolution since antifouling paint regulation. In: Proceedings of the Oceans '86 Organotin Symposium, Vol. 4, Washington, DC, 23–25 September 1986. Institute of Electrical and Electronics Engineers, New Jersey, pp. 1130–1134.

Alzieu, C. 1991. Environmental problems caused by TBT in France: assessment, regulations, prospects. *Mar. Environ. Res.*, **32**, 7–17.

Alzieu, C. and M. Heral. 1984. Ecotoxicological effects of organotin compounds on oyster culture. In: *Ecotoxicological Testing for the Marine Environment*, Vol. 2, G. Persoone, E. Jaspers and C. Claus (Eds). State Univ. Ghent and Inst. Mar. Scient. Res., Bredene, Belgium, pp. 187–196.

Alzieu, C. and J.E. Portmann. 1984. The effect of tributyltin on the culture of *C. gigas* and other species. Proceedings of the Fifteenth Annual Shellfish Conference, 15–16 May 1984. Shellfish Association, London, pp. 87–101.

Alzieu, C., Y. Thibaud, M. Héral and B. Boutier. 1980. Evaluation des risques dus à l'emploi des peintures anti-salissures dans les zones conchylicoles. *Rev. Trav. Inst. Pêches Marit.*, **44**, 301–349.

Alzieu, C., M. Héral, Y. Thibaud, M.J. Dardignac and M. Feuillet. 1982. Influence des peintures antisalissures à base d'organostanniques sur la calcification de la coquille de l'huître *Crassostrea gigas*. *Rev. Trav. Inst. Peches Marit.*, **45**, 101–116.

Alzieu, C., J. Sanjuan, J.P. Deltreil and M. Borel. 1986. Tin contamination in Arcachon Bay: effects on oyster shell anomalies. *Mar. Pollut. Bull.*, **17**, 494–498.

Alzieu, C., J. Sanjuan, P. Michel, M. Borel and J.P. Dreno. 1989. Monitoring and assessment of butyltins in Atlantic coastal waters. *Mar. Pollut. Bull.*, **20**, 22–26.

Alzieu, C., P. Michel, I. Tolosa, E. Bacci, L.D. Mee and J.W. Readman. 1991. Organotin compounds in the Mediterranean: a continuing cause for concern. *Mar. Environ. Res.*, **32**, 261–270.

Anderson, C.D. and R. Dalley. 1986. Use of organotins in antifouling paints. In: Proceedings of the Oceans '86 Organotin Symposium, Vol. 4, Washington, DC, 23–25 September 1986. Institute of Electrical and Electronic Engineers, New Jersey, pp. 1108–1113.

Anon. 1985. The Control of Pollution (Anti-Fouling Paints) Regulations 1985. Statutory Instrument 1985 No. 2011, Her Majesty's Stationery Office (HMSO), 3 pp.

Anon. 1986a. The Control of Pollution (Anti-Fouling Paints) (Amendment) Regulations 1986. Statutory Instrument 1986 No. 2300, HMSO, 2 pp.

Anon. 1986b. The Control of Pesticides Regulations 1986. Statutory Instrument 1986 No. 1510, HMSO, 11 pp.

Anon. 1987. The Control of Pollution (Anti-Fouling Paints and Treatments) Regulations 1987. Statutory Instrument 1987 No. 783, HMSO, 3 pp.

Anon. 1989. TBT clean-up scheme. *Mar. Pollut. Bull.*, **20**, 361–362.

Bailey, S.K. and I.M. Davies. 1988a. Tributyltin contamination in the Firth of Forth (1975–1987). *Sci. Total Environ.*, **76**, 185–192.

Bailey, S.K. and I.M. Davies. 1988b. Tributyltin contamination around an oil terminal in Sullom Voe (Shetland). *Environ. Pollut.*, **55**, 161–172.

Bailey, S.K. and I.M. Davies. 1989. The effects of tributyltin on dogwhelks (*Nucella lapillus*) from

Scottish coastal waters. *J. Mar. Biol. Ass. UK*, **69**, 335–354.

Bailey, S.K. and I.M. Davies. 1991. Continuing impact of TBT, previously used in mariculture, on dogwhelk (*Nucella lapillus* L.) populations in a Scottish sea loch. *Mar. Environ. Res.*, **32**, 187–199.

Balls, P.W. 1987. Tributyltin (TBT) in the waters of a Scottish sea loch arising from the use of antifoulant treated netting by salmon farms. *Aquaculture*, **65**, 227–237.

Beaumont, A.R. and M.D. Budd. 1984. High mortality of the larvae of the common mussel at low concentrations of tributyltin. *Mar. Pollut. Bull.*, **15**, 402–405.

Beaumont, A.R. and P.B. Newman. 1986. Low levels of tributyltin reduce growth of marine microalgae. *Mar. Pollut. Bull.* **17**, 457–461.

Bressa, G. and L. Cima. 1986. Contaminazione ambientale da stannorganic. Quali rischi? *Inquinamento*, **28**, 52–55.

Bryan, G.W., P.E. Gibbs, L.G. Hummerstone and G.R. Burt. 1986. The decline of the gastropod *Nucella lapillus* around south-west England: evidence for the effect of tributyltin from antifouling paints. *J. Mar. Biol. Ass. UK*, **66**, 611–640.

Bryan, G.W., P.E. Gibbs, G.R. Burt and L.G. Hummerstone. 1987. The effects of tributyltin (TBT) accumulation on adult dog-whelks, *Nucella lapillus*: long term field and laboratory experiments. *J. Mar. Biol. Ass. UK*, **67**, 525–544.

Bryan, G.W., P.E. Gibbs and G.R. Burt. 1988. A comparison of the effectiveness of tri-n-butyltin chloride and five other organotin compounds in promoting the development of imposex in the dog-whelk, *Nucella lapillus*. *J. Mar. Biol. Ass. UK*, **68**, 733–744.

Caithness, the Earl of. 1989. Dangerous substances in water: control. Written answer, 10 April 1989, House of Lords Official Report (Hansard), 506, cols 121–123. Related News Release No. 194, Department of the Environment, 10 April 1989.

Champ, M.A. and W.L. Pugh. 1987. Tributyltin antifouling paints: introduction and overview. In: Proceedings of the Oceans '87 International Organotin Symposium, Vol. 4, Halifax, Nova Scotia, 29 September–1 October 1987, Institute of Electrical and Electronics Engineers, New Jersey, pp. 1296–1308.

Cleary, J.J. 1991. Organotin in the marine surface microlayer and sub-surface waters of south-west England: relation to toxicity thresholds and the U.K. Environmental Quality Standard. *Mar. Environ. Res.*, **32**, 213–222.

Cleary, J.J. and A.R.D Stebbing. 1985. Organotin and total tin in coastal waters of southwest England. *Mar. Pollut. Bull.*, **16**, 350–355.

Cleary, J.J. and A.R.D. Stebbing. 1987a. Organotin in the surface microlayer and subsurface waters of southwest England. *Mar. Pollut. Bull.*, **18**, 238–246.

Cleary, J.J. and A.R.D. Stebbing. 1987b. Organotins in the water column – enhancement in the surface microlayer. In: Proceedings of the Oceans '87 International Organotin Symposium, Vol. 4, Halifax, Nova Scotia, 29 September–1 October 1987. Institute of Electrical and Electronics Engineers, New Jersey, pp.1405–1410.

Commission of the European Communities. 1988. Proposal for a Council Directive amending for the eighth time Directive 76/769/EEC on the approximation of laws, regulations and administrative provisions of the Member States relating to restrictions on the marketing and use of certain dangerous substances and preparations. COM(88) 7 final. *Official Journal of the European Communities*, 16 February 1988, No. C 43/9–12.

Commission of the European Communities. 1989. Amendment to the proposal for a Council Directive amending for the eighth time Directive 76/769/EEC on the approximation of the laws, regulations and administrative provisions of the Member States relating to restrictions on the marketing and use of certain dangerous substances and preparations. COM(89) 316 final – SYN 119. *Official Journal of the European Communities*, 28 July 1989, No. C 191/6–8.

Commission of the European Communities. 1990. Council Directive of 21 December 1989 amending for the eighth time Directive 76/769/EEC on the approximation of the laws, regulations and administrative provisions of the Member States relating to restrictions on the marketing and use of certain dangerous substances and preparations. Directive 89/677/EEC *Official Journal of the European Communities*, 30 December 1989, No. L 398/19–24.

Davies, I.M. and S.K. Bailey. 1991. The impact of tributyltin from large vessels on dogwhelk (*Nucella lapillus*) populations around Scottish oil ports. *Mar. Environ. Res.*, **32**, 201–211.

Davies, I.M. and J.C. McKie. 1987. Accumulation of total tin and tributyltin in muscle tissue of farmed Atlantic salmon. *Mar. Pollut. Bull.*, **18**, 405–407.

Davies, I.M., J. Drinkwater, J.C. McKie and P. Balls. 1987a. Effects of the use of tributyltin antifoulants in mariculture. In: Proceedings of the Oceans' 87 International Organotin Symposium, Vol. 4, Halifax, Nova Scotia, 29 September–1 October 1987. Institute of Electrical and Electronics Engineers, New Jersey, pp. 1477–1481.

Davies, I.M., S.K. Bailey and D.C. Moore. 1987b. Tributyltin in Scottish sea lochs, as indicated by degree of imposex in the dogwhelk, *Nucella lapillus* (L). *Mar. Pollut. Bull.*, **18**, 400–404.

Department of the Marine, Ireland. 1987. Restriction of Use of Organotin Antifouling Compounds Bye-law No. 657, 1987. Government Publications Sales Office, Dublin.

Dixon, T. 1989. Early response to TBT ban. *Mar. Pollut. Bull.*, **20**, 2.

Duff, A. 1987. TBT ban. *Mar. Pollut. Bull.*, **18**, 146.

Gibbs, P.E. and G.W. Bryan. 1986. Reproductive failure in populations of the dog-whelk, *Nucella lapillus*, caused by imposex induced by tributyltin from antifouling paints. *J. Mar. Biol. Ass. UK*, **66**, 767–777.

Gibbs, P.E. and G.W. Bryan. 1987. TBT paints and the demise of the dog-whelk, *Nucella lapillus* (Gastropoda). In: Proceedings of the Oceans '87 International Organotin Symposium, Vol. 4, Halifax, Nova Scotia, 29 September–1 October 1987. Institute of Electrical and Electronics Engineers, New Jersey, pp 1482–1487.

Gibbs, P.E., G.W. Bryan, P.L. Pascoe and G.R. Burt. 1987. The use of the dogwhelk, *Nucella lapillus*, as an indicator of tributyltin (TBT) contamination. *J. Mar. Biol. Ass. UK*, **67**, 507–523.

Gibbs, P.E., P.L. Pascoe and G.R. Burt. 1988. Sex changes in the female dog-whelk, *Nucella lapillus*, induced by tributyltin from antifouling paints. *J. Mar. Biol. Ass. UK*, **68**, 715–731.

Gummer, J. 1987. Anti-fouling compounds. Written answer, 3 February 1987. House of Commons Official Report (Hansard), 109, col 574. Related Press Release No. 21, Ministry of Agriculture, Fisheries and Food, 3 February 1987.

Harris, J.R.W., C.C. Hamlin and A.R.D. Stebbing. 1991. A simulation study of the effectiveness of legislation and improved dockyard practice in reducing TBT concentrations in the Tamar Estuary. *Mar. Environ. Res.*, **32**, 279–292.

Hayter, S. 1988. Survey of dogwhelk (*Nucella lapillus*) populations around the United Kingdom. Second interim report, December 1988, Marine Conservation Society and Institute of Biology, 52 pp.

Health and Safety Executive. 1987. Food and Environment Protection Act 1985: pesticides. *The London Gazette*, 26 June 1987, 50977, 8196–8201.

Holdgate, M.W. 1979. *A Perspective of Environmental Pollution*. Cambridge University Press.

House of Lords. 1989. Anti-fouling paints. Select Committee on the European Communities, Session 1988–89 4th report, 7 February 1989, HL Paper 23, HMSO, 10 pp.

Langston, W.J., G.R. Burt and Z. Mingjiang. 1987. Tin and organotin in water, sediments and benthic organisms of Poole Harbour. *Mar. Pollut. Bull.*, **18**, 634–639.

Linden, O. 1987. The scope of the organotin issue in Scandinavia. In: Proceedings of the Oceans '87 International Organotin Symposium, Vol.4, Halifax, Nova Scotia, 29 September–1 October 1987. Institute of Electrical and Electronics Engineers, New Jersey, pp. 1320–1323.

Ludgate, J.W. 1987. The economic and technical impact of TBT legislation on the USA marine industry. In: Proceedings of the Oceans '87 International Organotin Symposium, Vol. 4, Halifax, Nova Scotia, 29 September–1 October 1987. Institute of Electrical and Electronics Engineers, New Jersey, pp. 1309–1313.

Ministry of Agriculture, Fisheries and Food. 1985. Pesticides: implementing Part III of the Food and Environment Protection Act 1985. Consultative document, November 1985, Ministry of Agriculture, Fisheries and Food, 62 pp.

Ministry of Agriculture, Fisheries and Food. 1986. Consultation paper on the application of Part III of the Food and Environment Protection Act 1985 to marine antifouling treatments, 11 April 1986, Ministry of Agriculture, Fisheries and Food, 3 pp.

Ministry of Agriculture, Fisheries and Food/Health and Safety Executive. 1989. Pesticides 1989: pesticides approved under the Control of Pesticides Regulations 1986. Reference Book 500, HMSO, 407 pp.

Ministry of the Environment, France. 1982a. Utilisation des peintures marines antisalissures, Order of 19 January 1982. Journal Officiel de la République Française, 27 January 1982, 1056 N.C.

Ministry of the Environment, France. 1982b. Utilisation des peintures marines antisalissures, Order of 3 May 1982. Journal Officiel de la République Française, 13 May 1982, 4562 N.C.

Ministry of the Environment, France. 1982c. Décret No. 82-782 du 14 septembre 1982 relatif à l'utilisation des peintures marines antisalissures. Journal Officiel de la République Française, 16 September 1982, 2792.

Ministry of the Environment, France. 1982d. Conditions d'étiquetage des peintures marines antisalissures, Order of 21 September 1982. Journal Officiel de la République Française, 10 October 1982, 9112.

Ministry of the Environment, France. 1985a. Décret No. 85-233 du 12 fevrier 1985 relatif à l'utilisation des peintures marine antisalissures. Journal Officiel de la République Française, 18 February 1985, 2181–2182.

Ministry of the Environment, France. 1985b. Arrêté du 2 mai 1985 relatif aux conditions d'étiquetage des peintures marines antisalissures. Journal Officiel de la République Française, 11 May 1985, 5346.

Ministry of the Environment, France. 1987. Decret No 87-181 du 10 mars 1987 modifiant le decret no. 85–233 du 12 février 1985 relatif à l'utilisation des peintures marines antisalissures. Journal Officiel de la République Française, 20 mars 1987, 3177.

Organisation for Economic Co-operation and Development, 1988. Legislative and regulatory developments; dangerous substances and preparations. OECD Chemicals Programme News Sheet No. 3, September–December 1988.

Paris Commission. 1988. Ninth Annual Report on the Activities of the Paris Commission. Paris Commission, London.

Paris Commission. 1989. Tenth Annual Report on the Activities of the Paris Commission. Paris Commission, London.

Ritsema, R., R.W.P.M. Laane and O.F.X. Donard. 1991. Butyltins in marine waters of the Netherlands in 1988 and 1989: concentrations and effects. *Mar. Environ. Res.*, **32**, 243–260.

Royal Yachting Association. 1986. The marine environment: 'Don't Foul Things Up': guidelines for the antifouling user. Leaflet published by Royal Yachting Association, 6 pp.

Rumbold, A. 1986. Anti-fouling paint. Written answer, 27 February 1986. House of Commons Official Report (Hansard), 92, cols 669–670.

Side, J. 1987. Organotins – not so good relations. *Mar. Pollut. Bull.*, **18**, 205–206.

Spence, S.K., P.V. Horsman and S.J. Hawkins. 1987. Survey of dog whelk (*Nucella lapillus*) populations around the United Kingdom. Interim report, September 1987, Marine Conservation Society and Institute of Biology, 33 pp.

Stebbing, A.R.D. 1985. Organotins and water quality – some lessons to be learned. *Mar. Pollut. Bull.*, **16**, 383–390.

Stebbing, A.R.D. and J.R.W. Harris. 1988. The role of biological monitoring. In: *Pollution of the North Sea – An Assessment*, W. Salomons, B.L. Bayne, E.K. Duursma and U. Forstner (Eds). Springer-Verlag, Berlin, pp. 655–665.

Thain, J.E. 1983. The acute toxicity of bis (tributyl tin) oxide to the adults and larvae of some marine organisms. International Council for the Exploration of the Sea, Marine Environmental Quality Committee, Paper CM 1983/E:13, 5 pp.

Thain, J.E. 1986. Toxicity of TBT to bivalves: effects on reproduction, growth and survival. In: Proceedings of the Oceans '86 Organotin Symposium, Vol. 4, Washington, DC, 23–25 September 1986. Institute of Electrical and Electronics Engineers, New Jersey, pp. 1306–1313.

Thain, J.E. and M.J. Waldock. 1986. The impact of tributyltin (TBT) antifouling paints on molluscan fisheries. *Water Sci. Technol.*, **18**, 193–202.

Thompson, D. 1987. Tributyltin. Written answer, 23 April 1987. House of Commons Official Report (Hansard), 114, col. 697.

UK Department of the Environment. 1982. Monitoring the marine environment – into the eighties: the third report of the Marine Pollution Monitoring Management Group 1979–1981. DOE Pollution Report 14, 42 pp.

UK Department of the Environment. 1985a. The control of use of marine antifouling paints containing organo-tin compounds: draft regulations. Consultation paper, Department of the Environment, February 1985, 4 pp.

UK Department of the Environment. 1985b. Organotin use in antifouling paint – note of a meeting held to consider the Government's monitoring and research programme and the need for coordination of related work: 20 September 1985.

UK Department of the Environment. 1986a. Organotin in antifouling paints: environmental considerations. DOE Pollution Paper 25, HMSO, 82 pp.

UK Department of the Environment. 1986b. Organotin use in antifouling paint – note of a meeting held to review preliminary monitoring results to determine the need for future publicity and to assess the likelihood for future control of organotin use: 4 November 1986.

UK Department of the Environment. 1987a. TBT antifouling paint – the Broads problem – how you can help. Information note prepared January 1987.

UK Department of the Environment. 1987b. Shipyards and the marine environment: guidelines for applying and removing antifouling paint. Leaflet published by the Department of the Environment, 8 pp.

UK Department of the Environment. 1988. 'Don't Foul Things Up': guidelines for the antifouling user (revised 1988). Leaflet published by the Department of the Environment, 4 pp.

UK Department of the Environment/Welsh Office. 1986. Control of Pollution (Anti-fouling Paints) Regulations 1985. Joint Circular, 7 March 1986, 7/86 (DOE), 14/86 (Welsh Office), 4 pp.

UK Department of the Environment/Welsh Office. 1987a. Control of Pollution (Anti-Fouling Paints) (Amendments) Regulations 1986. Joint Circular, 24 February 1987, 3/87 (DOE), 9/87 (Welsh Office), 4 pp.

UK Department of the Environment/Welsh Office. 1987b. Control of Pollution (Antifouling Paints and Treatments) Regulations 1987. Joint Circular, 12 June 1987, 19/87 (DOE), 32/87 (Welsh Office), HMSO, 3 pp. Revisions to Annex listing paints and treatments which conform to the Regulations issued 1 May 1988, 1 March 1989, 1 November 1989.

UK Department of the Environment/Welsh Office. 1988. Inputs of dangerous substances to water: proposals for a unified system of control. The Government's consultative proposals for tighter controls over the most dangerous substances entering the aquatic environment ('the Red List'). Consultation Paper, July 1988, Department of the Environment and Welsh Office, 35 pp.

UK Department of the Environment/Welsh Office. 1989. Water and the environment: the implementation of European Community directives on pollution caused by certain dangerous substances discharged into the aquatic environment. Joint Circular, 30 March 1989, 7/89 (DOE), 16/89 (Welsh Office), HMSO, 86 pp.

United Nations Environment Programme. 1988. Assessment of organotin compounds as marine pollutants and proposed measures for the Mediterranean. UNEP(OCA)/MED WG. 1/7, 1 April 1988, UNEP, Athens, 62 pp.

Vrijhof, H. 1985. Organotin compounds and international treaties on the pollution of water by dangerous substances: black or grey list substances? *Sci. Total Environ.*, **43**, 221–231.

Waite, M.E., K.E. Evans, J.E. Thain and M.J. Waldock. 1989. Organotin concentrations in the Rivers Bure and Yare, Norfolk Broads, England. *Appl. Organometal. Chem.*, **3**, 383–391.

Waite, M.E., M.J. Waldock, J.E. Thain, D.J. Smith and S.M. Milton. 1991. Reductions in TBT concentrations in U.K. estuaries following legislation in 1986 and 1987. *Mar. Environ. Res.*, **32**, 89–111.

Waldegrave, W. 1985a. Paints (damage to marine life). Written answer, 13 February 1985. House of Commons Official Report (Hansard), 73, col. 175. Related Press Notice No. 61, Department of the Environment, 13 February 1985.

Waldegrave, W. 1985b. Anti-fouling paint. Written answer, 24 July 1985. House of Commons Official Report (Hansard), 83, cols 550–552. Related Press Notice No. 373, Department of the Environment, 24 July 1985.

Waldegrave, W. 1987a. Anti-Fouling Paint. Written answer, 13 January 1987. House of Commons Official Report (Hansard), 108, cols 143–144. Related News Release No. 15, Department of the Environment, 13 January 1987.

Waldegrave, W. 1987b. Anti-Fouling Paint. Written answer, 24 February 1987. House of Commons Official Report (Hansard), 111, cols 202–203. Related News Release No. 97, Department of the Environment, 24 February 1987.

Waldock, M.J. 1986. TBT in UK estuaries, 1982–86. Evaluation of the environmental problem. In: Proceedings of the Oceans '86 Organotin Symposium, Vol. 4, Washington, DC, 23–25 September 1986. Institute of Electrical and Electronics Engineers, New Jersey, pp. 1324–1330.

Waldock, M.J. and J.E. Thain. 1983. Shell thickening in *Crassostrea gigas*: organotin antifouling or sediment induced. *Mar. Pollut. Bull.*, **14**, 411–415.

Waldock, M.J., J.E. Thain and M.E. Waite. 1987a. The distribution and potential toxic effects of TBT in UK estuaries during 1986. *Appl. Organometal. Chem.*, **1**, 287–301.

Waldock, M.J., M.E. Waite and J.E. Thain. 1987b. Changes in concentrations of organotins in U.K. rivers and estuaries following legislation in 1986. In: Proceedings of the Oceans '87 International Organotin Symposium, Vol. 4, Halifax, Nova Scotia, 29 September–1 October 1987. Institute of Electrical and Electronics Engineers, New Jersey, pp. 1352–1356.

Waldock, M.J., M.E. Waite and J.E. Thain. 1988. Inputs of TBT to the marine environment from shipping activity in the U.K. *Environ. Technol. Lett.* **9**, 999–1010.

Zabel, T.F., J. Seager and S.D. Oakley. 1988. Proposed environmental quality standards for List II substances in water: organotins. Technical Report 255, Water Research Centre, 73 pp.

Zuolian, C. and A. Jensen. 1989. Accumulation of organic and inorganic tin in blue mussel, *Mytilus edulis*, under natural conditions. *Mar. Pollut. Bull.* **20**, 281–286.

REGULATORY POLICIES AND STRATEGIES FOR ORGANOTIN COMPOUNDS

3

Michael A. Champ[1] *and Terry L. Wade*[2]

[1]*Texas Engineering Experiment Station, Washington, DC Office, Texas A&M University System, 4601 North Fairfax Drive, Suite 1130, Arlington, Virginia 22203, USA*

[2]*Geochemical & Environmental Research Group, Texas A&M University, College Station, Texas 77895, USA*

Organotin. Edited by M.A. Champ and P.F. Seligman. Published in 1996 by Chapman & Hall, London. ISBN 0 412 58240 6

ABSTRACT

An historical overview of the use of organotin compounds as biocides in antifouling boat-bottom paints as well as a wide range of national and international regulatory and legislative options for policy and decision makers is presented and discussed in this chapter. The discussion includes the US Antifouling Paint Control Act of 1988 as well as international regulatory policies and practices from the United Kingdom, France, Switzerland, Germany, Japan, the Commission of the European Communities, and international conventions. The impact of present regulatory policies and practices are reviewed in accordance with economic and environmental costs and benefits, such as the effectiveness of regulations by developed countries at reducing local environmental concentrations as well as the shift of organotin-related environmental hazards (and associated economic loss of shipyard business in developed countries) to lesser developed countries (due to inexpensive labor and less stringent environmental regulations), which could be referred to as the transfer of contamination to those countries least able to deal with it. The following regulatory options are recommended as the most promising: (1) limit the release rate of organotin compound(s) used in copolymer antifouling paints to the adjacent water column and national water quality standards; (2) restrict the use of organotin compounds on mariculture structures, piers, and utility cooling water intake and discharge pipes; (3) develop a special antifouling coating R&D user's port and harbor dockage fee(s); and (4) institute a vessel certification system by the International Maritime Organization for vessels using toxic antifouling coatings to provide a source of funding for the development of national research and development (R&D) programs for the most cost-effective handling, collection, storage, treatment, processing, and disposal of hazardous and toxic antifouling paint wastes from ships, drydocks, and contaminated bottom sediments.

3.1. INTRODUCTION

The purpose of this chapter is to provide an historical overview (in conjunction with Chapter 2 of this volume by Abel, of the regulatory policies and practices in the United States, as compared with France, the United Kingdom, Switzerland, Germany, and Japan, as well as the Commission of European Communities (CEC) and other international bodies governing regulation of the use of organotin compounds as biocides in antifouling boat-bottom paints. Following this overview, a wide range of regulatory options available to policy and decision makers for consideration in developing future organotin compound policies at national, regional, and global levels are presented and discussed.

The impact of regulatory policies and practices is reviewed in accordance with economic and environmental costs and benefits. A comparative review of historical data is conducted to determine the effectiveness of regulations by developed countries at reducing local environmental concentrations. In addition a global perspective of how regulatory practices of developed countries can impact lesser developed countries is delineated in which the international marketplace brings about a shift of the market and application of organotin antifouling paints to countries where cheaper labor and less stringent environmental regulations bring about the transfer of contamination to those countries least able to deal with it.

For the purposes of this chapter and book, 'organotin' or 'organotin compound' is defined as any compound of tin used as a biocide in antifouling paint, which includes three tributyltin (TBT) compounds: bis(tributyltin) oxide, tributyltin fluoride, and tributyltin methacrylate. In the United States, these organotin compounds are the only currently active antifoulant registrations for organotins, and they are commonly referred to as butyltins.

3.2. BRIEF HISTORICAL OVERVIEW OF ORGANOTIN COMPOUND REGULATORY AND POLICY ISSUES IN THE UNITED STATES

There are few historical reviews or overviews on US policy related to organotins; most have one of two perspectives (user or environmental). To get a United States historical perspective (and the political sense of events), the reader must read papers such as Bruce Reid's articles at the *Daily Press/The Times-Herald* of Newport News, Virginia (Reid, 1986), the Congressional Hearing Records (US Congress, 1986, 1987a,b), Weis and Cole (1989), reports from NAVSEA (1984, 1986), and EPA Position Documents (PD/1, PD2/3) (US EPA, 1985, 1987), and the EPA Report to Congress (US EPA, 1991).

The widespread use of organotin-based antifouling boat-bottom paints began in the early 1970s. In 1974, oyster growers first reported the occurrence of abnormal shell growth in *Crassostrea gigas*, the Pacific oyster, along the east coast of England (Key *et al.*, 1976). It wasn't until the mid 1980s, however, that researchers in France (first) and the United Kingdom began to suggest that the use of TBT in antifouling paints was adversely impacting a number of marine species other than fouling organisms, including economically important species such as oysters (Alzieu *et al.*, 1980, 1981–1982, 1982, 1986, 1987, 1989; Alzieu, 1981, 1986, 1991; Thain, 1983, 1986; Thain and Waldock, 1983, 1985; Waldock and Miller, 1983; Waldock and Thain, 1983; Waldock *et al.*, 1983; Alzieu and Portman, 1984; UK DoE, 1986c; Waldock, 1986; Thain *et al.*, 1987; Waldock *et al.*, 1987a,b; Abel, Chapter 2; His, Chapter 12; and Waldock *et al.*, Chapter 11; and references therein). The effects on untargeted species and the difficulty in delineating cause-and-effect relationships attracted international concern (Stebbing, 1985).

Subsequently, in the United States, the use of organotin antifouling paints became very controversial and attracted public attention when the US Navy issued its Environmental Assessment (NAVSEA, 1984) on the use of organotin antifouling paints in compliance with the National Environmental Policy Act (NEPA). The US Navy assessment determined that the fleet-wide (600 ships) use of organotin antifouling paints would not have significant adverse environmental consequences in Navy harbors [Interim Finding of Non Significant Impact (FONSI)] and would save them over $100 million per year in fuel avoidance costs (US Navy, 1986). In addition, the Navy estimated that if its entire fleet were to be painted with TBT copolymer paints, only 268 kg y^{-1} of TBT would be added to the environment, which at that time represented less than 2.0% of the environmental loading (Eastin, 1987). Controversy first developed in the US press (Reid, 1986). It was spearheaded by scientists and environmentalists concerned that the impact of TBT on oysters in France and England could occur in the United States and particularly in the high oyster production areas of lower Chesapeake Bay, which were in the proximity of large naval facilities.

The US Navy's proposal met with resistance even though it called for using the safer (slow release) copolymer (for discussion of release rates, see Chapter 1, this volume) organotin paints, which were widely used by both commercial and pleasure craft at that time. Immediately following the statement of the then Secretary of the Navy, John F. Lehman, that the Navy would paint up to 50 ships with TBT in FY 86, Senator Paul S. Trible from Virginia introduced language into the Report of the Continuing Resolution for FY 86's Federal Budget that restricted the Navy's use of organotin compounds in antifouling paints until the US EPA approved of such paints. Navy advisors to Congress were comfortable with this language since the Navy had planned to use paint containing low-release-rate copolymer TBT.

Subsequently, a group of US scientists led by Robert J. Huggett of the Virginia Institute of Marine Science (located in the lower Chesapeake Bay near large naval shipyards and facilities and commercial floating and drydock facilities) and Edward D. Goldberg of the Scripps Institution of Oceanography became concerned about the impact of TBT on US oyster production. Attention was focused on results of studies in the United Kingdom and France and on preliminary studies conducted at the University of Maryland and at the Virginia Institute of Marine Science that suggested that larvae and eggs of *Crassostrea virginica* (the main oyster of Chesapeake Bay) were more sensitive than adults and that these results were similar to the results of Thain (1983) in the United Kingdom.

Also, there were a series of scientific questions which at that time could not be answered: (1) what concentrations of organotin compounds in the environment were causing the demise of oyster populations (acute and chronic dose exposure levels, for TBT and its metabolites); (2) what were the degradation rates, fate, and behavior of TBT and its metabolites in the environment; (3) what were the release rates of TBT in different types of antifoulant paints and the processes that affected these release rates; and (4) what was the distribution of TBT and its metabolites in the environment and the processes that influenced these distributions (Champ and Pugh, 1987).

Dr Huggett summarized the political issue when he noted that 'the Chesapeake Bay is the home of the largest naval facility in the United States,' and since the bay produces about a fourth of the nation's oysters, he adds, 'anything that might threaten those resources is looked at with a very callused eye' (Reid, 1986). Prompted by such concerns, the US Congress withheld funds for the Navy to paint its ships with TBT-based antifouling paintings, and stipulated that EPA must first complete a study of TBT's effects on the marine environment.

On 8 January 1986, the EPA announced and issued the initiation of a Special Review of all registered pesticide products containing TBT compounds (EPA, 1985) used as additives in antifouling paints. The DCI required the submission of product chemistry, ecological effects, environmental fate, TBT paint release rate, worker exposure, quantitative usage and application, and efficacy data. The decision to initiate the Special Review was triggered when EPA determined that the pesticide use of these compounds resulted in TBT exposure to nontarget aquatic organisms at concentrations resulting in acute and chronic toxicity, and that when applied as antifoulant paint, would meet or exceed the risk criteria.

Edward D. Goldberg, the keynote speaker at the 6th International Ocean Disposal Symposium (21–25 April 1986) at the Assilomar Conference Center in Pacific Grove, California, pointed out that 'TBT was perhaps the most toxic substance ever deliberately introduced to the marine environment by mankind.' He was concerned about the obvious parallel between the impact of TBT upon nontarget organisms and that of DDT in the

1960s (Goldberg, keynote address, unpublished manuscript; also see Goldberg, 1986).

At the 1986 International Organotin Symposium in Washington, DC (23–25 September 1986), the Chemistry Session Chairman, Frederick E. Brinckman of the National Institute of Standards and Technology (formerly the National Bureau of Standards), addressed the need to consider research on organotins as a 'new lost leader' for environmental studies because of their unique nature and properties as new high-technology chemicals. Dr Brinckman realized that studies of organotins would provide a future look at whole new lines of high-technology toxic compounds. As such, they would offer researchers an advance look at many future environmental dilemmas.

In oversight hearings on 30 September 1986 of the Committee on Merchant Marine and Fisheries of the US House of Representatives, the US Congress learned that it would take a year or more for the information requested by the DCs to be collected and analyzed fully by EPA before the decision of the Special Review could be completed. Subsequently, key members of Congress determined that the regulation of organotin compounds could not wait and decided to legislate the regulation of organotin compounds by United States law. Additional hearings were held on 29 April 1987 (Committee on Environment and Public Works, US Senate) and 8 July 1987 (Committee on Merchant Marine and Fisheries, US House of Representatives; see US Congress, 1986, 1987a,b).

The US coastal states concerned about impacts to commercial marine resources (Maine, New York, New Jersey, Delaware, Maryland, Virginia, Florida, California, Oregon, and Washington) followed closely the organotin issues through their representatives in Congress, their local water quality boards or natural resource agencies, and interested marine researchers in their states. By mid-1987, most coastal states were planning or had implemented restrictions on the use of organotins. Virginia was among the first to become concerned and implemented a regulatory strategy developed by the Virginia Water Control Board after which the subsequent federal law was modeled. One of the areas that states could regulate organotin usage was linked to the ambient water quality concentrations of organotins through State Environmental Quality Standards. Virginia initially accepted an EPA advisory allowable level of 10 ng l^{-1} for salt water; however, immediately upon passage of the Organotin Antifouling Paint Control Act (OAPCA), Virginia Water Control Board indicated that it thought that the level should be reduced to 1.0 ng l^{-1} (Commonwealth of Virginia, 1988) and was subsequently followed by the State of California with a level of 6.0 ng l^{-1}. Virginia also passed a state law that set the release rate at not greater than 4 $\mu g\ cm^{-2}\ d^{-1}$ (US Congress, 1987a,b). It should be noted that this tremendous interest and activity at the state level really gave Congress the support needed to enable it to draft US national legislation.

The United States Federal laws and regulations concerning the use (and subsequent disposal) of organotin compounds as additives or biocides in antifouling boat-bottom paints are coded in P.L. 100-333, the Organotin Antifouling Paint Control Act of 1988 (see section 3.4.8 of this chapter) (US Congress, 1988). The legislation was signed into United States law by President Reagan on 16 June 1988. It should be noted that the organotins are the only chemical compounds for which environmental legislation in the United States has been enacted by name. The earlier additional US laws that are related to the use, application, environmental and human health risks, monitoring, and disposal of organotin compounds are discussed in sections 3.4.1 to 3.4.7. One provision of OAPCA was to require the United States Navy to monitor US harbors in which navy vessels coated with organotin antifouling paints were home ported. These monitoring activities were focused in Pearl Harbor, Hawaii; San Diego, California and Norfolk, Virginia (see Grovhoug, 1992; Grovhoug *et al.*, Chapter 25).

The purpose of the Antifouling Paint Control Act of 1988 was to reduce immediately the quantities of organotins entering the waters of the United States. In the Act, there are two permanent sections: the 25-m size requirement and the prohibition of retail sale of TBT antifouling paint additives. The release rate portion of the bill had a duration time period that would remain in effect until a final decision of the Administrator of the EPA regarding continued registration of TBT as an ingredient in antifouling paints was made.

The following sections present in greater detail the regulatory and legislative actions associated with the use of organotin compounds in antifouling paints in the United States, other nations, and international conventions.

3.3. US ENVIRONMENTAL PROTECTION AGENCY REGULATORY ACTIONS

On 8 January 1986, the EPA announced the initiation of a *Special Review* of all registered pesticide products containing TBT compounds used as additives in antifouling paints applied to boat and ship hulls to inhibit the growth of certain aquatic organisms (US EPA, 1986a,b). The EPA issued the Data Call-in Notice (DCI), under the authority of Section 3(c) (2) (B) of the Federal Insecticide, Fungicide, and Rodenticide Act (FIFRA), to all registrants of TBT antifoulant products. The DCI required the submission of product chemistry, ecological effects, environmental fate, TBT paint release rate, worker exposure, quantitative usage and application, and efficacy data (US EPA, 1986b). When organotins were first registered in the US (in some cases 10–15 y earlier), the initial registration data for TBT did not evaluate the toxicity of TBT to oysters (or even molluscs), which was unknown at that time (US EPA, 1983).

In initiating the Special Review for TBT (US EPA, 1986a), the EPA identified significant gaps in the technical information needed to support the registration of organotin compounds used in antifoulant paints. To obtain the information necessary to assess the risks and benefits, EPA issued a Data Call-in Notice (DCI) to the registrants levying extensive data requirements to continue their registration pursuant to EPA's authority under the Federal Insecticide, Fungicide, and Rodenticide Act (FIFRA) (US EPA, 1986b). Under FIFRA, EPA registers pesticides for use on the basis of scientific data adequate to demonstrate with appropriate safety factors and precautions that their use following label directions will not result in unreasonable adverse effects or risk to human health or the environment. If new or additional scientific evidence is introduced that causes EPA to question the safety of a registered pesticide, the agency can invoke the Special Review process under FIFRA.

The Special Review process is a comprehensive and intensive scientific and economic assessment of the risks and benefits of a pesticide to determine if a pesticide should be registered by EPA for specified uses. This process provides an opportunity for all interested parties [pesticide manufacturers (registrants), the user community, research organizations, environmental groups and the general public] to provide EPA with data, information, and comments on the chemical(s) under consideration for reregistration. The decision to initiate the Special Review was triggered when EPA determined that the pesticide use of these compounds resulted in TBT exposure to nontarget aquatic organisms at concentrations (measured in parts per trillion for some molluscs) resulting in acute and chronic toxic effects, and when applied as antifoulant paint, met or exceeded the risk criteria as described in 40 CFR 162.11. A review of the information used to make this decision was published in the 'Tributyltin Support Document (PD-1)' (US EPA, 1985).

Under the Special Review, the EPA requested data from organotin registrants under several areas including: (1) chemical release rate studies of TBT from antifoulant paints; (2) product chemistry; (3) ecological effects, environmental fate; (4) worker exposure; (5) quantitative usage; (6) efficacy of TBT products, (7) specific toxicity tests with a wide range of organisms; and (8) specific environmental monitoring data. For the release rate DCI, which was issued in July 1986, a protocol for testing and chemically measuring release rates was developed in cooperation with the American Society for Testing and Materials (ASTM). The protocol provides information on daily release rates and 6-week cumulative release rates for different TBT paint products. Based on the rate of compliance with this DCI, about 300 of the 359 registrations for TBT were voluntarily canceled by companies within several months. Nevertheless, it is anticipated that it could take several years for the information requested by the DCIs to be collected and analyzed by EPA before the decision of the Special Review could be completed.

The regulatory and policy options being considered by the US EPA were discussed by Dr John Moore, the Assistant Administrator for Pesticides and Toxic Substances, in his presentation before the Subcommittee on Environmental Protection, of the Committee on Environment and Public Works of the US Senate, 29 April 1987. Dr Moore suggested a two-phase regulatory process, based upon results of initial monitoring programs carried out in selected estuaries and on the information obtained from the EPA DCI notices on release-rate test data. Dr Moore reported that EPA would be in a position to make a preliminary regulatory decision by Fall 1987. At the Hearing, Dr Moore felt that it would be 'premature to speculate on specifics' but that 'these measures will most likely involve the initiation of actions to restrict or cancel, the registrations of organotin antifouling paints, based on release-rate or other criteria as appropriate.'

At that time Dr Moore felt that as data on the long-term ecological effects and environmental fates became available over the next 4 y, and as additional monitoring data are accumulated, 'the Agency might refine its interim decision, including the imposition of more stringent restrictions, if necessary.' Dr Moore also discussed the progress EPA's Office of Water was having in developing water quality criteria with regard to TBT, which are developed under Section 304(a) of the Clean Water Act. Under this Congressional mandate, the EPA has to follow a prescribed set of guidelines in deriving criteria. These guidelines establish a minimum required data set, and define the methods for assessing the data. (Generally, it requires about 2 y or more to develop a criteria document.)

A water quality advisory was issued by EPA in October 1987 (EPA, 1987d), through the Criteria and Standards Division of the Office of Water Regulations and Standards. Water quality advisories have been developed as an interim vehicle for transmitting the best available scientific information concerning the aquatic life and human health effects of selected chemicals for which information is needed quickly but for which sufficient data, resources, or time are not available to allow derivation of national ambient water quality criteria. The TBT aquatic life advisor issued by EPA (1987d) reports that

> if the measured or estimated ambient concentration of tributyltin (TBT) exceeds 26 ng l^{-1} in fresh water or 10 ng l^{-1} in salt water, the discharger, after consultation with an appropriate regulatory agency, should evaluate the available exposure and effect data and complete one or more of the following options within a reasonable period of time:

a. Obtain additional measurements of the ambient concentration.
b. Improve the estimate of the ambient concentration.
c. Obtain additional laboratory and/or field data on the effects of TBT on aquatic organisms and their uses so that a revised, and usually higher, aquatic life advisory concentration or a water quality criterion can be derived.
d. Conduct appropriate toxicity tests on the effluent.
e. Reduce the ambient concentration of TBT to an acceptable level.

In October 1987, EPA issued a regulatory proposal (PD 2/3), based on a review of the risks and benefits of tributyltins (TBTs), to reduce loading of TBT without producing economic loss or harm to users. In this preliminary action, EPA proposed limiting TBT use to products: (1) with a maximum 14-d cumulative release rate of 168 $\mu g\ cm^{-2}$ of TBT (short term); (2) whose labels prohibit use on non-aluminum-hulled vessels less than 65 feet (19.8 m) in length; (3) classified as restricted use pesticides; and (4) in compliance with application, removal, and disposal requirements (US EPA, 1987a,b).

At the same time, EPA also issued in the *Federal Register* a partial conclusion of the Special Review on 4 October 1988 (53 FR 39022). This regulatory announcement incorporated the restrictions of the Organotin Antifouling Paint Control Act (OAPCA) of 1988 (see section 3.4.8), and reclassified most of TBT antifouling paint products as restricted-use pesticides, limiting sale to certified commercial applicators trained specifically in proper application, removal, and disposal. The action also limits the use of TBT to certified commercial applicators or to persons under the direct supervision of an on-site certified commercial applicator. These steps were taken to minimize environmental and human health risks.

In issuing this regulatory action, EPA also complied with the mandate of Congress to determine if these regulatory and legislative actions were adequate in reducing TBT concentrations in the aquatic environment and in protecting nontarget species. The *Federal Register* notice also reported that EPA might at a later date issue a final determination regarding TBT release rates, which would replace OAPCA's interim release rate restriction, if data submitted by registrants indicate a need for further reduction in loading to the environment, if a lower efficacious release rate becomes available, or if a more definitive standard method to measure release rates is developed resulting in greater precision or sensitivity.

In this decision, EPA indicated that available data did not warrant a more restrictive action at that time. Nevertheless, the agency needed additional data and information for use in the possible refinement of its risk/benefit assessment and issued a second DCI requiring registrants to conduct and report the results of a 3–5 y monitoring study in US waters. The purpose of this monitoring requirement was to provide EPA with information on the extent and levels of TBT in the aquatic environment, the impact of TBT on selected indicator organisms in situ, and the effect of the regulatory actions taken to date on TBT levels adjacent to harbors, drydocks, marinas, and other sensitive aquatic areas.

As a public information and risk communication action, EPA's Office of Policy Analysis of the Office of Policy, Planning and Evaluation, and the National Marine Pollution Program Office of the National Oceanic and Atmospheric Administration (NOAA) co-funded an assessment (Champ and Bleil, 1988b,c,d) to prepare and evaluate different public information leaflets (environmental and human health risks) as public self-regulatory actions) for distribution with the assistance of the Coastal States Organization.

From 30 November to 2 December 1988, EPA chaired a Tributyltin Workshop sponsored by the Organization for Economic Cooperation and Development (OECD) in Paris. The workshop focused on environmental monitoring, analysis of TBT residues in the water column, sediment, and animal tissues, and release rate testing. A draft document for monitoring guidance of protocol was also reviewed with the aim of its being issued later in the year as a DCI. At the end of the conference, EPA agreed that the United States would take the lead in the development of certified standards for chemical analysis of TBT in tissue, sediment, and water, and the organization of intercalibration exercises.

The Monitoring/Efficacy Data Call-in Notice was issued on 22 March 1989 to all registrants. The monitoring program was required to analyze for tributyltins, mono- and dibutyltin, and tetrabutyltin, a contaminant of TBT, and to create an historical database of levels in water, tissue (quarterly), and sediment (semiannual) in the vicinity of drydocks, shipyards, commercial harbors and ports, and marinas. A mussel watch monitoring program was also included similar to NOAA's National Status & Trends Program (NS&T) but in the areas recommended above. There also was a requirement for registrants to evaluate the effects of tributyltin concentrations on appropriate sensitive biological indicator species 5 y following the monitoring program with *Crassostrea gigas* (Pacific oyster) even though it is an introduced species (from Japan) and not native to the United States, and *Nucella lapillus* (dogwhelk snail).

The registrants' consortium lobbied hard for a less intensive and expensive study than EPA had proposed. In addition to reduced costs, they wanted to rely on the results of navy harbor monitoring studies in San Diego, Pearl Harbor, and Norfolk to fulfill EPA's requirement for monitoring in the Pacific Southwest, Hawaii, and Atlantic Northeast. EPA subsequently incorporated some of the modifications requested to reduce the cost of the monitoring program while maintaining its scientific integrity. Nevertheless, the Atochem North American and Sherex Chemical Company filed a complaint in the US District Court of the District of Columbia on 26 November 1990 challenging EPA's authority to require the monitoring data under FIFRA. On 12 March 1991, the Court rendered a decision in favor of EPA's authority to impose the aquatic monitoring requirement. The DCI was amended to reflect the time delays, and the scoping – preliminary survey was initiated in May 1991 and the long-term monitoring program in October 1991 (EPA, 1991).

In following up on the above requirements, it should be noted that a preliminary US field survey (sponsored by a consortium of TBT manufacturers) for tributyltin in aquatic environments was undertaken by Little (1991), in which over 800 samples were collected and analyzed.

On 1 July 1994, the EPA released a press advisory announcing the voluntary cancellation of tributyltin fluoride (TBTF) by the sole registrant of technical TBTF. EPA determined that the sole registrant had failed to meet its obligation of reregistration of its TBTF products by its failure to meet certain data requirements of the October 1991 DCI.

3.4. UNITED STATES LAWS

The use, application, and disposal of organotin compounds in antifouling paints also fall under the jurisdiction of the following United States laws.

3.4.1. FEDERAL INSECTICIDE, FUNGICIDE, AND RODENTICIDE ACT (FIFRA)

This act provides for the identification of environmental problems, the setting of water quality standards for toxic substances, and the monitoring of environmental concentrations of toxicants for the purpose of regulating toxic substances. Authority to develop and implement regulations under this act is given to EPA. A pesticide product may be sold or distributed in the United States only if it is registered or exempted from registration under the Federal Insecticide, Fungicide, and Rodenticide Act (FIFRA), as amended (US Congress, 1972c). Before a product can be registered, 'it must be shown that it can be used without "unreasonable adverse effects on the environment" (FIFRA Section 3(c)(5)), that is, without causing any unreasonable risk to man or the environment, taking into account the economic, social, and environmental costs and benefits of the use of the pesticide' (FIFRA Section 2(bb)).

When evidence is identified that causes EPA to question the safety of a registered pesticide (or group of pesticides), that pesticide may then be reviewed under the EPA's Special Review Process. The Special Review (RPAR – Rebuttable Presumptions Against Registration) Process (40 CFR 162.11) provides a mechanism through which EPA gathers risk and benefit information about pesticides that appear to pose unreasonable risks of adverse effects to human health or the environment.

Most of the early scientific data and information about the toxic effects from organotin antifouling paints were from UK scientists at the Ministry of Agriculture, Food and Fisheries (MAFF) fisheries research laboratories. Papers were written for and presented at regional meetings of the International Council for the Exploration of the Sea (ICES) and were not readily available to US scientists in the general scientific literature. Michael G. Norton, originally from the MAFF laboratory at Burnham-on-Crouch, where most of the early UK research was conducted, in 1986 was on assignment as the First Secretary for Science at the British Embassy in Washington, DC. Dr Norton shared with the first author of this chapter (who at that time was serving as a senior science advisor in EPA's Office of Policy, Planning and Evaluation) the early findings of the MAFF researchers. Subsequently, Champ (1985, 1986a,b,c,d) reviewed these ICES papers, navy studies and a review of US analytical capabilities for organotins. The long-term decline of oyster production in Chesapeake Bay, combined with the finding of deformed oysters in Coos Bay, Oregon (Wolniakowski *et al.*, 1987), led EPA to conclude that the use of TBT antifouling paints might result in exposure of certain nontarget marine organisms to TBT concentrations. Consequently, these activities convinced the agency to initiate the Special Review Process through the Office of Pesticides to collect and assess additional data.

In the Special Review Process, EPA, through the issuance of notices and support documents, publicly establishes its position and invites pesticide registrants, Federal and state agencies, user and environmental groups, and any other interested persons to participate in the agency's review process. If EPA determines that the risks appear to outweigh the benefits, then it can initiate action under FIFRA to cancel, suspend, and/or modify the terms and conditions of a given pesticides registration.

The Special Review Process is an intensive scientific risk assessment and benefit analysis (which can utilize the expertise of 100 or more of EPA staff and take up to 3–7 y (or more) depending upon available data) that culminates in a decision on the future registration of the uses of the pesticide being reviewed. In initiating the Special Review, EPA selected to regulate the use of organotin compounds in antifouling paints under the legislative mandates of FIFRA.

3.4.2. TOXIC SUBSTANCES CONTROL ACT (TSCA)

The Toxic Substances Control Act (TSCA), P.L. 94-469, establishes responsibilities for definition of problems caused by the presence of toxic substances in the environment, and for setting acceptable levels or standards to insure maintenance of acceptable water quality (US Congress, 1974). Monitoring environmental levels, drafting of regulations concerning use of the toxic substance, and the enforcement of these regulations is also provided by TSCA. Under TSCA, EPA has the authority to regulate the use of organotin compounds in the environment. For the past decade, EPA has utilized FIFRA as the preferred authority for the regulation of organotins.

3.4.3. CLEAN WATER ACT (CWA) AND AMENDMENTS

The Federal Water Pollution Control Act of 1972 as amended by the Clean Water Act of 1977 (P.L. 92-500 and P.L. 95-217) is known as the Clean Water Act (CWA) (US Congress, 1972a). The objective of this act is to restore and maintain the chemical, physical, and biological integrity of the nation's waters. Regarding control of marine pollution, the act provides for a water quality surveillance system for monitoring the quality of navigable waters including the contiguous zone (Section 104[a][5]). Although the major responsibility under this act is the Administrator of the EPA, the act expressly encourages other Federal, State, and local agencies to cooperate in establishing national programs for the prevention, reduction, and elimination of pollution. This includes research, investigations, experiments, training, demonstrations, surveys, and studies relating to the causes, effects, extent, prevention, reduction, and elimination of pollution (Section 104[a][1]). The National Oceanic and Atmospheric Administration (NOAA) is directed to establish, equip, and maintain a water quality surveillance system for the purpose of monitoring the quality of navigable waters, the contiguous zone, and the oceans.

Section 303 also directs NOAA to engage in activities responsive to the establishment of water quality standards, and effluent discharge standards in Section 307(a). The act directs the Administrator of the EPA to identify and quantify the distribution of in-place toxic contaminants with emphasis on toxic substances in harbors and navigable waterways. The EPA Administrator is also authorized, acting through the Secretary of the Army, to make contracts for the removal and appropriate disposal of such materials from critical port and harbor areas (Section 115). Therefore, in the future, should the concentrations of organotin compounds increase in estuarine bottom sediments faster than they degrade, it may become necessary to implement the regulatory process as developed for the disposal of contaminated dredged materials.

3.4.4. OCCUPATIONAL SAFETY AND HEALTH ACT (OSHA)

With regard to organotins, the focus of the Occupational Safety and Health Act (OSHA) (P.L. 91-596; US Congress, 1976) would be on the health and safety of shipyard and marina workers and 'do-it-yourself' boat owners exposed to organotin compounds during application or removal of antifouling paints. Research in the area of worker protection is being conducted by the Navy and by the paint companies. Evaluation of protection strategies to insure the safety of workers during the handling and use of TBT antifouling paints has been conducted by the US Navy, which includes hull sealing, mobile spray paint enclosures, paint spray collection mechanisms, and drydock clean-up means. Investigations have been conducted by the US Navy on disposal alternatives for spent paint film and grit resulting from the removal of spent paint films (Adema and Schatzberg, 1983; Schatzberg, 1987).

Worker exposure and TBT body burden studies have been conducted by M&T Chemical Corporation. Alternative worker protection coverings were also investigated in the study, which was conducted in response to the DCI from EPA under FIFRA rather than under OSHA. The OSHA regulations for permissible paint exposure currently exist, and no suggestions have been made that these regulations would be evaluated or revised until the final conclusion of the EPA Special Review.

3.4.5. MARINE PROTECTION, RESEARCH AND SANCTUARIES ACT (MPRSA)

The Marine Protection, Research, and Sanctuaries Act of 1972 is commonly called the 'Ocean Dumping Act' (P.L. 92-532; US Congress, (1972b). In this Act, Congress declared that it is the policy of the United States to regulate the dumping of all types of materials into ocean waters that would adversely affect human health, welfare, or amenities, or the marine environment, ecological systems, or economic potentialities. The dumping of wastes or toxic materials at sea is strictly regulated by the US EPA and the London Dumping Convention of which the US is a Contracting Party. Regarding organotin compounds, this Act is concerned with the disposal of contaminated dredged materials from waterways, harbors, or marinas that would be contaminated with toxic materials. Under Title II, Section 201, NOAA, in coordination with the US Coast Guard, and EPA are responsible for a comprehensive and continuing program of monitoring and research regarding the effects of the dumping of material into ocean waters.

Under Section 202, NOAA is also responsible for a comprehensive and continuing program of research and monitoring with respect to the possible long-range effects from exposure to contaminants. This act also assigns the Army Corps of Engineers (COE) with the authority to issue permits for the transportation of dredged materials for the purpose of disposal at sea (ocean dumping), 'where it has been determined that the dumping will not unreasonably degrade or endanger human health, welfare or amenities, or the marine environment, ecological systems or economic potential.' In order to establish the potential environmental impact or risk from disposal of contaminated sediments, the Chief of Engineers is authorized to conduct a comprehensive program of research, study, and experimentation relating to dredged material. The Ocean Assessments Division (OAD) in NOAA has been designated this responsibility, and studies were initiated in 1987 to monitor organotin environmental concentrations in bottom sediments by the Status and Trends Program.

3.4.6. NATIONAL OCEAN POLLUTION PLANNING ACT (NOPPA)

The National Ocean Pollution Planning Act (NOPPA), P.L. 95-273, and amendments directs the Administrator of NOAA, in consultation with the Director of the Office of Science and Technology Policy of the Executive Office of the President, and other appropriate Federal officials to prepare and update biennially a comprehensive 5-y plan (Section 4) for the overall federal effort in ocean pollution research and development and monitoring (US Congress, 1978). It further directs the Administrator of NOAA to establish within NOAA a comprehensive, coordinated, and effective ocean pollution research and development and monitoring program consistent with the 5-y plan (Section 5) and to provide financial assistance for research and development and monitoring projects or activities that are needed to meet priorities of the 5-y plan if these are not being adequately addressed by any Federal department, agency, or instrumentality (Section 6). Finally, the Act directs that the Administrator of NOAA shall ensure that results, findings, and information regarding federal ocean pollution research and development and monitoring programs be disseminated in a timely manner and useful form to federal and non-federal user groups having an interest in such information (Section 8).

3.4.7. NATIONAL ENVIRONMENTAL POLICY ACT (NEPA)

The National Environmental Policy Act (NEPA), P.L. 91-190, declared a national policy that encourages productive and enjoyable harmony between humans and the environment; to promote efforts that will prevent or eliminate damage to the environment and biosphere and stimulate the health and welfare of humans; to enrich the understanding of the ecological systems and natural resources important to the nation; and established the Council on Environmental Quality (CEQ) (US Congress, 1969). This act requires that all agencies of the federal government shall utilize a systematic, interdisciplinary approach that will insure the integrated use of environmental, social, and economic information in planning and decision-making to assess the impact of human activities on the environment; that unquantified environmental amenities and values may be given appropriate consideration in decision-making along with economic and technical considerations; and that a detailed statement of the action, commonly referred to as an Environmental Impact Statement (EIS), which details the environmental impact of the proposed action, including identifying alternatives to the proposed action, and identification of any irreversible and irretrievable commitments of resources.

In 1986 the US Navy prepared an Environmental Assessment (EA), which is the first step in fulfilling the requirements of NEPA, and issued an Interim Finding of No Significant Impact (Interim FONSI) for the proposed fleet-wide implementation of organotin antifouling hull paints (US Navy, 1986; see also Bailey, 1986). This decision was based on 10 y of navy studies and comprehensive environmental assessment (NAVSEA, 1984). When the Navy issued its Interim FONSI, it made commitments to monitor initial implementation operations, continue environmental studies, and reassess in 1988 its decision to proceed with full fleetwide implementation (NAVSEA, 1986).

3.4.8. THE ANTIFOULING PAINT CONTROL ACT OF 1988

Organotin is the only chemical compound regulated by law in the United States in which environmental legislation has been enacted solely for the chemical by name: the Organotin Antifouling Paint Control Act of 1988 (33 USC 2401) (US Congress, 1988). An historical perspective is presented below on how this law came into being.

The purpose of the Antifouling Paint Control Act of 1988 was 'to protect the aquatic environment by reducing immediately the quantities of organotin entering the waters of the United States.' In the Act, signed into law by President Reagan, there are two permanent sections, the 25-m size requirement, and the prohibition of retail sale of TBT antifouling paint additives. The release rate portion of the bill had a duration time period that would be in effect until a final decision of the administrator of the EPA regarding continued registration of TBT as an ingredient in antifouling paints takes effect.

It is interesting to note that in the United States the major stimulus for development of this act was initially proposed actions (US Navy proposal to use organotin antifouling paint for the fleet) for which the act subsequently did not regulate due to its exclusion of vessels greater than 25 m in length. In 1984 the US Navy prepared an 'Environmental Assessment on Fleetwide Use of Organotin Antifouling Paint' (NAVSEA, 1984) in compliance with the National Environmental Policy Act. In 1986, the US Navy issued a Finding of Non-Significant Impact (Interim FONSI) after determining that the fleetwide (600 ships) use of organotin antifouling paints would not have significant adverse environmental consequences in navy harbors (US Navy, 1986). These actions stimulated the development of a major controversy over concern that the impact of TBT on oysters that occurred in France and England could occur in the United States and particularly in the high oyster production areas of lower Chesapeake Bay, which were in the proximity of large naval facilities. Also, a series of scientific questions were raised by scientists who participated in the international organotin symposia (United States, Canada, and Monaco) held in 1986, 1987 and 1988, which at that time could not be answered. Additional symposia were held in 1989 and 1990 (see Proceedings of National and International Organotin Symposia in References).

Immediately following Navy Secretary John F. Lehman's statement that the Navy would paint up to 50 ships with TBT in FY 86, Senator Paul S. Trible from Virginia introduced on November 12 1985 the following language into the Report of the Continuing Resolution for FY 86's Federal Budget: that 'none of the funds appropriated by this joint resolution may be obligated or expended to carry out a program to paint any naval vessel with paint known by the trade name of Organotin or with any other paint containing the chemical compound tributyltin, until such time as the EPA certifies to the Department of Defense that whatever toxicity is generated by organotin paints as included in Navy specifications does not pose an unacceptable hazard to the marine environment.' Navy advisors to Congress were comfortable with this restriction since the Navy had planned to use the low-release-rate copolymer TBT paints, and figured that EPA would be able to approve some of these paints quickly.

However, Senator Trible's wording required EPA to certify that specific TBT paint formulations that the Navy wanted to use did not pose *an unacceptable hazard to the marine environment*. This, in essence, required EPA to analyze all of the toxicological and environmental data available from the DCIs and all other sources and to complete its Special Review Study (which could take several years) prior to giving the Navy permission to use a specific paint formulation. For fiscal years 1987 and 1988, this language was placed in the actual legislation of each year's Continuing Resolution. For more information on the concerns of Congress during this period, refer to the following documents: (1) the Hearing Record of the House Committee on Merchant Marine and Fisheries, 30 September 1986 (US Congress, 1986); (2) the Senate Congressional Record dated 2 February 1987 – S1481 (US Congress, 1987a); (3) the Senate Hearing Record of the Committee on Environment and Public Works, 29 April, 1987 (US Congress, 1987b); and (4) the Hearing Record of the House Committee on Merchant Marine and Fisheries 8 July 1987 (US Congress, 1987c).

Senator William Cohen of Maine joined Senator Trible in pushing through these amendments to the Continuing Resolution Bills. Subsequently, Senators Trible and Cohen began to draft comprehensive legislation, first as S. 428 (2 February 1987). The exact language in this bill was also introduced in the House by Congressman Bateman (VA) as H.R. 1046 on 9 February 1987. Senate Hearings were held for Bill S. 428, by the Committee on Environment and Public Works (US Congress, 1987b); However, the bill was never marked up. Senators Trible and Cohen on 15 October 1987 introduced Senate Bill S. 1788 the 'Tributyltin-Based Antifouling Paint Control Act of 1987', US Senate 1988 (which subsequently became the Senate legislative vehicle. The bill was cosponsored by Senators Mitchell and Chatee and focused on regulating the release rate of TBT from the paint to the water column in an attempt to attain some determined minimal-effect level. In discussions during the drafting of this bill, the TBT release rates proposed varied from 0.5 (±20%) $\mu g\ cm^{-2}\ d^{-1}$ of wetted hull surface area to 5.0 (±20%) μg

cm^{-2} d^{-1} due to standardization of the release rate testing protocol.

A new bill was also introduced in the House on 29 April 1987 by Mr Jones and subsequently amended by Mr Studds entitled 'the Organotin Antifouling Paint Control Act of 1987.' This bill (H.R. 2210) had the following requirements: a release rate of 'not more than 5.0 $\mu g/cm^2/d$, and prohibited use of organotin antifouling paints on any vessel less than 65 feet (19.8 meters) in length, except for aluminum hull vessels.' This bill became the House legislative vehicle, and began to be compared to the Senate legislative vehicle (S. 1788).

The following is a chronology of events associated with the passage of these bills:

9 November 1987, House passed H.R. 2210 and sent to the Senate;
12 December 1987, Senate passed H.R. 2210 with S. 1788 language inserted into the bill in place of H.R. 2210 language;
Staff negotiations for agreement on 4.0 release rate (rather than 3.0 – Senate, or 5.0 – House), 25 m length prohibition (rather than House 65 feet), 6 months for sale of existing stocks, and 1 y for application of existing stocks;
18 April 1988, Senate Amended H.R. 2210 and passed the bill back to the House, adding a section requiring a Water Quality Criteria Document from EPA by 30 March 1989;
24 May 1988, House agrees to Senate Amendment of H.R. 2210 and sends the bill to the President for signature into law; and
16 June 1988, President Reagan signs the bill into United States law as Public Law 100-333: The 'Organotin Antifouling Paint Control Act of 1988.'

The following paragraphs summarize highlights of key sections and identify specific regulatory aspects of the act.

Sec. 3. Definitions: The act defines the term 'organotin' as 'any compound of tin used as a biocide in an antifouling paint.' Organotins defined in this way specifically includes the three (TBT) compounds: bis(tributyltin) oxide, tributyltin fluoride, and tributyltin methacrylate. These are the only three organotin compounds regulated by EPA as active antifoulant registrations for organotins. This definition also includes the degradation products from TBT dibutyltin (DBT) and monobutyltin (MBT). The term 'release rate' is a measurement of the rate at which organotin is released from an antifouling paint over the long term, as determined by EPA, using either the American Society for Testing and Materials (ASTM) standard test method, which the EPA required in its 29 July 1986 DCI notice on tributyltin compounds used in antifouling paints or any similar test method specified by EPA. The term 'vessel' includes every description of water craft or other artificial contrivance used, or capable of being used, as a means of transportation on water.

Sec. 4. Prohibitions: 'No person in any State may apply to a vessel that is less than 25 meters in length an antifouling paint containing organotin' with the following exceptions: '(1) the aluminum hull of a vessel that is less that 25 meters in length; and (2) the outboard motor or lower drive unit of a vessel that is less than 25 meters in length.'

No person in any State may: (1) sell or deliver to, or purchase or receive from, another person an antifouling paint containing organotin; or (2) apply to a vessel an antifouling paint containing organotin; unless the antifouling paint is certified by the Administrator [of EPA] as being a qualified antifouling paint containing organotin, and (3) sell or deliver to, or purchase or receive from, another person at retail any substance containing organotin for the purpose of adding such substance to paint to create an antifouling paint.

Sec. 6. Certification: The Administrator shall certify each antifouling paint containing organotin that the Administrator has determined has a release rate of not more than

4.0 $\mu g\ cm^{-2}\ d^{-1}$ on the basis of the information submitted to EPA in response to its DCI notices.

Sec. 7. Monitoring and Research of Ecological Effects: Section 7(a) of the act requires that the Administrator of the EPA, in consultation with the Under Secretary of Commerce for Oceans and Atmosphere (NOAA), monitor the concentrations of organotin in the water column, sediments, and aquatic organisms of representative estuaries and near-coastal waters in the United States. These monitoring programs are to remain in effect for 10 y following the date of enactment of the act. In addition, Section 7(b) of the act requires the Secretary of the Navy to provide for periodic monitoring (not less than quarterly), of waters serving as the home port for any navy vessel coated with an antifouling paint containing organotin compounds to determine the concentrations of organotins in the water column, sediments, and aquatic organisms of such waters. The act also requires that the US Navy continue existing navy programs evaluating the laboratory toxicity and environmental risks associated with the use of antifouling paints containing organotin.

To the extent practicable, the EPA shall assist states in monitoring waters in such states for the presence of organotin as a requirement to initiate a nationwide tributyltin monitoring program and in analyzing samples taken during such monitoring. The Secretary of the Navy is required to submit annual reports of its findings to the EPA and to each state in which a navy home port is monitored. The EPA is required to submit annual reports to Congress that detail the results of all of these monitoring programs. A special report to Congress by EPA (due 5 y from the enactment) is mandated to assess the effectiveness of existing laws and regulations concerning organotin compounds in ensuring protection of human health and the environment, with recommendations for additional measures to protect human health and the environment.

Sec. 8. Alternative Antifoulant Research: The Navy and EPA shall conduct research into chemical and non-chemical alternatives to organotin antifouling paints with a report to Congress at the end of a 4-y period of their findings.

Sec. 9. Water Quality Criteria Document: Not later than 30 March 1989 the Administrator shall issue a final water quality criteria document concerning organotin compounds pursuant to Section 304(a) of the Federal Water Pollution Control Act.

Sec. 10. Penalties: Any person violating Sections 4 or 5 shall be assessed a civil penalty of not more than $5000 for each offense.

Sec. 11. Other Authorities; State Laws: Nothing in this act shall limit or prevent EPA from establishing a lower permissible release rate for organotin under other authorities.

Sec. 12. Effective Dates; Use of Existing Stocks: Existing (on-the-shelf) stocks of antifouling paints containing organotin and organotin additives that exist before the date of the enactment of this act may be sold, delivered, or purchased up until 180 d of the enactment of this act. Existing stocks of any antifouling paints containing organotin and organotin additives that exist before the date of the enactment of this act may be applied for up to a time period not to exceed 1 y. In reality, this last section delays the effective starting date of the act for 1 y for the prohibitions that enact restrictions for specific release rates and boat lengths.

Although OAPCA and subsequent US EPA regulations allowed use of TBT coatings by large vessels, the US Navy in 1989 decided not to use organotin coatings because of environmental concerns and the uncertain regulatory future at the state and regional levels.

3.5. INTERNATIONAL REGULATORY POLICIES AND PRACTICES

International regulations are briefly summarized in the following sections; greater detail is presented in individual chapters prepared by Abel (Chapter 2), Abel *et al.* (1986, 1987); Stebbing (1985); and Alzieu (1986, 1991), Anonymous (1994), Hunt (1994), Evans *et al.* (1995a), WWFN (1995).

3.5.1. FRANCE

France was the first country to regulate the use of organotin antifouling paints in an attempt to reduce environmental concentrations. On 19 January 1982, the French Ministry of Environment announced a temporary 2-y ban on TBT paint containing more than 3% (by weight) organotin for the protection of hulls of boats of less than 25 metric tonnes, for both the Atlantic coasts and the English Channel, following the official recommendation issued by the 'Evaluation Committee on Ecotoxicity of Chemical Substances.' The decree of 16 September 1982 extended the ban to the whole coastal area and to all organotin paints, beginning 1 October, 1982. These regulations also only allow the application of antifouling paints containing organotin to hulls of all boats and marine craft having an overall length of greater than 25 m. Hulls made of aluminum or aluminum alloys were exempted from the ban. This extension was effective through 12 February 1987 and banned the application of antifouling paints containing organotin on vessels less than 25 m in length (Alzieu, 1986, 1991).

Alzieu *et al.* (1980, 1981–1982) and Alzieu and Portman (1984) have reported both the concentrations of organotins and anomalies in oysters in Arcachon Bay since 1982. Concentrations of organotins in marina waters in Arcachon Bay decreased slowly between July 1982 and November 1985 as the ban became effective. This decrease in organotin contamination in Arcachon Bay has paralleled considerable improvement in oyster culture; the settling and survival of oyster spats was reported satisfactory in 1982 and 'superabundant' in the following years. The number of incidences of oyster shell thickening and abnormalities was drastically reduced in 1983; however, while 40% of the oysters near marinas in 1985 still demonstrated characteristics of chambering, the effects on growth were less noticeable (Alzieu, 1986; Alzieu *et al.*, 1986). Alzieu *et al.* (1989) continued to measure concentrations of organotins in seawater samples in 1986–1987, and expanded the sampling areas to northern sites on the French Atlantic coast, including marinas and oyster culture areas [Anse de Camaret (Brest), marinas at La Trinte and La Rochelle]. They found that organotin concentrations ranged from <2 to 1500 ng l^{-1} and that there was good agreement between TBT levels in seawater and the frequency of oyster shell anomalies in Arcachon Bay.

Sarradin *et al.* (1994) analyzed 14 sediment cores in Arcachon Bay in 1990 and found widely varying concentrations and considerable amounts of butyltin compounds in bottom sediments, even though concentrations in the water have been below 5 ng l^{-1} since 1987.

3.5.2. UNITED KINGDOM

This section is a brief summary of the UK regulatory actions; for a more complete discussion, see Abel (Chapter 2). The first regulatory action in the United Kingdom to reduce the environmental impact of organotin compounds from antifouling paint was announced by the Environment Minister in Parliament on 24 July 1985. The action consisted of the following five steps: (1) develop regulations to control the retail sale of the most damaging organotin-containing paints [beginning 1 January 1986, they intended to ban the use of 'free association' organotin-based paints by small boat owners, and to set the maximum levels for the organotin content of 'copolymer' paints]; (2) establish a notification scheme for all new antifouling agents; (3) develop guidelines for the cleaning and painting of boats coated with antifoulants; (4)

propose the establishment of a provisional ambient environmental quality target (EQT) for the concentration of tributyltin in water [20 ng l^{-1} was proposed as the UK's EQT]; and (5) coordinate and further develop organotin monitoring and research programs so that the government could assess the effectiveness of these regulatory actions at a later date.

The first legislation to control the retail sale of organotin-based antifoulant paints was introduced on 18 December 1985 under the Control of Pollution (Anti-Fouling Paints) Regulations of 1985, which came into force on 13 January 1986. These regulations were developed under Sections 100 and 104(1) of the Control of Pollution Act of 1974. They prohibited the retail sale of antifouling paints containing organotin compounds if (1) the total concentration of tin in dried copolymer paints exceeded 7.5% (by weight) of tin, or (2) the total concentration of tin in other non-copolymer [free association] paints exceeded 2.5% (by weight) of tin [The Control of Pollution (Anti-Fouling Paints) Regulations (UK DoE, 1986a,b)]. These regulatory actions were enacted with the provision that they would be reviewed with the interim results of the comprehensive scientific studies that were being carried out by both government and non-government laboratories, which included studies on the distribution, fate, and effects of TBT in the environment and laboratory toxicity studies.

During the 1986 boating season, a detailed monitoring program was conducted by the Ministry of Agriculture, Fisheries and Food (MAFF) (as proposed above) in oyster cultivation areas of nine estuaries of England and Wales (see Waite *et al.*, 1991; Dowson *et al.*, 1993; Waite *et al.*, Chapter 27). These studies found that TBT concentrations in the water column still exceeded 20 ng l^{-1} (the UK-proposed EQT) by up to ten times in six of the nine estuaries. An in-depth review of these data and additional data provided by other UK researchers led to the conclusion that the proposed EQT was not being met in areas of high boat densities. Also, there were measurable environmental problems resulting from the use of TBT as an antifoulant on nets and pens used in salmon cultivation in Scotland (see Davies *et al.*, 1986, 1987; Davies and McKie, 1987), and significant findings of imposex (sex changes of females to males) in dogwhelk populations (see Gibbs and Bryan, Chapter 13). The finding of these high levels following regulatory actions led the DoE in July of 1987 to initiate the total ban on the use of TBT in antifoulant paints on small boats and to revise the previous regulatory actions, because these results suggested that the existing controls were not effective enough in reducing TBT concentrations in the water column to acceptable levels to protect sensitive species.

The UK DoE subsequently lowered the TBT water quality standard from 20.0 ng l^{-1} to 2.0 ng l^{-1} (Abel, Chapter 2). These new regulations, introduced in January 1987, reduced the maximum allowable tin content of copolymer paints from 7.5% to 5.5% through the Control of Pollution Act of (COPA) of 1986, which amended the Control of Pollution Act of 1985 (the Anti-fouling Paints Regulations of 1985) (UK DoE, 1986a,b, 1987). These prohibited the retail sale and the supply for retail sale of antifouling paints containing a triorganotin compound as well as the wholesale and retail sale of anti-fouling treatments containing such a compound. The ban also did not make any exceptions to accommodate vessels with aluminum hulls, outboard drives, parts or fittings, as have US regulatory strategies. The DoE expected the paint companies to provide suitable products in the short term, and in fact both International Paints and Blakes are now marketing paints in the United Kingdom which they say are suitable for this use (AL27 and Lynx).

These regulations came into force on 28 May 1987 (The Control of Pollution (Anti-Fouling Paints and Treatments) Regulations, 1987 – Statutory Instruments No. 783, 1987).

It also should be noted that the control of pesticide regulatory actions in the United Kingdom shifted from the DoE to MAFF on 1 July 1987 through powers conferred to MAFF by sections 16(2) and 24(3) of the UK Food and Environmental Protection Act of 1985 and Regulation 5 of the Control of Pesticides Regulations 1986, as reflected in the Statutory Instruments No. 1510 (the Control of Pesticides Regulations, 1986).

The UK government also enacted the Food and Environment Protection Act (FEPA) to ensure that in the future all antifouling agents of any kind would be screened in the same way as other pesticides under provision of Part III. This was coordinated with the Control of Pesticides Regulations of 1986, which provided for the statutory screening of antifouling paints beginning 1 July 1987. These regulations prohibit the advertisement, sale, supply, storage, or use of any pesticide – including antifouling paints and treatments – unless approved by Ministers (see Abel *et al.*, 1986, 1987, for an in-depth historical perspective). The DoE has also revised (January 1988) its 'Don't Foul Things Up' leaflet to draw again the attention of boaters to the likely environmental impact from organotin in antifouling paints in 1988 when they expect a large amount of TBT paint to be removed from small vessels following the ban (January 1988).

3.5.3. SWITZERLAND AND GERMANY

Both Switzerland and Germany have banned all use of TBT in antifouling paints in freshwater environments. In the Federal Republic of Germany, the following regulations for organotin compounds are already in force:

1. a ban on its use for boats less than 25 m long;
2. a ban on retail sale;
3. a ban on its use on structures for mariculture;
4. a TBT limit of 3.8% (by weight) in copolymeric paints; and
5. regulation for the safe disposal of antifouling paints after removal (MEPC 30/20/2; IMO, 1990).

3.5.4. COMMISSION OF THE EUROPEAN COMMUNITIES

On 1 February 1988 the Commission of the European Communities proposed an amendment for Council Directive 76/769/EEC restricting the marketing and use of certain dangerous substances and preparations (COM(88) 7 Final – Brussels). The proposal lists 'organostannic compounds' and restricts their use as substances and constituents of preparations intended for use to prevent the fouling by microorganisms, plants or animals of (a) the hulls of boats of an overall length, as defined by ISO 8666, of less than 25 m, and (b) cages, floats, nets, and any other appliances or equipment used for fish or shellfish farming (see Davies *et al.*, 1986, 1987; Davies and McKie, 1987), and may be sold only to professional users in packaging of a capacity of not less than 20 liters.

3.5.5. JAPAN

Monitoring studies in Japan in the late 1980s found that a 'biologically significant' amount of organotin compounds derived from antifouling paints had been released to the marine environment with high residues in fish that ranged from 0.06 to 0.75 ng l^{-1} TBT and 0.03 to 2.6 ng l^{-1} TPT (triphenyltin), giving some concern for future human health affects. Also concentrations in bird tissues were found to range from 0.03 to 0.05 ng l^{-1} TPT (MEPC 30/WP.1). In 1990, given these findings and the results of laboratory and field studies, 7 TPT compounds (January), and 13 TBT compounds (September) were designed as Class II Specified Chemical

Substances. Subsequently the production, import, and use of these compounds has come under the domestic law concerning the examination and regulation of manufacture of chemical substances. Japanese government ministries have introduced domestic countermeasures to prohibit the application of TPT antifouling paints on all vessels including boats, ships, and marine structures. Regarding TBT antifouling paints, the following restrictions came into force in July 1990 (MEPC 30/WP.1; IMO, 1990):

(a) TBT antifouling paints shall not be applied to non-aluminum hulled vessels engaged in domestic voyages as well as on non-aluminum hulled vessels engaged in international voyages with a drydocking interval of approximately 1 y; and
(b) TBT antifouling paints shall not be applied to hulls, other than shell plating between the load line and the bile keel, of vessels engaged in international voyages with a drydocking interval of longer than 1 y. Shell plating between the load line and bilge keel of such vessels may be painted with antifouling paints containing a low percentage of TBT compounds.

At the 30th Session of the Marine Environmental Pollution Committee (MEPC) of the International Maritime Organization (IMO), the Japanese delegation indicated that it felt that the above interim measures were insufficient and that 'a total ban on the use of TBT antifouling paints on all vessels including vessels engaged in international voyages should be introduced as soon as possible as an international agreement.' They also noted that the primary argument for impeding a total ban was the lack of an alternative equivalent. The Japanese delegation raised the point that the alternative copper could provide acceptable control of fouling for more than 2 y, and that this should be satisfactory because most commercial vessels are drydocked for inspection every 2 y (MEPC 30/WP.1; IMO, 1990).

3.5.6. INTERNATIONAL CONVENTIONS

Most of the international conventions since 1986 have developed advisory positions or recommendations for contracting parties regarding the utilization of organotin compounds as biocides in antifouling paints, covering manufacture, registration, sale, application, removal, and disposal.

3.5.6a. The London Convention

The Marine Environmental Protection Committee (MEPC) of the IMO (IMO serves as the Secretariat of the London Dumping Convention) has for several years reviewed the position of organotin compounds in its lists of hazardous substances and collected information on the effects of organotin compounds on the marine environment and human health. Concern had been expressed within the Consultative Meeting of Contracting Parties to the London Dumping Convention (now referred to as the London Convention).

At its 29th session (on 27 April 1990), the MEPC reviewed the actions taken by the Consultative Meetings of the Contracting Parties to the London Convention. The MEPC for some years has reviewed the position of organotin compounds in its lists of hazardous substances and collected information on the effects of organotin compounds on the marine environment and human health. Particular concern had been raised within the MEPC of the potential hazards caused by disposal at sea of dredged material from areas such as marinas and dockyards containing high levels of organotin compounds (MEPC 29/15, MEPC 29/22; IMO, 1990). The MEPC identified a series of options for the regulation of antifouling paints considered within various national and regional bodies as follows:

1. the total ban of organotins in antifouling paints;

2. the regulation of the use of organotin compounds by the length of vessels, such as prohibition on vessels of less than 25 m in length;
3. the limitation of the amount of organotin compounds (on a percentage basis) in the paint;
4. the limitation of the release rate of organotin compounds from the paint to the adjacent water column; and
5. the development and enactment of local, national, or regional water quality standards.

At MEPC 29, the delegation from the United States agreed to prepare a draft resolution on the use of TBT compounds in antifouling paints for ships for consideration at MEPC 30. The delegation expressed its view that recommendations from the Committee were needed concerning the use of organotin compounds taking into account the hazards of these substances to the marine environment and human health.

Subsequently, the Third International Organotin Symposium, of which IMO was a co-sponsor, was held in Monaco (17–20 April 1990). A special policy and regulatory session was chaired at the symposium by the first author of this chapter and he presented a conceptual *list of regulatory requirements* that national and international bodies should consider in developing national and international regulations for the use of organotin compounds in antifouling paints:

1. Implement no- or low-cost regulatory requirements.
2. Implement fee schedules. Biocide producer pays all registration fees. Benefited user pays user benefit fee as an environmental degradation fee.
3. Create an environmental degradation fund from user benefit fees to support regional research, monitoring, and mediation activities. To be coordinated by a national research review panel.
4. Implement limited cost (<10%) bureaucratic and administrative management structures to manage these funds and activities.
5. Require all international vessels (as part of the ship's registration papers) to have certified and duly recorded the following specific data related to the use of organotin compounds in antifouling paints: the specific type, composition, release rate, and quantity of organotin utilized.

In this session at Monaco, the first author identified the following comprehensive range of regulatory options that could be considered for regulating the use of organotin compounds in antifouling boat-bottom paints:

1. Place a total ban on the use of organotin compounds in antifoulant paints.
2. Regulate the use of organotin compounds by the length of vessels, such as prohibition on vessels of less than 25 m in length with approval on all aluminum hull vessels; ban use on non-commercial or recreational vessels of any length.
3. Limit the amount of organotins (on a percentage basis) in a specific paint formulation.
4. Limit the release rate of organotins from antifouling paints to the adjacent water column.
5. Regulate the application/removal of antifouling paints that utilize organotins to trained and certified applicators.
6. Regulate the removal, containment, clean-up, and disposal of antifouling paints which contain organotins which are removed from vessels in drydock facilities.
7. Regulate the discharge rates of organotins in discharge waters from drydock facilities by standard prevention practices and clean-up procedures.
8. Regulate the dockage time of large vessels >25 m) that utilize organotin-based antifouling paints to specific time periods

with limited excess at-anchor time in harbors.

9. Require foreign vessels utilizing organotin-based antifouling paints in harbors to pay an environmental degradation fee (US $1200 d^{-1} or $50 h^{-1} for anchoring time in estuaries or ports.
10. Publish self-regulatory strategies for small boat owners who had painted their boat with organotin-based antifouling paints within the last 5 y.

Participants at Monaco felt that many of the above suggestions were either impracticable (e.g. environmental charges for use of the paints, restriction on the amount of time spent in waterways) or not relevant to IMO (such as a ban on the use of organotin on vessels of less than 25 m). Two measures discussed attracted the greatest support:

1. Regulate maximum release rate of organotins from antifouling paints.
2. Encourage member countries to introduce carefully controlled drydocking practices (MEPC 30/20: IMO, 1990).

Without any UN authority or mandate, the first author of this chapter and the symposium participants reworked the above suggestions into a series of proposed recommendations to be reported to MEPC by the UK delegate Dr Michael Waldock for MEPC consideration in developing recommendations for IMO for the global regulation of organotin compounds (MEPC 30/20; IMO, 1990: Mee and Fowler, 1991):

1. to establish a lower limit for vessel hull release rates for tributyltin compounds used in antifouling paints levels of micrograms per square centimeter per day from vessel hulls, including an agreement on a standard method of measurement;
2. to establish uniform industrial process instructions and regulations for the application, removal, and disposal of all organotin-based antifouling paints, residues, and discharges that have developed from their application in boatyards or drydocks, including certification of operators for application and removal;
3. to establish public information leaflets to serve a self-regulatory information strategy for small boat owners, who have previously painted their vessel with organotin-based antifouling paints;
4. to establish an IMO record system to register and certify every vessel according to type of antifouling paint used.

Also at the 30th session of MEPC, a series of overview papers were presented by invited experts along with position papers and review comments prepared by delegates to the MEPC. At the 29th session of MEPC, the United States volunteered to prepare a draft resolution on measures to control the use of tributyltin compounds in antifouling paints. The US delegation's draft resolution (MEPC 30/20/1; IMO, 1990) recognized that there are a number of different measures and approaches that can be employed to control the use of tributyltin compounds in antifouling paints in order to reduce or eliminate potential adverse impacts on the marine environment. The following MEPC Resolution (MEPC.45(30)) was adopted considering all of the above suggestions on 16 November 1990 (IMO, 1990):

(a) to recommend that governments adopt and promote effective measures within their jurisdictions to control the potential for adverse impacts to the marine environment associated with the use of tributyltin compounds in antifouling paints, and as an interim measure specifically consider actions as follows:
   (i) to eliminate the use of antifouling paints containing tributyltin compounds on non-aluminum-hulled vessels of <25 m in length;
   (ii) to eliminate the use of antifouling paints containing tributyltin com-

pounds that have an average release rate >4 $\mu g\ cm^{-2}\ d^{-1}$;

(iii) to develop guidance in sound management practice applicable to ship maintenance and construction facilities to eliminate the introduction of tributyltin compounds into the marine environment as a result of painting, paint removal, cleaning, sandblasting, or waste disposal operations, or runoff from such facilities;

(iv) to encourage development of alternatives to antifouling paints containing tributyltin compounds, giving due regard to any potential environmental hazards that might be posed by such alternative formulations; and

(v) to engage in monitoring to evaluate the effectiveness of control measures adopted and provide for sharing such data with other interested parties.

(b) to consider appropriate ways toward the possible total prohibition in the future of the use of tributyltin compounds in antifouling paints for ships.

3.5.6b. The Paris Commission

The Paris Commission deals with land-based sources of pollution to the North East Atlantic Ocean under the auspices of the Paris Convention. The Convention recommended in 1987 that contracting parties should take effective action to eliminate pollution by TBT of the inshore areas within the convention. One of the key restrictions recommended was that restrictions should be considered on the use of organotins on seagoing vessels. This recommendation was debated in 1988, and the Commission concluded that for economic reasons a ban on seagoing vessels was not achievable. Contracting parties agreed, however, 'to develop procedures and technologies aimed at a reduction of the amount of organotins released from boat yards and dry docks due to sand-blasting, dust, paint chips, over spray, etc. and to implement them in the near future' (MEPC 30/INF.5; IMO, 1990).

3.5.6c. The Barcelona Convention

In 1989, the contracting parties to the Barcelona Convention (for protection of the Mediterranean Sea against pollution) approved a restriction on large vessels. At that time, they also agreed to develop a code of practice to minimize the contamination in the vicinity of boatyards and drydocks to reduce contamination from removal of spent antifouling paints and application of fresh ones. For the Mediterranean Sea, comprehensive assessments of organotin compounds have been prepared by United Nations organizations: United Nations Environmental Program (UNEP) and the Food and Agriculture Organization (FAO) in cooperation with the World Health Organization (WHO) and the International Atomic Energy Agency (IAEA) to support the Mediterranean Action Plan (MEPC 29/15/1), MEPC 29/INF.19). The data and information from these assessments have led to a set of recommendations on organotin compounds that was adopted by the Sixth Ordinary Meeting of the contracting parties of the Barcelona Convention:

1. as from 1 July 1991 not to allow the use in the marine environment of preparations containing organotin compounds intended for the prevention of fouling by micro-organisms, plants or animals;
2. on hulls of boats having an overall length (as defined by the International Standards Organization (ISO) Standards No. 8666) of >25 m; and
3. on all structures, equipment, or apparatus used in mariculture.

In addition, 'Contracting Parties not having access to substitute products for organotin compounds by 1 July 1991 would be free to make an exception for a period not exceeding two years, after having so informed the Secretariat.' A recommendation was also made 'that a code of practice be developed to minimize the contamination of the marine environment in the vicinity of boatyards, drydocks, etc., where ships are cleaned of old anti-fouling paint and subsequently repainted' (MEPC 29/22; IMO, 1990).

3.5.6d. The Helsinki Commission

The Helsinki Commission (protection of the Baltic Sea) recommended in 1988 that their Contracting Parties should consider the need for restrictions on the use of organotin compounds on seagoing vessels (MEPC 30/INF.5; IMO, 1990).

3.6. IMPACT OF CURRENT REGULATORY POLICIES AND PRACTICES

The impact of current regulatory policies and practices can be assessed in two ways: (1) loss of economic and environmental benefits from the use of organotin compounds in antifouling coatings, and (2) decline in environmental concentrations following national regulatory and legislative actions.

3.6.1. ECONOMIC AND ENVIRONMENTAL BENEFITS

Fouling creates roughness on vessel hulls due to the growth of aquatic plants and animals. This roughness increases turbulent flow and drag, reducing vessel speed per unit of energy consumption (Milne, 1990a). A 10-μm increase in average hull roughness creates an increase in fuel consumption of between 0.3 and 1%. Fuel is the largest single cost in operating a ship. For bulk carriers, fuel costs can be 50% of the total vessel operating costs. In 1985/86, the fuel bill for the *Queen Elizabeth II* was $17 million.

Over the past 2000 y, naval fleets with superior hull antifoulings have often proved more effective in combat. Some examples include:

1. Nelson's victory over the French fleet at Trafalgar in 1805. The British fleet was 'copper bottomed' and foul free; the French fleet was heavily fouled and hence less maneuverable.
2. In World War II, US naval antifouling technology was more effective at controlling fouling than that used by the Japanese. This advantage provided the US fleet with a significantly greater fuel efficiency and subsequent operating range over the Japanese fleet.
3. During the Falklands War of 1982, the cruise liner *Queen Elizabeth II* was converted to troopship status in days. Thanks to her organotin copolymer antifouling bottom paint, she required virtually no hull coating work prior to dispatch to the Falklands, and arrived there ahead of schedule.

The benefits to navies of using TBT copolymers include not only increased combat performance but also the following:

1. Extended in-service deployment periods of 5–7 y between drydockings versus 24–30 months at present. Improved ship's operating readiness, which is a critical factor in a time of national emergency, enabling ships to be available on short notice for deployment with clean, foul-free hulls, without requiring drydocking to remove fouling or to repaint hulls.
2. Increased operating range, which is important in distant tropical waters such as the Indian Ocean, Persian Gulf, and South Pacific.
3. Maintenance of top vessel speed capabilities and lower fuel consumption during extended high-speed operations, such as the 40 knots needed for the launching of aircraft from aircraft carriers.
4. Elimination of costly and time-consuming underwater hull cleaning to remove fouling during deployment. Copolymers self 'polish' and 'smooth,' providing the possibility of reducing underwater hull noise.

5. Application to underwater advanced sonar and electronic communication and defense systems.

In 1985, the US Navy calculated that if the entire fleet (600 ships) were to be painted with TBT antifoulant paints, the fuel avoidance costs (extra consumption) would exceed $120 million annually (calculated with fuel costing approximately $18/barrel) (NAVSEA, 1986). Because of improved copper-based antifouling coatings, more recent estimates have reduced this cost avoidance estimate. Also, additional costs that are difficult to estimate because they vary significantly for different oceans are costs from operational activities for fouling reduction such as increased underwater cleaning and drydock costs for repainting every 18–30 months for non-organotin-based paints.

The use of tributyltin antifouling paints on commercial ships, fishing vessels, and private boats in the United States could add another $300–400 million (or 2 billion gallons) in fuel savings annually. Moreover, these estimated cost savings do not include the savings from decreased wear on propulsion machinery and down time for hull scraping, cleaning, and painting resulting from the use of organotin paints.

In the United States, the regulation of organotin compounds in antifouling paints was projected by M&T Chemicals, Inc. and International Paint Company (major US manufacturer of organotin antifouling paints) to have the following additional negative impacts: (1) deepsea vessels would go to foreign shipyards for painting; (2) higher antifoulant protection costs to vessel owners; (3) higher transportation costs; (4) domestic vessels would have a dramatic increase in operating costs; (5) severe hardship to US shipyards (125 000 workers); (6) no new construction (non-military); (7) maintenance and repair declining now; (8) TBT ban would push many shipyards over the edge, and foreign vessels and shipyards would capture market; (9) more than 70% of the world's fleet uses organotin copolymers; (10) national defense and military preparedness and be reduced; (11) extended drydock intervals; and (12) TBT-painted hulls would still be in US waters (Gibbons, 1986; Ludgate, 1987).

At the 30th session of the Marine Environmental Protection Committee (MEPC) of the IMO, A. Milne of Courtaulds NCT, presented a paper entitled: 'Cost/benefit analysis of SPC organo-tin antifoulings' (Milne, 1990b). In his study he considered the vessel as an industrial plant, and that time in drydock and associated delays constituted expensive 'down time' for loss of vessel revenue. His analysis was framed around the following categories: direct fuel savings (1976–86), extension of drydocking interval, improved plant utilization, capital savings, and antifoulings and the environment. The results of his study are presented below:

1. The marine transport industry burns 184 million tonnes of fuel per annum; at $100 per tonne, the fuel bill is 18.4×10^9 US dollars.
2. The cost of not having fouling protection was approximately 40% or 72 million tonnes of oil per year. It should be noted that this is greater than 60% of the North Sea oil production.
3. The cost of fouling failure for oil tankers was estimated to be $500 000 per 200 000 dry weight tonnage (DWT) vessel per annum, estimating this failure to occur beyond 14 months.
4. Self-polishing antifouling copolymers of TBT introduced in 1974 were estimated to provide the world fleet with an improvement in fuel efficiency of 2% with a 'very conservative' estimate of 2% savings from fouling for a total of 4% in power and fuel equivalent to 7.2 million tonnes of fuel or $\$0.7 \times 10^9$ saved annually.

In terms of antifouling performance in the 1970s, Milne reported that the drydocking intervals were: industry demand = 30 months, achieved = 18 months, and guaranteed = 12 months. The self-polishing antifouling copolymers of TBT (for a sample of over 4000 vessels per annum) by 1986 had shifted the mean docking interval to 27 months. The tonnage docked per annum was estimated to be 280 × 10^6 DWT. The mean cost was estimated at $10 per DWT. The calculated savings in drydock fees were $20.2 × 10^6 y^{-1}. His calculations for improved plant utilization were $409 × 10^6 y^{-1}. Capital savings were estimated to be $500 × 10^6 y^{-1}. The sum of these gave an estimate of $2449 × 10^6 y^{-1} in total savings to the world commercial fleet (over 6000 tankers).

In addition the use of organotin-based antifouling coatings provided the following environmental benefits: a reduction of 23 million tonnes per year of greenhouse gases and a reduction of 580 000 tonnes per year in acid rain. Milne (1990b) concluded that it was based on the above figures that the environmental impact of the continued use of organotins in antifouling paints needed to be assessed.

A brochure published in 1992 by the Organotin Environmental Programme Association (ORTEP) in the Netherlands and the Marine Painting Forum in the United Kingdom has summarized and updated a number of technical papers presented to the IMO MEPC Committee meeting in November 1990 (MEPC 30) as organized by the European Chemical Industry Council (CEFIC). This document revised Milne's calculated cost savings using current fuel prices and operating practices and added an estimate of $1000 million dollars in cost savings due to indirect savings giving a total estimate of 2.7 billion dollars per year of 'significant' economic benefits to the marine industry from the use of TBT copolymer antifouling paints.

3.6.2. SHIFT OF APPLICATION TO LESSER DEVELOPED COUNTRIES

A consideration that should not be omitted here involves the forces (economics and regulations) that drive international maritime companies to look for cheap labor and cheap environmental laws in lesser developed countries for painting their vessels with organotin antifouling paints. The length exclusion (>25 m) allows for the use of organotin compounds by large oceangoing vessels and gives the world's maritime fleet significant economic benefits. The regulatory logic for this exclusion is that they spend most of their time at sea (except when anchored in estuaries awaiting port space or goods, and/or at the loading dock) and do not contribute significantly to the critical environmental concentrations of organotin compounds in estuaries or near coastal waters where sensitive species of molluscs reside.

Environmental scientists in lesser developed countries have begun to find similar deformities in oysters as seen in the United Kingdom, France, and United States and realized that there has been a large increase in the number of vessels being painted with organotin-based antifouling paints in local shipyards due to cheap labor and lack of environmental regulations for organotins in their country. US Navy studies at Pearl Harbor, Hawaii, conducted during painting and release of ships from drydocks, found that, with appropriate environmental management practices, drydock effluents could be maintained at the low nanogram-per-liter levels with costs that were estimated to be $27/m^2 (see Adema *et al.*, 1988). For simulation of effectiveness of improved dockyard practice see Harris *et al.* (1991). In essence, economics and regulation in the developed countries have shifted environmental problems to the countries least able to address them (Isensee *et al.*, 1994; Ten Hallers-Tjabbes, 1994; Abd-Allah, 1995; Evans *et al.*, 1995b; Hardiman and Pearson, 1995).

3.6.3. REDUCTION IN ENVIRONMENTAL CONCENTRATIONS

In the United States, since the passage of the Antifouling Paint Control Act of 1988, the environmental concentrations of organotin compounds have declined (US EPA, 1991; Valkirs *et al.*, 1991; Wade *et al.*, 1991; Huggett *et al.*, 1992). The results of these studies are summarized below.

In the United States, the NOAA National Status and Trends Program funded the collection of a national (and standardized) prelegislative or regulatory data set for TBT concentrations for organotins in bivalves (oysters and mussels) and adjacent bottom sediments for multiple estuaries or near coastal waters (Wade *et al.*, 1988a,b, 1990). Of the sites sampled, they found that 97% of bivalves (oysters and mussels) and 75% of sediment from these sites were contaminated with butyltins. In these studies, mussels and oysters were collected from 36 US coastal sites (with three stations at each site) distributed on the Atlantic, Gulf of Mexico, and Pacific coasts, including one site from Hawaii. The average total butyltin concentration in bivalves (640 ng g^{-1} as Sn) was 18 times higher than the average concentration in sediment (36 ng g^{-1} as Sn). Butyltin concentrations in sediment samples ranged from <5 to 282 ng g^{-1} Sn. The most toxic butyltin, TBT, accounted on the average for 74% (in bivalves) and 80% (in sediments) of the total butyltins present (Wade *et al.*, 1988a,b, 1990):

Subsequent studies by Wade *et al.* (1991) measuring the concentrations of total butyltins and tributyltin in oysters from 59 sites in the Gulf of Mexico indicate that the concentrations have decreased at 85% of the sites after passage of the Organotin Paint Control Act of 1988. At 5% of the sites there was no measurable decrease in butyltin concentrations and 10% of the sites had increased in concentration between 1989, 1990 and 1991. (Figure 3.1(a) plots the available data for concentrations of TBT concentrations of TBT in oysters for 1989, 1990 and 1991. Figure 3.1(b) plots percent change for the same period). In spite of the decline in TBT concentrations in oysters at many Gulf of Mexico sites, the TBT concentrations in adjacent bottom sediments have not shared the same correlation.

In 1992, 169 sediment samples were collected by the Geochemical and Environmental Research Group (GERG) as part of the US EPA's Environmental Monitoring and Assessment – Estuarine (EMAP-E) Program in the Louisiana Providence in the Gulf of Mexico. These bottom sediment samples were analyzed for butyltins (Wade *et al.*, in preparation). EMAP-E uses a probability-based sampling strategy, which allows for the estimation of the percent of the resource contaminated by measured contaminants in a region. An initial statistical summary of these data estimated that tributyltin would be present (>1 ng l^{-1}) in 42 ± 11% of the estuarine sediments from the Louisiana Providence, which extends from Anclate, Florida, to Rio Grande River on the Texas–Mexico border (Macauley *et al.*, 1994). Also, sediment with TBT concentrations estimated to be greater than 5 ng l^{-1}, the concentration of TBT that is considered to cause toxic effects to benthic organisms, was 3 ± 3% for the entire region. If the data separate out the estimate for the large tidal rivers, however (i.e. Mississippi River), the distribution of this high level of contamination increases to 20 ± 4%. This suggests that even though butyltin concentrations have decreased in bivalves from these same coastal waters, bioavailable concentrations of butyltins may still be present in bottom sediments, which may act as a long-term source of organotins for many years. See also Sarradin *et al.* (1994) and Steur-Lauridsen and Dahl (1995).

During the mid-1980s, several studies of organotins in San Diego Bay reported finding surfacewater concentrations in excess of several hundred nanograms per liter in some areas associated with marinas in San Diego

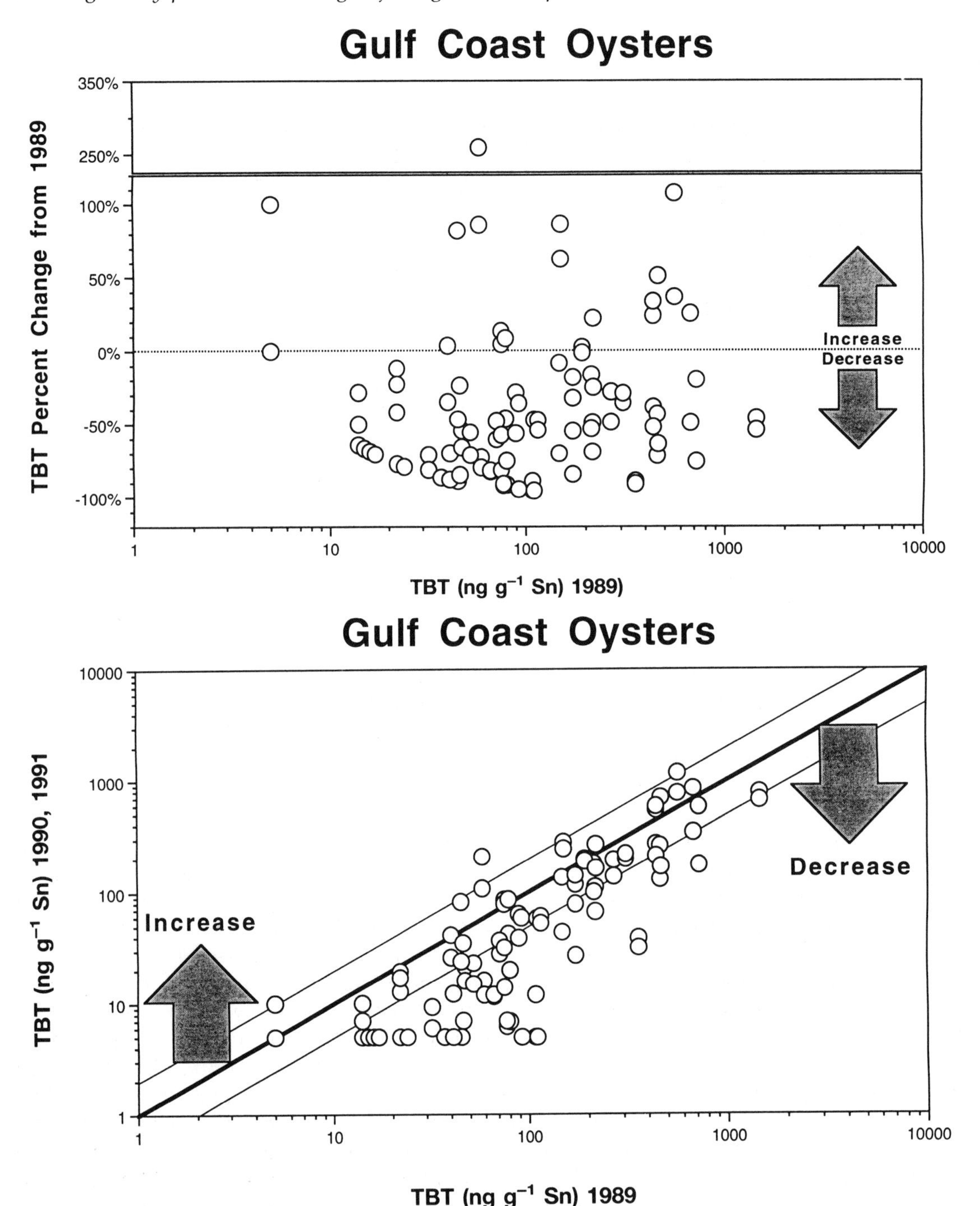

Figure 3.1 (a) Oyster (*Crassostrea virginica*) tissue concentrations (ng g^{-1} Sn) of total butyltins and tributyltins from 73 sites in the Gulf of Mexico for 1989, 1990 and 1991; (b) a plot of percent change in oyster (*Crassostrea virginica*) tissue TBT concentrations (ng g^{-1} Sn) from 73 sites in the Gulf of Mexico for 1989, 1990, and 1991. Modified from Wade *et al.* (1991) and Garcia-Romero *et al.* (1993).

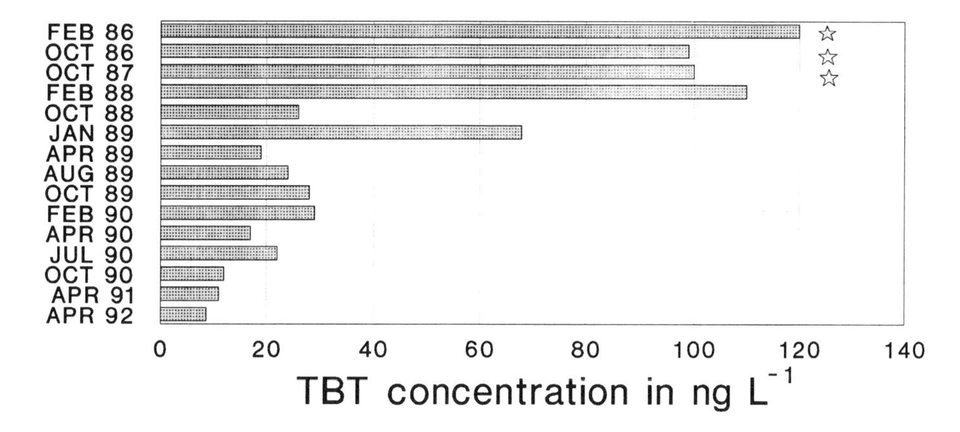

Figure 3.2 Surfacewater TBT concentrations (ng l^{-1}) from San Diego Bay for 1986 to 1992. Modified and updated from Valkirs *et al.* (1991).

Bay (Clavell *et al.*, 1986; Grovhoug and Seligman, 1986; Grovhoug *et al.*, 1986, Chapter 25; Seligman *et al.*, 1986a,b, 1989, 1990; Stang *et al.*, 1989; Valkirs *et al.*, 1986a,b, 1987). Recent studies of tributyltin in San Diego Bay, California (Valkirs *et al.*, 1991) however, have found that mean concentrations of TBT in surface water have significantly decreased in San Diego Bay following legislative restriction on the use of organotin antifouling paints in California (Fig. 3.2). Surfacewater concentrations in three of the four regions studied in San Diego Bay are below the 6 ng l^{-1} water quality criteria concentration adopted by the State of California (in January 1988) for the protection of marine species. Regression analysis of the San Diego data suggests that surfacewater concentrations would decrease by 50% in 8–24 months. It was found that TBT concentrations in sediment in San Diego Bay did not reflect recent decreases in values in the water column and were variable among stations over time, and that tissue concentrations in *Mytilus edulis* have generally declined in San Diego Bay since February 1988 (significantly since April and July 1990) (Valkirs *et al.*, 1991; Grovhoug, 1992).

Similar findings have been reported for the Chesapeake Bay by US EPA (1987c) and Huggett *et al.* (1992) for the Hampton, Virginia, area of the bay. Surfacewater samples analyzed after the passage of the Organotin Antifouling Paint Control Act (OAPCA) of 1988 in marinas and yacht clubs (Fig. 3.3) indicated that TBT concentrations had significantly decreased when compared with results of earlier studies by Huggett (1986, 1987); Huggett *et al.*, (1986); Hall (1986, 1988); Hall *et al.* (1986, 1987); and US EPA (1987c).

In the United Kingdom, during 1985, the UK government took action under the Control of Pollution Act of 1974 to regulate the use of TBT antifouling paints on small vessels and set an environmental quality target (EQT) concentration for TBT at 20 ng d^{-1} m^{-3} (see Abel, Chapter 2 and UK DoE, 1986, for further regulatory discussion of

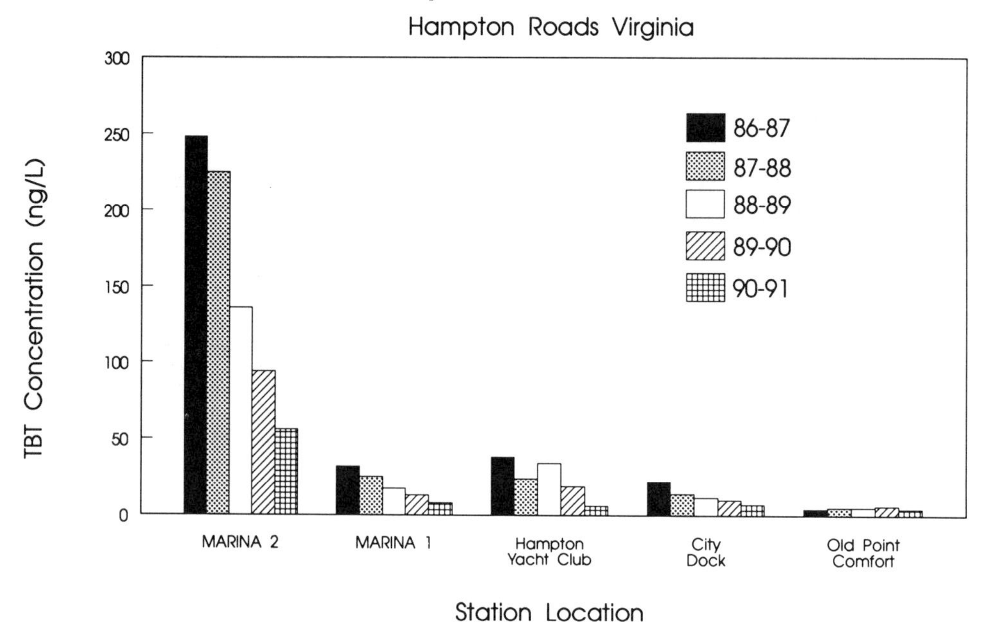

Figure 3.3 Surface water TBT concentrations (ng l^{-1}) from marinas and yacht clubs in the Hampton, Virginia, area of Chesapeake Bay (from Huggett *et al.*, 1992).

these deliberations). Subsequent studies by UK researchers during the next summer (Cleary and Stebbing, 1985; Waldock, 1986) found that, as in the past several years, organotin concentrations increased in the spring with the launching of yachts, usually followed by a secondary peak in later summer or autumn associated with repainting or hosing-off activities, and that concentrations declined during the winter. These studies suggested that the EQT needed to be reduced by a factor of ten to achieve environmental protection. As a result of these studies, in February 1987 the UK government announced its intention to introduce further controls under the Control of Pollution Act, which included a complete ban on retail sale of TBT antifouling paint formulations and a complete ban on the sale of products containing TBT used to treat fish-farm cages.

Studies subsequent to this second regulatory action carried out by researchers at MAFF have found significant concentrations of TBT in harbors and at anchorages in a study that focused on large vessel contributions. They have also found that drydocking practices result in discharges of hazardous concentrations of TBT (Waldock *et al.*, 1988). However, they have found reductions in concentrations of organotins in estuarine surface water and sediment in England and Wales following the 1987 legislation (see Waldock *et al.*, 1987a,b; Waite *et al.*, 1991, and Chapter 27; Evans *et al.*, 1994; Law *et al.*, 1994; Matthiessen *et al.*, 1995). However, they found that TBT concentrations in surface water in many areas exceeded the new EQT of 2 ng l^{-1}, and that studies in new marinas suggested that the higher than expected concentrations may have resulted from illegal use of TBT by boat owners. Drydocks in these studies were also singled out as a major source of TBT to estuaries (Anonymous, 1994).

A new set of environmental data that does

not share the above similarities (decline in organotin concentration since enactment of regulatory actions) has been published by Hardy and Cleary (1992) in a study of TBT concentrations in the surface microlayer from the German Bight to the North Sea. These studies developed as part of the 1990 Bremerhaven Workshop on Biological Effects of Contaminants (Stebbing and Dethlefsen, 1992). The workshop was a field research project with a fleet of seven research vessels and over 100 scientists conducting in-situ studies. Hardy and Cleary (1992) collected samples of sea-surface microlayer and near-surface bulkwater at five stations up to 200 km offshore in the North Sea. They found that a zone of toxic surface water, with contaminant levels exceeding UK water quality standards (EQS) extended from 100 to 200 km offshore. TBT concentrations in the surface microlayer (> 20 ng l^{-1}) were 10 times higher than needed to induce imposex in dogwhelks. With the high occurrence of fish egg and larval fish abnormalities found in this region by Dethlefsen *et al.* (1985), Hardy and Cleary concluded that these high levels of microlayer contaminants could pose a threat to fisheries recruitment in the North Sea. Stebbing realized that this was the first time toxic concentrations of any contaminant have been found in the open ocean, with the implication that this type of pollution is from oceangoing ships and may be occurring in oceans throughout the world (Coghlan, 1990). Similar results could be expected in highly productive estuarine waters.

3.7. CONCLUSIONS AND RECOMMENDATIONS

The history of strategies for regulating antifouling coatings that contain organotin compounds (as reviewed in preceding sections) is an excellent example of how well-intentioned policy and regulatory strategies responding to urgent concerns perceived by the public often fall short of achieving long-term goals. The long-term economic and environmental public goals should be that vessels (regardless of length) need effective antifouling coating technologies and that this effectiveness must not impact nontarget organisms. The regulation of antifouling coatings is a genuine public policy concern because they influence the price that the public pays for vessel-shipped common goods, food, and energy, due to increased cost of maritime commercial shipping.

Most effective antifouling coatings today contain toxic additive substances known as biocides. Organotin compounds have been found to be the most effective biocides developed to date; however, as engineered today, they are too effective because they also endanger nontarget organisms. The challenge is essentially a matter of engineering a means of reducing or controlling the scope of their effectivenes (Champ and Lowenstein, 1987). The shift from high-release-rate paints such as free-association (see Chapter 1 for definition of terms) to copolymer-based paints (and the development of self-polishing copolymer paints) so as to lower the concentrations of organotins in the environment is a step in the right direction. There are other additional technological advances that should be explored, however, in developing an economically and environmentally sound regulatory strategy. For a national regulatory strategy to be supportive of the creation of high-technology chemicals and products, the strategy must include the promotion of continued research and development to push these compounds to additional refinements that enhance environmental attributes and improve competitiveness in the global marketplace. It is easy for regulation to inhibit economic development of products and technologies.

The current regulatory strategies for organotin have several major shortcomings. First, the regulatory strategies of developed nations unfortunately focus on short-term national self interests and may not represent a 'think globally, act locally' philosophy. The principal regulatory approach is to reduce organotin concentration in the local environment by reducing the concentration in the paint (or in the release rate) and in the concentrations discharged to the environment from shipyards. Regulatory bodies of developed nations, in setting regulatory environmental concentrations (water quality standards) to protect local coastal waters, are, in effect, encouraging shipping companies to take their antifouling repainting business abroad at the economic loss of domestic shipyards. US and European shipyards cannot effectively compete in the lesser developed non-environmentally regulated marketplace, if, in addition to high labor and operational costs, they must also shoulder the expense of waste treatment and disposal of antifouling residues from removal of spent antifouling paints to achieve a regulated discharge (environmental water quality standard) level to protect local waters. Consequently, large vessel owners can enjoy the double cost benefit of being able to have their vessel painted by cheap labor without having to be responsible for environmental degradation and human health hazards (externalities) in lesser developed countries.

In the United States and United Kingdom, regulatory strategy has reduced local environmental concentrations; however, the strategy has shifted the shipyard impact (from repainting large vessels) to environmentally unregulated countries. This strategy may seem appropriate locally, but in fact it is both bad for the environment and bad for American and European businesses. In essence, the developed nations' regulatory strategies have addressed only a small portion of the antifouling problem, that is, removal of the use on small vessels of organotin antifouling paints from shallow estuaries where shellfish reside and shipyards discharge to local waters.

The ultimate long-term solution to the problem of antifouling coating technology is to come up with effective regulatory strategies that promote the development of new and advanced antifouling coating technologies that are not toxic to nontarget organisms and are inexpensive to treat or degrade in shipyard waste treatment systems as well as public education, clean-up, and monitoring efforts. In the short term, however, regulatory policies must focus on developing processes or mechanisms by which the user who benefits from being able to use a specific technology pays user fees for that benefit and these fees are used to support the development of future technological advances in those antifouling coatings that provide greater public benefits. Some may ask, why not let the forces of the marketplace drive the development of these technologies? The problem is that the uneven environmental regulation (developed versus lesser developed nations) that occurs in global markets defeats the driving forces in the marketplace for better products (supply and demand).

A low- or no-cost way to address this regulatory problem (which also would reduce the shift to lesser developed nations) would be to develop a system of user fees directed at both coating manufacturers and ship owners/operators. Currently the direct 'regulatory registration' costs (paint pesticide regulation costs and perhaps toxicity monitoring studies) for organotins in antifoulant paints is passed on to the user at the time of sale. A second operational 'exposure' fee for indirect costs could be associated with each vessel and assessed based on ship size, specific type of organotin paint used, and amount of time spent anchored or docked in harbors or nearshore waters and collected each time a ship docks in a port (similar to an airport passenger tax, except that it would have a sliding scale, perhaps from 1 to 10 certified by IMO as part of the ship's papers.

This 'polluter pays' free-market approach would encourage organotin antifoulant users to 'run the numbers' and decide whether it makes financial sense for them to use organotins. These collected fees could be placed in a restricted environmental and technology development fund for financing future R&D activities related to product advancement. This approach would discourage low-technology applications by having vessel owners pay higher fees for them than for environmentally preferred, higher technology, environmentally benign antifouling coatings. This approach also addresses a weakness of FIFRA in the United States, whereby industry provides EPA with the regulatory data and information (DCI) to support their decision-making. This amounts to allowing the 'fox to guard the chicken house.' To avoid conflict of interest and insure scientific objectivity and integrity, regulatory data and information should instead come from independently funded third-party studies.

It is interesting to note that there has been a stepped-up research and development program within the world's navies to develop new types of antifouling coatings that do not contain organotins and provide perhaps 24–30 months of protection. In reinventing a new technology, however, the effort to make an excellent one better has been abandoned. Thus, the developed nations' organotin regulatory strategy fails to recognize that antifouling coatings are needed and that, at present, commercially viable non-toxic alternatives are not available. By banning the organotin-based antifouling paints, policy makers have ignored the vast advances in the development of antifouling technology made over the past several decades.

Finally, the organotin ban has sent the wrong signal to the general public as to the extent of the existing organotin contamination problem. The current public perception is that the problem has been solved. Consequently, vital research and development funds have been redirected toward 'higher priority' problems, leaving the antifouling problem to linger on. For example, the valuable research into fate and effects in the US, which could enhance our understanding of TBT processes and release rates, has been discontinued and a wide range of scientific questions have been left unresolved (see Champ and Seligman, Chapter 29).

Has the regulatory strategy of developed nations served 'to protect the aquatic environment by reducing immediately the quantities of organotin entering the waters of the United States' as required by the Antifouling Paint Control Act of 1988? The answer is yes, the environmental concentrations of organotins in surface waters and bivalves have declined in both United States and European coastal waters. But, on a global level, considerable negative environmental impacts stemming from the use of organotin compounds may be fully realized by the end of this century. The impact of the high concentrations of organotins 200 nautical miles offshore in the surface microlayer of the North Sea (Harding and Cleary, 1992) and the residual concentrations and bioavailibility of 'historic' organotins in river and estuarine sediments (Macauley, *et al.* 1994) may bring about a serious rethinking of current regulatory strategies.

The effect of the passage of the Antifouling Paint Control Act of 1988 in the United States was to close the door on future US organotin research and development as funded by US federal agencies and industries. This was both shortsighted and foolish: without a comprehensive understanding of the environmental mechanisms and processes that control organotin distribution, critical pathways of toxicity, and biological uptake mechanisms, chemists cannot continue to develop the high technology that created tributyltin antifouling paints to make them more effective and safer. The regulatory strategies reviewed in this chapter have failed to address the above mentioned aspects of the antifouling coatings problem.

ACKNOWLEDGMENTS

The authors thank Joyce G. Nuttall for her contributions to the technical editing of this manuscript and the entire book. In addition, the following are to be thanked for their contributions to this chapter: Linda J. Hogg, David F. Bleil, Manfried Nauke, Robert Abel, Lee R. Crockett, Sandra Panem, Peter F. Seligman, W. Lawrence Pugh, Michael H. Salazar, Linda K. Vlier, Janet Anderson, Jill Bloom, Judith Koontz, Judith S. Weis, Krystyna U. Wolniakowski, Ronald B. Landy, and Benjamin W. Patton.

REFERENCES

Abd-Allah, A.M.A. 1995. Occurrence of organotin compounds in water and biota from Alexandria harbours. *Chemosphere*, **30**(4), 707–715.

Abel, R., N.J. King, J.L. Vosser and T.G. Wilkinson. 1986. The control of organotin use in antifouling paint – the UK's basis for action. In: Proceedings of the Oceans '86 Organotin Symposium, Vol. 4, Marine Technology Society, Washington, DC, pp. 1314–1323.

Abel, R., R.A. Hathaway, N.J. King, J.L. Vosser and A.G. Wilkinson. 1987. Assessment and regulatory actions for TBT in the UK. In: Proceedings of the Oceans '87 International Organotin Symposium, Vol. 4, Marine Technology Society, Washington, DC, pp. 1314–1319.

Adema, C.M. and P. Schatzberg. 1983. Organotin antifouling paints and the environment drydock phase. *Navy Eng. J.*, **96**(3), 209–217.

Adema, C.M., W.M. Thomas, Jr. and S.R. Mangum. 1988. Butyltin releases to harbor water from ship painting in a dry dock. In: Proceedings, Oceans '88 Conference, Vol. 4, Marine Technology Society, Washington, DC, pp. 1656–1667.

Alzieu, Cl. 1986. TBT detrimental effects on oyster culture in France – evolution since antifouling paint regulation. In: Proceedings of the Ocean '86 Organotin Symposium, Vol. 4, Marine Technology Society, Washington, DC, pp. 1130–1134.

Alzieu, Cl. 1991. Environmental problems caused by TBT in France: assessment, regulations, prospects. *Mar. Environ. Res.*, **32**, 7–17.

Alzieu, Cl. and J.E. Portman. 1984. The effect of tributyltin on the culture of *Crassostrea gigas* and other species. Proceedings of the 50th Annual Shellfish Conference, 15–16 May, Shellfish Association, London, pp. 87–104.

Alzieu, Cl., Y. Thibaud, M. Herall and B. Boutier. 1980. Evaluation des risques dus à l'emploi des peintures anti-salissures dans les zones conchylicoles. *Rev. Trav. Inst. Pêches Marit.*, **44**, 301–349.

Alzieu, Cl., M. Heral, Y. Thibaud, M.J. Dardignac and M. Feuillet. 1981–1982. Influence des peintures anti-salissures à base d'organostanniques sur la calcification de la coquille de l'huître *Crassostrea gigas*. *Rev. Trav. Inst. Pêches Marit.*, **45**, 101–116.

Alzieu, Cl., J. Sanjuan, J.P. Deltreil and M. Borel. 1986. Tin contamination in Arcachon Bay: effects on oyster shell anomalies. *Mar. Pollut. Bull.*, **17**, 494–498.

Alzieu, Cl., G. Barbier and J. Sanjuan. 1987. Evolution des teneurs en cuivre des huîtres du bassin d'Arcachon: influence de la legislation sur les peintures anti-salissures. *Oceanol. Acta*, **10**, 463–468.

Alzieu, Cl., J. Sanjuan, P. Michell, M. Borel and J.P. Dreno. 1989. Monitoring and assessment of butyltins in Atlantic coastal waters. *Mar. Pollut. Bull.*, **20**(1), 22–26.

Anonymous. 1994. TBT on the way out. *Mar. Pollut. Bull.*, **28**(9), 519.

Bailey, W.A. 1986. Assessing impacts of organotin paint use. In: Proceedings of the Oceans '86 Organotin Symposium, Vol. 4, Marine Technology Society, Washington, DC, pp. 1101–1107.

Champ, M.A. 1985 (8 January). Summary of background information on the toxicity of tributyltin used in antifouling boat paints to oyster larvae from studies conducted in the United Kingdom. US EPA/OPPE Science Group, 722 Jackson Place, Washington, DC, 7 pp.

Champ, M.A. 1986a (10 January). Proposal for an interagency (hosted by NAVY/NOAA) workshop on tributyltin compounds (held 3–5 June 1986 at the US Naval Academy). US EPA/OPPE Science Group, 722 Jackson Place, Washington, DC, 3 pp.

Champ, M.A. 1986b. Tributyltin data collection needs. US EPA/OPPE Science Group, 722 Jackson Place, Washington, DC, 10 pp.

Champ, M.A. 1986c. Monitoring prospectus for tributyltins. US EPA/OPPE Science Group, 722 Jackson Place, Washington, DC, 9 pp.

Champ, M.A. 1986d. Introduction and overview. In: Proceedings of the Oceans '86 Organotin Symposium, Vol. 4, Marine Technology Society, Washington, DC, pp. i–vii.

Champ, M.A. and D.F. Bleil. 1988a. Research needs concerning organotin compounds used in antifouling paints in coastal environments.

NOAA Technical Report Published by the Office of the Chief Scientist, National Ocean Pollution Office. 5 parts plus appendices.

Champ, M.A. and D.F. Bleil. 1988b. Safer use of boat bottom paints – public health risks. US EPA risk communication leaflet. EPA Office of Policy Analysis, Washington, DC.

Champ, M.A. and D.F. Bleil. 1988c. Safer use of boat bottom paints – environmental risks. US EPA risk communication leaflet. EPA Office of Policy Analysis, Washington, DC.

Champ, M.A. and D.F. Bleil. 1988d. Communication of chemical risks from antifouling paints to the boating public: relative effectiveness of risk communications that appeal to public health concerns versus environmental concerns. Science Applications International Company Technical Report. EPA Contract No. 68-02-4210. USEPA Office of Policy Analysis, Washington, DC, 62 pp.

Champ, M.A. and F.L. Lowenstein. 1987. TBT: the dilemma of hi-technology antifouling paints. *Oceanus*, **30**(3), 69–77.

Champ, M.A. and L.W. Pugh. 1987. Tributyltin antifouling paints: introduction and overview. In: Proceedings of the Oceans '87 International Organotin Symposium, Vol. 4, Marine Technology Society, Washington, DC, pp. 1296–1308.

Clavell, D., P.F. Seligman and P.M. Stang. 1986. Automated analysis of organotin compounds: a method for monitoring butyltins in the marine environment. In: Proceedings of the Oceans '86 Organotin Symposium, Vol. 4, Marine Technology Society, Washington, DC, pp. 1152–1154.

Cleary, J.J. and A.R.D. Stebbing. 1985. Organotin and total tin in coastal waters of southwest England. *Mar. Pollut. Bull.*, **16**(9), 350–355.

Coghlan, A. 1990. Lethal paint makes for the open sea. *New Scientist*, 8 December, p. 16.

Commonwealth of Virginia. 1988. State Water Control Board Proceedings, 27–28 June. Richmond, Virginia.

Davies, I.M. and J.C. McKie. 1987. Accumulation of total tin and tributyltin in muscle tissue of farmed Atlantic salmon. *Mar. Pollut. Bull.*, **18**(7), 405–407.

Davies, I.M., J.C. McKie and J.D. Paul. 1986. Accumulation of tin and tributyltin from antifouling paint by cultivated scallops (*Pectin maximus*) and Pacific oysters (*Crassostrea gigas*). *Aquaculture*, **55**(2), 103–114.

Davies, I.M., J. Drinkwater, J.C. McKie and P. Balls. 1987. Effects of the use of tributyltin antifoulants in mariculture. In: Proceedings of the Oceans '87 International Organotin Symposium, Vol. 4, Marine Technology Society, Washington, DC, pp. 1477–1481.

Dethlefsen, V., P. Cameron and H. von Westernhagen. 1985. Untersuchungen uber die Haufigkeit von Missbildungen in Fischembryonen der sudlichen Nordsee. *Inf. Fischwirtsch*, **32**, 22–27.

Dowson, P.H., J.M. Bubb and J.N. Lester. 1993a. Temporal distribution of organotins in the aquatic environment: five years after the 1987 UK retail ban on TBT based antifouling paints. *Mar. Pollut. Bull.*, **26**(9), 487–494.

Dowson, P.H., J.M. Bubb and J.N. Lester. 1993b. Depositional profiles and relationships between organotin compounds in freshwater and estuarine sediment cores. *Environ. Monitoring and Assessment*, **28**(2), 145–160.

Eastin, K.E. 1987. Tributyltin paint – the navy perspective. *Sea Technol.*, **28**(3), 69.

Evans, S.M., S.T. Hawkins, J. Porter and A.M. Samosir. 1994. Recovery of dogwhelk populations on the Isle of Cumbrae, Scotland, following legislation limiting the use of TBT as an antifoulant. *Mar. Pollut. Bull.*, **28**(1), 15–17.

Evans, S.M., T. Leksono and P.D. McKinnel. 1995a. Tributyltin pollution: a diminishing problem following legislation limiting the use of TBT-based anti-fouling paints. *Mar. Pollut. Bull.*, **30**(1), 14–21.

Evans, S.M., M. Dawson, J. Day, C.L.J. Frid, M.E. Gill, L.A. Pattisina and J. Porter. 1995b. Domestic waste and TBT pollution in coastal areas of Ambon Island (Eastern Indonesia). *Mar. Pollut. Bull.*, **30**(2), 109–115.

Garcia-Romero, B., T.L. Wade, G.G. Salata and J.M. Brooks. 1993. Butyltin concentrations in oysters from the Gulf of Mexico from 1989 to 1991. *Environ. Poll.*, **81**, 103–111.

Gibbons, T.J. 1986. Testimony of International Paint Company for the House Committee on Merchant Marine and Fisheries Hearing. In: *Hearing Record*, 30 September 1986. Serial No. 99-49 (65-830-0). US Government Printing Office. Washington, DC 20402, pp. 35–60.

Goldberg, E.D. 1986. TBT: an experimental dilemma. *Environment*, **22**, 17–20, 42–44.

Grovhoug, J.G. 1992. Evaluation of sediment contamination in Pearl Harbor. Naval Command, Control and Ocean Surveillance Center, RDT and E Division. (NRAD-TR-1502), San Diego, California. 83 pp.

Grovhoug, J.G. and P.F. Seligman. 1986. Navy monitoring of butyltins in U.S. harbors and estuaries, San Diego Bay: case study. In: Interagency Workshop on Aquatic Monitoring and Analysis for Organotin. NOAA/NMPPO, Rockville, MD, pp. 33–38.

Grovhoug, J.G., P.F. Seligman, G. Vafa and R.L. Fransham. 1986. Baseline measurements of butyltin in U.S. harbors and estuaries. In: Proceedings of the Oceans '86 Organotin Symposium, Vol. 4, Marine Technology Society, Washington, DC, pp. 1283–1288.

Hall, L.W. 1986. Monitoring organotin concentrations in Maryland waters of Chesapeake Bay. In: Interagency Workshop on Aquatic Monitoring and Analysis for Organotin. NOAA/NMPPO. Rockville, MD, pp. 27–28.

Hall, L.W. Jr, 1988. Tributyltin environmental studies in Chesapeake Bay. *Mar. Pollut. Bull.*, **19**, 431–438.

Hall, L.W., M.J. Lenkevich, W.S. Hall, A.E. Pinkney and S.J. Bushong. 1986. Monitoring organotin concentrations in Maryland waters of Chesapeake Bay. In: Proceedings of the Oceans '86 Organotin Symposium, Vol. 4, Marine Technology Society, Washington, DC, pp. 1275–1279.

Hall, L.W. Jr, M.J. Lenkevich, W.S. Hall, A.E. Pinkney and S.J. Bushong. 1987. Evaluation of butyltin compounds in Maryland waters of Chesapeake Bay. *Mar. Pollut. Bull.*, **18**, 78–83.

Hardiman, S. and B. Pearson. 1995. Heavy metals, TBT and DDT in the Sydney rock oyster (*Saccostrea commercialis*) sample from the Hawkesbury River Estuary, NSW, Australia. *Mar. Pollut. Bull.*, **30**(8), 563–567.

Hardy, J.T. and J. Cleary. 1992. Surface microlayer contamination and toxicity in the German Bight. *Marine Ecology Progress Series*, **91**, 203–210.

Harris, J.R.W., C.C. Hamlin and A.R.D. Stebbing. 1991. A simulation study of the effectiveness of legislation and improved dockyard practice in reducing TBT concentrations in the Tamar Estuary. *Mar. Environ. Res.*, **32**, 279–292.

Huggett, R.J. 1986. Monitoring tributyltin in Southern Chesapeake Bay. In: Proceedings of the Interagency Workshop on Aquatic Monitoring and Analysis for Organotin. Sponsored by NOAA/NMPPO, Rockville, MD, pp. 29–30.

Huggett, R.J. 1987. Statement for Senate Hearing. The effects of the chemical tributyltin (TBT) on the marine environment. Hearing Record. Senate Subcommittee on Environmental Protection of the Committee on Environment and Public Works, 29 April. S. HGR 100-89. US Government Printing Office. 73-832. Washington, DC, pp. 23–28 (oral); pp. 68–74 (written).

Huggett, R.J., M.A. Unger and D.J. Westbrook. 1986. Organotin concentrations in southern Chesapeake Bay. In: Proceedings of the Oceans '86 Organotin Symposium, Vol. 4, Marine Technology Society, Washington, DC, pp. 1262–1265.

Huggett, R.J., M.A. Unger, P.F. Seligman and A.O. Valkirs. 1992. The marine biocide tributyltin. *Environ. Sci. Technol.*, **26**(2), 232–237.

Hunt, J. 1994. Science and policy in North Sea pollution. Lancaster University. 436 pp.

International Maritime Organization. 1990. MEPC (Marine Environmental Protection Committee) of the International Maritime Organization (IMO). Background papers and meeting notes. MEPC 29th and 30th Sessions. IMO, London.

Isensee, J., B. Watermann and H.-D. Berger. 1994. Emissions of antifouling-biocides into the North Sea – an estimation. *Deutsche Hydrographische Zeltschrift*, **46**(4), 355–364.

Key, D., R.S. Nunny, P.E. Davidson and M.A. Leonard. 1976. Abnormal shell growth in the Pacific oyster *Crassostrea gigas*. Some preliminary results from experiments undertaken in 1975. ICES C.M. Paper (K11), 7 pp.

Law, R.J., M.J. Waldock, C.R. Allchin, R.E. Laslett and K.J. Bailey. 1994. Contaminants in seawater around England and Wales: results from monitoring surveys, 1990–1992. *Mar. Pollut. Bull.*, **28**(11), 668–675.

Little, A.D. 1991. Preliminary field survey completed for national monitoring program for tributyltin (TBT) in Aquatic Environments. *Marine Science Update*, **1**(2).

Ludgate, J.W. 1987. Testimony of International Paint (U.S.A.) Inc. for the U.S. House of Representatives Committee on Merchant Marine and Fisheries Hearing. In: *Hearing Record*, 8 July 1987. Serial No. 100-28 (78-297). US Government Printing Office, Washington, DC, pp. 73–86.

Macauley, J.M., J.K. Summers, P.T. Heitmuller, U.D. Engle, G.T. Brooks, M. Babikow and A.M. Adams. 1994. Annual statistical summary: EMAP – estuaries Louisianian Providence 1993. US EPA, Office of Research & Development, Environmental Research Laboratories. Gulf Breeze, FL. EPA/620/R-94/002, 82 pp.

Matthiessen, P., R. Waldock, J.E. Thain, M.E. Waite and S. Scropte-Howe. 1995. Changes in periwinkle (*Littorina littorea*) populations following the ban on TBT-based antifoulings on small boats in the United Kingdom. *Ectoxicol. Environ. Safety*, **30**(2), 180–194.

Mee, L.D. and S.W. Fowler (Guest Editors). 1991. Editorial. *Mar. Environ. Res.*, **32**, 1–5.

Milne, A. 1990a. Roughness and drag from the marine chemist's viewpoint. Proceedings of the International Workshop on Marine Roughness and Drag. Royal Institution of Naval Architects, London.

Milne, A. 1990b. Cost/benefit analysis of SPC organo-tin antifoulings. In: International Maritime Organization (IMO) Marine Environmental Protection Committee (MEPC) Uses of Tributyltin compounds in Anti-fouling Paints for Ships. Document MEPC 30/INF.16. IMO, London, pp. 7–9.

Minchin, D., J. Oehlmann, C.B. Duggan, E. Stroben and M. Keatinge. 1995. Marine TBT antifouling contamination in Ireland, following legislation in 1987. *Mar. Pollut. Bull.*, **30**(10), 633–639.

NAVSEA (US Naval Sea Systems Command). 1984. Environmental assessment of fleetwide use of organotin antifouling paint. NAVSEA, Washington, DC, 128 pp.

NAVSEA (US Naval Sea Systems Command). 1986. Organotin antifouling paint: US Navy's needs, benefits, and ecological research. A report to Congress. NAVSEA, Washington, DC, 38 pp + appendix.

Reid, B. 1986. A chemical on trial. *Daily Press/The Times-Herald*, Newport News, VA (reprinted section – collection of articles).

Sarradin, P.-M., A. Astruc, R. Sabrier and M. Astruc. 1994. Survey of butyltin compounds in Archon Bay sediments. *Mar. Pollut. Bull.*, **28**(10), 621–628.

Schatzberg, P. 1987. Organotin antifouling paints and the US Navy: a historical perspective. In: Proceedings of the Oceans '87 International Organotin Symposium, Vol. 4, Marine Technology Society, Washington, DC, pp. 1324–1333.

Seligman, P.F., A.O. Valkirs and R.F. Lee. 1986a. Degradation of tributyltin in San Diego Bay, California waters. *Environ. Sci. Technol.*, **20**, 1229–1234.

Seligman, P.F., J.G. Grouvhoug and K. Richter, 1986b. Measurement of butylins in San Diego Bay, CA: monitoring strategy. In: Proceedings of the Oceans' 86 Organotin Symposium, Vol. 4, Marine Technology Society, Washington, DC, pp. 1289–1296.

Seligman, P.F., C.M. Adema, P.M. Stang, A.O. Valkirs and J.G. Grovhoug. 1987. Monitoring and prediction of tributyltin in the Elizabeth River and Hampton Roads, Virginia. In: Proceedings of the Oceans '87 International Organotin Symposium, Vol. 4, Marine Technology Society, Washington, DC, pp. 1357–1363.

Seligman, P.F., J.G. Grovhoug, R.L. Fransham, B. Davidson and A.O. Valkirs. 1990. US Navy Statutory Monitoring of Tributyltin in Selected US Harbors. Annual Report: 1989. Naval Ocean Systems Center Technical Report No. 1346. Naval Ocean Systems Center, San Diego, CA, 32 pp.

Stang, C.M., D.R. Bower and P.F. Seligman. 1989. Stratification and tributyltin variability in San Diego Bay. *Appl. Organometal. Chem.*, **3**, 105–114.

Stebbing, A.R.D. and V. Dethlefsen. 1992. Introduction to the Bremerhaven workshop on biological effects of contaminants. *Marine Ecology Progress Series*, **91**, 1–8.

Stebbing, A.R.D. 1985. Organotins and water quality – some lessons to be learned. *Mar. Pollut. Bull.*, **16**(10), 383–390.

Steur-Lauridsen, F. and B. Dahl. 1995. Source of organotin at a marine water/sediment interface – a field study. *Chemosphere*, **30**(5), 831–845.

Ten Hallers-Tjabbes, C.C. 1994. TBT in the open sea: a case for a total ban on the use of TBT antifouling paint. *North Sea Monitor*, **12**(3), 12–14.

Ten Hallers-Tjabbes and J.P. Boon. 1995. Whelks (*Buccinum undatum* L.), dogwhelks (*Nucella lapillus* L.) and TBT – a cause for confusion. *Mar. Pollut. Bull.*, **30**(10), 675–676.

Thain, J.E. 1983. The acute toxicity of bis (tributyl tin) oxides to the adults and larvae of some marine organisms. ICES C.M. Paper (E13), 5 pp.

Thain, J.E. 1986. Toxicity of tributyltin to bivalves: effects on reproduction, growth and survival. In: Proceedings of the Oceans '86 Organotin Symposium, Vol. 4, Marine Technology Society, Washington, DC, pp. 1306–1313.

Thain, J.E. and M.J. Waldock. 1983. The effect of suspended sediment and bis (tributyl tin) oxide on the growth of *Crassostrea gigas* spat. ICES C.M. Paper (E10), 11 pp.

Thain, J.E. and M.J. Waldock. 1985. The growth of bivalve spat exposed to organotin leachates from antifouling paints. ICES C.M. Paper (E28), 10 pp.

Thain, J.E., M.J. Waldock and M.E. Waite. 1987. Toxicity and degradation studies of tributyltin (TBT) and dibutyltin (DBTA) in the aquatic environment. In: Proceedings of the Oceans '87 International Organotin Symposium, Vol. 4, Marine Technology Society, Washington, DC, pp. 1398–1404.

UK DoE (Department of Environment). 1986a. The Control of Pollution Act (COPA) of 1985, the Control of Pollution Act of 1985 (the Anti-fouling Paints Regulations, of 1985, Statutory Instruments No. 2011, No. 2300, 1986). Joint Circular, 7 March 1986, 7/86. DoE/HMSO, London, 4 pp.

UK DoE (Department of Environment). 1986b. The Control of Pollution Act (COPA) of 1986, which amended the Control of Pollution Act of 1985 (the Anti-fouling Paints Regulations of 1986). Joint Circular, 24 February 1987, 3/87, DoE/HMSO, London, 4 pp.

UK DoE (Department of the Environment) Central Directorate of Environmental Protection. 1986c. Organotin in antifouling paints, environmental considerations. HMSO, London, England. Pollution Paper No. 25, 82 pp.

UK DoE (Department of Environment). 1987. The Control of Pollution (Antifouling Paints and Treatments) Regulations 1987. Statutory Instruments No. 783, 1987). Joint Circular, 12 June 1987, 19/87. Revision to Annex listing paints and treatment which conform to Regulations issued 5 January 1988, 3 January 1989, 11 January 1989. DoE/HMSO, London, 3 pp.

US Congress. 1969. The National Environmental Policy Act (NEPA), 91-190, In: 42 US Code 4321 *et seq*. US Code Government Printing Office, Washington, DC.

US Congress. 1972a. The Federal Water Pollution Control Act of 1972 as amended by the Clean Water Act of 1977 (P.L. 92-500 and P.L. 95-217) are known as the Clean Water Act (CWA). In: 86 Stat. 816.33U.S.C. 1251 *et seq* as amended. US Code. US Government Printing Office, Washington, DC.

US Congress. 1972b. The Marine Protection, Research, and Sanctuaries Act of 1972 (the Ocean Dumping Act). P.L. 92-532. In: 16 US Code 1431 *et seq* and 33 US Code 1401 *et seq*. US Government Printing Office, Washington, DC.

US Congress. 1972c. Federal insecticide, Fungicide, and Rodenticide Act (FIFRA), P.L. 92-516. In: 86 Stat. 975. 7 US Code 136 *et seq*. as amended. US Government Printing Office, Washington, DC.

US Congress. 1974. Toxic Substances Control Act (TSCA). P.L. 94–469. In: 100 Stat. 2989. 15 US Code 6A *et seq*. as amended. US Government Printing Office, Washington, DC.

U.S. Congress. 1976. Occupational Safety and Health Act (OSHA). P.L. 91-596. In: 42 US Code 6A *et seq*. as amended. US Government Printing Office, Washington, DC.

US Congress. 1978. The National Ocean Pollution Planning Act of 1978 (NOPPA). P.L. 95-273. In: 33 US Code 1701 *et seq*. US Government Printing Office, Washington, DC.

US Congress. 1986. The Hearing Record for the Oversight Hearings on Tributyltin in the Marine Environment. Committee on Merchant Marine and Fisheries. US House of Representatives. Ninety-Ninth Congress, 30 September, 1986. H. Hrg. Serial 99-49. (65-830-0). US Government Printing Office, Washington, DC, 117 pp.

US Congress. 1987a. The Senate Congressional Record, 2 February, 1987 – S1481. US Government Printing Office, Washington, DC.

US Congress. 1987b. The Senate Hearing Record for the Effects of the Chemical Tributyltin (TBT) on the Marine Environment. Committee on Environment and Public Works. US Senate, 29 April, 1987. S. Hrg. Serial 100-89 (73-832). US Government Printing Office, Washington, DC, 189 pp.

US Congress. 1987c. The Senate Hearing Record for Antifouling Paints for HR 1046 and HR 2210. Committee on Merchant Marine and Fisheries. US House of Representatives. One Hundredth Congress, 8 July, 1987. Serial No. 100-28 (78-297). US Government Printing Office, Washington, DC, 234 pp.

US Congress. 1988. The 'Antifouling Paint Control Act of 1988'. The Organotin Antifouling Paint Control Act of 1988 Public Law 100-333 (33 USC 2401).

US Environmental Protection Agency. 1983. Alkyltin compounds: responses to the interagency testing committee. *Federal Register*, **48**(217), 51361–51368.

US Environmental Protection Agency. 1985. Tributyltin support document. Position Document 1. EPA/OPP (Office of Pesticide Programs), Washington, DC, 46 pp.

US Environmental Protection Agency. 1986a. Initiation of a special review of certain pesticide products containing tributyltin used as antifoul-

ants. Notice of Availability of Support Document. *Federal Register*, **51**(5), 778–779.
US Environmental Protection Agency OPP (Office of Pesticide Programs). 1986b. Data call-in notice for data to support the continued registration of pesticide products containing tributyltin active ingredients used as paint antifoulants. 28 pp + ASTM/EPA Standard Test Method for Organotin Release Rates of Antifouling Coating Systems in Sea Water. 10 pp + EPA Supplement to the Method. OPTS/OPP, Washington, DC.
US Environmental Protection Agency 1987a. Tributyltin technical support document. Position Document 2/3. Office of Pesticide Programs, Washington, DC, 156 pp.
US Environmental Protection Agency. 1987b. Preliminary determination to cancel certain registrations of tributyltin products used as antifoulants unless terms and conditions of registration are modified. Technical Support Document and Draft Notice of Intent to Cancel. OPP-3000/49A. Federal Regulations L3273–9, 43 pp.
US Environmental Protection Agency. 1987c. Survey of tributyltin and dibutyltin concentrations at selected harbors in Chesapeake Bay – Final Report. Chesapeake Bay Program/TRS 14/87. Annapolis, MD. 58 pp + appendix 40 pp.
US Environmental Protection Agency Criteria and Standards Division. 1987d. Water quality advisory for tributyltin. Office of Water Regulations and Standards, Washington, DC, 23 pp.
US Environmental Protection Agency 1991. Congressional Report on Environmental Monitoring of Organotin. Office of Pesticide Programs, Washington, DC 23 pp.
US Navy. 1986. Interim FONSI. *Federal Register*, **50** (120), 25748.
US Senate. 1988. Organotin Antifouling Paint Control Act of 1987 Senate Amendments. Congressional Record – Senate. US Government Printing Office, Washington, DC. 18 April S. 4204–4211.
Valkirs, A.O., P.F. Seligman, P.M. Stang, V. Homer, S.H. Lieberman, G. Vafa and C.A. Dooley. 1986a. Measurement of butyltin compounds in San Diego Bay. *Mar. Pollut. Bull.*, **17**(7), 319–324.
Valkirs, A.O., P.F. Seligman and R.F. Lee. 1986b. Butyltin partitioning in marine waters and sediments. In: Proceedings in the Oceans '86 Organotin Symposium, Vol. 4, Marine Technology Society, Washington, DC, pp. 1165–1170.
Valkirs, A.O., M.O. Stallard and P.F. Seligman. 1987. Butyltin partitioning in marine waters. Proceedings of the Oceans '87 International Organotin Symposium, Vol. 4, Marine Technology Society, Washington, DC, pp. 1375–1380.
Valkirs, A.O., B. Davidson, L.L. Kear, R.L. Fransham, J.G. Grovhoug and P.F. Seligman. 1991. Long-term monitoring of tributyltin in San Diego Bay, California. *Mar. Environ. Res.*, **32**, 151–167.
Wade, T.L., B. Garcia-Romero and J.M. Brooks. 1988a. Tributyltin analyses in association with NOAA's National Status and Trends Mussel Watch Program. In: Proceedings of the Oceans '88 Organotin Symposium, Vol. 4, Marine Technology Society, Washington, DC, pp. 1198–1201.
Wade, T.L., B. Garcia-Romero and J.M. Brooks. 1988b. Tributyltin contamination in bivalves from United States coastal estuaries. *Environ. Sci. Technol.*, **22**, 1488–1493.
Wade, T.L., B. Garcia-Romero and J.M. Brooks. 1990. Butyltins in sediments and bivalves from United States coastal areas. *Chemosphere*, **20**(6), 647–662.
Wade, T.L., B. Garcia-Romero and J.M. Brooks. 1991. Oysters as biomonitors of butyltins in the Gulf of Mexico. *Mar. Environ. Res.*, **32**, 233–241.
Waite, M.E., M.J. Waldock, J.E. Thain, D.J. Smith and S.M. Milton. 1991. Reductions in TBT concentrations in UK estuaries following legislation in 1986 and 1987. *Mar. Environ. Res.*, **32**, 89–112.
Waldock, M.J. 1986. Tributyltin in UK estuaries, 1982–86: evaluation of the environmental problem. In: Proceedings of the Oceans '86 Organotin Symposium, Vol. 4, Marine Technology Society, Washington, DC, pp. 1324–1330.
Waldock, M.J. and D. Miller. 1983. The acute toxicity of bis(tributyl tin) oxides to the adults and larvae of some marine organisms. ICES C.M. Paper (E13), 5 pp
Waldock, M.J. and J.E. Thain. 1983. Shell thickening in *Crassostrea gigas:* organotin antifouling or sediment induced? *Mar. Pollut. Bull.*, **14**(11), 411–415.
Waldock, M.J., J.E. Thain and D. Miller. 1983. The accumulation and depuration of bis(tributyl tin) oxide in oysters: a comparison between the Pacific oyster (*Crassostrea gigas*) and the Euro-

pean flat oyster (*Ostrea edulis*). ICES C.M. Paper (E52), 9 pp.

Waldock, M.J., J.E. Thain and M.E. Waite. 1987a. The distribution and potential toxic effects of TBT in UK estuaries during 1986. *Appl. Organometal. Chem.*, **1**, 287–301.

Waldock, M.J., M.E. Waite and J.E. Thain. 1987b. Changes in concentrations of organotins in UK rivers and estuaries following legislation in 1986. In: Proceedings of the Oceans '87 International Organotin Symposium, Vol. 4, Marine Technology Society, Washington, DC, pp. 1352–1356.

Waldock, M.J., M.E. Waite and J.E. Thain. 1988. Inputs of TBT to the marine environment from shipping activity in the UK. *Environ. Technol. Lett.* **9**, 999–1010.

Weis, J.S. and L.A. Cole. 1989. Tributyltin and public policy. *Environ. Impact Assess. Rev.*, **9**, 33–47.

Wolniakowski, K., M.D. Stephenson and G. Ichikawa 1987. Tributyltin concentrations and Pacific oyster deformations in Coos Bay, Oregon. In: Proceedings of the Oceans '87 International Organotin Symposium, Vol. 4, Marine Technology Society, Washington, DC, pp. 1438–1442.

WWFN (World Wide Fund for Nature). 1995. Marine pollution by triorganotins. Marine Update. **21**, 4 pp.

PURGE-TRAP METHOD FOR DETERMINATION OF TRIBUTYLTIN BY ATOMIC ABSORPTION SPECTROMETRY

4

James H. Weber[1], *Cesar Clavell*[2], *Martha O. Stallard*[3] and *Aldis O. Valkirs*[3]

[1]*University of New Hampshire, Chemistry Department, Parsons Hall, Durham, New Hampshire 03824-3598, USA*
[2]*Naval Command Control and Ocean Surveillance Center, RDT&E Division, Environmental Sciences (Code 52), San Diego, California 92152-5000, USA*
[3]*Computer Sciences Corporation, 4045 Hancock Street, San Diego, California 92110-5164, USA*

Organotin. Edited by M.A. Champ and P.F. Seligman. Published in 1996 by Chapman & Hall, London. ISBN 0 412 58240 6

ABSTRACT

This chapter describes direct determination of sub-nanogram amounts of methyl- and butyltin compounds, particularly tributyltin, in aqueous environmental samples, plant extracts, and shellfish extracts by two similar methods. Both procedures use purge-and-trap methods following derivatization with sodium borohydride that allows concentration of organotin hydrides in a trap at liquid nitrogen temperature (−196°C). The organotin hydride species are eluted from the trap and are atomized in a quartz burner mounted in an atomic absorption spectrophotometer. The chapter discusses advantages and disadvantages of the two methods. Automation of one method permits near-real-time, aboard-ship determinations of tributyltin in seawater under in-situ conditions.

4.1. INTRODUCTION

The hydride generation technique is a method of converting organotin ions into volatile organotin hydrides that can be purged from solution, concentrated in a trap, and quantitated by atomic absorption spectrometry. About 10 y ago two groups published hydride generation methods for determination of organotin compounds in the aquatic environment. Braman and Tompkins (1979) detected inorganic tin and methyltin compounds in rainwater, urine, ocean, estuarine, and freshwater samples by flame emission spectrometry of the Sn–H bond. Hodge *et al.* (1979), in contrast, used atomic absorption spectrometry for detection. They measured inorganic tin as well as methyl-, ethyl-, butyl-, and phenyltin compounds in freshwater and seawater environments. These two papers are an important part of the foundation for the methods described below.

This chapter emphasizes *experimental* details of the hydride generation method for determination of tributyltin and other organotin compounds at the nanogram and sub-nanogram levels in environmental matrices such as shellfish, plants, and water. Two major advantages of the hydride generation technique compared with competing techniques are a one-step concentration method without extractions, and sufficient sensitivity for most environmental samples without pretreatment. The goal of this chapter along with its references is to explain clearly how to determine low concentrations of organotin compounds, especially tributyltin, in environmental samples by the hydride generation approach. The three main sections of the chapter are the non-automated approach of the University of New Hampshire and of the Naval Oceans Systems Center (NOSC) groups, and automation of the method by the NOSC group.

4.2. BASIC EXPERIMENTAL TECHNIQUE (UNIVERSITY OF NEW HAMPSHIRE)

Butyltin hydrides were formed in a hydride generation flask, purged on to a cold trap, and detected with a Perkin–Elmer Model 503 atomic absorption spectrophotometer (AAS) using a quartz furnace. After initial development by Donard *et al.* (1986), these procedures have evolved from research of several members of the Weber group mentioned in the references. Thus, details are from several publications, but information in this section is from Randall *et al.* (1986) unless stated otherwise.

4.2.1. APPARATUS

4.2.1a. Reaction vessel, transfer lines, cryotrap, and packing

The hydride generation flask (5 × 7 cm) is connected to a 25-cm Teflon PTFE transfer line. The inner tube [0.3 cm outside diameter (o.d.), 0.1 cm wall] was inserted into the Tygon outer tube (0.5 cm o.d., 0.05 cm wall), which was wrapped with Nichrome wire (32 gauge, 0.35 ohm cm^{-1} resistance) and insulated with Teflon tape. The other end of the transfer line was connected to a U-shaped trap [35 cm, 0.4 cm inside diameter (i.d.) that contained 2.5 g Chromosorb G AW-DMCS (45/60 mesh) coated with 3% SP-2100]. The trap was prepared from Cole Parmer Type T-6407-44 Teflon TFE tubing, and was wrapped with Nichrome wire (26 gauge, 0.088 ohm cm^{-1} resistance). In more recent work (Francois and Weber, 1988; Francois *et al.*, 1989) the packing material was silanized. Ten microliters of 5% dimethyldichlorosilane in toluene was injected on the head of the column (not connected to the furnace) and the column was flushed with helium for 5 min. This treatment was repeated three times, twice at 100°C and the third time at 200°C.

4.2.1b. Quartz furnace

The trap was connected to the quartz furnace (Anderson Glassblowing, Fitzwilliam, New Hampshire), which has inlets for the helium purge gas, hydrogen, and oxygen (Donard *et al.*, 1986). The furnace was treated with an aqueous hydrofluoric acid–nitric acid mixture, rinsed, and annealed at 1100°C (Hatfield, 1987). The major helium inlet was wrapped with 26-gauge Nichrome wire. The body of the furnace was wrapped with a doubled strand of 26-gauge Nichrome wire. In contrast to our previous work (Donard *et al.*, 1986), the quartz furnace was wrapped with materials from Wale Apparatus Company (Hellertown, Pennsylvania) to eliminate use of asbestos. Woven quartz tape was used between the furnace and Nichrome wire, and Waletex high-temperature glass tape was used as a heat insulator. The latter tape is somewhat stiff and is easier to handle after wetting. The furnace was mounted on a custom-made stainless steel frame placed on the AAS burner head.

Teflon-to-Teflon and Teflon-to-quartz connections were made with Omnifit Teflon variable-bore connectors (Rainin Instrument Corp.). Power to the transfer line, trap, furnace inlet, and furnace was supplied by individual Variacs.

4.2.1c. Atomic absorption detection

The Perkin–Elmer 503 AAS was fitted with a Westinghouse electrodeless discharge lamp (EDL) for tin, operated on continuous mode by a Westinghouse EDL power supply. The output signal (1 V full deflection) was amplified 10-fold and filtered at 1.3 Hz by a low-pass filter. (In current experiments the output signal is not amplified or filtered.) The output signal was integrated by a Hewlett-Packard Model 3392 A integrator.

4.2.1d. Operating parameters

Operating parameters determined by Simplex optimization were gas flow rates (ml min^{-1}): helium (400), oxygen (21), and hydrogen (833). Hydrogen and helium flow rates were controlled by Cole Parmer flow meters and oxygen flow rate by a precise Brooks flow meter. The tin EDL was used at 10 W in continuous mode. The AAS was operated at 224.61 nm with 1 mm slit and 0.3 s integration repeat mode. Variac outputs were set to attain temperatures of 95–150°C (hot to the touch) for the transfer line (25 V), 20–200°C for the trap (0–30 V), and 750°C (19 V) for the furnace. The Variac setting for the furnace inlet was 8 V. Temperatures were measured with a Chromel/Alumel thermocouple.

4.2.2. REAGENTS AND STANDARDS

The water used in all experiments was deionized twice and then distilled through a Corning Mega-pure still. All glassware was soaked in 7% nitric acid (diluted 9:1 from Fisher 70% nitric acid) for at least 12 h and rinsed with water. Aqueous 6% sodium borohydride was prepared from Aldrich sodium borohydride. The solutions were filtered through a Nuclepore filter (0.2 μm pore size), put in a glass flask, and refrigerated for at least 12 h to reduce the tin blank. Methanol was Fisher certified ACS spectroanalyzed grade.

Mono- and dibutyltin chlorides, tetrabutyltin, and triethyltin bromide were obtained from Alfa Ventron. Tributyltin hydride and tributyltin chloride were purchased from Aldrich. These chemicals were used without additional purification. Standards were made by diluting 100 $\mu g\ ml^{-1}$ (as Sn) methanolic stock solutions with methanol (Han and Weber, 1988). Stock solutions were stored in 10-ml Hypovials sealed with crimp-on Teflon-lined septa. All solutions were stored in the dark at 4°C. Stock solutions were stable over several months. Standards were injected into the hydride generator by gas chromatographic microliter syringes.

4.2.3. OPERATING PROCEDURE

Samples and the internal standard (triethyltin bromide) were introduced into the hydride generation flask containing 40 ml of water and 0.5 ml of glacial acetic acid. The solution was stirred using a magnetic stirrer while 2.0 ml of 6% sodium borohydride was added via the septum and the solution purged for 3 min (Francois and Weber, 1988; Francois *et al.*, 1989). The hydride formation step for tributyltin was quantitative as demonstrated by 100% ± 5% recovery of tributyltin chloride relative to pre-formed tributyltin hydride. The Dewar flask containing liquid nitrogen was removed from the trap. The trap was left unheated for a time, and then heated with a Variac program that gave good peak separation. This procedure is effective for a mixture of the six ionic methyl- and butyltin compounds, tetramethyltin, and tetrabutyltin (Donard *et al.*, 1986; Francois and Weber, 1988; Han and Weber, 1988).

An alternative procedure can be used for measurement of tributyltin (TBT) alone (Francois and Weber, 1988; Francois *et al.*, 1989). Tributyltin hydride can be trapped on a 30-cm trap containing 0.5 g of packing material at room temperature. (Details of the trap were described in section 4.2.1a, Reaction vessel, transfer lines, cryotrap, and packing.) Tributyltin hydride is eluted by using a helium flow of 200 cm min^{-1} when the trap is heated to ~180°C.

These basic procedures preceded by appropriate extraction steps give excellent results from environmental matrices such as estuarine water (Donard *et al.*, 1986), oysters (Han and Weber, 1988), and eelgrass (Francois and Weber, 1988; Francois *et al.*, 1989). However, it was necessary to purge off the extractant dichloromethane by using a room-temperature trap (Francois *et al.*, 1989) since dichloromethane burned in the quartz furnace and also prevented tributyltin hydride formation (Francois and Weber, 1988).

Interferences have not usually been a problem. Donard *et al.* (1986) showed that sodium chloride, fulvic acid, EDTA, and iron (III) did not interfere, but that sulfide ion could interfere. The latter could be removed as hydrogen sulfide by purging acidic solutions. High percent recoveries from oyster (Han and Weber, 1988) and eelgrass (Francois and Weber, 1988) matrices confirm a lack of interferences in these samples.

4.2.4. CALIBRATION AND LIMITS OF DETECTION

Analytical results are quantitated on the basis of calibration curves of analytes and the triethyltin bromide internal standard. First, calibration curves were made from standard additions in pure water (Randall *et al.*, 1986) or in an experimental matrix such as an oyster extract (Han and Weber, 1988). The internal standard, triethyltin bromide, was added to all samples. The amount of analyte in the sample was then determined by comparing the calibration curve slopes of the internal standard and the analyte.

Limit of detection (LOD) for each organotin compound was defined as background signal plus three standard deviations in the matrix being studied. In all cases LOD values are as Sn. Among inorganic tin and methyl- and butyltin compounds, LOD was highest for inorganic tin because it was a contaminant in the reagents used. The LOD was always higher in environmental matrices than in pure water. For example, LOD for TBT was 45 pg in pure water (Randall *et al.*, 1986), 2.5 ng in the extract from 0.1-g oyster samples (Han and Weber, 1988), and 1.0 ng in the extract from 40-mg eel grass samples (Francois and Weber, 1988). Larger oyster or eelgrass samples would, of course, decrease LOD in their extracts. LOD of mono- and dibutyltin was lower in all matrices studied (Randall *et al.*, 1986; Francois and Weber, 1988; Han and Weber, 1988).

4.3. BASIC EXPERIMENTAL TECHNIQUE (NAVAL OCEANS SYSTEMS CENTER)

The hydride generation purge-and-trap system described is similar to the experimental technique of the University of New Hampshire described in section 4.2. Initial development followed from procedures published by Hodge *et al.* (1979). Additional developments by Valkirs *et al.* (1985, 1987) permitted determination of tributyltin in seawater at low environmental concentrations. The system and procedures described in this chapter were developed by Stallard *et al.* (1989) unless otherwise referenced.

4.3.1. APPARATUS

4.3.1a. Reaction vessel and transfer lines

A modified 500-ml gas washing bottle, which has an injection port on its side and inlet and outlet ports at its top (Fig. 4.1, part B), is used. Teflon tape is wrapped around 3.2-mm o.d. Teflon tubing to form a secure seal when the tubing is threaded through the outlet port and pulled snug. The other end of the tubing line is pressure fitted into the end of a 3-mm i.d. glass U-shaped tube that functions as the cryotrap. Improved sensitivity was achieved by avoiding the use of Swagelok fittings, which may form dead spaces in the transfer line system and attenuate transport of the analyte to the detector. It is also essential to avoid dead space in the gas washing bottle by filling the bottle to the 500-ml mark ~4 cm from the neck of the bottle). A large decrease in sensitivity of twofold or more occurs if a small sample volume with a large dead space is used. Filling the bottle closer to the outlet port junction results in excess transfer of water vapor to the outlet line. The tall, cylindrical shape of the modified gas washing bottle coupled with the use of a Kimax 12C frit to disperse bubbles formed by the helium carrier gas has provided the most efficient combination for purging the evolved methyl- and butyltin hydrides from the sample. Coarser frits are less efficient due to larger bubble formation, and finer frits have caused too much back pressure, which causes problems in controlling the helium flow.

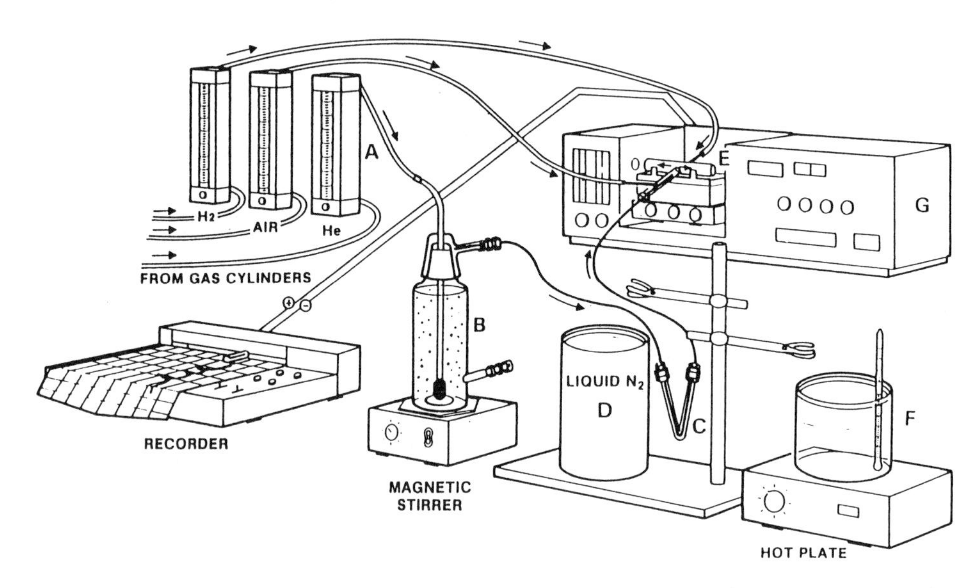

Figure 4.1 Hydride generation/atomic absorption spectrometry system for measurement of organotin species: (A) helium carrier gas line; (B) hydride generator, modified gas washing bottle; (C) 3% OV-1 Supelcoport packed glass trap; (D) liquid nitrogen for cryogenic cooling; (E) quartz furnace; (F) high-temperature silicone oil bath; and (G) atomic absorption spectrometer.

Larger bottles were fabricated to permit larger sample volumes for analysis, and thus an increased amount of analyte at the detector. Larger reaction vessels do, however, cause more formation of water vapor resulting in unpredictable signal attenuation.

The Teflon transfer line from the cryotrap to the quartz furnace, which is 3.2-mm o.d. Teflon as well, transports tin hydrides evolved from the cryotrap to the quartz furnace. This line is pressure fitted into the exit end of the cryotrap.

4.3.1b. Cryotrap and packing material

The cryotrap is a 3-mm i.d. glass U-shaped tube with side arms of 6.5–7.5 cm. Silanizing the trap is essential to avoid active sites, which may bind the hydride species formed and cause peak reduction or tailing. Proper silanization requires a thorough cleaning by soaking in hot 2% RBS-35 critical cleaner (VWR Scientific Co.), rinsing with distilled water, and drying. The traps are then washed with hexane and methanol and dried in an oven. While still warm, the traps are filled with silanizing fluid (Supelco Silon CT) and left for at least 30 min to dry. The traps are thoroughly flushed with hexane and then methanol, and conditioned in an oven at 200°C for 10–15 min. Twenty to thirty milligrams of 3% OV-1 on Supelcoport 80/100 mesh is added to each trap. The packing material is loosely settled and held in place by a very small silanized glass wool plug placed at each end of the trap immediately in contact with the packing material. Compressing the packing material should be avoided as this may result in broad peaks with poor definition. The 20–30 mg of chromatographic packing is adequate, and larger amounts tend to result in broad peaks. We have also investigated use of Teflon traps with the same internal diameter as the glass traps. The Teflon traps gave similar results to the silanized glass traps; however, Teflon does not conduct heat well, and water vapor con-

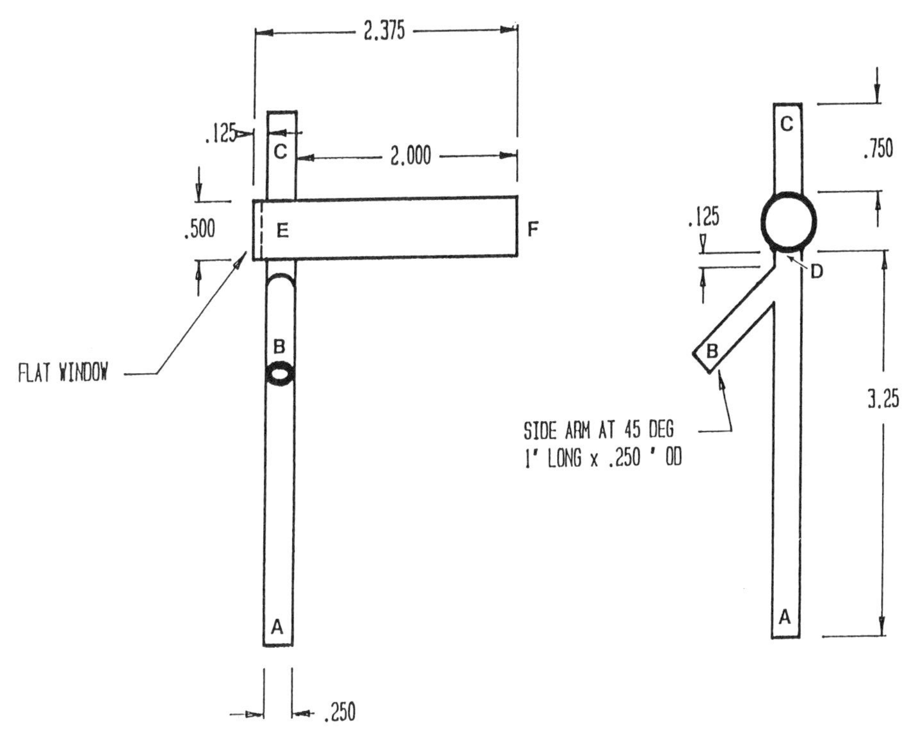

Figure 4.2 Closed-end quartz burner (dimensions in inches; 1 in = 2.54 cm): (A) sample entry; (B) air entry; (C) hydrogen entry; (D) position of glass capillary tip in Teflon tube introducing tin hydrides into burner; (E) hottest area of burner, 500°C; and (F) open end of burner, 400°C. US patent number 4 913 648.

densed in the ends of the trap resulting in greatly reduced sensitivity after a few runs.

4.3.1c. Quartz furnace

Several furnace designs have been reported in the literature in analytical systems similar to the NOSC model (Braman and Tompkins, 1979; Hodge *et al.*, 1979; Andreae, 1981; Donard *et al.*, 1986). The diameter and length of the burner cylinder; orientation of oxidizer, fuel, and carrier gas lines; and single open-ended or double open-ended cylinders are all important factors. The present furnace was fabricated by closing the end of the furnace facing the hollow cathode tube in the AAS, moving all gas entries off center near the closed end, and lengthening the main burner cylinder. Details of this system (Fig. 4.2) were recently reported (Stallard *et al.*, 1989). The position of the tube that introduces the tin hydrides influenced sensitivity. A 1.5-cm-long glass capillary was forced into the end of the Teflon tube from the cryotrap to prevent the end from burning where it is in close proximity to the flame. This line is threaded through a drilled-out Swagelok reduction fitting at the sample entry port of the quartz furnace (Fig. 4.2, part A). The use of a Swagelok fitting at this junction presents no problems with dead space formation as the fitting simply acts to hold in place the line passing through it. The tip of the glass capillary tube was positioned at the edge of the main cylinder of the quartz furnace (Fig. 4.2, part D). This introduces the evolved tin hydrides as a small jet into the hottest part of the furnace (Fig. 4.2, part E).

External heating of the quartz furnace with a Nichrome wire coil did not seem to improve sensitivity and resulted in a longer time for stabilization of the system after it was turned on. We believe the improved sensitivity demonstrated with the single open-ended burner may result from concentrating the tin hydrides formed in a small area for a longer time period, which resulted in more tin atoms in the light path and therefore increased light absorption. The rear of the burner where the hydride species are introduced is also the hottest part of the burner. During operation, and particularly before trapped hydride species are purged into the furnace, the flame should be checked and re-ignited if necessary.

The quartz furnace is mounted in the AAS by wrapping the main cylinder with a 2-mm-thick piece of Fiberfrax ceramic fiber (Lab Safety Supply, Janesville, Wisconsin). The insulated furnace cylinder is then mounted on a custom-machined adjustable aluminium frame placed on the AAS burner head. The burner head is then positioned with the AAS adjustments to permit maximum light transmission through the flame in the center of the furnace.

It is very important that the furnace be carefully constructed. Considerable differences in sensitivity were observed among the 12 furnaces that were developed. It is especially important that the flat quartz window be perfectly clear and free of imperfections, bubbles, or scratches. The sample, hydrogen, and air entry ports should be carefully connected to the main furnace cylinder so that the space around area E (Fig. 4.2) is minimal.

4.3.1d. Atomic absorption detector

The volatile methyl- and butyltin hydrides formed by sodium borohydride derivatization are detected with a single-beam AAS (Buck Scientific Model 200) at 286.3 nm. The 224.6-nm line gave somewhat better sensitivity but considerably more baseline noise (Valkirs *et al.*, 1985). Since very small peaks were difficult to resolve accurately, the 286.3-nm wavelength was selected for routine use. A tin-specific hollow cathode lamp set at a lamp current of 8 mA was the light source. Highest sensitivity occurred when the gas flow rate was set at 220 ml min^{-1} for hydrogen and 140 ml min^{-1} for air. Helium carrier gas was set at 50–60 ml min^{-1}. Background correction was not used since the derivatization purge-and-trap collection effectively removes the tin hydride species from the sample matrix.

4.3.2. REAGENTS AND STANDARDS

Volumetric glassware, the reaction vessel, and polycarbonate bottles used for sample collection and storage were cleaned by soaking for several days in 2% RBS-35 critical cleaning solution. We have found this simple procedure is effective in removing even nanogram amounts of TBT from polycarbonate sample bottles. Polycarbonate plastic does not adsorb organotin compounds (Dooley and Homer, 1983; Valkirs *et al.*, 1986; Carter *et al.*, 1989).

Sodium borohydride (98%) from J.T. Baker was dissolved in 18 megaohm deionized water containing 1% sodium hydroxide (AR, Mallinckrodt Inc.). A 4% $NaBH_4$ solution was prepared fresh daily. Baker analyzed reagent-grade acetic acid was dissolved in 18 megaohm deionized water to make a 10% solution. UPC-grade helium, hydrogen, and zero-grade air were used. Mono-, di-, and tributyltin chlorides were used as received from Aldrich (Milwaukee, Wisconsin). Ethyl alcohol (USP, 95%) was obtained from US Industrial Chemical (Anaheim, California). Seawater for blanks and sample dilution was taken directly from the seawater flume at the Scripps Institution of Oceanography pier in La Jolla, California.

Primary standards were prepared bimonthly by weighing mono-, di-, and tributyltin chlorides into 50-ml volumetric flasks and diluting with ethanol. All concentrations in this section are reported *as butyltin cations*. Each primary standard contained ~0.1 mg ml^{-1} of a butyltin chloride. Primary standards were stored at 4°C in the dark. A secondary working standard with all three butyltin species added was prepared every other day by diluting 100 μl of each of the primary standards to 100 ml, resulting in an approximate concentration of 0.1 ng μl^{-1}.

4.3.3. OPERATING PROCEDURE

4.3.3a. Nature of sample

The NOSC system has been used primarily for determinations in seawater samples. A typical seawater sample with a salinity of 32–33 g kg^{-1} is adjusted to approximately pH 5.5 with 0.5 ml of 10% acetic acid: prior to addition of sodium borohydride. The initial pH and buffering capacity of the sample are important since over-acidification results in excess hydrogen liberation and inconsistent results. Estuarine samples with low salinities (10–20 g kg^{-1}) may be analyzed by dilution of the sample with seawater. If the TBT concentration of such a sample is <0.5 ng l^{-1}, it may be necessary to modify the acetic acid: sodium borohydride ratio without dilution. Routine analysis of freshwater samples has not been attempted because of difficulties that occurred with low-salinity estuarine samples. However, buffering of the samples and an appropriate acetic acid:sodium borohydride ratio may allow determinations in low-salinity samples.

Other samples, such as marine sediments, are amenable to direct sodium borohydride derivatization as well. It is possible to place a wet sediment sample (50–200 mg wet wt) in the reaction vessel with a 500-ml volume of blank seawater and proceed as with the seawater analysis. Larger sample weights may cause analytical interferences due to the often complex nature of marine sediments, particularly those from yacht harbors and marine-repair facilities where high fuel and oil concentrations are often encountered. With samples from such areas, and with marine sediments in general, it is advisable to verify the absence of significant matrix effects by addition of standards to the sample. Repetitive analysis of marine sediment samples has indicated that the method's precision is comparable to that of analysis of water samples (Valkirs *et al.*, 1985).

Some success was achieved with sodium borohydride derivatization of tissue extracts (Dooley and Vafa, 1986). Dilution of the extract and a sodium hydroxide wash of the extract were necessary to minimize matrix interferences. Other methods have proven effective for determination of TBT in tissues from marine species. Consequently, we have not pursued application of the sodium borohydride derivatization process for tissue analysis.

4.3.3b. Purging and trapping

The purging and trapping of tin hydrides is initiated by acidification of the bulk seawater sample by addition of 500 μl of 10% acetic acid to the sample contained in the reaction vessel. Regardless of the sample volume used, the final volume is always brought to 500 ml with blank seawater when sample volumes of <500 ml are used, resulting in an initial sample pH of ~5.5. The glass cryotrap is placed in a liquid nitrogen bath at a level just above the packing material, and 5 ml of 4% sodium borohydride solution is injected into the reaction vessel. Volatile organotin hydrides formed are purged from solution by the helium carrier gas and collected on the chromatographic packing in the cryotrap. A 5-min purge-and-trap cycle is sufficient for quantitative analysis of the butyltin hydrides present. During purging the solution is stirred by a Teflon stirring bar.

4.3.3c. Detection and integration

After the purging-and-trapping cycle, the trap is removed from the liquid nitrogen bath so it may warm to room temperature. During this time stannane, methyltin hydrides, and monobutyltin hydride evolve from the trap in order from lowest to highest boiling point. After the appearance of monobutyltin hydride, the trap is placed in a 50°C water bath, resulting in rapid evolution of dibutyltin as a sharp peak. Tributyltin hydride is evolved from the trap by placing it in a silicone oil bath at 180°C (Fig. 4.1, part F).

The methyl- and butyltin hydrides are carried to the quartz burner through a Teflon transfer line and atomized in a hydrogen–air flame. Absorbance areas are recorded on a Shimadzu CR3A Chromatopac recorder. After each sample analysis, the exit transfer line is removed from the cryotrap at the pressure fit junction, and the cryotrap is purged with helium for 8 min to remove water vapor collected during the purging-and-trapping cycle. Simultaneously, the reaction vessel is emptied and an empty, clean, and dry vessel is put in place. This permits only dry helium carrier gas to pass through the system. After the 8-min period for purging of water vapor, the exit transfer line is also briefly purged with helium to remove any water vapor. The system is reassembled as before with a clean reaction vessel containing a new sample, and the purge-and-trap cycle is repeated. We have found that three analytical cycles per hour are possible. Attempts to process more than three cycles per hour result in excess retention of water vapor and signal attenuation.

4.3.3d. Interferences

In the routine analysis of seawater we have seldom encountered signal attenuation due to sample matrix effects. Standard additions are made to subsets of samples collected during monitoring surveys to assess matrix effects. In 66 of 76 samples (87%) analyzed from Norfolk, Virginia; Pearl Harbor, Hawaii; and San Diego Bay, California, the recovery of tributyltin added (1–6 ng) exceeded 75%. Previous comparisons of TBT values calculated from calibration curves and by the method of standard additions indicated that values calculated from calibration curves agreed well with those determined by standard additions (Valkirs *et al.*, 1985). Elements such as copper, silver, gold, and iron may interfere with tin hydride generation by consuming the sodium borohydride reductant, but this interference is not serious if excess sodium borohydride is used (Van Loon, 1980).

The presence of volatile compounds such as diesel fuel at high concentrations has resulted in severe signal attenuation (Valkirs *et al.*, 1985). It is possible that at very high concentrations such compounds may create active sites in the transfer lines and trap, thus attenuating the signal. Andreae (1981) noted that adsorption of butyltin hydrides to internal surfaces may be the most persistent difficulty with their determination by the hydride derivatization method. Attempts to determine butyltin compounds from areas where pollutants such as fuel oils are present should be supported by assessments of matrix effects by standard additions.

Early experiments in our laboratory indicated that sulfide ion might interfere with measurement of TBT in marine water and sediment samples. For example, the presence of 1.6 mg l^{-1} sulfide in seawater samples spiked with TBT (0.15 $\mu g\ l^{-1}$) resulted in ~50% loss of TBT measured after a 24-h exposure period. Marine sediment from a local yacht basin exposed to 16.2 mg l^{-1} sulfide for 2 min exhibited a 58% decrease in the TBT peak compared with that in the original sediment (Valkirs *et al.*, 1985). Calibration curves produced from sodium borohydride reduction of tributyltin chloride and tributyltin sulfide standards in ethanol gave comparable slopes, suggesting that the reduced amount of TBT measured in environmental matrices in the presence of sulfide ion was due to interaction of sulfide with the samples (Valkirs *et al.*, 1985). Thus some possible interference with sodium borohydride derivatization might be expected in environmental samples, particularly marine sediments, where high sulfide concentrations exist.

4.3.4. CALIBRATION AND LIMITS OF DETECTION

A standard curve was prepared daily by adding 10, 25, 50, or 100 μl of the secondary standard to 500 ml of blank seawater. Offshore seawater from the Scripps Institution of Oceanography was used as a reference blank solution. A standard was repeated after every sixth sample. Standard curves typically range from 1 to 10 ng at the detector. Occasional standard additions were made to samples to confirm the absence of matrix effects.

A solution prepared for an interlaboratory TBT determination by the National Bureau of Standards (NBS) was a useful reference material. The initial TBT concentration was 1.42 mg l^{-1}. One milliliter of the NBS solution was diluted to 100 ml with ethanol or distilled water to form a secondary reference. Daily analysis of a 250 μl aliquot of the secondary reference added to blank seawater (3.6 ng TBT at the detector) for several months by five analysts operating four different instruments gave consistent results. The relative standard deviation was 11–18% of the mean value. This variability is similar to that reported in an interlaboratory comparison of tributyltin determinations (Valkirs *et al.*, 1987). Our experience with analysis of an initially sterile TBT solution prepared at a high concentration suggests that dilution and analysis over time can provide a satisfactory reference analysis procedure for determining analytical accuracy. Additional investigation is necessary to determine the stability of the stock TBT solution.

Detection limits have recently been greatly improved by a modified burner design and silanization of the cryotrap (Stallard *et al.*, 1989). Current detection limits calculated from three standard deviations of five replicate analyses of seawater solutions spiked with 0.5 ng each of mono-, di-, and tributyltin are 120, 80, and 180 pg, respectively, as butyltin cations. Detection limits for TBT have ranged from 180 to 80 pg. This range reflects sensitivity differences in the five systems we use for monitoring butyltin compounds in environmental samples (Stallard *et al.*, 1989). Overall this system has been used to measure several thousand environmental water samples from diverse harbors and estuaries (Grovhoug *et al.*, Chapter 25; Huggett *et al.*, Chapter 24).

4.4. AUTOMATION OF SYSTEM (NAVAL OCEANS SYSTEM CENTER)

Because of the large numbers of samples required for monitoring, an automated analytical system was developed at NOSC. The instrument is comprised of three main subsections: the system controller, the analytical section, and the detector section (Fig. 4.3). The controller and detector are packaged as one unit, and the analytical section is a separate module. The analytical and detection system is as described above for the NOSC method.

4.4.1. SYSTEM CONTROLLER

The controller consists of a board-level IBM-AT compatible computer (International Business Machines; Diversified Technologies, Model 902), which supports a 3.5-in floppy disk and a hard drive (Conner, CP-340). The system display is a 640 × 200 LC flat panel with integral RGB controller (Sharp, model QA7LCS02). Interface electronics for I/O control are controlled by a half-size IBM-bus card (Strawberry Tree Computer, model PC JR). There are eight channels of analog input, with software-programmable ranges of ±25 mV to ±10 V, multiplexed to a 12-bit analog-to-digital converter. Dynamic range changing accommodates the large variation in signal

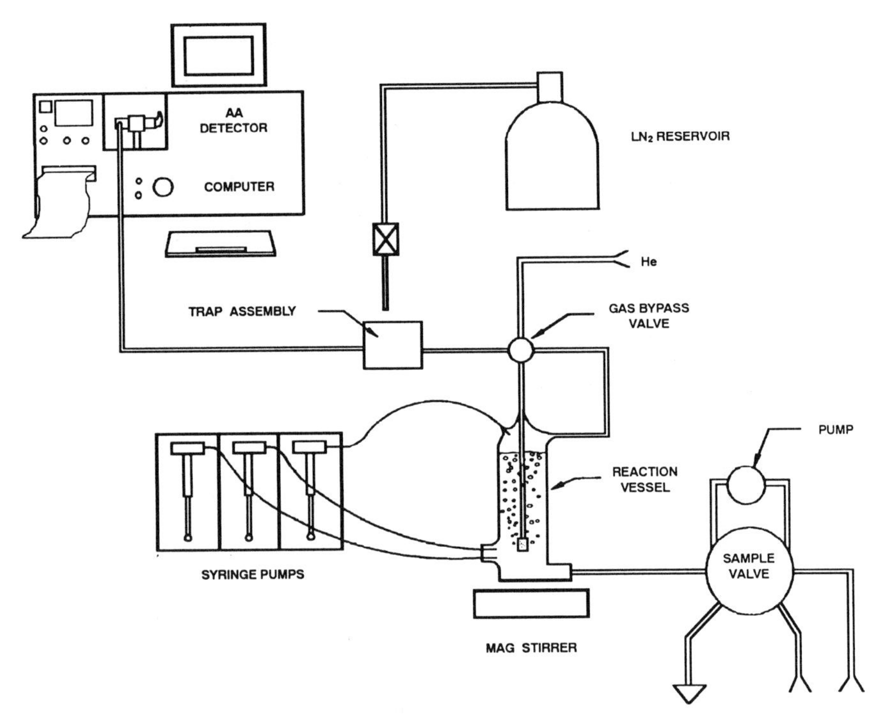

Figure 4.3 Diagrammatic representation of automated system.

level input from the atomic absorption detector, the temperature thermocouple, and the liquid nitrogen level sensor. There are 12 bidirectional digital I/O lines available with eight configured as outputs and four as inputs. The output lines control the mechanical components via solid-state relays mounted on an external termination board. The input lines monitor limit switches and status bits.

Software is written in the PolyFORTH (Forth Inc.) version of the Forth language, which provides a multiuser, multitasking environment. The system program is comprised of several 'tasks' that can run simultaneously, permitting input or retrieval of parameters or data at any time. This feature enables the user to review system status and alter parameters; to input free-form notes relating to the current analysis; and to adjust previous run data, with a peak editor function, during an analysis cycle without interrupting the analysis process. User interaction with the system is through a menu-driven interface implemented as function key selections with pop-up windows. All data collected are automatically archived in 720 kilobyte files, representing 70 analysis runs, to facilitate back-up on 3.5-in floppy disks. The data for each individual analysis include a complete record of all system parameters, user-entered notes, standard curve values, and the raw binary peak values. This storage scheme allows an analysis to be retrieved by its run number or date and the raw data reprocessed at any future date. Utility routines provide data file manipulation and cataloging functions.

Troubleshooting of the instrument is facilitated by a resident diagnostic software task. A diagnostic session can be initiated by either the system itself, upon an internal fault indication, or by the operator. The goal of the diagnostic process is to lead the user, via a

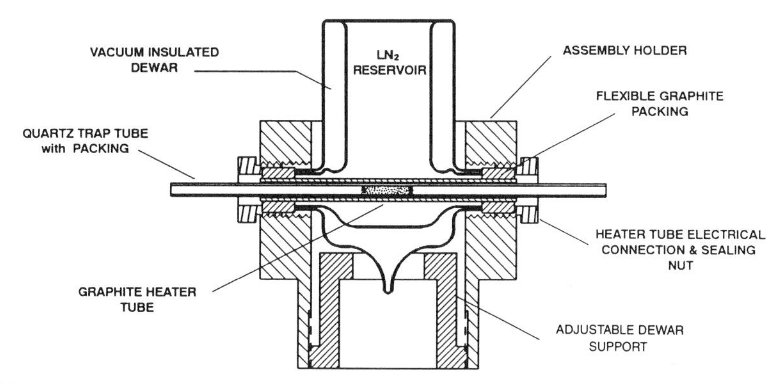

Figure 4.4 Diagrammatic representation of variable-temperature trap assembly.

question–test–result style interaction, to determine the cause of the fault and then to provide the necessary information to correct it, if possible. Information and advice given can be tailored to the expected technical proficiency of the operator.

4.4.2. ANALYTICAL SECTION

The analytical unit contains hardware to perform automatically the three functions of sample handling, additions of reagents and standards, and concentration and separation of methyl- and butyltin hydrides.

4.4.2a. Sample handling

Samples can be pumped from any source into the 500-ml custom-made, glass reaction vessel. Following analysis, the system can be instructed to flush the reaction vessel, using an alternate water source or the sample source, and then refill the vessel with a fresh sample.

4.4.2b. Addition of reagents and standards

Additions to the reaction vessel are made by a set of three computer-controlled syringe pumps (CAVRO Inc.). The pumps feature interchangeable syringe bodies and have a delivery capability of 0.001 full scale.

4.4.2c. Concentration and separation of methyl- and butyltin hydrides

Trapping and concentrating the volatile hydride reaction products are accomplished with a patented variable-temperature trap assembly (Fig. 4.4). The computer-controlled trap and its liquid nitrogen source can provide temperature sequencing of up to five selectable set points, where the temperature can be held (±2°C) for a specified time. The trap temperature range is −200 to +180°C. Heating rates of 8–23°C s^{-1} are possible with a pulsed 500-W heater. The trap temperature is monitored by a custom-type K thermocouple (3.2 mm diameter × 9.5 cm long, Therm Southwest, #TJ-36-CA316SS-010G-3.75/MC) inserted coaxially in the trap tube. Trap tubes are made of either FEP tubing (11.4 cm long × 4.7 mm o.d. × 2.5 mm i.d.) or quartz tubing (11.4 cm long × 4.2 mm o.d. × 2.9 mm i.d.) packed with 1.5 cm of chromatographic support material (3% OV-1 on Supelcoport, 80/100 mesh) centered in the tube. The trap tube is inserted coaxially in a heater consisting of a tightly fitting, 7.6-cm-long graphite tube connected to a 7 V, 200 W toroidal transformer. The heater is controlled by a software algorithm that pulses the power to the tube (+500 W peak) in a proportional manner to avoid excessive set point overshoot. The trap assembly (trap tube and heater) is inserted into a vacuum-insulated quartz reservoir that is filled with liquid nitrogen to maintain the temperature of the trap tube at −196°C during the concentration period. A 300-ohm, 0.125-W, 10% carbon resistor serves as a level sensor for control of the liquid nitrogen addition valve. The resistor is wired as one leg of a voltage divider with a 12-V input and generates approximately a 2-V output differential when immersed. The components comprising the analytical subsystem are mounted on a lightweight base and communicate with the computer–detector unit via a multiconductor cable. All component parts and tubing that contact the sample stream are constructed from either glass or Teflon.

The system can operate continuously (processing 3–4 samples per hour) or on an individual sample basis. To begin an analysis run, a sample is pumped into a 500-ml reaction vessel and 0.5 ml of 10% acetic acid is added. Simultaneously, the trap is filled with liquid nitrogen. When the trap has cooled to −150°C, the gas bypass valve directs helium gas (55 ml min^{-1}) into the reaction vessel and then 5 ml of 4% (wt/v) sodium borohydride in 1% NaOH is added. The helium is bubbled through the sample for 5 min purging the methyl- and butyltin hydrides from the reservoir and concentrating them in the cold trap. After the concentration period, the gas valve redirects the helium to the trap, bypassing the reaction vessel. This is followed by the simultaneous initiation of the heating program, the acquisition of data from the AAS, and the real-time plotting of the raw data. The trap temperature is raised in discrete steps, with a holding time between steps, yielding well-resolved chromatograms (Fig. 4.3). The trap is maintained at the final temperature of 180°C for 6 min to remove all traces of moisture. During this time, the computer processes the data and outputs an annotated plot with temperature vs. millivolt scales, peak areas and concentrations for each component, and any notes input by the user. If the automatic peak detection has incorrectly identified the limits of a peak, the operator may call up the peak editor, correct the limits, and then reprocess the data to generate a new plot. The multitasking capability of the software allows all these functions to be performed concurrently.

4.4.3. DETECTOR SYSTEM

The detector is based on a Buck Instruments Model 200 AAS. The single-beam grating instrument was modified by removing all nonessential features and repackaging the components in a custom enclosure. The new chassis incorporates a one-piece, vibration-damped optical bench mounted on a stiff, boxed frame. Sensitivity for the butyltin and arsenic compounds of interest is in the sub-nanogram per liter range. The compact size, light weight, simplicity, and apparent ruggedness of this unit facilitate its use in a transportable system.

4.5. CONCLUSIONS

We described two procedures for organotin determinations by hydride derivatization with purge-and-trap collection and atomic absorption detection. Both methods have similar detection limits, and each has advantages and difficulties. Only the NOSC method has been automated.

The NOSC method uses a fairly large sample volume (500 ml), rather high pH (5.5), and a small trap with very little packing material. This method uses relatively low gas flow rates. It also uses an unheated, closed-end quartz furnace and relies upon small helium bubbles to purge tin hydrides from solution. It is very sensitive to changes in pH, and this can cause difficulties in determining organotin compounds in freshwater or estuarine samples of low salinity. The small, simple trap is easy to change and can be used for >100 runs if it is carefully dried between runs. The manually heated trap requires close operator attention. The automated version of the method allows measurement of a large number of samples aboard ship.

The University of New Hampshire method uses a fairly small volume of sample, a low pH, and an electrothermally heated trap with a fairly large mass of packing, and a heated quartz furnace. The trap and furnace can be used hundreds of times without deterioration. This method depends mainly on evolved hydrogen to purge tin hydrides from solution. The method is equally effective in samples of low and high salinity and for extracts of sediments, shellfish, and plants. It is possible to determine sub-nanogram quantities of methyl- and butyltin compounds in the same run.

ACKNOWLEDGMENTS

JHW thanks the National Science Foundation for financial support from Grant NSF CES 86-12972, and his co-workers Dr Olivier Donard and Dr Louise Randall for developing the University of New Hampshire method for determination of methyl- and butyltin compounds. Work of the Naval Oceans Systems Center, San Diego, California, was sponsored by the Office of the Chief of Naval Research under the Navy Energy Research and Development Program.

REFERENCES

Andreae, M.O. 1981. The determination of the chemical species of some of the 'hydride elements' (arsenic, antimony, tin and germanium) in seawater: methodology and results, In: NATO Symposium on Trace Metals in Sea Water, Erice, Sicily, 30 March–3 April 1981.

Braman, R.S. and M.A. Tompkins. 1979. Separation and determination of nanogram amounts of inorganic tin and methyltin compounds in the environment. *Anal. Chem.*, **51**, 12–19.

Carter, R.J., N.J. Turoczy and A.M. Bond. 1989. Container adsorption of tributyltin (TBT) compounds: implications for environmental analysis. *Environ. Sci. Technol.*, **23**, 615–617.

Donard, O.F.X., S. Rapsomanikis and J.H. Weber. 1986. Speciation of inorganic tin and alkyltin compounds by atomic absorption spectrometry using electrothermal quartz furnace after hydride generation. *Anal. Chem.*, **58**, 772–777.

Dooley, C.A. and V. Homer. 1983. Organotin compounds in the marine environment: uptake and sorption behavior. Naval Ocean Systems Center Technical Report #917, San Diego, California, 19 pp.

Dooley, C.A. and G. Vafa. 1986. Butyltin compounds and their measurement in oyster tissues. Proceedings of the Oceans, 86 Organotin Symposium, Vol. 4, Marine Technology Society, Washington, DC, 23–25 September 1986, pp. 1171–1176.

Francois, R. and J.H. Weber. 1988. Speciation of methyltin and butyltin compounds in eel grass (*Zostera marina* L.) from the Great Bay Estuary (NH). *Mar. Chem.*, **25**, 279–289.

Francois, R., F.T. Short and J.H. Weber. 1989. Accumulation and persistence of tributyltin in eelgrass (*Zostera marina* L.) tissue. *Environ. Sci. Technol.*, **23**, 191–196.

Han, J.S. and J.H. Weber. 1988. Speciation of methyl- and butyltin compounds and inorganic tin in oysters by hydride generation atomic absorption spectrometry. *Anal. Chem.*, **60**, 316–319.

Hatfield, D.B. 1987. Electrically heated quartz atomization cell for hydride generation–atomic absorption spectrophotometry. *Anal. Chem.*, **59**, 1887–1888.

Hodge, V.F., S.L. Seidel and E.D. Goldberg. 1979. Determination of tin(IV) and organotin compounds in natural waters, coastal sediments and macro algae by atomic absorption spectrometry. *Anal. Chem.*, **51**, 1256–1259.

Randall, L., O.F.X. Donard and J.H. Weber. 1986. Speciation of *n*-butyltin compounds by atomic absorption spectrometry with an electrothermal quartz furnace after hydride generation. *Anal. Chim. Acta*, **184**, 197–203.

Stallard, M.O., S.Y. Cola and C.A. Dooley. 1989. Optimization of butyltin measurements for seawater, tissue, and marine sediment samples. *Appl. Organomet. Chem.*, **3**, 105–114.

Valkirs, A.O., P.F. Seligman, G. Vafa, P.M. Stang, V. Homer and S.H. Lieberman. 1985. Speciation of butyltins and methyltins in seawater and marine sediments by hydride derivatization and atomic absorption detection. Naval Ocean Systems Center Technical Report #TR 1037, San Diego, California, 25 pp.

Valkirs, A.O., P.F. Seligman and R.F. Lee. 1986. Butyltin partitioning in marine waters and sediments. Proceedings of the Oceans, 86 Organotin Symposium, Vol. 4, Marine Technology Society, Washington DC, 23–25 September 1986, pp. 1165–1170.

Valkirs, A.O., P.F. Seligman, G.J. Olson, F.E. Brinckman, C.L. Matthias and J.M. Bellama. 1987. Di- and tributyltin species in marine and estuarine waters. Inter-laboratory comparison of two ultratrace analytical methods employing hydride generation and atomic absorption or flame photometric detection. *Analyst*, **112**, 17–20.

Van Loon, J.C. 1980. *Analytical Atomic Absorption Spectroscopy – Selected Methods*. Academic Press, Orlando, Florida, 337 pp.

GAS CHROMATOGRAPHIC DETERMINATION OF BUTYLTIN COMPOUNDS IN WATER USING HYDRIDE DERIVATIZATION

5

Cheryl L. Matthias

Mine Safety Appliances Co., 38 Loveton Circle, Sparks, Maryland, MD, USA

Organotin. Edited by M.A. Champ and P.F. Seligman. Published in 1996 by Chapman & Hall, London. ISBN 0 412 58240 6

ABSTRACT

The marine antifouling agent tributyltin and its degradation products were determined in estuarine water by in-situ hydride derivatization using aqueous sodium borohydride with simultaneous extraction into dichloromethane. The butyltin hydrides were determined using gas chromatography with tin-selective flame photometric detection. The detection limit for tributyltin in a 200-ml water sample is 5 ng l^{-1} (0.017 nM) by using this method. An aqueous tributyltin solution prepared by the National Institute of Technology and Standards (formerly the National Bureau of Standards) for a 35-laboratory comparison was analyzed by the new method. The result obtained by the new method is within 4% of the mean value of the determination for the 35 laboratories and within 9% of the value determined by neutron activation analysis. Fourteen marine and estuarine water samples were analyzed for di- and tributyltin in an interlaboratory exercise. For tributyltin, agreement between the new method and an existing one was within 20% for 9 of the 14 samples.

5.1. INTRODUCTION

The analytical method described here was developed for the determination of the antifouling agent tributyltin and the related butyltin compounds. The method takes advantage of the ease of derivatization of *in situ* hydride generation methods and permits the use of commercially available gas chromatographic equipment. Commercial equipment eliminates the in-house fabrication of transfer lines often required in purge-and-trap methods and thereby improves system reliability. Use of a tin-selective flame photometric detector (FPD) eliminates 'clean-up' of the extracted samples.

5.2. GAS CHROMATOGRAPHY OF BUTYLTIN HYDRIDES

5.2.1. BACKGROUND

Modern flame photometric detectors (FPD) for use in gas chromatography are based on the design of Brody and Chaney (1966), who detected chemiluminescence from phosphorus and sulfur compounds in a hydrogen-rich diffusion flame. The flame both decomposes (burns) the sample and generates the optical emission. The first application of an FPD to the determination of organotin compounds was reported by Aue and Flinn (1977), who optimized flame conditions for detection of tetraalkyltins. They detected tin emissions in an 'open' configuration without any wavelength discrimination and with an optical interference filter centered on the 610 nm emission attributed to SnH (Dagnall *et al.*, 1968). They reported detection of 2×10^{-13} g of tetrapropyltin. Maguire and Huneault (1981) detected pentyl-derivatives of methyl and butyltin species using the 'open' configuration. Jackson *et al.* (1982) used a modified commercial FPD detector with flame gases optimized for tin detection to determine methyltin hydrides, and mono- and dibutyltin hydrides. The optical emission was monitored with a 600-nm 'cut-on' optical filter (band-pass 600–2000 nm) to transmit the SnH emission lines. The work described here is the first to determine butyltin hydrides in organic solvent solutions using commercially available gas chromatographic–flame photometric detection (GC–FPD) systems.

5.2.2. MATERIALS AND METHODS

5.2.2a. Gas chromatography system

A Hewlett-Packard (HP) (Avondale, Pennsylvania) Model 5730A gas chromatograph equipped with an HP flame photometric detector (Model 18764A) was used. The chromatographic separation was carried out on a 2-mm inner diameter × 6-foot (1.83-m) glass column packed with 1.5% OV-101 (liquid methyl silicone) on Chromosorb G HP (Varian, Sunnyvale, California). Nitrogen gas (zero grade) carrier flowed at a measured rate of 20 ml min^{-1}. A hydrogen-rich flame was sustained with hydrogen flowing at 150 ml min^{-1}, air at 50 ml min^{-1}, and oxygen at 5 ml min^{-1}. The signal output was recorded on an integrator plotter (HP Model 3390A). Initial column temperature was 23°C. After an initial 2-min hold, the column was heated at the rate of 32°C min^{-1} to a final temperature of 180°C. The detector temperature was maintained at 200°C and the injection port at 150°C. The GC–FPD was routinely equipped with a 600-nm 'cut-on' optical interference filter with a band pass of 600–2000 nm (Ditric Optical, Inc., Hudson, Massachusetts) to monitor SnH molecular emission.

5.2.2b. Reagents and standard solutions

Spectrograde dichloromethane was obtained from Fisher Scientific (Silver Spring, Maryland) and from Burdick and Jackson Laboratories, Inc. (Muskegon, Michigan). Spectrograde methanol was obtained from Fisher. Sodium borohydride (sodium tetrahydroborate) was obtained from Aldrich Chemical Co. (Milwaukee, Wisconsin). Deionized water of 15–18 MΩ resistivity was obtained using a Milli-Q reagent-grade water system (Millipore Corp., Bedford, Massachusetts). A fresh solution of sodium borohydride (4% wt/v) was prepared daily in deionized water.

Mono-, di-, and tributyltin chlorides and tetrabutyltin were obtained from Alfa Chemical (Milwaukee, Wisconsin). Dipropyltin dichloride for use as an internal standard was obtained from Organomet, Inc. (East Hampstead, New Hampshire). These compounds were all greater than 95% purity and were used as received without further purification. All organotin concentrations reported in this paper are as the fully dissociated anion, such as Bu_3Sn^+. Stock solutions of butyltin and propyltin chlorides were prepared gravimetrically in spectrograde methanol at ~2000 ng μl^{-1} (~6–11 mM). These stock solutions are stable for at least three months at room temperature in borosilicate glass bottles if protected from light. Working dilutions of these stock solutions were prepared daily.

5.2.3. RESULTS AND DISCUSSION

5.2.3a. Detector selectivity

While the earliest tin FPDs were used without wavelength discrimination, in environmental samples there is a possibility of interference from sulfur or phosphorus species, which have complex emission spectra in the 350–450 nm and 500–575 nm regions, respectively (Brody and Chaney, 1966). While tin has emission in the 405 nm region, monitoring the red emissions centered around 609.5 nm reduces the probability of S or P interferences. The 600–2000 nm wide band-pass cut-on filter provides sensitive tin-selective detection with minimal interference from sulfur or phosphorus emissions. More tin-specific detection can be achieved by using a narrow band-pass filter of 610 nm (band width 10 nm). This filter is centered on the most intense peak of the SnH emission spectrum, which is found at 609.5 nm, and yields higher Sn specificity with a concomitant reduction in sensitivity. The narrow band-pass filter reduces the detector's response to about one-third that of the wide band filter.

The flame photometric detector, which monitors molecular emission, is element selective, not element specific, regardless of the wavelength monitored. The possibility therefore exists that in environmental samples some interfering species, which may produce an FPD response in the spectral region being monitored, will co-elute with the desired analyte. A simultaneous dual detector would be very useful in addressing this problem. By monitoring two different wavelength regions simultaneously, it is possible to determine if the observed peak is indeed tin or is some non-tin-containing compound with a similar chromatographic retention time. Alternatively, repeat analysis of the suspect peak with a sulfur- or phosphorus-selective (394 nm and 526 nm, respectively) filter can be used to aid in peak identification, if needed. The potential for co-eluting spectral interference is much higher when studying methyltin compounds than butyltins because there are several ubiquitous environmental compounds in the same molecular weight and boiling point range that could cause an FPD response, including dimethyldisulfide and methylgermanium species (Andreae and Byrd, 1984). There are fewer possible interferants in the same molecular weight range as TBT, and interferants have not been a problem in TBT analysis of water samples.

5.2.3b. Detector stability

The sensitivity (detector response) of the tin-selective FPD has been reported to be quite dependent on the 'history' of the detector (Maguire and Tkacz, 1983) because the detector is easily fouled by large amounts of tin compounds or by other chemicals such as tropolone, which is used to improve butyltin extraction in some analytical methods. In this work, a causal relationship between detector history and sensitivity has not been observed; however, day-to-day variations in the detector response were observed, due to aging of detector O-rings and other causes. In order to monitor detector response, a quality control standard solution of tetrabutyltin (TTBT) in dichloromethane was prepared at a concentration of 0.3 ng μl^{-1} (0.86 μM) and stored in the dark in a bottle equipped with a 'Mininert' cap. This cap permits syringe withdrawals of the test solution through a rubber diaphragm, which can be resealed with an airtight valve, thus preventing evaporation of the solvent. Three replicate 5-μl injections (1.5 ng TTBT) of this solution were analyzed by GC–FPD (isothermal at 170°C) each day prior to analytical runs in order to evaluate the condition of the detector. In order to insure adequate detector sensitivity for analysis of low-concentration environmental samples, a detector response of at least 6×10^4 area units was defined as the minimum acceptable detector response. If the detector sensitivity to the test TTBT solution fell below 6×10^4 area units, the instrument was not used for quantitative analysis of environmental samples until sensitivity was restored. This precaution insured that no samples were reported to be below the detection limit in TBT when in fact the system did not have adequate sensitivity on the day of the analysis. The detector response to the test solution was often significantly above the required 6×10^4 area units, occasionally by as much as a factor of two.

The run-to-run reproducibility of the detector was evaluated by making seven replicate injections of a dichloromethane solution of the hydride-derivatized butyltins. The percent relative standard deviation for MBT was 12%, for DBT 9%, for TBT 15%, and for TTBT 9% for full-scale peaks.

Maguire and Tkacz (1983) reported the formation of substantial amounts of a white powder, presumed to be SnO, in the FPD used for tin determination. The presence of this powder lowered detector response, and the powder had to be removed from the

Table 5.1 Minimum detectable amounts of butyltins

Species	*Butyltin cation injected (ng)*	*Sn injected (ng)*
$BuSn^{3+}$	0.26	0.17
Bu_2Sn^{2+}	0.22	0.11
Bu_3Sn^{+}	0.20	0.08
Bu_4Sn	0.80	0.28

detector assembly by mechanical means on a regular basis. Formation of significant amounts of white powder was not noted in this work, although occasionally the quartz heat shields of the burner were noted to be plated with an intractable white material that could not be removed by organic solvents, acids, or mechanical abrasion. Detector stability was improved by keeping the flame burning constantly. This resulted in longer intervals between detector maintenance, particularly O-ring replacement. Maguire and Tkacz (1983) also reported improved day-to-day stability when the detector was constantly burning.

5.2.3c. Detector sensitivity

The 'sensitivity' of the detector indicates the detector response per unit of analyte. As discussed in the preceding section, detector sensitivity varied considerably from day to day, so determination of absolute detection limits for the GC–FPD depended on the detector sensitivity on the day of the experiment. Under the defined minimum acceptable detector conditions, using the packed column and temperature program described above, the detection limits for the butyltin compounds, at the 95% confidence interval as defined by Parris *et al.* (1977), are as shown in Table 5.1. These limits are higher than those of Aue and Flinn (1977) who reported 10^{-12}–10^{-13} g (as Sn) as the detection limit for their system operating in the 'open' configuration. They reported that the detector was ~100 times less sensitive when operated with the 610-nm optical filter. The detection limits determined in this work are in accord with those findings.

5.2.3d. Chromatographic separation

The use of an element-selective detector simplifies the task of obtaining satisfactory chromatographic resolution by limiting the number of chromatographic peaks produced. In the case of butyltin compounds, a simple 6-ft (1.83-m) packed column provided sufficient resolution to achieve the required separation (Fig. 5.1). A 10-m 'megabore' (530-μm i.d.) fused silica column, HP-1 (Hewlett Packard), was also evaluated for butyltin separation.

There was baseline resolution for di-, tri-, and tetrabutyltin on both columns. However, the megabore capillary column did not satisfactorily resolve monobutyltin hydride from the solvent front. It may be possible to resolve this peak with a sub-ambient oven temperature, but this was not possible on the available instrumentation. The capillary col-

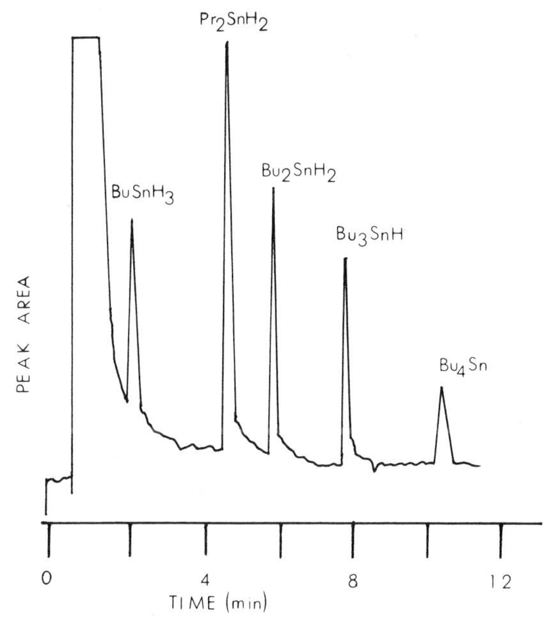

Figure 5.1 Typical chromatogram showing $Bu_nSnH_{(4-n)}$ (n = 1–4) and Pr_2SnH_2.

umn does permit more rapid analysis than the packed column. The HP-1 column can be used with a carrier gas flow of 20 ml min^{-1}; thus, no make-up gas is required for optimum detector performance. A disadvantage of this column for environmental work is that it may become fouled by non-volatile components of the sample, but the use of an off-column (splitless) injection technique with a disposable gas liner in the injection port to trap non-volatile components minimizes this problem.

Both packed column and megabore capillary columns produced peaks with a small degree of tailing. This tailing has been attributed to detector design (Aue and Flinn, 1977) rather than column characteristics and does not hinder separation or quantitation.

5.3. SAMPLE PREPARATION

5.3.1. BACKGROUND

In the determination of ultra-trace concentrations of pollutants in environmental samples, successful chromatographic separation and selective and sensitive detector response are often only half the problem. The other half of the problem is the sample preparation to minimize matrix effects, to preconcentrate the sample, and to derivatize the analyte. Typically, TBT sample preparation has been done either using modifications of the hydride generation and purge-and-trap system of Braman and Thompkins (1979) and others or by solvent extraction followed by alkylation using a Grignard reagent (Unger *et al.*, 1986) The approach taken in this work combines *in situ* hydride derivatization with solvent extraction according to the reaction scheme:

$$Bu_3Sn^+_{(aq)} \xrightarrow{BH_4^-} Bu_3SnH_{(aq)} \xrightarrow{CH_2Cl_2} Bu_3SnH_{(org)}$$

The hydride derivatization–extraction system is optimized for the analyte of most interest, the bioactive agent TBT. The other butyltin species are also determined, but not necessarily optimally.

5.3.2. MATERIALS AND METHODS

5.3.2a. Gas chromatography systems

The gas chromatographic system and reagents used were as described previously.

5.3.2b. Reagents and standards

All glassware was cleaned prior to use by washing with laboratory detergent followed by 12–24 h of leaching with 10% nitric acid. The clean glassware was then rinsed with copious amounts of deionized water.

For analysis of water samples with typical TBT concentrations [<250 ng l^{-1} (0.86 nM)], a sample volume of 200 ml is necessary in order to achieve a method detection limit of 5 ng l^{-1} (0.017 nM). For samples with higher TBT concentration, such as samples from some marinas, microlayer samples, or samples from some toxicological or degradation studies, a sample volume of 50–100 ml is sufficient. For samples of 200 ml, the extraction-derivatization procedures were carried out in a 250-ml straight-sided separatory funnel equipped with a Teflon®-lined (E.I. Du Pont, Wilmington, Delaware) screw cap and a Teflon stopcock. These funnels were made to our specification for this work by Wheaton Scientific (Millville, New Jersey), based on the design of a 125-ml funnel available as a standard item from Wheaton. For smaller samples, the 125-ml capacity funnels were used. These funnels can be secured to a wrist-action shaker (Burrell, Pittsburgh, Pennsylvania) to allow mechanization of a portion of the analytical workup. The procedure outlined below is that for 200-ml samples. For smaller samples, volumes of solvents and reagents are reduced proportionally.

5.3.2c. Analysis procedure for TBT in water

To a 200-ml sample of unfiltered water in a 250-ml separatory funnel, 50 μl of 0.5 ng μl^{-1} (2.4 μM) Pr_2Sn^{2+} in deionized water is added as an internal standard. The samples are then shaken and the internal standards are allowed to equilibrate with the samples for 10–15 min. Hydride derivatization is achieved by the addition of 3 ml of 4% (wt:v) aqueous sodium borohydride (sodium tetrahydroborate). Six milliliters of dichloromethane is then added, the funnel is capped and shaken by hand for ~30 s and vented. The funnel is then secured to the wrist action shaker and shaken (240 stroke min^{-1}) for 10 min. After shaking, the layers are allowed to separate and the lower, organic layer is removed, either to a Reactivial (Wheaton Scientific, Millville, New Jersey) or to a 15-ml borosilicate glass centrifuge tube (Corning Glass, Corning, New York). An additional 3 ml of dichloromethane is added and the shaking step is repeated. The second organic portion is combined with the first for a total solvent volume of ~5–6 ml. (Dichloromethane has an aqueous solubility of ~1.5 ml l^{-1}.) In some estuarine water samples, the dichloromethane forms an emulsion with algae and other materials present in the water. This dichloromethane emulsion is considerably heavier than water and can be removed easily from the separatory funnel. The emulsion is then broken by centrifugation at 700 × *G* for 5–10 min using a benchtop centrifuge. The organic (bottom) layer is removed to a glass Reactivial, using a Pasteur pipet. The dichloromethane is then evaporated under a gentle stream of dry air to 50–200 μl as required. Generally 5-μl portions of the concentrated sample are injected onto the GC column, although sample sizes of up to 10 μl can be used.

5.3.3. RESULTS AND DISCUSSION

5.3.3a. Solvent choice and hydride derivatization

Dichloromethane is often the solvent of choice for extraction of organic compounds from natural waters and is effective in extracting organotins from tissues (Seligman, 1984). Dichloromethane is sufficiently volatile to permit easy sample concentration by evaporation, but it is not difficult to prevent evaporation to dryness. For liquid–liquid extraction techniques, a heavier-than-water solvent simplifies removal of the organic layer and eliminates the need for transfer of the aqueous portion to a second separatory funnel if multiple extractions are required. For this application, dichloromethane has the disadvantage of producing a response from the FPD (unlike non-chlorinated solvents such as hexane). This is not a severe problem, since the solvent front is separated from the earliest eluting peak, butyltin trihydride, on the packed column.

Hydride derivatization is required to produce volatile butyltin derivatives for GC analysis, by converting the aquated cations and hydroxides of mono- and dibutyltin to non-polar, covalent species that are easily extracted into the organic phase.

5.3.3b. Efficiency of TBT extraction using the simultaneous system

Determination of the recovery of TBT by this method is complicated by the simultaneous derivatization–extraction procedure, which necessitates determination both of completeness of chemical reaction and of extraction efficiency. Preliminary determination of TBT extraction efficiency from deionized water into dichloromethane [TBT concentration ~3000 ng l^{-1} (10 nM)] was made by tin-specific graphite furnace atomic absorption analysis of the water sample before and after solvent extraction. TBT extraction was ~95% after a single extraction with dichloromethane without hydride derivatization. This recovery is in good agreement with Meinema and coworkers (1978) who reported quantitative recovery of TBT into benzene, cyclohexane, and chloroform without the use of tropolone complexation as is required for extraction of mono- and dibutyltins. Tributyltin hydride (TBTH) of >95% purity is commercially available (Alfa Products, Danvers, Massachusetts), and was used to construct a calibration curve of peak area vs. nanograms of TBTH. The efficiency of the new method for determination of tributyltin was determined by spiking at 2000 ng l^{-1} (6.9 nM) with tributyltin a 100-ml sample of Chesapeake Bay water (CBI station 858C) of low intrinsic TBT concentration and performing the simultaneous hydride derivatization–extraction. This concentration was chosen because it is sufficiently low to be realistic for environmental samples and sufficiently high that no evaporation of the solvent was required in order to produce a dichloromethane solution of TBTH of sufficient concentration to yield a GC–FPD peak. Evaporation of the organic layer would have complicated the system further by introducing the possibility of analyte loss due to volatilization. The amount of tributyltin recovered (determined from the TBTH calibration curve) was 112% ± 10% for five replicates.

5.3.3c. Analyte losses due to solvent evaporation

Analyte loss on solvent evaporation was investigated using the dichloromethane extract of borohydride-treated Chesapeake Bay water spiked with 1000 ng l^{-1} (5.6–2.9 nM) $Bu_nSn^{(4-n)+}$ (n = 1–4). One milliliter (1.0 ml) of the extract was evaporated under a gentle stream of gas to ~50 μl and then rediluted back to 1.0 ml with fresh dichloromethane. GC–FPD analysis of the samples was done before and after evaporation, and losses were determined by changes in peak area. Comparisons were done between dry air and dry nitrogen and between 22°C (ambient temperature) and 0°C (ice water bath). Under all conditions minimal losses (0–12%) on evaporation were noted for di-, tri-, and tetrabutyltin; and approximately 50% loss was noted for monobutyltin. Dry air and ambient temperature were used for all subsequent analysis.

5.3.3d. TBT stability

The possibility that the analytical workup, in particular the derivatization procedure or the chromatography, might induce either degradation or rearrangement reactions is always of concern when working with organotin compounds (Burns *et al.*, 1980). The absence of a TBT standard reference material has been a limitation in evaluating analytical methods for this problem. A research reference material of aqueous tributyltin was prepared chromatographically at the National Institute of Technology and Standards (formerly the National Bureau of Standards; Blair *et al.*, 1986) at 2 ng μl^{-1} (6.9 μM) concentration. Six randomly chosen bottles of this material were analyzed repeatedly at concentrations ranging from milligrams per liter to <50 ng l^{-1} with no evidence of any degradation or rearrangement. Spikes of mono- and di- and tetrabutyltin added to dilutions of the reference material indicate that in solutions of TBT concentration of 2 μg l^{-1} (6.9 μM), detection of <3% conversion of tributyltin to dibutyltin or tetrabutyltin would be readily achieved.

5.3.3e. Quantitation, detection limits, and reproducibility

Quantitation was done by preparing a series of butyltin-spiked water samples of salinity approximating the salinity of the environmental samples. For estuarine samples, artificial seawater diluted to 10–25‰ salinity was usually satisfactory. The butyltin spikes generally ranged from 0 to 400 ng l^{-1} (400 ng l^{-1} = 2.2 nM to 1.2 nM, depending on butyltin species) in each butyltin species to be determined, this being the most common concentration range found in the environment. A known amount of internal standard dipropyltin was then added to the spiked samples. Generally 50 μl of 0.5 ng μl^{-1} dipropyltin in water was added to a 200-ml water sample to give a final concentration of 125 ng l^{-1} (0.60 nM) of internal standard. These spiked samples were then subjected to the usual analytical procedure, and the relative peak areas of the butyltin analytes were determined as the ratio of the butyltin peak area to the peak area of the internal standard. Calibration curves were then plotted as relative peak area vs. concentration (ng l^{-1}) of butyltin analyte.

Statement of the limits of detection for ultra-trace analysis methods is a topic of considerable controversy. The usual definition of limit of detection – that concentration that gives an instrument signal significantly different from the 'blank' or 'background' signal (Miller and Miller, 1984) – leaves considerable margin for interpretation by the analyst. In this work the criterion used for the limit of detection is the analyte concentration giving a signal equal to the blank plus three standard deviations of the noise (as recommended by the American Chemical Society, 1980). For water samples of 200 ml volume, in which the dichloromethane extract is concentrated to 50 μl and 10 μl is injected for GC analysis, the limit of detection was determined to be 5 ng l^{-1} (0.021–0.014 nM) for di-, tri-, and tetrabutyltin and ~20–25 ng l^{-1} (0.11–0.14 nM) for monobutyltin. These limits may vary depending on the initial sample volume, the degree of preconcentration, and the volume injected for GC analysis. Because of the variability in response of the FPD detector, the detector response was monitored daily. The detection limits reported here are defined at the minimum acceptable detector response. If the detector response to the test sample fell below the value defined in section 5.2.3b, Detector stability, the instrument was deemed unreliable for determination of TBT in environmental samples and was not used for that purpose until sufficient detector response was restored, usually by cleaning and refurbishing the detector. The detector response was often above the defined lower limit, and in these cases the limiting value of 5 ng l^{-1} (0.017 nM) for tributyltin often produces a signal significantly larger than three times baseline noise, and often approaches the requirement for limit of quantitation [viz. blank plus 10 times the standard deviation of the noise; American Chemical Society (1980)].

There have been few problems with maintaining a very low blank level for all butyltin species, as long as scrupulous care is taken in washing and acid leaching of glassware. In the earliest work, polypropylene centrifuge tubes rather than borosilicate glass tubes were used. The use of polypropylene tubes was discontinued when it was found that the plastic tubes were causing a large, broad GC peak of retention time similar to that of tetrabutyltin. Subsequently, only glass or polycarbonate vessels were used.

Reproducibility of the analytical workup was determined at several concentration levels using the packed column, and at 20 ng l^{-1} (0.07 nM) using the 520-μm fused silica column. Reproducibility of the method at 2000 ng l^{-1} (6.8 nM) was determined using spiked Chesapeake Bay water (CBI site 858). Reproducibility at 100 ng l^{-1} (0.3 nM) and 20 ng l^{-1} (0.07 nM) was determined in artificial seawater. The results of these determinations are shown in Table 5.2. For tributyltin at 2000 ng l^{-1} (6.8 nM) the reproducibility is 11% relative standard deviation (n = 6), and at 20 ng l^{-1} (0.07 nM) reproducibility is 30% on the packed column and 22% on the 530-μm column (n = 5).

5.3.3f. Effect of pH on analysis

In order to determine the optimum pH for hydride derivatization and to determine

Table 5.2 Reproducibility of butyltin analysis[a]

	Relative standard deviation (%)			
	2000 ng l^{-1}[b] *(5.8–11 nM) (n = 6)*	*100 ng l^{-1}*[c] *(0.43–0.34 nM) (n = 5)*	*20 ng l^{-1}*[c] *(0.085–0.069 nM) (n = 5)*	
Butyltin species	*packed column*	*packed column*	*packed column*	*530μm column*
MBT	9	—[d]	—	—
DBT	15	3	20	22
TBT	11	17	30	22
TTBT	11	—	—	—

[a]Percent relative standard deviation.
[b]Spiked Chesapeake Bay water.
[c]Spiked artificial seawater.
[d]Dash indicates not determined.

whether natural variations in pH found in environmental samples would adversely affect the analytical scheme, the four butyltins and dipropyltin were studied in deionized water solutions of 2500 ng l^{-1} (14 nM–7 nM) of MBT, DBT, TBT, TTBT, and DPT. Duplicate 100-ml samples were adjusted to pH 1.0, 2.7, 4.0, 7.0, 9.0, 11.0, and 13.0 using HCl and NaOH. These samples were treated with the simultaneous hydride derivatization–extraction procedure. Solvent evaporation was not required prior to GC–FPD analysis.

For the four species that undergo hydride derivatization (mono-, di-, and tributyltin and dipropyltin) there is no difference in peak area over the pH range 4.0–11.0. There was a significant difference at the lowest and highest pH (1.0 and 13.0). The pH of water samples from the environment will normally be within 4.0–11.0, and water samples were not pH adjusted prior to analysis.

5.3.3g. Effect of salinity on analysis

The effect of variation in salinity on the analytical scheme was assessed using artificial seawater diluted with deionized water to one-fourth, one-half, and three-fourths and full-salinity seawater (35‰). These diluted seawater samples and deionized water were spiked with the four butyltins and dipropyltin at 2500 ng l^{-1} (14–7 nM). Duplicate 100-ml samples were analyzed as described for the pH experiments.

The resulting peak areas as a function of salinity show poor correlation between peak area and salinity for all species except tetrabutyltin. The correlation coefficient (r) of the regression lines for MBT, DBT, TBT, and DPT ranges from 0.05 for DPT to 0.49 for MBT, indicating essentially no correlation between salinity and peak area for these species.

5.3.3h. Stability of butyltin hydrides

Organotin hydrides are subject to decomposition by oxidation. Although there was no apparent loss of analyte during a normal working day, there was concern that the butyltin hydrides required analysis immediately after derivatization. In order to evaluate the stability of the derivatized analytes in a situation in which the analysis could not be completed on the same day as the samples were prepared, five replicate dichloromethane extracts of the hydride-derivatized butyltins, dipropyltin, and TTBT were prepared at a final concentration of ~0.2 ng μl^{-1} (1–0.6 μM) of each species of organotin in the extract. This is a concentration comparable to that found in extracts of environmental samples. These samples were stored in 5-ml Reactivials with tightly closed Teflon-lined caps, in a freezer at −20°C. The samples were chromatographed prior to storage ($t = 0$) and after 21, 45, and 69 h of storage. After 69 h there were no detectable changes in the concentration of any of the organotin species.

5.3.4. VALIDATION OF METHOD

5.3.4a. Analysis of NBS TBT research reference material

The simultaneous hydride-extraction method was used to determine the concentration of TBT in deionized water solution in a research reference material prepared by NBS for an interlaboratory comparison of TBT analytical methods (Blair *et al.*, 1986). The aqueous solution of the reference material was diluted 1:1000 in deionized water and analyzed by the procedure as described in section 5.3.2c, Analytical procedure for TBT in water. The concentration of TBT ion in the reference material was determined to be 2.34 ± 0.07 mg l^{-1} (ppm) (8.04 ± 0.24 μM) as Bu_3Sn^+ for 10 determinations. The mean value determined by the 35 laboratories participating in the intercomparison was 2.44 ± 0.12 mg l^{-1} (ppm) (8.38 ± 0.41 μM). The value determined by neutron activation analysis was 2.58 ± 0.12 mg l^{-1} (ppm) (8.87 ± 0.41 μM).

5.3.4b. Laboratory intercomparison

Comparison of results obtained using a newly developed method with those obtained by an accepted method is a commonly used approach to validate the accuracy of the newly developed method, particularly in the absence of a standard reference material (SRM) for the analyte in question, as is the case for TBT. This method has been compared with the hydride generation, purge-and-trap system (Valkirs *et al.*, 1987) with good agreement between the two approaches. In 6 of 14 samples collected on the east and west coast of the United States and an estuary in the United Kingdom, the TBT concentrations determined by the two methods were within 20% of their mean concentration. For DBT, only four of the values varied by more than 20% of the mean values. The RSDs for the replicate analysis were also similar for the two methods, being in the range of 11–15% for the two tin species.

5.4. CONCLUSIONS

A new method for the determination of TBT and other butyltins in water employing hydride derivatization with solvent extraction and gas chromatography with tin-selective flame photometric detection has been shown to be sensitive, selective, and unaffected by salinity or pH, and to give results comparable to those of other analytical methods. The new method is fast and sensitive, and requires no sample clean-up step; it uses commercially available instrumentation, eliminating the need for in-house fabrication of transfer lines, thus increasing reliability.

ACKNOWLEDGMENTS

The author thanks the David Taylor Naval Ship Research and Development Center for financial support of the experimental work described herein. The author also thanks P. Schatzberg, J. Bellama, G. Olson, F. Brinckman, and W. Blair for their generous assistance.

REFERENCES

American Chemical Society. 1980. Committee on Environmental Improvement, Subcommittee on Environmental Analytical Chemistry. Guidelines for data acquisition and data quality evaluation in environmental chemistry. *Anal. Chem.*, **52**, 2242–2249.

Andreae, M.O. and J.T. Byrd. 1984. Determination of tin and methyltin species by hydride generation and detection with graphite-furnace atomic absorption or flame emission spectrometry. *Anal. Chim. Acta*, **156**, 147–157.

Aue, W.A. and C.G. Flinn. 1977. A photometric tin detector for gas chromatography. *J. Chromatogr.*, **142**, 145–154.

Blair, W.R., G.J. Olson, F.E. Brinckman, R.C. Paule and D.A. Becker. 1986. An International Butyltin Measurement Methods Intercomparison: Sample Preparation and Results of Analyses. National Bureau of Standards Internal Report 86-3321 prepared for the Office of Naval Research, Arlington, Virginia, 29 pp.

Braman, R.S. and M.A. Tompkins. 1979. Separation and determination of nanogram amounts of inorganic tin and methyltin compounds in the environment. *Anal. Chem.*, **51**, 12–19.

Brody, S.S. and J.E. Chaney. 1966. Flame photometric detector. *J. Chromatogr.*, **4**, 42–45.

Burns, D.T., F. Glockling and M. Harriott. 1980. Comparative assessment of gas–liquid chromatography and high-performance liquid chromatography for the separation of tin tetraalkyls and alkyltin halides. *J. Chromatogr.*, **200**, 305–308.

Dagnall, R.M., K.C. Thompson and T.S. West. 1968. Molecular-emission spectroscopy in cool flames. Part III. The emission of tin in diffusion flames. *Analyst*, **93**, 518–521.

Jackson, J.A., W.R. Blair, F.E. Brinckman and W.P. Iverson. 1982. Gas-chromatographic speciation of methylstannanes in the Chesapeake Bay using purge and trap sampling with a tin-selective detector. *Environ. Sci. Technol.*, **16**, 110–119.

Maguire, R.J. and H. Huneault. 1981. Determination of butyltin species in water by gas chromatography with flame photometric detection. *J. Chromatogr.*, **209**, 458–462.

Maguire, R.J. and R.J. Tkacz. 1983. Analysis of butyltin compounds by gas chromatography: comparison of flame photometric and atomic absorption spectrophotometric detectors. *J. Chromatogr.* **268**, 99–101.

Miller, J.C. and J.N. Miller. 1984. *Statistics for Analytical Chemistry*. John Wiley, New York, pp. 97–100.

Meinema, H.A., T. Burger-Wiersma, G. Versluis-de Haan and E.C. Gevers. 1978. Determination of trace amounts of butyltins compounds in aqueous systems by gas chromatography/mass spectrometry. *Environ. Sci. Technol.* **12**, 288–293.

Parris, G.E., W.R. Blair and F.E. Brinckman. 1977. Chemical and physical considerations in the use of atomic absorption detectors coupled with a gas chromatograph for determination of trace organometallic gases. *Anal. Chem.*, **49**, 378–386.

Seligman, P.F. 1984. Fate and Effects of Organotin Antifouling Leachates in the Marine Environment. Progress report prepared for the Energy Research and Development Office, David W. Taylor Ship Research and Development Center, Bethesda, Maryland, pp. 1–47.

Unger, M.A., W.G. MacIntyre, J. Greaves and R.J. Huggett. 1986. GC determination of butyltins in natural waters by flame photometric detection of hexyl derivatives with mass spectrometric confirmation. *Chemosphere*, **15**, 461–470.

Valkirs, A.O., P.F. Seligman, G.J. Olson, F.E. Brinckman, C.L. Matthias and J.M. Bellama. 1987. Di- and tri-tributyltin species in marine and estuarine waters. Interlaboratory comparison of two ultra-trace analytical methods employing hydride generation and atomic absorption or flame photometric detection. *Analyst*, **112**, 17–21.

GRIGNARD DERIVATIZATION AND MASS SPECTROMETRY AS TECHNIQUES FOR THE ANALYSIS OF BUTYLTINS IN ENVIRONMENTAL SAMPLES

6

M.A. Unger[1], *J. Greaves*[2] *and R.J. Huggett*[1]

[1]*Virginia Institute of Marine Science, School of Marine Science, College of William and Mary, Gloucester Point, Virginia 23062, USA*

[2]*Department of Chemistry, University of California, Irvine, California 92717, USA*

ABSTRACT

When analyzing environmental samples for tributyltin (TBT) and its breakdown products, it is important to speciate the various organotins that may be present in the samples. Most commonly used methods require a derivatization step to form volatile organotin hydrides or tetraalkyltin derivatives. The tetraalkyltins formed by reaction with Grignard reagents are stable compounds that are readily separated by gas chromatography and produce characteristic mass spectra that are easily interpreted. Various types of Grignard reagents and solvents have been used in the published methods for the analysis of TBT in environmental samples. When the longer chain (pentyl and hexyl) derivatives are formed, sample extracts can be concentrated to small volumes without significant losses from volatilization.

Gas chromatography with modified flame photometric detection has been shown to be a sensitive and somewhat selective technique for quantifying butyltins. However, the complexity of environmental extracts means that the only way to identify butyltins positively is

Organotin. Edited by M.A. Champ and P.F. Seligman. Published in 1996 by Chapman & Hall, London. ISBN 0 412 58240 6

with mass spectrometry. Gas chromatography–mass spectrometry, in combination with either electron ionization or positive chemical ionization, has been the most commonly used method. Qualitative identification of butyltins has usually been accomplished by full scanning and is facilitated by the characteristic isotopic pattern displayed by tin (10 stable isotopes). Selected ion monitoring has been the technique used most often for quantification by mass spectrometry. Mass spectrometry–mass spectrometry, alternative ionization methods, and stable isotope monitoring are all techniques that have also been applied to the analysis of butyltins by mass spectrometry.

6.1. INTRODUCTION

It is important to differentiate between tributyltin (TBT) and the various butyltin species present as degradation products when analyzing environmental samples. Some investigators have quantified TBT by atomic absorption spectrophotometric analysis of organic extracts of the samples. Since such extracts will likely contain some dibutyltin (DBT), monobutyltin (MBT), and other tin–organic complexes, methods without an organotin speciation technique cannot be used to determine TBT concentrations accurately. This was shown by Dooley and Vafa (1986) in an examination of organic extracts of bivalve tissue.

Methods that separate TBT from other organotins typically utilize gas chromatography (GC) or a temperature gradient desorption from a purge-and-trap apparatus as the separation techniques. These separation techniques are usually followed by flame photometric or mass spectrometric detection. A recently developed method avoids the use of chromatography completely by utilizing the highly selective technique of mass spectrometry–mass spectrometry (MS–MS) (Siu *et al.*, 1989; Siu and Berman, Chapter 8).

Most commonly used methods require a derivatization step to form either volatile hydrides or tetraalkyl derivatives of the butyltins. Some researchers have found that by doping polar gas chromatography columns with HCl they can form butyltin chlorides in situ. These butyltin chlorides can be effectively separated, provided that doping with HCl is continued. This chapter will concentrate on the advantages and disadvantages of using Grignard reagents to form tetraalkyltin derivatives prior to GC, and will discuss the use of mass spectrometry for butyltin analysis.

6.2. GRIGNARD DERIVATIZATION OF BUTYLTIN COMPOUNDS

A Grignard reagent is an organometallic compound that can be represented by the formula RMgBr, where R can be any alkyl group. These organometallic compounds behave as strong bases and are attracted to electron-deficient centers. Commercial Grignard reagents are usually dissolved in ether and are sealed under an inert atmosphere. Exposure to water or the atmosphere can result in spontaneous reaction. The reaction with water, shown in Eq. 1, is exothermic and typical for these highly reactive compounds.

$$\mathrm{RMgBr + HOH \rightarrow RH + Mg(OH)Br} \quad 1$$

For this reason, vessels containing Grignard reagents should be purged with nitrogen whenever open to the atmosphere. Exposure to the atmosphere will reduce the effective concentration of the Grignard reagent in the solution, which may cause low recoveries during derivatization reactions.

When added to organic extracts containing butyltin compounds, Grignard reagents react with the butyltins by adding alkyl groups to form tetraalkyltin derivatives. The following reactions (Eqs 2, 3, and 4) illustrate the derivatization of TBT, DBT, and MBT, respectively, by n-hexylMgBr:

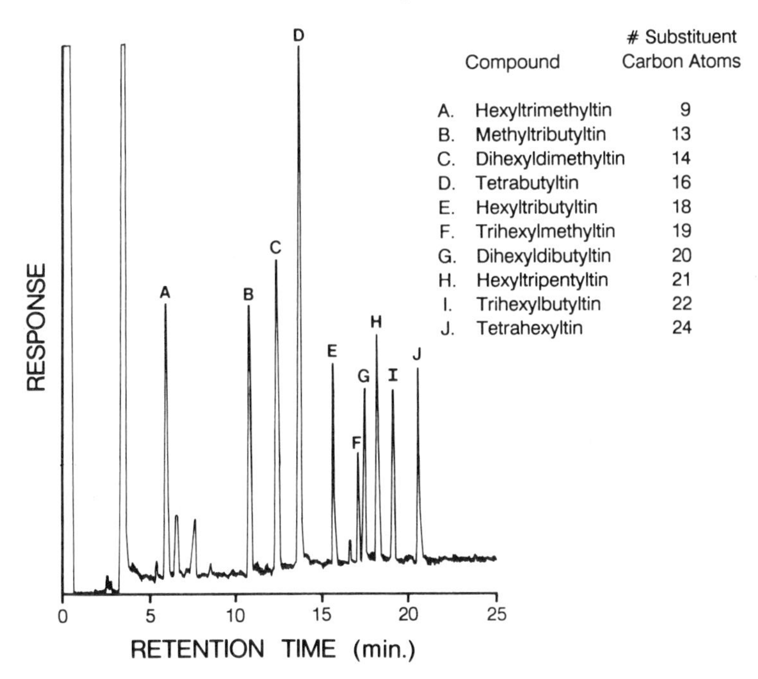

Figure 6.1 GC–FPD chromatogram of tetraalkyltin derivatives (adapted from Unger *et al.*, 1986).

$$(\text{n-butyl})_3\text{Sn}^+ + \text{n-hexylMgBr} \rightarrow (\text{n-hexyl})(\text{n-butyl})_3\text{Sn} \quad 2$$

$$(\text{n-butyl})_2\text{Sn}^{2+} + 2\ \text{n-hexylMgBr} \rightarrow (\text{n-hexyl})_2(\text{n-butyl})_2\text{Sn} \quad 3$$

$$(\text{n-butyl})\text{Sn}^{3+} + 3\ \text{n-hexylMgBr} \rightarrow (\text{n-hexyl})_3(\text{n-butyl})\text{Sn} \quad 4$$

The resulting tetraalkyltins are stable compounds that can be easily separated by gas chromatography. Figure 6.1 shows a chromatogram of the hexyl derivatives of butyltins. Other tetraalkyltins are present in the chromatogram and illustrate the relative retention times of these compounds. The total number of carbon atoms in the chains substituted onto the tin atom is also given in the figure. The relative retention times of the derivatized compounds are directly correlated with the number of carbon atoms present.

Unlike hydride derivatives of butyltins, which may degrade in hours or days, the tetraalkyl derivatives formed with Grignard reagents are stable for extended periods of time. Huggett *et al.* (1986) repeatedly analyzed hexyl derivatized sample extracts over a 10-week period and found no significant loss of butyltins.

Many different Grignard reagents have been used for the derivatization of organotin compounds. Derivatives formed by reaction with methyl and ethyl Grignards are considerably more volatile than those formed with larger alkyl (pentyl or hexyl) groups. Mueller (1984) warned that when concentrating samples containing methyltributyltin, great caution must be used to prevent loss of the analyte through volatilization.

The reactions discussed above are performed in the organic extracts of samples. Because of the reactive nature and range of Grignard reagents, available choices of extraction solvents and internal standards are important and require discussion. Figure 6.2 shows a flowchart for the general analytical procedure that is typical for the analysis of butyltins in environmental samples. Variations in extraction procedures, internal standards, Grignard reagents, and detection

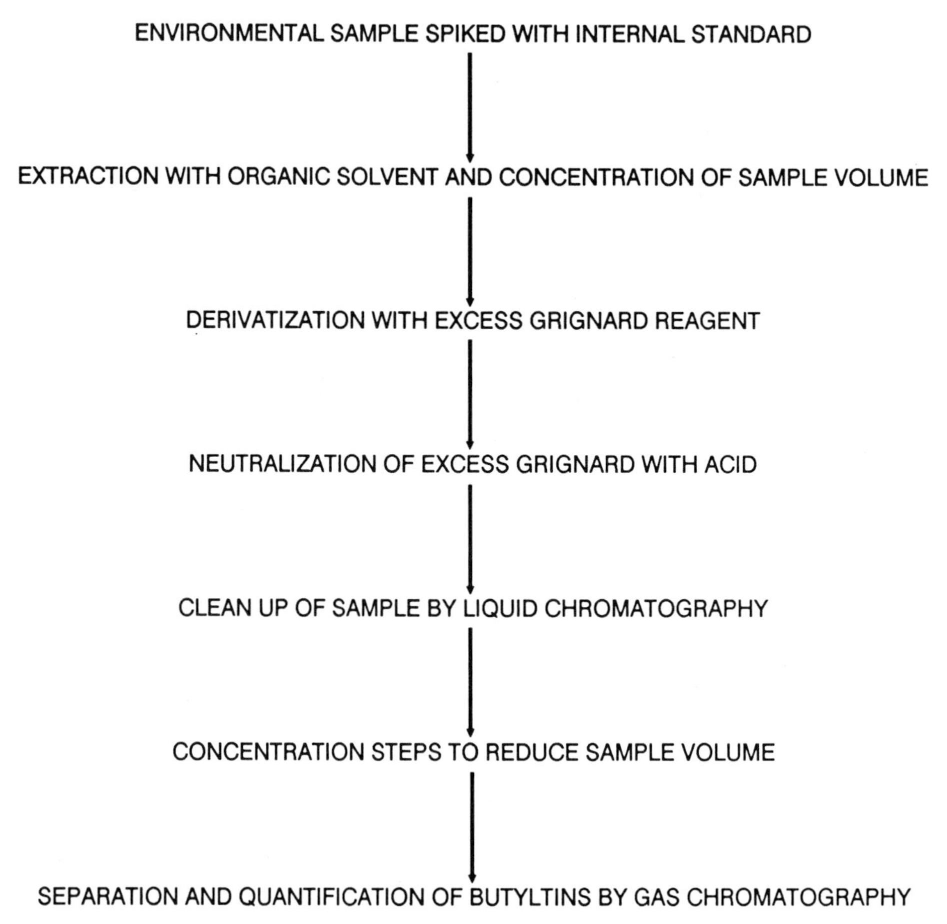

Figure 6.2 Flowchart of the general analytical procedure for the analysis of butyltins in environmental samples via solvent extraction, Grignard derivatization, and chromatographic analysis.

techniques have resulted in many different methods being published for the analysis of TBT and its degradation products. A summary of these methods is given in Table 6.1. Many of these methods are modifications of those published earlier or are applications of previous methods to a new sample matrix. The authors have tried to include all currently published techniques to show the variety of modifications and applications for this type of butyltin analysis.

A number of different solvents have been shown to be satisfactory for the extraction of TBT from water samples. DBT and MBT are more difficult to extract quantitatively due to their more ionic nature. For this reason, most methods use tropolone, added to the solvent, as a complexing agent to increase the recovery of these compounds. Solid-phase extraction procedures have also been reported. Mueller (1987) found that tropolone-treated C-18 packed columns will successfully remove butyltins from environmental water samples. The butyltins are then eluted from the cartridges with diethyl ether and are ready for derivatization with Grignard reagent. Matthias *et al.* (1987) also reported on the use of C-18 bonded silica for extracting butyltins. These authors had good success when butyltins were at high concentrations,

Table 6.1 Methods of butyltin analysis with GC or MS analysis. Grignard derivatization reagents are indicated as appropriate

Sample matrix	*Extraction solvent*	*Grignard reagent*	*Internal standard*	*Detector used for quantification*[a]	*Reference*
Standards	Ethyl ether	Methylmagnesium bromide	Phenylcyclohexane	GC/TCD	Steinmeyer *et al.* (1965)
Water	Benzene/tropolone Chloroform/tropolone Methylene chloride/ tropolone	Methylmagnesium bromide n-Butylmagnesium bromide	Hexyldibutylmethyltin	GC/MS	Meinema *et al.* (1978)
Water	Benzene/tropolone	n-Pentylmagnesium bromide	None	GC/FID (MS confirmation)	Boettner *et al.* (1981)
Water	Benzene/tropolone	n-Pentylmagnesium bromide	None	GC/FPD (MS confirmation)	Maguire and Huneault (1981)
Sediment	Benzene/tropolone	n-Pentylmagnesium bromide	None	GC/FPD (MS confirmation)	Maguire (1984)
Water, sediment	Solid phase, pentane	Methylmagnesium bromide	Dihexyldibutyltin	GC/FPD (MS confirmation)	Mueller (1984)
Water	Hexane/tropolone	n-Hexylmagnesium bromide	Tripentyltin chloride	GC/FPD (MS confirmation)	Unger *et al.* (1986)
Water, sewage sludge, sediment	Tropolone/C_{18} cartridge Ether/tropolone	Ethylmagnesium bromide	Dihexyldibutyltin	GC/FPD GC/MS	Mueller (1987)
Bivalve tissue, sediment	Hexane	n-Hexylmagnesium bromide	Tripentyltin chloride	GC/FPD	Rice *et al.* (1987)
Finfish tissue	Hexane	n-Pentylmagnesium bromide	Tetrabutyltin	GC/AA	Short (1987)
Finfish	Hexane/ethyl ether	Not applicable	(Standard addition)	GC/ECD	Takeuchi *et al.* (1987)
Water	Hexane/tropolone bromide	n-Hexylmagnesium chloride	Tripentyltin	GC/MS	Greaves and Unger (1988)
Bivalve tissue	Methylene chloride tropolone/hexane	n-Hexylmagnesium bromide	Tripropyltin chloride	GC/FPD GC/MS	Wade *et al.* (1988)
Bivalve tissue	Benzene/tropolone	n-Pentylmagnesium bromide	Tetrapentyltin	GC/AA	Short and Sharp (1989)
Bivalve tissue, sediment	Methylene chloride, hexane	n-Hexylmagnesium bromide	Tripentyltin bromide Tripropyltin chloride	GC/FPD (MS confirmation)	Stallard *et al.* (1989)
Bivalve tissue	Methylene chloride	n-Hexylmagnesium bromide	Tripentyltin chloride	GC/FID	Page (1989)
Sediment	n-Butanol, methanol/ iso-octane	Not applicable	(Standard addition)	MS/MS	Siu *et al.* (1989)

[a]Abbreviations: AA – atomic absorption spectrophotmeter; ECD – electron capture detector; FID – flame ionization detector; FPD – flame photometric detector; GC – gas chromatography; MS – mass spectrometer; MS–MS – mass spectrometry–mass spectrometry; TCD – thermal conductivity detector.

but found reduced recoveries at the low nanogram-per-liter concentrations typical of environmental samples.

An internal standard should be used to quantify reliably low concentrations of butyltins in environmental samples. The internal standard should have similar chemical properties to the analyte so that relative recoveries stay constant throughout the analytical procedure. When hexyl derivatization is used, tripentyltin chloride has proven to be an excellent internal standard. While not commercially available, it can be synthesized through a series of reactions (Unger *et al.*, 1986). Stallard *et al.* (1989) prepared tripentyltin bromide for use as an internal standard via bromination of tetrapentyltin. Holland (1987) reported on techniques for the direct synthesis of triorganotin compounds. This method may simplify the preparation of tripentyltin.

One of the biggest problems encountered when analyzing environmental samples for low concentrations of TBT (e.g. nanogram per liter) is sample contamination by impure Grignard reagents. Contamination of Grignard reagents may occur when chemical manufacturers use the same inert atmosphere apparatus for synthesizing butyltin compounds and preparing Grignard reagents. This is a common problem with some commercial sources, and therefore each batch of Grignard reagent should be checked for contamination. To overcome these contamination problems, Grignard reagents may be prepared as needed by the reaction of magnesium metal with the appropriate alkyl bromide. To assure that the product will be free of organotins, the reagents should be checked for butyltin contamination prior to use.

Despite the problems encountered with contaminated reagents, methods utilizing Grignard derivatization have proven reliable and adaptable for a wide variety of sample types. When the longer chain Grignard reagents are used, sample extracts can be concentrated to small volumes without significant losses of the derivatives due to volatilization. The tetraalkyltin compounds produced by these methods are thermally stable compounds that are easily separated by GC and produce characteristic mass spectra.

6.3. MASS SPECTROMETRY OF BUTYLTINS AND RELATED COMPOUNDS

The use of mass spectrometry (MS) for research on organotins, and in particular the alkyltins, can be divided into two time periods. Prior to 1984 most work was concerned with the chemistry of organometallics, although some environmentally oriented papers were published. Since 1984 there has been an increase in the number of papers that report the use of MS in the analysis of TBT in environmental samples. The mass spectrometer is the only detector that can be used to give positive identification of butyltins, as well as allowing quantification, in environmental samples. Because of the stringent requirements for low detection limits in environmental samples, a variety of sample introduction techniques and ionization methods have been investigated.

Gielen and Mayence (1968) were among the first to report mass spectra of alkyltins including the tetraalkyltins and trialkyltin halides. The spectra were obtained by electron ionization (EI, 70-eV electron energy) and showed extensive fragmentation. Cleavage of the Sn–C bond was the predominant process. Molecular ions, if present, were small (<4%), and the base peak corresponded to the loss of an alkyl group in the case of the trialkyltin halides. For the tetraalkyltins the base peak was less consistent but usually resulted from the loss of multiple alkyl groups. The same authors (Gielen and Mayence, 1972), working with mixed alkyl substituted tins, determined that the favored cleavage was of the largest substituent. Reduction of the electron energy to 7.5 eV improved the intensity of the molecular ion but failed to increase it above 10%.

Positive chemical ionization (PCI) was applied to a variety of organotins by Fish *et al.* (1974). They found that although the spectra were simplified, frequently to one or two clusters of ions, PCI with methane as the reagent gas failed to yield protonated molecular or C_2H_5 adduct ions. Further, MS studies using field desorption and field ionization (Weber *et al.*, 1980) gave spectra in which the molecular ions predominated. In the same paper, collision activation dissociation, using argon as the collision gas, was used to fragment the molecular ion. The resulting spectra showed one major fragment corresponding to the loss of an alkyl group followed by minor peaks attributable to further alkyl group removal. Somewhat surprisingly, fast atom bombardment MS (Greaves, 1989) did not result in observation of a molecular ion. Spectra obtained were similar to those obtained by PCI. Examples of mass spectra of alkyltins are shown in Fig. 6.3. Figures 6.3A and 6.3B are spectra of tetrabutyltin. Figure 6.3A exemplifies the extensive fragmentation observed for EI, and Fig. 6.3B is the much simpler PCI spectrum. Figure 6.3C is a PCI spectrum of dihexyldibutyltin. It illustrates the favored loss of the heavier alkyl group to form *m/z* 319. Molecular ions are not observed in Fig. 6.3A, B, or C. In all monostannic spectra the typical tin isotopic pattern is observed. The masses of the main tin isotopes and their relative abundances are 116 (14.3%), 117 (7.6%), 118 (24.0%), 119 (8.6%), 120 (32.9%), 122 (4.7%), and 124 (5.9%). The isotopic pattern becomes very complex if more than one tin atom is present in the molecule, as is the case for bis TBT-oxide illustrated by the PCI spectrum shown in Fig. 6.3D. It is observed that this compound does produce a protonated molecular ion cluster at *m/z* 597. None of the tetraalkyltins gives useful spectra in negative chemical ionization with methane as the reagent gas.

The earliest reported environmental application of MS related to TBT was that of Meinema *et al.* (1978). These workers used selected ion monitoring (SIM) of methylated extracts of fresh water and seawater spiked with butyltins at the 1 $\mu g\ l^{-1}$ Sn level. The ions monitored were *m/z* 135 $[SnMe]^+$ and *m/z* 193 $[SnBuMeH]^+$. Maguire and Huneault (1981) used MS to confirm the identities of the tetraalkyl derivatives that they had prepared, using pentylmagnesium bromide, to improve the chromatography of TBT, DBT, and MBT. Both groups used EI. In both papers the authors noted the value of the multiple isotopes of tin as a diagnostic tool for identification of organotins.

The pace of publication on TBT has increased since the early 1980s. However, MS is still used less frequently than GC with flame photometric detection (FPD). Matthias *et al.* (1986), Unger *et al.* (1986), Mueller (1987), Takeuchi *et al.* (1987), Sasaki *et al.* (1988), and Wade *et al.* (1988) have all reported use of MS in the detection of butyltins in a variety of environmental samples.

Unger *et al.* (1986) reported on extracts of water samples that were analyzed by GC–FPD and GC–MS, the latter using PCI with methane as the reagent gas. The advantage of using PCI was illustrated because the reduced fragmentation resulted in the ion current being concentrated in the ion clusters, $[M\text{-hexyl}]^+$ and $[M\text{-butyl}]^+$. This enabled an unambiguous spectrum to be obtained from an extract of a water sample containing TBT at a concentration of 8.6 ng l^{-1} Sn. The improvement in sensitivity of PCI over EI was also noted by Mueller (1987) in a report on butyl- and phenyltins in water, lake sediment, and sewage sludge from Switzerland. Wade *et al.* (1988) used EI–GC–MS in both the full scanning and selected ion monitoring modes to confirm the identities of TBT, DBT, and MBT in the shellfish, *Mytilus edulis*, *Crassostrea virginica*, and *Ostrea sandwichensis*. This group noted that in EI–GC–MS there were many other compounds present

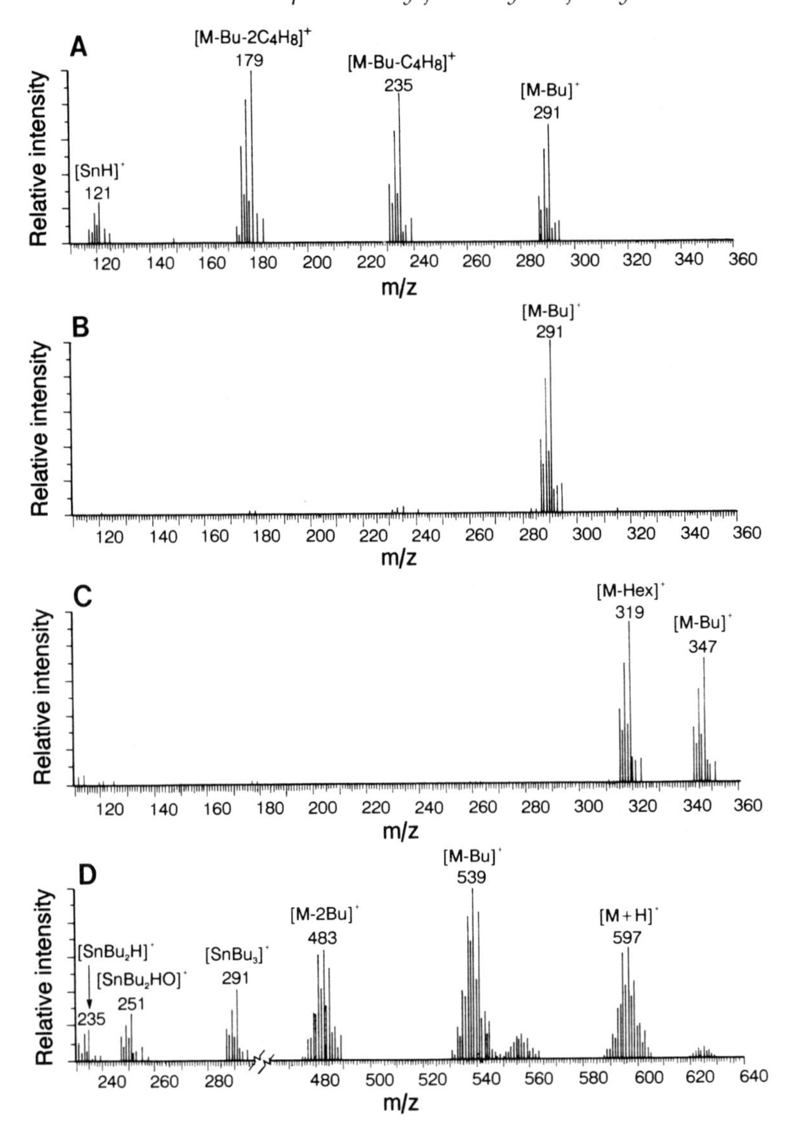

Figure 6.3 Mass spectra of butyltins: (A) EI spectrum of tetrabutyltin; (B) PCI spectrum of tetrabutyltin; (C) PCI spectrum of dihexyldibutyltin; and (D) PCI spectrum of TBTO.

that reduced the efficiency of the method. The SIM was conducted on *m/z* 121, which is the $[SnH]^+$ ion of ^{12}Sn, the major tin isotope. PCI–GC–MS was used by Espourteille (1988) for confirmation of the presence of TBT in *Crassostrea virginica*.

Matthias *et al*. (1986), working with a water sample from a drydock, identified tetrabutyltin using EI–GC–MS. They argued that this may indicate a redistribution reaction either between 2 TBT^+ ions or 1 DBT^{2+} and 1 TBT^+. They also suggested that the same compound was present in a microlayer sample from Baltimore Harbor but were unable to confirm it by GC–MS. Analysis of bis-tributyltin oxide (TBTO) in fish (yellowtail, *Seciola quinqueradiata*) was undertaken by Takeuchi *et al.* (1987). These researchers extracted the alkyltin as TBTCl and then chromatographed the compound without derivatization. In order to obtain good gas chromatographic performance for the chloride, the column must be continuously doped with HCl in acetone. Mass spectrometric confirmation of the presence of TBTCl used EI and full scanning. Sasaki *et al.* (1988), working on yellowtail as well as other fish species, used both full scanning and selected ion monitoring EI–MS to identify and quantify the methyl and pentyl derivatives of TBT and DBT. It is of interest to note that when TBT is analyzed as TBTCl it becomes feasible to use electron capture detectors for GC detection. On the same lines it would be expected that negative chemical ionization MS may be a suitable detection method for the TBTCl. However, to the authors' knowledge there are no reports in the literature on the use of this ionization technique for assaying TBTCl.

Taking advantage of the added sensitivity and selectivity of PCI (methane-reagent gas) and SIM, Greaves and Unger (1988) have reported a selected ion monitoring assay for TBT, DBT, and MBT, as their hexyl derivatives, in water using m/z 319 and m/z 347 – the former for TBT and the latter for DBT, MBT, and the internal standard, tripentyltin. The detection limit of this method is $\sim$1 ng l^{-1}. The authors evaluated the PCI–GC–MS method with respect to GC–FPD and showed that the two techniques compared very favorably with each other. Greaves (1986) demonstrated the applicability of SIM for monitoring butyltins in sediment. Figure 6.4 illustrates the use of PCI SIM for analysis of water (Part A) and sediment (Part B).

MS, with its ability to distinguish individual masses, enables the use of stable isotope tracers. In recent work (Meyers-Schulte and Dooley, 1989; Testa and Dooley, 1989) the synthesis of ^{124}Sn-labeled TBT has been reported. This compound can be traced using GC–MS and can therefore be used to follow the behavior of TBT in natural systems, even when TBT having the normal isotope distribution is present. In addition, because the ion current is concentrated in a single isotope rather than the 10 isotopes normally present, the detection limits for TBT are improved by about a factor of three.

Mass spectrometry will sometimes detect organic compounds that may coelute with butyltins. For this reason, it has been criticized as not being as specific a detection technique as atomic absorption or FPD modified for tin detection. Careful selection of target ions when using SIM will help decrease the likelihood that interfering compounds will be detected. Another technique, presaged by the work of Weber *et al.* (1980), is the use of MS–MS as recently reported by Siu *et al.* (1989). In this technique an ionspray interface was coupled to an atmospheric pressure ionization MS–MS system. The authors used a simple extraction procedure requiring only sonication of a 4-g sediment sample with n-butanol or an HCl/methanol/iso-octane combination. The resulting extract is directly injected into the ionspray interface without further purification or derivatization, and TBT is measured using selected reaction monitoring of the parent/daughter pair of m/z 291:m/z 179. The detection limit for this technique is $\sim$5 pg Sn pure compound or -0.2 μg g^{-1} Sn in sediment. It should be emphasized that this methodology required no sample clean-up after extraction, no derivatization, and no chromatographic separation of the extract. Siu *et al.* (1989) note that under these conditions SIM is ineffective and only selected reaction monitoring MS–MS is capable of providing the required specificity.

6.4. SUMMARY

This chapter has summarized the use of Grignard reagents for the derivatization of butyltins and the application of mass spectrometry to the analysis of these compounds. The tetraalkyltin derivatives formed in the Grignard reaction are stable and readily

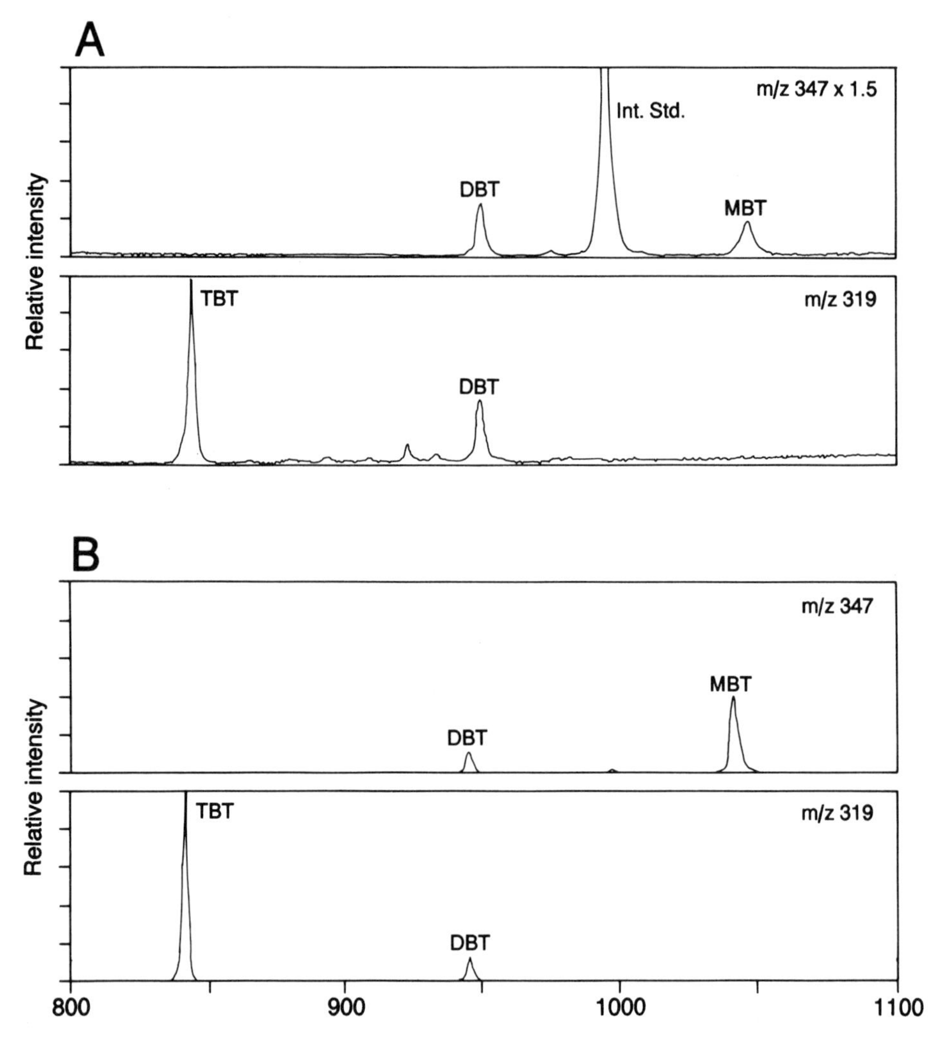

Figure 6.4 SIM chromatograms (PCI) of environmentally contaminated samples: (A) water and (B) sediment.

amenable to GC analysis. A variety of detectors have been used, in particular FPD and MS. The latter is an expensive and specialized technique. Its place alongside GC–FPD in the analysis of organotins is, however, assured because of the versatility of the instrumentation. This allows a wide variety of experiments to be conducted. Information gathered may include identification, quantification, and isotopically labeled tracer behavior.

ACKNOWLEDGEMENT

This chapter is VIMS contribution number 1619.

REFERENCES

Boettner, E.A., G.L. Ball, Z.H. Hollingsworth and R. Aquino. 1981. Organic and Organotin Compounds Leached from PVC and CPVC Pipe. Report No. EPA–600/1–81–062, prepared for Office of Research and Development, US Environmental Protection Agency, Cincinnati, Ohio, 102 pp.

Dooley, C.A. and G. Vafa. 1986. Butyltin compounds and their measurement in oyster tissue. In: Proceedings of the Oceans '86 Organotin Symposium, Vol. 4, 23–25 September 1986, Institute of Electrical and Electronics Engineers and the Marine Technology Society, Washington, DC, pp. 1171–1176.

Espourteille, F.A. 1988. An Assessment of Tributyltin Contamination in Sediments and Shellfish in the Chesapeake Bay. Masters Thesis. Virginia Institute of Marine Science, College of William and Mary, Gloucester Point, Virginia, 78 pp.

Fish, R.H., R.L. Holmstead and J.E. Casida. 1974. Chemical tonization mass spectrometry of organotin compounds. *Tetrahedron Lett.*, **14**, 1303–1306.

Gielen, M. and G. Mayence. 1968. Organometallic compounds. VII. Electron impact fragmentation of trialkyltin halides. *J. Organomet. Chem.*, **12**, 363–368.

Gielen, M. and G. Mayence. 1972. Organometallic compounds. XXXVII. Mass spectra of mixed tetraalkyltins and of trimethyltin halides. *J. Organomet. Chem.* **46**, 281–288.

Greaves, J. 1986. Gas chromatography studies on butyltins and their analysis in environmental samples. 34th Conference on Mass Spectrometry, American Society of Mass Spectrometry and Allied Topics, Cincinnati, Ohio, 8–13 June 1986.

Greaves, J. 1989. Alkyltins: good for antifouling paint, not so good for oysters. Miami Beach, Florida. 37th Conference on Mass Spectrometry, American Society for Mass Spectrometry and Allied Topics, 21–26 May 1989.

Greaves, J. and M.A. Unger. 1988. A selected ion monitoring assay for tributyltin and its degradation products. *Biomed. Environ. Mass Spectrom.*, **15**, 565–569.

Holland, F.S. 1987. The direct synthesis of triorganotin compounds: process and reaction mechanism. *Appl. Organometal. Chem.*, **1**, 449–458.

Huggett, R.J., M.A. Unger and D.J. Westbrook. 1986. Organotin concentrations in the southern Chesapeake Bay. In: Proceedings of the Oceans '86 Organotin Symposium, Vol. 4, 23–25 September 1986, Institute of Electrical and Electronics Engineers and the Marine Technology Society, Washington, DC, pp. 1262–1265.

Maguire, R.J. 1984. Butyltin compounds and inorganic tin in sediments in Ontario. *Environ. Sci. Technol.* **18**, 291.

Maguire, R.J. and H. Huneault. 1981. Determination of butyltin species in water by gas chromatography with flame photometric detection. *J. Chromatogr.*, **209**, 463–466.

Matthias, C.L., J.M. Bellama, G.J. Olson and F.E. Brinckman. 1986. Comprehensive method for determination of aquatic butyltin and butylmethyltin species at ultratrace levels using simultaneous hybridization/extraction with gas chromatography–flame photometric detection. *Environ. Sci. Technol.*, **20**, 609–615.

Matthias, C.L., J.M. Bellama and F.E. Brinckman. 1987. Determination of tributyltin in estuarine water using bonded C-18 silica solid phase extraction, hydride derivatization and GC–FPD. In: Proceedings of the Oceans '87 International Organotin Symposium, Vol. 4, 28 September–1 October 1987, Institute of Electrical and Electronics Engineers and the Marine Technology Society, Halifax, Nova Scotia, pp. 1344–1347.

Meinema, H.A., T. Burger-Wiersma, G. Verslurs-de Haan and E. Ch. Gevers. 1978. Determination of trace amounts of butyltin compounds in aqueous systems by gas chromatography–mass spectrometry. *Environ. Sci. Technol.*, **12**, 288–293.

Meyers-Schulte, K.J. and C.A. Dooley. 1989. The use of ^{124}Sn as a tracer of tributyltin degradation. Society of Environmental Toxicology and Chemistry. 30 October–2 November 1989, 10th Annual Meeting, Toronto, Canada.

Mueller, M.D. 1984. Tributyltin detection at trace levels in water and sediments using GC with flame-photometric detection and GC–MS. *Fresen. Z. Anal. Chem.*, **317**, 32.

Mueller, M.D. 1987. Comprehensive trace level determination of organotin compounds in environmental samples using high-resolution gas-chromatography with flame photometric detection. *Anal. Chem.*, **59**, 617–623.

Page, D.S. 1989. An analytical method for butyltin species in shellfish. *Mar. Pollut. Bull.*, **20**, 129–133.

Rice, C.D., F.A. Espourteille and R.J. Huggett. 1987. Analysis of tributyltin in estuarine sediments and oyster tissue (*Crassostrea virginica*). *Appl. Organometal. Chem.*, **1**, 541–544.

Sasaki, K., T. Ishizaka, T. Suzuki and Y. Saito. 1988. Determination of tri-n-butyltin and di-n-butyltin compounds in fish by gas chromatography with flame photometric detection. *J. Assoc. Off. Anal. Chem.*, **71**, 360–363.

Short, J.W. 1987. Measuring tri-n-butyltin in salmon by atomic absorption: analysis with and without gas chromatography. *Bull. Environ. Contam. Toxicol.*, **39**, 412–416.

Short, J.W. and J.L. Sharp. 1989. Tributyltin in bay mussels (*Mytilus edulis*) of the Pacific coast of the United States. *Environ. Sci. Technol.*, **23**, 740–743.

Siu, K.W.M., G.J. Gardner and S.S. Berman. 1989. Ionspray mass spectrometry/mass spectrometry: quantitation of tributyltin in a sediment reference material for trace metals. *Anal. Chem.*, **61**, 2320–2322.

Stallard, M.O., S.Y. Cola and C.A. Dooley. 1989. Optimization of butyltin measurements for seawater, tissue, and marine sediment samples. *Appl. Organometal. Chem.*, **3**, 105–114.

Steinmeyer, R.D., A.F. Fentiman and E.J. Kahler. 1965. Analysis of alkyltin bromides by gas liquid chromatography. *Anal. Chem.*, **37**, 520–523.

Takeuchi, M., K. Mizuish, H. Yamanobe and Y. Watanabe. 1987. Determination of tributyltin compounds in fish and shell-fish by GC using electron capture detector. *Bunseki Kagaku*, **36**, 138–142.

Testa, J.P. and C.A. Dooley. 1989. ^{124}Sn-labelled tetra-n-butyltin and tri-n-butyltin bromide. *J. Labelled Compd Rad.*, **27**, 753–761.

Unger, M.A., W.G. MacIntyre, J. Greaves and R.J. Huggett. 1986. GC determination of butyltins in natural waters by flame photometric detection of hexyl derivatives with mass spectrometric confirmation. *Chemosphere*, **15**, 461–470.

Wade, T.L., B. Garcia-Romero and J.M. Brooks. 1988. Tributyltin contamination in bivalves from United States coastal estuaries. *Environ. Sci. Technol.*, **22**, 1488–1493.

Weber, R., F. Visel and K. Levsen. 1980. Field ionisation and field desorption for collisional activation mass spectrometric determination of organic tin compounds. *Anal. Chem.*, **52**, 2299–2304.

ANALYTICAL METHODS FOR TRIBUTYLTIN IN SEDIMENTS AND TISSUES

7

R.J. Huggett[1], *F.A. Espourteille*[2], *C.D. Rice*[3], *M.A. Unger*[1], *C.A. Dooley*[4], *and S.Y. Cola*[5]

[1]*Virginia Institute of Marine Science, Schools of Marine Science, College of William and Mary, Gloucester Point, Virginia 23062, USA*
[2]*Rutgers University, New Brunswick, New Jersey 07102, USA*
[3]*Department of Anatomy, Medical College of Virginia, Richmond, Virginia 23298, USA*
[4]*Naval Command Control and Ocean Surveillance Center, RDT&E Division, Environmental Sciences (Code 52), San Diego, California 92152-5000, USA*
[5]*Computer Sciences Corporation, 4045 Hancock Street, San Diego, California 92110-5164, USA*

ABSTRACT

Analytical methodologies have been developed over the last decade to quantify tributyltin in sediments and tissues. The methods involve extracting the substance from solid matrices with either an organic solvent (sometimes acidified) or an acid. Tributyltin (TBT) is then converted to a more volatile compound, either an alkyl or hydride derivative to facilitate separation from potentially interfering compounds and subsequent detection by gas chromatography, atomic absorption spectrophotometry, or mass spectrometry. Detection levels reported vary by more than an order of magnitude, but accurate quantitation at concentrations <10 ng g^{-1} appears possible. Experiments conducted with oysters (*Crassostrea virginica*), exposed to TBT in the environment indicate that natural variability in TBT concentrations is such that ten replicates, were required to assess con-

Organotin. Edited by M.A. Champ and P.F. Seligman. Published in 1996 by Chapman & Hall, London. ISBN 0 412 58240 6

tamination at a station within a 95% degree of confidence.

7.1. INTRODUCTION

Tributyltin (TBT) contamination in lake, riverine, and coastal waters has caused concern about the potential contamination of the underlying sediments and the biota that inhabit the systems. The data set for TBT in solid matrices is small compared with that in water. The reason is simply that quantitative analyses of TBT in sediments and tissues are much more difficult than in the aqueous phase. The substance is difficult to remove from the solids without chemically altering it, and other substances extracted at the same time can interfere with derivatization, detection, and quantification. This chapter is a summary of representative analytical methodologies developed to quantify TBT in sediment and tissues. In addition, data are presented on the variability of TBT concentrations in sediments and oysters collected from the same locations since it is an important aspect of the total analytical scheme.

7.2. ANALYTICAL METHODS

Numerous investigators have developed and used procedures to estimate concentrations of TBT in sediment and tissue with varying degrees of success. Some of the early attempts relied on extracting the sample with an organic solvent and acid, adjusting pH, and analyzing by atomic absorption spectrophotometry (AA). While these methods did yield crude estimates of TBT levels, more accurate and precise methodologies have been developed. Some of the later analytical schemes are discussed in this section. All of these methods involve converting TBT to a more volatile compound, either an alkyl derivative or the hydride.

A method to determine TBT in sediments and macroalgae was reported by Hodge *et al.* (1979). In this procedure the materials are totally digested with HNO_3, $HClO_4$, and HF–HCl. The butyltins are converted to their hydrides and cryogenically trapped. The organotin hydrides are separated by their boiling points and detected by AA.

Maguire (1984) and Mueller (1984) reported two techniques for quantifying TBT in sediments via derivatization. Maguire refluxed air-dried sediments with benzene and tropolone to remove TBT followed by conversion to pentyltributyltin using n-pentylmagnesium bromide. After deactivation of the excess Grignard reagent, the solution was cleaned by florisil column chromatography and analyzed by gas chromatography (GC) with flame photometric detection (FPD).

Mueller (1984) acidified wet sediment with HCl to a pH of 2 and extracted organotins with diethylether. The ether phase was dried over anhydrous calcium chloride. Methylmagnesium chloride was used to produce methyltributyltin derivatives. Excess Grignard reagent was removed with HCl and the resulting aqueous phase was washed with ether. The ether phase was again dried over anhydrous calcium chloride and cleaned by silica gel column chromatography with pentane elution. Methyltributyltin was quantified by GC–FPD and gas chromatography–mass spectrometry (GC–MS). This procedure was reported to have a detection limit of 0.5 ng g^{-1}.

Tsuda *et al.* (1986) extracted wet sediment with hexane following acid leaching with HCl. After centrifugation, aliquots of the organic layer were evaporated to near dryness, and residues were dissolved in ethanol prior to hydride derivatization. Cleanup was achieved by column chromatography using silica gel, and GC with electron capture (EC) was employed for quantitation. They achieved a 0.5–1.0 ng g^{-1} detection limit for sediment with this procedure. Their methodology for tissues closely followed that for sediments with the exception that sodium chloride and ethyl acetate were also employed in the extraction step. Their reported detection limit was 1.0–2.0 ng g^{-1} in a 1-ml final volume.

Tributyltin has been extracted from freeze-dried sediments with a solution of calcium chloride and HCl (Randall *et al.*, 1986). Following conversion to the hydride, TBT concentrations were determined by AA using tetramethyltin as an internal standard. The detection limit was reported to be 0.6 ng g^{-1} (dry wt) [*sic*].

Cooney *et al.* (1988) refluxed wet sediment to which triethyltin bromide had been added as an internal standard with acidic methanol. After cooling and centrifugation, the organotins were derivatized to the hydrides. The volatile hydrides were trapped cryogenically. Quantitation was achieved by eluting the hydrides into an AA furnace by heating. They reported a limit of detection of 0.4 ng of organotin per gram of sediment.

Ashby and Craig (1989) mixed wet sediment with distilled water and added tripropyltin chloride as an internal standard. The mixture was acidified with HCl and allowed to sit overnight. Tropolone in dichloromethane was then added and the liquid separated by vacuum filtration. Uncomplexed tropolone was removed by adding ferrous sulfate, and the organic layer was separated and evaporated to dryness. The organotin–tropolone complexes were redissolved in ethanol and derivatized to the hydrogen and ethyl derivatives. Quantitation was achieved by GC–AA and GC–MS. They reported a 0.7 ng for 2 g sediment detection limit for the hydrides and 1.2 ng for 2 g sediment for the ethyl derivatives.

Rice *et al.* (1987) reported that TBT in sediment and tissues could be removed by Soxhlet-extracting samples that had been desiccated with anhydrous sodium sulfate and precipitated silica. The extracted TBT and an internal standard (tripentyltin chloride, which had been added before desiccation) were converted to their hexyl derivatives by hexylmagnesium bromide. The derivatized extract was cleaned by fluorisil column chromatography and TBT was quantified by GC–FPD. Detection levels were in the low nanogram per gram range.

Takami *et al.* (1987) extracted TBT from tissue with HCl and ethanol in a separatory funnel. The extract was paper filtered, washed further with ethanol, and reduced in volume by evaporation. Sodium chloride and HCl were added prior to extraction with ethyl acetate hexane. A cation exchange resin clean-up cartridge was used to trap the concentrated ethanol extracts of TBT. Hydride derivatization was achieved on-column by passing sodium borohydride through the cartridge. The hydrides were removed from the column with ethanol–hexane and washed with hexane–water, and the hexane layer was separated and reduced in volume prior to TBT quantitation by GC–MS. The detection limit was reported as 8 ng g^{-1} for a 10-g sample.

Short (1987) utilized an HCl digestion to remove TBT from tissue. The digest was extracted with hexane. He reported a screening step whereby the hexane was evaporated to dryness and redissolved in nitric and acidic acids followed by AA analyses. A more precise and accurate TBT estimate was achieved by converting the TBT to the pentyl derivatives and analyzing them by GC–AA after silica gel clean-up. The reported detection limits were approximately 5 ng g^{-1} for the screening method and 15 ng g^{-1} for derivatization.

Sasaki *et al.* (1988) extracted TBT from tissues by homogenizing with methanol prior to extraction with a combination of diethylether and hexane to which NaCl and HCl had been added. The organic phase was removed and dried with anhydrous sodium sulfate. The TBT was converted to the ethyl derivative, cleaned by silica gel column chromatography, and quantified by GC–FPD at a detection limit of near 5 ng g^{-1}.

A methodology for shellfish tissues was reported by Rapsomanikis and Harrison (1988). After freeze drying, TBT was extracted by either methylene chloride and tropolone or HCl. Hydride derivatization was followed by purge-and-trap GC and quartz furnace AA with a reported 3.5 ng for a 1.58 g limit of detection.

Wade *et al.* (1988) utilized sodium sulfate drying followed by methylene chloride–tropolone extraction to remove TBT and an internal standard, tripropyltin chloride, from tissues. They made hexyl derivatives and quantified by GC–FPD and GC–MS. Detection limits were reported to be 5 ng g^{-1}.

Stallard *et al.* (1989) published a TBT methodology for sediments and tissues that employs extraction of a wet sample with HCl and methylene chloride. After a solvent exchange to hexane, tripentyltin bromide and/or tripropyltin chloride was added as an internal standard(s). The organotins were converted to their hexyl derivatives before quantitation with GC–FPD. The absolute detection limit for sediments and tissues was reported to be 0.1 ng for TBT.

A method for determining TBT in plant tissues was reported by Francois and Weber (1988). Plant material was ground in liquid nitrogen and extracted with methylene chloride and methanol. TBT was derivatized to the hydride, purged, and trapped on silanized Chromosorb®. The TBT hydride was stripped from the column and analyzed by AA. A detection limit was reported of 1 ng TBT in 100 mg eelgrass tissue (wet wt).

7.3. TBT NATURAL VARIABILITY

Sediments or individual animals taken from the same location will often contain contaminant levels that differ by factors of two or more among samples due to the different microenvironments around the sample. Physical differences (e.g. grain size, organic carbon content, and currents) and physiological differences (e.g. sex, maturity, and reproductive stage) play important roles. For this reason, it is essential to have some estimate of the range of concentrations expected at a location to assist in sampling design and interpretation of data.

Experiments were conducted on sediments and oysters *Crassostrea virginica* from the Chesapeake Bay to shed light on this phenomenon (Espourteille, 1988). Five replicate sediment samples were collected from the same location in the Lafayette River while the vessel was 'swinging on its anchor' (Fig. 7.1). Sediments were collected with a Ponar Grab, and only the top 2 cm of the undisturbed sediment column were collected. Samples were stored in the dark while in the field. Once returned to laboratory (in <8 h), samples were kept at 12°C for no longer than 3 d before being analyzed by the method of Rice *et al.* (1987).

Estimates were made of natural variability in oysters by collecting individual oysters by dredge from each of two beds. One was located in the James River (19 replicates) and the other was from the relatively rural Rappahannock River (20 replicates) (Fig. 7.1). Samples were kept on ice until returned to the laboratory where they were either processed immediately or frozen at −20°C until analysis. TBT tissue burdens were determined in individuals (Rice *et al.*, 1987).

The sediment samples had a mean TBT concentration of 43 ppb. The standard error of the mean was ±3.5 ppb (Table 7.1).

Oysters collected from a given location displayed considerable variability in TBT concentration (Table 7.2). Mean and standard error for TBT from the James River was 310 ± 16 ppb. Similar results were obtained for Rappahannock River oysters. Mean and standard error for TBT was 310 ± 18 ppb.

These results were used to determine the sample size (n) required to assess TBT contamination at a station within a given degree of confidence. According to Zar (1984) the following equation allows for the determination of the sample size needed to estimate the population mean with the desired precision:

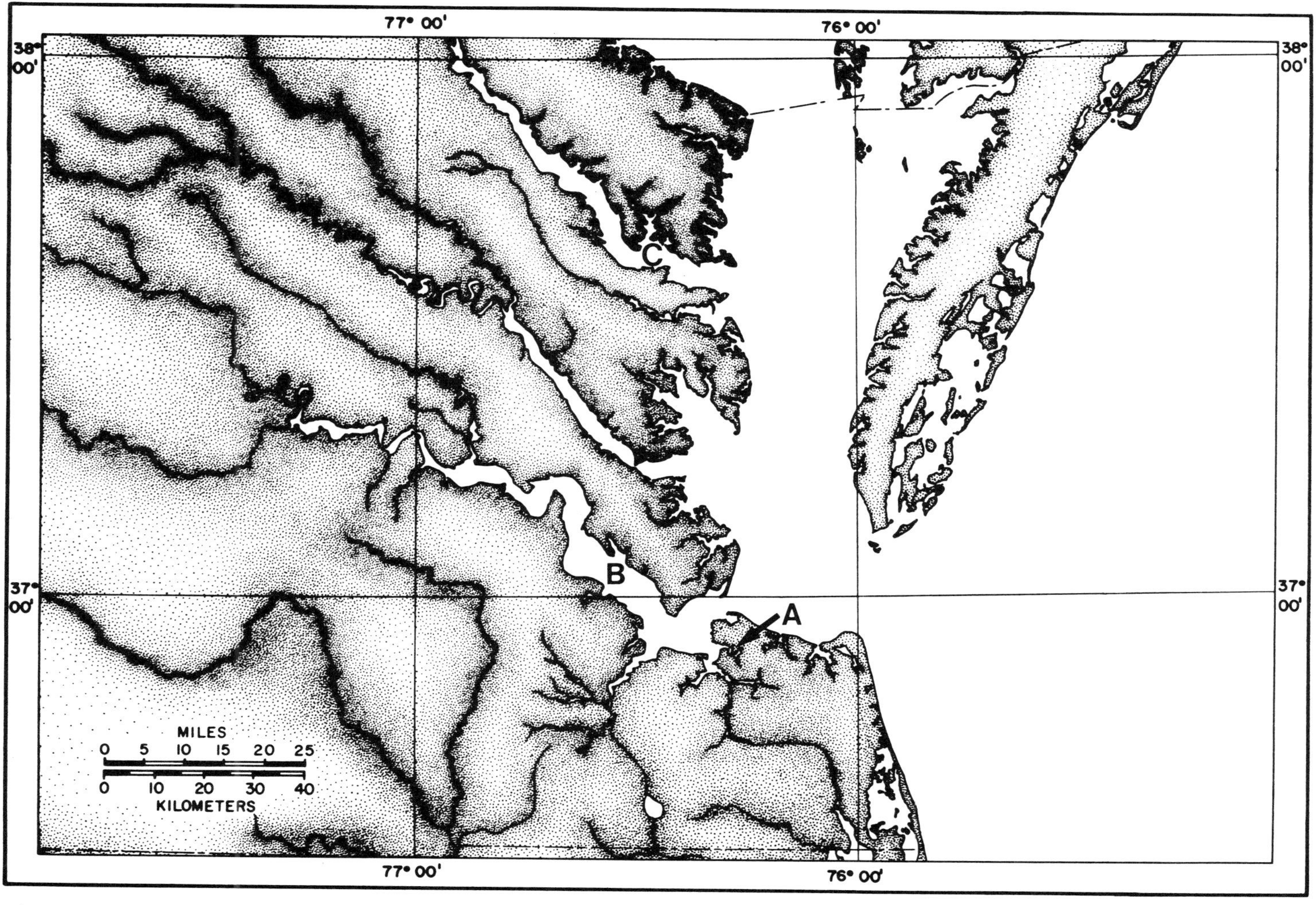

Figure 7.1 Map of Chesapeake Bay with sediment and oyster sampling stations. A = Lafayette River, B = James River, Warwick Station, C = Rappahannock River, Corrotoman Station.

Table 7.1 Natural variability in tributyltin among sediments from the same station (Lafayette River, Marker 13)

	TBT+[a]
Grab: a	37
Grab: b	53
Grab: c	41
Grab: d	50
Grab: e	34
Average TBT+ concentration	43
Standard deviation	±8.2
Standard error of the mean	±3.5

[a]Concentrations are in micrograms per kilogram (dry wt).

Table 7.2 Natural variability of tributyltin in oysters[a]

	TBT concentration	
Oyster sample no.	*James River (Warwick Station)*	*Rappahannock River (Corrotoman Station)*
1	270	270
2	260	200
3	190	250
4	460	350
5	250	310
6	310	350
7	350	310
8	230	230
9	330	220
10	320	240
11	370	280
12	400	250
13	430	350
14	280	550
15	240	370
16	340	350
17	330	450
18	290	300
19	290	310
20	—	320
Average value per oyster	310	310
Standard deviation	70	82
Standard error of the mean	16	18

[a]Concentrations are in micrograms per kilogram (dry wt).

$$N = \frac{(S^2 T^2_{a(2),\,(n-1)}\, F_{b(1),\,(n-1,\,v)}}{d^2}$$

where N is the sample size; S^2 is the variance of the oysters sampled, which is an estimate of the population variances; $T_{a(2),\,(n-1)}$ is the two-tailed critical value of Student's T, a being the confidence interval with n–1 degrees of freedom; $F_{b(1),\,(n-1,\,v)}$ is the one-tail critical value for the F distribution, b being the assurance that the confidence interval will be no larger than specified with n–1 and v degrees of freedom; d is the half width of the desired confidence interval, and it was set to equal the standard deviation, which for Table 7.2 is equal to 70 ppb. It was decided to set the confidence level (a) at 95%. The power of the test was set at 90%, which specifies a 90% probability that the confidence interval will be no larger than specified (b). For the two oyster populations used in this study, n equals 10.

7.4. CONCLUSION

Much progress has been made in precisely and accurately analyzing sediments and tissues for TBT. Concentrations in the low nanogram per gram range can now be determined routinely.

Naturally exposed sediments and oysters (i.e. not exposed under laboratory conditions) exhibit considerable variability in TBT concentrations. This variability must be taken into account when designing monitoring programs or surveys to assess spatial or temporal distributions of TBT in sediments or biota. An inadequate number of replicates collected per station will result in sampling 'errors' that will overwhelm chemical analytical errors. Natural variability should be considered part of the total analytical scheme.

ACKNOWLEDGMENTS

This work was supported by the Commonwealth of Virginia and the US Navy, and is VIMS contribution number 1705.

REFERENCES

Ashby, J.R. and P.J. Craig. 1989. New method for the production of volatile organometallic species for analysis from the environment; some butyltin levels in U.K. sediments. *Sci. Total Environ.*, **78**, 219–232.

Cooney, J.J., A.T. Kronick, G.J. Olson, W.R. Blair and F.E. Brinckman (Eds) 1988. A modified method for quantifying methyl and butyltins in estuarine sediments. *Chemosphere*, **17**, 1795–1802.

Espourteille, F.A. 1988. An Assessment of Tributyltin Contamination in Sediments and Shellfish in the Chesapeake Bay. Masters Thesis, College of William and Mary. 78 pp.

Francois, R. and J.H. Weber. 1988. Speciation of methyltin and butyltin compounds in eelgrass (*Zostera marina*) leaf tissue from the Great Bay Estuary (NH). *Marine Chem.*, **25**, 279–289.

Hodge, V.F., S.L. Seidel and E.D. Goldberg. 1979. Determination of tin(iv) and organotin compounds in neutral waters, coastal sediments and macro algae by atomic absorption spectrometry. *Anal. Chem.*, **51**, 1256–1259.

Maguire, R.J. 1984. Butyltin compounds and inorganic tin in sediments in Ontario. *Environ. Sci. Technol.*, **18**, 291–294.

Mueller, M.D. 1984. Tributyltin detection at trace levels in water and sediments using GC with flame-photometric detection and GC–MS. *Fresen. Z. Anal. Chem.*, **317**, 32–36.

Randall, L., J.S. Han and J.H. Weber. 1986. Determination of inorganic tin, methyltin and butyltin compounds in sediments. *Environ. Technol. Lett.*, **7**, 571–576.

Rapsomanikis, S. and R.M. Harrison. 1988. Speciation of butyltin compounds in oyster samples. *Appl. Organometal. Chem.*, **2**, 151–157.

Rice, C.D., F.A. Espourteille and R.J. Huggett. 1987. Analysis of tributyltin in estuarine sediments and oyster tissue, *Crassostrea virginica*. *Appl. Organometal. Chem.*, **1**, 541–544.

Sasaki, K., T. Suzuki and Y. Saito. 1988. Determination of tri-n-butyltin and di-n-butyltin compounds in yellowtails. *Bull. Environ. Contam. Toxicol.*, **41**, 888–893.

Short, J.W. 1987. Measuring tri-n-butyltin in salmon by atomic absorption: analysis with and without gas chromatography. *Bull. Environ. Contam. Toxicol.*, **39**, 412–416.

Stallard, M.O., S.Y. Cola and C.A. Dooley. 1989. Optimization of butyltin measurements for seawater, tissue and marine sediment samples. *Appl. Organometal. Chem.*, **3**, 105–113.

Takami, K., H. Yamamoto, T. Okumura, A. Sugimae and M. Nakamoto. 1987. Application of 'clean-up' cartridge for gas chromatographic determination of di- and tri-n-butyltin in fish. *Anal. Sci.*, **3**, 63–67.

Tsuda, T., H. Nakanishi, T. Morita and J. Takebayashi. 1986. Simultaneous gas chromatographic determination of dibutyltin and tributyltin compounds in biological and sediment samples. *J. Assoc. Off. Anal. Chem.*, **69**, 981–984.

Wade, T.L., B. Garcia-Romero and J.M. Brooks. 1988. Tributyltin contamination in bivalves form the United States coastal estuaries. *Environ. Sci. Technol.*, **22**, 1488–1493.

Zar, J.H. 1984. *Biostatistical Analysis*, 2nd Edn, Prentice Hall, Englewood, California.

COMPARISON OF METHODOLOGIES FOR BUTYLTIN SPECIES DETERMINATION

8

K.W.M. Siu and S.S. Berman

Institute for Environmental Chemistry, National Research Council of Canada, Montreal Road, Ottawa, Ontario K1A 0R9, Canada

ABSTRACT

This chapter reviews, evaluates, and critically compares the major techniques for the determination of butyltin species; discusses an analytical protocol for good laboratory practice; and presents results of several interlaboratory comparisons, including the certification of PACS-1 (the world's first certified reference material for butyltins). The certified concentrations for tributyl-, dibutyl-, and monobutyltin are 1.21 ± 0.24, 1.14 ± 0.20, and 0.28 ± 0.17 mg Sn kg^{-1} of dry sediment, respectively. The uncertainties are 95% confidence intervals for the analysis of an individual subsample.

8.1. INTRODUCTION

A variety of analytical methodologies have been developed over the last 15 y or so for the determination of butyltin species. This chapter aims to discuss, group, and compare some of these analytical methods as well as to point out the advantages and disadvantages of various techniques. Needless to say, information of this type can be very subjective

Organotin. Edited by M.A. Champ and P.F. Seligman. Published in 1996 by Chapman & Hall, London. ISBN 0 412 58240 6

in nature and unavoidably reflects the personal experiences and biases of the authors.

8.2. ANALYTICAL CHEMISTRY OF BUTYLTINS

Tributyltin (TBT) and its degradation products, dibutyltin (DBT), monobutyltin (MBT), and inorganic tin, exhibit very different toxicological properties. Consequently, it is imperative that analytical techniques used for butyltin determinations be able to differentiate the various species. Some of these techniques are described in the preceding chapters on butyltin determination (Chapters 4, 5, 6, and 7).

Butyltins are usually extracted from environmental samples as halides or tropolone complexes by using solvent or solid-phase extraction. They may be derivatized into hydrides (e.g. Donard *et al.*, 1986; Stallard *et al.*, 1989; Weber *et al.*, Chapter 4) or tetraalkylated forms (e.g. Maguire and Huneault, 1981) for subsequent analysis. Alternatively, for the analysis of waters, the preconcentration and derivatization step may be combined in a single hydride generation procedure (Matthias *et al.*, 1986; Matthias, Chapter 5).

The majority of measurement techniques are based on the selective detection of tin such as flame photometric detection (FPD) (e.g. Maguire and Huneault, 1981), inductively coupled plasma–atomic emission spectrometry (ICP–AES) (Suyani *et al.*, 1989a), inductively coupled plasma–mass spectrometry (ICP–MS) (e.g. McLaren *et al.*, 1990), atomic absorption spectrometry (AAS) (e.g. Chau *et al.*, 1982), or the individual butyltin species such as electron impact mass spectrometry (EIMS) (e.g. Meinema *et al.*, 1978), and ionspray mass spectrometry/mass spectrometry (ISMS/MS) (Siu *et al.*, 1989b). Selectivity in detection is important as environmental samples have complex matrices.

To differentiate the butyltin species, a separation procedure such as gas or liquid chromatography, or a selective extraction is employed. Chromatography is far more efficient and hence more popular. Various chromatographic modes have been combined with an array of selective detection methods to produce a matrix of hyphenated techniques for butyltin species determination. These include strong cation exchange (SCX)–ICP–MS (McLaren *et al.*, 1990), ion pairing (IP)–ICP–MS (Suyani *et al.*, 1989a), SCX–AAS (Ebdon *et al.*, 1985), gas chromatography (GC)–FPD (e.g. Maguire and Huneault, 1981), and GC–AAS (e.g. Chau *et al.*, 1982).

8.2.1. SAMPLE TREATMENT

Numerous procedures have been designed for various sample types. Each of these procedures is usually divisible into three sections: solvent extraction, derivatization, and clean-up.

Except for methodologies involving hydride generation AAS, all procedures require extraction of the butyltin species into an organic solvent with or without the addition of tropolone. Sometimes an acid (usually hydrochloric or hydrobromic) or a salt (sodium chloride or bromide), or both, is added to release the butyltin species as halides and to facilitate extraction. Several solvents [e.g. dichloromethane (e.g. Wade *et al.*, 1988; Stallard *et al.*, 1989), chloroform (Ebdon *et al.*, 1985), hexane (e.g. Unger *et al.*, 1986), toluene (e.g. Uhler, personal communication), benzene (e.g. Maguire and Huneault, 1981), dichloromethane–chloroform (Sullivan *et al.*, 1988), and hexane–isobutyl acetate (Siu *et al.*, 1989a)] have been reported to extract TBT, DBT, and MBT efficiently, usually with the aid of tropolone, from various matrices. This multiplicity of choice can be confusing and does not reflect the difficulty in achieving quantitative or near-quantitative extraction of the individual butyltin species. It is important to remember that a published extraction and clean-up method may only be applicable to the particular type of sample such as mussel, oyster, or sediment that was analyzed during the course of that work. Extrapolation of extraction efficiencies to other sample types should be done with caution, and it is always advisable to verify the recoveries of the butyltin species whenever a different biological species or sediment type is encountered.

Our experience indicates that hexane and toluene may be the best solvents for overall good performance. Toluene is also championed by another group (Uhler, personal communication). Solid-phase extraction with tropolone-loaded reversed-phase materials, usually octadecyl-bonded silica gel, has been used for extracting and preconcentrating butyltin species from waters (Matthias *et al.*, 1987). This technique is attractive and has been proven effective in the extraction of trace hydrophobic compounds in waters. Nevertheless, degradation of butyltins has been observed on some reversed-phase materials (Siu and McLaren, personal communication), and caution should be exercised in their use.

The butyltin species are usually derivatized into an inert form for determination after solvent extraction. The type of derivatization performed is normally dictated by the analytical technique subsequently employed. For liquid chromatography, no derivatization is necessary. For gas chromatography, the derivatization may be hydride generation, alkylation, or halide conversion (see section 8.2.2, Gas chromatography, for details).

Sample clean-up is normally performed after derivatization for sediments and biological materials. This is necessary to ensure good chromatography and reasonable column life. This step may be omitted when the sample matrix is relatively simple or when the butyltin concentrations are relatively high and only a small quantity of sample material is needed for analysis. One of the most effective clean-up methods is the use of silica gel, sometimes in conjunction with alumina and sodium sulphate (e.g. Wade *et al.*, 1988) in a clean-up column. The retention characteristics of silica gel are highly dependent on its moisture content. In some recipes, silica gel with a controlled moisture content is utilized (Tsuda *et al.*, 1986) while in others anhydrous silica gel is preferred (e.g. Wade *et al.*, 1988). In either case, careful regulation of the water content is essential.

No clean-up is necessary for water samples, which are usually handled by hydride generation. In this case, the butyltin hydrides are formed and purged from the sample, obviating the clean-up step (e.g. Donard *et al.*, 1986).

8.2.2. GAS CHROMATOGRAPHY

The butyltin species may be separated and eluted as hydrides, tetraalkylated forms, or halides after appropriate derivatization and clean-up. For routine analysis, derivatization to hydrides or tetraalkylated forms is probably advisable because of the relatively higher stability of these forms. The halide derivatives require the addition of hydrogen halide into the carrier gas stream to ensure quantitative elution (Aue *et al.*, 1989). The butyltin derivatives, whether they are hydrides, tetraalkyls, or halides, are effectively separated from one another on common non-polar or moderately polar liquid phases. Both packed and capillary columns have been successfully used. When analyzing environmental samples, it is often difficult to resolve the butyltin species completely from the matrix components. This, however, usually poses little problem when tin-selective detection, such as FPD or AAS, is employed. The quality of the GC separation becomes more important when a less-selective means of detection, such as EIMS, electron capture detection (ECD), or flame ionization detection (FID), is used.

In the case of FPD, detection is based on tin molecular emission in a hydrogen–air flame. The band most commonly selected is centered at ~610 nm and is attributed to SnH emission in the gas phase (Pearce and Gaydon, 1963). A broader emission band, 100–1000 times more sensitive, centered at ~390 nm, and attributed to a quartz surface-induced tin luminescence, has been observed (Flinn and Aue, 1980). This emission, however, has not found any application to the analysis of environmental extracts due to its temperamental nature and susceptibility to poisoning (Aue and Flinn, 1977). For AAS, tin atomic absorption is usually monitored via the 286.3- or 224.6-nm line after atomization in a hydrogen–air flame using a quartz burner designed to maximize light absorption. An electrodeless discharge lamp is usually used in place of the conventional, but less intense, hollow cathode lamp for tin (Donard *et al.*, 1986).

The butyltin species do yield rather characteristic mass spectrometric fragments (the isotopic pattern of tin is usually unmistakable) under electron impact (Meinema *et al.*, 1978) and chemical ionization (Unger *et al.*, 1986), but unfortunately their *m/z* values are rather small and hence rather common. As a result, observations of non-tin peaks and high background are possible during selected ion monitoring of these fragments in environmental sample analysis. Both ECD and FID are non-selective. The electron capture detector is more sensitive to the tin derivatives (e.g. Hattori *et al.*, 1984).

There is an apparent trend in the choice of tin-selective detection according to the type of derivatization employed: FPD is favored when tetraalkylation is used, while AAS is preferred when hydride generation is involved [although examples of opposite pairing are quite common (Chau *et al.*, 1982; Matthias *et al.*, 1986)]. This trend is likely historical as hydride generation is a long-established technique in AAS, and alkylation is a proven derivatization procedure in gas chromatography. To some, the practice of gas chromatography in the hydride derivatization of butyltin species for AAS may not be immediately apparent because of the possible absence of a gas chromatograph. The trapping of the butyltin hydrides and their sequential volatilization are, of course, in GC terminology, cold trapping and temperature programming, respectively.

Hydride generation has an advantage over alkylation for dilute aqueous samples such as waters. Hydride generation produces a large volume of hydrogen as a byproduct, which facilitates the purging of butyltin hydrides from a large volume of sample. Also, the derivatization and preconcentration of butyltin species are achieved simultaneously (e.g. Donard *et al.*, 1986; Valkirs *et al.*, 1987; Stallard *et al.*, 1989). In alkylation, the butyltin species have to be extracted and preconcentrated into a water-immiscible solvent and then alkylated with a Grignard reagent. Extraction may be performed by using a solvent (e.g. Maguire and Huneault, 1981; Wade *et al.*, 1988) or a solid-phase cartridge (Matthias *et al.*, 1987). In either case, the process is rather time consuming.

Contrarily, alkylation is often simpler and preferred for sediments and biological tissues since solvent extraction with a water-immiscible solvent is mandatory for these samples. The pentyl- and hexyl- derivatives are considered to be most suitable since their volatility is high enough for GC yet low enough to prevent losses during solvent evaporation (Unger *et al.*, 1986, and Chapter 6). (Caution should be exercised in using commercial Grignard reagents as they are occasionally contaminated with organotins.) Hydride generation has enjoyed limited success in the analyses of sediments and tissues (Randall *et al.*, 1986a; Francois and Weber, 1988; Han and Weber, 1988). Non-quantitative generation of the hydride derivatives due to matrix interference has been reported (Uhler and Durell, personal communication).

Simultaneous hydride generation and extraction into dichloromethane for GC–FPD has been described (Matthias *et al.*, 1986, and Chapter 5). The use of the halide derivatives is increasing, partly as a result of the attraction of doing away with a discrete derivatization step (Junk and Richard, 1987; Krull *et al.*, 1989; Siu *et al.*, 1989a). Although butyltin species are usually extracted as halides or tropolone complexes, the actual species matter little as any non-halides are derivatized on column into halides by the hydrogen halide that is added into the carrier gas stream. The hydrogen halide may be administered either as a solution (e.g. in methanol), which volatilizes in the injection port, or as a gas mixture. This addition is imperative for good chromatography even when butyltin halides are injected, as it discourages their on-column decomposition by shifting the equilibrium in favor of the halides (Aue *et al.*, 1989).

8.2.3. LIQUID CHROMATOGRAPHY

Liquid chromatography is the second (to GC) most popular means for separating the butyltin species. The two most widely used modes are ion-exchange and ion-pairing chromatography. The butyltin species are sufficiently ionic (ionic character increases from tributyltin to monobutyltin) to be partly ionized in protic solvents, and hence may be separated by ion-exchange or ion-pairing chromatography. The separation is relatively simple because tributyl-, dibutyl-, and monobutyltin ions are singly, doubly, and triply charged, respectively. For strong cation exchange, the eluant is usually a citrate buffer (Jewett and Brinckman, 1981) while for ion-pairing chromatography, an alkyl sulfate or sulfonate solution (Suyani *et al.*, 1989a,b) is typically employed. Detection is usually by ICP–MS (Suyani *et al.*, 1989b; McLaren *et al.*, 1990), ICP–AES (Suyani *et al.*, 1989a), flame AAS (Ebdon *et al.*, 1985), or graphite furnace AAS (Jewett and Brinckman, 1981) with ICP–MS being the choice (albeit expensive) means of detection with respect to sensitivity, ease of interfacing, and reliability.

Compared with GC, liquid chromatography (LC) is attractive in that no derivatization is necessary. For LC separation, the extraction procedure used is important. Extraction solvents such as 1-butanol (Epler *et al.*, 1988), hexane (McLaren *et al.*, 1990), and chloroform (Ebdon *et al.*, 1985) are compatible with the subsequent strong cation exchange separation, whereas complexometric extractants, such as tropolone, are not. Tropolone forms relatively strong complexes with butyltins and thus interferes with their ion exchange separation (Siu and McLaren, personal communication). Of the few LC studies that reported the analysis of real samples, only one (McLaren *et al.*, 1990) involved tributyl- as well as dibutyltin. Exclusion of tropolone makes it impossible to extract quantitatively all three butyltin species. Complexometric extractants are expected to exert a similar effect on ion-pairing separation. A further difficulty in this type of separation is the potential risk of butyltin degradation, which has been observed on certain reversed-phase materials (Siu and McLaren, personal communication).

The ICP mass spectrometer is virtually the ideal LC detector for butyltins. When the mass spectrometer monitors the tin isotopic cluster for quantification, it is almost tin specific. Interferences may arise from ionized oxides and molecular species that have the same *m/z* as tin, but these are quite rare. The technique is sensitive and tolerates the use of organic solvents (up to 80% or higher) and buffer salts [up to 0.3 M of ammonium citrate has been used (McLaren *et al.*, 1990)]. This means that ion-exchange and ion-pairing separation schemes designed for other detection systems may be adapted with little or no modification for ICP–MS. Another noteworthy mass spectrometric detector is the tandem mass spectrometer designed for atmospheric pressure ionization. This system, when coupled to ionspray or electrospray, becomes a tributyltin-specific detector that requires no prior separation and derivatization (Siu *et al.*, 1989b). These powerful techniques have been employed for the determination of TBT in a certified reference material (PACS-1, National Research Council of Canada). The obvious drawbacks of these mass spectrometric detectors are their high capital cost and the relatively skilled personnel required to operate and maintain them, which put them out of reach of most monitoring laboratories. ICP–AES (Suyani *et al.*, 1989a) and flame AAS (Ebdon *et al.*, 1985) usually do not have sufficient sensitivity for environmental sample analysis.

8.2.4. FIGURES OF MERIT

It is often difficult, even impossible, to compare directly figures of merit from various studies due to different definitions used, conditions of measurement, latitude allowed, and incomplete descriptions. Absolute detection limits for the above methods are typically between 10 and 100 pg Sn for environmental samples, while relative detection limits are usually ~1–10 ng g^{-1} Sn for sediments and biological materials and ~1 ng l^{-1} Sn for waters. Detector linearity is rarely a problem. The detectors mentioned above have typical linear ranges of a few decades, and usually only trace concentrations of butyltins are encountered in environmental analysis.

8.3. ANALYTICAL PROTOCOL

It is not sufficient in a discussion of this nature to focus just on techniques *per se* without mentioning analytical protocol. Experience has shown that when a set of data generated by several laboratories using comparable techniques is examined, the largest source of bias is usually within the laboratory itself (see section 8.4, Interlaboratory comparisons). In an analysis as labor intensive as butyltin species determination, it is not difficult to see why. The sample treatment is tedious; any small changes in conditions, some of which may not be operator controllable, may affect sample recovery. In addition, methodologies developed for one biological species or sediment type may not be directly transferable to another.

An analytical protocol for butyltin determination has been advocated (Uhler and Durell, personal communication), most of which is common sense for good laboratory practice. The following details are observed in our laboratory and are deemed worthy of discussion.

1. *Internal standard*. A tin compound similar in nature to the butyltin species [for example, tripropyltin chloride for GC–FPD (e.g. Wade *et al.*, 1988) and triethyltin bromide for AAS (Randall *et al.*, 1986b)] should be added to the sample prior to extraction as an internal standard to compensate for variation in solvent and sample volume, and to a smaller extent for the possible partial loss of the butyltin species. This last compensation may be inexact because TBT, DBT, and MBT have quite different chemical properties. Tripropyltin bromide and triethyltin bromide are good internal standards only for TBT.
2. *Percent recovery*. This should always be assessed as even samples of the same type (sediment, fish, or shellfish) may exhibit quite different butyltin recoveries. Ideally, recoveries should be quantitative. This may not be attainable because the butyltin species have different chemistries, and lengthy sample treatment procedures are employed thus increasing the possibilities of losses. Recoveries higher than 80% may be deemed acceptable in many instances.
3. *Standard additions*. Known quantities of TBT, DBT and MBT are added to a fraction of the samples together with the internal standard prior to sample treatment. An assay based on the non-spiked and the spiked set compensates for losses of the butyltin species provided the degree of loss is the same for the sample and the spiked sample. A comparison of the results obtained via standard additions and a standard calibration reveals immediately any change in percent recoveries. Although standard additions increase the

number of analyses and reduce the number of samples that can be processed per day, it is recommended that they be performed as much as possible for quality assurance. Even so, the results of standard additions are not foolproof. The technique is only valid provided the butyltins in the spike and the sample are in the same chemical forms, or it has been previously established that the method of standard additions provides reliable results for that type of sample. The butyltin spikes in most analyses are added as the chlorides while the butyltin chemical forms in real samples are unknown. It is generally assumed that, provided sufficient equilibration time is allowed, the butyltins in the spike will be bound by the sample matrix and converted to the same chemical forms as the native butyltins. However, this has never been proven and the time frame necessary is unknown. If standard additions demonstrate low recoveries, there are surely problems. However, full recoveries of the additions are not necessarily proof that all of the natural species has been recovered.

4. *Blank.* Blanks must be carried throughout the complete analytical procedure. This is essential laboratory practice. Although contamination of common laboratory reagents by butyltins is rare, commercial Grignard reagents may sometimes contain significant concentrations of various butyltin species.
5. *Calibration of standards.* Commercial butyltin standards are usually impure and should be standardized against a tin solution (usually inorganic) of accurately known concentration. The species purity of these standards should also be verified.
6. *Interspecies conversion.* Although interspecies conversion among TBT, DBT, MBT, and inorganic tin during sample treatment and analysis has not been documented, it is a possibility in view of the chemistry of butyltin compounds (Neumann, 1970). This may be checked by running known quantities of TBT, DBT, and MBT spiked separately to butyltin-free samples through the sample treatment and analysis to ascertain any degradation or interspecies conversion.
7. *Certified reference materials.* A basic step in the development and verification of any analytical method is the analysis of valid certified reference materials (CRMs). Unfortunately, these are often not available during the initial phases of research and development for a particular analytical problem. The only available CRM for TBT, DBT, and MBT is a marine harbor sediment (PACS-1) prepared by the National Research Council of Canada. A biological CRM is currently under preparation by the National Institute for Environmental Studies, Japan (sea bass) and should become available in the not-too-distant future. The availability of this new CRM will facilitate the verification of new analytical methodologies and laboratory practices.

8.4. INTERLABORATORY COMPARISONS

We know of six interlaboratory exercises (more than two participating laboratories) that have been conducted for the butyltin species in the past few years. The results of some of these, at the time of writing, are still being evaluated. These intercomparisons were designed to test the participants' abilities to determine the species in a particular matrix, in some cases to arrive at consensus values for the butyltin concentrations, and to evaluate particular methods or techniques. Unfortunately, results of these studies have not been published in the open literature and thus are generally not well known. Also, most of these studies involved too few laboratories with the resulting inability, in general, to assess differences among techniques or even individual laboratory performance.

The detailed results of four of these studies have been made available to us:

1. Study 1 was a round-robin measurement of the concentration of a tributyltin solution in water that was coordinated by the US National Institute of Standards and Technology (formerly the National Bureau of Standards) and concluded in early 1986 (Blair *et al.*, 1986).
2. Study 2 was an intercomparison exercise for organotins in an oyster sample coordinated by Institut français de recherche pour l'exploitation de la mer, and concluded in 1987 (Michel, 1987).
3. Study 3 was an intercomparison of butyltin determinations in mussel tissue and sediments coordinated by Moss Landing Marine Laboratories, United States, and reported in 1987 (Stephenson *et al.*, 1987).
4. Study 4 was the certification of the certified marine sediment PACS-1 coordinated by the National Research Council of Canada and concluded in 1988 (National Research Council of Canada, 1989).

Study 1 involved the analysis of a synthetic solution of TBT in water and had the largest number of participants at 32. The design of studies 2, 3, and 4 was very similar. Ten or fewer laboratories were involved in each study. The participants were asked to analyze the environmental samples for TBT, DBT, and MBT (study 2 asked for total tin as well).

Of the four studies, interlaboratory agreement was the highest in study 1 with a standard deviation of the mean of 5% (after elimination of results from three laboratories). The good agreement was probably due to the presence of one butyltin species, TBT, in the solution, which permitted the great majority of the laboratories to use techniques for determination of total tin for their measurements. The various analytical methodologies employed were classified into seven categories (Table 8.1), which included popular speciation techniques such as hydride

Table 8.1 Analytical methodologies employed in study 1

Method	*Extractant*	*Derivatization*	*Separation*	*Detection*
A	None	None or $KMnO_4$, H_2SO_4	None	Graphite furnace AAS
B	Methyl-isobutyl ketone, cyclohexane, toluene or dichloromethane with tropolone and HCl or HBr	None	None	Graphite furnace AAS
C	None or acetic acid	Hydride	None or desorption	Quartz furnace AAS
D	Benzene with tropolone or hexane with acetic acid	CH_3MgCl or bomb digestion	None or GC	Quartz furnace or other AAS
E	Benzene, pentane, dichloromethane or hexane; some with tropolone and HCl or HBr	Hydride or Grignard	Desorption or GC	FPD
F	Cyclohexane or hexane with HCl	None	Desorption or GC	ECD
G	Chloroform or dichloromethane	Dithiol, Grignard or H_2O_2–H_2SO_4	None or GC	Colorimetry or MS

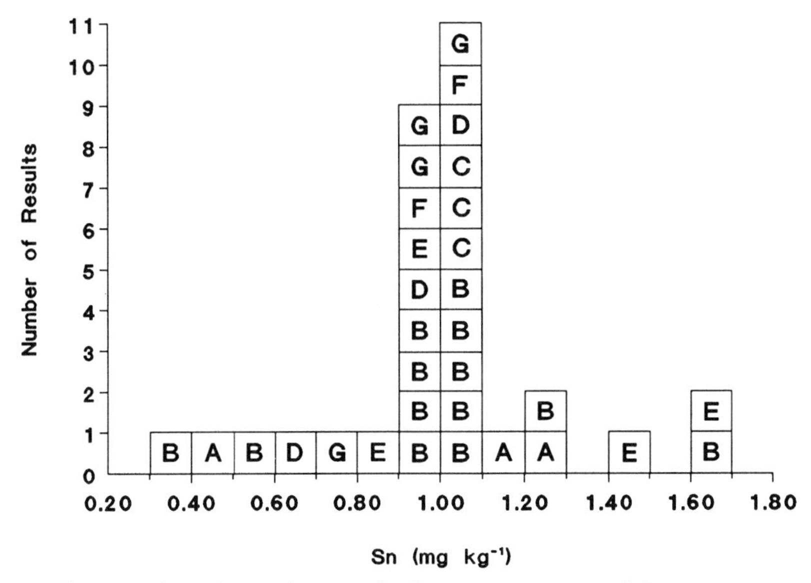

Figure 8.1 Distribution of results in the round-robin measurement of the concentration of tributyltin solution in water (Blair *et al.*, 1986). Letters refer to methods listed in Table 8.1.

Table 8.2 Methodologies employed for the certification of butyltins in PACS-1

Laboratory	*Extraction*	*Derivatization*	*Technique*
1	HCl–hexane–iso-butyl acetate	None	HPLC–ICP–MS
2	HCl–hexane–iso-butyl acetate	Chloride	GC–FPD
3	HCl–toluene–iso-butyl acetate–tropolone	Chloride	GC–FPD
4	1-butanol or HCl–iso-octane	None	ISMS/MS
5	Dichloromethane–tropolone	Pentylation or hexylation	GC–FPD
6	HCl–dichloromethane	Pentylation	GC–FPD
7	HCl–hexane–tropolone	Hexylation	GC–EIMS
8	NaCl–sodium benzoate–KI–benzene–tropolone	Methylation	GC–AAS
9	Solvent	Hexylation	GC–FPD
10	Solvent	Pentylation	GC–FPD

generation AAS and alkylation followed by GC–FPD. The results are summarized in Fig. 8.1. The letters correspond to the methods summarized in Table 8.1. No inherent bias among the methods is evident, and any one of the seven methods employed can apparently yield an accurate total tin value for the single species in aqueous solution. The real differences appear to lie in the diverse abilities of different laboratories.

Studies 2, 3, and 4 were similar in nature and yielded similar results. Most of the following discussion is based on data in study 4, which is best known to us. Six institutions participated in the certification process for the butyltin species in the marine sediment reference material PACS-1. The participants were asked to employ their method of choice for determining TBT, DBT, and MBT and report their concentrations as tin on a dry-weight basis. One institution submitted four sets of data for TBT using four independent techniques; another submitted two sets making a total of ten data sets for TBT. For DBT and MBT, there were nine and six data sets, respectively. The methodolo-

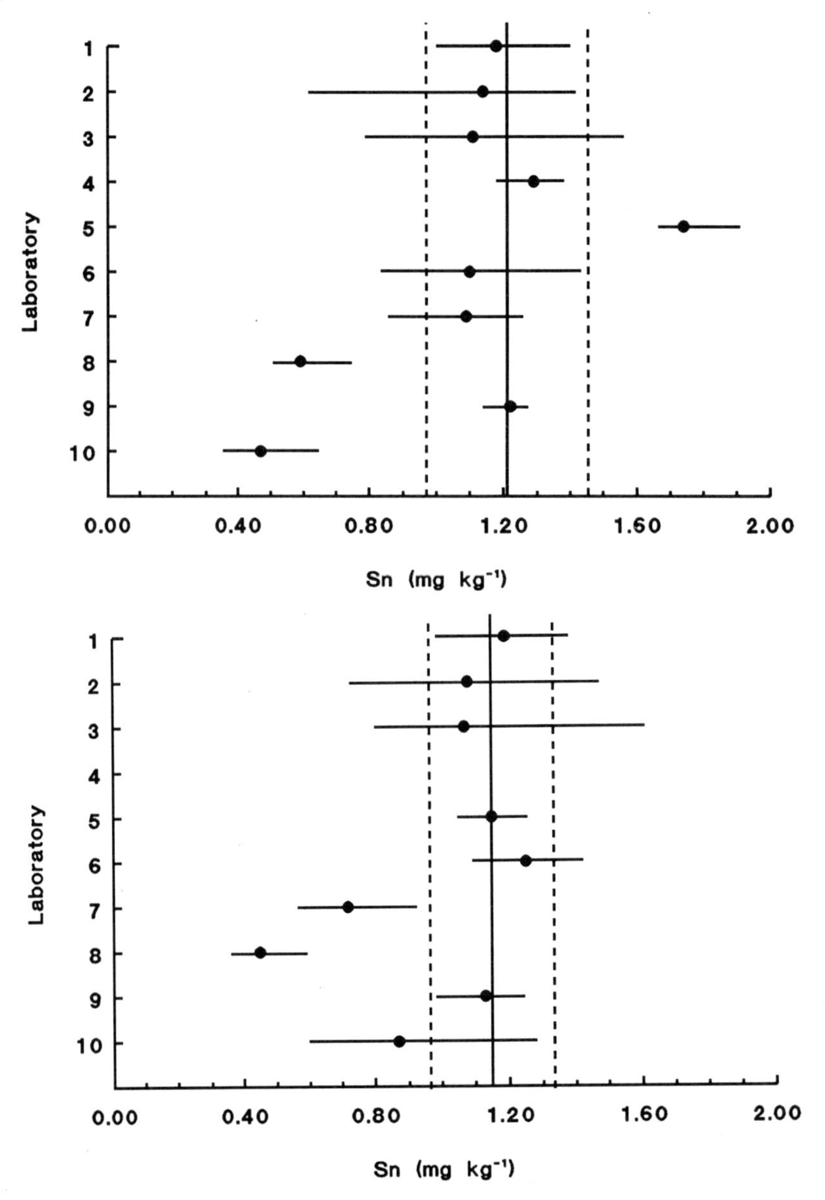

Figure 8.2 Laboratory performance in the certification of (a) tributyltin, and (b) dibutyltin in PACS-1 (National Research Council of Canada, 1989). Methodologies are detailed in Table 8.2. The solid vertical lines represent the certified values. The dotted vertical lines show the 95% confidence intervals.

gies employed in the ten data sets are summarized in Table 8.2. For TBT, there were at least nine different extraction methods and five different detection techniques used, with derivatization followed by GC–FPD being the most popular technique (six data sets).

Results for TBT and DBT determination are summarized in Fig. 8.2, which shows the range and the mean for each data set as well as the certified values and their uncertainties. Three sets of data for each of TBT and DBT, and two sets for MBT, were deemed outliers and were not used in the final calculations. Also, some individual results from accepted data sets were rejected. The concentrations for the butyltin species in PACS-1 were calculated to be TBT, 1.21 ± 0.24; DBT, 1.14 ± 0.20; and MBT, 0.28 ± 0.17 mg kg^{-1} Sn in dry sediment. The uncertainties are 95% confidence intervals for the analysis of an individual subsample. Due to the diversity of methods used and the relatively small uncertainties, these concentrations are now regarded as the certified values for the butyltin species in PACS-1. They are slightly, but not significantly, different from the original values issued by the National Research Council of Canada in June 1989. The change is due to the availability of new data after the original values were issued.

Studies 2 and 3 yield the same conclusions as study 4. The majority of the participants, usually using different methodologies, were able to determine the individual butyltin species to a reasonable degree of agreement. Also, there were no discernible differences in results due to methodology. However, a much larger population of participants would be needed to truly test the latter conclusion. As in study 1, the principal factor contributing to bias and imprecision seems to be individual laboratory performance.

ACKNOWLEDGMENT

We greatly appreciate the cooperation of W.R. Blair, the US National Institute of Standards and Technology; P. Michel, Institut français de recherche pour l'exploitation de la mer, France; P. Quevauviller, Community Bureau of References, Belgium; A.D. Uhler, Battelle Ocean Sciences, United States; and L. Pugh, US National Oceanic and Atmospheric Administration, in sharing their data and experience with us. Unfortunately, one set of data arrived too late to be included. We are grateful to W.R. Blair and A.D. Uhler for their comments on this chapter.

REFERENCES

Aue, W.A. and C.G. Flinn. 1977. A photometric tin detector for gas chromatography. *J. Chromatogr.*, **142**, 145–154.

Aue, W.A., B.J. Flinn, C.G. Flinn, V. Paramasigamani and K.A. Russell. 1989. Transformation and transmission of organotin compounds inside a gas chromatograph. *Can. J. Chem.*, **67**, 402–410.

Blair, W.R., G.J. Olson, F.E. Brinckman, R.C. Paule and D.A. Becker. 1986. An International Butyltin Measurement Methods Intercomparison: Sample Preparation and Results of Analyses. NBSIR 86-3321. Report prepared for the Office of Naval Research, Arlington, Virginia, 31 pp.

Chau, Y.K., P.T.S. Wong and G.A. Bengert. 1982. Determination of methyltin(IV) and tin(IV) species in water by gas chromatography/atomic absorption spectrophotometry. *Anal. Chem.*, **54**, 246–249.

Donard, O.F.X., S. Rapsomanikis and J.H. Weber. 1986. Speciation of inorganic tin and alkyltin compounds by atomic absorption spectrometry using electrothermal quartz furnace after hydride generation. *Anal. Chem.*, **58**, 772–777.

Ebdon, L., S.J. Hill and P. Jones. 1985. Speciation of tin in natural waters using coupled high-performance liquid chromatography–flame atomic absorption spectrometry. *Analyst*, **110**, 515–517.

Epler, K.S., T.C. O'Haver, G.C. Turk and W.A. MacCrehan. 1988. Laser-enhanced ionization as a selective detector for the liquid chromatographic determination of alkyltins in sediment. *Anal. Chem.*, **60**, 2062–2066.

Flinn, C.G. and W.A. Aue. 1980. Surface luminescence in the detection of organotin compounds following gas chromatography. *Can. J. Spectrosc.*, **25**, 141–148.

François, R. and J.H. Weber. 1988. Speciation of methylin and butyltin compounds in eel grass (*Zostera marina* L.) from the Great Bay Estuary (NH). *Mar. Chem.*, **25**, 279–289.

Han, J.S. and J.H. Weber. 1988. Speciation of methyl- and butyltin compounds and inorganic tin in oysters by hydride generation atomic absorption spectrometry. *Anal. Chem.*, **59**, 1887–1888.

Hattori, Y., A. Kobayashi, S. Takemoto, K. Takami, Y. Kuge, A. Sugimae and M. Nakamoto. 1984. Determination of trialkyltin, dialkyltin, and triphenyltin compounds in environmental water and sediments. *J. Chromatogr.*, **315**, 341–349.

Jewett, K.L. and F.E. Brinckman. 1981. Speciation of trace di- and triorganotins in water by ion-exchange HPLC–GFAA. *J. Chromatogr. Sci.*, **19**, 583–593.

Junk, G.A. and J.J. Richard. 1987. Solid phase extraction, GC separation and EC detection of tributyltin chloride. *Chemosphere*, **16**, 61–68.

Krull, I.S., K.W. Panaro, J. Noonan and D. Erickson. 1989. The determination of organotins (TBT) in fish and shellfish via gas chromatography–flame photometric detection and direct current plasma emission spectroscopy (GC–FPD/DCP). *Appl. Organometal. Chem.*, **3**, 295–308.

McLaren, J.W., K.W.M. Siu, J.W. Lam, S.N. Willie, P.S. Maxwell, A. Palepu, M. Koether and S.S. Berman. 1990. Applications of ICP–MS in marine analytical chemistry. *Fresen. Z. Anal. Chem.*, **337**, 721–728.

Maguire, R.J. and H. Huneault. 1981. Determination of butyltin species in water by gas chromatography with flame photometric detection. *J. Chromatogr.*, **209**, 458–462.

Matthias, C.L., J.M. Bellama, G.J. Olson and F.E. Brinckman. 1986. Comprehensive method for determination of aquatic butyltin and butylmethyltin species at ultratrace levels using simultaneous hydridization/extraction with gas chromatography–flame photometric detection. *Environ. Sci. Technol.*, **20**, 609–615.

Matthias, C.L., J.M. Bellama and F.E. Brinckman. 1987. Determination of tributyltin in estuarine water using bonded C-18 silica solid phase extraction, hydride derivatization and GC–FPD. In: Proceedings of the Oceans '87 International Organotin Symposium, Vol. 4, Halifax, Nova Scotia, 28 September–1 October 1987, pp. 1344–1347.

Meinema, H.A., T. Burger-Wiersma, G. Versluis-de Haan and E.C. Gevers. 1978. Determination of trace amounts of butyltin compounds in aqueous systems by gas chromatography/mass spectrometry. *Anal. Chem.*, **12**, 288–293.

Michel, P. 1987. Organotin Intercomparison Exercise on Oyster Sample. Report prepared for the participants. Institut français de recherche pour l'exploitation de la mer, France, 8 pp.

National Research Council of Canada. 1989. Marine Analytical Chemistry Standards Program Update, National Research Council of Canada, June, 2 pp.

Neumann, W.P. 1970. *The Organic Chemistry of Tin*, John Wiley & Sons, London, 282 pp.

Pearce, R.W.B. and A.G. Gaydon. 1963. *The Identification of Molecular Spectra*, Chapman & Hall, London, 289 pp.

Randall, L., J.S. Han and J.H. Weber. 1986a. Determination of inorganic tin, methyltin and butyltin compounds in sediments. *Environ. Technol. Lett.*, **7**, 571–576.

Randall, L., O.F.X. Donard and J.H. Weber. 1986b. Speciation of n-butyltin compounds by atomic absorption spectrometry with an electrothermal quartz furnace after hydride generation. *Anal. Chim. Acta*, **184**, 197–203.

Siu, K.W.M., P.S. Maxwell and S.S. Berman. 1989a. Extraction of butyltin species and their gas chromatographic determination as chlorides in a sediment certified reference material for trace metals, PACS-1. *J. Chromatogr.*, **475**, 373–379.

Siu, K.W.M., G.J. Gardner and S.S. Berman. 1989b. Ionspray mass spectrometry/mass spectrometry: quantitation of tributyltin in a sediment reference material for trace metals. *Anal. Chem.*, **61**, 2320–2322.

Stallard, M.O., S.Y. Cola and C.A. Dooley. 1989. Optimization of butyltin measurements for seawater, tissue and marine sediment samples. *Appl. Organometal. Chem.*, **3**, 105–113.

Stephenson, M.D., D.R. Smith, L.W. Hall, Jr, W.E. Johnson, P. Michel, J. Short, M. Waldock, R.J. Huggett, P. Seligman and S. Kola. 1987. An international intercomparison of butyltin determinations in mussel tissue and sediments. In: Proceedings of the Oceans '87 International Organotin Symposium, Halifax, Nova Scotia, Vol. 4, 28 September–1 October 1987, pp. 1334–1338.

Sullivan, J.J., J.D. Torkelson, M.M. Wekell, T.A. Hollingworth, W.L. Saxton, G.A. Miller, K.W. Panaro and A.D. Uhler. 1988. Determination of tri-n-butyltin and di-n-butyltin in fish as hydride derivatives by reaction gas chromatography. *Anal. Chem.*, **60**, 626–630.

Suyani, H., J. Creed, T. Davidson and J. Caruso. 1989a. Inductively coupled plasma mass spectrometry and atomic emission spectrometry coupled to high-performance liquid chromatography for speciation and detection of organotin compounds. *J. Chromatogr. Sci.*, **27**, 139–143.

Suyani, H., D. Heitkemper, J. Creed and J.

Caruso. 1989b. Inductively coupled plasma mass spectrometry as a detector for micellar liquid chromatography: speciation of alkyltin compounds. *Appl. Spectrosc.*, **43**, 962–967.

Tsuda, T., H. Nakanishi, T. Morita and J. Takebayashi. 1986. Simultaneous gas chromatographic determination of dibutyltin and tributyltin compounds in biological and sediment samples. *J. Assoc. Off. Anal. Chem.*, **69**, 981–984.

Unger, M.A., W.G. MacIntyre, J. Greaves and R.J. Huggett. 1986. GC determination of butyltins in natural waters by flame photometric detection of hexyl derivatives with mass spectrometric confirmation. *Chemosphere*, **15**, 461–470.

Valkirs, A.O., P.F. Seligman, G.J. Olson, F.E. Brinckman, C.L. Matthias and J.M. Bellama. 1987. Di- and tributyltin species in marine and estuarine waters. Inter-laboratory comparison of two ultratrace analytical methods employing hydride generation and atomic absorption or flame photometric detection. *Analyst*, **112**, 17–21.

Wade, T.L., B. Garcia-Romero and J.M. Brooks. 1988. Tributyltin contamination in bivalves from United States coastal estuaries. *Environ. Sci. Technol.*, **22**, 1488–1493.

A REVIEW OF ACUTE EFFECTS OF TRIBUTYLTIN COMPOUNDS ON AQUATIC BIOTA

9

Lenwood W. Hall, Jr[1] *and Steven J. Bushong*[2]

[1]*The University of Maryland, Agricultural Experiment Station, Wye Research and Education Center, Box 169, Queenstown, Maryland 21658, USA*

[2]*Holme, Roberts & Owen LLC, 401 Pearl St, Suite 400, Boulder, Colorado 80302, USA*

ABSTRACT

The objective of this chapter was to collect, synthesize, and interpret data on acute tributyltin (TBT) toxicity for both freshwater and saltwater organisms. Survival of test organisms (LC_{50}) was the most frequently used endpoint for these experiments, although other sublethal parameters were also evaluated. Tributyltin toxicity data were evaluated for 29 freshwater species and 56 saltwater species. Most of the freshwater data and approximately half of the saltwater data were generated from experiments using nominal concentrations of TBT. Acute toxicity data generated from studies using nominal concentrations are suspect because TBT is hydrophobic and tends to adsorb to most contact material. Tributyltin also degrades in solution over time. The most sensitive freshwater species tested was the coelenterate *Hydra* sp. [96-h LC_{50} = 0.5 $\mu g\ l^{-1}$ tributyltin oxide (TBTO)]. The bluegill (*Lepomis macrochirus*) was the most resistant freshwater species with a 96-h LC_{50} of 240 $\mu g\ l^{-1}$ TBTO. Copepods and mysids were the most sensitive saltwater organisms with acute effects reported at TBT concentrations of 0.4–0.5 $\mu g\ l^{-1}$. Adult oysters were reported to be the most resistant saltwater species with 96-h LC_{50} values >200 $\mu g\ l^{-1}$ TBT.

Organotin. Edited by M.A. Champ and P.F. Seligman. Published in 1996 by Chapman & Hall, London. ISBN 0 412 58240 6

9.1. INTRODUCTION

Organotin compounds have developed into important industrial commodities in the past 30 y because of their wide variety of uses. Major organic applications of organotin compounds are as polyvinyl chloride stabilizers, industrial catalysts, industrial and agricultural biocides, wood preserving agents, and antifouling agents (Piver, 1973; Zuckerman *et al.*, 1978; World Health Organization, 1980; Hall and Pinkney, 1985). Major organotin compounds released to the environment are those of triphenyltin, tricyclohexyltin, di-n-octyltin, di-n-butyltin, dimethyltin, and tri-n-butyltin (TBT) (Maguire, 1987). Triphenyltin and tricyclohexyltin species are agricultural pesticides. Di-n-octyltin compounds are used as stabilizers in some food wrappings. Di-n-butyltin compounds are used as catalysts in a number of industrial processes. Tri-n-butyltin compounds are used as slimicides in cooling water, as lumber preservatives, and as antifouling agents in paints for watercraft and docks. TBT-based antifouling paints are widely used on both recreational and commercial watercraft because of their excellent antifouling properties, long lifetime, and lack of corrosion problems.

TBT is an effective biocide because it is toxic to target fouling organisms at low concentrations. Unfortunately, these same toxic properties of TBT cause adverse effects to non-target organisms. Current interest in the toxicity of TBT to aquatic non-target organisms has resulted in numerous studies designed to evaluate the adverse effects of this biocide. The following reviews have summarized some of these TBT toxicity data: Hall and Pinkney, 1985; Laughlin and Linden, 1985; Thompson *et al.*, 1985; Oceans, 1986, 1987; Hall, 1988; Nicklin and Robson, 1988; Eisler, 1989. The quality of toxicity studies designed to evaluate the effects of TBT on aquatic biota has greatly improved in recent years due to the refinement of analytical methods used to measure TBT in aquatic media. Accurate toxicity experiments require the measurement of TBT in the test chambers (containing the test organisms) during the experiments. Continuous-flow experiments are the best and most accurate estimate of toxicity because they allow a more complete assessment of the relationship between exposure duration and effect. Continuous-flow tests result in more uniform and stable test conditions. Waste products of test organisms are also removed during continuous-flow tests.

Both acute and chronic laboratory toxicity tests are used for evaluating the effects of TBT on aquatic biota. This chapter will emphasize acute TBT toxicity studies; Laughlin *et al.* (Chapter 10) will focus on chronic toxicity studies. Acute tests are usually <14 d (depending on the life cycle of the organism) and in most cases these are 4-d experiments. The most common acute toxicity test is the acute lethality test where death of the organism is the final endpoint. Acute tests provide a practical means to determine the following under a set of given test conditions: (1) derive estimates of the upper limit of concentrations that produce toxic effects; (2) evaluate the relative toxicity of large numbers of test materials; (3) assess the relative sensitivity of different aquatic organisms or life stages to test materials; (4) determine the relationship of water quality to the toxicity of the test material; (5) develop an understanding of the concentration–dose and exposure pathways relationship between test organisms and exposure systems and the significance of duration of exposure to the test material (Macek *et al.*, 1978). Acute toxicity data are important in the regulatory process because these data are used by the US Environmental Protection Agency to develop water quality criteria for toxic chemicals.

The specific objective of this review chapter is to collect, synthesize, and interpret recent data on acute TBT toxicity for both freshwater and saltwater organisms. Toxicity data for freshwater and saltwater organisms are presented by trophic level: phytoplankton, zooplankton, macroinvertebrate, fish, and amphibian. Each paper has been objectively evaluated and presented in a tabular format using the following criteria: (1) test organism (common name and scientific name), (2) tributyltin compound, (3) result, (4) analytical method used for TBT analysis, (5) water quality data, (6) type of exposure, and (7) reference. Numerous papers were screened for these tables; however, if adequate information was not reported, then these references were omitted.

9.2. DATA ON ACUTE TRIBUTYLTIN TOXICITY IN FRESH WATER

Data on acute TBT toxicity were available for 29 species of freshwater organisms (Table 9.1). These included algae, a cyanobacterium, plankton, macroinvertebrates, fish, and amphibians. Most of the data on TBT toxicity in Table 9.1 were generated from experiments before 1984 where nominal concentrations were used to calculate acute toxicity values.

Data on acute toxicity of TBT were available for two freshwater species of phytoplankton, one natural mixed community of phytoplankton, and one cyanobacterium species. Wong *et al.* (1982) reported 4-h IC_{50} (concentration inhibiting primary productivity by 50%) values ranging from 16 to 20 $\mu g l^{-1}$ tributyltin oxide (TBTO) for the algae *Scenedesmus quadricauda* and *Ankistrodesmus falcatus*. These investigators also reported a lower 4-h IC_{50} value of 3 $\mu g l^{-1}$ TBTO for algal communities from Lake Ontario. In the same study, a 4-h IC_{50} value of 13 $\mu g l^{-1}$ TBTO was reported for the cyanobacterium *Anabaena flosaquae*.

Data on acute TBT toxicity were available for three freshwater species of zooplankton. A 48-h LC_{50} value of 1.67 $\mu g l^{-1}$ TBTO was reported for the water flea *Daphnia magna* (M & T Chemicals Co., 1976a). Meador (1986) reported 96– and 120-h LC_{50} values of 5.9 and 3.4 $\mu g l^{-1}$ tributyltin chloride (TBTCl), respectively, for *D. magna*. This investigator also reported that significantly lower concentrations of TBTCl (0.5 $\mu g l^{-1}$) caused significant alteration of photobehavior in *Daphnia* after 96–120 h of exposure. Brooke *et al.* (1986) reported a 48-h EC_{50} value of 4.3 $\mu g\ l^{-1}$ TBT for *D. magna*. Floch et al. (1964) reported 24-h and 72-h LC_{100} values of 120 and 60 $\mu g\ l^{-1}$ TBTO, respectively, for the water flea *Daphnia longispina*. These investigators also reported 23-h and 72-h LC_{50} values of 300 and 150 $\mu g\ l^{-1}$ tributyltin acetate (TBTA), respectively, for the same species. The ostracod *Cypridopsis hartwigi* was reported to be more resistant to TBT than *Daphnia* at exposures of <48 h (Floch *et al.*, 1964). These investigators reported 24- and 48-h LC_{100} values of 4000 and 2000 $\mu g\ l^{-1}$ TBTO, respectively, for this species. A 96-h LC_{100} of 120 $\mu g\ l^{-1}$ TBTO was also reported by these investigators for the ostracod. Floch *et al.* (1964) reported 24-, 48-, and 96-h LC_{100} values of 2000, 1000, and 150 $\mu g\ l^{-1}$ TBTA, respectively, for the ostracod *Cyprinopsis hartwigi*. The 4-d LC_{100} values for this species were similar for TBTO and TBTA.

Brooke *et al.* (1986) reported the following 96-h EC_{50} values for various freshwater invertebrates: 0.5 $\mu g\ l^{-1}$ TBT for the coelenterate *Hydra* sp.; 3.7 $\mu g\ l^{-1}$ TBT for the amphipod *Gammarus pseudolimnaeus*; 5.4 $\mu g\ l^{-1}$ TBT for the annelid *Lumbriculus variegatus*; and 10.2 $\mu g\ l^{-1}$ TBT for mosquito larvae *Culex* sp. The 96-h EC_{50} value of 0.5 $\mu g\ l^{-1}$ for the *Hydra* is the lowest value for acute toxicity reported for any freshwater species in Table 9.1. All data generated in the Brooke *et al.* (1986) study were from tests with measured concentrations of TBT.

Data on acute TBT toxicity were available for nine species of freshwater snails. Ritchie *et al.* (1964) reported 24-h LC_{50} values ranging from 40 to 50 $\mu g\ l^{-1}$ TBTO for the snail *Australorbis glabratus*. Frick and de Jimenez (1964) reported 24-h LC_{50} and LC_{100} values of 31 and 200 $\mu g\ l^{-1}$ TBTA, respectively, for *A. glabratus* eggs. Twenty-four-hour LC_{50} and LC_{100} values of 85 and 200 $\mu g\ l^{-1}$ TBTA, respectively, were reported for adult *A. glabratus*.

Table 9.1 Data for acute toxicity of TBT for freshwater organisms. (The following abbreviations are used for tributyltin (TBT) compounds: TBTO = bis(tributyltin) oxide; TBTCl = tributyltin chloride; TBTF = tributyltin fluoride; $TBTMSO_3$ = tributyltin methanesulfonate; TBTS = tributyltin sulfide; TBTA = tributyltin acetate.)

Test organism	*TBT compound*	*Result*	*Analytical method*	*Water quality data*[1]	*Type of exposure*	*Reference*
Alga: *Scenedesmus quadricauda*	TBTO in acetone	4-h IC_{50} = 16 μg l^{-1}	Nominal concentration	T = 20°C	Static	Wong *et al.*, 1982
Alga: *Ankistrodesmus falcatus*		4-h IC_{50} = 20 μg l^{-1}				Wong *et al.*, 1982
Cyanobacterium: *Anabaena flosaquae*		4-h IC_{50} = 13 μg l^{-1}				Wong *et al.*, 1982
Lake Ontario algae (natural community)		4-h IC_{50} = 3 μg l^{-1}				Wong *et al.*, 1982
Water flea: *Daphnia magna*	TBTO	48-h LC_{50} = 1.67 μg l^{-1}	Nominal concentration	Not reported	Static	M & T Chemicals Co., 1976a
Water flea: *Daphnia magna*	TBTO	48-h EC_{50} = 70 μg l^{-1} (immobilization)	Nominal concentration	Not reported	Not reported	Foster, 1981
Water flea: *Daphnia longispina*	TBTO	24-h LC_{100} = 120 μg l^{-1}	Nominal concentration	Not reported	Static	Floch *et al.*, 1964
		3-d LC_{100} = 60 μg l^{-1}	Nominal concentration	Not reported	Static	Floch *et al.*, 1964
	TBTA	23-h LC_{100} = 300 μg l^{-1}	Nominal concentration	Not reported	Static	Floch *et al.*, 1964
		3-d LC_{100} = 150 μg l^{-1}	Nominal concentration	Not reported	Static	Floch *et al.*, 1964
Ostracod: *Cypridopsis hartwigi*	TBTO	24-h LC_{100} = 4000 μg l^{-1}	Nominal concentration	Not reported	Static	Floch *et al.*, 1964
		48-h LC_{100} = 2000 μg l^{-1}	Nominal concentration	Not reported	Static	Floch *et al.*, 1964
		4-d LC_{50} = 120 μg l^{-1}	Nominal concentration	Not reported	Static	Floch *et al.*, 1964
	TBTA	24-h LC_{100} = 2000 μg l^{-1}	Nominal concentration	Not reported	Static	Floch *et al.*, 1964
		48-h LC_{100} = 1000 μg l^{-1}	Nominal concentration	Not reported	Static	Floch *et al.*, 1964
		4-d LC_{100} = 150 μg l^{-1}	Nominal concentration	Not reported	Static	Floch *et al.*, 1964
Ostracod: *Cypridopsis hartwigi*	TBTO	24-h LC_{100} = 4000 μg l^{-1}	Nominal concentration	pH = 7	Static	Deschiens and Floch, 1968
Water flea: *Daphnia longispina*	TBTO	24-h LC_{100} = 120 μg l^{-1}	Nominal concentration	pH = 7	Static	Deschiens and Floch, 1968
Water flea: *Daphnia magna*	TBTO	48-h EC_{50} = 4.3 μg l^{-1}	Matthias *et al.*, 1986	T = 22°C Hardness = 52 mg l^{-1} $CaCO_3$	Continuous flow	Brooke *et al.*, 1986
Water flea: *Daphnia magna* (adult)	TBTCl	96-h LC_{50} = 5.9 μg l^{-1} 120-h LC_{50} = 3.4 μg l^{-1} 0.5 μg l^{-1} caused significant alteration of photobehavior after 96–120 h of exposure	Nominal concentration	T = 21°C		Meador, 1986
Snail: *Australorbis glabratus*	TBTO	24-h exposure with 24-h recovery LC_{50} = 40; 50 μg l^{-1} (TBTO from two suppliers)	Nominal concentration	T = 25°C	Static	Ritchie *et al.*, 1964

Snail: *Australorbis glabratus*	TBTA; in methanol	24-h exposure followed by 24-h recovery Eggs (1–6 h old): LC_{50} = 31 μg l^{-1} LC_{100} = 200 μg l^{-1} Mature snails: LC_{50} = 85 μg l^{-1} LC_{100} = 200 μg l^{-1}	Nominal concentration	Dechlorinated tap water	Static	Frick and de Jimenez, 1964
Snail: *Australorbis glabratus*	TBTO	6-h LC_{50} = 410 μg l^{-1} 6-h LC_{100} = 2000 μg l^{-1}	Not reported	Not reported	Static	Seiffer and Schoof, 1967
	TBTA in 1.5% ethanol	6-h LC_{50} = 290 μg l^{-1} 6-h LC_{100} = 1000 μg l^{-1}	Nominal concentration			Seiffer and Schoof, 1967
Snail: *Australorbis glabratus* (adults)	TBTA in alcohol	24-h exposure followed by 72-h recovery: LC_{50} = 170 μg l^{-1} LC_{90} = 300 μg l^{-1}	Nominal concentration	T = 23 ± 2°C	Static	Camey and Paulini, 1964
Snail: *Australorbis glabratus* and *Bulinus contortus*	TBTO	24-h exposure followed by 72-h recovery: LC_{100} = 75 μg l^{-1}	Nominal concentration		Static	Floch *et al.*, 1964
		48-h exposure followed by 72-h recovery:				
	TBTO	LC_{100} = 50 μg l^{-1}				
	TBTA	LC_{100} = 150 μg l^{-1}				
		5-d exposure followed by 72-h recovery:				
	TBTO	LC_{100} = 30 μg l^{-1}				
	TBTA	LC_{100} = 30 μg l^{-1}				
Annelid: *Lumbriculus variegatus*	TBTO	96-h EC_{50} = 5.4 μg l^{-1}	Matthias *et al.*, 1986	T = 18°C Hardness = 52 mg l^{-1} $CaCO_3$	Continuous flow	Brooke *et al.*, 1986
Coelenterate: *Hydra* sp.	TBTO	96-h EC_{50} = 0.5 μg l^{-1}	Matthias *et al.*, 1986	T = 26°C Hardness = 51 mg l^{-1} $CaCO_3$	Static	Brooke *et al.*, 1986
Amphipod: *Gammarus pseudolimnaeus*	TBTO	96-h LC_{50} = 3.7 μg l^{-1}	Matthias *et al.*, 1986	T = 18°C Hardness = 52 mg l^{-1} $CaCO_3$	Continuous flow	Brooke *et al.*, 1986
Mosquito larvae: *Culex* sp.	TBTO	96-h EC_{50} = 10.2 μg l^{-1}	Matthias *et al.*, 1986	T = 17°C Hardness = 52 mg l^{-1} $CaCO_3$	Static	Brooke *et al.*, 1986
Snails: *Biomphalaria glabratus*, *Biomphalaria camerounensis*, *Bulinus contortus* (adults)	TBTO	24-h exposure followed by 72-h recovery: LC_{100} = 75 μg l^{-1}	Nominal concentration	pH = 7	Static	Deschiens and Floch, 1968
		3-d exposure followed by 3-d recovery:				
	TBTO	LC_{100} = 50 μg l^{-1}				
Eggs	TBTO	24-h LC_{50} = 100 μg l^{-1}				Deschiens and Floch, 1968

Table 9.1 Continued

Test organism	*TBT compound*	*Result*	*Analytical method*	*Water quality data*[1]	*Type of exposure*	*Reference*
Snail: *Biomphalaria glabrata* (adults)	TBTO	2–5 d LC_{100} = 100 μg l^{-1}	Nominal concentration	Not reported	Static renewal	Ritchie *et al.*, 1974
Snail: *Biomphalaria glabrata* (adults)	^{14}C-TBTO in slow-release pellets	After 96 and 120 h in 60 μg l^{-1} head–foot tissue had BCF values of 28.6 and 23.7, respectively; at 3 μg l^{-1} 120-h BCF was 47.5; ^{14}C-TBTO was found first in head–foot tissue and later in reproductive and digestive organs	Liquid scintillation counter	Not reported	Static	Allen *et al.*, 1980
		Exposure to 60 μg l^{-1} resulted in rapid cessation of browsing activity and mobility; retraction also occurred; animals died within a few days				
Snail: *Biomphalaria glabrata*	TBTO in slow-release pellets	50% mortality at 5400 μg l^{-1} after 6 d; 50% mortality at 2300 μg l^{-1} after 5.2 d	Nominal concentration	Not reported	Static	Cardarelli and Evans, 1980
Snail: *Biomphalaria glabrata*	$TBTMSO_3$; in dimethyl sulfoxide	24-h exposure followed by 48-h recovery: LC_{50} = 10–50 μg l^{-1}	Nominal concentration	Not reported	Static	Smith *et al.*, 1979
	$TBTMSO_3$	LC_{50} = 25–50 μg l^{-1}				
Snail: *Biomphalaria glabrata*	TBTO	24-h exposure followed by 144-h recovery: LC_{50} = 50–100 μg l^{-1}	Nominal concentration	pH = 5.5	Static	Hopf *et al.*, 1967
	TBTA in Tween 80	24-h exposure followed 48-h recovery: LC_{50} = 100–300 μg l^{-1}	Nominal concentration	pH = 5.5	Static	
Snails: *Biomphalaria camerounensis* and *Lymnaea* sp.	TBTO	100% mortality occurred in 4–7 d at 15 μg l^{-1}	Nominal concentration	T = 23°C pH = 6.4	Field tests: experiments conducted in fish culture ponds	Deschiens *et al.*, 1966a

Snail: *Biomphalaria glabrata*	TBTO	100% mortality in 6 d at 15 $\mu g\ l^{-1}$; 100% mortality in 4 d at 30 $\mu g\ l^{-1}$	Nominal concentration	Not reported	Field tests: experiments conducted in fish culture ponds	Deschiens *et al.*, 1966b
Snail: *Biomphalaria sudanica*	TBTA	24-h exposure followed by 48-h recovery:	Nominal concentration	T = 22–24°C pH = 7.5–7.8	Static	Webbe and Sturrock, 1964
2–3-d-old eggs		LC_{100} = 250 $\mu g\ l^{-1}$				
4–6 weeks old		LC_{50} = 14 $\mu g\ l^{-1}$				
9–11 weeks old		LC_{50} = 34 $\mu g\ l^{-1}$				
Snail: *Bulinus nasutus*	TBTA	LC_{100} = 250 $\mu g\ l^{-1}$				Webbe and Sturrock, 1964
2–3-d-old eggs		LC_{50} = 150 $\mu g\ l^{-1}$				
5.5–8 weeks old		LC_{50} = 33 $\mu\ l^{-1}$				
11–32 weeks old						
Asiatic clam: *Corbicula fluminera* larvae	TBTO	24-h LC_{50} = 2100 $\mu g\ l^{-1}$	Nominal concentration	Not reported	Not reported	Foster, 1981
Rainbow trout: *Salmo gairdneri* yolk sac fry	TBTCl	10–12-d LC_{100} = 5 $\mu g\ l^{-1}$	Nominal concentration	T = 13–15°C DO = 9 ± 1 mg l^{-1} pH = 6.8 Hardness = 94–102 mg l^{-1} $CaCO_3$	Continuous flow	Seinen *et al.*, 1981
	TBTO	24-h LC_{50} = 28 $\mu g\ l^{-1}$ 48-h LC_{50} = 21 $\mu g\ l^{-1}$	Nominal concentration	T = 18°C Hardness = 250 mg l^{-1} $CaCO_3$	24-h static renewal	Alabaster, 1969
Bluegill: *Lepomis macrochirus*	TBTO	96-h LC_{50} = 240 $\mu g\ l^{-1}$	Nominal concentration	Not reported	Not reported	Foster, 1981
Guppy: *Lebistes reticulatus*	TBTO	7-d LC_{100} = 60 $\mu g\ l^{-1}$	Nominal concentration	Not reported	Static	Cardarelli, 1973
	TBTF	7-d LC_{100} = 30 $\mu g\ l^{-1}$	Nominal concentration	Not reported	Static	Cardarelli, 1973
Goldfish: *Carassius auratus*	TBTO	24-h LC_{100} = 75 $\mu g\ l^{-1}$	Nominal concentration	Not reported	Static	Floch *et al.*, 1964; Deschiens and Floch, 1968
	TBTA	24-h LC_{100} = 75 $\mu g\ l^{-1}$				
Guppy: *Lebistes reticulatas*	TBTO	24-h LC_{100} = 75 $\mu g\ l^{-1}$				
	TBTA	24-h LC_{100} = 75 $\mu g\ l^{-1}$ 48-h LC_{100} = 30 $\mu g\ l^{-1}$				
Cichlids: *Tilapia nilotica Hemichromis* sp. (adults and juveniles)	TBTO	Application of 45 $\mu g\ l^{-1}$ caused 70% mortality after 24–48 h	Nominal concentration	T = 23°C pH = 6.4	Field experiments conducted in fish culture ponds	Deschiens *et al.*, 1966a
Cichlids: *Tilapia nilotica* (adults and juveniles)	TBTO	Application of 15 $\mu g\ l^{-1}$ caused no mortality after 6 d; 100% mortality occurred at 30 $\mu g\ l^{-1}$ after 4 d	Nominal concentration	Not reported	Field experiments conducted in fish culture ponds	Deschiens *et al.*, 1966b

Table 9.1 Continued

Test organism	*TBT compound*	*Result*	*Analytical method*	*Water quality data*[1]	*Type of exposure*	*Reference*
Rainbow trout: *Salmo gairdneri*	TBTO in acetone	24-h EC_{50} = 31µg l^{-1} (loss of positive rheotaxis). An increase of the packed cell volume and hemoglobin concentration occurred after a 21.5-h exposure to 53 µg l^{-1}; a 10-min exposure to 5850 µg l^{-1} caused gill epithelium to separate from the basement membrane and pillar cells; flattening of bile-duct columnar epithelial cells and separation from connective tissue occurred after 5-d exposure to 1.7 µg l^{-1}; destruction of corneal epithelium occurred after 7-d exposure to 1.7 µg l^{-1}	Not reported	T = 16°C DO = 7.6 mg l^{-1} pH = 7.6	Continuous flow	Chliamovitch and Kuhn, 1977
Cichlid: *Tilapia rendalli*		24-h EC_{50} = 53 µg l^{-1} (loss of positive rheotaxis)		T = 25°C pH = 7.9 DO = 7.1 mg l^{-1}		Chliamovitch and Kuhn, 1977
Goldfish: *Carassius auratus*	TBTA in acetone	24-h LC_{100} = 400 µg l^{-1}	Nominal concentration	T = 21°C pH = 6.8	Static	Gras and Rioux, 1965

American eel: *Anguilla anguilla* (young)	TBTA in acetone	24-h LC_{100} = 400 μg l^{-1}	Nominal concentration	T = 25°C pH = 6.8	Static	Gras and Rioux, 1965
Bluegill: *Lepomis macrochirus*	TBTO	96-h LC_{50} = 7.6 μg l^{-1}	Nominal concentration	Not reported	Static	M & T Chemicals Co., 1976b
Rainbow trout: *Salmo gairdneri*	TBTO	96-h LC_{50} = 6.9 μg l^{-1}	Nominal concentration	Not reported	Static	M & T Chemicals Co., 1978
Channel catfish: *Ictalurus punctatus*	TBTO	96-h LC_{50} = 12 μg l^{-1}	Nominal concentration	Not reported	Static	M & T Chemicals Co., 1976c
Rainbow trout: *Salmo gairdneri*	TBT	96-h LC_{50} = 3.4 μg l^{-1}	Maguire *et al.*, 1982	T = 15.5°C DO = 8.9 mg l^{-1} pH = 7.7	Continuous flow	Martin *et al.*, 1991
Lake trout *Salvelinus namaycush*	TBT	96-h LC_{50} = 12.6 μg l^{-1}				Martin *et al.*, 1991
Rainbow trout (juvenile) *Salmo gairdneri*	TBTO	96-h LC_{50} = 3.9 μg l^{-1} TBT	Matthias *et al.*, 1986	T = 16°C Hardnesss = 51 mg l^{-1} $CaCO_3$	Continuous flow	Brooke *et al.*, 1986
Fathead minnow (juvenile) *Pimephales promelas*	TBTO	96-h LC_{50} = 2.6 μg l^{-1} TBT	Matthias *et al.*, 1986	T = 24°C Hardness = 52 mg l^{-1} $CaCO_3$		Brooke *et al.*, 1986
Channel catfish (juvenile) *Ictalurus punctatus*	TBTO	96-h LC_{50} = 5.5 μg l^{-1} TBT		T = 18°C Hardness = 52 mg l^{-1} $CaCO_3$		Brooke *et al.*, 1986
European frog: *Rana temporaria* (post-gastrula egg)	TBTO in acetone	40% mortality occurred after 5-d exposure to 30 μg l^{-1};	Nominal concentration	Not reported	Static renewal	Laughlin and Linden, 1982
	TBTF in acetone	50% mortality occurred after 5-d exposure to 30 μg l^{-1}; % body water decreased at 30 μg l^{-1}				
European frog: *Rana temporaria* (tadpole)	TBTO	24-h LC_{100} = 75 μg l^{-1}	Nominal concentration	pH = 7	Static	Deschiens and Floch 1968

[1]Hardness is a measure of the buffering capacity of water in milligrams per litre of $CaCO_3$.

glabratus. Seiffer and Schoof (1967) reported 6-h LC_{50} values of 410 and 290 μg l^{-1} for TBTO and TBTA, respectively, for this snail species. Camey and Paulini (1964) reported a 24-h LC_{50} of 170 μg l^{-1} TBTA for *A. glabratus*. Floch *et al*. (1964) reported 24-, 48-, and 120-h LC_{100} values of 75, 50, and 30 μg l^{-1} TBTO. All exposures were followed by a 72-h recovery period.

Deschiens and Floch (1968) reported 24-h LC_{100} values of 75 μg l^{-1} TBTO for three snail species (*Biomphalaria glabratus, B. camerounensis*, and *Bulinus contortus*). Seventy-two-hour LC_{100} values of 50 μg l^{-1} TBTO were also reported for these species. Ritchie *et al*. (1974) reported 2–5-d LC_{100} values of 100 μg l^{-1} TBTO for the snail *B. glabratus*.

Allen *et al*. (1980) reported bioconcentration factors (BCFs) of 28.6 and 23.7 in the head–foot tissue of the snail *B. glabrata* after 96 and 120 h of exposure, respectively, to 60 μg l^{-1} TBTO. Rapid cessation of browsing activity and mobility and retraction also occurred at 60 μg l^{-1} TBTO. A BCF of 47.5 was reported in this species after 120-h exposures to 3 μg l^{-1} TBTO. Allen *et al*. (1980) also reported that ^{14}C TBTO was first found in the head–foot tissue and later in the reproductive and digestive organs. Cardarelli and Evans (1980) reported 50% mortality of *B. glabrata* at 5400 μg l^{-1} TBTO after 6 d. Twenty-four-hour LC_{50} values of 10–50 μg l^{-1} $TBTMSO_3$ (in dimethyl sulfoxide), 50–100 μg l^{-1} TBTO, and 100–300 μg l^{-1} TBTA were reported for the snail *B. glabrata* (Hopf *et al*., 1967; Smith *et al*., 1979).

Deschiens *et al*. (1966a) reported 100% mortality in 4–7 d at 15 μg l^{-1} TBTO for the snails *B. camerounensis* and *Lymnaea* sp. These field experiments were conducted in fish culture ponds. These same investigators conducted similar experiments with *B. glabrata* and reported 100% mortality in 6 d at 15 μg l^{-1} TBTO (Deschiens *et al*., 1966b). This pond water likely contained naturally occurring organic material, bacteria, phytoplankton, and particulate material that may have absorbed or degraded the nominal TBT test solution used in these experiments.

Toxicity studies conducted by Webbe and Sturrock (1964) were designed to evaluate the effects of snail age on the toxicity of TBT. Twenty-four-hour LC_{100} values of 250 μg l^{-1} TBTA were reported for 2–3 d old *B. sudanica* eggs. Twenty-four-hour LC_{50} values ranging from 14 to 34 μg l^{-1} TBTA were reported for 4- to 11-week-old *B. sudanica*. These investigators also reported a 24-h LC_{100} of 250 μg l^{-1} TBTA for 2- to 3-d-old *B. nasutus* eggs. Twenty-four-hour LC_{50} values ranging from 33 to 150 μg l^{-1} were reported for 5.5–32 week old *B. nasutus*.

One study was conducted to evaluate the acute effects of TBTO on the larvae of the Asiatic clam *Corbicula fluminera*. Foster (1981) reported a 24-h LC_{50} of 2100 μg l^{-1} TBTO for the larval stage of this clam species. This is the highest LC_{50} reported from a 24-h test with any freshwater organism listed in Table 9.1. The study has limitations that must be considered when evaluating the data. Nominal concentrations were used for evaluating the 24-h LC_{50}; therefore, much lower concentrations may have been causing an effect. Failure to report the type of exposure also makes these data difficult to interpret.

Data on acute TBT toxicity were available for 11 species of freshwater fish. Five separate studies were conducted with rainbow trout (*Salmo gairdneri*). Ninety-six-hour LC_{50} values ranging from 3.4 to 3.9 μg l^{-1} TBT were reported for juvenile and adult *S. gairdneri* (Brooke *et al*., 1986; Martin et al., 1991). M & T Chemicals Co. (1978) reported a 96-h LC_{50} of 6.9 μg l^{-1} TBTO for this species. In longer acute experiments, a 10–12-d LC_{100} of 5 μg l^{-1} TBTCl was reported for rainbow trout (Seinen *et al*., 1981). Chliamovitch and Kuhn (1977) reported a 24-h EC_{50} of 31 μg l^{-1} TBTO (loss of positive rheotaxis) for rainbow trout. These investigators also reported an increase in the packed cell volume and hemoglobin concentration after a 21.5-h exposure to 53 μg l^{-1} TBTO. A 10-min exposure to 5850 μg l^{-1}

TBTO caused gill epithelium to separate from the basement membrane and pillar cells. In rainbow trout flattening of columnar epithelial cells of the bile duct and separation from connective tissue occurred after a 5-d exposure to 1.7 $\mu g\ l^{-1}$ TBTO. Destruction of corneal epithelium occurred after a 7-d exposure to 1.7 $\mu g\ l^{-1}$ TBTO.

Two studies of acute TBT toxicity were conducted with bluegill (*Lepomis macrochirus*). Nominal TBTO concentrations were used in both studies. M & T Chemicals Co. (1976b) reported a 96-h LC_{50} of 7.6 $\mu g\ l^{-1}$ TBTO for the bluegill. Foster (1981) reported a 96-h LC_{50} of 240 $\mu g\ l^{-1}$ TBTO for the same species. The 30-fold difference in toxicity between the two studies may be caused by differences in age, sex, genotypes, acclimation of test species, or health of test species. The 96-h LC_{50} of 240 $\mu g\ l^{-1}$ reported by Foster (1981) is generally higher than acute values reported for other fish species listed in Table 9.1.

Cardarelli (1973) reported 7-d LC_{100} values of 60 and 30 $\mu g\ l^{-1}$ TBTO and tributyltin fluoride (TBTF), respectively, for the guppy *Lebistes reticulatus*. Other investigators have reported two 24-h and 48-h LC_{100} values of 75 $\mu g\ l^{-1}$ TBTO, 75 $\mu g\ l^{-1}$ TBTA, and 30 $\mu g\ l^{-1}$ TBTA for this species (Floch *et al.*, 1964; Deschiens and Floch, 1968). Twenty-four-hour LC_{100} values of 75 $\mu g\ l^{-1}$ TBTO and TBTA were reported for the goldfish *Carassius auratus* (Floch *et al.*, 1964; Deschiens and Floch, 1968). In another study, a much higher 24-h LC_{100} value of 400 $\mu g\ l^{-1}$ TBTA was reported for this species (Gras and Rioux, 1965).

Deschiens *et al.* (1966a) reported 70% mortality for the cichlids *Tilapia nilotica* and *Hemichromis* sp. after 24–48-h exposure to 45 $\mu g\ l^{-1}$ TBTO. TBTO concentrations of 30 $\mu g\ l^{-1}$ caused 100% mortality to *T. nilotica* after 4 d of exposure (Deschiens *et al.*, 1966b). Behavioral modification (loss of positive rheotaxis) was reported for the cichlid *T. rendalli* after 24 h of exposure to 53 $\mu g\ l^{-1}$ TBTO (Chliamovitch and Kuhn, 1977).

Two studies on acute TBT toxicity were conducted with channel catfish (*Ictalurus punctatus*). Brooke *et al.* (1986) reported a 96-h LC_{50} of 5.5 $\mu g\ l^{-1}$ TBTO for juvenile channel catfish. A higher 96-h LC_{50} of 12 $\mu g\ l^{-1}$ TBTO was reported for adult channel catfish (M & T Chemicals Co., 1976c).

Data for acute TBT toxicity were available for three other species of freshwater fish. Gras and Rioux (1965) reported a 24-h LC_{100} value of 400 $\mu g\ l^{-1}$ TBTA for young American eels (*Anguilla anguilla*). Ninety-six-hour LC_{50} values of 12.6 $\mu g\ l^{-1}$ TBT and 2.6 $\mu g\ l^{-1}$ TBTO were reported for lake trout (*Salvelinus namaycush*) and fathead minnow (*Pimephales promelas*), respectively (Brooke *et al.*, 1986; Martin et al., 1991).

Two studies on acute TBT toxicity were conducted with the European frog (*Rana temporaria*). Laughlin and Linden (1982) reported 50% mortality to European frog post-gastrula eggs after 5-d exposures to 30 $\mu g\ l^{-1}$ TBTF. These investigators also reported that the percentage of body water decreased at these conditions. Deschiens and Floch (1968) reported a 24-h LC_{100} value of 75 $\mu g\ l^{-1}$ for *R. temporaria* tadpoles.

9.3. DATA ON ACUTE TRIBUTYLTIN TOXICITY IN SALTWATER

Data for acute TBT toxicity have been reported for 56 species of saltwater organisms including bacteria, plankton, crustaceans, polychaetes, molluscs, and fish (Table 9.2). Over 50% of the experiments reported used nominal test concentrations, and many experiments had no renewal of test water. The tests encompassed different life stages and used either lethality or a variety of sublethal endpoints. The most common experiments were short-term ($\leq$96-h) tests using mortality as a measure of toxicity. Tests up to 15 d in duration were included in this review. Some experiments $\leq$15 d in duration would be considered chronic tests for organisms with short life cycles (e.g. phytoplankton or copepods).

Table 9.2 TBT acute toxicity data for saltwater organisms (the following abbreviations are used for TBT compounds: TBTO = bis(tributyltin)oxide; TBTBr = tributyltin bromide; TBTCl = tributyltin chloride; TBTF = tributyltin fluoride; TBTS = tributyltin sulfide; and TBTA = tributyltin acetate. The following abbreviations were used for analytical methods: AAS = atomic absorption spectrometry; AAS–GF = atomic absorption spectrometry with graphite furnace; GC = gas chromatography; and ICAP = inductively coupled argon plasma. SAL is an abbreviation for salinity.)

Test organism	*TBT compound*	*Result*	*Analytical method*	*Water quality data*	*Type of exposure*	*Reference*
Luminescent bacterium: *Photobacterium phosphoreum*	TBTCl in ethanol	5-min EC_{50} = 0.06 μM (20 μg l^{-1} TBTCl) 15-min EC_{50} = 0.02 μM (7 μg l^{-1} TBTCl)	Not reported	T = 15°C SAL = 2% NaCl	Static	Dooley and Kenis, 1987
	TBTBr in ethanol	5-min EC_{50} = 0.13 μM (48 μg l^{-1} TBTBr) 15-min EC_{50} = 0.06 μM (22 μg l^{-1} TBTBr) (EC_{50} = 50% reduction in light output by bacterium)				
Alga: *Skeletonema costatum*	TBTO (BioMet Red in acetone)	14-d EC_{50}>0.125<0.250 μg l^{-1} (based on dry cell wt)	Not reported	T = 20°C	Static	M & T Chemicals Co., 1981a
Alga: *Skeletonema costatum*	TBTO (alkyl–source in acetone)	14-d EC_{50} = 0.064 μg l^{-1} (based on dry cell wt)				M & T Chemicals Co., 1981a
Alga: *Skeletonema costatum*	TBTO in acetic acid or ethanol	5-d exposure; algistatic concentrations (cell division ceased but population recovered upon transfer to uncontaminated medium) = 1–18 μg l^{-1}	Not reported	Not reported	Static	Thain, 1983
Alga: *Tetraselmin suecica*		5-d exposure; algistatic concentrations = 560–1000 μg l^{-1}				Thain, 1983
Alga: *Skeletonema costatum*	TBTCl in acetone	5-d EC_{50}<1.0 μg l^{-1} (cell density and chlorophyll) 1.68 μg l^{-1} immediately reduced photosynthesis by 50%, the 24-h photosynthetic EC_{50} = 1.22 μg l^{-1}	Nominal concentrations	T = 20 ± 4°C SAL = 20‰	Static	Ho, 1984

Alga: *Prorocentrum mariae-bebouriae*		5-d EC_{50} = 2.41 μg l^{-1} (cell density) 5-d EC_{50} = 2.62 μg l^{-1} (chlorophyll) 2.73 μg l^{-1} immediately reduced photosynthesis by 50%, the 24-h photosynthetic EC_{50} = 1.81 μg l^{-1}				Ho, 1984
Alga: *Isochrysis galbana*		5-d EC_{50} = 9.58 μg l^{-1} (cell density) 5-d EC_{50} = 1.14 μg l^{-1} (chlorophyll) 10.23 μg l^{-1} immediate reduced photosynthesis 50%, the 24-h photosynthetic EC_{50} = 4.69 μg l^{-1}				Ho, 1984
Natural assemblages of York River phytoplankton:						Ho, 1984
Surface water phytoplankton		Photosynthetic 2-h EC_{50} = 5.99–7.26 μg l^{-1}				
Deepwater phytoplankton		Photosynthetic 2-h EC_{50} 5.62–7.57 μg l^{-1}				
Alga: *Skeletonema costatum*	TBTO in acetone	72-h EC_{50} = 0.33 μg l^{-1} (cell density)	Nominal concentrations (stocks measured by ICAP spectro-photometry)	T = 20 ± 0.5°C SAL = 30‰	Static	Walsh *et al.*, 1985
	TBTA in acetone	72-h EC_{50} = 0.36 μg l^{-1}				
	TBTCl in acetone	72-h EC_{50} = 0.36 μg l^{-1}				
	TBTF in acetone	72-h EC_{50} = 0.25–0.5 μg l^{-1}				
Alga: *Thalassiosira pseudonana*	TBTO in acetone	72-h EC_{50} = 1.03 μg l^{-1}				Walsh *et al.*, 1985
	TBTA in acetone	72-h EC_{50} = 1.28 μg l^{-1}				
Alga: *Gymnodinium splendens*	TBTO	72-h exposures to ≥1.5 μg l^{-1} killed cultures	Nominal concentrations (stocks measured by AAS–GF)	T = 18°C	Static	Salazar, 1985
Alga: *Dunaliella* sp.		1.5 μg l^{-1} significantly reduced growth in 72 h				Salazar, 1985

Table 9.2 Continued

Test organism	*TBT compound*	*Result*	*Analytical method*	*Water quality data*	*Type of exposure*	*Reference*
Alga: *Phaeodactylum tricornutum*		No observable effect at 1.5–6.0 μg l^{-1} after 72 h				Salazar, 1985
Alga: *Skeletonema costatum* *Pavlova lutheri* *Dunaliella tertiolecta*	TBTO in glacial acetic acid	All cultures dead in 48 h at 5 μg l^{-1} TBTO. Growth significantly reduced in all species after 15 d at 0.1 μg l^{-1} TBTO	Nominal concentrations (initial test solutions confirmed by AAS; one measured after experiment showed a 50% loss of TBT over time)	SAL = 34–40‰	Static	Beaumont and Newman, 1986
Hydroid: *Campanularia flexuosa*	TBTF in acetone	Growth stimulation (hormesis) occurred at 0.01 and 0.1 μg l^{-1} after 11 d: 100% inhibition of growth occurred at 1.0 μg l^{-1} after 11 d.	Nominal concentrations	T = 20°C SAL = 35.1‰	Static renewal	Stebbing, 1981
Harpacticoid: *Nitrocra spinipes*	TBTO in acetone TBTF in acetone	96-h LC_{50} = 2 μg l^{-1} 96-h LC_{50} = 2 μg l^{-1}	Nominal concentrations	T = 21 ± 1°C SAL = 7‰	Static	Linden *et al.*, 1979
Copepod: *Acartia tonsa*	TBTO in acetone	96-h LC_{50} = 1 μg l^{-1} 144-h EC_{50} = 0.4 μg l^{-1} (moribundity and mortality)	AAS–GF	T = 20 ± 0.5°C	Static renewal	U'ren, 1983
Copepod: *Acartia tonsa* (subadult) Copepod: *Eurytemora affinis* (subadult)	TBTCl	48-h LC_{50} = 1.1 μg l^{-1} TBT Repeated experiments: 48-h LC_{50} = 1.4 μg l^{-1} TBT 72-h LC_{50} = 0.6 μg l^{-1} TBT 48-h LC_{50} = 2.5 μg l^{-1} TBT 72-h LC_{50} = 0.5 μg l^{-1} TBT	Matthias *et al.*, 1986	T = 20 ± 1°C SAL = 10–12‰	Continuous flow	Bushong *et al.*, 1988
Copepod: *Eurytemora affinis* (eggs→adults →nauplii)	TBTCl	0.088 μg l^{-1} TBT significantly reduced survival of nauplii within 6 d of a 13-d test. In a second experiment, survival was significantly reduced at 0.224 μg l^{-1} TBT after 13 d but no significant effects were reported at ≤0.1 μg l^{-1}	Matthias *et al.*, 1986	T = 17–20°C SAL = 9.9–10.7‰ (exp 1) SAL = 13.2–16.1‰ (exp 2)	Continuous flow	Hall *et al.*, 1988

Copepod: *Acartia tonsa*	TBTO	100% mortality at 1 $\mu g\ l^{-1}$ within 24 h	Measurements made periodically at some treatments (Valkirs *et al.*, 1986)	T = 13–14°C SAL = 33–34‰	Static	Salazar and Salazar, 1989
Copepod: *Acartia tonsa*	TBTO in glacial acetic acid	Small but significant reductions in egg production after 120 h at concentrations $\geq 0.01\ \mu g\ l^{-1}$	Nominal	T = 18°C SAL = 28‰	Static	Johansen and Mohlenberg, 1987
Copepod: *Acartia tonsa* (nauplii→subadult/adult)	TBTCl	Repeated 6-d experiments; (measured response is survival) LOEC = 0.023 $\mu g\ l^{-1}$ TBT NOEC = 0.012 $\mu g\ l^{-1}$ TBT Chronic value = 0.017 $\mu g\ l^{-1}$ TBT LOEC = 0.024 $\mu g\ l^{-1}$ TBT NOEC = 0.01 $\mu g\ l^{-1}$ TBT Chronic value = 0.016 $\mu g\ l^{-1}$ TBT	Unger *et al.*, 1986	T = 20–21.4°C SAL = 10–12‰	Continuous flow	Bushong *et al.*, 1990
Shrimp: *Crangon crangon*	TBTO in acetic acid or ethanol	Adult: 96-h LC_{50} = 41 $\mu g\ l^{-1}$ Larvae: 96-h LC_{50} = 2 $\mu g\ l^{-1}$	Not reported	Not reported	Static renewal	Thain, 1983
Crab: *Carcinus maenas* (larvae)		96-h LC_{50} = 10 $\mu g\ l^{-1}$				Thain, 1983
American lobster: *Homarus americanus* (larvae)	TBTO in acetone	24-h LC_{100} = 20 $\mu g\ l^{-1}$ 6-d LC_{100} = 5 $\mu g\ l^{-1}$	Nominal concentrations	T = 20°C SAL = 32‰	Static renewal	Laughlin and French, 1980
Shore crab: *Hemigrapsus nudus* (larvae)	TBTO in acetone	2-d LC_{100} = 500 $\mu g\ l^{-1}$ Estimated 50% mortality at 25 $\mu g\ l^{-1}$ for 6.2 d		T = 15°C SAL = 32‰		Laughlin and French, 1980
Amphipod: *Orchestia traskiana*	TBTO in acetone	Lethality for 9-d exposure: 47% @ 6 $\mu g\ l^{-1}$ 80% @ 10 $\mu g\ l^{-1}$ 93% @ 15 $\mu g\ l^{-1}$	GC	T = 18°C SAL = 30‰	Static renewal	Laughlin *et al.*, 1982
	TBTF in acetone	Lethality for 9-d exposure: <50% @ 6 $\mu g\ l^{-1}$ 87% @ 10 $\mu g\ l^{-1}$ 100% @ 150 $\mu g\ l^{-1}$				
Baltic amphipod: *Gammarus oceanicus*	TBTF-based paint panels (two types)	5-d LC_{100} = 5 $\mu g\ l^{-1}$ (as TBTO equivalent); 85% mortality occurred from 6-d exposure to 3 $\mu g\ l^{-1}$ (as TBTO equivalent)		SAL = 7‰	Static renewal	Laughlin *et al.*, 1982

Table 9.2 Continued

Test organism	*TBT compound*	*Result*	*Analytical method*	*Water quality data*	*Type of exposure*	*Reference*
Mud crab: *Rhithropanopeus harrisii* zoeae	TBTO in acetone	15-d exposure: 63% mortality and delayed metamorphosis occurred at 25 μg l^{-1}; decreased weight occurred at ≥15 μg l^{-1}	Nominal concentrations	T = 25°C SAL = 15‰	Static renewal	Laughlin *et al.*, 1983
	TBTS in acetone	15-d exposure: 22% mortality at 22 μg l^{-1} delayed metamorphosis occurred and decreased weight occurred at ≥20 μg l^{-1}; 74% mortality occurred at 30 μg l^{-1}				
Baltic amphipod: *Gammarus oceanicus* (females with embryos in marsupia)	TBTO in acetone	Exposure to 3 μg l^{-1} resulted in 100% mortality in adults in 16 d (TBTO)	GC	T = 10–15°C	Static renewal	Laughlin *et al.*, 1984
Amphipod: *Gammarus* sp.	TBTCl	Larvae: 96-h LC_{50} = 1.3 μg l^{-1} Adults: 96-h LC_{50} = 5.3 μg l^{-1}	Matthias *et al.*, 1986	T = 20 ± 1°C SAL = 10–12‰	Continuous flow	Bushong *et al.*, 1988
Grass shrimp: *Palaemonetes* sp.		40% mortality at 31 μg l^{-1} after 96 h; 96-h LC_{50} >31 μg l^{-1}				Bushong *et al.*, 1988
Grass shrimp: *Palaemonetes pugio*	TBTO	Avoidance did not occur at total organic Sn concentrations ranging from 2.3 to 30 μg l^{-1} (5.6 to 75 μg l^{-1} TBTO)	AAS–GF	T = 22–27.5°C SAL = 9.9–11.2‰	Continuous flow	Pinkney *et al.*, 1985
Grass shrimp: *Palaemonetes pugio*	TBTO in acetone in triethylene glycol	96-h LC_{50} = 20 μg l^{-1}	Nominal concentrations	T = 22 ± 1°C or 25 ± 1°C	Continuous flow	Clark *et al.*, 1987
Amphioxus: *Branchiostoma caribaeum*		100% mortality at 10 μg l^{-1} after 9 h				Clark *et al.*, 1987
Fiddler crab: *Uca pugilator*	TBTO in acetone	1-week exposure to 5.0μg l^{-1} TBTO during the sensitive period of limb regeneration caused increased malformation of limbs	Nominal concentrations	T = 23–25°C SAL = 25‰	Static renewal	Weis and Kim, 1988

Pink shrimp: *Penaeus duoarum* (juvenile)	TBTO in acetone	96-h LC_{50} = 7.6–9.1 μg l^{-1}	Nominal concentrations	T = 21–22°C SAL = 23–24‰	Static	M & T Chemicals Co., 1981b
Mysid: *Mysidopsis bahia* (juvenile)	TBTO (Bio MetRed in dimethyl-formamide)	96-h LC_{50} = 8 μg l^{-1}	Nominal concentrations	T = 22°C SAL = 27‰	Static	M & T Chemicals Co., 1981c
Mysid: *Acanthomysis sculpta*	TBT leachate from panels	Juveniles: 96-h LC_{50} = 0.61 μg l^{-1} TBT Adults: 96-h LC_{50} = 1.68 μg l^{-1} TBT	Valkirs *et al.*, 1986	Not reported	Continuous flow	Valkirs *et al.*, 1985
Mysid: *Acanthomysis sculpta* (juveniles)	TBT leachate from panels	96-h LC_{50} = 0.42 μg l^{-1} TBT	Valkirs *et al.*, 1986	T = 18 ± 1°C	Static renewal	Davidson *et al.*, 1986
Mysid: *Acanthomysis sculpta* (juveniles)	TBT leachate from panels	30% survival after 10 d at 0.61 μg l^{-1} TBT	Valkirs *et al.*, 1986	T = 14.1–15.3°C SAL = 32–33‰	Continuous flow	Salazar and Salazar, 1985
Mysid: *Mysidopsis bahia*	TBTCl in triethylene glycol	≤1-d-old: 96-h LC_{50} = 1.1 μg l^{-1} 5-d-old: 96-h LC_{50} = 2.0 μg l^{-1} 10-d-old: 96-h LC_{50} = 2.2 μg l^{-1}	Matthias *et al.*, 1986 (modified)	T = 25 ± 1°C SAL = 19–22.3‰	Continuous flow	Goodman *et al.*, 1988
Mysid: *Metamysidopsis elongata*	TBTO or TBT leachate from panels	Juveniles: 7-d LC_{50}<1.0 μg l^{-1} Adults: 6-d LC_{50} ≈ 2.0 μg l^{-1}	Measurements made periodically at some treatments (Valkirs *et al.*, 1986)	T = 13–14°C SAL = 33–34‰	Static Static and static renewal	Salazar and Salazar, 1989
Polychaete worm: *Neanthes arenaceodentata*	TBTO	Juveniles: 96-h LC_{50} ≈ 7 μg l^{-1} Adults: 96-h LC_{50} ≈ 20 μg l^{-1}			Static	Salazar and Salazar, 1989
Lugworm: *Arenicola cristata* (larvae)	TBTO in acetone	100% mortality at 4 μg l^{-1} after 96 h. No effect at 2 μg l^{-1} after 168 h	Nominal concentrations (stocks measured by ICAP spectro-photometer)	T = 20 ± 0.5°C SAL = 28‰	Static	Walsh *et al.*, 1986
	TBTA in acetone	100% mortality at 10 μg l^{-1} after 96 h. Abnormal development at 5 μg l^{-1} after 96 h. In 168 h, 100% mortality at 5 μg l^{-1} and no effect at 2.5 μg l^{-1}				
Snail: *Neritina* sp.	TBTA	6-d LC_{100} = 10 μg l^{-1}	Not reported	SAL = 5, 10, 15, 20, 25, 30, or 35‰	Static	Good and Dundee, 1981

Table 9.2 Continued

Test organism	*TBT compound*	*Result*	*Analytical method*	*Water quality data*	*Type of exposure*	*Reference*
Clam: *Rangia cuneata*	TBTA	5-d LC_{100} = 1000 μg l^{-1}; symptoms observed in clams exposed to 40–100 μg l^{-1} included partial retraction of the siphons, excess mucus, and swollen, rubbery siphons and mantle edges	Spectrophotometric (λ = 450 nm)	Not reported	Static	Good *et al.*, 1980
European oyster: *Ostrea edulis*	TBTO in acetic acid or ethanol	96-h LC_{50} = 210 μg l^{-1}	Not reported	Not reported	Static renewal	Thain, 1983
Common mussel: *Mytilus edulis*		Larvae: 48-h LC_{50} = 2.3 μg l^{-1} Adult: 96-h LC_{50} = 38 μg l^{-1}				Thain, 1983
Pacific oyster: *Crassostrea gigas*		Larvae: 48-h LC_{50} = 1.6 μg l^{-1} Adult: 96-h LC_{50} = 290 μg l^{-1}				Thain, 1983
Eastern oyster: *Crassostrea virginica*	TBTO	96-h LC_{50} = 560–1000 μg l^{-1}	Nominal concentrations	Not reported	Static	M & T Chemicals Co., 1976c
Eastern oyster *Crassostrea virginica* (embryo)	TBTO in 1:1 mixture of acetone and methanol	48-h EC_{50} = 0.9 μg l^{-1}	Nominal concentrations (actual measured concentrations similar to nominal: AAS–GF, however, data are reported as nominal concentrations)	T = 20 ± 1°C SAL = 22‰	Static	M & T Chemicals Co., 1977
Pacific oyster: *Crassostrea gigas*	TBTA	Embryo: incomplete formation of larvae at 1 μg l^{-1} in 24 h; mortality after 48 h Larvae: ≈50% dead at 1 μg l^{-1} after 96–120 h	Nominal concentrations	Not reported	Static	Robert and His, 1981
Mussel: *Mytilus galloprovincialis*		Embryo: 96-h LC_{50} = 1.0 μg l^{-1} Larvae: 120-h LC_{50} = 5.0 μg l^{-1}				Robert and His, 1981

Pacific oyster *Crassostrea gigas* (spat)	TBTA	0.1 μg l^{-1} caused slow growth and 100% mortality after 12 d; 0.05 μg l^{-1} caused slow growth and high mortality by day 10	Nominal concentrations	Not reported	Static	His and Robert, 1985
European oyster: *Ostrea edulis* (spat)	TBTO in glacial acetic acid	Growth rate severely curtailed after 10 d at 0.06 μg l^{-1}	Nominal concentrations	T = 20 ± 0.5°C SAL = 30‰	Static renewal	Thain and Waldock, 1985
Pacific oyster: *Crassostrea gigas* (spat)	TBTO in glacial acetic acid	Significant decrease in O_2 consumption and feeding rate after 14 d at 0.05 μg l^{-1} Hypoxia compensation significantly affected at 0.01 μg l^{-1}	Nominal concentrations	T = 10°C	Not reported	Lawler and Aldrich, 1987
Eastern oyster: *Crassostrea virginica*	TBTCl in acetone in glacial acetic acid	Embryo: 48-h LC_{50} = 0.71–1.3 μg l^{-1} TBTCl Larvae: 48-h LC_{50} = 3.96 μg l^{-1} TBTCl	Unger *et al.*, 1986	T = 20–24°C SAL = 18–22‰	Static renewal	Roberts, 1987
Hardshell clam: *Mercenaria mercenaria*		Embryo: 48-h LC_{50} = 1.13 μg l^{-1} TBTCl. Delayed development observed at 0.77 μg l^{-1} TBTCl Larvae: 48-h LC_{50} = 1.65 μg l^{-1} TBTCl				Roberts, 1987
Hardshell clam: *Mercenaria mercenaria*	TBTO in acetone	24–48 h old larvae: 100% mortality in 48 h at ≥2.5 μg l^{-1} and 100% mortality at 1 μg l^{-1} in 7 d. Reduced survival and growth inhibition at 0.6 μg l^{-1} in 8 d.	Valkirs *et al.*, 1986	SAL = 33‰	Static renewal	Laughlin *et al.*, 1989
		Post larvae (4 weeks old): no apparent effect on survival in 10 d at concentrations up to 10 μg l^{-1}			Continuous flow	

Table 9.2 Continued

Test organism	*TBT compound*	*Result*	*Analytical method*	*Water quality data*	*Type of exposure*	*Reference*
Hardshell clam: *Mercenaria mercenaria* (embryo→larvae)	TBTO in acetone	No dose-dependent mortality observed after 14 d at concentrations of 0.01–0.5 $\mu g\ l^{-1}$. Significant reductions in growth reported at all concentrations in 14 d.	Nominal concentrations (concentrations were measured: Valkirs *et al.*, 1986; however, measurements showed significant reductions in TBT over time and nominal concentrations are reported)	T = 25°C SAL = 32‰	Static renewal	Laughlin *et al.*, 1988
Hardshell clam: *Mercenaria mercenaria* (larvae)	TBTCl	96-h $EC_{50} = 0.016\ \mu g\ l^{-1}$	Nominal concentrations	Not reported	Static	Becerra-Huencho, 1984
Common mussel: *Mytilus edulis*	TBTO	10-d $LC_{50} \approx 8 \mu g\ l^{-1}$ TBTO	Measurements made periodically at some treatments: Valkirs *et al.*, 1986)	T = 13–14°C SAL = 33–34‰	Static	Salazar and Salazar, 1989
Clam: *Protothaca staminea*		10-d $LC_{50} \approx 100–120\ \mu g\ l^{-1}$ TBTO				Salazar and Salazar, 1989
Common mussel: *Mytilus edulis* (larvae)	TBTO in glacial acetic acid	≈50% larvae dead after 15 d at 0.1 $\mu g\ l^{-1}$ TBTO and growth reduced relative to controls	Nominal concentrations (initial test solutions measured by AAS)	T = 15 ± 1°C SAL = 33 ± 2‰	Static renewal	Beaumont and Budd, 1984
Common mussel: *Mytilus edulis* (post-fertilized eggs)	TBTO in glacial acetic acid	1.0 $\mu g\ l^{-1}$ TBT caused significant reduction in the percentage of normal larvae produced after 3 d	Nominal concentrations	T = 15 ± 1°C SAL = 2‰	Not reported	Beaumont *et al.*, in press
Common mussel: *Mytilus edulis* (juveniles, 5–8 mm)	TBTO in glacial acetic acid	Significant reduction in length growth rate occurred at concentrations ≥0.4 $\mu g\ l^{-1}$ TBTO in 7 d. No observed effect at 0.1 $\mu g\ l^{-1}$	Nominal concentrations	T = 8°C SAL = 33.7‰	Continuous flow	Stromgren and Bongard, 1987

Common mussel: *Mytilus edulis* (larvae 12 h old)	TBTO in acetone	No genotoxicity, as measured by sister chromatid exchange and chromosomal aberrations, was measured after 24-h exposures to 0.05–1.0 μg l^{-1} TBTO. A dose-dependent reduction in survival and development rate was observed at 0.05–5.0 μg l^{-1} TBTO in 96 h	Nominal concentrations	T = 14°C	Static	Dixon and Prosser, 1986
Bleak (fish): *Alburnus alburnus*	TBTO in acetone TBTF in acetone	96-h LC_{50} = 15 μg l^{-1} 96-h LC_{50} = 6–8 μg l^{-1}	Nominal concentrations	T = 10°C SAL = 7‰	Static	Linden *et al.*, 1979
Mosquitofish: *Gambusia affinis*	TBTA	Concentrations of 200–1000 μg l^{-1} caused death of all fish species after exposure from 20 to 180 min; complete mortality occurred for all species at 100 μg l^{-1} after 24-h exposure	Spectrophotometric (λ = 450 nm)	Not reported	Static	Good *et al.*, 1980
Sailfin molly: *Poecilia latipinna*						Good *et al.*, 1980
Gulf killifish: *Fundulus grandis*						Good *et al.*, 1980
Poacher: *Agonus catophractus*	TBTO in acetic acid or ethanol	96-h LC_{50} = 16 μg l^{-1}	Not reported	Not reported	Static renewal	Thain, 1983
Sole: *Solea solea*		Larvae: 96-h LC_{50} = 2 μg l^{-1} Adult: 96-h LC_{50} = 36 μg l^{-1}				Thain, 1983
Common mummichog: *Fundulus heteroclitus*	TBTO	96-h LC_{50} = 24 μg l^{-1}	Nominal concentrations	Not reported	Static	M & T Chemicals Co., 1976c
Sheepshead minnow: *Cyprinodon variegatus* (juvenile)	TBTO in acetone	96-h LC_{50} = 13–17 μg l^{-1}	Nominal concentrations	T = 22°C SAL = 20‰	Static	M & T Chemicals Co., 1979
Sheepshead minnow: *Cyprinodon variegatus*	TBTO in 1:1 (v/v) methanol and acetone	4-d LC_{50} = 1.5–3.2 μg l^{-1} Sn (3.7–7.9 μg l^{-1} TBTO) 7-d LC_{50} = 1.8 μg l^{-1} Sn (4.5 μg l^{-1} TBTO) 14-d LC_{50} = 1 μg l^{-1} Sn (2.5 μg l^{-1} TBTO)	AAS–GF	T = 29 ± 1°C SAL = 28–32‰	Continuous flow	Ward *et al.*, 1981

Table 9.2 Continued

Test organism	*TBT compound*	*Result*	*Analytical method*	*Water quality data*	*Type of exposure*	*Reference*
Common mummichog: *Fundulus heteroclitus*	TBTCl	Larvae (33–38 d): 96-h LC_{50} = 23.4 μg l^{-1} TBT; Subadult: 96-h LC_{50} = 23.8 μg l^{-1} TBT	Matthias *et al.*, 1986	T = 20 ± 1°C SAL = 10–12‰	Continuous flow	Bushong *et al.*, 1988
Sheepshead minnow: *Cyprinodon variegatus* (subadult)		96-h LC_{50} = 25.9 μg l^{-1} TBT				Bushong *et al.*, 1988
Atlantic menhaden: *Brevoortia tyrannus* (juvenile)		Repeated experiments: 96-h LC_{50} = 5.2 μg l^{-1} TBT 96-h LC_{50} = 4.5 μg l^{-1} TBT 96-h LC_{50} = 3.0 μg l^{-1} TBT				Bushong *et al.*, 1988
Inland silverside: *Menidia beryllina* (larvae 4–10 d)						Bushong *et al.*, 1988
Atlantic silverside: *Menidia menidia* (subadult)		96-h LC_{50} = 8.9 μg l^{-1} TBT				Bushong *et al.*, 1988
Mummichog: *Fundulus heteroclitus*	TBTO	Avoidance occurred at total organic Sn concentrations of 3.7 μg l^{-1} (9 μg l^{-1} TBTO)	AAS–GF	T = 22–36°C SAL = 9.9–11.2‰	Continuous flow	Pinkney *et al.*, 1985
Striped bass: *Morone saxatilis* (juvenile)	TBTO	Avoidance occurred at 25 μg l^{-1} total organic Sn (63 μg l^{-1} TBTO)	AAS–GF	T = 24–28.5°C SAL = 9–11‰	Continuous flow	Hall *et al.*, 1984
Atlantic menhaden: *Brevoortia tyrannus* (juveniles)	TBTO	Avoidance occurred at total organic Sn concentrations of 6 μg l^{-1} (15 μg l^{-1} TBTO)				Hall *et al.*, 1984
Striped bass: *Morone saxatilis* (larvae)	TBT leachate from panels	Concentrations ≥1.5 μg l^{-1} TBT caused 100% mortality in 5–6 d; concentrations ≥0.77 μg l^{-1} TBT significantly reduced survival in 6 d; 6-d exposure to 0.067 μg l^{-1} TBT decreased body depth of larvae	Valkirs *et al.*, 1986	T = 18–20°C SAL = 1.1–3.0	Continuous flow	Pinkney *et al.*, 1990

Common mummichog: *Fundulus heteroclitus*	TBTO in acetone	96-h LC_{50} = 17.2 µg l^{-1} TBT	AAS–GF	T = 23–25°C SAL = 15‰	Continuous flow	Pinkney *et al.*, 1989b
Striped bass: *Morone saxatilis*	TBT leachate from panels	Gil Na^+K^+ATPase was significantly increased at 0.1 µg l^{-1} after 14-d exposure	Valkirs *et al.*, 1986	Not reported	Continuous flow	Pinkney *et al.*, 1989a
	TBTCl	Significant inhibition of Na^+K^+ATPase at 106 µg l^{-1} and inhibition of Mg^{++}ATPase at 53 µg l^{-1} in vitro	Nominal concentrations (stocks measured with AAS–GF)	T = 25°C	Static (in vitro)	
Common mummichog: *Fundulus heteroclitus*	TBTO	Significant inhibition of Na^+K^+ATPase at 25.3 µg l^{-1} and inhibition of Mg^{++}ATPase at 5.1 µg l^{-1} in gill tissue (in vitro)				Pinkney *et al.*, 1989a
Chinook salmon: *Oncorhynchus tshawytscha* (juveniles)	TBTO in glacial acetic acid	96-h LC_{50} = 1.5 µg l^{-1} TBT	AAS	T = 4 ± 1°C SAL = 28‰	Static	Short and Thrower, 1986
Flatfish: *Citharichthys stigmaeus*	TBTO	96-h LC_{50} ≈19 µg l^{-1} TBTO 14-d LC_{50} ≈5–9 µg l^{-1} TBTO	Measurements made periodically at some treatments: (Valkirs *et al.*, 1986)	T = 13–14°C SAL = 33–34‰	Static	Salazar and Salazar, 1989
California grunion: *Leuresthes tenuis*	TBT leachate from panels	Larvae: No significant effect on survival after 7 d at concentrations of 0.04–1.65 µg l^{-1}	Hodge *et al.*, 1979 (modified)	Not reported	Continuous flow	Newton *et al.*, 1985
		Embryo: 0.14–1.72 µg l^{-1} enhanced hatch success and stimulated growth after 10-d exposure. No adverse effects		T = 20.5°C		
		Pre-fertilized in 74 µg l^{-1} reduced gametes and hatching success by ≈50%				

These data were included in this review because they enhanced the understanding of the organism and its response to TBT.

Data for acute TBT toxicity were available for one species of bacterium (*Photobacterium bacterium*). Fifteen-minute EC_{50} values of 7 and 22 $\mu g\ l^{-1}$ for TBTCl and TBTBr, respectively, were obtained in these Microtox experiments (Dooley and Kenis, 1987). These Microtox data are useful for ranking the sensitivity of an organism to various compounds, but have limitations when comparisons are attempted with other standard toxicity tests.

Data for acute TBT toxicity are presented for ten species of marine algae and a natural phytoplankton assemblage. All algal tests reported used static conditions, and either nominal concentrations or no analytical method were reported. The accuracy of TBT exposure concentrations causing effects is therefore suspect. The most intensively studied algal species was the marine diatom *Skeletonema costatum* with nine separate toxicity tests reported in Table 9.2. In general, the varying endpoints and test conditions gave comparable results for *S. costatum* in short-term experiments with EC_{50} values typically $<1.0\ \mu g\ l^{-1}$. Walsh *et al.* (1985) reported a 72-h EC_{50} of 0.33 $\mu g\ l^{-1}$ TBTO. Studies of longer duration, although not necessarily acute exposures, did provide evidence of effects at lower concentrations. M & T Chemicals Co. (1981a) reported a 14-d EC_{50} of 0.064 $\mu g\ l^{-1}$ TBTO, and Beaumont and Newman (1986) reported significant decreases in growth after 15 d at 0.1 $\mu g\ l^{-1}$ for *S. costatum*.

Studies on other algal species demonstrated considerable variability in TBT sensitivity; however, most algal species appeared to be less sensitive than *S. costatum*. Results varied from a 72-h EC_{50} of 1.0 $\mu g\ l^{-1}$ for *Thalassiosira pseudonana* (Walsh *et al.*, 1985) to no observable effect at 6.0 $\mu g\ l^{-1}$ after 72 h with *P. tricornutum* (Salazar, 1985). Thain (1983) reported a 5-d algistatic concentration of 560–1000 $\mu g\ l^{-1}$ for *T. suecica*, ~500 times greater than the algistatic concentration reported for *S. costatum*. Thain (1983) showed that algae recovered from acute TBT exposure after being transferred to uncontaminated media. In a study of longer duration (a chronic study), Beaumont and Newman (1986) reported reduced growth after 15 d at 0.1 $\mu g\ l^{-1}$ in *Pavola lutheri* and *Dunaliella tertiolecta*.

One study was conducted with natural phytoplankton assemblages from the York River. In this experiment, Ho (1984) reported a photosynthetic 2-h EC_{50} ranging from 5.62 to 7.57 $\mu g\ l^{-1}$ TBTCl with no significant differences between surface water and deep-water plankton assemblages.

Toxicity data on three copepod species, *E. affinis*, *A. tonsa*, and *N. spinipes*, are presented in Table 9.2. Comparable short-term toxicity data from three studies have been reported for adult or subadult *A. tonsa*: 48-h LC_{50} of 1.1 $\mu g\ l^{-1}$ TBT (Bushong *et al.*, 1988), 96-h LC_{50} of 1 $\mu g\ l^{-1}$ TBTO (U'ren, 1983), and 100% mortality at 1 $\mu g\ l^{-1}$ TBTO in 24 h (Salazar and Salazar, 1989). A 48-h LC_{50} of 1.4–2.5 $\mu g\ l^{-1}$ TBT and 72-h LC_{50} of 0.5–0.6 $\mu g\ l^{-1}$ TBT were reported for *E. affinis* (Bushong *et al.*, 1988). *N. spinipes* is the most resistant copepod species studied with a 96-h LC_{50} of 2.0 $\mu g\ l^{-1}$ TBTO (Linden *et al.*, 1979).

Experiments on the young life stages of copepods (nauplii) or studies of longer duration (chronic studies) using sublethal endpoints demonstrate sensitivity to TBT at very low concentrations. Copepods have a short life cycle; therefore, 6-d exposures are classified as chronic studies. Hall *et al.* (1988) reported significantly reduced survival of *E. affinis* nauplii within 6 d at 0.088 $\mu g\ l^{-1}$ TBT. U'ren (1983) reported a 6-d EC_{50} of 0.4 $\mu g\ l^{-1}$ TBTO for adult *A. tonsa*, and Johansen and Mohlenberg (1987) reported reduced egg production at 0.01 $\mu g\ l^{-1}$ TBTO in adults exposed for 5 d. Bushong *et al.* (1990) reported significant mortality of *A. tonsa* nauplii after 6 d at a measured concentration of 0.023 $\mu g\ l^{-1}$ TBT and determined the

NOEC (no observed effect concentration) to be 0.010 μg l^{-1} TBT. Although the Bushong *et al.* (1990) experiments were not acute experiments, they do demonstrate mortality of *A. tonsa* at low nanogram per liter concentrations in a relatively short time period.

Four crab species were evaluated in tests for acute TBT toxicity using both lethal and a variety of sublethal endpoints. In tests with larval crabs, *C. maenas* (Thain, 1983), *H. nudus* (Laughlin and French, 1980), and *R. harrisii* (Laughlin *et al.*, 1983), lethal and sublethal effects were reported at concentrations ranging from 10 to 25 μg l^{-1} TBTO. Larval lobsters (*H. americanus*) appear to be somewhat more sensitive than crabs with a 6-d LC_{100} of 5 μg l^{-1} TBTO (Laughlin and French, 1980). Exposure of adult fiddler crabs (*U. pugilator*) for 1 week at 5.0 μg l^{-1} TBTO (nominal concentrations) caused increased malformation of regenerated limbs (Weis and Kim, 1988).

Data for acute TBT toxicity were reported for three amphipod species. An approximate LC_{50} of 6 μg l^{-1} TBTO was reported for *O. traskiana* (Laughlin *et al.*, 1982). An LC_{50} value for *G. oceanicus* appeared to be <3 μg l^{-1} TBTO (Laughlin *et al.*, 1982, 1984); 96-h LC_{50} values of 1.3 and 5.3 μg l^{-1} TBT were reported for larvae and adult *Gammarus sp.*, respectively (Bushong *et al.*, 1988).

The mysids were one of the most intensely studied Crustacea, as 10 studies covering different life stages of three species were reported. Three early life stages of *M. bahia* (≤1 d, 5 d, and 10 d old mysids) had 96-h LC_{50} values ranging from 1.1 to 2.2 μg l^{-1} TBTCl (Goodman *et al.*, 1988). M & T Chemicals Co. (1981c) reported a 96-h LC_{50} of 8 μg l^{-1} TBTO for juvenile *M. bahia*. Two studies on the mysid *A. sculpta* demonstrated similar 96-h LC_{50} values of 0.61 μg l^{-1} TBT (Valkirs *et al.*, 1985) and 0.42 μg l^{-1} TBT (Davidson *et al.*, 1986) for juveniles. Salazar and Salazar (1985) found 30% survival after 10 d at 0.6 μg l^{-1} with juvenile *A. sculpta*. Valkirs *et al.* (1985) found adult *A. sculpta* to be somewhat more resistant than juveniles (96-h LC_{50} of 1.7 μg l^{-1} TBT). The mysid *M. elongata* appears equally as sensitive to TBT as *M. bahia* and *A. sculpta*. Salazar and Salazar (1989) reported a 7-d LC_{50} <1.0 μl^{-1} for juveniles and a 6-d LC_{50} ≈2.0 μg l^{-1} for adult *M. elongata*.

The shrimp *C. crangon* demonstrated a 20-fold difference in sensitivity between larvae (96-h LC_{50} of 2 μg l^{-1} TBT)) and adults (96-h LC_{50} of 41 μg l^{-1} TBTO) (Thain, 1983). Studies on the grass shrimp (*P. pugio*) demonstrate that it is a relatively resistant crustacean with 96-h LC_{50} values of 20 μg l^{-1} TBTO (Clark *et al.*, 1987) and >31 μg l^{-1} TBT (Bushong *et al.*, 1988). Pinkney *et al.* (1985) found that *P. pugio* did not avoid concentrations as high as 75 μg l^{-1} TBTO in avoidance experiments.

Acute toxicity data on two species of polychaetes are reported. Walsh *et al.* (1986) reported 100% mortality in larval lugworms (*A. cristata*) after 96-h exposure to 4 μg l^{-1} TBTO. Abnormal development was not reported at concentrations that did not also affect survival. Adult *N. arenaceodentata* were approximately three times more resistant (96-h LC_{50} of 20 μg l^{-1} TBTO) than the juveniles (96-h LC_{50} of 7 μg l^{-1} TBTO) (Salazar and Salazar, 1989).

Nine species of the Mollusca phylum, primarily bivalves, are presented in Table 9.2. In general, the adults are quite resistant to TBT while larval stages appear to be very sensitive. One species of the class Gastropoda was tested in an acute exposure. In this test, Good and Dundee (1981) reported an LC_{100} of 10 μg l^{-1} for *Neritina* sp.

Data for acute TBT toxicity are presented for three oyster species. A 96-h LC_{50} of 210 μg l^{-1} TBTO was reported for adult European oyster (*O. edulis*) (Thain, 1983). Thain and Waldock (1985) reported severe growth inhibition of larval *O. edulis* after 10-d exposures to 0.06 μg l^{-1} TBTO.

A 96-h LC_{50} of 290 μg l^{-1} TBTO was reported for adult Pacific oysters (*C. gigas*) (Thain, 1983). Larval *C. gigas* were more sensitive than adults, with acute toxicity reported at near 1 μg l^{-1} [48-h LC_{50} of 1.6 μg l^{-1} TBTO (Thain, 1983); and 96-h LC_{50} ≈1.0 μg l^{-1} TBTA (Robert and His, 1981)]. Studies of longer duration and with sublethal endpoints reported much lower effect levels. His and Robert (1985), Alzieu (1991), and His (Chapter 12) found slow growth and high mortality in larval *C. gigas* at 0.05 μg l^{-1} TBTA after 10 d. In 14-d studies with larval *C. gigas*, Lawler and Aldrich (1987) reported a decrease in O_2 consumption and feeding rates at 0.05 μg l^{-1} TBTO and found hypoxia compensation significantly affected at 0.01 μg l^{-1}.

Adult eastern oysters (*C. virginica*) were resistant to acute TBT exposure [96-h LC_{50} of 560–1000 μg l^{-1} (M & T Chemicals Co., 1976c)]. Roberts (1987) reported a 48-h LC_{50} of 0.71–1.3 μg l^{-1} TBTCl and a 48-h LC_{50} of 3.96 μg l^{-1} TBTCl for *C. virginica* embryos and larvae, respectively. M & T Chemicals Co. (1977) reported a comparable 48-h EC_{50} value for embryo *C. virginica* of 0.9 μg l^{-1} TBTO.

Seven TBT toxicity studies were reported for the mussel *M. edulis*. Thain (1983) reported a 96-h LC_{50} of 38 μg l^{-1} TBTO for adult *M. edulis*. In a longer study, Salazar and Salazar (1989) reported a 10-d LC_{50} of 8 μg l^{-1} TBTO for adult *M. edulis*. Stromgren and Bongard (1987) found significant reductions in the growth of juvenile *M. edulis* after 7 d at 0.4 μg l^{-1} TBTO, and Thain (1983) reported a 48-h LC_{50} of 2.3 μg l^{-1} TBTO for larval *M. edulis*. In a 15-d study, Beaumont and Budd (1984) found high mortality and significant reductions in growth of larval *M. edulis* at 0.1 μg l^{-1} TBTO. No genotoxicity was reported after 24-h exposures at 0.05–1.0 μg l^{-1} TBTO in larval *M. edulis* (Dixon and Prosser, 1986). In a study with the mussel *M. galloprovincialis*, Robert and His (1981) reported a 96-h LC_{50} of 1.0 μg l^{-1} TBTA for embryos and a 120-h LC_{50} of 5.0 μg l^{-1} TBTA for larvae.

Toxicity data for three species of clams are presented in Table 9.2. Adult *P. staminea* and *R. cuneata* were both very resistant to TBT exposure with a 10-d LC_{50} ≈100–120 μg l^{-1} TBTO (Salazar and Salazar, 1989) and a 5-d LC_{100} of 1000 μg l^{-1} TBTA (Good *et al.*, 1980), respectively. Good *et al.* (1980) also reported observing sublethal effects at lower concentrations (40–100 μg l^{-1}) in *R. cuneata*. Several studies were conducted that evaluated young life stages of the hardshell clam *M. mercenaria*. Roberts (1987) reported a 48-h LC_{50} of 1.13 μg l^{-1} TBTCl for embryos and a 48-h LC_{50} of 1.65 μg l^{-1} TBTCl for larval *M. mercenaria*. In these studies Roberts (1987) also reported delayed development of embryos at 0.77 μg l^{-1} TBTCl. Laughlin *et al.* (1989) found results similar to Roberts (1987) in acute experiments with larval *M. mercenaria*, but in experiments with older post larvae, Laughlin *et al.* (1989) observed no effects after 10 d at concentrations up to 10 μg l^{-1}. Becerra-Huencho (1984) reported an acute toxicity value for larval *M. mercenaria* (96-h EC_{50} of 0.016 μg l^{-1} TBTCl) noticeably lower than found by Roberts (1987) and Laughlin *et al.* (1988). Data from the Becerra-Huencho (1984) study resulted from experiments where TBT was not measured in test containers. These unpublished data have limited use because they are two orders of magnitude lower than data from two other published studies. In a 14-d study, Laughlin *et al.* (1988) reported reduced growth in embryos of *M. mercenaria* at TBT concentrations ≥0.01 μg l^{-1} TBTO.

Acute toxicity data covering 15 species of saltwater fish are presented. In general, fish are less sensitive to TBT in acute exposures than sensitive invertebrates such as copepods. The most sensitive fish species reported were juvenile Chinook salmon (*O. tshawytscha*) with a 96-h LC_{50} of 1.5 μg l^{-1} TBT (Short and Thrower, 1986), larval sole (*S. solea*) with a 96-h LC_{50} of 2 μg l^{-1} TBTO (Thain, 1983), and larval inland silversides (*M. beryllina*) with a 96-h LC_{50} of 3.0 μg l^{-1} TBT (Bushong *et al.*, 1988). The most resistant fish species reported was adult sole (*S. solea*) with a 96-h LC_{50} of 36 μg l^{-1} TBTO (Thain, 1983).

Some investigators have reported increased sensitivity to TBT with increased exposure time. In studies with the flatfish *S. stigmaeus*, Salazar and Salazar (1989) reported a 4-d LC_{50} ≈19 μg l^{-1} TBTO and a 14-d LC_{50} ≈5–9 μg l^{-1} TBTO. Ward *et al.* (1981) reported a 4-d LC_{50} of 3.7–7.9 μg l^{-1} TBTO and a 14-d LC_{50} of 2.5 μg l^{-1} TBTO for the sheepshead minnow (*C. variegatus*). Bushong *et al.* (1988) with a 96-h LC_{50} of 25.9 μg l^{-1} TBTO and M & T Chemicals Co. (1979) with a 96-h LC_{50} of 13–17 μg l^{-1} TBTO reported higher acute toxicity values for *C. variegatus* than Ward *et al.* (1981). These differences may in part be due to differences in temperature and salinity of the test water.

A number of the studies presented in this review examined sublethal responses of fish to acute TBT exposures. Pinkney *et al.* (1985) reported avoidance behavior at concentrations of 9.0 μg l^{-1} TBTO by the mummichog (*F. heteroclitus*). Inhibition of Mg^{++}ATPase was reported in gill tissue (in vitro) of *F. heteroclitus* at 5.1 μg l^{-1} TBTO (Pinkney *et al.*, 1989a). Acute 96-h LC_{50} values for *F. heteroclitus* range from 17.2 μg l^{-1} TBT (Pinkney *et al.*, 1989b) to 24 μg l^{-1} TBTO (M & T Chemicals Co., 1976c) and 23.8 μg l^{-1} TBT (Bushong *et al.*, 1988).

Pinkney *et al.* (1990) reported significant mortality of larval striped bass (*M. saxatilis*) at 0.77 μg l^{-1} TBT in 6 d. Six-day exposures to 0.067 μg l^{-1} TBT decreased body depth of larval *M. saxatilis* (Pinkney *et al.*, 1990). In another study, Pinkney *et al.* (1989a) found increased gill Na^+K^+ATPase in striped bass exposed to 0.1 μg l^{-1} for 14 d. Other researchers have found sublethal effects in *M. saxatilis* at TBT exposure exceeding lethal doses (Hall *et al.*, 1984; Pinkney *et al.*, 1989a).

Hall *et al.* (1984) reported avoidance behavior in Atlantic menhaden (*B. tyrannus*) at concentrations of 15 μg l^{-1} TBTO. This value is approximately three times greater than the 96-h LC_{50} reported for *B. tyrannus* by Bushong *et al.* (1988).

In a series of experiments with the California grunion (*L. tenuis*), Newton *et al.* (1985) found no effect on survival of larvae at 0.04–1.65 μg l^{-1} TBT in 7 d. Newton *et al.* (1985) reported enhanced hatching success and growth of embryo *L. tenuis* after 10-d exposures to 0.14–1.72 μg l^{-1} TBT, indicating a possible hormestic effect. Reduced hatching success of *L. tenuis* was reported when prefertilized gametes were exposed at 74 μg l^{-1} TBT.

9.4. SUMMARY

The preceding sections have provided a review of available data on acute tributyltin toxicity for both freshwater and saltwater organisms. The following summaries were determined for each water type after an interpretation of the literature.

9.4.1. FRESH WATER

Most of the data on TBT toxicity in fresh water were generated from experiments where nominal concentrations were used to calculate toxicity values. Survival of test organisms was the most frequently used endpoint in these acute tests although other endpoints such as behavioral modification, BCF determinations, and histological changes have been reported. Twenty-nine species of freshwater organisms tested were found to have a wide range of sensitivity to TBT after acute exposures. *Hydra* were the most sensitive species tested (in standard 96-h tests) with a 96-h EC_{50} of 0.5 μg l^{-1} TBTO. Bluegills were the most resistant species tested with a 96-h LC_{50} of 240 μg l^{-1} TBTO.

Four-hour IC_{50} values ranged from 3 to 20 μg l^{-1} TBTO for two freshwater algae species, one cyanobacterium, and a mixed algae community. *Daphnia* were the most sensitive zooplankton tested with a 48-h LC_{50} of 1.67 μg l^{-1} TBTO; the ostracod was one of the most resistant zooplankton species tested (4-d LC_{50} of 120 μg l^{-1} TBTO). Two experiments to determine acute TBT toxicity with invertebrates were designed to evaluate parameters other than mortality. Significant alteration of photobehavior occurred with *Daphnia* after 96–120 h of exposure to 0.5 μg l^{-1} TBTCl. Bioconcentration factors of 23.7–28.6 were reported in the head and foot area of snails after acute exposures to TBTO.

Data on acute TBT toxicity were available for 11 fish species. Ninety-six-hour LC_{50} data were similar for most species as values ranged from 2.6 to 12.6 μg l^{-1} TBT. Loss of positive rheotaxis (a behavioral modification) was evaluated in acute tests with two fish species; however, the concentrations of TBT that caused this behavioral modification were similar to concentrations causing mortality.

9.4.2. SALTWATER

The review covered 56 species of saltwater organisms that demonstrated a wide range of sensitivity to TBT. The most sensitive organisms were copepods and mysids with acute effects reported as low as 0.4–0.5 μg l^{-1} TBT. Bivalve embryos were also sensitive as acute effects were reported at ~1.0 μg l^{-1}. Chronic effects have been reported at much lower concentrations (0.023 μg l^{-1} TBT) with copepods. The most resistant organisms were adult oysters with 96-h LC_{50} values >200 μg l^{-1}. In many instances, large differences in sensitivity were reported between larvae and adults of a species. All tests were ≤15 d in duration and used either mortality or a variety of sublethal parameters as an endpoint. Over half of the experiments reported in this review used nominal test concentrations, and many experiments had no renewal of test water.

The marine diatom *S. costatum* was the most intensely studied phytoplankton species and one of the most sensitive to TBT (72-h EC_{50} of 0.33 μg l^{-1} TBTO). The most resistant algal species were *P. tricornutum* (no effect after 72 h at 6.0 μg l^{-1} TBTO) and *T. suecica* (5-d algistatic concentration of 560–1000 μg l^{-1} TBTO). Concentrations of TBT as high as 560–1000 μg l^{-1} are rarely found in the environments; therefore, these values have limited value. Studies of longer duration demonstrated enhanced sensitivity of phytoplankton to TBT. Reduced growth was reported in three phytoplankton species at 0.064–0.1 μg l^{-1} after 14–15 d.

Crustaceans demonstrated a wide range in sensitivity to TBT. The most sensitive crustaceans were the copepods and mysids. Copepods such as *A. tonsa* (48-h LC_{50} of 1.1 μg l^{-1} TBT) and *E. affinis* (72-h LC_{50} of 0.5 μg l^{-1} TBT) were particularly sensitive to TBT. Sublethal effects (reduced egg production) were observed in *A. tonsa* adults after 5-d exposures to 0.01 μg l^{-1} TBTO, and significant mortality was observed in *A. tonsa* nauplii after 6-d exposures to 0.023 μg l^{-1} TBT. Mysids were also sensitive to TBT exposure. The range of 4- to 7-d LC_{50} values for early life stages or juveniles of three mysid species (*A. sculpta, M. elongata,* and *M. bahia*) was 0.42–2.2 μg l^{-1} TBT. The most resistant crustacean tested was the adult shrimp *C. crangon* with a 96-h LC_{50} of 41 μg l^{-1} TBTO. Larval *C. crangon* were 20 times more sensitive than adult *C. crangon*.

In general, bivalve adults were very resistant to TBT whereas larval bivalves were very sensitive. Adult bivalves may have been much more resistant to TBT than larvae because this bivalve life stage can close its valves for 4 d with no intake of toxicant. Therefore, tests for acute TBT toxicity with adult oysters have limited value. The 96-h LC_{50} values for adult oysters (*O. edulis, C. gigas,* and *C. virginica*) ranged from 210 to 560 μg l^{-1}. Short-term toxicity tests with larvae and embryos of oysters produced much lower LC_{50} values between 1 and 2 μg l^{-1}. Sublethal effects were observed at 0.01–0.05 μg l^{-1} in longer studies with larval oysters. Similar observations were made on mussels with growth inhibition reported in larval mussels (*M. edulis*) at 0.1 μg l^{-1}.

Fish species were generally less sensitive to TBT than bivalves and crustaceans. The 96-h LC_{50} values ranged from 1.5 $\mu g\ l^{-1}$ TBT for juvenile Chinook salmon (*O. tshawtscha*) to 36 $\mu g\ l^{-1}$ TBTO for adult sole (*S. solea*). In many instances, larvae were more sensitive than adults of the same species, and longer test durations produced lower levels for a toxic effect. Various investigators reported that sublethal parameters could be used to detect effects of TBT at lower, nonlethal concentrations. A TBT concentration of 0.1 $\mu g\ l^{-1}$ caused increased gill $Na^{+}K^{+}$ATPase, and a concentration of 0.067 $\mu g\ l^{-1}$ decreased body depth of larval striped bass (*M. saxatilis*).

9.5. CONCLUSIONS AND RECOMMENDATIONS

Acute toxicity data for both freshwater and saltwater organisms have been reported for numerous species. Most of these data were generated from experiments where nominal concentrations of TBT were used to calculate values for acute toxicity. Only nine freshwater species have been tested using measured concentrations of TBT. Approximately three times as many saltwater species have been tested in acute experiments using measured TBT concentrations. For many of the earlier studies, analytical methods to measure TBT in the parts per trillion range (the toxicity range for TBT) were not available for use in toxicity studies. Failure to measure TBT in test containers may result in inaccurate data, because TBT adsorbs to the test container or degrades in solution over time. Measurement of TBT in test containers must be conducted to insure that accurate concentrations were reported in stock solutions used for calculating acute toxicity values. Data from acute studies where TBT was not measured in test containers (nominal concentrations) should not be used in the policy or regulatory process. Toxicity data generated from studies using nominal concentrations provide only estimates of toxic concentrations of TBT.

The following criteria are recommended for acute studies of TBT toxicity that are used to develop water quality criteria or regulations: (1) TBT concentrations should be measured at least every 24 h during exposure using acceptable analytical methods with detection limits of least 0.010 g l^{-1} TBT; (2) continuous flow conditions should be used; (3) all contact material used in tests (e.g. test containers, hoses, and head tanks) should be polycarbonate or Teflon to minimize absorption problems; and (4) basic parameters for water quality [temperature, dissolved oxygen, pH, conductivity, salinity (saltwater studies), and hardness (freshwater studies)] should be measured every 24 h during tests, and water quality should simulate natural conditions of the test species. Other acceptable procedures as described in Stephen *et al.* (1985) should also be followed for acute tests.

Future studies of acute TBT toxicity should focus on sensitive life stages of species found in the natural environment where TBT may exist. These acute studies are needed primarily for freshwater species. Although acute laboratory studies of toxicity do have value in the regulatory process, long-term chronic studies with sensitive life stages of resident species are more useful and provide a more realistic assessment of environmental effects of TBT. These type studies are recommended for both freshwater and saltwater species using the experimental design criteria previously described for acute studies. Tributyltin concentrations used in these chronic studies should reflect those concentrations found in the environment.

ACKNOWLEDGEMENTS

MAES Contribution Number 9114/MAES Scientific Article Number A7792.

REFERENCES

Alabaster, J.S. 1969. Survival of fish in 164 herbicides, insecticides, fungicides, wetting agents and miscellaneous substances. *Int. Pesticide Control*, **11**, 29–35.

Allen, A.J., B.M. Quitter and C.M. Radick. 1980. The biocidal mechanism of controlled release bis (tri-n-butyltin) oxide in *Biomphalaria glabrata*. In: *Controlled Release of Bioactive Materials*, R.W. Baker (Ed.). Symposium of the International Meeting, Controlled Release Society, 1979. Academic Press, New York, pp. 399–413.

Beaumont, A.R. and M.D. Budd. 1984. High mortality of the larvae of the common mussel at low concentrations of tributyltin. *Mar. Pollut. Bull.*, **15**, 402–405.

Beaumont, A.R. and P.B. Newman. 1986. Low levels of tributyltin reduce growth of marine micro-algae. *Mar. Pollut. Bull.*, **17**, 457–461.

Beaumont, A.R., P.B. Newman and J. Smith. (in press). Some effects of tributyltin from antifouling paints on early development and veliger larvae of the mussel *Mytilus edulis*. *Malacologia*.

Becerra-Huencho, R.M. 1984. The Effect of Organotin and Copper Sulfate on the Late Development and Presettlement Behavior of the Hard Clam *Mercenaria mercenaria*. Masters Thesis, Chesapeake Biological Lab., University of Maryland, Solomons Island, Maryland, 83 pp.

Brooke, L.T., D.J. Call, S.H. Poirier, T.P. Markee, C.A. Lindberg, D.J. McCauley and P.G. Simmonson. 1986. Acute Toxicity and Chronic Effects of Bis(tri-n-butyltin) Oxide to Several Species of Freshwater Organisms. Report to Battelle Memorial Research Institute, Columbus, Ohio, 20 pp.

Bushong, S.J., L.W. Hall, Jr, W.S. Hall, W.E. Johnson and R.L. Herman. 1988. Acute toxicity of tributyltin to selected Chesapeake Bay fish and invertebrates. *Water Res.*, **22**, 1027–1032.

Bushong, S.J., M.C. Ziegenfuss, M.A. Unger and L.W. Hall, Jr. 1990. Chronic tributyltin toxicity experiments with the Chesapeake Bay copepod, *Acartia tonsa*. *Environ. Toxicol. Chem.*, **9**, 359–366.

Camey, T. and E. Paulini. 1964. Molluscicidal activity of some organotin compounds. *Rev. Bras. Malariol. Doencas Trop.*, **16**, 487–491.

Cardarelli, N.F. 1973. Development and Testing of Molluscicidal and Cercariacidal formulations. Annual Report of Creative Biology Laboratory to World Health Organization, Norton, Ohio, 38 pp.

Cardarelli, N.F. and W. Evans. 1980. Chemodynamics and environmental toxicology of controlled release organotin molluscicides. In: *Controlled Release of Bioactive Materials*, N.F. Cardarelli, W. Evans and R.W. Baker (Eds). 6th International Meeting, Controlled Release Society, 1979. Academic Press, New York, pp. 357–385.

Chliamovitch, Y.P. and C. Kuhn. 1977. Behavioural, hematological and histological studies on acute toxicity of bis(tri-n-butyltin) oxide on *Salmo gairdneri* Richardson and *Tilapia rendalli* Boulenger. *J. Fish Biol.*, **10**, 575–585.

Clark, J.R., J.M. Patrick, Jr, J.C. Moore and E.M. Lores. 1987. Waterborne and sediment-source toxicities of six organic chemicals to grass shrimp (*Palaemonetes pugio*) and amphioxus (*Branchiostoma caribaeum*). *Arch. Environ. Con. Tox.*, **16**, 401–407.

Davidson, B.M., A.O. Valkirs and P.F. Seligman. 1986. Acute and chronic effects of tributyltin on the mysid *Acanthomysis sculpta* (Crustacea, Mysidacea). In: Proceedings of the Oceans '86 Organotin Symposium, Vol. 4, 25–27 September 1986, Washington, DC, pp. 1219–1225.

Deschiens, R. and H.A. Floch. 1968. Action biologique comparéi de 6 molluscicides chimiques daus le cadre de la prophylaxie des bilharzioses, conclusions pratiques. *Bull. Soc. Pathol. Exotique*, **61**, 640–650.

Deschiens, R., H. Brottes and L. Mvogo. 1966a. Application sur le terrain au Cameroun dans la prophylaxie des bilharzioses de l'action molluscicide de l'oxyde de tributylétain. *Bull. Soc. Pathol. Exotique*, **59**, 968–973.

Deschiens, R., H. Brottes and L. Mvogo. 1966b. Contrôle sur la terrain des propriétés molluscicides de l'oxyde de tributyl-étain (prophylaxis des bilharzioses). *Bull. Soc. Pathol. Exotique*, **59**, 231–240.

Dixon, D.R. and H. Prosser. 1986. An investigation of the genotoxic effects of an organotin antifouling compound (bis(tributyltin)oxide) on the chromosomes of the edible mussel, *Mytilus edulis*. *Aquat. Toxicol.*, **8**, 185–195.

Dooley, C.A. and P. Kenis. 1987. Response of bioluminescent bacteria to alkyltin compounds. In: Proceedings of the Oceans '87 International Organotin Symposium, Vol. 4, September 28–1 October 1987, Halifax, Nova Scotia, pp. 1517–1524.

Eisler, R. 1989. Tin Hazards to Fish, Wildlife, and Invertebrates: A Synoptic Review. Report US Fish and Wildlife Service 85 (1.15), Laurel, Maryland, 83 pp.

Floch, H., R. Deschiens and T. Floch. 1964. Sur les propriétés molluscicides de l'oxyde et de l'acetate de tributyl-étain (prophylaxie des bilharzioses). *Bull. Soc. Pathol. Exotique*, **57**, 454–465.

Foster, R.B. 1981. Use of Asiatic Clam Larvae in Aquatic Hazard Evaluation. In: *Ecological Assessments of Effluent Impacts on Communities of Indigenous Aquatic Organisms*, J.M. Bates and C.I. Weber (Eds). American Society for Testing and Materials, Philadelphia, pp. 280–288.

Frick, L.R. and W.Q. de Jimenez. 1964. Molluscicidal qualities of three organo-tin compounds revealed by 6-hour and 24-hour exposures against representative stages and sizes of *Australorbis glabratus*. *Bull. World Health Organiz.*, **31**, 429–431.

Good, M.L. and D.S. Dundee. 1981. Final Report on Chemical, Physical, and Biological Investigations of Marine Antifouling Coatings, Contract No. DO-AO1-78-00-3049, Report to US Maritime Administration, Dept of Commerce, Washington, DC, 44 pp.

Good, M.L., D.S. Dundee and G. Swindler. 1980. Bioassays and environmental effects of organotin marine antifoulants. In: *Controlled Release of Bioactive Materials*, R. Baker (Ed.). Academic Press, New York, pp. 387–397.

Goodman, L.R., G.M. Cripe, P.H. Moody and D.G. Halsell. 1988. Acute toxicity of malathion, tetrabromobisphenol-A and tributyltin chloride to mysids (*Mysidopsis bahia*) of three ages. *Bull. Environ. Contam. Toxicol.* **41**, 746–753.

Gras, G. and J.A. Rioux. 1965. Relation entre la structure chimique et l'activité insecticide des composes organiques de l'étain (essai sur les larves de *Culex pipiens* L.) *Arch. Inst. Pasteur Tunis*, **42**, 9–22.

Hall, L.W., Jr. 1988. Tributyltin environmental studies. *Mar. Pollut. Bull.* **19**, 431–438.

Hall, L.W., Jr and A.E. Pinkney. 1985. Acute and sublethal effects of organotin compounds on aquatic biota: an interpretative literature evaluation. *CRC Crit. Rev. Toxicol.*, **14**, 159–209.

Hall, L.W., Jr, A.E. Pinkney, S. Zeger, D.T. Burton and M.J. Lenkenvich. 1984. Behavioral-responses to two estuarine fish species subjected to bis (tri-n-butyltin) oxide. *Water Resources Bull.* **20**, 235–239.

Hall, L.W., Jr, S.J. Bushong, W.S. Hall and W.E. Johnson. 1988. Acute and chronic effects of tributyltin on a Chesapeake Bay copepod. *Environ. Toxicol. Chem.* **7**, 41–46.

His, E. and R. Robert. 1985. Developpement des velígères de *Crassostrea gigas* dans le bassin d'Arcachon études sur les mortalités larvaires. *Rev. Trav. Inst. Peches Marit.*, **47**, 63–88.

Ho, S.L. 1984. Evaluation of Carbon-14 Uptake Algal Toxicity Assay and Its Application in Field Assessment of Tributyltin Chloride and Chlorinated Sewage Toxicities. Master's Thesis, The College of William and Mary, Williamsburg, Virginia, 108 pp.

Hodge, V.F., S.L. Seidel and E.D. Goldberg. 1979. Determination of tin (IV) and organotin compounds in natural waters, coastal sediments and macroalgae by atomic absorption spectrometry. *Anal. Chem.*, **51**, 1256–1259.

Hopf, H.S., J. Duncan, J.S.S. Beesley, D.J. Webley and R.F. Sturrock. 1967. Molluscicidal properties of organotin and organolead compounds with particular reference to triphenyllead acetate. *Bull. World Health Organiz.*, **36**, 955–961.

Johansen, K. and F. Mohlenberg. 1987. Impairment of egg production in *Acartia tonsa* exposed to tributyltin oxide. *Ophelia*, **27**, 137–141.

Laughlin, R.B. and W.J. French. 1980. Comparative study of the acute toxicity of a homologous series of trialkyltins to larval shore crabs, *Hemigrapsus nudus* and lobster, *Homarus americanus*. *Bull. Environ. Contam. Toxicol.*, **25**, 802–809.

Laughlin, R.B. and O. Linden. 1982. Sublethal responses of the tadpoles of the European frog *Rana temporaria* to two tributyltin compounds. *Bull. Environ. Contam. Toxicol.*, **28**, 494–499.

Laughlin, R.B. and O. Linden. 1985. Fate and effects of organotin compounds. *Ambio*, **14**, 88–94.

Laughlin, R.B., O. Linden and H.E. Guard. 1982. Acute toxicity of tributyltins and tributyltin leachates from marine antibiofouling paints. *Bull. Liaison Comité Int. Permanent Rech. Preservation Matoriaux Milieu Mar. (COIPM)*, **13**, 3–20.

Laughlin, R., W. French and H.E. Guard. 1983. Acute and sublethal toxicity of tributyltin oxide (TBTO) and its putative environmental product, tributyltin sulfide (TBTS) to zoeal mud crabs, *Rhithropanopeus harrisii*. *Water Air Soil Pollut.*, **20**, 69–79.

Laughlin, R.B., K. Nordlund and O. Linden. 1984. Long-term effects of tributyltin compounds on the Baltic amphipod *Gammarus oceanicus*. *Mar. Environ. Res.*, **12**, 243–271.

Laughlin, R.B., R. Gustafson and P. Pendoley. 1988. Chronic embryo–larval toxicity of tributyltin (TBT) to the hard shell clam *Mercenaria mercenaria*. *Mar. Ecol.*, **48**, 29–36.

Laughlin, R.B., R.G. Gustafson and P. Pendoley. 1989. Acute toxicity of tributyltin (TBT) to early life history stages of the hard shell clam,

Mercenaria mercenaria. Bull. Environ. Contam. Toxicol., **42**, 352–358.

Lawler, I.F. and J.C. Aldrich. 1987. Sublethal effects of bis(tri-n-butyltin) oxide on *Crassostrea gigas* spat. *Mar. Pollut. Bull.*, **18**, 274–278.

Linden, E., B.E. Bengtsson, O. Svanberg and G. Sundstrom. 1979. The acute toxicity of 78 chemicals and pesticide formulations against two brackish water organisms, the bleak (*Alburnus alburnus*) and the harpacticoid (*Nitocra spinipes*). *Chemosphere*, **11**, 843–851.

M & T Chemicals Company. 1976a. Acute Toxicity of Tributyltin Oxide to *Daphnia magna*. Unpublished study, EPA Accession No. 136469.

M & T Chemicals Company. 1976b. Acute Toxicity of Tri-n-butyltin Oxide to Bluegill (*Lepomis macrochirus*). Unpublished study, EPA Accession No. 136471.

M & T Chemicals Company. 1976c. Acute Toxicity of Tri-n-butyltin Oxide to Channel Catfish (*Ictalurus punctatus*), the Freshwater Clam (*Elliptio complanatus*), the Common Mummichog (*Fundulus heteroclitus*), and the American Oyster (*Crassostrea virginica*). Unpublished study, EPA Accession No. 136470.

M & T Chemicals Company. 1977. Toxicity of Tri-n-butyltin Oxide (TBTO) to Embryos of Eastern Oysters (*Crassostrea virginica*). Report submitted by EG & G Bionomics to M & T Chemicals Co., Rahway, New Jersey.

M & T Chemicals Company. 1978. The Toxicity of Bis(tri-n-butyltin Oxide) (TBTO) to Rainbow Trout (*Salmo gairdneri*). Unpublished study. EPA Accession No. 106966.

M & T Chemicals Company. 1979. Acute Toxicity of Three Samples of TBTO (Tributyltin Oxide) to Juvenile Sheepshead Minnows (*Cyprinodon variegatus*). Report No. BP-79-9-151 submitted by EG & G Bionomics to M & T Chemicals Co., Rahway, New Jersey.

M & T Chemicals Company. 1981a. Comparative Toxicity of Tri-n-butyltin Oxide (TBTO) Produced by Two Different Chemical Processes to the Marine Alga *Skeletonema costatum*. Report No. BP-81-6-109 submitted by EG & G Bionomics to M & T Chemicals Co., Rahway, New Jersey.

M & T Chemicals Company. 1981b. Comparative Toxicity of Tri-n-butyltin Oxide (TBTO) Produced by Two Different Chemical Processes to Pink Shrimp (*Penaeus duorarum*). Report No BP-81-4-55 submitted by EG & G Bionomics to M & T Chemicals Co., Rahway, New Jersey.

M & T Chemicals Company. 1981c. Acute Toxicity of BioMet* 204 Red to Mysid Shrimp (*Mysidopsis bahia*). Report No. BP-81-2-15 submitted by EG & G Bionomics to M & T Chemicals Co., Rahway, New Jersey.

Macek, K., W. Birge, F.L. Mayer, A.L. Buikema, Jr and A.W. Maki. 1978. Toxicological effects. In *Estimating the Hazard of Chemical Substances to Aquatic Life*, J. Cairns, K.L. Dickson and A.E. Maki (Eds) American Society for Testing and Materials, Philadelphia, pp. 27–32.

Maguire, J.R. 1987. Environmental aspects of tributyltin. *Appl. Organometal. Chem.*, **1**, 475–498.

Maguire, J.R., Y.K. Chan, G.A. Bengert, E.J. Hale, P.T.S. Wong and O. Kramer. 1982. Occurrence of organotin compounds in Ontario lakes and rivers. *Environ. Sci. Technol.*, **16**, 698–702.

Martin, R.C., D.G. Dixon, R.J. Maguire, P.V. Hodson and R.J. Tkacz. 1991. Acute toxicity, uptake, depuration and tissue distribution of tri-n-butyltin in rainbow trout, *Salmo gairdneri*. *Aquatic Toxicol.* **15**, 37–52.

Matthias, C.C., J. Bellama, G. Olson and F. Brinckman. 1986. Comprehensive method for the determination of aquatic butyltin and butylmethyltin species at ultratrace levels using simultaneous hybridization/extraction with gas chromatographic–flame photometric detection. *Environ. Sci. Technol.*, **20**, 609–615.

Meador, J.P. 1986. An analysis of photobehavior of *Daphnia magna* exposed to tributyltin. In: Proceedings of the Oceans '86 Organotin Symposium, Vol. 4, 25–27 September 1986, Marine Technology Society, Washington, DC, pp. 1213–1218.

Newton, F., A. Thum, B. Davidson, A. Valkirs and P. Seligman. 1985. Effects on the Growth and Survival of Eggs and Embryos of the California Grunion (*Leuresthes tenuis*) Exposed to Trace Levels of Tributyltin. Technical Report 1040, Naval Ocean Systems Center, San Diego, California, 15 pp.

Nicklin, S. and M.W. Robson. 1988. Organotins: toxicology and biological effects. *Appl. Organometal. Chem.*, **2**, 487–508.

Oceans. 1986. Proceedings of the Organotin Symposium, Vol. 4, 25–27 September 1986, Marine Technology Society and IEEE Ocean Engineering Society, Washington, DC, 229 pp.

Oceans. 1987. Proceedings of the International Organotin Symposium, Vol. 4, 28 September–1 October 1987, Marine Technology Society and

IEEE Ocean Engineering Society, Washington, DC, 229 pp.

Pinkney, A.E., L.W. Hall, Jr, M.J. Lenkevich and D.T. Burton. 1985. Comparison of avoidance responses of an estuarine fish, *Fundulus heteroclitus* and a crustacean, *Palaemonetes pugio*, to bis-(tri-n-butyltin) oxide. *Water, Air Soil Pollut.*, **25**, 33–40.

Pinkney, A.E., D.A. Wright, M.A. Jepson and D.W. Towle. 1989a. Effects of tributyltin compounds on ionic regulation and gill ATPase activity in estuarine fish. *Comparative Biochem. Physiol.*, **92C**, 125–129.

Pinkney, A.E., D.A. Wright and G.M. Hughes. 1989b. A morphometric study of the effects of tributyltin compounds on the gills of the mummichog, *Fundulus heteroclitus*. *J. Fish Biol.*, **34**, 665–677.

Pinkney, A.E., L.L. Matteson and D.A. Wright. 1990. Effects of tributyltin on survival, growth, morphometry and RNA–DNA ratio of larval striped bass, *Morone saxatilis*. *Arch. Environ. Contam. Toxicol.*, **19**, 235–241.

Piver, W.T. 1983. Organotin compounds: industrial applications and biological investigations. *Environ. Health Perspective*, **4**, 61–79.

Ritchie, L.S., L.A. Berrios-Duran, L.P. Frick and I. Fox. 1964. Molluscicidal time–concentration relationships of organotin compounds. *Bull. World Health Organiz.*, **31**, 147–149.

Ritchie, L.S., V.A. Lopez and J.M. Cora. 1974. Prolonged applications of an organotin against *Biomphalaria glabrata* and *Schistosoma mansoni*. In: *Molluscicides in Schistosomiasis Control*, T.C. Cheng (Ed.). Academic Press, New York, pp. 77–88.

Robert, R. and E. His. 1981. Action de l'acetate de tributyl-étain sur les oeufs et les larves de deux mollusques d'interêt commercial: *Crassostrea gigas* (Thunberg) et *Mytilus galloprovincialis* (Lamark). Internat. Council Explor. Sea Paper, CM1981:F42, 16 pp.

Roberts, M.H., Jr. 1987. Acute toxicity of tributyltin chloride to embryos and larvae of two bivalve mollusks, *Crassostrea virginica* and *Mercenaria mercenaria*. *Bull. Environ. Contam. Toxicol.*, **39**, 1012–1019.

Salazar, S.M. 1985. The Effects of Bis(tri-n-butyltin) Oxide on Three Species of Marine Phytoplankton. Technical Report 1039, Naval Ocean Systems Center, San Diego, California, 16 pp.

Salazar, M.H. and S.M. Salazar. 1985. The effects of sediment on the survival of mysids exposed to organotins. In: Proceedings of the 11th US/Japan Experts Meeting, Management of Bottom Sediments Containing Toxic Substances, 4–6, November, Seattle, Washington, pp. 176–192.

Salazar, M.H. and S.M. Salazar. 1989. Acute Effects of (Bis)tributyltin Oxide on Marine Organisms, Summary of Work Performed 1981–1983. Draft Report, Naval Ocean Systems Center, San Diego, California, 88 pp.

Seiffer, E.A. and H.F. Schoof. 1967. Tests of 15 experimental molluscicides against *Australorbis glabratus*. *Public Health Report*, **82**, 833–839.

Seinen, W., T. Helder, H. Vernij, A. Penninks and P. Leeuwangh. 1981. Short term toxicity of tri-n-butyltinchloride in rainbow trout (*Salmo gairdneri* Richardson) yolk sac fry. *Sci. Total Environ.*, **19**, 155–166.

Short, J.W. and F.P. Thrower. 1986. Tri-n-butyltin caused mortality of Chinook salmon, *Oncorhynchus tshawytscha*, on transfer to a TBT-treated marine net pen. In: Proceedings of the Oceans '86 Organotin Symposium, Vol. 4, Washington, DC, pp. 1201–1205,

Smith, P.J., A.J. Crowe, V.G. Kumar Das and J. Duncan. 1979. Structure–activity relationships for some organotin molluscicides. *Pesticide Sci.*, **10**, 419–422.

Stebbing, A.R.D. 1981. Hormesis stimulation of colony growth in *Campanularia flexuosa* (Hydrozoa) by copper, cadmium and other toxicants. *Aquatic Toxicol.*, **1**, 227–238.

Stephen, C.E., D.I. Mount, D.H. Hansen, J.H. Gentile, G.A. Chapman and W.A. Brungs. 1985. Guidelines for Deriving Numerical Water Quality Criteria for the Protection of Aquatic Organisms and their Uses. Report, US Environmental Agency, Duluth, Minnesota, 90 pp.

Stromgren, T. and T. Bongard. 1987. The effect of tributyltin oxide on growth of *Mytilus edulis*. *Mar. Pollut. Bull.*, **18**, 30–31.

Thain, J.E. 1983. The Acute Toxicity of Bis (Tributyltin) Oxide to the Adults and Larvae of Some Marine Organisms, International Council for the Exploration of the Sea, Copenhagen, Paper CM1983/E:13, 5 pp.

Thain, J.E. and M.J. Waldock. 1985. The Growth of Bivalve Spats Exposed to Organotin Leachates from Antifouling Paints. International Council for the Exploration of the Sea, Copenhagen, Paper CM1985/E:28, 6 pp.

Thompson, J.A., M.G. Sheffer, R.C. Pierce, Y.K. Chau, J.J. Cooney, W.R. Cullen and R.J. Maguire. 1985. *Organotin Compounds in the Aquatic Environment: Scientific Criteria for Assessing their Effects on Environmental Quality*. National Resource Council, Canada. Publication NRCC 22494. NRCC/CNRC, Ottawa, 284 pp.

Unger, M.A., W.G. MacIntyre, J. Greaves and R.J. Huggett. 1986. GC determination of butyltins in natural waters by flame photometric detection of hexylderivatives with mass spectrometric confirmation. *Chemosphere*, **15**, 461–470.

U'ren, S.C. 1983. Acute toxicity of bis(tributyltin) oxide to a marine copepod. *Mar. Pollut. Bull.*, **14**, 303–306.

Valkirs, A., B. Davidson and P. Seligman. 1985. Sublethal Growth Effects and Mortality to Marine Bivalves and Fish from Long-term Exposure to Tributyltin. Technical Report 1042, Naval Ocean Systems Center, San Diego, California, 35 pp.

Valkirs, A.O., P.F. Seligman, P.M. Stang, V. Homer, S.H. Lieberman, S.H. Vafa and C.A. Dooley. 1986. Measurements of butyltin compounds in San Diego Bay. *Mar. Pollut. Bull.*, **17**, 319–324.

Walsh, G.E., L.L. McLaughlan, E.M. Lores, M.K. Louie and C.H. Deans. 1985. Effects of organotins on growth and survival of two marine diatoms, *Skeletonema costatum* and *Thalassiosira pseudonana*. *Chemosphere*, **14**, 383–392.

Walsh, G.E., M.K. Louie, L.L. McLaughlan and E.M. Lores. 1986. Lugworm (*Arenicola cristata*) larvae in toxicity tests: survival and development when exposed to organotins. *Environ. Toxicol. Chem.*, **5**, 749–754.

Ward, G.S., G.C. Cramm, P.R. Parrish, H. Trachman and A. Slesinger. 1981. Bioaccumulation and chronic toxicity of bis(tributyltin) oxide (TBTO): tests with a saltwater fish. In *Aquatic Toxicology and Hazard Assessment: Fourth Conf.*, D.R. Branson and K.L. Dickson (Eds), ASTM STP 737, American Society for Testing and Materials, Philadelphia, pp. 183–200.

Webbe, G. and R.F. Sturrock. 1964. Laboratory tests of some new molluscicides in Tanganyika. *Ann. Trop. Med. Parasitol.*, **58**, 234–239.

Weis, J.S. and K. Kim. 1988. Tributyltin is a teratogen in producing deformities in limbs of the fiddler crab, *Uca pugilator*. *Arch. Environ. Contam. Toxicol.*, **17**, 583–589.

Wong, P.T.S., Y.K. Chau, O. Kramar and G.A. Bengert. 1982. Structure–toxicity relationship of tin compounds on algae. *Can. J. Fish. Aquat. Sci.*, **39**, 483–488.

World Health Organization. 1980. Tin and organotin compounds: a preliminary review. In: *Environmental Health Criteria 15*, World Health Organization, Task Group on Environmental Health Aspects of Tin and Organotin Compounds, Geneva, Switzerland, 109 pp.

Zuckerman, J.J., R.P. Reisdorf, H.V. Ellis, III and R.R. Wilkinson. 1978. Organotins in biology and the environment. In: *Organo Metals and Organometalloids, Occurrence and Fate in the Environment*, F.E. Brinckman and J.M. Bellama (Eds). American Chemical Society Symposium Series, 82, Washington, DC, pp. 388–424.

EXPERIMENTAL STUDIES OF CHRONIC TOXICITY OF TRIBUTYLTIN COMPOUNDS

10

Roy B. Laughlin, Jr,[1] *John Thain,*[2] *Brad Davidson,*[3] *Aldis O. Valkirs*[3] *and Frederick C. Newton, III*[4]

[1]*Azurea, Inc., PO Box 561178, Rockledge, Florida 32955, USA*
[2]*Ministry of Agriculture, Fisheries and Food, Fisheries Laboratory, Remembrance Avenue, Burnham-on-Crouch, Essex CM0 8HA, UK*
[3]*Naval Ocean Systems Center, Environmental Services Branch, Code 522, San Diego, California 92152-5000, USA*
[4]*Kinnetic Laboratories, Inc., 5225 Avenida Encinas, Suite H, Carlsbad, California 92008, USA*

Organotin. Edited by M.A. Champ and P.F. Seligman. Published in 1996 by Chapman & Hall, London. ISBN 0 412 58240 6

ABSTRACT

This chapter examines chronic toxicity of TBT through a discussion of five comprehensive studies performed during the last 8 y. The studies share the following characteristics. They are experimental studies conducted under laboratory conditions and designed to demonstrate a causal relationship between TBT exposure and a sublethal response, which was usually reduction in growth or abnormal development; biological endpoints, such as metamorphosis or hatching, usually defined the period of observation; and experiments lasted more than 96 h in all cases. The studies are characterized as a progression of exposure levels that began above 1 $\mu g\ l^{-1}$, but dropped to 1–10 ng l^{-1}, concomitant with increasingly sensitive chemical analysis methods for TBT. The lowest levels tested are similar to those found in many nearshore and estuarine habitats. Test organisms include developmental stages of fish (*Leuresthes tenuis*), decapod crustaceans (*Rhithropanopeus harrisii*), mysid shrimp (*Acanthomysis sculpta*), oysters (*Crassostrea gigas*), and clams (*Mercenaria mercenaria*). Results of these studies demonstrate a range of taxonomically correlated sensitivity to TBT of two to three orders of magnitude. Fish and crab larvae were most tolerant, with exposures in excess of 1 $\mu g\ l^{-1}$ causing notable reductions in growth or survival. Mysids formed an intermediate group, with a sensitivity about one order of magnitude less. Clams and oysters are the most sensitive group, showing significant sublethal effects in TBT concentrations of 10–50 ng l^{-1}. The taxonomic pattern of sensitivity depends, at least partially, on possession of metabolic pathways to rid tissue of TBT. Studies with molluscs clearly show a causal relationship between TBT exposure and adverse sublethal effects that is consistent with observations made in field studies.

10.1. INTRODUCTION

Nearly all experimental studies of chronic toxicity of TBT to marine organisms were conducted within the past 15 y. In the earliest studies, investigators were more concerned with the effectiveness of TBT compounds for controlling snail vectors of schistosomiasis (Ritchie *et al.*, 1974) than for their environmental toxicity to nontarget species, although such concerns had been noted (Deschiens *et al.*, 1966). One of the earliest studies of nontarget organisms with chronic exposure to TBT was conducted by Seinen *et al.* (1981) with rainbow trout (*Salmo gairdneri*) yolk sac larvae. In their experiments, fish were exposed for 110 d to a nominal concentration of 1 $\mu g\ l^{-1}$ tributyltin chloride. The concentration ranges these investigators tested were similar to those that might be expected after application to control schistosomiasis vectors and are much higher than those expected (and subsequently found) for TBT released from antifouling coatings. These studies clearly suggested, and subsequent experience showed, that nontarget species were adversely affected, sometimes severely, by exposure to dissolved TBT.

The discovery of the chronic effects of TBT compounds followed a typical pattern in that TBT use became firmly established prior to recognition of its harmful exposure to nontarget species (Alzieu *et al.*, 1980; Alzieu, 1991; His, Chapter 12). Field studies in France provided compelling evidence of a role for TBT in the decline of the oyster (*Crassostrea gigas*) fishery. Reasonable skepticism remained, however, until experiments were performed demonstrating a causal role for TBT producing shell abnormalities and poor growth. TBT was shown to dominate environmental and genetic factors. Description of the sensitivity of *C. gigas* to TBT was serendipitous but added impetus to support for experimental studies of other nontarget species. Few people felt that it was appropriate to wait until new problems occurred elsewhere as use of TBT in antifouling paints gained acceptance in, for example, US markets.

10.2. COMPREHENSIVE STUDIES OF SPECIES

Early studies on the chronic effects of TBT performed in the laboratory (Laughlin and French, 1981; Laughlin *et al.*, 1982, 1984) showed that TBT is a slow-acting toxin whose toxicity was significantly underestimated by short-term tests. In spite of such limitations, some of these studies demonstrated adverse effects due to exposure concentrations well below 1 $\mu g\ l^{-1}$, below limits of analytical detection for TBT available at the time. For these reasons, chronic tests became an important method for measuring the true effect of TBT on nontarget organisms.

Chronic toxicity tests imply exposures lasting longer than the 24–96 h required by standard protocols of acute tests. Chronic toxicity tests are best defined not by the duration but rather by an interval that continues until a biologically defined endpoint. For example, larval studies may last <2 weeks for many species, but are operationally defined by the normal interval during embryogenesis, ending with achievement of some distinct morphological stage, frequently following metamorphosis. The unquestioned validity of chronic tests with larvae rests upon the occurrence of cellular growth, division, and differentiation during a relatively short, but significant, part of the life history of the organism. Other chronic test designs may require much longer periods of exposure and observation. Those involving reproduction and assessment of gamete–embryo viability may require weeks or months during the normal course of reproductive events and provide another approach to testing effects on cellular growth and differentiation.

In this chapter we will discuss several experimental chronic studies that have successfully met analytical and biological challenges posed by TBT. Our criterion of selection was primarily that a relevant biological endpoint other than death was utilized in exposures lasting longer than 4 d. These studies are not necessarily identical protocols performed in different species. They are similar in that they were conducted with early developmental stages of marine organisms, lasted for the duration of a developmental period, and examined sublethal responses of exposure to TBT. These studies were attempts to identify sensitive species and to describe responses to TBT that were particularly acute or diagnostic of TBT exposure.

10.2.1. LIFE CYCLE TESTS ON MYSID SHRIMP

Mysid shrimp have been used as model species in life-cycle tests because generation times of many species are short, frequently only several weeks. The mysid *Acanthomysis sculpta* was used for experiments described in this section. It has been used previously in bioassays (Tatem and Portzer, 1985; Salazar and Salazar, Chapter 15) and was shown to compare favorably with the more familiar Atlantic–Gulf of Mexico estuarine mysid, *Mysidopsis bahia*. Experiments presented here describe three separate investigations, one each on long-term mortality, growth during chronic exposure, and effects on larval production and viability following long-term exposure of the gravid females.

10.2.1a. Materials and methods

During exposure to TBT, mysids were kept in 500-ml polycarbonate cylinders suspended in aquaria. Flowing seawater (200 ml min^{-1}) entered at the top. The bottom of each cylinder was covered with 202-μm Nitex mesh. In the long-term mortality and growth experiments, up to 15 mysids were initially put into each cylinder. In the larval production experiment, only one female was placed into each of 14 cylinders. Daily, mysids were fed at least 30 *Artemia salina* nauplii per mysid, a ration at or above maintenance levels. Water quality was measured daily during flowthrough chronic tests and was consistent with ambient offshore water.

Tributyltin was introduced into the exposure system as a leachate from panels coated with antifouling paint (International Paint Corporation SPC 954). The system delivered five separate dilutions of the test solution released from the panels plus a seawater control. TBT concentrations during tests reported here were from 30 to 600 ng l^{-1}, varying slightly among the different tests since measured release rates from the same set of panels varied over time. Specific concentrations are given in the results of each experiment. Analyses of TBT concentrations in exposure tanks were performed using hydride derivatization followed by atomic absorption detection (Valkirs *et al.*, 1985).

Length and weight were determined at different intervals given in figures and tables for each experiment and are for living mysids only. Selected individuals were anesthetized with MS-222 (tricaine methanesulfonate) to stop movement but avoid shrinkage and distortion from immediate preservation. Mysids were then measured with a Bausch & Lomb dissecting microscope fitted with an ocular micrometer. Within 8 h following length measurement, the mysids were freeze dried and weighed on a Cahn gram electrobalance. Additional details of these experiments are given in Davidson *et al.* (1986).

10.2.1b. Results

During the long-term mortality test, most deaths occurred during the first 22 d, when mysids were in juvenile and subadult instars (Fig. 10.1). All mysids exposed to 480 ng l^{-1} TBT died within 7 d, but, on day 22, mortality in the other test concentrations (in 380 ng l^{-1} TBT and below) was not dose dependent, and differences were not statistically significant ($p > 0.05$). Similar trends were evident on day 41, the last day on which observations were made prior to release of young by females. Thus, through day 41, the 'no

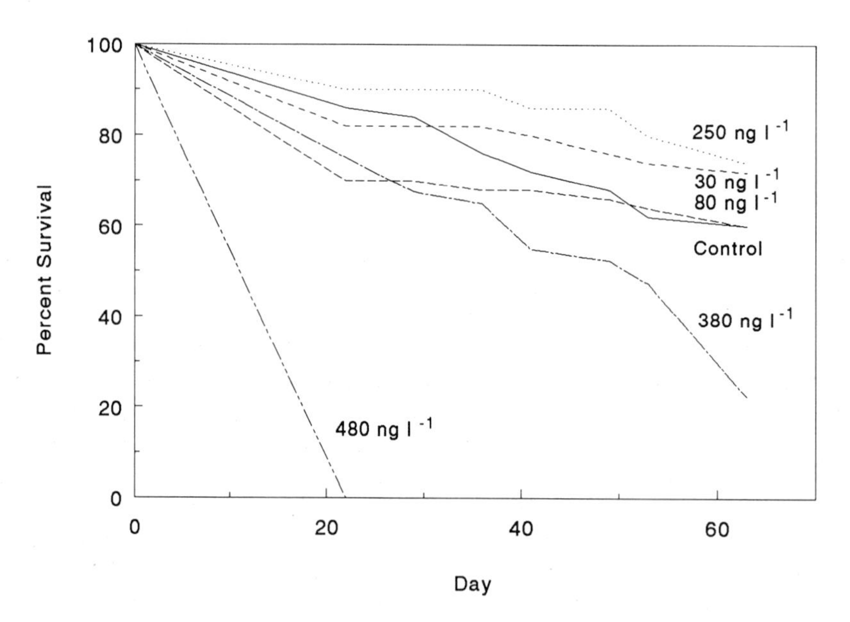

Figure 10.1 Long-term mortality of *A. sculpta* exposed for up to 63 d to TBT leachates from painted panels. The percent of individuals surviving after days of exposure to various TBT concentrations is presented. All mysids in the 480 ng l^{-1} concentration died within 7 d.

observable effects concentration' (NOEC) with respect to mortality for this experiment was 380 ng l^{-1}.

The experiment ended on day 63. At this time, mean survival in 380 ng l^{-1} TBT was only 22% compared with 60% in controls and 80 ng l^{-1} TBT, the next closest groups. This difference was statistically significant (p <0.05 ANOVA followed by Duncan's Multiple Range Test). Among the other groups, 63-d cumulative mortality was not dose dependent. The 68-d NOEC for mortality was 250 ng l^{-1}. The decrease in the NOEC value after females began releasing juveniles from their brood pouch suggests these mysids may become more sensitive to TBT during this process.

Effects of TBT on growth of mysids were measured during four separate trials, including the bioassay just discussed. These trials lasted from 21 to 63 d. Data are grouped according to trial, age (in days) of the mysids, and sex, since males tended to be smaller than females at maturity (Fig. 10.2).

Sexually mature females in the higher dose levels (<300 ng l^{-1}) tended to exhibit less growth than their counterparts in the lower dose levels and less than sexually mature males in all dose levels. For example, in trial 4 (Fig. 10.2), means for both length and weight of females were significantly lower in the 380 ng l^{-1} exposure than in any of the lower levels (p <0.05, ANOVA followed by Duncan's Multiple Range Test). In trial 1 females showed a significant decrease in growth compared with controls at day 39 in all dose levels. Differences of mean weights and lengths of males, although obvious, were not statistically significant. In trial 2, females displayed a significant decrease in both length and weight in concentrations of 490 ng l^{-1} TBT. Again no statistically significant differences occurred in corresponding males as a function of exposure to TBT. In addition to reductions in growth of females, TBT appears to reduce growth and development of juveniles and subadult *A. sculpta* exposed to concentrations >430 ng l^{-1} TBT (final chronic value). A significant difference in mean length was detected on day 14 (trial 1, Fig. 10.2) where mean lengths of 4.07 and 3.99 mm in 190 and 330 ng l^{-1} TBT, respectively, were significantly less than the mean length, 4.37 mm, measured for controls. Data from day 24 of the same trial also showed a significantly lower mean length of 6.73 mm in 70 ng l^{-1} TBT compared with 7.50 mm mean length in controls. During the second trial, a significant difference in mean length of mysids in 600 ng l^{-1} TBT of 2.96 mm, and that of controls of 3.59 mm, was also observed. Notably, in the third trial, which tested TBT concentrations only to 310 ng l^{-1} for 21 d, no statistically significant effects on growth of subadults was observed.

During reproduction success tests, females in all but the highest TBT concentrations, 520 ng l^{-1}, produced brood pouches or young. Among mysids in controls and the remaining four exposure levels, no significant differences occurred with respect to the number of juveniles released per individual female, numbers of individuals in unhatched broods (compared with those holding broods), or number of days from hatching of a female to the release of her juveniles. However, the number of females that actually released viable juveniles differed significantly as a function of TBT exposure (p <0.05; $\chi^2 = 8.91$). The mean number of larvae released by mysids in 190 and 330 ng l^{-1} was much less than the average released by controls, 30, or 90 ng l^{-1} TBT exposure groups (Table 10.1). Upon visual examination of mysids that did not release juveniles, we found that many of the brood pouches were empty or contained dead (undeveloped) eggs. We suspect that females holding no eggs had produced, then aborted them shortly afterwards, as all females had been observed holding eggs at some time during the experiments. To summarize briefly, if females produced viable larvae, the number and their development rate were not greatly influenced by TBT

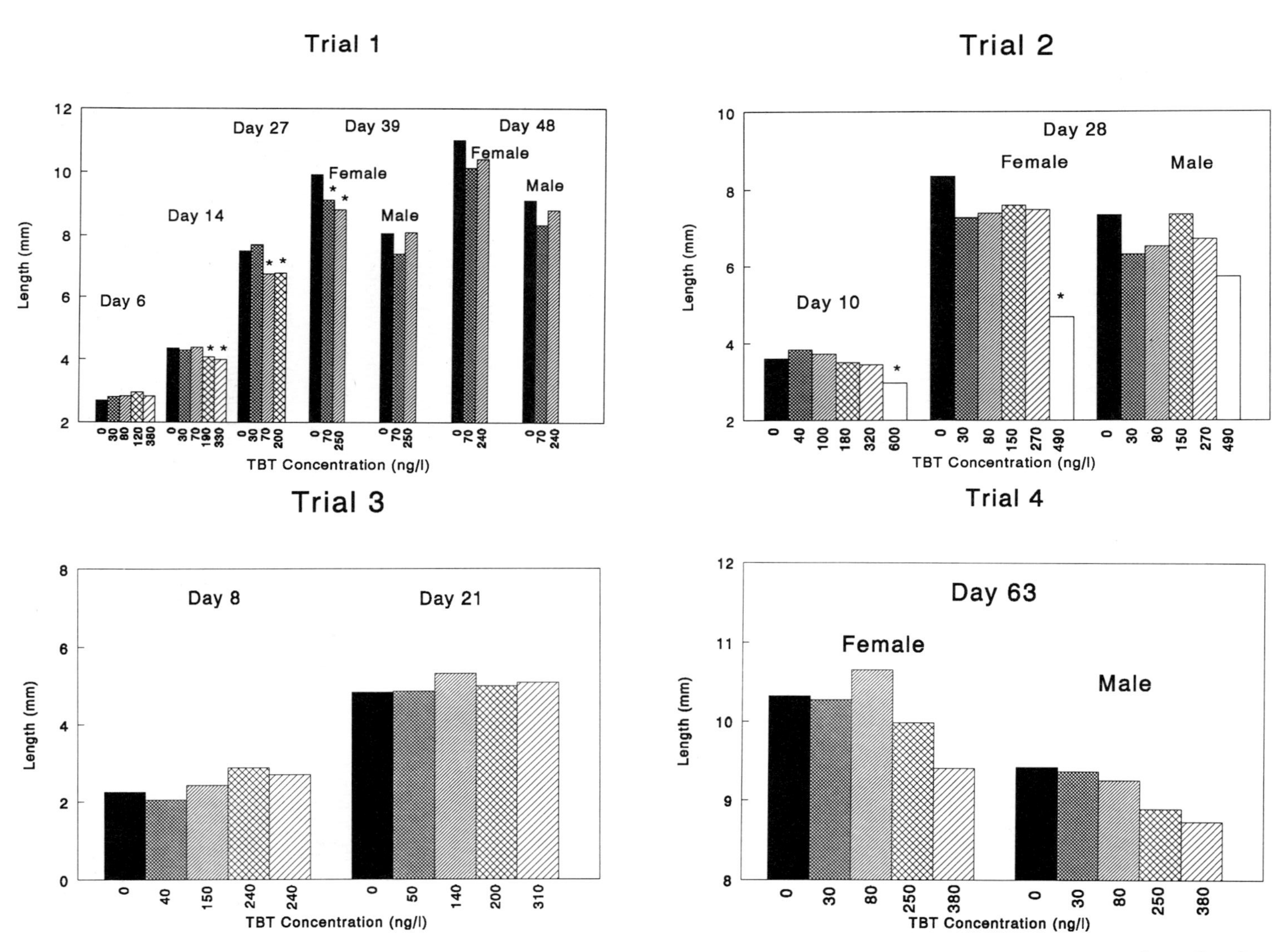

Figure 10.2 Results of *Acanthomysis sculpta* growth tests for trials 1–4. Tributyltin concentrations were measured on the day growth observations were made. Males and females were measured as separate groups after sexual maturity. Bars denoted with an asterisk (*) are statistically different within a group from those not denoted.

Table 10.1 Number of viable juveniles released from female *Acanthomysis sculpta* exposed to TBT for the duration of the mothers' life cycle

	Dose (ng l^{-1})					
Female ID number	*Control*	*30*	*90*	*190*	*330*	*520*
1	10	17	13	0	0	—[a]
2	10	0	6	0	0	—
3	0	11	10	0	0	—
	17	0	14	13	0	—
5	0	2	0	0	0	—
6	5	0	7	0	0	—
7	0	6	10	0	0	—
8	0	17	20	14	0	—
9	0	1	5	7	0	—
10	0	0	10	0	0	—
11	18	10	0	0	0	—
12	0	21	12	0	0	—
13	8	19	0	0	0	—
14	*[b]	20	*	*	0	—

[a]Dash indicates individual dead before sexual maturity.
[b]Asterisk indicates data not collected.

exposure; however, TBT exposure reduced the number of females that successfully produced any larvae.

We compared sensitivities to disruption by TBT of the three processes we measured, survival, growth, and reproduction. The 'no observable effects concentration' (NOEC) has been used as a standard chronic value for parameters tested but has been referred to by several terms including 'maximum acceptable toxicant concentration' (MATC) (Mount and Stephan, 1967) and 'no observable effects level' (NOEL) (Payne and Hall, 1979). Stephen *et al.* (1985) further defined a 'chronic value' as the geometric mean of the lower and upper chronic limits of a chronic test. The lower limit referenced here is the highest concentration of a toxicant tested that did not show an adverse effect (the actual definition of a NOEC). The upper limit is the lowest concentration tested that caused an adverse effect. Following this procedure, we calculated a final chronic value with respect to the three indices; long-term mortality, growth, and reproduction, measured during chronic bioassays (Table 10.2). For long-term mortality, the NOEC and final chronic value were 250 and 300 ng l^{-1}, respectively. Values for growth calculated for subadults (NOEC, 320 ng l^{-1}; final chronic value, 430 ng l^{-1}) were greater than those calculated for adult females (NOEC, 250 ng l^{-1} TBT; final chronic value, 310 ng l^{-1} TBT). Corresponding values for reproduction were lowest of all with a NOEC of 90 ng l^{-1} TBT and a final chronic value of 130 ng l^{-1} TBT.

Table 10.2 Summary of chronic values of TBT (ng l^{-1}) in mysid shrimp *Acanthomysis sculpta*

Test	*Lower limit (NOEC)*	*Upper limit*	*Final chronic value*
Long-term mortality	250	380	300
Growth			
Subadults	320	600	430
Adult females	250	380	310
Reproduction	90	190	130

10.2.2. EFFECTS OF TBT ON GRUNION EGGS AND EMBRYOS

Toxicological studies on the California grunion (*Leuresthes tenuis*) were conducted to provide information about the tolerance of fish gametes and embryos to TBT exposure. We selected this species for several reasons: (a) eggs, which are deposited in sediments, develop in an environmental compartment potentially enriched by TBT or other environmental contaminants; (b) embryos are easily cultured in the laboratory (Ehrlich and Farris, 1971; May, 1971); and (c) large numbers of gametes are routinely available from March through August. This section describes experiments to study effects of TBT on gametes and also on survival and development of fertilized eggs reared and exposed to dissolved TBT in the laboratory.

10.2.2a. Materials and methods

Ripe adult grunion were collected during the first night of their bimonthly spawning run on Ocean Beach (San Diego, California, USA) from 2230 to 2345 hours, 1 June 1984. Eggs from live, spawning females were stripped directly into three 250-ml polycarbonate canisters containing 25 ml of 0, 10, or 74 μg l^{-1} TBT in filtered seawater. Then milt from three males was added immediately to fertilize the eggs since the chorion hardens after extrusion so that sperm cannot penetrate it (Thum and Ehrlich, 1979). Following the fertilization procedure, eggs were strained onto 0.5 mm Nitex mesh and rinsed with a TBT–seawater solution of the same concentration to which the eggs were exposed during fertilization. At this point, all eggs exposed to the same TBT concentration were combined into 25 ml of TBT–seawater solution and transported to the laboratory, care being taken to maintain an ambient temperature of 19.5°C.

Upon arrival at the laboratory, randomly selected eggs from each field-dose level were mixed with coarse silica sand (1–2 mm) at a ratio of ~1000 eggs to 150 ml sand. Eggs in

Table 10.3 Distribution of replicate incubators (three per exposure combination) within exposure combination levels for *Leuresthes tenuis*

Gamete exposure (*μg l^{-1}*)	*Mean exposure concentration for embryos (ng l^{-1})*				
	0	*140*	*330*	*900*	*1700*
0	3	3	3	3	3
10	3	–[a]	–	–	3
74	3	–	–	–	3

[a]Dash indicates no testing at that exposure combination.

sand were placed into an incubation apparatus modeled after that of Ehrlich and Farris (1971). A total of 30 of these incubators containing grunion eggs were exposed to TBT in the same flowthrough aquarium bioassay system used for mysids (Davidson *et al.*, 1986). Flow rates of seawater containing TBT were 180 ml min^{-1}.

This experiment may be viewed as two separate dosing protocols. In the first, fertilization and early embryological development occurred in a seawater control, 10, or 74 μg l^{-1} TBT. Use of high concentrations simulates expected interstitial water concentrations. In the second experiment, fertilized eggs not previously exposed to TBT were exposed for the duration of embryonic development until hatching to TBT concentrations of 0, 140, 330, 900, or 1700 ng l^{-1}. In addition, embryos obtained by fertilizing eggs in 10 or 74 μg l^{-1} TBT were exposed to 0 or 1700 ng l^{-1} TBT for the remainder of larval development. Each exposure group consisted of three replicates (Table 10.3).

After 10 d of incubation at an average water temperature of 20.5°C, the grunion eggs were ready to hatch. Hatching was initiated by depositing eggs and sand from an incubation apparatus into a 3-l glass crystallizing dish containing 2 l of 18°C seawater and gently swirling with a plastic spatula. The eggs, initially suspended in the water column by stirring, settled to the center of the container on top of the sand and began to hatch within

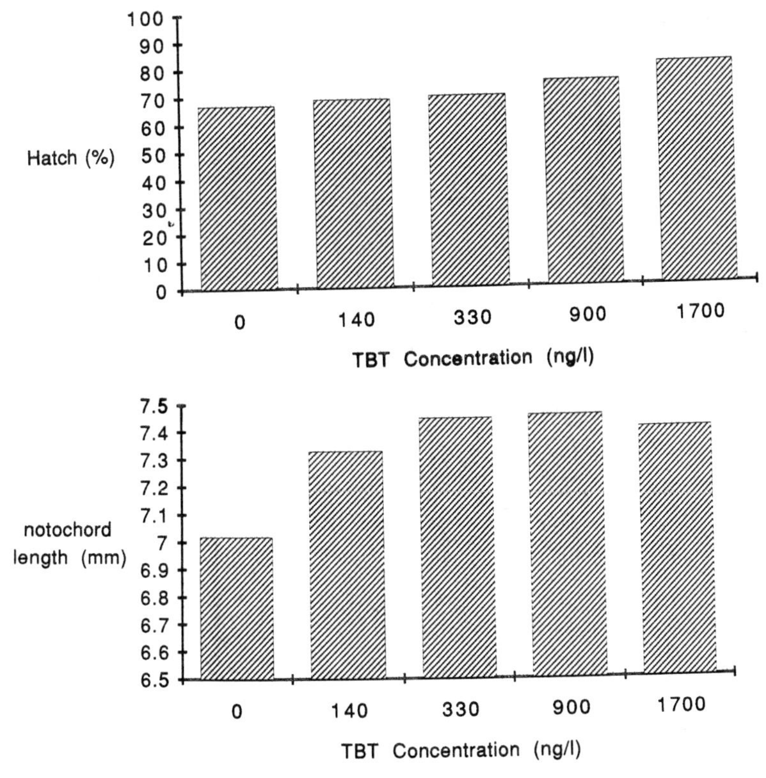

Figure 10.3 *Leuresthes tenuis*. Effects of TBT on hatching success and notochord length of grunion exposed during embryogenesis in the eggs. $n = 10$.

2 min. Hatching continued for 15 min, and was terminated by addition of an anesthetic. Unhatched eggs were removed and divided into groups (fertilized:unhatched and unfertilized:opaque), counted, and preserved. Larvae were also counted and preserved. Ten randomly selected larvae from each replicate were measured for notochord length and examined for anomalous development.

10.2.2b. Results

Mean hatching success of grunion exposed to TBT during embryonic development in the eggs was 67% in controls compared with 81% in 1700 ng l^{-1} TBT, values considered to indicate good hatching success for this species (Fig. 10.3). Hatching success was consistently exposure dependent; that is, increasing exposure to TBT during larval development increased hatching success by 13% in TBT concentrations from 140 to 1700 ng l^{-1}. This difference was not statistically significant ($p > 0.05$).

Mean notochord length at hatch was lowest in controls, 7.02 mm, consistently increasing with increasing exposure to 7.46 mm in 900 ng l^{-1} TBT, then declining to 7.41 mm in 1700 ng l^{-1} TBT (Fig. 10.3). These differences were marginally significant ($p = 0.0685$).

When exposure of grunion eggs to TBT began during fertilization and continued through embryogenesis, effects on hatching success and notochord length were different from the effects when exposure occurred only during embryogenesis. The percentage of unfertilized eggs was not significantly affected by TBT exposure (Fig. 10.4). Hatching success, however, displayed an unusual

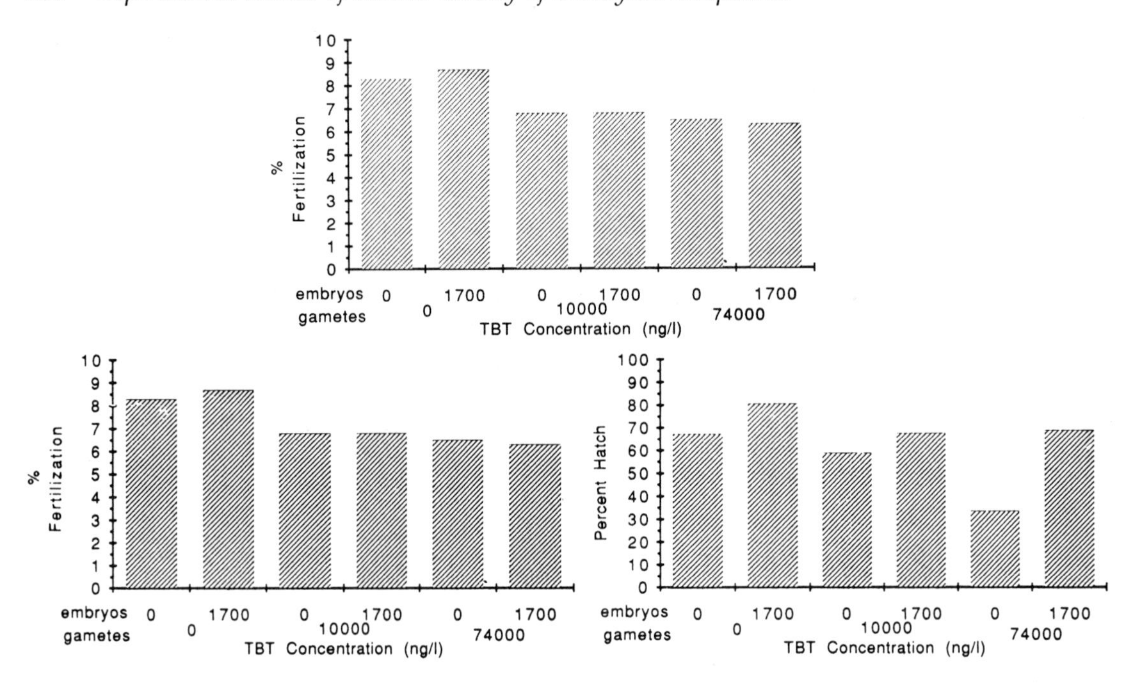

Figure 10.4 *Leuresthes tenuis*. Effects of TBT on growth and hatching success of embryos when exposure began with gametes. The number of measurements of notochord length in each group was ten. Percent hatch or unfertilized is based on three groups in each concentration combination.

pattern. Grunion exposed only at fertilization had generally lower hatching success than those exposed for the duration of development. Highest hatching success occurred in the group exposed to 1700 ng l^{-1} during embryogenesis. The difference in the highest exposure (1700 ng l^{-1}) was statistically significant (two-way ANOVA, $p = 0.005$ for percent hatch). Analysis using the Student–Newman–Keuls multiple range test (SNK) indicated three groups consisting of (1) those exposed only as embryos to 1700 ng l^{-1}, (2) those exposed to 74 μg l^{-1} gametes – 1700 ng l^{-1} embryos, and (3) all other groups (Table 10.4).

Effects of TBT exposure on notochord length were similar to those observed for hatching success. All fish exposed to TBT had notochord lengths longer than those of controls. Experimental fish exposed only after fertilization were larger than those also exposed as gametes before fertilization (Fig. 10.4). Controls had the smallest notochord. The SNK test showed four groups in the data (Table 10.5).

Table 10.4 SNK rating ($p = 0.05$) of percent hatch results for *Leuresthes tenuis*. Due to a significant interaction, the main effects have been combined. Means with the same grouping letter are not significantly different

Grouping	*Mean*	*N*	*Interaction*
A	80.77	3	0 ng l^{-1}–1700 ng l^{-1}
B	68.62	3	74 μg l^{-1}–1700 ng l^{-1}
B	67.61	3	10 μg l^{-1}–1700 ng l^{-1}
B	67.44	3	0 ng l^{-1}–0 ng l^{-1}
B	58.85	3	10 μg l^{-1}–0 ng l^{-1}
C	33.42	3	74 μg l^{-1}–0 ng l^{-1}

10.2.3. EFFECTS OF TBT ON MUD CRAB LARVAE

Like mysid shrimp, mud crab larvae (*Rhithropanopeus harrisii*, zoeae and megalops), have

Table 10.5 SNK rating (p = 0.05) of notochord length at hatch for *Leuresthes tenuis*. Due to a significant interaction, the main effects have been combined. Means with the same grouping letter are not significantly different

Grouping	*Mean*	*N*	*Interaction*
A	7.41	30	0 ng l^{-1}–1700 ng l^{-1}
BA	7.32	30	10 μg l^{-1}–1700 ng l^{-1}
BC	7.24	30	74 μg l^{-1}–1700 ng l^{-1}
BC	7.23	30	10 μg l^{-1}–0 ng l^{-1}
C	7.12	30	74 μg l^{-1}–0 ng l^{-1}
D	7.02	30	0 ng l^{-1}–0 ng l^{-1}

become a commonly used representative species for laboratory studies involving developmental stages of crustaceans. Unlike many of the other species discussed in this chapter, mud crabs hatch as zoeae, a relatively advanced larval stage, before their use in bioassays. Because the developmental paradigm is highly structured with regard to the number of stages, morphology, and the usual duration of zoeal development, a number of parameters exist that can be measured or observed to provide an index of response. This species is very hardy in laboratory culture, facilitating robust statistical analysis of treatment effects.

In this section, our purpose is to present a comparison of the acute and chronic toxicity of TBT to two geographically an isolated populations of *R. harrisii*, one from Florida and the other from California (USA). Additional studies, published elsewhere, suggest that population differences are produced primarily by interaction of temperature and salinity, dominant physical factors in the environment during embryogenesis (Laughlin and French, 1989a,b).

10.2.3a. Materials and methods

Ovigerous mud crabs, *Rhithropanopeus harrisii*, were collected either from Petaluma River (California, USA) or Sykes Creek, a part of the Indian River Lagoon system (Florida, USA). Petaluma River is a tidal creek in a temperate climatic zone while the Sykes Creek collection site is without tides and in a subtropical area. In general, the Petaluma River population experiences a wider range of temperature and salinity in its habitat during the year, including the reproductive season (summer), but the Florida population experiences higher temperatures overall (Laughlin and French, 1989b). We selected crabs whose eggs had eyespots and had thus completed most embryogenesis under ambient conditions in the field.

Until their eggs hatched, usually within 3–5 d, female crabs were maintained in the laboratory in 25°C, 25‰ salinity made by diluting sand-filtered natural seawater with deionized water. Exposure to TBT began within 18 h after hatching. Three groups of 10 zoeae were put into 10-cm-diameter finger bowls containing 50 ml seawater. We replicated this procedure for three hatches, giving a total of 90 larvae exposed to each TBT concentration. Seawater solutions of toxicant were changed each day and freshly hatched *Artemia* nauplii were added as food for the crab larvae. We continued exposure until all larvae in a bowl died or metamorphosed to megalops. We collected megalops within 24 h after metamorphosis, rinsed them with deionized water to remove superficial sea salts, lyophilized, and weighed them individually to the nearest 0.1 μg.

We exposed mud crab larvae to bis(tri-n-butyl)tin oxide (Cahn-Ventron, purity: +97%) by making a series of stock solutions in acetone of sufficient strength so that 100 μl of stock per liter of seawater yielded the desired exposure concentration in the bowls. Tributyltin concentrations were from 5 to 40 μg l^{-1}. We included both a seawater and an acetone control (100 μl acetone per liter of seawater) in all experiments.

We calculated LC_{50} values for the duration of zoeal development using the Proc Probit subroutine in SAS (Statistical Analysis System Institute, 1982). We used regression analysis to further examine exposure–response relationships for survival, duration of zoeal development, and megalopal dry weight. Our procedures are similar to those used by Bookhout *et al.* (1980). Survival data were transformed by arcsin $\sqrt{x}$, where x is the fractional survival of each treatment group (Steel and Torrie, 1960).

10.2.3b. Results

The LC_{50} value for the Florida population of *R. harrisii* was higher than for the population from California, 33.6 μg l^{-1} (95% CI = 22.05–59.25) and 13.0 μg l^{-1} (95% CI = 12.5–15.0), respectively (Fig. 10.5). These differences are statistically significant ($p < 0.05$).

Regression of transformed survival on TBT exposure concentration shows that Florida zoeae were much less sensitive to increasing TBT concentration compared with those from California (Fig. 10.6a) The slope of the line for California zoeae is 2.5 times greater than that for the Florida population (Table 10.6).

The Florida population was also less sensitive than the California population to effects of TBT on duration of zoeal development (Fig. 10.6b). Larvae from California crabs

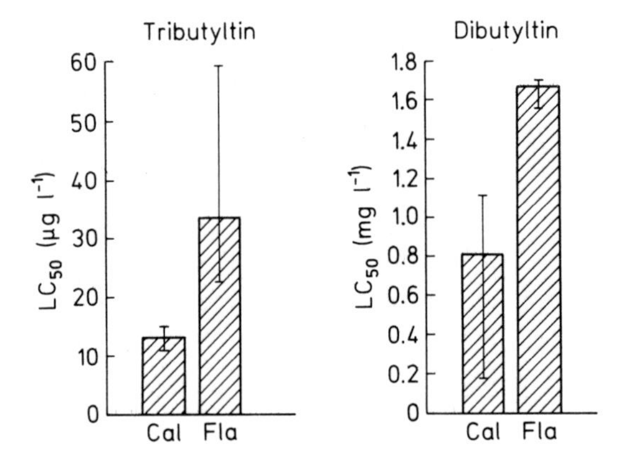

Figure 10.5 *Rhithropanopeus harrisii.* Comparison of LC_{50} value for zoeae of crabs from Florida (Fla) or California (Cal) exposed to TBT for the duration of zoeal development. Error bars indicated 95% confidence intervals about means.

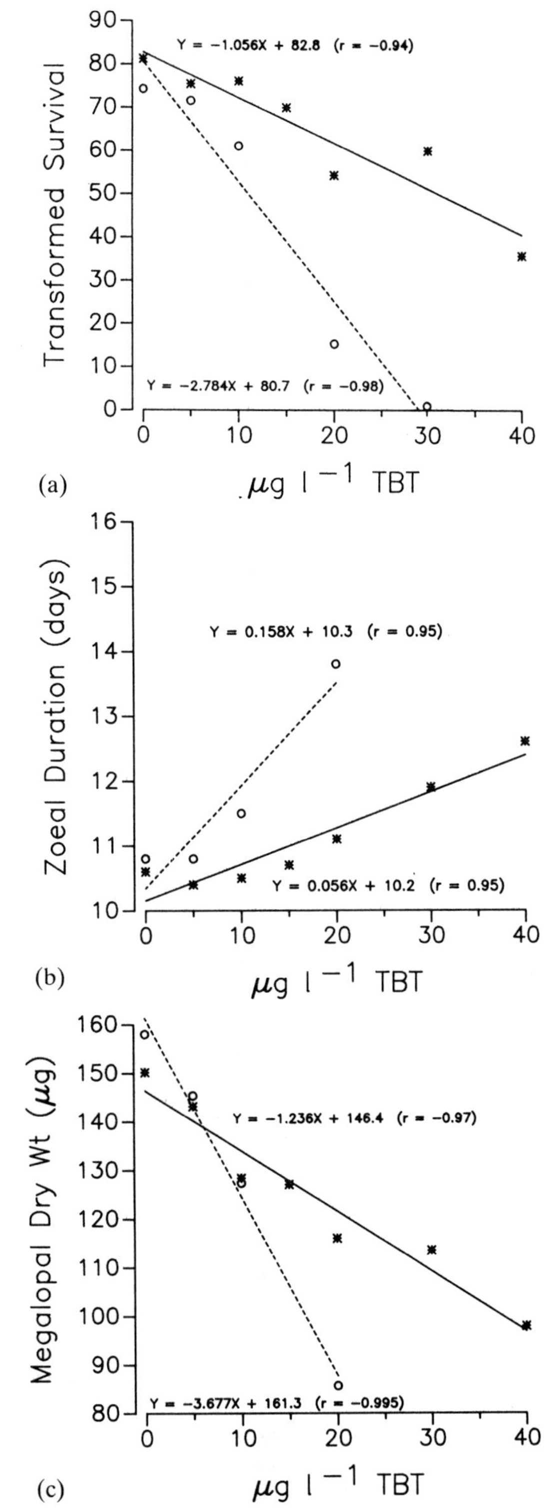

consistently required more time, on average, to reach the megalops stage and were more sensitive to increasing TBT concentrations. That is, the lines for the two populations are not parallel; the one for the California population has a significantly steeper slope (Table 10.6).

Megalopal dry weights were influenced by TBT exposure in a pattern similar to that observed previously. Florida crabs were much less sensitive to increasing TBT concentrations (Fig. 10.6c). Mean control weights of the Florida larvae were slightly less than those from California. This, and the greater decrease in megalopal dry weights in response to increasing TBT concentrations, led to a 2.9 times greater sensitivity in the California population. Again, this difference is statistically significant (Table 10.6).

To summarize briefly, exposure to TBT during the interval of development from hatching to metamorphosis reduces zoeal survival and increases the duration of zoeal development; the resulting megalops is smaller than controls in a dose-dependent relationship. Zoeae from Florida females were, however, less sensitive to the effects of TBT than were those from California crabs.

10.2.4. EFFECTS OF TBT ON OYSTER GROWTH

Sensitivity of the Pacific oyster (*Crassostrea gigas*) to tributyltin leached from antifouling paints is the primary example motivating recent attention to TBT as an environmental issue (Alzieu *et al.*, 1980; Alzieu, 1991; His, Chapter 12). Even though results of field work showed highly significant correlations between TBT exposure and abnormal settlement and growth of oysters, reasonable skepticism remained until experimental studies were reported that showed causality of TBT, not other environmental factors. Demonstration of exposure dependence within environmental concentration ranges was one of the most important contributions of these experimental studies. In this section, we will emphasize experiments that show nonlethal responses to chronic exposure to TBT. We selected several specific experiments for discussion here from among those performed during a multiyear program. Their results are representative of those obtained from the other experiments even though the procedures and techniques (tin analysis protocols, for example) were modified and improved during the course of the work. In addition to *C. gigas*, references to other oyster species will be made when appropriate.

Figure 10.6 *Rhithropanopeus harrisii.* (a) Comparison by regression analysis of survival to metamorphosis of zoeae of crabs from Florida (*) or California (o) exposed to TBT. Data were transformed for analysis to arcsin $\sqrt{x}$ where x is fractional survival. (b) Comparison of duration of zoeal development of zoeae. (c) Comparison of megalopal dry weights of megalops.

10.2.4a. Materials and methods

These experiments were conducted at the Fisheries Laboratory at Burnham on Crouch, England. Hatchery-reared oyster spat, *Crassostrea gigas*, were acclimated over a 3-d period to 18°C in the laboratory after being reared in 12°C. Following temperature acclimation, oysters were randomly subdivided into groups of 10 animals for assignment to experimental groups. Individual length and weight of these experimental oysters were determined. In addition, 10 oysters from the original lot of hatchery-reared spat were sacrificed for determinations of TBT content, condition index [(wet meat weight/internal shell volume) × 100], and shell thickness index (after Alzieu *et al.*, 1980).

Spat were held under experimental conditions in glass containers holding 20 l of seawater (29–32 ‰ salinity). Seawater was passed through a Gelman pleated cartridge to remove particles >1 μm diameter. Seawater in exposure containers was changed daily,

Table 10.6 Statistics of regression analysis of survival, duration of zoeal development, and megalopal dry weights of *Rhithropanopeus harrisii* larvae exposed to TBT. Error estimates are ±1 standard error. Survival values were transformed by $\sqrt{x}$, where x is fractional survival

Statistic	*California*	*Florida*
Survival		
Slope	−1.06±0.18 (p<0.001)	−2.8±0.40 (p<0.001)
Intercept	82.8 ±3.80 (p<0.001)	80.7±6.20 (p<0.001)
r^2	0.877	0.950
Zoeal duration		
Slope	0.06±0.01 (p<0.001)	0.16±0.04 (p<0.001)
Intercept	10.2 ±0.20 (p<0.001)	10.3 ±0.50 (p<0.001)
r^2	0.901	0.903
Megalopal weight		
Slope	−1.23±0.13 (p<0.001)	−3.68±0.26 (p<0.001)
Intercept	146.4 ±2.90 (p<001)	161.3 ±2.80 (p<0.001)
r^2	0.845	0.990

maintained at 18 ± 1°C and agitated to keep sediment and microalgae, *Tetraselmis suecica* and *Skeletonema costatum* (food), in suspension. Microalgae suspensions were added twice daily at concentrations that prevented growth limitation due to insufficient food.

Experimental exposure continued for 56 d. At weekly intervals, oysters in each group were measured for increases in wet weight. At the end of the experiment, remaining oysters were collected, measured, and sacrificed to collect tissues to be analyzed for TBT content. At this time measurements to determine a condition index [(wet meat weight/ internal shell volume) × 100] were also made.

Treatments to examine the role of TBT and sediment as factors causing abnormal growth of oysters were also conducted. Stock solutions of TBT were prepared in glacial acetic acid and aliquots (<150 μl) were added to the experimental tanks to yield nominal low (0.2 μg l^{-1}) and high (2.0 μg l^{-1}) TBT concentrations. After 1 h, measured concentrations in eight trials were within the range of 0.15–0.25 and 1.6–2.0 μg l^{-1} TBT in the low and high treatments, respectively. In tanks containing TBT dissolved in seawater, up to 10% of the TBT was lost from solution within 24 h; while in tanks containing sediment along with TBT, over 60% of the TBT was still in solution after 24 h. No TBT was detected in River Crouch seawater used in these experiments.

Sediment concentrations tested ranged from 30 to 75 mg l^{-1} TBT and were based on an analysis of sediment concentrations of TBT in the River Crouch (Sheldon, 1968). One group of oysters was reared in either high or low TBT concentrations with 30 mg l^{-1} Crouch sediment. In addition, other groups were reared in sediment concentrations of 30–75 mg l^{-1} without TBT, to test the importance of sediment as a factor causing abnormal growth.

Tributyltin in water and tissues was measured using acid–solvent extraction with identification of TBT compounds by flameless atomic absorption spectrophotometry (AAS). Additional details of the methods and procedures are given in Waldock and Thain (1983).

In a second set of experiments, oyster spat were exposed in the laboratory to TBT concentrations between 2 and 200 ng l^{-1} as well as a seawater control. Exposures lasted for 49 d. Rearing conditions were similar to those of previous experiments (Thain *et al.*, 1987). The purpose of these experiments was to characterize the abnormal shell morphology

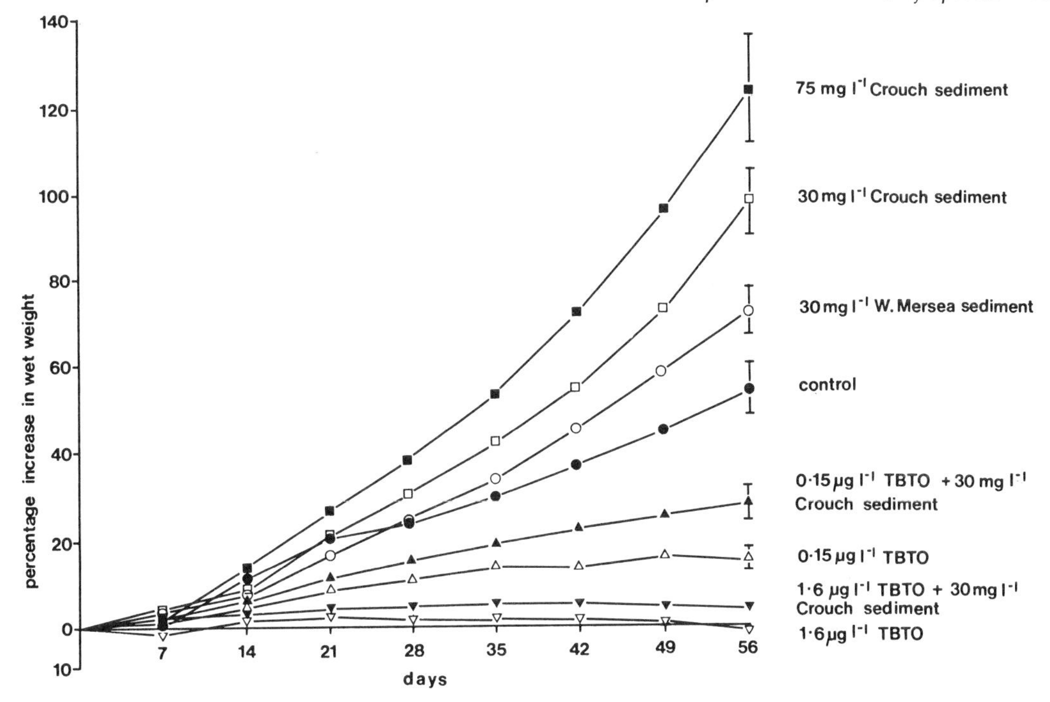

Figure 10.7 *Crassostrea gigas*. Wet weight of oyster spat exposed to TBT, sediment, or TBT–sediment combinations during a 56-d experiment. Error bars shown for day 56 are ± one standard error.

and to determine a threshold of the exposure–response curve for the effect of TBT on abnormal shell development.

10.2.4b. Results

Mortality of oysters during the 56-d observation period was minimal. Only two oysters exposed to 1.6 $\mu g\ l^{-1}$ dissolved TBT, one oyster exposed to 1.6 $\mu g\ l^{-1}$ dissolved TBT with 30 mg l^{-1} Crouch sediment, and one oyster exposed to 0.15 $\mu g\ l^{-1}$ dissolved TBT died. The effects of TBT at levels tested here were dominated by obvious sublethal responses.

TBT exposure, but not exposure to suspended sediment, markedly reduced growth of oyster spat (Fig. 10.7). Spat exposed to 1.6 $\mu g\ l^{-1}$ dissolved TBT hardly increased in size during the entire 56-d observation period; those in 1.6 $\mu g\ l^{-1}$ dissolved TBT with 30 mg l^{-1} Crouch sediment showed ~ 4.4% increase in wet weight, a minor value compared with controls (54% increase) or those reared in 0 $\mu g\ l^{-1}$ dissolved TBT with 75 mg l^{-1} Crouch sediment (126% increase in wet weight). Growth of oysters in 0.15 $\mu g\ l^{-1}$ TBT was only 16% and slightly higher, 28% in 0.15 $\mu g\ l^{-1}$ TBT with 30 mg l^{-1} Crouch sediment. Intermediate sediment concentrations, without TBT, caused oysters to increase their wet weight more rapidly than controls without sediment in their tanks.

Reductions in wet weight gain were generally accompanied by increases in shell height (Table 10.7). Oysters exposed to sediment alone were longer and markedly heavier than control spat. Those exposed to low TBT concentrations (0.15 $\mu g\ l^{-1}$) and separately to West Mersea sediment lengthened less than expected from weight gain. In these groups, shell thickening accounted for most of the weight gain (Table 10.7). In 1.6 $\mu g\ l^{-1}$ dissolved TBT, the spat did not thicken, but

Table 10.7 *Crassostrea gigas* biological data and tributyltin concentrations after a 56-d growth experiment

Treatment	*Wet weight increase (%)*	*Length increase (%)*	*Length/thickness ratio of upper value*	*Condition index[a]*	*Tributyltin content ($\mu g\ g^{-1}$) wet tissue*
Day 0	—[b]	—	53.0	66	<0.08
Control (filtered seawater)	53.7	4.7	26.9	62	<0.08
1.6 $\mu g\ l^{-1}$ TBTO	−1.1	−0.1	51.6	23	3.70
1.6 $\mu g\ l^{-1}$ TBTO + 30 mg l^{-1} Crouch sediment	4.4	0.0	33.1	36	4.89
0.15 $\mu g\ l^{-1}$ TBTO	16.1	−0.2	15.1	51	1.71
0.15 $\mu g\ l^{-1}$ TBTO + 30 mg l^{-1} Crouch sediment	28.2	1.4	6.2	55	1.30
30 mg l^{-1} West Mersea sediment	72.5	7.8	10.2	64	0.59
30 mg l^{-1} Crouch sediment	99.4	22.4	23.0	63	0.09
75 mg l^{-1} Crouch sediment	125.7	27.3	28.5	61	<0.08

[a]Condition index = $\frac{\text{Wet net weight}}{\text{Internal shell volume}} \times 100$.
[b]Dash indicates no prior value for calculation.

they also failed to grow and lost condition. Presence of sediment in 1.6 $\mu g\ l^{-1}$ TBT modified the pattern somewhat, but the effect was small.

After 2 months in experimental treatments, the condition index of spat held in control tanks and in tanks containing West Mersea sediment and 30 mg l^{-1} and 75 mg l^{-1} Crouch sediments was similar to the condition index of spat at the beginning of the experiment (Table 10.7). Oysters exposed to 0.15 $\mu g\ l^{-1}$ dissolved TBT displayed a poor condition index while those in 1.6 $\mu g\ l^{-1}$ dissolved TBT showed a markedly poor condition index and wasting.

Results of the experiment to determine an exposure–response relationship are notable. Slight chambering of oyster shells occurs in all *C. gigas*, even those in controls and 2 ng l^{-1} TBT. A valve from a control oyster and one exposed to TBT (Fig. 10.8) illustrates this point. Oysters exposed to TBT concentrations >2 ng l^{-1} displayed progressive increases in chambering or balling. Oysters exposed to 100 ng l^{-1} TBT and above exhibited severely abnormal shell morphology (Fig. 10.9). These oysters are not harvestable as a commercial crop.

10.2.5. EFFECTS OF TBT ON CLAM EMBRYOS AND LARVAE

On a monetary value basis, hard shell clams (*Mercenaria mercenaria*) support one of the most important Atlantic estuarine fisheries. An unpublished LC_{50} value (96 h) of 15 ng l^{-1} for toxicity of TBT to clam larvae (Bacerra-Huencho, 1984) provided considerable incentive for experimental determinations of chronic toxicity in order to avoid repetition of the *C. gigas* experience. Experiments initiated with clams had the following goals: (a) to test TBT concentrations between 10 and 500 ng l^{-1}, those most commonly encountered in estuarine waters receiving inputs from antifouling coatings; (b) to continue exposure during larval development through metamorphosis to the pediveliger stage; and (c) to determine effects on growth and growth rates as well as survival.

10.2.5a. Materials and methods

Ripe female clams (*Mercenaria mercenaria*) were collected from the Indian River Lagoon (Florida, USA). We obtained gametes from several adults induced to spawn by cyclical exposure to temperatures of 20 and 30°C,

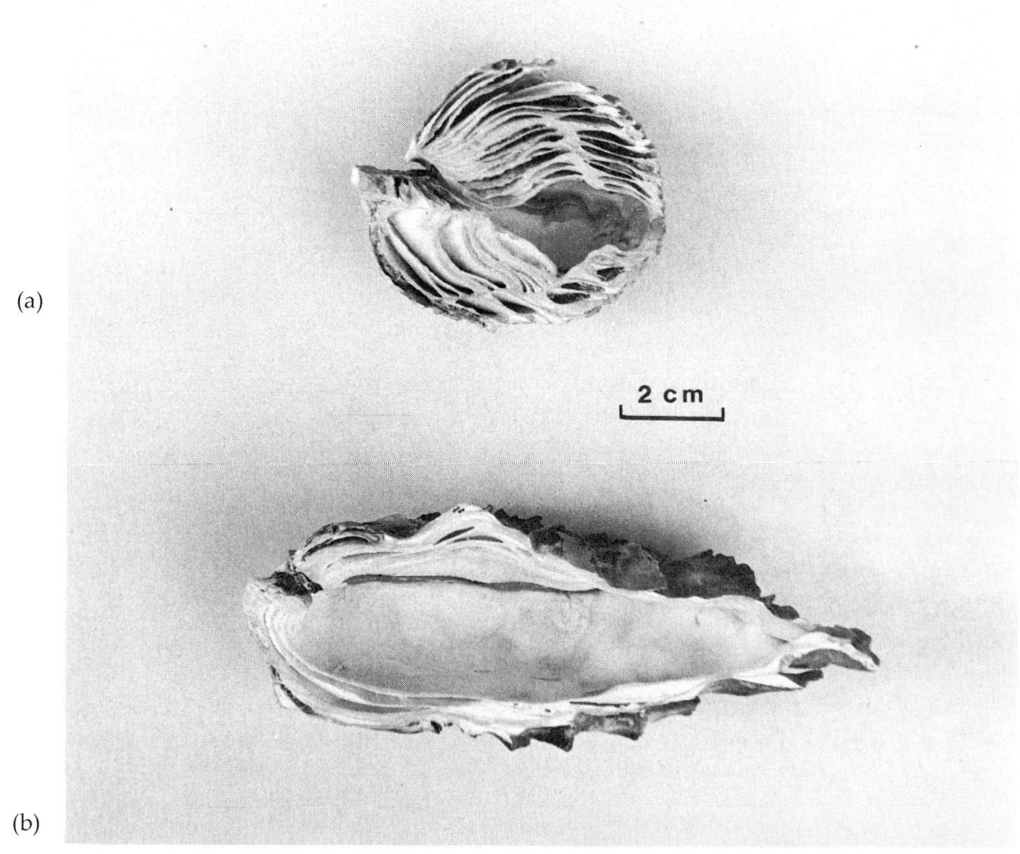

Figure 10.8 *Crassostrea gigas*. Photograph of sectioned values of oysters to show chambering (a) and its relationship to abnormal shell growth, termed 'balling' (b).

each cycle lasting 2 h. Exposure to TBT commenced within 4 h after fertilization. At this time, ~150 gametes were transferred into finger bowls containing 50 ml seawater–TBT exposure solution. Exposure solutions were made from a series of TBT-in-acetone stock solutions of sufficient strength so that 10 μl of stock solution yielded the desired aqueous TBT concentrations. Exposure concentrations were 10, 25, 50, 100, 250 or 500 ng l^{-1} TBT. A seawater and an acetone control were also tested. Two different exposure protocols were employed. The first group was exposed for the entire 14-d period. In the second group, exposures to TBT lasted 5 d, followed by continued observation of the clam larvae while they were reared in uncontaminated seawater. We designated this the 'recovery' group. Exposure was static with daily renewal. Initial and 24-h TBT concentrations in exposure containers were determined analytically using the borohydride reduction method of Valkirs *et al.* (1985, 1987).

Each day after the second exposure, the contents of each bowl were strained through Nitex mesh and put into bowls containing freshly prepared exposure solutions and microalgae (*Isochrysis galbana*, 40 000 cells ml^{-1}) as food. At times indicated in the figures, clam larvae in one bowl in each concentration of TBT were collected, the number of living bivalves was counted, and

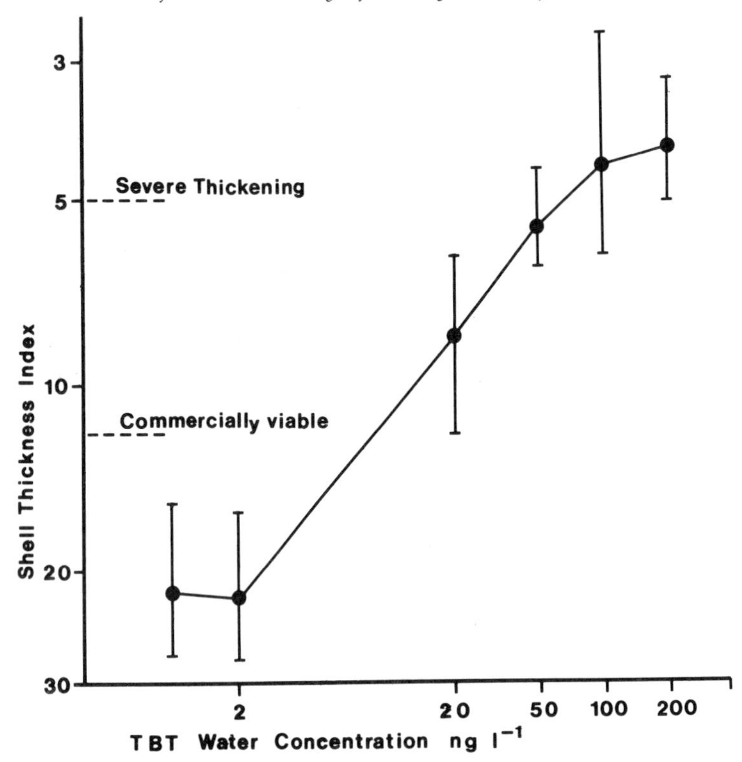

Figure 10.9 *Crassostrea gigas*. Exposure–response relationship for shell thickness index in oysters exposed to TBT concentrations up to 200 ng l^{-1}.

the valve length was determined. We replicated experiments three times with larvae from different spawns. Additional experimental details are given in Laughlin *et al.* (1988).

10.2.5b. Results

Survival of clam larvae exposed for 14 d to TBT concentrations from 10 to 500 ng l^{-1} was not significantly different, statistically, from controls (Fig. 10.10). Mean survival of groups of clam larvae exposed for 5 d and then given an 8-d recovery period also did not differ significantly from controls (Fig. 10.11). In both groups, mean survival declined throughout the 14-d observation period. Statistical analysis indicated that this daily attrition was the only significant factor in the pattern of survival; neither exposure to TBT nor duration of exposure significantly correlated with survival (Table 10.8).

Growth of control clams was rapid. Mean valve lengths increased from 100 μm on the first day to between 210 and 250 μm on day 14 (Fig. 10.12). Numbers for day 14 indicate the range of means for three hatches tested. Exposure to TBT significantly reduced growth of clam larvae regardless of the duration of exposure (Fig. 10.13). Cumulative growth on day 14 was less than controls for all groups exposed to TBT. Clam larvae exposed to 250 or 500 ng l^{-1} TBT showed almost no growth and likely would not have survived another week. Exposure to TBT for 5 d produced similar reductions in cumulative growth on day 14, although 'recovery' groups showed slightly but statistically insignificantly greater valve length compared with

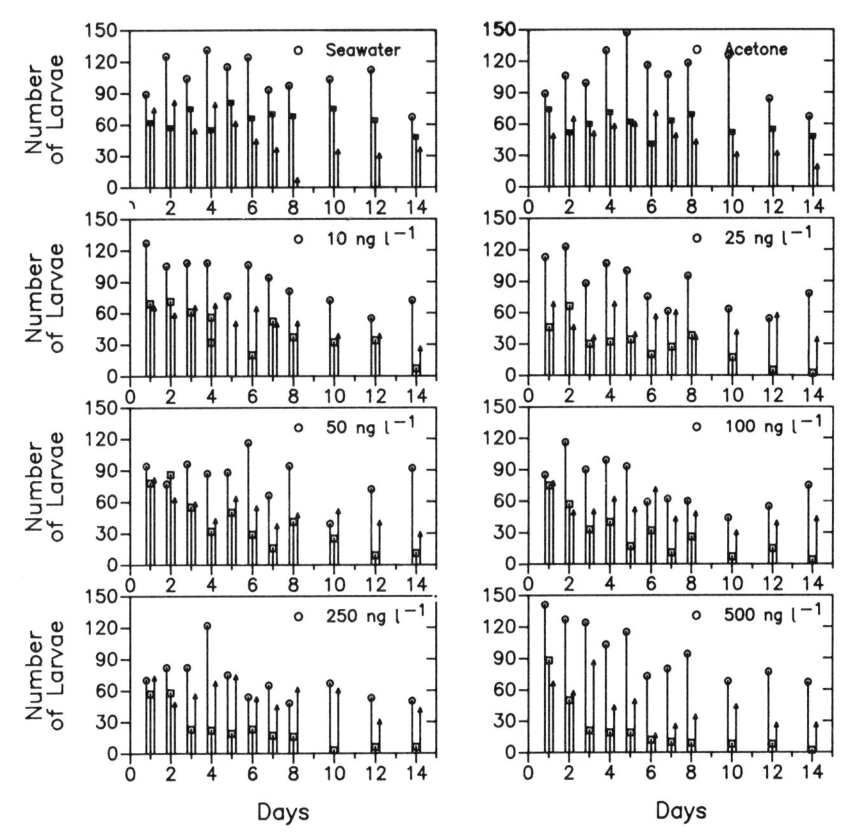

Figure 10.10 *Mercenaria mercenaria*. Survival of clam larvae exposed for 14 d to TBT. Each symbol designates one of three replicates.

their respective 'continuous' exposure group (Figs 10.12, 10.13; Table 10.9).

Mean daily growth rates of controls were between 5 and 15 μm d^{-1} (Fig. 10.14). Exposure to TBT significantly reduced growth rates in a dose-dependent fashion so that in 250 and 500 ng l^{-1} TBT, daily growth rates were generally <5 μm d^{-1}. Growth rates of TBT-exposed larvae usually declined throughout the course of the experiment, probably correlated with incipient mortality in the highest TBT concentrations (250 and 500 ng l^{-1}). Differences in growth rate were dose-dependent and statistically significant (Table 10.10).

Clam larvae exposed to TBT displayed less color, particularly in the gut tissues, which were yellow and dense in controls. We did not observe any shell abnormalities in *M. mercenaria* of the type displayed by *Crassostrea gigas*.

10.3. DISCUSSION

In 1980, French scientists attributed the decline of their oyster fishery, which was based on *Crassostrea gigas*, to TBT released from antifouling coatings on small pleasure craft moored and used in areas where oysters were also grown (Alzieu *et al*., 1980). Geographical occurrence of shell abnormalities correlated with proximity of boats and the use of TBT antifouling treatment. Analysis of water and oyster tissues for total tin suggested that organotins caused the abnormalities. Evidence from field studies, although

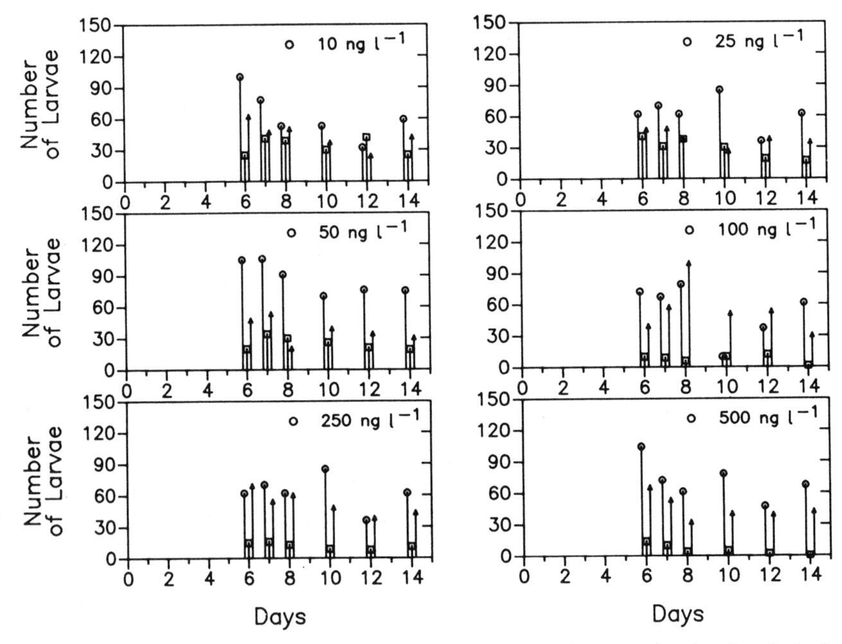

Figure 10.11 *Mercenaria mercenaria*. Survival of clam larvae exposed to TBT for the first 5 d of development, then kept in uncontaminated seawater for the next 9 d. Counting of veligers began after TBT exposure ended. Each symbol designates one of three replicates and corresponds with usage in Fig. 10.5.

Table 10.8 Analysis of variance of survival for clam (*Mercenaria mercenaria*) larvae exposed to TBT. The dependent variable is the number of living clam larvae

Source	*df*	*SS*	*MS*	*F*	*p* >*F*
TBT concentration	1	946	946	1.31	0.2533
Day	1	44 831	44 831	62.11	0.0001
Duration[a]	1	4	4	0.01	0.9427
Error	302	217 990			
Total	305	263 771			
$R^2 = 0.174$					

[a]Duration refers to 5- or 14-d exposure to TBT; it is treated as a class variable.

compelling, was largely circumstantial. Experimental exposures in the field and laboratory provided causal evidence for the role of TBT. Corroboration of the French work by Waldock and Thain (1983), discussed in this chapter, convinced most people that TBT from antifouling coatings was responsible for the severe effects of TBT on *C. gigas*. Experimental work has shown TBT concentrations causing these effects are <100 ng l^{-1} when exposures continued for weeks or months.

The value of studies of chronic exposure to tributyltin compounds became apparent by comparisons of results of toxicity bioassays lasting 5 d and those that continued even several days longer (Laughlin *et al.*, 1982). Short-term tests seriously underestimate the toxicity of these compounds. Until 1984, chemical analysis of TBT at 1–250 ng l^{-1}

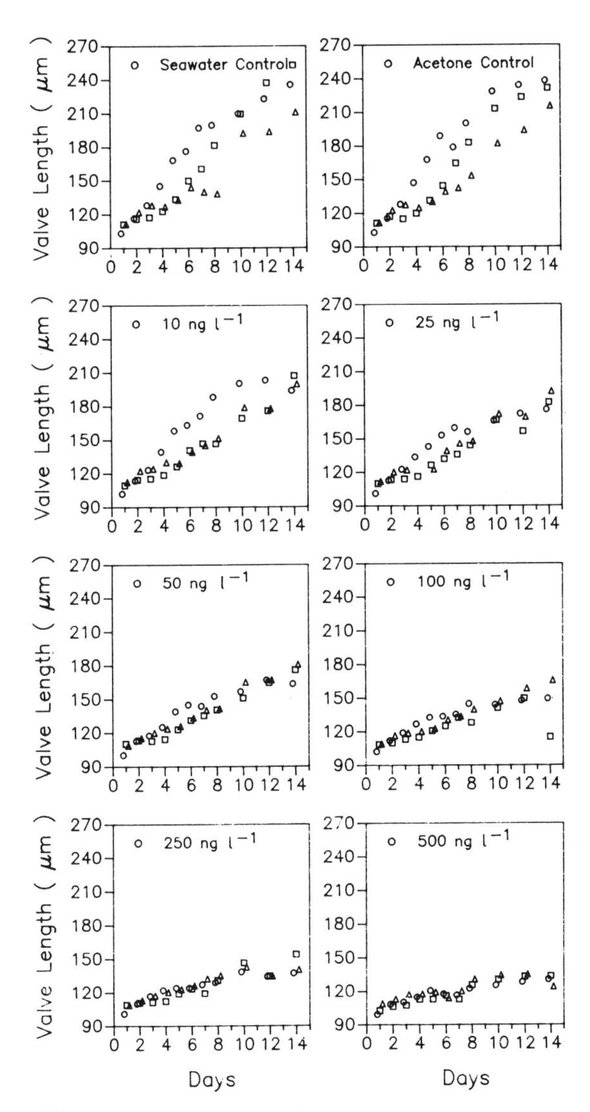

Figure 10.12 *Mercenaria mercenaria*. Growth of clam larvae exposed to TBT for 14 d. Each symbol represents one of three separate groups of clam larvae. Each point is the mean of up to 25 larvae. Points are offset for clarity.

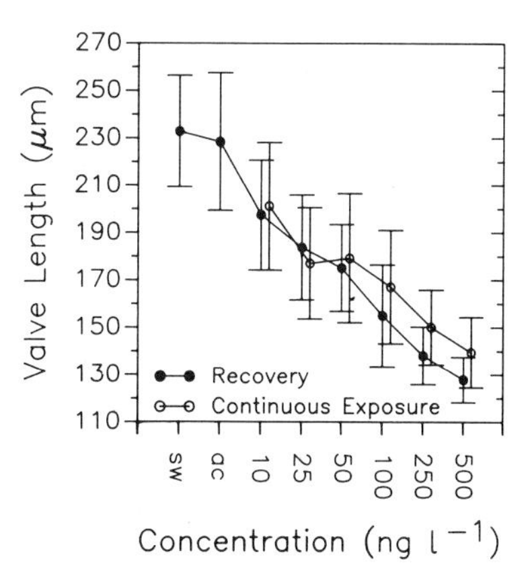

Figure 10.13 *Mercenaria mercenaria*. Mean valve lengths of clam larvae on day 14 for those exposed continuously or for 5 d followed by 9 d recovery. Points for each line are offset slightly for clarity, relative to the *x*-axis scale. Error bars are ±1 standard deviation of the mean.

concentrations, corresponding to environmental levels, was not a routine procedure useful in support of replicated bioassays (Jackson *et al.*, 1982; Matthias *et al.*, 1986; Unger *et al.*, 1986). Thus, chronic bioassays during the past decade are characterized not only by an increasing variety of species tested, but also by declining concentrations of TBT tested. The lowest TBT level has dropped over three orders of magnitude, from 1 μg l^{-1} to near 1 ng l^{-1}, and has approached levels commonly encountered in many contaminated habitats. Because measured chemical concentrations in contemporary laboratory bioassays and the field are similar, results of laboratory studies of TBT are in close agreement with conclusions based upon correlations found in field studies linking TBT exposure with adverse effects on bivalve molluscs, in particular. We can be confident that species found to be sensitive or resistant to TBT toxicity in laboratory studies are similarly at risk in the field.

Studies discussed in this chapter show that taxa differ greatly in their sensitivity to TBT. Decapod crustaceans and fish show the greatest and roughly similar resistance. The model species discussed here, the mud crab (*Rhithropanopeus harrisii*) and the grunion (*Leuresthes tenuis*), tolerate concentrations of at least 10 μg l^{-1}. Mysid shrimp (*Acanthomysis sculpta*), which show marked adverse effects at concentrations between 100 and 1000 ng l^{-1},

Table 10.9 Analysis of growth of clam (*Mercenaria mercenaria*) larvae exposed to TBT

(A) Analysis of variance of valve length of clams on day 14. Data were transformed by $\log_{10}$

Source	*df*	*SS*	*MS*	*F*	p >*F*
Duration[a]	1	0.1357135	0.1357135	30.06	0.0001
Concentration	1	3.2057969	3.2057969	710.12	0.0001
Error	864	3.9004557			
Total	866	7.2419661			
$R^2 = 0.461$					

(B) Student–Newman–Keuls test. Means of two groups underscored by the same line are statistically indistinguishable. Abbreviations: SW, seawater controls; AC, acetone control

$\underline{\text{SW} \quad \text{AC}}$ $\underline{\text{10}}$ $\underline{\text{25} \quad \text{50}}$ $\underline{\text{100}}$ $\underline{\text{250}}$ $\underline{\text{500}}$

(C) *t*-test comparisons of valve length between controls and 10, 25, and 50 ng l^{-1} TBT exposures. C: Pooled seawater and acetone controls

Variable	n	t	*df*	p >t
SW	74	1.0665	112.5	0.2885
AC	67	7.6590		
C	141	7.6590	197	0.0001
10	58			
C	141	11.1083	195	0.0001
25	56			
C	141	11.4485	182	0.0001
50	43			

[a]'Duration' refers to 5- or 14-d exposure and is treated as a class variable.

fall into an intermediate class; and molluscs (*Crassostrea gigas* and *Mercenaria mercenaria*) compose a very sensitive class.

Species-specific differences may be due to several factors. Some taxa possess enzyme systems that degrade or metabolize TBT to less toxic or more easily excreted products (Lee, 1985; Chapter 18). Specific investigations of fish, crabs, and oysters show a descending order of ability to degrade TBT via the mixed function oxygenase system (Lee, 1985), strongly suggesting that enzymatic modification is a significant factor in the intoxication process of TBT. The role of metabolic modification would be particularly important in those species where TBT acts as a non-specific narcotic toxin. Even within a taxon, however, there is a range of sensitivity to TBT (Fig. 10.15), implying that some species have a specific point of TBT action. For example, not all bivalve molluscs are equally sensitive to TBT; and, in particular, post-settlement stages of the oyster *Crassostrea virginica* are relatively hardy to TBT exposure compared with the congeneric *C. gigas*.

The occurrence of imposex in some gastropod species (Smith, 1981; Bryan *et al.*, 1986; Gibbs *et al.*, 1987, 1991; Spooner *et al.*, 1991; Gibbs and Bryan, Chapter 13) is another example of apparently specific effects of TBT on sensitive species. In no case, however, has interference with the specific metabolic pathways leading to stereotypic responses to chronic exposure to TBT been completely

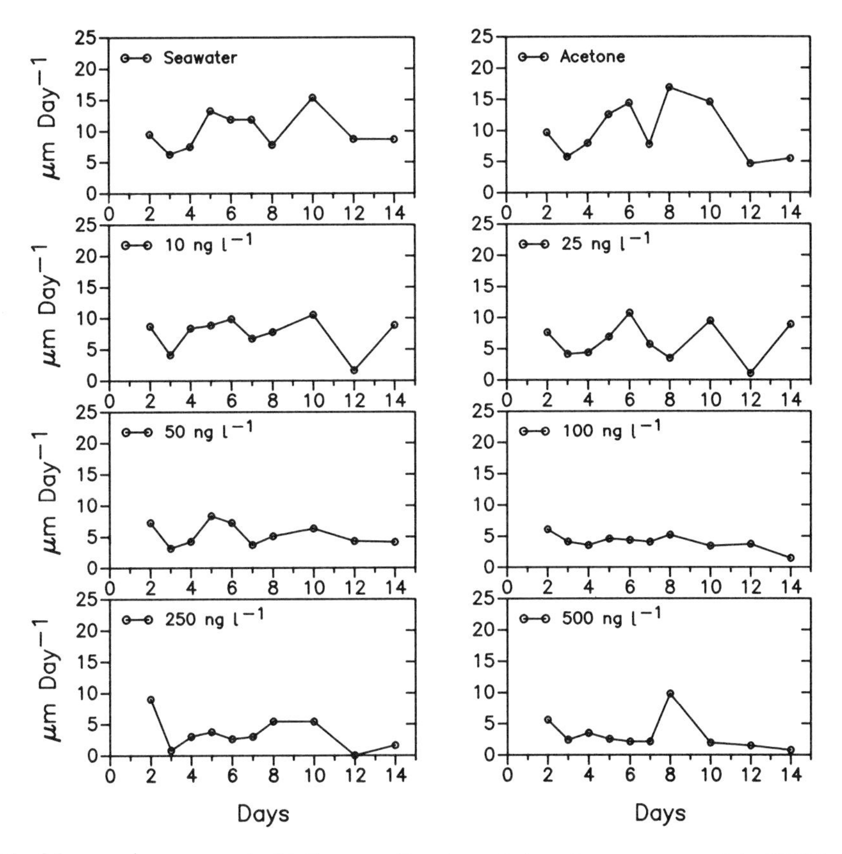

Figure 10.14 *Mercenaria mercenaria*. Daily growth rates of clam larvae exposed to TBT for 14 d. Each point is the mean of three replicates.

Table 10.10 Analysis of variance for data of daily growth rates of clam (*Mercenaria mercenaria*) larvae

Source	*df*	*SS*	*MS*	*F*	p >*F*
TBT concentration	1	840.78	840.78	30.81	0.001
Day	1	47.66	47.66	1.75	0.1876
Error	237	6468.22	27.29		
Total	239	7356.67			
r^2 = 0.121					

described. Such descriptions would be a significant addition to environmental toxicology. They should not, however, become the primary goal of further research on environmental toxicology since TBT appears to have a complex mode of action that is a function of exposure concentration, duration of exposure, and metabolic pathways in the organism (Laughlin, 1987). Mitochondria, for example, may be sites of one type of toxic action in which TBT acts as an ionophore to collapse the $Cl^-:OH^-$ gradient across mitochondrial membranes. Thus, a complete characterization of the chronic toxicity of TBT would

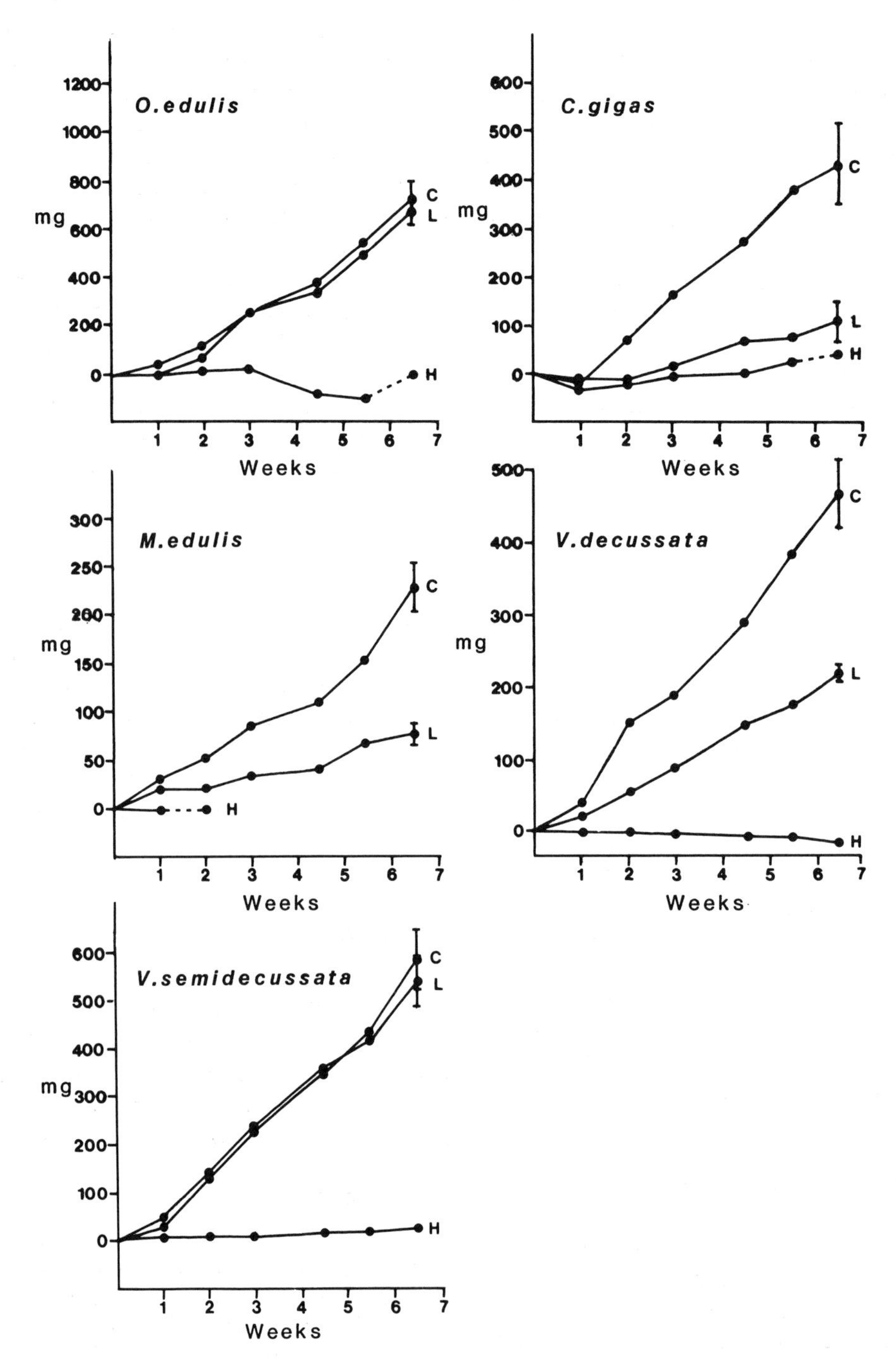

Figure 10.15 Increase in wet weight of bivalves exposed to tributyltin leachates. Symbols: C = control, L = 0.24 μg l^{-1}; H = 2.6 μg l^{-1} TBTO. Species are *Ostrea edulis, Crassostrea gigas, Mytilus edulis, Venerupis decussata, Venerupis semidecussata.*

include an understanding of the hierarchical relationship between exposure and site of action in both animals and plants.

The story of the chronic toxicity of tributyltin to marine organisms is a unique chapter in environmental toxicology. Tributyltin is the first anthropogenic chemical that has a primary use in the marine environment and that caused severe toxicity only to marine organisms. Other chemicals thought to cause adverse effects on marine invertebrates and fish find their way into the sea from uses on land (for example, chlorinated hydrocarbons, heavy metals, or petroleum). Because some molluscs demonstrate highly specific responses to very low TBT concentrations, unambiguous relationships between TBT and sublethal responses have been shown in both field studies and laboratory experiments. Field studies have been important to the TBT issue because they showed compelling correlations between TBT exposure and biological responses. Experimental laboratory studies have been as important because they showed the necessary causality between TBT exposure and the response attributed to it in the field. It was essential to distinguish experimentally between exposure to this chemical and to other environmental influences, such as silt load, temperature, salinity, nutritional factors, xenobiotics, and biological interactions, which are uncontrolled in field studies. It should be noted that no significant contradictions between results of field studies and laboratory experiments exist. Thus, experimental laboratory studies of chronic toxicity have played a crucial role in the recognition and resolution of the issue to the environmental toxicology of tributyltin.

Laboratory studies of the chronic toxicity of TBT to nontarget organisms have been challenging. The first studies conducted a decade ago indicated that sensitive species responded to exposures well below those detectable by analytical chemical techniques then available. In retrospect, the bioassays were surprisingly accurate, but, without the independent chemical analysis that subsequently became available, determination of a validated dose–response relationship was problematical. That bioassays showed significant sublethal responses when exposures were 10–500 ng l^{-1} was completely surprising initially. The bioassays gained considerable credibility after field studies demonstrated that TBT concentrations from antifouling paints were frequently in this range; and wild populations of molluscs, in particular, displayed sublethal responses to TBT. It is likely that restrictions on the use of TBT will reduce the potential for it to cause further environmental harm, and thus the topical importance of TBT as a research topic has decreased. In a larger perspective, however, these studies will retain their significance because they were the first rigorous investigations of agents in the marine environment with such high toxicity and which targeted specific species with distinctive responses such as shell abnormalities and imposex. Bioassay results spurred an appreciation of the need for more sensitive chemical analysis for these and other organometallic compounds. Clearly, these studies have contributed significantly to environmental studies.

ACKNOWLEDGMENTS

Research reviewed in this paper was supported by the Office of the Chief of Naval Research and the Ministry of Agriculture, Fisheries and Food. We thank the staff of the Center for Academic Publication, Florida Institute of Technology, for their assistance.

REFERENCES

Alzieu, C., Y. Thibaud, M. Héral and B. Boutier. 1980. Evaluation des risques dus à l'emploi des peintures antisalissures dans les zones conchylicoles. *Rev. Trav. Inst. Pêches Marit.*, **44**, 301–348.

Bacerra-Huencho, R. 1984. The Effects of Organotin and Copper Sulfate on the Metamorphosis of the Hard Clam *Mercenaria mercenaria*. Master's Thesis, University of Maryland, College Park.

Bookhout, C.G., J.D. Costlow, Jr and R. Monroe. 1980. Kepone effects on larval development of mud crab and blue crab. *Water Air Soil Pollut.*, **13**, 57–77.

Bryan, G.W., P.E. Gibbs, L.G. Hummerstone and G.R. Burt. 1986. The decline of the gastropod *Nucella lapillus* around southwest England: evidence for the effect of butyltin from antifouling paints. *J. Mar. Biol. Ass. UK*, **66**, 611–640.

Davidson, B.M., A.O. Valkirs and P.F. Seligman. 1986. Acute and chronic effects of tributyltin on the mysid *Acanthomysis sculpta* (Crustacea, Mysidacea). In: Proceedings of the Oceans '86 Organotin Symposium, Vol. 4, 25–27 September 1986, Washington, DC, pp. 1219–1225.

Deschiens, R., H. Brottes and L. Mvogo. 1966. Application sur le terrain, au Cameroun, dans la prophylaxie des bilharzioses de l'action molluscicide de l'oxyde de tributylétain. *Bull. Soc. Exotique*, **59**, 968–973.

Ehrlich, K.F. and D.A. Farris. 1971. Some influences of temperature on the development of grunion *Leuresthes tenuis* (Ayres). *Calif. Fish and Game*, **57**, 58–68.

Gibbs, P.E., G.W. Bryan, P.L. Pascoe and G.R. Burt. 1987. The use of the dog-whelk, *Nucella lapillus*, as an indicator of tributyltin (TBT) contamination (southwest England and Isles of Scilly). *J. Mar. Biol. Ass. UK*, **67**, 507–523.

Gibbs, P.E., G.W. Bryan and P.L. Pascoe. 1991. TBT-induced imposex in the dogwhelk, *Nucella lapillus*: geographical uniformity of the response and effects. *Mar. Environ. Res.*, **32**, 79–87.

Jackson, J.A., W.R. Blair, F.E. Brinckman and W.P. Iverson. 1982. Gas chromatographic speciation of methylstannanes in the Chesapeake Bay using purge and trap sampling with a tin selective detector. *Environ. Sci. Technol.*, **16**, 110–119.

Laughlin, R.B., Jr. 1987. Quantitative structure–activity studies of di- and triorganotin compounds. In: *Proceedings of the Second International Workshop on Quantitative Structure–Activity Relationships in Environmental Toxicology*, K.L.E. Kaiser (Ed.). Dordrecht, D. Reidel Publishing Company, pp. 189–206.

Laughlin, R.B., Jr and W.J. French. 1981. A comparative study of the acute toxicity of a homologous series of trialkyltins to larval shore crabs, *Hemigrapsus nudus*, and lobster, *Homarus americanus*. *Bull. Environ. Contam. Toxicol.*, **25**, 802–809.

Laughlin, R.B., Jr and W. French. 1989a. Interactions between temperature and salinity during brooding on subsequent zoeal development of the mud crab *Rhithropanopeus harrisii*. *Mar. Biol.*, **102**, 377–386.

Laughlin, R.B., Jr and W. French. 1989b. Differences in responses to factorial combinations of temperature and salinity by zoeae from two geographically isolated populations of the mud crab, *Rhithropanopeus harrisii*. *Mar. Biol.*, **102**, 387–396.

Laughlin, R.B., Jr and W. French. 1989c. Population-related toxicity responses to two butyltin compounds by zoeae of the mud crab, *Rhithropanopeus harrisii*. *Mar. Biol.*, **102**, 397–402

Laughlin, R.B., Jr, O. Linden and H.E. Guard. 1982. Acute toxicity of tributyltins and tributyltin leachate from marine antibiofouling paints. *Bulletin de liaison du Comité International Permanent pour la Recherche sur la Preservation des Materiaux en Milieu Marin*, **13**, 3–20.

Laughlin, R., K. Nordlund and O. Linden. 1984. Effects of tributyltin compounds on survival, growth and respiration rates of Baltic populations of *Gammarus oceanicus* (Crustacea: Amphipoda). *Mar. Environ. Res.*, **12**, 243–272.

Laughlin, R.B., Jr, R. Gustafson and P. Pendoley. 1988. Chronic embryo–larval toxicity of tributyltin (TBT) to the hard shell clam, *Mercenaria mercenaria*. *Mar. Ecol. – Prog. Ser.*, **48**, 29–36.

Lee, R.F. 1985. Metabolism of tributyltin oxide by crabs, oysters and fish. *Mar. Environ. Res.* **17**, 145–149.

Matthias, C., J.M. Bellama, G.J. Olson and F.E. Brinckman. 1986. A comprehensive method for the determination of aquatic butyltin and butylmethyltin species at ultra-trace levels using simultaneous hydridization/extraction with GC–FPD. *Environ. Sci. Technol.*, **20**, 609–615.

May, R.C. 1971. Effects of delayed feeding on larvae of the grunion, *Leuresthes tenuis* (Ayres). *Fish. Bull. – NOAA*, **69**, 411–423.

Mount, D.I. and C.E. Stephan. 1967. A method for establishing acceptable limits for fish – malathion and the butoxyethanol ester of 2,4-D. *Trans. Am. Fisheries Soc.*, **96**, 185–193.

Payne, A.G. and R.H. Hall. 1979. *Aquatic Toxicology: Second Conference, ASTM STP 667*. American Society for Testing and Materials, Philadelphia, pp. 171–180.

Ritchie, L.S., V.A. Lopez and J.M. Cora. 1974. Prolonged application of an organotin against

Biomphalaria glabrata and *Schistosoma mansoni*. In: *Molluscicides in Schistosomiasis Control*. T.C. Chang (Ed.). Academic Press, New York, pp. 77–88.

Salazar, M.H. and S.M. Salazar. 1985. Ecological Evaluation of Organotin-contaminated Sediment. Naval Ocean Systems Center Technical Report 1050. Naval Oceans System Center, San Diego, California.

Seinen, W., T. Helder, H. Vernij, A. Penninks and P. Leeuwangh. 1981. Short term toxicity of tri-n-butyltin chloride in rainbow trout (*Salmo gairdneri* Richardson) yolk sac fry. *Sci. Total Environ.*, **19**, 155–166.

Sheldon, R.W. 1968. Sedimentation in the estuary of the River Crouch, Essex, England. *Limnol. Oceanogr.*, **13**, 72–83.

Smith, B.S. 1981. Male characteristics in female and snails caused by antifouling bottom paints. *J. Appl. Toxicol.*, **1**, 22–25.

Spooner, N., P.E. Gibbs, G.W. Bryan and L.J. Coad. 1991. The effect of tributyltin upon steroid titres in the female dogwhelk, *Nucella lapidus*, and the development of imposex. *Mar. Environ. Res.*, **32**, 37–49.

Statistical Analysis System Institute, Inc. 1982. *SAS User's Guide, Version 5*. Statistical Analysis System Institute, Inc., Cary, North Carolina.

Steel, R.G.D. and J.A. Torrie. 1960. *Principles and Procedures of Statistics*. McGraw-Hill, New York.

Stephen, C.E., D.I. Mount, D.J. Hansen, J.H. Gentile, G.A. Chapman and W.A. Brungs. 1985. Guidelines for Deriving Numerical National Water Quality Criteria for the Protection of Aquatic Organisms and their Uses. US Environmental Protection Agency, Washington, DC.

Tatem, H.E. and A.S. Portzer. 1985. Culture and Toxicity Tests using Los Angeles District Bioassay Animals, *Acanthomysis* and *Neanthes*. Miscellaneous Paper EL-85–6, US Army Engineer Waterways Experiment Station, Vicksburg, Mississippi.

Thain, J.E., Waldock, M.J. and Waite, M.E. 1987. Toxicity and degradation studies of tributyltin (TBT) and dibutyltin (DBT) in the aquatic environment. In: Proceedings of the Oceans '87 International Organotin Symposium, Vol. 4, Halifax, Nova Scotia, Canada, 28 September–1 October 1987. Marine Technology Society, Washington, DC, pp. 1398–1404.

Thum, A.B. and K.F. Ehrlich. 1979. Ocean Sediment Study, Ontario, California. Lockheed Aircraft Service Company, Project Number 80011604, 91 pp.

Unger, M.A., W.G. MacIntyre, J. Greaves and R.J. Huggett. 1986. GC determination of butyltins in natural waters by flame photometric detection of hexyl derivatives with mass spectrometric confirmation. *Chemosphere (UK)*, **15**, 461–468.

Valkirs, A.O., P.F. Seligman, G. Vafa, P.M. Stang, V. Homer and S.H. Lieberman. 1985. Speciation of Butyltins and Methylbutyltins in Seawater and Marine Sediments by Hydride Derivatization and Atomic Absorption Detection. Technical Report 1087. Naval Oceans System Center, San Deigo, California.

Valkirs, A.O., P.F. Seligman, P.M. Stang, V. Homer, S.H. Lieberman, G. Vafa and C.A. Dooley. 1986. Measurement of butyltin compounds in San Diego Bay. *Mar. Pollut. Bull.*, **17**, 319–324.

Valkirs, A.P., P.F. Seligman, G.J. Olson, F.E. Brinckman, C.L. Mattias and J.M. Bellama. 1987. Di- and tributyltin species in marine and estuarine waters. Interlaboratory comparisons of two ultratrace analytical methods employing hydride generation and atomic absorption or flame photometric detection. *Analyst*, **112**, 17–21.

Waldock, M.J. and J.E. Thain. 1983. Shell thickening in *Crassostrea gigas*: organotin antifouling or sediment induced? *Mar. Pollut. Bull.*, **14**, 411–415.

AN ASSESSMENT OF THE VALUE OF SHELL THICKENING IN *CRASSOSTREA GIGAS* AS AN INDICATOR OF EXPOSURE TO TRIBUTYLTIN

11

M.J. Waldock,[1] J.E. Thain,[1] and M.E. Waite[2]

[1]*MAFF, Directorate of Fisheries Research, Fisheries Laboratory, Remembrance Avenue, Burnham-on-Crouch, Essex CM0 8HA, UK*

[2]*National Rivers Authority, South Western Region, Manley House, Kestrel Way, Exeter, Devon EX2 7LQ, UK*

ABSTRACT

Data from laboratory and field studies on the effects of tributyltin (TBT) on the oyster *Crassostrea gigas* are presented in order to provide a framework for the use of the species as a bioindicator when monitoring for environmental concentrations of TBT, and for monitoring for harmful effects of TBT on marine ecosystems. Concentrations of TBT in tissues have been demonstrated to reflect concentrations of TBT in the water column. Shell thickness index and accompanying loss of meat yield in the oysters have been shown to be related to tissue concentrations of TBT or concentrations of TBT in the water column in a readily predictable way. In field transplanting experiments, tissue concentrations

Organotin. Edited by M.A. Champ and P.F. Seligman. Published in 1996 by Chapman & Hall, London. ISBN 0 412 58240 6

of TBT >0.2–0.3 μg g^{-1} occurred concomitantly with a decrease in shell thickness index in the oysters, demonstrating that harm was occurring to a valuable marine resource. Deployment of *C. gigas* in the field has been shown to be a useful component in programs designed to monitor long-term changes in TBT concentrations. It is proposed that the presence of normally shaped *C. gigas* oysters yielding good meats are indicative of concentrations of TBT of <10 ng l^{-1}

11.1. INTRODUCTION

Before outlining the way in which bivalves may be used as indicator species, it is necessary to consider the objectives of the proposed monitoring program. There are many categories of monitoring programs in support of environmental quality objectives (EQOs) (see for example Holdgate, 1979, and Abel, Chapter 2), and there is often confusion about their respective aims.

The UK Department of the Environment has recently published environmental quality standards (EQSs) for TBT, based on the EQO of protecting aquatic organisms; the EQS for TBT in saltwater is 2 ng l^{-1} and for fresh water it is 20 ng l^{-1}. Since the standard is set in terms of TBT concentration in water, bivalves cannot be used for determination of compliance. As yet an acceptable daily intake (ADI) for TBT has not been set, and therefore measurement of TBT in seafood is not required to demonstrate compliance to an ADI.

Bivalves are useful for monitoring trends in environmental contamination. The use of the common mussel (*Mytilus edulis*) as a monitor for long-term changes in contaminant levels of a variety of metals, hydrocarbons, and organochlorine compounds has long been accepted and forms the basis of the 'mussel watch' program throughout a number of countries (Farrington *et al.*, 1983; Phillips and Segar, 1986). As shown below, both mussels and oysters are similarly suitable indicators of changes in environmental TBT contamination over a period of time. Furthermore, the Pacific oyster (*Crassostrea gigas*) may also be used as an indicator of environmental impact of TBT because the species is sensitive to TBT and, once above a threshold concentration, responds to the presence of the compound in a readily detectable and predictable way. It is important to draw this distinction between the use of an organism to determine environmental contamination (i.e. a measure of the concentration of TBT in the environment) and environmental impact [i.e., the presence of TBT as a pollutant (GESAMP definition) having harmful effects on biological systems].

Shell thickening in combination with reduced meat yields in Pacific oysters has been demonstrated to be symptomatic of exposure of these animals to low concentrations of TBT (>10 ng l^{-1}) in France, the United Kingdom, and North America. Despite the growing lists of researchers who have shown a positive correlation between increasing shell thickness and increased tissue concentrations of TBT (Alzieu *et al.*, 1986; Davies *et al.*, 1987; Minchin *et al.*, 1987), the evidence of cause and effect has been quoted as equivocal by others (Salazar and Champ, 1988). The ambiguity perhaps stems from the fact that although measurement of shell thickness in *C. gigas* is commonly used as an indicator of the presence of TBT in marine and estuarine environments, the protocols for deployment and limitations of the usefulness of the animal as a sentinel species have not been fully documented. This chapter will cover the following three areas:

1. It will highlight both the advantages and disadvantages of the use of bivalves as indicators of environmental contamination by TBT;
2. It will indicate the usefulness of the shell thickness as a measure of environmental impact.
3. It will clarify the manner in which oysters and mussels may be deployed in the field.

The relative merits of setting environmental quality standards in terms of water or in animal tissues are also considered.

The following discussion relates to long-term (year-to-year) trend monitoring and draws upon experience gained from the UK monitoring program funded by the UK Department of the Environment (DoE) and Ministry of Agriculture, Fisheries and Food (MAFF). The program was designed to measure long-term temporal changes in TBT concentrations following legislative measures. The strategy for the whole program is outlined in Waite *et al.* (Chapter 27) and relies on the measurement of TBT in water, indicator species, and sediments.

11.2. BIVALVES AS INDICATORS OF ENVIRONMENTAL CONTAMINATION

11.2.1. GENERAL STRATEGIES FOR MEASUREMENT OF ENVIRONMENTAL CONTAMINATION

A 'good' indicator of environmental contamination, in the context of long-term trend monitoring, is one that integrates the short-term temporal variability in concentrations of TBT. This avoids the expensive and laborious option of integrating small-scale heterogeneity by analyzing a large number of water samples. Purely mechanical devices, such as the sea star sampler (a continuous ion-exchange sampler), have been designed for this purpose (Schatzberg *et al.*, 1986), but the costs are high if several such devices are employed simultaneously.

Sediments form a repository for TBT and, in areas where they slowly accumulate, are potentially good indicators of changes in the degree of contamination. However, to date most monitoring effort in the United Kingdom has been centered on dynamic estuarine systems with mobile sediments, resulting in rapid changes of measured TBT concentrations at monitoring sites. Point sources of TBT from boatyards and harbors further complicate the distribution and give rise to contamination of sediments with particles of TBT-based paints. The degradation and bioavailability of TBT from paint particles is as yet unknown, and this makes the interpretation of trend data a difficult task.

Where animals are used as indicators of environmental contamination, time integration is not a problem since filter-feeding animals pass several liters of water across active gill surfaces each hour, and other routes of uptake to animal systems (e.g. from contaminated food) also contribute to the tissue concentration. Obviously such a system measures only the bioavailable fraction of the compound present in the environment and integrates that accumulation over time, representing a normalized dose rather than pulses. A further advantage of using a bioaccumulating species is that it may be easier and more cost effective to analyze for the higher tissue concentrations of TBT than the much lower corresponding concentrations in the water.

11.2.2. SUITABILITY OF *CRASSOSTREA GIGAS*

The sentinel organism should be selected carefully: tissue residues of TBT reflect not only environmental exposure but physiology (e.g. feeding strategy) and biochemistry (ability to metabolize the compound) of the animal. Ideally, the animal should accumulate TBT at a sufficiently slow rate to allow the tissue concentrations to reflect mean environmental concentrations occurring over a period of time and not respond rapidly to short-term fluctuations. The animal should be at a low trophic level to minimize complications with food webs, and it should be sessile. Bivalves are well suited in all of the above respects.

One drawback of using oysters (and indeed mussels) is that they are relatively sensitive to TBT, and therefore cannot be deployed in areas of high contaminant concentrations (greater than micrograms per liter) because they die. In order to establish the relationship between environmental contamination and tissue concentration, experiments are necessary to 'calibrate' the indicator species by characterizing the rates of accumulation and depuration of TBT in the animal under defined environmental conditions, both in time and across a large range of sublethal concentrations that are encountered in the environment.

11.2.3. CALIBRATION OF EXPOSURE CONCENTRATION AND TISSUE CONCENTRATION

The accumulation and depuration of TBT in *C. gigas* at two exposure concentrations (1.25 and 0.15 μg l^{-1}) was first determined in 1983 (Waldock *et al.*, 1983). Animals were exposed to tributyltin oxide (TBTO) in a continuous flow-through system for 21 d and allowed to depurate for a further 23 d. Concentrations of TBT in oyster flesh reached equilibrium with water concentrations after 2 weeks, and bioconcentration factors (BCFs) of 2000-fold and 6000-fold were recorded at the high and low concentrations, respectively. The depuration study indicated that the initial loss of TBT from the animals was rapid (approximately half the accumulated TBT was lost within 2 weeks), but thereafter in the surviving animals at the low concentration, TBT loss became much slower (see Fig. 11.1).

Tissue concentrations from controlled exposure experiments were also derived in a further experiment in 1983, which was designed to determine whether or not TBTO induced the thickening response in *C. gigas*,

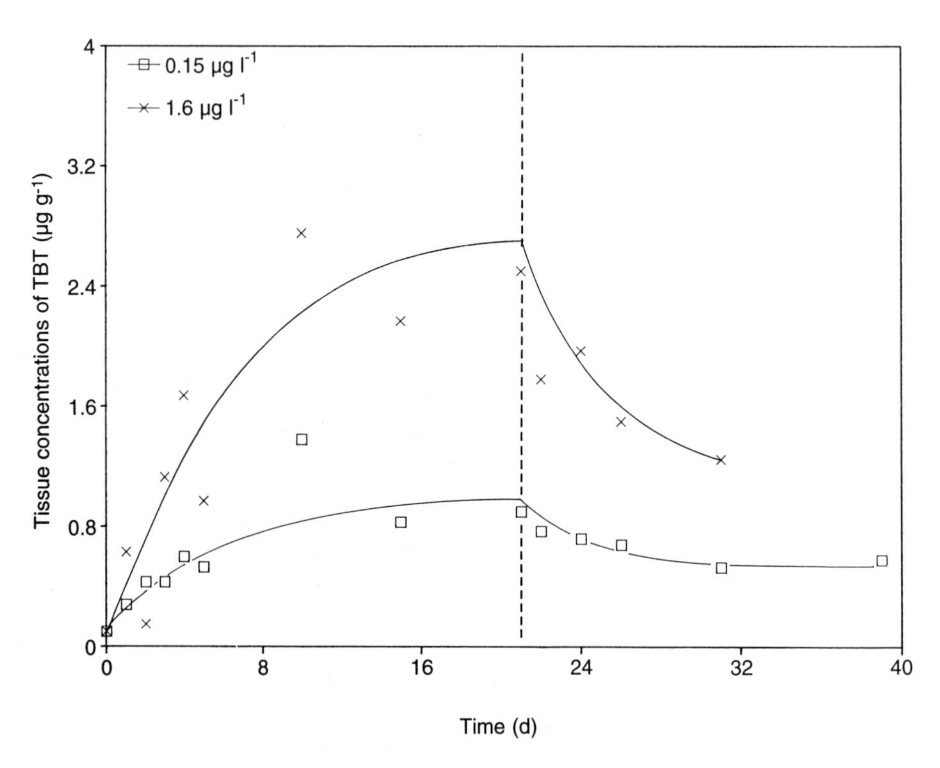

Figure 11.1 Accumulation and depuration of TBTO in *Crassostrea gigas*. Source Waldock *et al.* (1983). The oysters were held in settled seawater, temperature 13 ± 1°C, salinity 28–35 practical salinity units (psu). Each data point represents a single analysis of the combined tissue from three animals. All animals had died in the higher concentration by day 31.

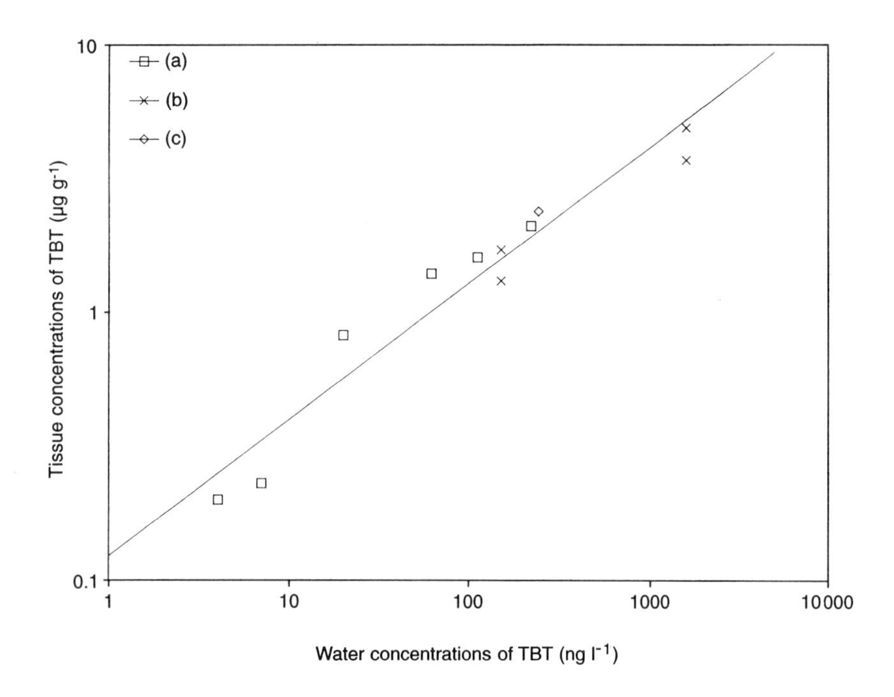

Figure 11.2 Concentrations of TBT in *Crassostrea gigas* flesh after exposure to TBT for at least 45 d under controlled laboratory conditions. (a) Static renewal holding conditions for 49 d in artificial seawater, 30 ± 1 psu, 21 ± 1°C. (b) Static renewal holding conditions for 56 d in settled seawater, 29–32 psu, 18 ± 1°C (Waldock and Thain, 1983). (c) Flowthrough holding conditions for 45 d, 28.5–34.2 psu, 20–21°C (Thain and Waldock, 1985).

reported by Waldock and Thain (1983). BCF values of 2000–3000-fold and 9000–11 000-fold were obtained from exposure to measured concentrations of 1.6 or 0.15 μg l^{-1} TBTO, respectively, under static renewal conditions. In 1985 Thain and Waldock exposed a variety of bivalve spat to TBT leachates in a flow-through experiment. A tissue concentration of 2.38 μg g^{-1} was recorded after a 45-d exposure to 240 ng l^{-1} TBT. More recently, tissue concentrations were measured at a further five water concentrations (Thain *et al.*, 1987) in order to derive a no-effect concentration for the thickening response. BCF values were 9500, 14 300, 22 400, 33 000, 41 000, and 50 000, respectively, for exposure to measured concentrations of 220, 112, 62, 20, 7, and 4 ng l^{-1} TBTO. The relationship for tissue concentration and exposure concentration for all three experiments is shown in Fig. 11.2. The data suggest that it should be relatively easy to predict exposure concentration from tissue concentrations. However, both the time taken to reach equilibrium and slow depuration rates must be given careful consideration when the animals are deployed in the field. Laboratory depuration rates suggest that the animals deployed during several weeks of high ambient TBT concentrations would partially reflect those concentrations for a period of several months even if ambient water concentrations fell rapidly.

In a similar series of experiments, accumulation in *M. edulis* was determined at four concentrations in a flow-through experiment (Fig. 11.3). BCF values for mussels were 6600,

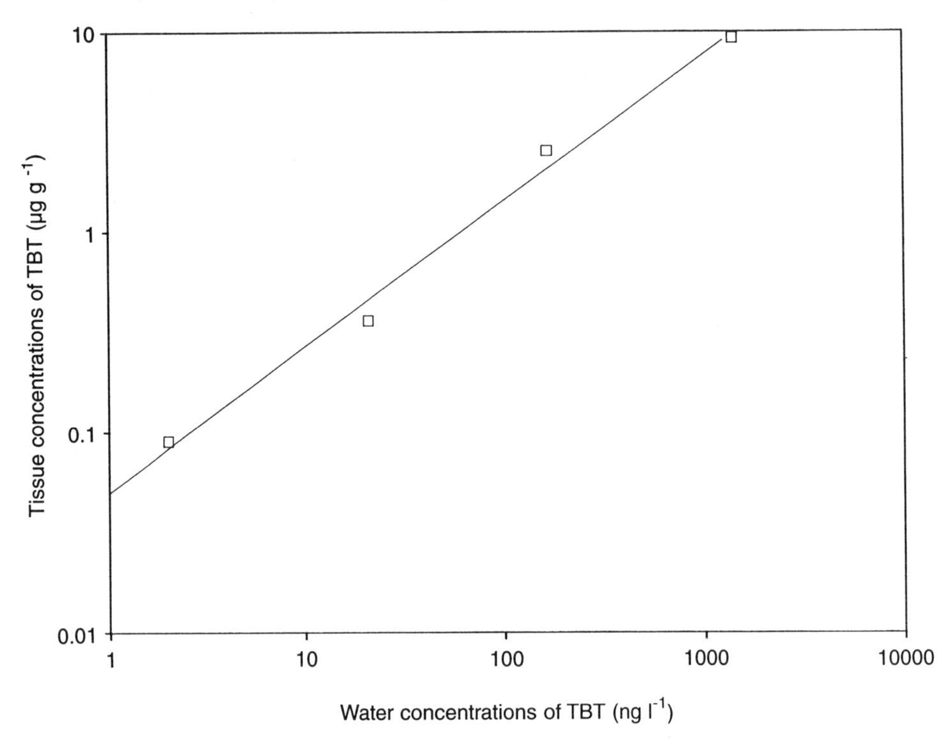

Figure 11.3 Concentrations of TBT in *Mytilus edulis* tissue after exposure to TBT for at least 21 d under laboratory (flowthrough holding) conditions.

15 400, 17 100, and 45 000 for exposure to measured concentrations of 1410, 165, 21, and 2 ng l^{-1} TBT. The BCF values fall within the same range as those reported by Salazar and Salazar (1988).

11.3. BIVALVES AS INDICATORS OF ENVIRONMENTAL IMPACT

11.3.1. REQUIREMENTS FOR GOOD INDICATOR SPECIES

In addition to being a good accumulator of TBT, reliable indicators of environmental impact need to demonstrate a readily detectable and quantifiable physiological response to a range of concentrations of the contaminant. The response should also be specific to TBT and ideally be graded in relation to the concentration of the compound, so that it is possible to gauge environmental concentrations of TBT from the severity of the effect. Practical considerations dictate that the study population should be reasonably isolated, in order to ensure that there is no import or export of less- or more-contaminated individuals.

11.3.2. SUITABILITY OF *CRASSOSTREA GIGAS*

The very specific responses of shell thickening, with associated loss of meat yield, in *Crassostrea gigas* make it an excellent candidate for monitoring environmental changes in TBT concentrations. *C. gigas* is reasonably sensitive to TBT with progressive shell thickening produced at concentrations greater than approximately 10 ng l^{-1} TBT (Thain *et al.*, 1987) and with reversion to normal shell growth in cleaner water. This species has shown a graded response in thickening away from emission sources of TBT, and to date none of the other toxic substances likely to have caused the effect in estuarine systems has been shown to induce this specific biological effect. Copper, for example, has little effect on growth of the oysters, and cages made from cupro-nickel mesh produce excellent oyster growth while preventing fouling from other species (Paul and Davies, 1986). Neff and Boehm (1985) reported high concentrations of aromatic hydrocarbons in *C. gigas* tissues (500 $\mu g\ g^{-1}$) after the *Amoco Cadiz* spill with no decrease in oyster quality. Furthermore, in field trials Key *et al.* (1976) measured residues of several possible causative agents of shell thickening (metals and pesticides) and found no correlation of their concentrations with growth performance. *C. gigas* also has the advantage of not being able to breed in the United Kingdom, and so there can be no inadvertent mixing of captive populations with wild stocks.

It should be noted that not all races of *C. gigas* oysters exhibit the thickening phenomenon. Okoshi *et al.* (1987) have recently reported that the Myagi race are particularly prone to producing chambers within the shell structure, whereas the Hiroshima race produce few if any chambers in the shell. However, the thickening malformation is not restricted to *Crassostrea* species; the Sydney rock oyster *Saccostrea commercialis* has also been reported to thicken in the presence of TBT (Batley *et al.*, 1989). Conversely, other *Crassostrea* species (notably *C. virginica*) have been shown not to thicken in the presence of TBT (Thain *et al.* 1987).

11.3.3. CALIBRATION OF EXPOSURE CONCENTRATION AND SHELL THICKNESS

The relationship between shell thickness index (length divided by the thickness of the upper shell valve) and exposure to TBT has been examined in two experiments (Waldock and Thain, 1983; Thain *et al.*, 1987).

Length in the above studies was determined by the distance from the umbo to the outer edge, along the mid-line of the shell; this distance is usually defined as shell height by the scientific community (Tebble, 1966), but as length by mariculturalists. Shell thickness is measured with calipers and is the distance from the inside to the outside of the upper valve in the center of the shell, at a point adjacent to the adductor muscle scar.

In laboratory studies at low water concentrations of TBT ($<7\ ng\ l^{-1}$), oyster growth was normal; at 20 $ng\ l^{-1}$ the oysters' shells thickened, and the amount of thickening increased with concentrations of up to 240 $ng\ l^{-1}$. At a concentration of 1.6 $\mu g\ l^{-1}$ no growth occurred, and therefore the oyster shells remained thin. Using these data together with data for the calibration of tissue residues to water concentrations, a curve can be produced to predict shell thickening behavior in the presence of TBT either in water or from tissue concentrations (Fig. 11.4a).

The relationship between TBT concentrations in water and production of thickened shells is not, however, a simple one. Two points about shell thickness should be emphasized. Firstly, all Pacific oysters grown in the United Kingdom are of the Miyagi race, which tend to produce cavities in the shell, resulting in thicker shells with increasing age. When grown in water with a low content of suspended solids under ideal conditions, a newly metamorphosed spat will have a thickness index in excess of 50; by the time the oyster reaches marketable size (>70 g in perhaps 2 y) the index could be as low as 12. Secondly, when grown in turbid water, Pacific oysters will produce a much heavier and therefore thicker shell. At the first marketable size the thickness index may be ~9, but as the oyster gets older indices of 6 may not be unusual.

All such normal oysters produce a large body cavity and yield good meats. Excessive shell thickening resulting from exposure to TBT is an extreme version of the natural

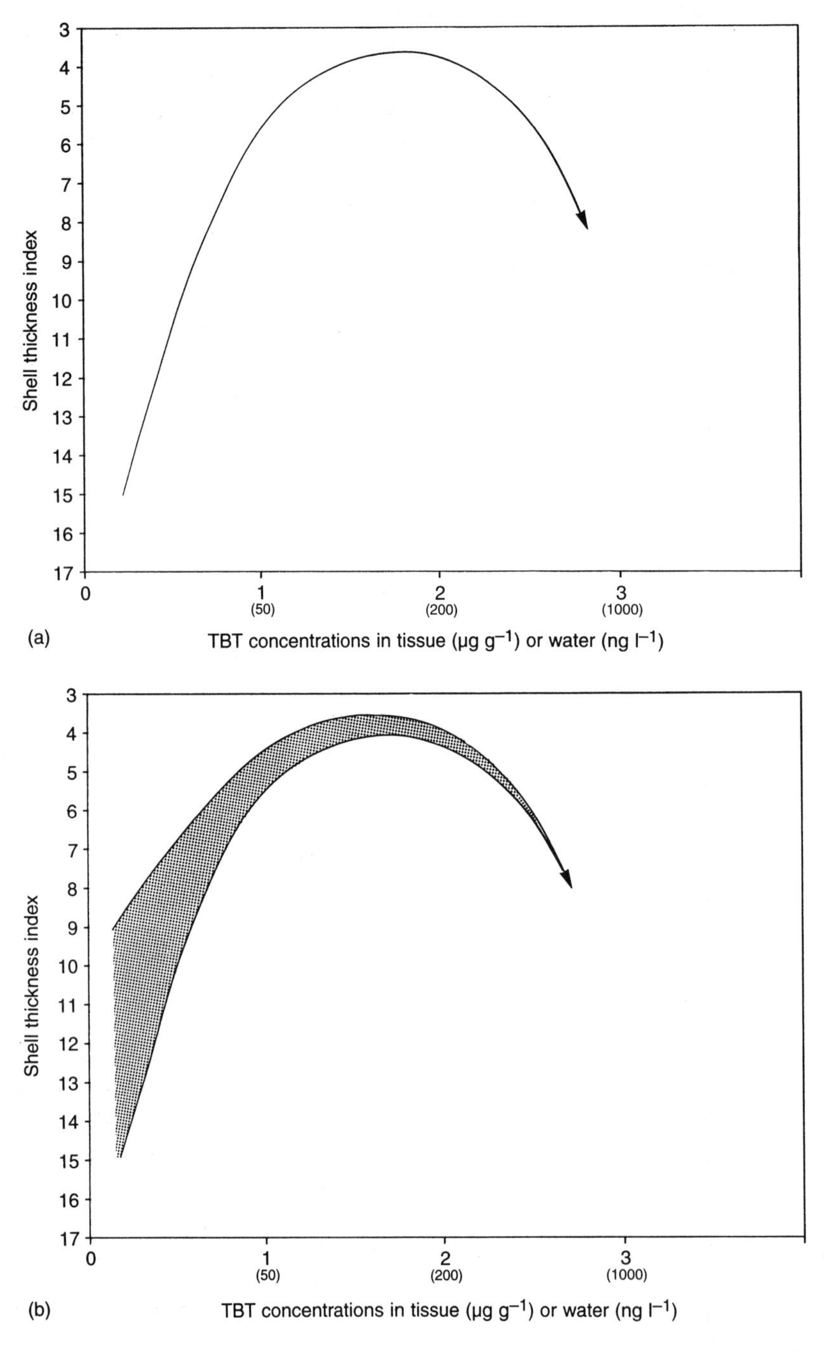

Figure 11.4 (a) The relationship of shell thickness index and TBT concentration in *Crassostrea gigas* in clear water. The diagram represents the likely progression of thickening if young oysters are exposed to TBT in their first growing season. (b) When exposed to water with a wide range of suspended solids regimes, thicker oysters occur in more turbid waters. As the concentration of TBT increases, the differences in thickness index become less marked.

thickening process, but in this case it is associated with a small body cavity and poor meat yields. Shell thickness indices <5 are indicative of 'ball-shaped oysters', which rarely contain sufficient meat to market.

The idealized curve for the shell thickening response to TBT shown in Fig. 11.4a would have to be modified if oysters were held in conditions of higher suspended particulate matter. Figure 11.4b represents the range of thickness indices that could occur if juveniles were held in a range of suspended particulate regimes before exposure to TBT. The diagram would have to be modified further if older animals were deployed, because the curve would be displaced to lower thickness indices. Animals held in the River Crouch for 3 y in the mid-1980s, for example, had index values as low as 2.3. The continuous nature of the thickening response makes it imperative to define carefully the exposure period when the animals are deployed as indicator species.

11.4. DEPLOYMENT OF BIVALVES IN THE FIELD

Field relaying trials by MAFF for oysters and mussels have been standardized on ~1-g animals. Juveniles have the advantage of fast growth rates, so that growth effects may be measured in relatively short periods, and yet 10–20 animals yield sufficient tissue for TBT analysis. In practice it was extremely difficult to obtain a single size class of oysters from the hatchery (MAFF Shellfish Cultivation Unit, Conwy, or Guernsey Sea Farms) at a particular time, and therefore, in the field-relaying trials carried out between 1983 and 1989, the initial weight of stocks supplied for the trials has varied from 0.3 g (spat produced early in the year) to 2.5 g (spat produced the previous autumn).

The method of deployment followed guidelines published by the MAFF Shellfish Cultivation Unit (Spencer and Gouch, 1978; Spencer *et al.*, 1985). Approximately 200 oysters or mussels were held in Netlon cages (100 × 50 × 8 cm, 6 mm mesh, Netlon Ltd, Blackburn) on trestles placed at low water spring tides on the foreshore. Stocking density did not exceed 1 g cm^{-2}, and the cages were cleaned of fouling organisms each month. Since a sample of 20 individuals was removed from the cage each month, the oysters rarely outgrew the space available. However, where necessary the animals were moved on to larger cages with a broader mesh (15-mm mesh). There were some exceptions to this strategy in areas where intertidal deployment was impossible. In such cases animals were kept fully submersed throughout the tidal cycle as this was expected to promote an acceptable growth rate.

Animals were normally relaid in April each year and samples taken at monthly intervals thereafter. The accumulation and depuration curves of animals exposed in the laboratory suggest that after 3 weeks the concentrations of TBT in bivalve tissue would be in equilibrium with TBT concentrations in water and should give a reliable indication of average water concentrations through to the autumn. In the autumn, when boats are removed from many UK estuaries, concentrations in the animals could reduce less quickly than those in the water column, and hence they are a less reliable indication of very recent exposure. Figure 11.5 shows data from the River Crouch, Essex, in 1986. There is a lag between highest water concentrations and highest tissue concentrations, which results from the inability of the animal to rapidly depurate TBT.

The seasonal nature of changes in TBT concentrations due to small boat activity in the United Kingdom is well documented (Waldock, 1986). In areas where TBT concentrations do not fluctuate widely, the bivalves may give a reliable indication of water concentrations for a longer period.

In the River Crouch, *C. gigas* were relaid at stations ~3 km apart along the river, whereas

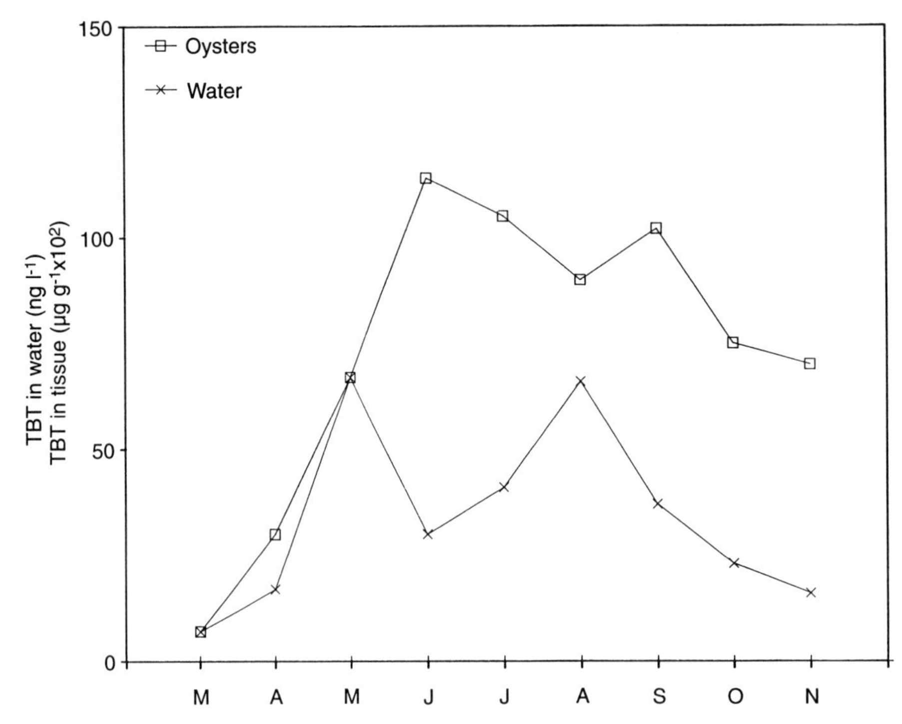

Figure 11.5 Concentrations of TBT in *Crassostrea gigas* and water in the Crouch at Burnham in 1986. Tissue data are based on a single analysis of a composite sample of 20 individuals. Water data are based on a mean of up to four determinations per month.

in other estuaries they were deployed at only one site. The former approach yields detailed spatial information on a graded response away from large numbers of boats, while the latter allows for a larger number of sites to be monitored.

At other monitoring sites in the United Kingdom, bivalves were generally placed in areas traditionally used for mariculture, to ensure that control measures were effective at the most potentially productive areas within estuaries. Additionally, in order to ensure that sites with a range of TBT concentrations were being monitored, one site was chosen to give a worst-case scenario (i.e. close to a marina) and one site was chosen as a control. However, due to the ubiquitous presence of TBT in English estuaries, the latter had to be re-designated as a 'low-level TBT site'.

11.5. SITE-SPECIFIC RESULTS

11.5.1. OYSTERS AND MUSSELS AS INDICATORS OF ENVIRONMENTAL CONTAMINATION

Results for tissue concentrations of TBT in relaid oysters at Burnham on Crouch (1986–1988) are shown plotted against measured water concentration of TBT in Fig. 11.6. The results generally conform to the relationship predicted by laboratory studies, despite the fact that relatively few water samples were taken.

Changes in TBT concentrations in tissues of oysters held at seven stations on the Crouch between 1987 and 1988 are shown in Fig. 11.7. The example demonstrates how oysters may be used to gain spatial information on the distribution of TBT concentrations within an estuary. Over a thousand

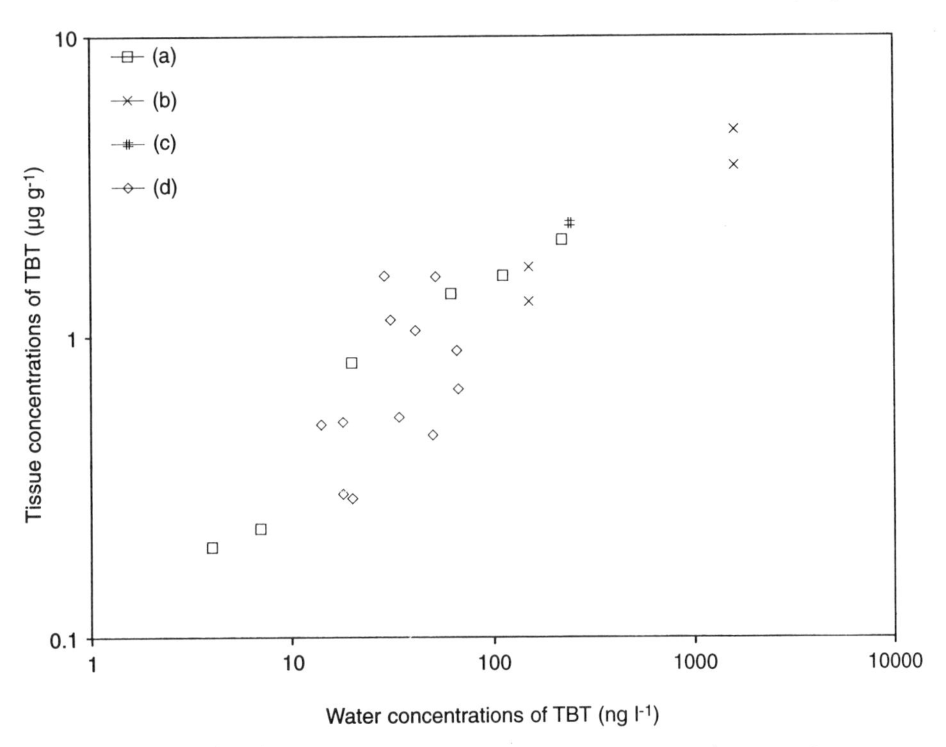

Figure 11.6 The relationship between TBT concentrations in water and oyster tissue measured at Burnham on Crouch, May–August 1986, 1987, and 1988. For calibration data for (a), (b), and (c) see Fig. 11.2. (d) Data from oysters deployed in the field.

boat moorings are situated midway down the estuary at Burnham 8 km from the open sea, with a few hundred moorings at the head of the estuary at Fambridge.

Single water samples taken at slack-water low tide show that the highest TBT concentrations occur at Burnham (Thain *et al.*, 1987). A combination of migration of this contaminated plume of water up the estuary on the flood tide and the additional inputs from marinas at Fambridge would account for the higher concentrations of TBT in tissue measured in animals at the top of the estuary. Toward the mouth of the estuary, concentrations of TBT are reduced due to water exchange with the open sea. This distribution of TBT in the estuary would require an intensive spatial and temporal program of water sampling to give data comparable to those obtained from oyster tissue analysis.

Following legislation in July 1987 that banned the use of TBT on small boats, concentrations of TBT in tissues decreased at all sites along the estuary. The spatial trend reflects that of 1987, but the temporal differences among the years are clear (Fig. 11.7). Data for TBT in the water column confirm the decrease in concentrations from 1987 to 1988.

The relationship between tissue concentrations of TBT in mussels and water in UK, US, and Danish field trials is shown in Fig. 11.8. Both Figs 11.6 and 11.8 demonstrate that the relationship between water and tissue concentration shows a well-defined positive correlation (R^2 = 0.78 and 0.79, respectively), despite the fact that the data used are based on limited measurements of water concentrations and incorporated year-to-year variability, different size classes of animals, different analytical methods, different expos-

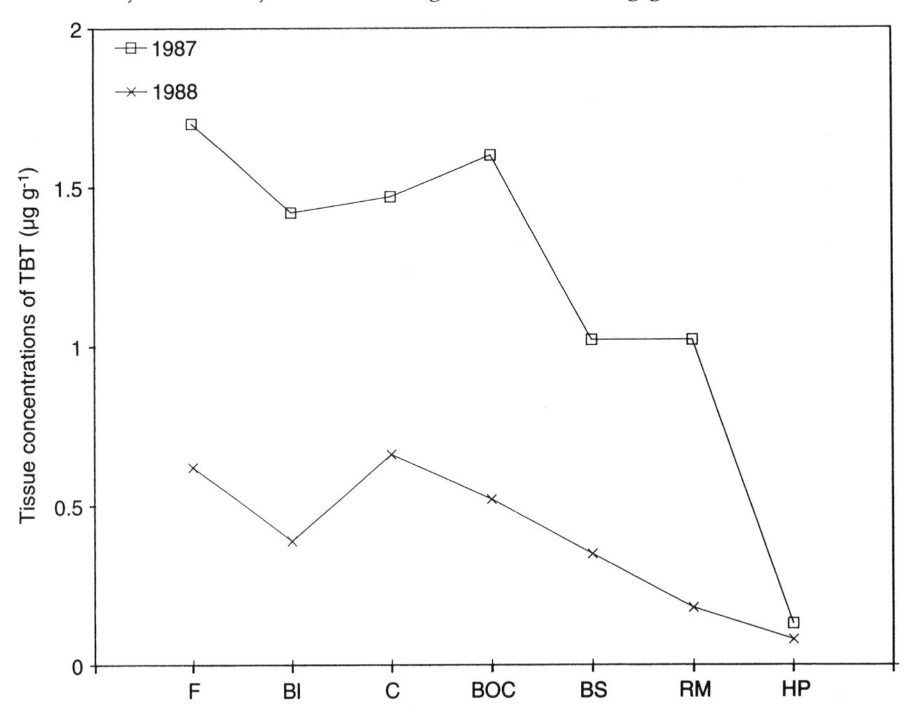

Figure 11.7 Tissue concentrations of TBT in *Crassostrea gigas* held in the River Crouch in 1987 and 1988 (F = Fambridge, BI = Bridgemarsh Island, C = Creeksea, BOC = Burnham on Crouch, BS = Bush Shore, RM = Roach Mouth, and HP = Holliwell Point). Fambridge represents the uppermost station at the head of the estuary, and Holliwell Point is at the mouth. The stations are ~3 km apart. Data points represent the mean of three determinations (in June, July, and August) of tissue concentration of TBT from a composite sample of 20 individuals (Table 11.1 shows typical ranges of TBT concentrations from June to August).

ure regimes, different water temperatures, and the contribution of other sources of TBT (i.e. food and sediment have not been considered).

If the data for bivalves illustrated here are used to predict water concentrations, the range of water concentrations associated with a particular tissue concentration could be large; for example, from Fig. 11.6, a tissue concentration of 1.0 $\mu g\ g^{-1}$ in oysters could be associated with exposure to between 15 and 160 ng l^{-1}. This large range of predicted concentrations stems principally from the poor data set for water. Variability in the bioavailable sediment-bound fraction of TBT also contributes to the uncertainty; for example, measured concentrations of TBT in surface sediment at Burnham for June–October 1986 varied over an order of magnitude. Until the relationship between bivalve tissue concentration and that in water, sediment, and food is better defined, oysters and mussels cannot be used as accurate predictors of water concentrations. Nevertheless, the consistency in the tissue concentration to exposure data generated in laboratory experiments for bivalves, the same consistency in results for animals deployed in the field over several years, and the gradation of tissue concentration from highly impacted to cleaner environments, suggest that the animals are a good measure of all sources of environmental

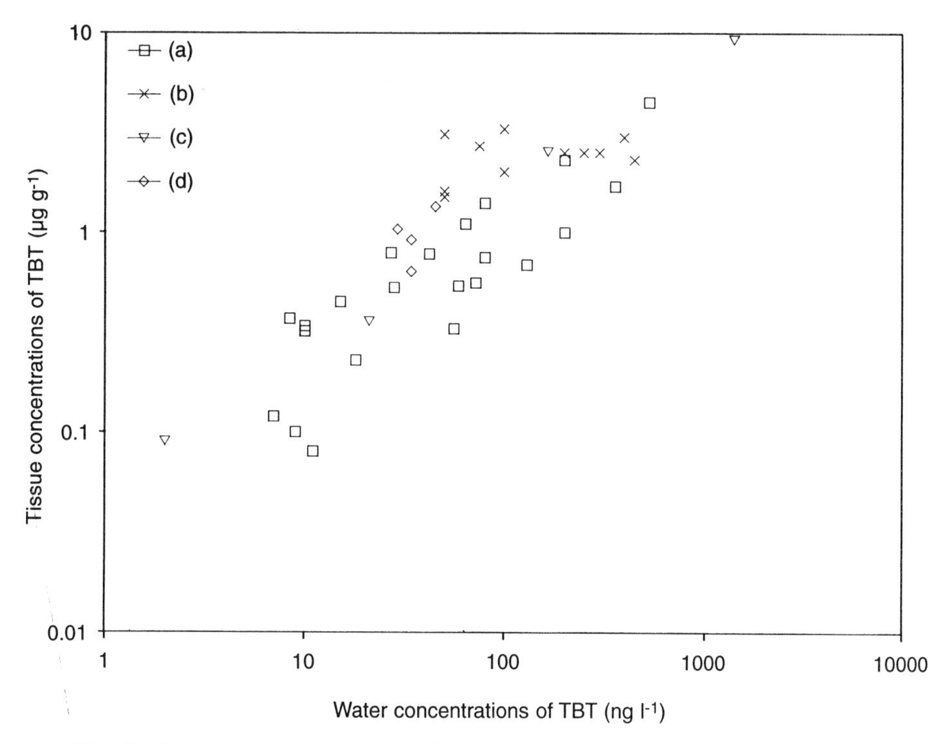

Figure 11.8 The field relationship between TBT concentrations in water and mussel tissue measured by US, UK, and Danish researchers. (a) Salazar and Salazar (1988), (b) Zuolian and Jensen (1989), (c) calibration (see Fig. 11.3), (d) Waldock *et al.* (unpublished).

contamination. Indeed they may be a much more reliable measure than analytical data derived from a few water samples.

Pacific oysters accumulated different amounts of TBT in tissues than did mussels when held at the same sites. Table 11.1 shows a comparison of tissue concentrations of oysters and mussels held in Essex under the same conditions at sites showing a range of contamination in water.

Mussels accumulated approximately half the tissue concentrations of TBT present in the oysters. Accumulation studies in this laboratory would suggest that if water were the only route of uptake, *M. edulis* should accumulate similar tissue concentrations at similar exposures to TBT in water at the range of values normally encountered in UK estuaries. The results are broadly in agreement with field results from Salazar and Salazar (1988) and Zuolian and Jensen (1989) who suggest BCF values for mussels of up to 44 000 and 60 000 respectively.

11.5.2 MEASUREMENT OF CHANGES IN OYSTER GROWTH

Comparison of year-to-year changes in bivalve growth is difficult. Yields of oysters in each growing season may be different due to variability in condition of the stocks, food availability, and water temperature. Stocks available from hatcheries may be of slightly different sizes; however, genetic variability is unlikely to be a problem since all *C. gigas* in the United Kingdom have been reared from the same race. Pacific oysters are also known to be heavier shelled in waters with a high load of suspended solids. Despite this, the relationship of excessive shell thickening and

Table 11.1 Concentrations of TBT in *Crassostrea gigas* and *Mytilus edulis* held at the same sites during 1987. Mean concentrations are from a composite sample of 20 individuals

	TBT concentration ($\mu g\ g^{-1}$ wet wt)		
Site and date	*Oyster*	*Mussel*	*Ratio mussel:oyster*
West Mersea			
June 1987	1.79	1.02	0.57
July 1987	2.75	0.74	0.27
Aug 1987	1.99	0.82	0.41
Sept 1987	2.33	0.79	0.34
Oct 1987	1.38	0.54	0.39
Roach Mouth			
May 1987	0.59	0.23	0.39
June 1987	1.14	0.62	0.54
July 1987	1.02	0.55	0.55
Aug 1987	0.79	0.41	0.52
Oct 1987	0.56	0.33	0.59
Goldhanger			
April 1987	0.22	0.12	0.55
June 1987	0.27	0.13	0.48
July 1987	0.19	0.12	0.63
Nov 1987	0.15	0.12	0.80

body burden of TBT appears to be well defined in the field and is consistent from year to year. Figure 11.9 summarizes 4 y of field data.

Since the thickening process is continuous, values of thickness index can be compared for only equivalent periods of exposure. Figure 11.9 depicts thickness values measured at the end of the summer, ~4 months after relaying small, and hence thin-shelled, individuals. At low ambient concentrations of TBT, the thickness index of the animal largely reflects the suspended solids regime of the site, animals generally having heavier shells in turbid waters (but still producing good meat yields). At higher TBT concentrations the animals become grossly thickened (i.e. have low values of thickness index) as growth rates reduce. At the very highest concentrations measured at marina sites, the oysters exhibit little or no growth at all, and therefore the thickness index does not decrease with time. However, animals held at Beaulieu within a marina on the south coast of Britain did survive and thicken even with high concentrations of TBT in the flesh (>5 $\mu g\ g^{-1}$). Although no explanation is presently available for the tenacious nature of these animals, J. Widdows (Plymouth Marine Laboratory, personal communication) has recently demonstrated antagonistic behavior between the toxic effects of TBT and the presence of hydrocarbons in mussels.

Shell thickening alone is not considered to be indicative of exposure to TBT. Loss in meat production is always associated with the malformed shells in regions of TBT exposure. Figure 11.10 shows the relationship between meat weight and TBT tissue concentrations of *Crassostrea gigas* deployed at 16 sites in the United Kingdom in 1986. Optimum meat yields were only achieved where TBT concentrations were low; and meat yield was severely affected at high TBT concentrations, despite the relatively short period of exposure. Deployment for longer periods amplifies such differences, and by December in the same year there was a 17-fold difference between meat weights at the best and worst sites (5.10 ± 1.47 g and 0.29 ± 0.2 g, respectively).

During the period from 1983 to 1987 there was little change in the growth characteristics of *C. gigas* at most of the monitoring sites. The small changes in TBT concentrations in the tissues suggest that marked improvements should not be expected, since concentrations were well above the thresholds at which the thickening response occurs. Best estimates of the concentration of TBT in oyster tissues that initiates the response in the field are ~0.2–0.3 $\mu g\ g^{-1}$. Oysters held in the River Teign (South Devon) generally accumulated TBT concentrations to this level and continued to grow normally. At Goldhanger on the River Blackwater (Essex) a local oyster grower had abandoned a site for

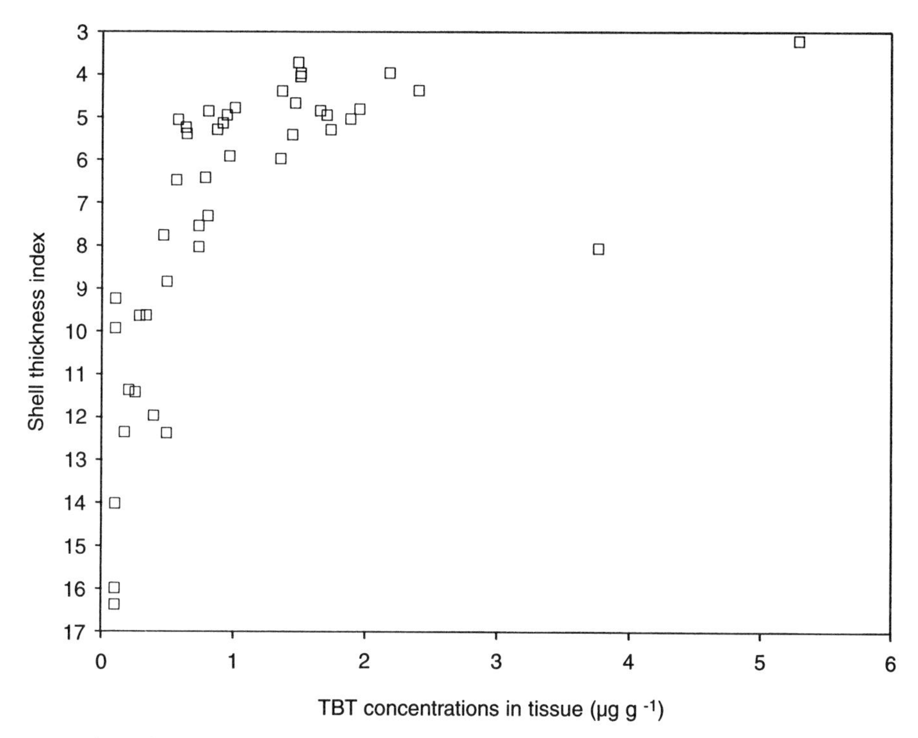

Figure 11.9 The relationship between TBT concentrations in tissue and shell thickness index measured in *Crassostrea gigas* deployed in the field in each year 1983 to 1987. Thickness index was measured in August. Each data point represents the mean of 20 individuals. Typical standard deviations would be 25% of the index value.

production of *C. gigas* in the late 1970s. In 1984 concentrations in mid-summer were 0.4 $\mu g\ g^{-1}$, and oysters failed to grow normally. Growth of the animals at the same site in 1985–1987 improved to the point of once again being commercially viable, and the business was restarted. TBT concentrations had decreased to ~0.2 $\mu g\ g^{-1}$.

During 1988, concentrations of TBT in oyster flesh showed a marked decrease at most sites compared with 1987 levels. They were also concurrent improvements in meat yields and increased shell thickness indices at less marginal sites. Table 11.2 shows comparative data for 1984 and 1988 at Creeksea (moderately contaminated) in the Crouch and the Teign (low TBT) in South Devon.

In order to improve the predictive ability of such bioindicators, further studies will be necessary to define the moderating effects of parameters not considered in detail here. Factors such as food availability, temperature, salinity, intertidal exposure, disease susceptibility, suspended particulate regimes, holding conditions, and other toxicant stress clearly have a role to play. Field transplanting experiments are fraught with difficulties in interpretation of such variables, and, as the list increases, the cost of monitoring additional determinands becomes prohibitive. Presently it would appear to be unwise to suggest that in the United Kingdom there are significant improvements in growth rates and thickness indices that are solely due to decreases in TBT concentrations following legislation. However, the trend for better growth performance is continuing in 1989, and for the first year of relaying trials,

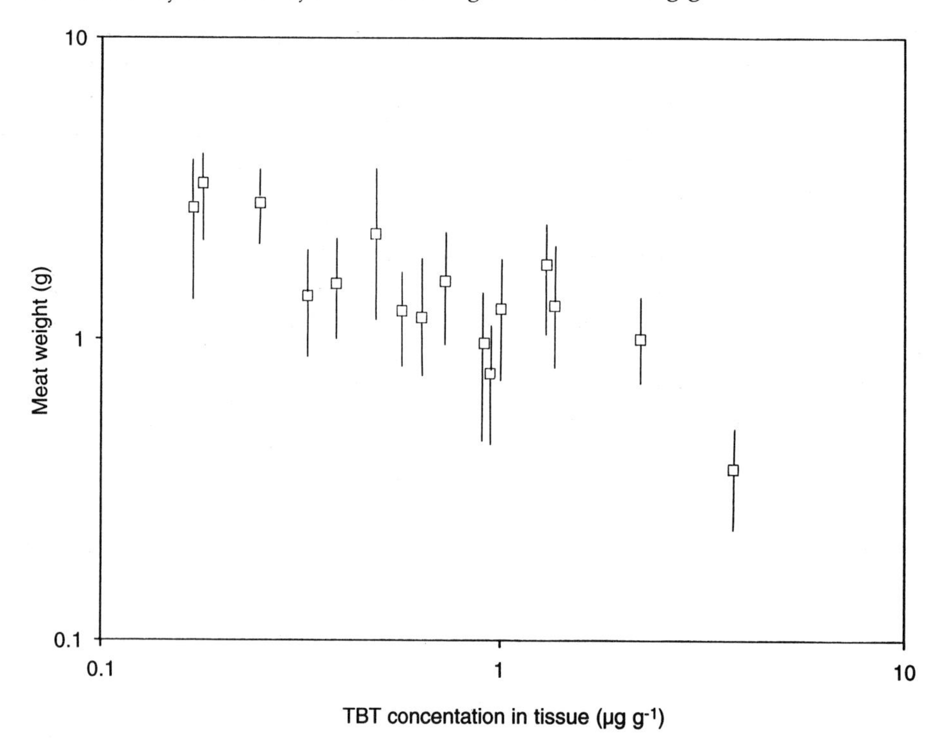

Figure 11.10 The relationship between meat yield and tissue concentration of TBT in *Crassostrea gigas* deployed at 16 sites in the United Kingdom during 1986. TBT concentrations are mean values for June–August, and meat weight was measured in August. Standard deviations are shown for meat weight determinations.

Table 11.2 Comparison of oyster growth and tissue concentrations of TBT in 1984 and 1988

Site	*Date*	*TBT concentration* ($\mu g\ g^{-1}$ *wet wt*)	*Whole weight (g)*	*Meat weight (g)*	*Shell thickness index*
Day 0	12 April 1984	<0.1	2.60 ± 0.87	0.68 ± 0.22	26.25 ± 11.14
	14 April 1988	0.01	2.37 ± 0.41	0.56 ± 0.11	36.98 ± 9.37
Creeksea	15 August 1984	1.5	4.54 ± 1.60	0.79 ± 0.33	4.95 ± 1.95
	11 August 1988	0.53	11.09 ± 2.07	2.23 ± 0.59	8.07 ± 2.32
Teign	28 August 1984	<0.1	15.46 ± 2.66	3.96 ± 1.05	9.42 ± 2.42
	30 August 1988	0.23	16.33 ± 2.95	5.22 ± 1.26	25.66 ± 6.37

animals held within a marina at Beaulieu have shown marked weight gain and normal shell growth in the period April–June 1989.

From our work it appears that shell thickness in *Crassostrea gigas* can be an indicator of environmental impact by TBT compounds, and this is corroborated by other researchers. For example, a considerable amount of field work has been carried out by Alzieu and his colleagues in France. Various indices associated with malformed oysters have been used as predictive tools in assessing the degree of contamination of French estuaries and bays by TBT. Early studies were carried out using the thickness index described above (Alzieu *et al.*, 1980, 1982; Alzieu, 1981). In addition to measurement of the upper shell valve, the presence of cavities in the lower valve has also been employed to indicate the most severe form of malformation (Alzieu *et al.*, 1986). The formation of a gel within the chambers has also been used to gauge temporal differences in exposure to TBT (M. Héral, personal communication). Following the ban on use of TBT on small boats in France (see His, Chapter 12), such indices have been measured to follow reductions in TBT concentrations and associated improvements in oyster growth. In areas where the ban has been rigidly enforced, such as in the Bay of Arcachon, there are no problems at present with oyster spatfall and abnormal growth (Alzieu *et al.*, 1989; C. Alzieu, personal communication).

Similar indices of shell thickness have also been used in laboratory or field studies in Scotland by Davies *et al.* (1987); Ireland by Minchin *et al.* (1987); the US by Wolniakowski *et al.* (1987); Canada, (E. Black, personal communication); and Portugal (H. Phelps, personal communication). In all cases TBT has been proposed as the causative agent.

Some researchers also suggest that shell thickening in mussels may act as an indication of environmental impact by TBT: Stephenson *et al.* (1986) have described mussel shell thickening close to harbors, and Page *et al.* (1989) have recently demonstrated that condition indices of mussels correlate well with TBT concentration and have found significant differences in shell thickness index for mussels with high and low concentrations of TBT in tissue. However, Salazar and Salazar (1988) suggest that growth in mussels at low concentrations of TBT (<200 ng l^{-1}) is predominantly controlled by other environmental factors. The use of mussels as bioindicators is discussed further in Laughlin *et al.*, Chapter 10.

11.6. CONCLUSIONS

MAFF studies to date show that the oyster *Crassostrea gigas* has a valuable role to play in monitoring programs designed to assess spatial and temporal trends in environmental concentrations of TBT. Controlled laboratory experiments show that the animals accumulate TBT from concentrations dissolved in water in a predictable way, with high BCF values at low concentrations and lower BCF values at high concentrations.

When bivalves are deployed in the field, exposure concentration cannot only be defined as TBT dissolved in water, since *C. gigas* can accumulate TBT from the particulate phase; and other routes of uptake such as sediment and food may affect tissue concentration. Nevertheless, the consistency in tissue concentrations in successive samples and gradation in concentration from clean to highly impacted areas point to the reliability of the animals as sentinel organisms.

Since the tissue concentration is less variable than values measured in spot samples of water and sediment, there is a strong case for setting EQS values in terms of tissue concentration of such species. Clearly there are advantages to proposing a shellfish standard; compliance could be demonstrated with relatively few samples per year, and all sources of bioavailable TBT are considered. The disadvantage of a shellfish standard would be that pulsed inputs of extremely high concentrations of TBT (e.g. from drydock discharges) might be missed if the animals closed their shell valves to minimize exposure.

All experience to date suggests that measurement of shell thickness index is a useful adjunct to trend monitoring programs, providing a low-cost indicator of environmental exposure to TBT. Cause-and-effect has been demonstrated, and the presence of normally shaped *C. gigas* (Miyagi race) that provide good meat yield is one way of demonstrating that ambient TBT concentrations are <10 ng l^{-1}. However, because of the difficulties in comparing thickness indices of different-aged stocks of oysters, it is not possible to use such parameters alone for predicting ambient TBT concentrations.

The UK temporal trend monitoring program will continue at least until 1992. In mid-1990, concentrations of TBT in water, bivalves, and sediments had continued to decrease. There has also been a concomitant increase in growth rates of *C. gigas* with thinner shelled individuals at more sites. The results confirm that the control measures taken by the UK government have been effective in protecting a valuable mariculture resource, and strengthen the arguments for the deployment of *C. gigas* oysters as bioindicator species.

REFERENCES

Alzieu, C. 1981. Evaluation des risques dus à l'emploi des peintures antisalissures dans les zones conchylicoles. Abnomalies de calcification. Thème surveillance continue et mecanismes d'action des pollutants. Addendum to the report of 5 June 1981. Report prepared by the Institut Scientifique et Technique des Pêches Maritimes, Nantes, 15 December 1981, 83 pp.

Alzieu, C., Y. Thibaud, M. Héral and B. Boutier. 1980. Evaluation des risques dus à l'emploi des peintures antisalissures dans les zones conchylicoles. *Rev. Trav. Inst. Pêches Marit.*, **44**, 301–349.

Alzieu, C., M. Héral, Y. Thibaud, M.J. Dardignac and M. Feuillet. 1982. Influence des peintures antisalissures à base d'organostaniques sur la calcification de la coquille de l'huître *Crassostrea gigas*. *Rev. Trav. Inst. Pêches Marit.*, **45**, 101–116.

Alzieu, C., J. Sanjuan, J.P. Deltreil and M. Borel. 1986. Tin contamination in Archachon Bay: effects on oyster shell anomalies. *Mar. Pollut. Bull.*, **17**, 494–498.

Alzieu, C., J. Sanjuan, P. Michel, M. Borel and J.P. Dreno. 1989. Monitoring and assessment of butyltins in Atlantic coastal waters. *Mar. Pollut. Bull.*, **20**, 22–26.

Batley, G.E., C. Fuhua, C.I. Brockbank and K.J. Flegg. 1989. Accumulation of tributyltin by the Sydney rock oyster *Saccostrea commercialis*. *Aust. J. Mar. Fresh. Res.*, **40**, 49–54.

Davies, I.M., J. Drinkwater, J.C. McKie and P. Balls. 1987. Effects of the use of tributyltin antifoulants in mariculture. In: Proceedings of Oceans '87 Conference and Exposition on Science and Engineering, Halifax, Nova Scotia, 28 September–1 October 1987, Institute of Electrical and Electronic Engineers, Piscataway, New Jersey, and Marine Technology Society, Washington, DC, pp. 1477–1481.

Farrington J.W., E.D. Goldberg, R.W. Risebrough, J.H. Martin and V.T. Bowen. 1983. US 'Mussel Watch' 1976–1978: an overview of the trace metal, DDE, PCB, hydrocarbon and artificial radio-nuclide data. *Environ. Sci. Technol.*, **17**, 490–496.

Holdgate, M.W. 1979. *A Perspective of Environmental Pollution*. Cambridge University Press, Cambridge, 278 pp.

Key, D., R.S. Nunny, P.E. Davidson and M.A. Leonard. 1976. Abnormal shell growth in the Pacific oyster *Crassostrea gigas*. Some preliminary results from experiments undertaken in 1975. International Council for the Exploration of the Sea, Copenhagen, CM 1976/K11, 12 pp (mimeo).

Minchin, D., C.B. Duggan and W. King. 1987. Possible effects of organotins on scallop recruitment. *Mar. Pollut. Bull.*, **18**, 604–608.

Neff, J.M. and P.D. Boehm. 1985. Petroleum contamination and biochemical alterations in oysters (*Crassostrea gigas*) and plaice (*Pleuronectes platessa*) from bays impacted by the *Amoco Cadiz* crude oil spill. *Mar. Environ. Res.*, **17**, 281–283.

Okoshi, J., K. Mori and T. Nomura. 1987. Characteristics of shell chamber formation between the two local races in the Japanese oyster *Crassostrea gigas*. *Aquaculture*, **67**, 313–320.

Page, D.S., E.S. Gilfillian, J. Foster and J. Widdows. 1989. Tributyltin in *Mytilus edulis* from coastal locations in Devon and Cornwall (UK) and Maine (US) and its effect on shell morphology. Fifth International Symposium on Responses of Marine Organisms to Pollutants, 12–14 April 1989, Plymouth Marine Laboratory, Plymouth, UK, unpaginated.

Paul, J.D. and I.M. Davies. 1986. Effects of copper and tin-based anti-fouling compounds on the growth of scallops (*Pecten maximus*) and oysters (*Crassostrea gigas*). *Aquaculture*, **54**, 191–203.

Phillips, D.J.H. and D.A. Segar. 1986. Use of bioindicators in monitoring conservative contaminants: programme design imperatives. *Mar. Pollut. Bull.*, **17**, 10–17.

Salazar, M.H. and M.A. Champ. 1988. Tributyltin and water quality: a question of environmental significance. In: Proceedings of Oceans '88 Conference, Maryland, 31 October–2 November 1988, Organotin Symposium. Institute of Electrical and Electronic Engineers, Piscataway, New Jersey and Marine Technology Society, Washington, DC, Vol. 4, pp. 1497–1506.

Salazar, M.H. and S.M. Salazar. 1988. Tributyltin and mussel growth in San Diego Bay. In: Proceedings of Oceans '88 Conference, Maryland, 31 October–2 November 1988, Organotin Symposium. Institute of Electrical and Electronic Engineers, Piscataway, New Jersey and Marine Technology Society, Washington, DC, Vol. 4, pp. 1188–1195.

Schatzberg, P., C.M. Adema, W.M. Thomas and S.R. Magnum. 1986. A time integrating, remotely moored, automated sampling and concentration system for aquatic butyltin monitoring. In: Proceedings of Oceans '86 Conference and Exposition on Science and Engineering, Washington, DC, 23–25 September 1986, Institute of Electrical and Electronic Engineers, Piscataway, New Jersey and Marine Technology Society, Washington, DC, Vol. 4, pp. 1155–1159.

Spencer, B.E. and C.J. Gouch. 1978. The growth and survival of experimental batches of hatchery-reared spat of *Ostrea edulis* L. and *Crassostrea gigas* Thunberg, using different methods of tray cultivation. *Aquaculture*, **13**, 293–312.

Spencer, B.E., C.J. Gouch and M.J. Thomas. 1985. A strategy for growing hatchery reared Pacific oysters (*Crassostrea gigas* Thunberg) to market size. Experiments and observations on costed small-scale trials. *Aquaculture*, **50**, 175–192.

Stephenson, M.D., D.R. Smith, J. Goetzl, G. Ichikawa and M. Martin. 1986. Growth abnormalities in mussels and oysters from areas with high levels of tributyltin in San Diego Bay. In: Proceedings of Oceans '86 Conference and Exposition on Science and Engineering, Washington, DC, 23–25 September 1986, Institute of Electrical and Electronic Engineers, Piscataway, New Jersey and Marine Technology Society, Washington, DC, Vol. 4, pp. 1246–1251.

Tebble, N. 1966. *British Bivalve Seashells. A Handbook for Identification*. British Museum, London/Alden Press, Oxford, 212 pp.

Thain, J.E. 1986. Toxicity of TBT to bivalves: effects on reproduction, growth and survival. In: Proceedings of Oceans '86 Conference and Exposition on Science and Engineering, Washington, DC, 23–25 September 1986, Institute of Electrical and Electronic Engineers, Piscataway, New Jersey and Marine Technology Society, Washington, DC, Vol. 4, pp. 1306–1313.

Thain, J.E. and M.J. Waldock. 1985. The growth of bivalve spat exposed to organotin leachates from antifouling paints. International Council for the Exploration of the Sea, Copenhagen, CM 1985/E29, 8 pp (mimeo).

Thain, J.E., M.J. Waldock. and M.E. Waite. 1987. Toxicity and degradation studies of tributyltin (TBT) and dibutyltin (DBT) in the aquatic environment. In: Proceedings of Oceans '87 Conference and Exposition on Science and Engineering, Halifax, Nova Scotia, 28 September–1 October 1987, Institute of Electrical and Electronic Engineers, Piscataway, New Jersey and Marine Technology Society, Washington, DC, Vol. 4, 1398–1404.

Waldock, M.J. 1986. TBT in UK estuaries 1982–1986. Evaluation of the environmental problem. In: Proceedings of Oceans '86 Conference and Exposition on Science and Engineering, Washington, DC, 23–25 September 1986, Institute of Electrical and Electronic Engineers, Piscataway, New Jersey and Marine Technology Society, Washington, DC, Vol. 4, pp. 1324–1330.

Waldock, M.J. and J.E. Thain. 1983. Shell thickening in *Crassostrea gigas* organotin antifouling or sediment induced? *Mar. Pollut. Bull.*, **14**, 411–415.

Waldock, M.J., J.E. Thain and D. Miller. 1983. The accumulation and depuration of bis(tributyltin) oxide in oysters: a comparison between the Pacific oyster *Crassostrea gigas* and the European flat oyster *Ostrea edulis*. International Council for the Exploration of the Sea, Copenhagen, CM 1983/E52, 9 pp (mimeo).

Wolniakowski, K.U., J. Grovhoug and K.E. Richter. 1987. Tributyltin concentrations and Pacific oyster deformations in Coos Bay, Oregon. In: Proceedings of Oceans '87 Conference and Exposition on Science and Engineering, Halifax, Nova Scotia, 28 September–1 October 1987, Institute of Electrical and Electronic Engineers, Piscataway, New Jersey and Marine Technology Society, Washington, DC, Vol. 4, pp. 1438–1442.

Zuolian, C. and A. Jensen. 1989. Accumulation of organic and inorganic tin in blue mussel, *Mytilus edulis*, under natural conditions. *Mar. Pollut. Bull.*, **20**, 281–286.

EMBRYOGENESIS AND LARVAL DEVELOPMENT IN *CRASSOSTREA GIGAS*: EXPERIMENTAL DATA AND FIELD OBSERVATIONS ON THE EFFECT OF TRIBUTYLTIN COMPOUNDS

12

E. His

Institut Français de Recherche pour l'Exploration de la Mer, Quai du Commandant Silhouette, 33120 Arcachon, France

ABSTRACT

The Bassin d'Arcachon, on the southwest Atlantic coast of France, is an important oyster-rearing area. From 1976 to 1981, the oyster industry was disturbed by shell abnormalities and by a drastic reduction of reproduction. As a consequence, the number of oyster farmers declined by 50%. The phenomena were consequent to a large increase in pleasure-craft activities in the bay; tributyltin compounds released from antifouling paints were suspected as the main cause of these abnormalities. The failure of larvae to survive the D larval stage was assessed experimentally by using the *Crassostrea gigas* embryo–

Organotin. Edited by M.A. Champ and P.F. Seligman. Published in 1996 by Chapman & Hall, London. ISBN 0 412 58240 6

larval bioassay. At first, it was established experimentally that the use of tributyltin compounds was dangerous in oyster-farming areas. Above 1 μg l^{-1} tributyltin (TBT) acetate, fertilized eggs cannot develop to the D larval stage; at 1 μg l^{-1} the D larval stage is reached, but all larvae are abnormal and die within a few days; from 0.5 to 0.05 μg l^{-1} abnormalities and mortalities are still considerable (>78% over a 12-d period), and larval growth is strongly affected. At 0.02 μg practically no action is observed, and this value seems to represent the threshold tolerance level of the larvae. On the other hand, neither low nor varying temperature could explain the lack of larval growth observed in the field; the failure in larval development could not be attributed to the action of TBT on the gamete viability from Arcachon adult oysters, or to its direct action on embryos and larvae. In 1981, D larvae isolated from the plankton of the Bassin d'Arcachon were reared in the laboratory, in seawater collected at the same place and time as the larvae. In the field, the stomach of the larvae remained uncolored (absence of food), and the larvae did not reach the early umbo stage; in the laboratory, larvae of the same brood stock fed normally (stomach well colored), and reached the early umbo and the umbo stage 12 d after the beginning of the experiments. From this it was supposed that D larvae in the field could not find the appropriate food required during the first days of their pelagic life; this was probably due to a disturbance in the development of the nanoplankton caused by the action of antifouling paints containing organotin compounds. Since the ban on TBT-based antifouling paints was put into effect in 1982, the Japanese oyster has reproduced on a commercial scale every year in the Bassin d'Arcachon.

12.1. INTRODUCTION

Early experiments on embryos and larvae of the Japanese oyster (*Crassostrea gigas*) (His and Robert, 1980) reported for the first time the possible deleterious effects of the presence of tributyltin on oyster-farming areas. To my knowledge, moreover, the only observations to date on the effects of this pollutant on oyster recruitment have been performed in the Bassin d'Arcachon, in southwest France (His and Robert, 1985; His *et al.*, 1986). In the present study, the Bassin d'Arcachon has been used as a model to assess the action of tributyltin released from antifouling paints on embryogenesis and larval development of *Crassostrea gigas*.

The Bassin d'Arcachon (Fig. 12.1), located along the Atlantic coast of France, is one of the greatest oyster-rearing areas in the world (Iversen, 1968). 'It is impossible to imagine any area better suited for oyster culture than this almost land-locked bay' (Yonge, 1960). The Pacific oyster was introduced on a commercial scale to the Bassin d'Arcachon in 1971, after the Portuguese oyster (*Crassostrea angulata*) had been completely destroyed in the French oyster-farming areas by two successive viral diseases (Combs, 1983). In the spring of 1971, large Japanese oysters were airshipped from British Columbia to serve as breeding stock in the Bassin d'Arcachon. Two months later, great quantities of spat had settled on collectors. Until 1972, spat from Japan and British Columbia continued to be introduced to complete the local recruitment. In the summers of 1971, 1973, 1975, and 1976, the Pacific oyster reproduced on a commercial scale. The highest level of production ever observed in the Bassin d'Arcachon was then 20 000 metric tons (t).

The Japanese oyster can reproduce from early June to early September, with an annual production of as high as 5×10^9 spats (His and Robert, 1985) and usually >12 000 t of marketable oysters. The economic base of oyster farming in the Bassin d'Arcachon is twofold: spat are produced and developed until harvested at marketable sizes; they are also sold to other French aquaculture centers such as Brittany, Normandy, and the

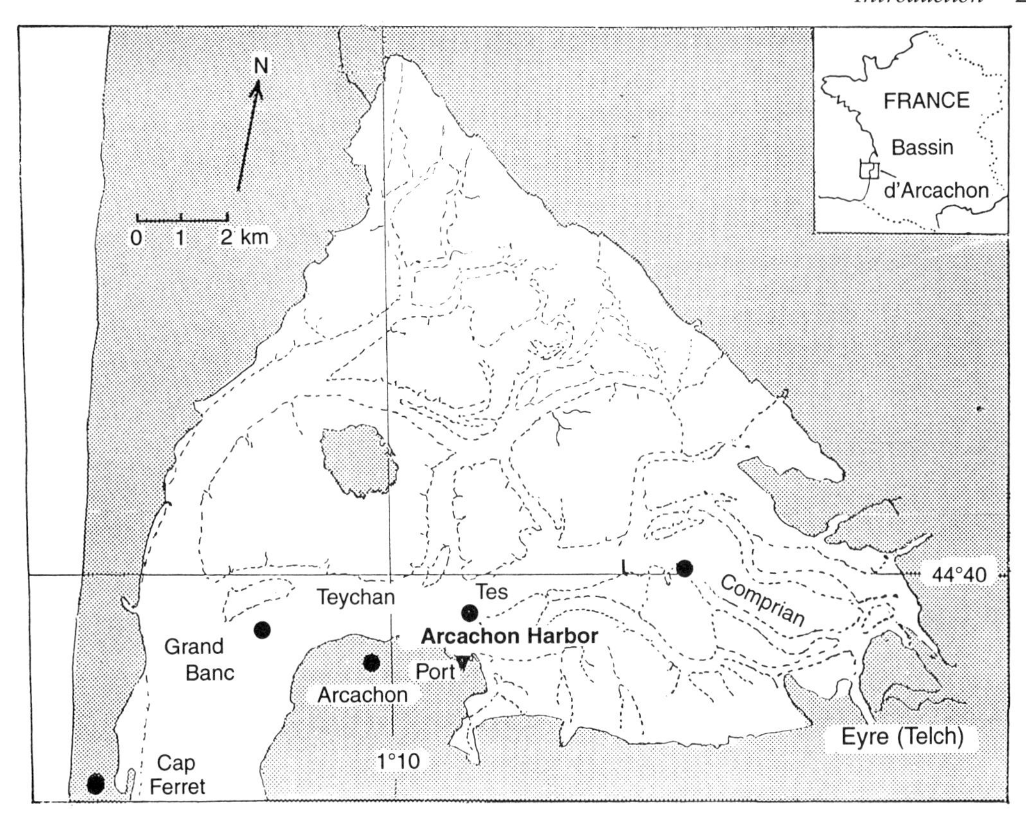

Figure 12.1 Locations of sampling stations in the Bassin d'Arcachon, France.

Mediterranean coast, where the Pacific oyster cannot reproduce.

As of 1976 and through 1981, the oyster industry was disturbed by two phenomena: poor growth and physical abnormalities. Shell malformations characterized by considerable thickening of the valve and the absence of fattening in autumn diminished the commercial value of marketable oysters. Moreover, settlement of spat either failed or was drastically reduced. As a consequence, the oyster farmers had either to buy spat in the Bassin de Marennes-Oléron (~100 km north of the bay), where the Japanese oyster continued to reproduce, or to import it again from Japan. The annual production progressively declined to the lowest level, <10 000 t in 1982, and the number of active oyster farmers was reduced by 50%. The abnormalities were still observed, even when the population of adult oysters was at its lowest; this fact clearly indicates that, in this case, overcrowding was not a cause of failure of the larvae to develop.

Anthropogenic activity was suspected as the major cause of the malformations and settlement failures. The same phenomena were observed near La Rochelle Harbor, which is situated ~150 km north of Arcachon. In both La Rochelle Harbor and the Bassin d'Arcachon, there is a concentration of pleasure craft, whose antifouling paints released organotins; discovery of this fact led to the study of the possible adverse effects of tributyltin compounds in oyster-growing areas.

Among various heavy metals tested on embryos and bivalve larvae (see Calabrese *et al.*, 1977; Martin *et al.*, 1981; Deslous-Paoli, 1982), tributyltin was rapidly shown to be one of the most highly toxic at levels far below those of any previously studied pollutant. As a consequence, it was established for the first time that the use of tributyltin paints, such as those used on pleasure craft, could be deleterious to oyster-farming areas (His and Robert, 1980). Subsequently, Alzieu *et al.* (1982) produced, in oysters experimentally exposed to tributyltin compounds, abnormalities similar to those found in field-reared oysters.

In the present study, the failure of oyster recruitment in the bay was studied experimentally by using the *Crassostrea gigas* embryo–larval bioassay to monitor the water quality in the bay. A second important aim was to demonstrate that results obtained in the laboratory have relevance to prevailing environmental conditions in the bay itself.

12.2. EXPERIMENTAL DATA: THE EFFECTS OF TRIBUTYLTIN ACETATE ON EMBRYOGENESIS AND LARVAL DEVELOPMENT IN *CRASSOSTREA GIGAS*

Adult oysters from an unpolluted area (Brittany) were collected and conditioned in the laboratory for 3 weeks at 20°C in a recirculating seawater system. Spawning was induced by thermal stimulation, and the eggs were fertilized in seawater containing the suspected pollutant. After incubation, 16 000 larvae were put into each of a series of 2-l hard glass beakers containing filtered (0.2 μm pore size) seawater and prepared at the desired test conditions. Duplicates were run for each set of experimental conditions. Natural seawater was used for the various treatment solutions; to avoid the presence of any pollutant, seawater was obtained in bulk 9 km offshore from the Cap Ferret peninsula, and its salinity was lowered to 28‰ by adding distilled water. This salinity is usually found in the central part of the bay in summer, when the oysters reproduce. The various tributyltin treatments were prepared by adding appropriate aliquots of a tributyltin acetate solution (50 mg l^{-1} in distilled water acidified by 2 ml of HCl), with the controls receiving the same quantity of acidified distilled water as the highest solution treatment. The concentration of tributyltin acetate in each treatment was not measured.

All experimental solutions were changed daily for the first 2 d after fertilization and then renewed at 48-h intervals; samples of the larvae were then taken over a 12-d period. Mortalities and abnormalities were assessed from a sample of 200 individuals from each beaker. Size was determined by measuring from photographs the height across the valve of 50 larvae from each culture. The mean valve height was calculated with 95% confidence limits. Measurements of growth rate were terminated when 50% mortality was reached. All larvae were reared at 24°C and fed on a mixed algal diet of *Isochrysis galbana* (or *Monochrysis lutheri*), *Chaetoceros calcitrans*, and *Tetraselmis suecica* (Helm and Millican, 1977).

When these experiments began, there were no data on the toxicity level of tributyltin compounds on bivalve embryos and larvae nor on the level of tributyltin contamination in the Bassin d'Arcachon (Alzieu *et al.*, 1980). Thus, in order to examine in detail the possible effects of TBT on *Crassostrea gigas* embryos and larvae, a first set of experiments was run with the range generally used in such experiments, 100–1 μg l^{-1} (His and Robert, 1980); a second set was run with the range of 5–1 μg l^{-1} (Robert and His, 1981); a third set of experiments with the range of 1–0.02 μg l^{-1} was necessary to reach levels at which there was practically no action on *C. gigas* larval growth (His *et al.*, 1983).

No straight-hinge larvae (D larvae) were observed at concentrations >1 μg l^{-1} TBT. At 3 and 5 μg l^{-1}, very irregularly shaped trochophores, which never developed to the straight-hinge stage, were observed. At 10 μg l^{-1}, only 80% of the eggs divided, and of the relatively few trochophores present all were abnormal. Above 10 μm l^{-1}, cleavage was

Table 12.1 Mean percentages of abnormalities of *Crassostrea gigas* larvae reared under different concentrations of tributyltin acetate

	Concentration of tributyltin acetate ($\mu g\ l^{-1}$)[a]						
Age (d)	*0*	*0.02*	*0.05*	*0.1*	*0.2*	*0.5*	*1*
1	3	4	14(v.m.)[b]	11(v.m.)	15(v.m.)	95(v.m.)	10(v.m.)
	(0.8)	(1)	(1.7)	(1.5)	(1.7)	(1.1)	(1.5)
2	3	4	5	11	10	18	100[c]
	(0.8)	(1)	(1)	(1.5)	(1.5)	(1.9)	
4	3	4	5	11	10	18	100[c]
	(0.8)	(1)	(1)	(1.5)	(1.5)	(1.9)	
6	3	4[d]	5[d]	11[c]	10[c]	18[c]	—[e]
	(0.8)	(1)	(1)	(1.5)	(1.5)	(1.9)	
8	3	4[d]	5[c]	11[c]	10[c]	—	—
	(0.8)	(1)	(1)	(1.5)	(1.5)		

[a]Numerals in parentheses are the standard error.
[b](v.m.) indicates abnormalities of the visceral mass, observed on day 1 only.
[c]Substantial disturbance of feeding activity occurred.
[d]Slight disturbance of feeding activity occurred.
[e]Dash indicates all larvae died.

progressively reduced with increasing concentrations (60% at 25 $\mu g\ l^{-1}$ and only 1% at 50 $\mu g\ l^{-1}$), while at 100 $\mu g\ l^{-1}$ the vitelline membrane failed to rise, suggesting eggs could not be fertilized normally. The embryonic development was inhibited if the eggs were exposed to 50 $\mu g\ l^{-1}$ tributyltin acetate for 30 min prior to fertilization.

At 1 $\mu g\ l^{-1}$, the D stage was reached within 24 h after fertilization, but all the larvae were of abnormal shape and size. The most easily and frequently detected abnormalities were irregular shell shapes, such as a concave hinge, irregular shell margins, and notching of the valves opposite the hinge; moreover, all the larvae died between days 4 and 6.

At lower concentrations (from 0.5 to 0.02 $\mu g\ l^{-1}$ TBT), D larvae were obtained 24 h after fertilization but from day 1 in each treatment, some larvae were characterized by the inability to withdraw all tissue into the shell, as described by Roberts (1987); this inability was observed on day 1 only and from day 2 until the end of the experiments, abnormalities were restricted to the shell or the velum. From day 2 until day 6, shell abnormalities varied from 18% (0.5 $\mu g\ l^{-1}$) to 5% (0.05 $\mu g\ l^{-1}$) and 4% (0.02 $\mu g\ l^{-1}$, with a value of 3% being observed in controls (Table 12.1); these values are significant at the 99.9% confidence limit. They remained unchanged in the surviving larvae through the end of the experiments.

Although the larvae were fed from day 1, at treatment with 1 $\mu g\ l^{-1}$ TBT they remained grayish in color, indicating considerable difficulty in food absorption (Table 12.1); inspection of individuals showed that their stomachs were empty in spite of the presence of algal food in the culture media. The phenomenon was observed from day 2 until larval death at day 6. It appeared later and later with decreased TBT concentration, appearing on day 4 at 0.5 and 0.2 $\mu g\ l^{-1}$ and on day 6 at 0.1 and 0.05 $\mu g\ l^{-1}$, and was only slightly observed on days 6 and 8 at 0.02 $\mu g\ l^{-1}$.

Mortality was <10% in control trials and at 0.02 $\mu g\ l^{-1}$ TBT (Table 12.2); above this concentration value, mortality was total or near 100% from day 8 to day 12.

Exposure to TBT reduced size and growth rates of exposed *Crassostrea gigas* larvae (Fig. 12.2). In spite of a slight difference, corresponding probably to the disturbance of the

Table 12.2 Mean cumulative mortality (%) of *Crassostrea gigas* larvae exposed to different concentrations of tributyltin acetate

	Concentration of tributyltin acetate ($\mu g\ l^{-1}$)[a]						
Age (d)	*0*	*0.02*	*0.05*	*0.1*	*0.2*	*0.5*	*1.0*
1	0	0	0	0	0	0	0
2	1	1	7	1	2	2	2
	(0.5)	(0.5)	(1.3)	(0.5)	(0.7)	(0.7)	(0.7)
4	3	4	8	1	3	3	25
	(0.8)	(1.0)	(1.3)	(0.5)	(0.8)	(0.8)	(2.1)
6	4	6	8	7	7	14	98
	(1.0)	(1.2)	(1.3)	(1.3)	(1.3)	(1.7)	(0.7)
8	8	8	14	13	15	90	—[b]
	(1.3)	(1.3)	(1.7)	(1.7)	(1.8)	(1.5)	
10	8	8	60	40	42	—	
	(1.3)	(1.3)	(2.4)	(2.4)	(2.4)		
12	8	8	78	82	99		
	(1.3)	(1.3)	(2.0)	(1.9)	(0.5)		

[a]Numerals in parentheses are the standard error.
[b]Dash indicates all larvae died.

feeding behavior noted above, treatments of 0.02 $\mu g\ l^{-1}$ TBT closely followed the pattern of the controls. Above this minimum concentration, however, larvae that remained active exhibited a sharp decrease in growth rate, the phenomenon increasing with the TBT concentration. Nevertheless, between 0.2 and 0.05 $\mu g\ l^{-1}$, some larvae that did not display disturbed feeding activity showed only a slight tendency to increase in size.

These results indicate that there is a relationship among TBT concentrations and abnormalities, mortalities, and shell growth. Table 12.3 sums up the effects of tributyltin acetate on embryogenesis and larval development of *C. gigas*.

In a complementary series of tests, initiation of TBT exposure began with 24-h-old D-stage larvae and continued for 12 d (Robert and His, 1981). Deleterious or lethal effects were shown. For example, all larvae died after an exposure to 5 $\mu g\ l^{-1}$ TBT, while on day 4 most larvae exhibited a large 'vacuolated' space within the shell (Roberts, 1987). Such abnormalities were observed at 3 $\mu g\ l^{-1}$ on day 8, and again all the larvae were dead at the end of the experiment. A 12-d exposure to 1 $\mu g\ l^{-1}$ severely affected the larval growth and also caused 100% mortality. Short-term toxicity tests with *C. gigas* larval stages (probably D larvae) were also performed by Thain (1983), who found a 2-d LC_{50} value (the concentration at which 50% of the population died in 48 h) of 1.6 $\mu g\ l^{-1}$. The acute toxicity of tributyltin chloride to embryos and D larvae of the American oyster (*Crassostrea virginica*) was determined by Roberts (1987). He found 2-d LC_{50} values of 1.30 $\mu g\ l^{-1}$ and 3.96 $\mu g\ l^{-1}$ for embryos and D larvae, respectively. He also concluded that at TBT concentrations of 0.77 $\mu g\ l^{-1}$ and higher, some larvae were abnormal (shell and soft tissue), but he did not enumerate the abnormal animals from the normal D larvae. Neither Thain for *C. gigas* nor Roberts for *C. virginica* indicated the influence of TBT on larval growth.

Anthropogenic toxic wastes are present in polluted areas at the time when gametes are released by the oviparous oysters and during subsequent fertilization. For this reason, toxicity experiments, carried out to determine the threshold level of effects of tributyltin compounds on oysters in shellfish farming

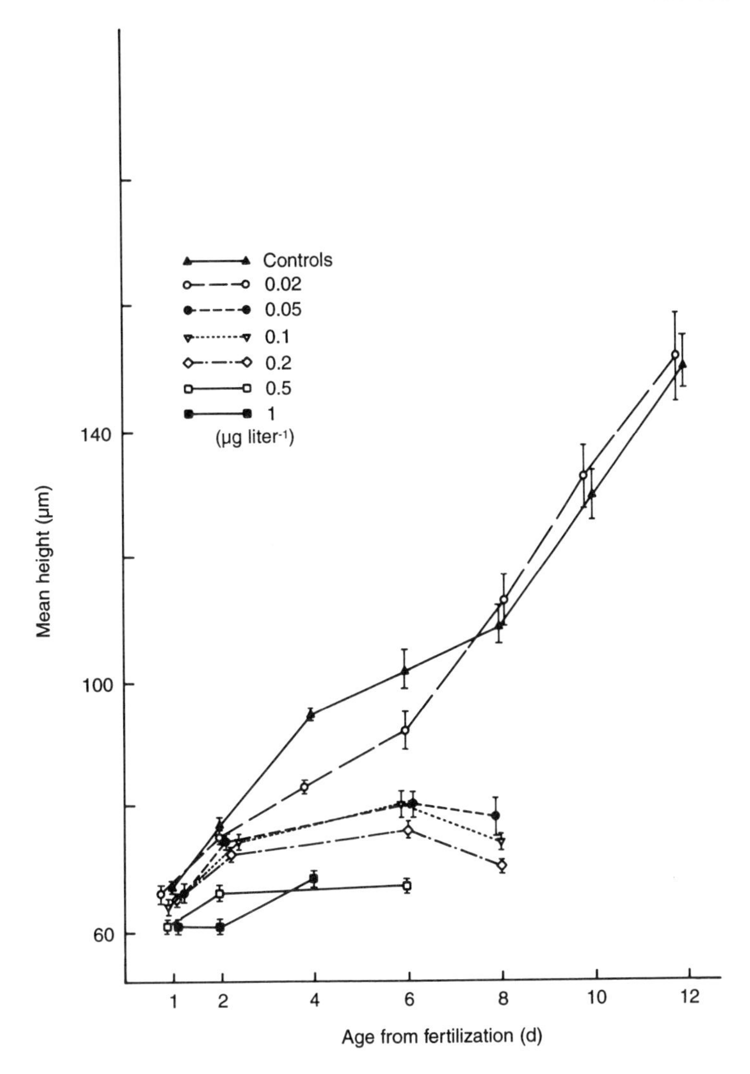

Figure 12.2 Mean shell height (±95% confidence limits) of *Crassostrea gigas* larvae exposed to tributyltin acetate.

areas, must be performed with fertilization occurring in the presence of the pollutant at the different concentrations tested.

Any reduction in larval growth prolongs the pelagic stage and thereby decreases the chances of larval survival (Calabrese *et al.*, 1977). So the observations of growth continued for 10–12 d after fertilization, a period of sufficient duration to measure even relatively small sublethal effects of TBT that are reflected in reduced growth. Reductions in growth were observed at TBT concentrations as low as 0.02 $\mu g\ l^{-1}$ from day 2 to day 6 (Fig. 12.2). For this reason, this value of 0.02 $\mu g\ l^{-1}$ TBT, much lower than that of Thain (1983) or Roberts (1987), seems to be closer to the threshold tolerance level of *C. gigas* larvae.

12.3. FIELD OBSERVATIONS

The reproductive patterns of *Crassostrea gigas* in the Bassin d'Arcachon have been described

Table 12.3 A summary of the effects of tributyltin acetate (TBT acetate) on embryogenesis and larval development of *Crassostrea gigas*

TBT acetate ($\mu g\ l^{-1}$)	*Effects on embryogenesis and larval development*
100	Inhibition of fertilization
50	Inhibition of cleavage of fertilized eggs
25	Partial (40%) inhibition of cleavage of fertilized eggs
10	No trochophores developed
3–5	No D larvae. All trochophores abnormal
1	Abnormal D larvae. No survival beyond day 6
0.5	Numerous abnormal D larvae. Feeding activity disturbed on day 6. Practically no shell growth. No survival beyond day 8.
0.2	Numerous normal D larvae. Feeding activity disturbed on day 6. Growth rate sharply decreased. No survival by day 12
0.1	Most larvae are normal. Feeding regime disturbed on day 6. Some growth from day 1 to day 6. Subtotal mortalities on day 12
0.05	Normal D larvae. Feeding regime disturbed on day 8. High mortality rate on day 10. Some growth from day 1 to day 6
0.02	Normal D larvae. Low mortality rate. Growth as in controls on day 12. Practically no action of TBT observed

in detail (His, 1976). Full ripeness is observed by the end of June, and the oysters reproduce until the beginning of September, generally with three or four spawning periods, at 3-week intervals. By recording the shell movements of oysters in the field, it was shown that spawning and hence fertilization occur principally in a 2-h period after high tide. On this basis, studies on the capacity of seawater to support normal embryogenesis and larval development have always involved sampling during the first 2 h of ebb tide.

Oyster spawning is sudden and generally simultaneous for most of the oyster beds in the bay; spawning streaks can be seen every year as large, whitish bands several meters long in the main channel of the bay (Teychan, Fig. 12.1). Therefore, individual larval broods may be studied closely; D larvae can be seen in the first 24 h in quantities of many hundred thousands per cubic meter. Larval growth in the bay is fast (Fig. 12.3, for instance 1975 and 1983) when the seawater temperature remains at 22°C and above. During the breeding season, the salinity is within the limits defined by Helm and Millican (1977) for maximum growth of *C. gigas* larvae (between 25 and 30‰). At 22°C and above, the very early umbo stage [see Quayle (1969) for the different stages in larval development] is reached on day 5–6 after spawning; the early umbo stage and the umbo stage are usually reached on days 8–10 and on days 12–15 respectively; the advanced umbo stage and the 'eyed' stage occur within 2 or 3 weeks, the pelagic life span being inversely related to seawater temperature; from many hundred thousands of larvae per cubic meter, many thousands of the two last stages per cubic meter are still alive in the plankton. Normally, by the end of the breeding season in September, from many hundreds to many thousands of spat have settled per tile collector, as a result of 'good environmental conditions' for oyster larvae.

When thunderstorms are frequent during the breeding season, the water temperature falls to <22°C, but the salinity remains within the limits previously defined by Helm and Millican (1977); growth rates decrease (see for instance 1974, Fig. 12.3), indicating 'bad environmental conditions' for oyster larvae. Although the different larval stages (early umbo and umbo stages in particular) and a few 'eyed stages' survive in the plankton, >100 spat per tile collector settle by September.

In July 1976, a heavy spawning occurred in the continental part of the bay. D larvae were very abundant (590 000 m^{-3}). They possessed

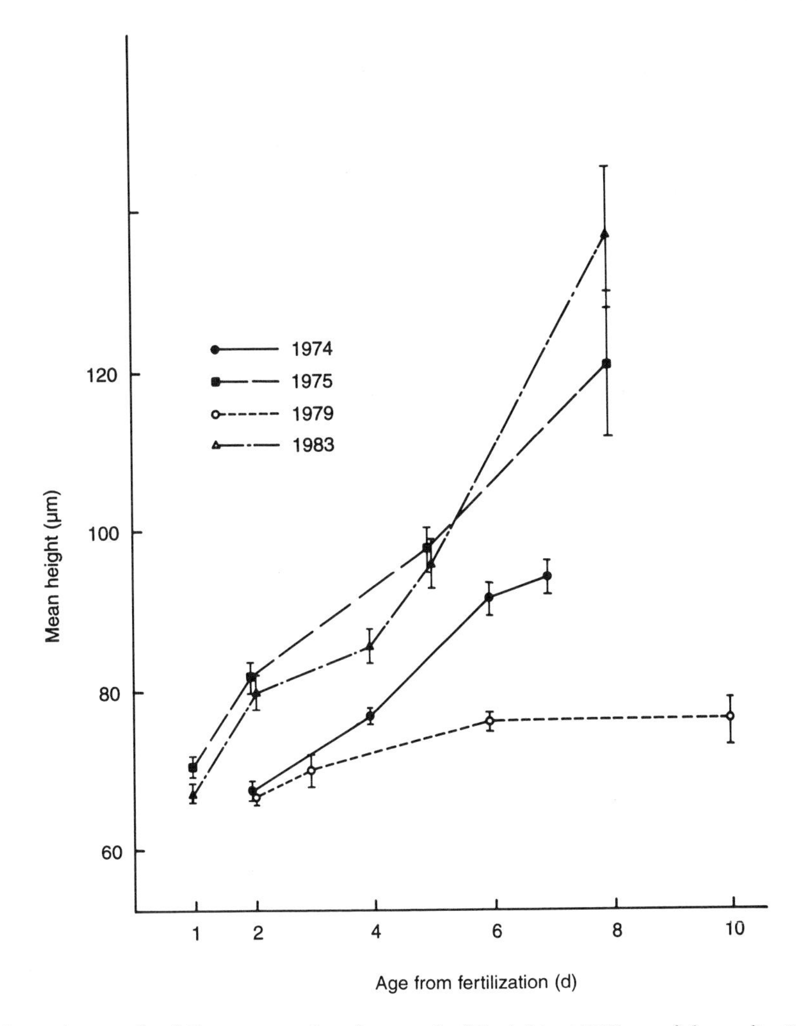

Figure 12.3 Larval growth of *Crassostrea gigas* (mean shell height, ±95% confidence limits) during 1975 and 1983 (favorable seawater thermal conditions), 1974 (low water temperatures), and 1979 (abnormal larval growth).

the characteristic shape and pearl-gray color, and did not exhibit any of the previously described abnormalities present in toxicity experiments, notably abnormalities of the shell or meat of the veligers, which are considered to characterize tributyltin contamination of seawater. Later on, practically no growth was observed in this brood; it disappeared progressively within a week. Only a few early umbo stages were evident even when water temperature was >23°C. This drastic and abnormal reduction in larval growth was observed over the entire bay from 1977 to 1981 (see for example 1979, Fig. 12.3).

The color of oyster larvae varies somewhat with the food, but healthy larvae are usually a yellow-brown color, which is especially noticeable in the area of the digestive diverticula, and which becomes apparent when the veligers are 2 or 3 d old. This color changes to brown and later to dark brown as the larvae grow to the pediveliger stage. In July 1976, the D larvae remained a very pale yellowish color, indicating a probable disruption in the absorption of food.

From 1977 to 1981 this particular phenomenon could be observed in samples from all over the bay. A very large difference was observed (Fig. 12.3) among larval growth during good spatting summers, that is, in good environmental conditions for oyster larvae (1975 and 1983 for example); during poor spatting summers resulting from bad hydroclimatic conditions (as in 1974, resulting from frequent thunderstorms); and in abnormal conditions during which oyster recruitment failed totally (as in 1979).

Observations of larvae collected from the plankton during the summer of 1980 did not reveal the presence of any disease-causing agent (Combs, personal communication).

The consequences of the abnormality in oyster growth and development were the failure of larvae to survive the umbo stage over several years and a consequent major disruption of oyster-farming activity in the bay. Two hypotheses formulated to explain the effects on oysters were the action of poor hydroclimatic conditions in the summers from 1977 to 1981 or the action of one or more pollutants present in the waters of the bay. Agricultural runoff that carried pesticides and herbicides from the booming corn-crop industry in the vicinity of the bay or from pine forest treatments was considered a possible source of pollution. Alternatively, the increase in pleasure-craft activities throughout the last decade and the concomitant increase in the use of antifouling paints could explain the larval abnormalities that were also observed near La Rochelle harbor.

With this in mind, experiments were performed with Pacific oyster larvae cultured under laboratory conditions, as suggested by Woelke (1967), to provide a biological measure of the quality of the water in the bay. Adult oysters were induced to spawn in the laboratory, and larvae were reared from the D stage to the umbo stage during a 10- or 12-d period, thereby including the period (first 8 d) of pelagic life during which abnormal development was observed in the field. As during the toxicity experiments, growth of larvae could be compared under various environmental conditions. In addition, trials were carried out with natural veligers. For the latter, larvae were isolated by sieving freshly sampled plankton through two mesh sieves of 150 and then 40 μm (pore size), followed by successive washings with 0.2 μm (pore size) filtered seawater. The first sieve retained the large zooplankton, principally copepods, while the second retained the D larvae, with phytoplankton and the seston being washed away. Using this procedure, several hundred thousand larvae could be isolated and reared under experimental conditions and their growth compared with that of the same larval stage, under conditions in the field. This approach led to a more clear understanding of the reasons why the Japanese oyster failed to reproduce in the Bassin d'Arcachon.

12.3.1. INFLUENCE OF WATER TEMPERATURE ON LARVAL DEVELOPMENT

Straight-hinge larvae were reared at a constant temperature of 18 or 24°C for a 12-d period. Mortality at 18°C was 50%, but decreased to only 10% at 24°C at the end of 12 d. Mean valve length of larvae increased by a factor of 2 at 18°C, but by a factor of 3.5 at 24°C (Fig. 12.4). More importantly, at 18°C development was much slower; nonetheless, at the end of 12 d, 52% of the larvae were in the early umbo and 8% in the umbo stage. At 24°C all the larvae were in the umbo stage.

Concurrent with the above trials, D larvae from the same brood were reared in the same seawater at temperatures that alternated between 24°C and 18°C at 2-d intervals; such a temperature differential is never observed during breeding seasons for the same larval cohort in the Bassin d'Arcachon. Final

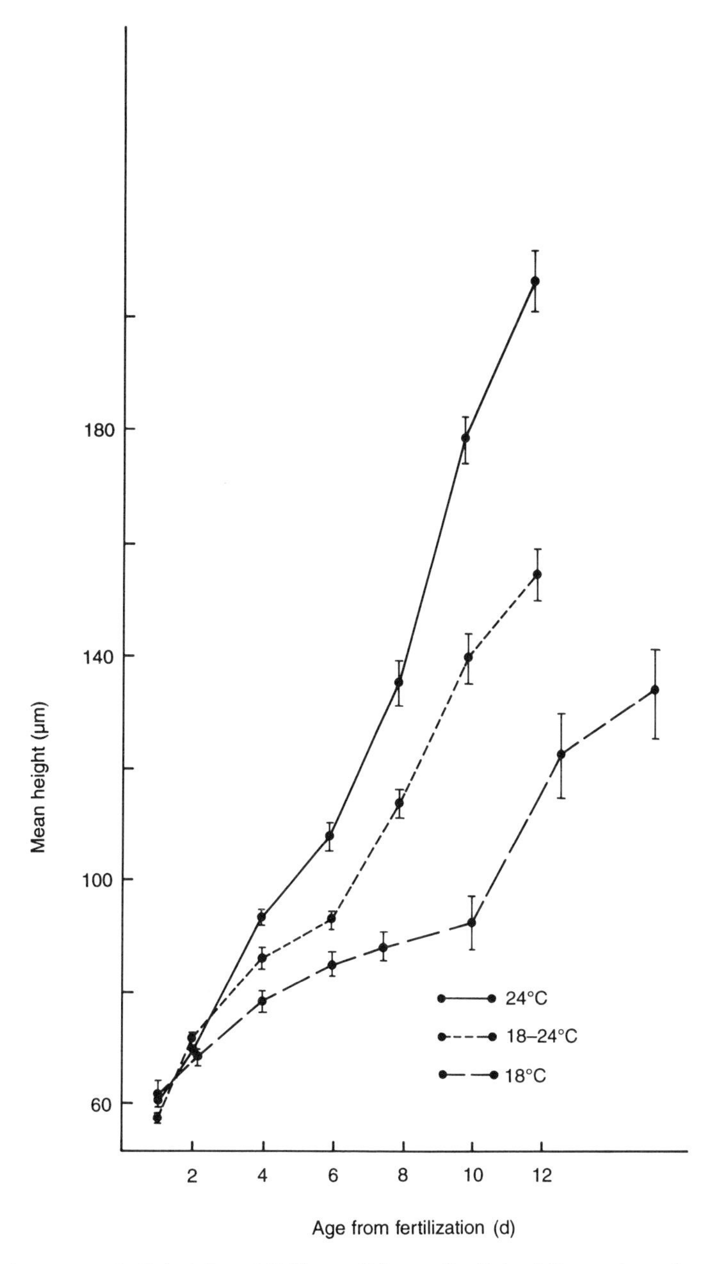

Figure 12.4 Growth (mean shell height, ±95% confidence limits) of *Crassostrea gigas* larvae reared at 18°C and 24°C, and with alternation between 18°C and 24°C.

mortality was 10%, as in controls. Larval growth was somewhat affected (Fig. 12.4); nonetheless, 60% of the larvae were in the early umbo stage and 40% in the umbo stage on day 12.

During these experiments, larvae were seen to feed normally and were normally colored, as in controls. Neither low nor varying temperatures could therefore explain the lack of larval growth in the field. This was further confirmed by observations in the Gironde Estuary and the Bassin de Marennes-Oléron, areas with reduced pleasure-craft activities as compared with the Bassin d'Arcachon, both situated ~100 km north of Arcachon, and where *C. gigas* reproduced normally during the time that abnormalities were observed in the Bassin d'Arcachon.

12.3.2. INFLUENCE OF WATER QUALITY ON LARVAL DEVELOPMENT

There were ~2540 oyster-farming boats and 5300 pleasure craft in the Bassin de Marennes-Oléron in 1980 as compared with 1000 oyster-farming boats and 20 000 pleasure craft in the Bassin d'Arcachon during the same year. The use of antifouling paints containing tributyltin could therefore be estimated at 2 t y^{-1} in the first case and at 30–35 t y^{-1} in the second (Alzieu *et al.*, 1980).

Furthermore, a comparison of the concentrations of metals in the seawater of the two areas gave the following results: in 1980, Hg was <0.7 μg l^{-1}, Cu varied from 0.28 to 12.6 μg l^{-1}, and Zn from 9.28 to 33.8 μg l^{-1} in the seawater of the Bassin d'Arcachon; at Marennes-Oléron Hg was ~1 μg l^{-1}, Cu varied from 2.71 to 5.2 μg l^{-1}, and Zn from 8.52 to 145.3 μg l^{-1}.

According to the data of MacInnes (1981) concerning the response of embryos of the American oyster to heavy metal mixtures, the abnormalities in larval growth of *C. gigas* could not be attributed to the presence of Cu, Hg, or Zn in the seawater of the bay (Alzieu *et al.*, 1980).

From 1980 to 1983, when experiments were terminated, current chemical methods were not sufficiently sensitive to measure concentrations of tributyltin that were shown experimentally to be toxic to embryos and larvae of the Pacific oyster in environmental samples. At that time the detection limit of TBT in seawater was 0.08 μg l^{-1} (Waldock and Thain, 1983). For this reason, bioassays were performed to compare experimental and field data, initially to determine whether the seawater was having a detrimental influence on adult oysters or directly on the larvae.

12.3.2a. Influence of water quality on gamete viability of Arcachon oysters

Shellfish, and oysters in particular, are known to take up pollutants from the seawater and to accumulate them to high concentrations, with consequent deleterious effects on the viability of their gametes (Bayne *et al.*, 1975). Exposure of adult oysters to heavy metals has been shown to create a stress on gametes that is severe enough to disturb embryonic development (Zarrogian and Morrisson, 1981). Alzieu *et al.* (1982) suggested that contamination of the gonad by tin could explain the absence of larval growth and the failure of spatting in certain oyster-farming areas such as the Bassin d'Arcachon.

Ripe oysters from Brittany (an unpolluted area) and from Arcachon were induced to spawn by thermal stimulation. Larvae from each group were reared for 12 d in seawater obtained off the shore of the Cap Ferret peninsula. Abnormalities, mortalities, and larval growth pattern were compared in two independent trials. D larvae from both Brittany and Arcachon exhibited low abnormality rates: 5% and 8% for Brittany larvae and 2% and 15% for Arcachon larvae. Mortalities also remained at a low level: 2% and 7% for Brittany larvae and 2% and 15% for Arcachon larvae. There were no differences in larval growth, early umbo larvae being observed as usual on day 6. Thus, the characteristic decrease in larval growth did not occur, and maturation to the pediveliger stage occurred within a 3-week period.

Effects of antifouling paint containing tributyltin compounds have also been studied on adult oysters, under natural field conditions in the Bassin d'Arcachon, after the ban on the use of antifouling paints containing tributyltin compounds (His and Robert,

1987). *Crassostrea gigas* 18-month-old oysters were reared for 13 months (September 1983 to October 1984) in wooden trays whose sides had been painted with organotin antifouling paint. When mature, contaminated and control oysters were induced to spawn in the laboratory, and larvae were reared under hatchery conditions. The contaminated oysters exhibited the abnormal thickening of the valves as described by Alzieu *et al.* (1982) and by Waldock and Thain (1983); the adult control oysters were normal as regards shell calcification (His and Robert, 1987). Nevertheless, it was shown that embryogenic and larval development from gametes of adult *C. gigas* exposed to organotin paint remained unaffected. While a slight decrease in larval growth was noticed (mean shell height of 149.0 ± 2.6 μm for larvae from contaminated oysters and 155.0 ± 2.7 μm for controls on day 11), the umbo stage was still observed on day 6. Inhibition of larval growth affecting veligers in the field did not occur. The results of these two sets of experiments are concordant. Larval development was not inhibited when parent oysters were selected from among either shell-thickened oysters contaminated in the bay, before the ban in January 1982 of organotin compounds in antifouling paints, or shell-thickened oysters obtained experimentally after the ban took effect.

The conclusion from these experiments, therefore, was that the failure in larval development in the Bassin d'Arcachon could not be attributed to the action of organotins on the viability of gametes from Arcachon adult oysters.

12.3.2b. Influence of water quality on the embryos and larvae of *Crassostrea gigas*

Ripe oysters were induced to spawn, and larvae were reared in seawater obtained from the main channel in the bay, at Le Banc, Le Tés, and Comprian (Fig. 12.1). Seawater for controls was again collected off the shore of the Cap Ferret peninsula. Larvae were laboratory reared for 12 d in April, June, and July 1981; in July a heavy spawning of oysters had just occurred in the bay when the experiments began, and in the field the growth of the uncolored larvae was strongly affected. By contrast, in the three sets of laboratory experiments, abnormalities and mortalities remained under 10%. Compared with controls, no decrease in larval growth was observed. By day 6, between 46 and 96% and, at day 12, 100% of the larvae had reached the umbo stage. In one trial, the larvae were reared until they had matured to the pediveliger stage.

In July, at the same time when larvae were laboratory reared in seawater from the bay, an equivalent cohort in the bay itself was found to have reduced survival, growth rate, and color of the veligers. It was therefore difficult to explain the differences between the two cohorts by the simple presence of one or several pollutants in solution in the seawater acting directly on the larvae.

Based upon the assumption that tributyltin compounds were the cause of larval growth inhibition in the Bassin d'Arcachon, it was decided to use seawater from Arcachon Harbor for rearing oyster larvae in the laboratory, as the presence of these compounds could be assumed to be even greater in the harbor water, where more than 1500 boats are moored year round. In such experiments, abnormalities were high (40%), and subsequent mortality was also high: 40% on day 12. The abnormal veligers exhibited poor growth, although normal veligers reached the umbo stage by the end of the experiments. In brief, the use of harbor water that was probably polluted did not produce the same abnormalities in larval growth and feeding as could be observed in the bay.

So, neither the viability of the gametes from adult oysters nor the direct action of some pollutants on embryos and larvae of the Japanese oyster could explain the failure of *C. gigas* larvae to survive and reach the umbo stage.

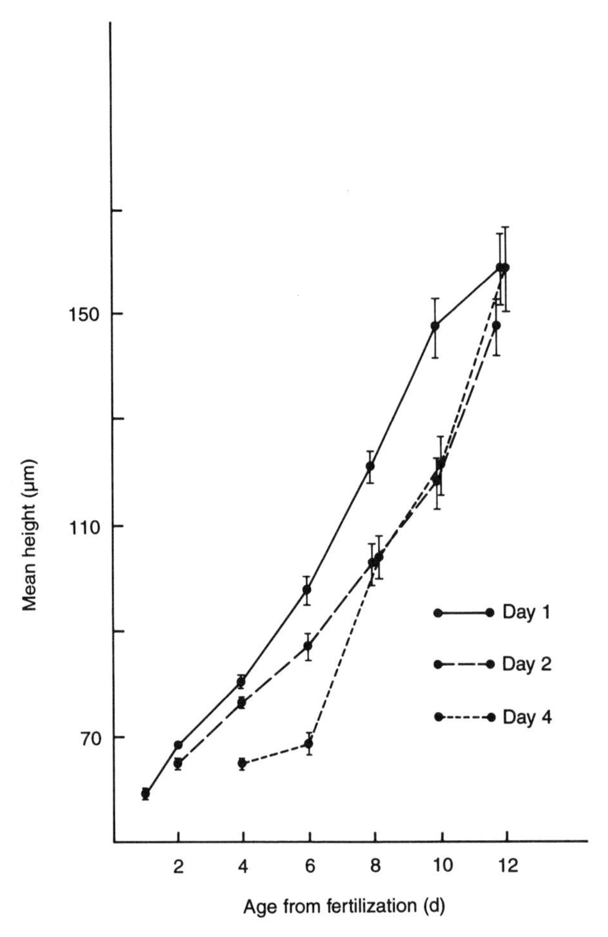

Figure 12.5 Growth of *Crassostrea gigas* larvae (mean shell height, ±95% confidence limits) collected from the Bassin d'Arcachon at days 1, 2, and 4 after spawning and reared in the laboratory.

12.3.3. EXPERIMENTS WITH LARVAE ISOLATED FROM THE BASSIN D'ARCACHON

In 1981, the Pacific oyster spawned for a second time on 29 July. On 30 and 31 July and 2 August, D larvae were isolated from plankton in the bay and were 1, 2, and 4 d old, respectively. The D larvae were reared in the laboratory in seawater collected from Le Tés at the same time as the veligers.

In the field, some larval growth was evident during the first 2 d (Fig. 12.5); this height increment corresponded to the endotrophic period, with the veliger reserves being used as a source of energy. Growth continues in *C. gigas* larvae during the subsequent exotrophic period only if nutritional sources are available (Gerdes, 1983). This did not seem to happen in the bay.

For the larvae isolated on day 1, abnormalities and mortalities remained at low levels throughout the experiments, with values of 2% and 5% on day 12, respectively; 35% of the veligers had reached the early umbo stage on day 6, 85% on day 8 while, as of day 12, 30% had reached the early umbo stage and 65% the umbo stage.

For the larvae isolated on day 2, abnormalities and mortalities were 2% and 6%, respectively, at the end of the experiments. Larval growth was practically the same as for 1-d larvae (Fig. 12.5) (i.e. 35% of the larvae had reached the early umbo stage and 60% the umbo stage on day 12).

On day 4, 6% mortality was observed at the time when the D larvae were isolated from the plankton; at the end of the rearing period, the figure was 26%. There were no early umbo larvae on day 6 (i.e. after 2 d in hatchery conditions). On day 8, however, 56% of the veligers were early umbo larvae. On day 12, the veligers that had survived were early umbo larvae (40%) and umbo larvae (52%). So, in spite of a 24-h delay (Fig. 12.5), the surviving larvae isolated from the field on day 4 resumed growing.

Within a few hours after transfer to the laboratory, visual inspection of the larvae under a compound microscope showed that most of their stomachs contained algal cells. This indicated that larvae from the bay, reared in seawater from the bay, were able to feed on appropriate external food sources, if present. The absence of any food in the gut of the equivalent larvae in the field suggested the absence of such an appropriate food source for the straight-hinge larvae of the Japanese oyster, *C. gigas*.

Toxicity experiments with tributyltin acetate had shown that concentrations as low as 0.05 $\mu g\ l^{-1}$ disturbed the feeding regime of the oyster larvae, whose gut remained empty even in the presence of appropriate algal food. From this it might be supposed that the abnormalities in the larval development observed in the Bassin d'Arcachon are explained by the lack of an appropriate food source, such as a specific nanoplankton, that is normally required by the D larvae during the first days of their pelagic life, but which is itself directly affected by a pollutant in the bay.

12.4. DISCUSSION

'One of the main causes of reduced oyster harvests has been, and continues to be, the failure over a several-year period of larvae to survive the planktonic stage of development' (Abbe, 1986). This was the case in the Bassin d'Arcachon, which is subject to high levels of pleasure-craft activity. It is now well established that embryos and larvae of bivalves, and those of the Japanese oyster in particular, are one of the marine organisms most sensitive to pollutants. A review of the effects of heavy metals on bivalve larvae is available in Martin *et al.* (1981); with the toxicity level expressed as 48-h EC_{50} (concentrations which caused 50% abnormal development in larvae during a 48-h period) the most toxic metals for *C. gigas* embryos and larvae were found to be Cu (5.3 $\mu g\ l^{-1}$), Hg (6.7 $\mu g\ l^{-1}$), and Ag (22 $\mu g\ l^{-1}$). *Mytilus edulis* embryos and larvae showed similar toxic responses to these metals and also to Pb, but *C. gigas* is more sensitive to Ni, Cd, and As by a factor of two Concerning the action of pollutants on *C. gigas* larval growth, copper chloride ($CuCl_2$) and copper sulfate ($CuSO_4$) have adverse effects above 10 $\mu g\ l^{-1}$ (Cu^{++} = 6.4 $\mu g\ l^{-1}$) and 25 $\mu g\ l^{-1}$ (Cu^{++} = 16 $\mu g\ l^{-1}$), respectively (His and Robert, 1981, 1982). The tributyltin acetate threshold level is 0.02 $\mu g\ l^{-1}$. Therefore, tributyltin compounds are markedly more toxic than the most toxic heavy metal (Cu^{++}). This conclusion concords with that of Thompson *et al.* (1985) who consider that the lowest concentration of tributyltin necessary to induce deleterious effects in many marine molluscs lies in the nanogram-per-liter range. For the Japanese oyster, the concentration limit at which spat develop shell abnormalities when exposed to organotin leachates from antifouling paints is below 0.08 $\mu g\ l^{-1}$ (Waldock and Thain, 1983). Thickening was especially evident at TBTO (bis tributyltin oxide) concentrations of 0.05, 0.02, and 0.01 μg l^{-1}; significant effects on oxygen consumption and feeding rate were found at 0.05 μg l^{-1}, and at 0.01 μg l^{-1} for the ability to compensate for hypoxia (Lawler and Aldrich, 1987). In *Mytilus edulis* veligers, which are more resistant than embryos, the toxicity in 7-d-old larvae lies far below 0.1 μg l^{-1} TBTO (Beaumont and Budd, 1984).

Imposex in species of stenoglossan marine snails was linked to the action of tributyltin compounds (Smith, 1981; Féral, 1982); high degrees of imposex could be induced in *Nucella lapillus* by 120-d exposure to water containing 0.02 μg l^{-1} tin leached from paint containing tributyltin compounds (Bryan *et al.*, 1986). These authors attributed the decline in recruitment of this marine gastropod in southeast England to contamination of the water by organotins. In the same way, Minchin *et al.* (1987) explained the decline in natural populations of the scallop *Pecten maximus* on the north coast of Ireland by a reduction in, or the failure of, settlement concomitant to the use of tributyltin net dips on salmonid farms.

In the Bassin d'Arcachon, the annual utilization of organotin-based antifouling paints, principally from Easter to July, could be estimated at 30–35 t (2 t at Marennes-Oléron, see Alzieu *et al.*, 1980). Abnormalities in both recruitment and shell thickening were shown to impede oyster-farming activity. As a result, the use of organotin antifouling paints on boats of <25 m (principally pleasure craft) was banned from January 1982.

In April 1982, approximately 9 months after the period of maximum use of antifouling paints, *C. gigas* embryos and larvae were reared in seawater collected from Arcachon Harbor at high tide, during ebb tide, and at low tide. No differences were observed among these oysters regarding abnormalities, mortalities, and larval growth, thus indicating a substantial improvement in water quality following the ban of organotin antifouling paints (His *et al.*, 1983). This was clearly confirmed in June 1982 when, after a preliminary spawning of the Japanese oyster in mid-June, healthy colored, early umbo larvae could be observed in the plankton, despite water temperatures close to 19°C. Hereafter the Pacific oyster recruitment has remained at a high annual level, and the incidence of shell abnormalities has declined drastically.

From January 1982 to November 1985, tin levels in the bay were 5–10 times lower than prior to the ban on organotin antifouling paints; and the decrease in shell anomalies 'seemed to be correlated with the decrease in tin contamination of the waters of the Bassin d'Arcachon' (Alzieu *et al.*, 1986). From this it was evident that the previous abnormalities in the development of embryos and larvae of *C. gigas* were linked, as has been suspected, to the use of organotin antifouling paints. Moreover, it was subsequently shown by larval bioassays that the freshwater runoff into the bay was itself free of pollutants that could have disturbed the reproduction of the oysters (His *et al.*, 1986).

The failure of the veligers to survive in the field may well be explained by an action of the tributyltins on the development of specific nanoplankton required by the D larvae. The lack of proper food is one of the major causes of larval mortality (Berg, 1971). Sublethal concentrations of pollutants can destroy one of the critical links in the larval food chain that could be replaced by microorganisms totally unsuitable for oyster larvae (Abbe, 1986). Thus, the action of toxicants could have been an indirect cause of failure of larvae to survive in the field at Arcachon.

Tributyltin compounds may inhibit primary production and cell multiplication of algae in natural phytoplankton of lake waters, and these pollutants may pose a threat to the survival of microalgae in certain polluted areas (Wong *et al.*, 1982). According to Walsh *et al.* (1985), tributyl- and triphenyltins are highly toxic to marine algae within 72 h, and the EC_{50} (concentration that inhibits growth by 50%) is lower than many LC_{50} values derived from animal tests reported in the literature. TBTO is algistatic with 5-d exposure at 1.0–18 μg l^{-1} to the diatom *Skeletonema costatum* (Thain, 1983). This diatom was found to be more sensitive by Beaumont and Newman (1986), who did not observe any growth with TBTO at a concentration of 0.1 μg l^{-1}; and a significant reduction in cell multiplication was evident in the two marine algae *Pavlova lutheri* and *Dunaliella tertiolecta*. It was also shown that the flagellate *Isochrysis galbana* and, more particularly, the diatom *Chaetoceros calcitrans* are more sensitive to the action of organotin paints than to tributyltin acetate (His *et al.*, 1986). With regard to shell abnormalities (Gendron, 1985), some synergistic action between the organotin salts in the paint and the different components of the paint, such as the binder, matrix, and solvents, may explain the more deleterious effects of the paint on cell multiplication of marine algae, which serve as a food source for *C. gigas* larvae.

The lack of proper food for oyster larvae could not be observed directly when the abnormalities affected the oyster recruitment, and thus must be regarded as a hypothesis. Studies on the phytoplankton of the Bassin d'Arcachon and the role of this phytoplankton in the feeding of oyster larvae started in spring 1982 (Maurer *et al.*, 1984), but during the following breeding season the abnormalities in larval development did not occur. Nevertheless, it was shown that phytoflagellates and diatoms such as Naviculaceae play an important role in the feeding of *C. gigas* larvae in the bay (His *et al.*, 1985; His and Robert, 1987). On the other hand, a particular phenomenon occurred for the first time in March 1976, that is to say during the first year the abnormalities in larval development were noticed: all the oysters in the bay were 'green gilled.' This bluish-green coloration, which is typical of fattening ponds ('claires') at Marennes-Oléron, has been shown to be due to the absorption of a pigment (marenin) from a very abundant and common diatom, *Navicula ostrearia*; this diatom has been shown to produce marenin under various unfavorable environmental conditions (Daste and Neuville, 1970). The phenomenon could be observed either on the whole bay or in its continental part every year in March or April, from 1976 to 1982. Curiously, it did not occur again in the following years, when the abnormalities in larval development had also disappeared.

Thus, if some disturbance could affect the phytoplankton in spring, before the period of maximum use of TBT, it is possible that the phytoplankton could also have been disturbed in summer, when pleasure-craft activities were very important.

12.5. CONCLUSION

The Bassin d'Arcachon is the first place in the world where the danger of the use of tributyltin in oyster-farming areas has been shown by toxicity experiments on embryos and larvae of the Japanese oyster, *Crassostrea gigas* (His and Robert, 1980).

In addition to the possible action on oyster recruitment, it was shown experimentally in 1981 that tributyltin pollution could also induce shell abnormalities in adult oysters (Alzieu *et al.*, 1981).

The present results concern the disruption of oyster recruitment in the bay. From 1976 to 1981, when these abnormalities were observed, chemical methods were not sensitive enough to measure the concentrations present in the bay. For this reason, treatments in the nominal range of 100–0.02 μg l^{-1} were used in tests. The higher concentration levels of treatment, habitually used in toxicity experiments with metal salts, have since been shown to have no relevance to the environment.

The 0.02 μg l^{-1} concentration represents the threshold level above which abnormalities in larval shell formation and feeding behavior, mortalities, and reduction of larval growth can be observed.

The larvae in the bay, which did not exhibit any abnormalities but in which growth rate was drastically reduced, did not reach the early umbo stage in great numbers; moreover, their pale yellowish color indicated a probable disturbance of their feeding mechanism.

The phenomena were studied by bioassays using embryos and larvae. It may be concluded that neither low nor varying temperature could explain the abnormalities observed. Furthermore, larvae from thick-shelled oysters collected from the bay prior to 1981, or oysters experimentally contaminated with TBT after the ban of TBT-based antifouling paints, grew to the umbo stage when reared in seawater from the bay, under experimental conditions.

Larval growth was not drastically reduced when larvae were reared in seawater from the probably highly TBT-polluted Arcachon Harbor. On the other hand, veligers isolated from the plankton and reared in seawater of the bay under laboratory conditions fed normally and grew to the early umbo and umbo stage, while in the field the veligers remained uncolored and did not grow.

Thus, during toxicity experiments, D larvae could not feed in the presence of an algal diet at concentrations of 0.05 $\mu g\ l^{-1}$ and above, but larvae from the field fed normally in the presence of the same algal diet. This observation led to the supposition that the abnormalities observed in the Bassin d'Arcachon could be better explained by the lack of proper food for D larvae, rather than by direct action of any one pollutant on either the adult oysters or the larvae.

This hypothesis could not be verified. Nevertheless, studies on the phytoplankton of the bay showed that diatoms play an important role in the feeding of oyster larvae; the presence of 'green-gilled' oysters in spring, from 1976 to 1981 concurrent with the presence of abnormalities in larval development, indicates that unfavorable conditions for some diatoms were present in the bay itself.

A few months after the bans on the use of TBT-based antifouling paints in January 1982, the Japanese oyster was again able to reproduce at Arcachon. Since the ban, the bay is the only oyster-farming area where *C. gigas* has reproduced every year on a commercial scale, with from 300 to 400 spat (1988 for example) to 5000 and 7000 spat per tile collector (1985 and 1989 for example) at the end of the breeding season. The annual production of commercially sized oysters has again reached 15 000 t.

The decrease in contamination of the bay has thus been shown to correspond with the decrease in the incidence and extent of shell abnormalities (Alzieu *et al.*, 1989; Alzieu, 1991). In addition, since last year, oyster farmers have once again observed the presence of the winkle (*Ocenebra erinacea*), which had practically disappeared from this region.

REFERENCES

Abbe, G.R. 1986. A review of some factors that limit oyster recruitment in Chesapeake Bay. *Am. Malacol. Bull.*, Special Edition 3, 59–70.

Alzieu, C. 1991. Environmental problems caused by TBT in France: assessment, regulations, prospects. *Mar. Environ. Res.*, **32**, 7–17.

Alzieu, C., Y. Thibaud, M. Héral and B. Boutier. 1980. Evaluation des risques dus à l'emploi des peintures antisalissures dans les zones conchylicoles. *Rev. Trav. Inst. Pêches Marit.*, **44**, 301–348.

Alzieu, C., M. Héral, Y. Thibaud, M.J. Dardignac and M. Feuillet. 1982. Influence des peintures antisalissures à base d'organostanniques sur la calcification de la coquille de l'huître *Crassostrea gigas*. *Rev. Trav. Inst. Pêches Marit.*, **45**, 100–116.

Alzieu, C., J. Sanjuan, J.P. Deltreil and M. Borel. 1986. Tin contamination in Arcachon Bay: effects on oyster shell anomalies. *Mar. Pollut. Bull.*, **17**, 494–498.

Alzieu, C., J. Sanjuan, P. Michel, M. Borel and J.P. Dréno. 1989. Monitoring and assessment of butyltins in Atlantic coastal waters. *Mar. Pollut. Bull.*, **20**, 22–26.

Bayne, B.L., P.A. Gabbott and J. Widdows. 1975. Some effects of stress in the adult on the eggs and larvae of *Mytilus edulis*. *J. Mar. Biol. Ass. UK*, **55**, 675–689.

Beaumont, A.R. and M.D. Budd. 1984. High mortality of the larvae of the common mussel at low concentrations of tributyltin. *Mar. Pollut. Bull.*, **15**, 402–405.

Beaumont, A.R. and P.B. Newman. 1986. Low levels of tributyltin reduce growth of marine algae. *Mar. Pollut. Bull.*, **17**, 457–461.

Berg, C.J. 1971. Review of the possible causes of mortality of oyster larvae of the genus *Crassostrea* in Tomales Bay, California. *Fish and Game*, **57**, 69–75.

Bryan, G.W., P.E. Gibbs, L.G. Hummerstone and G.R. Burt. 1986. The decline of the gastropod *Nucella lapillus* around south-west England: evidence for the effect of tributyltin from antifouling paints. *J. Mar. Biol. Ass. UK*, **66**, 611–640.

Calabrese, A., R.S. Collier, D.A. Nelson and J.R. MacInnes. 1973. The toxicity of heavy metals to embryos of the American oyster *Crassostrea virginica*. *Mar. Biol.*, **18**, 162–166.

Calabrese, A., J.R. MacInnes, D.A. Nelson and J.E. Miller. 1977. Survival and growth of bivalve larvae under heavy-metal stress. *Mar. Biol.*, **41**, 179–184.

Combs, M. 1983. Recherches Histologiques et Cytologiques sur les Infestations Intracellulaires des Mollusques Bivalves. Thèse de Doctorat és Sciences Naturelles, Université de Montpellier, France, 128 pp.

Combs, M., J.R. Bonami, C. Vago and A. Campillo. 1976. Une virose de l'huitre portugaise (*Crassostrea angulata* Lmk). *C. R. Acad. Sci. Paris*, **282 D**, 1991–1993.

Daste, Ph. and D. Neuville. 1970. Recherches sur le verdissement des huîtres en claires. I – Rappel technologique et revue critique des travaux antérieurs. *La Pêche maritime*, **1112**, 1–8.

Deslous-Paoli, J.M. 1982. Toxicité des éléments métalliques dissous pour les larves d'organismes marins: données bibliographiques. *Rev. Trav. Inst. Pêches Marit.*, **45**(1), 73–83.

Féral, C. 1982. Etude Expérimentale des Mécanismes Assurant l'Apparition, le Maintien et le Cycle d'un Tractus Génital Mâle Externe Chez les Femelles de *Nucella lapillus* (L), *Nassarius reticulatus* (L), *Ocenebra erinacea*, Mollusques Néogastéropodes Gonochoriques. Thèse de Doctorat és Sciences Naturelles, Université de Caen, France, 175 pp.

Gendron, F. 1985. Recherches sur la Toxicité des Peintures Antisalissures à Base d'Organostanniques et de l'Oxyde de Tributylétain, vis-à-vis de l'Huître *Crassostrea gigas*. Thèse de Doctorat és Sciences Naturelles, Université d'Aix-Marseille, France, 138 pp.

Gerdes, D. 1983. The Pacific oyster, *Crassostrea gigas*. Part I. Feeding behaviour of larvae and adults. *Aquaculture*, **31**, 195–219.

Helm, M.M. and P.F. Millican. 1977. Experiments in the hatchery rearing of Pacific oyster larvae (*Crassostrea gigas* Thunberg). *Aquaculture*, **11**, 1–12.

His, E. 1976. Contribution à l'Étude Biologique de l'Huître dans le Bassin d'Arcachon. Activité Valvaire de *Crassostrea angulata* et de *Crassostrea gigas;* Application à l'Étude de la Reproduction de l'Huître Japonaise. Thèse de Doctorat és Sciences Biologiques, Université de Bordeaux I, France, 63 pp.

His, E. and R. Robert. 1980. Action d'un sel organométallique, l'acétate de tributyl-étain sur les oeufs et les larves D de *Crassostrea gigas* (Thunberg). *Intern. Counc. Explor. Sea Comm. Meet. (Mariculture Comm.)*, F, **27**, 1–10.

His, E. and R. Robert. 1981. Effects of copper chloride on the eggs and D larvae of *Crassostrea gigas* (Thunberg). Preliminary results. *Intern. Counc. Explor. Sea Comm. Meet. (Mariculture Comm.)*, F, **43**, 1–14.

His, E. and R. Robert. 1982. Le danger des traitements par le sulfate de cuivre en zone conchylicole: toxicité vis à vis des oeufs et des jeunes larves de *Crassostrea gigas*. *Rev. Trav. Inst. Pêches Marit.*, **45**, 117–125.

His, E. and R. Robert. 1985. Développement des véligères de *Crassostrea gigas* dans le Bassin d'Arcachon. Etudes sur les mortalités larvaires. *Rev. Trav. Inst. Pêches Marit.*, **47**, 63–88.

His, E. and R. Robert. 1987a. Comparative effects of two antifouling paints on the oyster *Crassostrea gigas*. *Mar. Biol.*, **95**, 83–86.

His, E. and R. Robert. 1987b. L'isolement des véligères de *Crassostrea gigas*: un nouveau mode d'investigation sur la biologie larvaire. *Haliotis*, **16**, 573–575.

His, E., D. Maurer and R. Robert. 1983. Estimation de la teneur en acétate de tributylétain dans l'eau de mer par une méthode biologique. *J. Mollusc. Stud.*, **12A**, 60–68.

His, E., R. Robert and M.J. Chrétiennot-Dinet. 1985. Nouvelle méthode pour étudier la nutrition de jeunes larves de *Crassostrea gigas* (Thunberg) en milieu naturel. Premières données expérimentales. *C.R. Acad. Sci. Paris.* **300** (3), No. 8, 319–321.

His, E., D. Maurer and R. Robert. 1986. Observations complémentaires sur les causes possibles des anomalies de la reproduction de *Crassostrea gigas* (Thunberg) dans le Bassin d'Arcachon. *Rev. Trav. Inst. Pêches Marit.*, **48**, 45–54.

Iversen, E.S. (1968). *Farming the Edge of the Sea*. Fishing News Ltd, London, 301 pp.

Lawler, I.F. and J.C. Aldrich. (1987). Sublethal effects of bis(tri-n-butyltin) oxide on *Crassostrea gigas* spat. *Mar. Pollut. Bull.*, **18**, 274–278.

MacInnes, J. R. (1981). Response of embryos of the American oyster, *Crassostrea virginica* to heavy metal mixtures. *Mar. Environ. Res.*, **4**, 217–227.

Martin, M., K.E. Osborn, P. Billig and N. Glickstein. (1981). Toxicities of ten metals to *Crassostrea gigas* and *Mytilus edulis* embryos and *Cancer magister* larvae. *Mar. Pollut. Bull.*, **12**, 305–308.

Maurer, D., E. His and R. Robert. (1984). Observations sur le phytoplancton du Bassin d'Arcachon en période estivale. Rôle potentiel dans la nutrition des larves de *Crassostrea gigas*. *Intern. Counc. Explor. Sea Comm. Meet. (Biol. Oceanogr. Comm.)*, L, **14**, 1–12.

Minchin, D., C.B. Duggan and W. King. (1987). Possible effects of organotins on scallop recruitment. *Mar. Pollut. Bull.*, **18**, 604–608.

Quayle, D.B. (1969). Pacific oyster culture in British Columbia. *Bull. Fish. Res. Bd Can.*, **169**, 1–192.

Robert, R. and E. His. (1981). Action de l'acétate de tributyle étain sur les oeufs et les larves D de deux mollusques d'intérêt commercial: *Crassostrea gigas* (Thunberg) et *Mytilus galloprovincialis* (Lmk). *Intern. Counc. Explor. Sea Comm. Meet. (Mariculture Comm.)*, F, **42**, 1–15.

Roberts, M.H. (1987). Acute toxicity of tributyltin chloride to embryos and larvae of two bivalve mollusks, *Crassostrea virginica* and *Mercenaria mercenaria*. *Bull. Environ. Contam. Toxicol.*, **39**, 1012–1019.

Smith, B. (1981). Tributyltin compounds induce male characteristics in female mud snails *Nassarius = Ilyanassa obsoleta*. *J. Appl. Toxicol.*, **1**, 141–144.

Thain, J.E. 1983. The acute toxicity of bis(tributyltin) oxide to the adults and larvae of some marine invertebrates. *Intern. Counc. Explor. Sea (Marine Environment. Quality Committee)*, E, **13**, 1–5.

Thompson, J.A., M.G. Sheffer, R.C. Pierce, Y.K. Chau, J. Cooney, W.R. Cullen and R.J. Maguire. 1985. Organotin Compounds in the Environment: Scientific Criteria for Assessing Their Effects on Environmental Quality. National Research Council of Canada, NRCC Associate Committee on Scientific Criteria for Environmental Quality, Subcommittee on Water, Publication No. NRCC 22494 of the Environmental Secretariat, Ottawa, Canada, 284 pp.

Waldock, M.J. and J.E. Thain. 1983. Shell thickening in *Crassostrea gigas*: organotin antifouling or sediment induced? *Mar. Pollut. Bull.*, **14**, 411–415.

Walsh, G.E., L.L. McLaughan, E.M. Lores, M.K. Louie and C.H. Deans. 1985. Effects of organotins on growth and survival of two marine diatoms *Skeletonema costatum* and *Thalassiosira pseudonana*. *Chemosphere*, **14**, 383–392.

Woelke, C.E. 1967. Measurement of water quality with the Pacific oyster embryo bioassay. In: *Water Quality Criteria*, American Society for Testing and Materials, Philadelphia, pp. 112–120.

Wong, P.T.S., Y.K. Chau, O. Kramar and G.A. Benger, 1982. Structure–toxicity relationship of tin compounds on algae. *Can. J. Fish. Aquat. Sci.*, **39**, 483–488.

Yonge, C.M. 1960. *Oysters*. Collins Clear-Type Press, London, 209 pp.

Zarrogian, G.E. and G. Morrisson. 1981. Effects of cadmium body burdens in adult *Crassostrea virginica* and on fecundity and viability of larvae. *Bull. Environ. Contam. Toxicol.*, **27**, 344–348.

REPRODUCTIVE FAILURE IN THE GASTROPOD *NUCELLA LAPILLUS* ASSOCIATED WITH IMPOSEX CAUSED BY TRIBUTYLTIN POLLUTION: A REVIEW

13

P.E. Gibbs and G.W. Bryan
Plymouth Marine Laboratory, Citadel Hill, Plymouth, Devon PL1 2PB, UK

ABSTRACT

Most stenoglossan gastropods are gonochoristic (i.e. the sexes are separate). During the last two decades, the phenomenon of 'imposex,' the development of male sex organs on the female, has become increasingly prevalent, to the extent that >40 species worldwide are now known to exhibit the syndrome. Initial evidence linking imposex to the presence of the leachates of marine antifouling paints containing tributyltin (TBT) as a biocide was provided by studies of the American mudsnail *Ilyanassa obsoleta*, but no deleterious effect on its reproductive biology was detected. However, recent

Organotin. Edited by M.A. Champ and P.F. Seligman. Published in 1996 by Chapman & Hall, London. ISBN 0 412 58240 6

investigations of imposex in the European dogwhelk (*Nucella lapillus*) have demonstrated that in this species the effects of TBT can be profound, since breeding can be inhibited causing populations to decline and eventually disappear. This chapter summarizes the evidence that TBT pollution is responsible for the disappearance of *N. lapillus* in areas close to centers of boating activity.

Two methods of measuring the intensity of imposex in *N. lapillus* are described, namely the relative penis size (RPS) index and the vas deferens sequence (VDS) index: these indices provide indications of the relative development of the female penis and of the associated vas deferens. When fully developed, vas deferens tissue blocks the oviduct preventing the release of egg capsules, thus rendering the female sterile. Field surveys of much of the UK coastline demonstrate that the intensity of imposex, as measured by both indices, increases markedly with proximity to sources of TBT such as harbors and marinas. Close to sources, females are sterile, breeding activity has ceased, and populations are declining or have disappeared. Transplantations of animals from 'clean' sites to such contaminated areas promote imposex in the transplanted animals. Laboratory experiments in which animals were reared from the hatchling stage to maturity at 2 y of age show that at an ambient TBT concentration in water of 1–2 ng l^{-1} Sn imposex is fully developed; at $\geq$3 ng l^{-1} Sn all females are sterilized. At higher concentrations ($\geq$10 ng l^{-1} Sn) oogenesis is suppressed and is supplanted by spermatogenesis. Laboratory and field data indicate that imposex in *N. lapillus* is initiated at an ambient water TBT concentration of $<$1 ng l^{-1} Sn. The high sensitivity of this response provides one explanation for the fact that imposex is found throughout the species' geographic range except in remote areas (e.g. parts of northwest Scotland). In those areas where TBT pollution exceeds 2 ng l^{-1} Sn, the sterilizing effect of imposex is apparent in the lack of breeding activity and dwindling population numbers.

13.1. INTRODUCTION

The development of male sex organs by the female of gonochoristic stenoglossan gastropods is a widespread phenomenon that appears to have become more prevalent with the increasing usage, during the last two decades, of tributyltin (TBT)-containing antifouling paints, especially on small pleasure craft. The term 'imposex' was coined by Smith (1971) to describe the superimposition of male sex characteristics, including a penis and a vas deferens, on to the female. By 1990, $\sim$40 species were known to exhibit imposex, but the syndrome has been investigated in detail in only two species: the American mudsnail, *Ilyanassa obsoleta* (Say), and the dogwhelk, *Nucella lapillus* (L.). Studies of *I. obsoleta* by Smith (1980, 1981a,b,c) established that imposex can be induced by TBT and by leachates of TBT-containing paints, but noted that there was no obvious effect on the reproductive capacity and general ecology of affected populations. In the case of *N. lapillus*, however, imposex may greatly modify the female genital tract to the extent that populations fail to breed, decline, and eventually disappear (Gibbs and Bryan, 1987). The evidence gained so far suggests that this sequence of events has occurred in those areas where TBT contamination levels have exceeded 2 ng l^{-1} Sn for prolonged periods. The usage of antifouling paints containing triorganotins as a biocide has been restricted by UK government legislation since July 1987. The restrictions also apply to the use of these compounds in aquaculture.

Essentially, *N. lapillus* is an intertidal species found on rocky shores along both the east and the west sides of the North Atlantic (Russia to Portugal and Newfoundland to New York). Many aspects of the general

Figure 13.1 (A) *Nucella lapillus* (Bude population). (B) Cluster of egg capsules with emerging juveniles (several arrowed). Scale lines: (A) 10 mm; (B) 5 mm.

biology of this carnivore have been investigated (summarized by Crothers, 1985); in seeking an explanation of the decline of *N. lapillus* in relation to TBT contamination, some features of its biology need to be highlighted. Probably the most important of these is the fact that *N. lapillus* has very limited powers of dispersion since the adults remain on the same stretch of shore throughout their life, and there is no planktonic phase during its larval development. Larval development takes place within a durable capsule attached to the substratum, the juveniles emerging as miniature adults after a period of 3–4 months (Fig. 13.1). The life span of this species is known to exceed 6 y (Feare, 1970a) but could extend to a decade or more. Thus populations remain extant long after recruitment has ceased.

13.2. METHODS

13.2.1. TIDAL TANK EXPERIMENTS

Four 200-l tanks were continuously supplied with laboratory circulation seawater (pumped from Plymouth Sound) at a rate of ~0.6 1 min^{-1}; to produce a tidal effect the tanks were emptied every 12 h and took 5–6 h to refill. Initially, ~180 animals were placed in each tank and were confined in plastic cages containing rocks with food (barnacles or mussels) from a clean site. The TBT concentration in the laboratory water was ~1.5 ng l^{-1} Sn; this background level was enhanced by flowing the water supplied to each of the three experimental tanks through separate mixing chambers in which the ends of rods coated with a TBT-copolymer antifouling paint were immersed. Concentrations of TBT (as tin) were controlled by changing the depth of the rods in the mixing chambers (see Bryan *et al.*, 1987; Gibbs *et al.*, 1988).

13.2.2. MEASUREMENT OF IMPOSEX

In both sexes, the length of the penis was measured to the nearest 0.1 mm, and the relative penis size (RPS) index for the sample was determined as the cube of the mean length of the female penis expressed as a percentage of the cube of the mean length of the male penis. Structural changes in the development of the vas deferens were assessed using the vas deferens sequence (VDS) index (see section 13.3.1). Standard histological techniques were employed; details are given in Gibbs *et al.* (1988).

13.2.3. ANALYTICAL METHODS

The determination of TBT in tissues was based on the method of Ward *et al.* (1981) and described by Bryan *et al.* (1986). An aliquot of a homogenate prepared from frozen *N. lapillus* was dispersed in concentrated hydrochloric acid and then shaken with hexane to extract TBT and dibutyltin (DBT) compounds. The extract was analyzed for tin by graphite-furnace atomic absorption spectrophotometry before (TBT + DBT) and after DBT had been removed by shaking with 1N sodium hydroxide solution. Other aliquots of homogenate spiked with TBT or DBT were used as standards.

Replicate determinations (n = 5) gave coefficients of variation of 7% for both compounds at a concentration of ~0.25 μg g^{-1} Sn. Although not totally specific, this method separates TBT from tin added to homogenates as stannous chloride or as the halides of mono- and dibutyltin, di- and trimethyltin, and triphenyltin; however, coextraction of tripropyltin and some tetrabutyltin was observed (Bryan *et al.*, 1987, 1988). The detection limit for TBT was 0.01–0.02 μg g^{-1} Sn in dry tissue.

Water samples collected below the surface in 1-l glass bottles were acidified with 5 ml of concentrated hydrochloric acid and analyzed for TBT using a similar procedure to that outlined above. The detection limit was ~0.5 ng l^{-1} Sn.

Concentrations of TBT in this chapter are expressed as tin; these can be converted to concentrations of the TBT ion by multiplying by 2.44.

13.3. IMPOSEX IN *NUCELLA LAPILLUS*

13.3.1. DEVELOPMENT AND OCCURRENCE

Imposex was discovered in *N. lapillus* by Blaber (1970) who observed penis-like outgrowths on females in Plymouth Sound populations during the winter of 1969–70. No earlier observation of the syndrome has been found in the literature on this extensively investigated species. In this respect, the studies of Pelseneer (1926) and Moore (1938) can be cited here as relevant examples: both involved examinations of large samples (n >1000) of *N. lapillus* (at Wimereux, near Boulogne, in 1923–24 and in Plymouth Sound in 1935) and, using presence or absence of a penis to distinguish the sexes, both found females to predominate in adult populations. Clearly, imposex was not present in these populations. In the absence of any evidence to the contrary, it is concluded that the syndrome is not a natural condition but one that has arisen as a response to recently introduced pollution.

Since the study of Blaber, the incidence and intensity of imposex in *N. lapillus* have increased markedly in the same Plymouth populations (Bryan *et al.*, 1986), and it would appear that this trend has been a general one, extending to all UK populations except those in remote areas (cf. Bailey and Davies, 1988a). In populations inhabiting sheltered inlets used for recreational boating, such as those found along the southwest coast of England (Helford, Fal, Tamar, and Dart estuaries), the

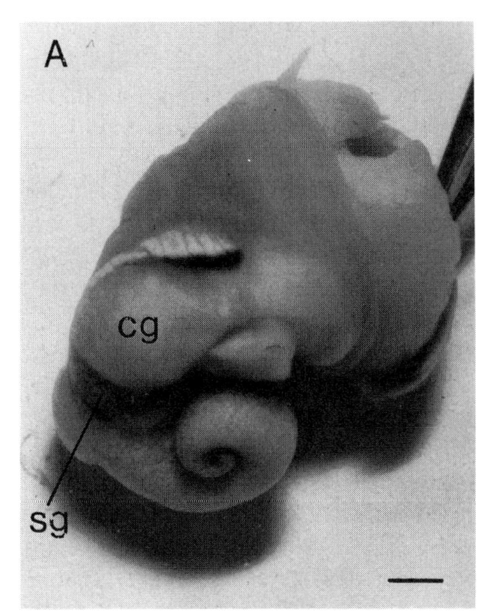

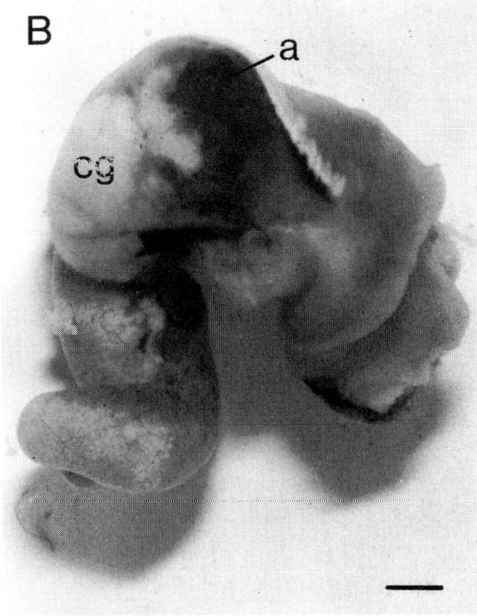

Figure 13.2 Dorsal views of adult female *N. lapillus* with shell removed showing typical appearance of specimens from (A) North Cornwall ('clean' site) and (B) Plymouth Sound (high TBT-contaminated site). A mass of aborted capsules (a) is visible through the distended capsule gland (cg) in (B). The sperm-ingesting gland (sg), used as a diagnostic feature of a female, is indicated in (A). Scale lines: 2 mm.

size of the penis on females is much the same as that of the male and thus other characteristics, such as the presence of a sperm-ingesting gland in females, must be used to separate the sexes (Fig. 13.2).

In *N. lapillus* the pallial portion of the male gonoduct is composed of three parts: (1) a glandular prostate region that is joined to (2) the vas deferens, a narrow duct crossing the floor of the mantle cavity, that continues as (3) the penial duct extending the length of the penis, a prominent organ sited on the head behind the right tentacle. The distal half of the male tract (2 and 3) appears to be derived from the primitive seminal groove found in other gastropods, a closed duct having been formed by the edges of the infolded epithelium becoming fused (Fretter, 1941). The initial formation of this structure signals the onset of imposex in female *N. lapillus*, commencing at a point close to the genital papilla. Development of this proximal portion of the vas deferens (Fig. 13.3, stage 1) is followed by initial penis formation (stage 2) with subsequent enlargement to form the penis proper (stage 3). From the base of the penis the distal section of the vas deferens appears, and this section eventually fuses with that developing proximally (stage 4). Most populations comprise females exhibiting imposex either in the early (1 and 2) or intermediate (3 and 4) stages, and breeding in these populations appears to be unaffected because the female oviduct opening (vulva) remains unobstructed, permitting copulation and capsule expulsion (Gibbs *et al.*, 1988). However, the females in populations close to boating centers exhibit more advanced stages of imposex manifest by an overgrowth of the genital papilla by vas deferens tissue thus occluding the vulva (stage 5). Internally, a prostate gland forms that displaces the capsule gland (Fig. 13.4). Such blockage of the oviduct prevents the expulsion of egg capsules but does not inhibit their manufacture (Gibbs and Bryan, 1986); consequently, aborted capsules accumulate

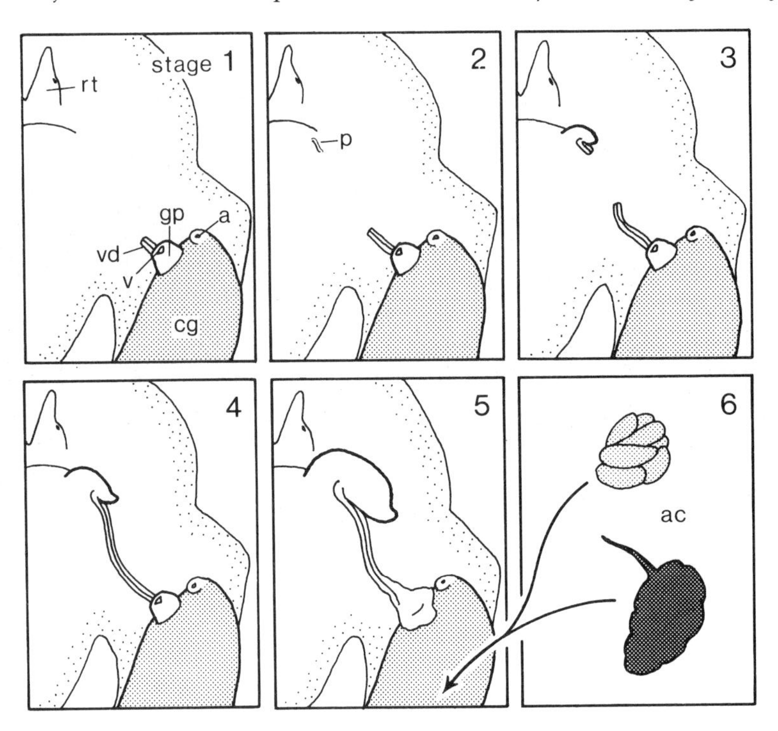

Figure 13.3 Six stages used to rank the development of imposex in *N. lapillus* by the vas deferens sequence (VDS) method. Abbreviations: a, anus; ac, aborted capsule masses; cg, capsule gland; gp, genital papilla; p, penis; rt, right tentacle; v, vulva; vd, vas deferens.

within the capsule gland (stage 6). Masses of aborted capsules are often visible through the oviduct wall (Figs. 13.2 and 13.5) and in extreme cases become so large as to rupture the distended oviduct, the extruded mass becoming fused to the shell (Fig. 13.5). Premature death of a female so maimed is assumed.

The degree of imposex development in a population can be expressed in a number of ways. The simplest is the percentage of penis-bearing females within the population, but this measure is appropriate only when imposex is in its initial phases: often virtually all females over long stretches of coast are penis-bearing and thus percentages have little meaning. The sizes of female penes within any one population are usually found to be remarkably consistent, and the mean length provides a useful parameter. However, adult *N. lapillus* vary considerably among populations in body size and therefore penis size. For comparisons among populations, the RPS index (see section 13.2, Methods), which compares the size of the female penis with that of the male, has proved a reliable measure (Gibbs *et al.*, 1987). This index provides a relatively simple and quick method of estimating the status of imposex, particularly in its intermediate stages (3 and 4), but provides no indication of the most significant effect, that is, the sterility it causes when at an advanced stage. The VDS index, based on six stages (Fig. 13.3), was devised to account for any reduction in the reproductive capacity of a population. The VDS index of a population is the mean of those stages in a sample; females lacking any visible sign of imposex are ranked as stage 0 and included in the calculation. An index

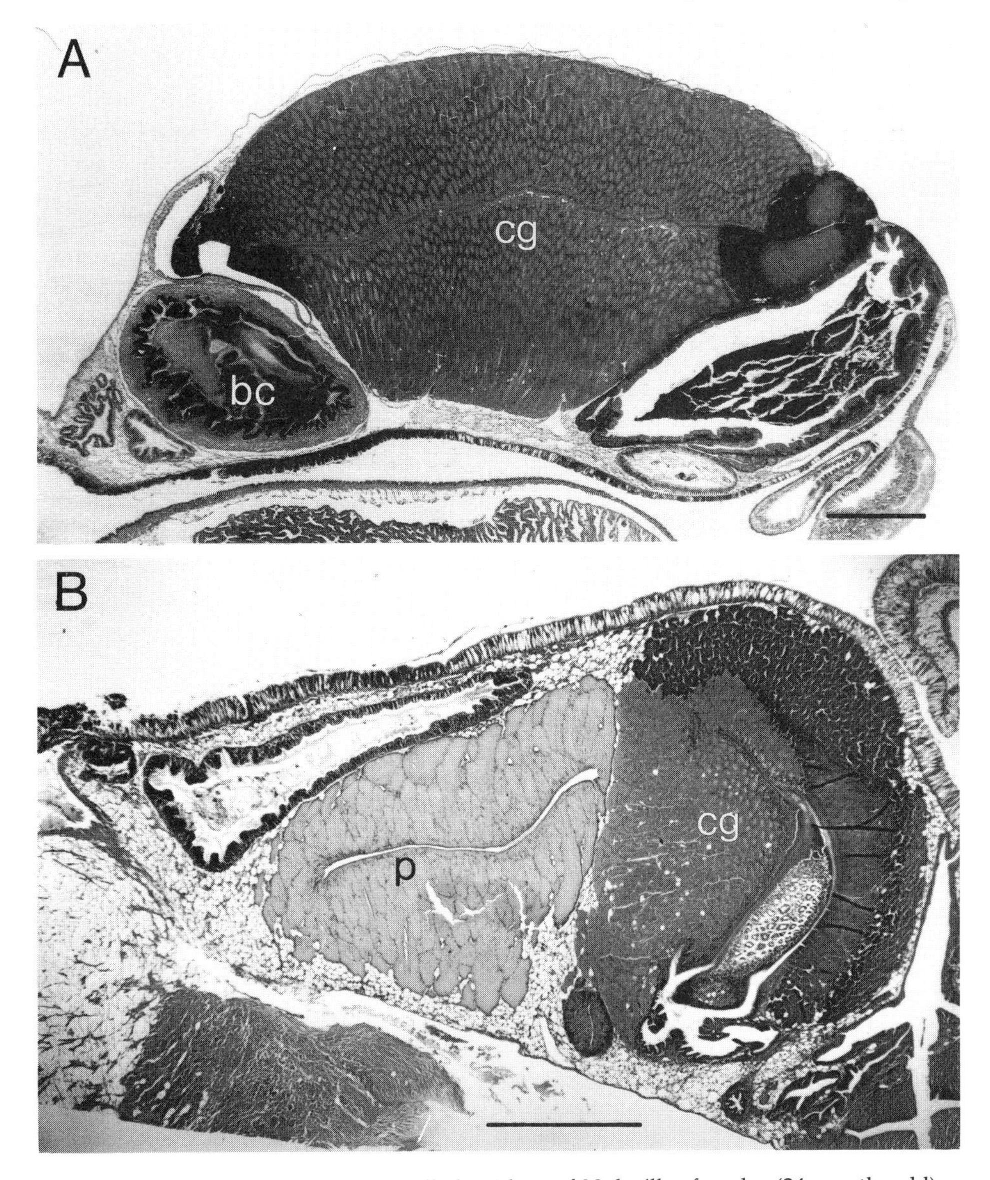

Figure 13.4 Longitudinal sections of the pallial oviduct of *N. lapillus* females (24 months old) reared at (A) 1–2 ng l^{-1} Sn, and (B) 20 ng l^{-1} Sn. The bursa copulatrix (bc) in (A) is filled with darkly staining spermatozoa. In (B) the bursa copulatrix is replaced by the imposed prostate (p), which displaces the capsule gland (cg) posteriorly. Scale lines: 2 mm.

above 4.0 indicates the presence of sterile females within the population and a reduction of its reproductive capacity.

RPS indices have been calculated for many populations around the United Kingdom, including those in remote Scottish lochs used for fish farming (see Bryan *et al.*, 1986; Davies *et al.*, 1987; Bailey and Davies, 1988b, 1989, 1991). The values obtained during surveys of the west Cornwall region are shown in Fig. 13.6; they illustrate the general trend for RPS indices to increase from <5% in remote areas

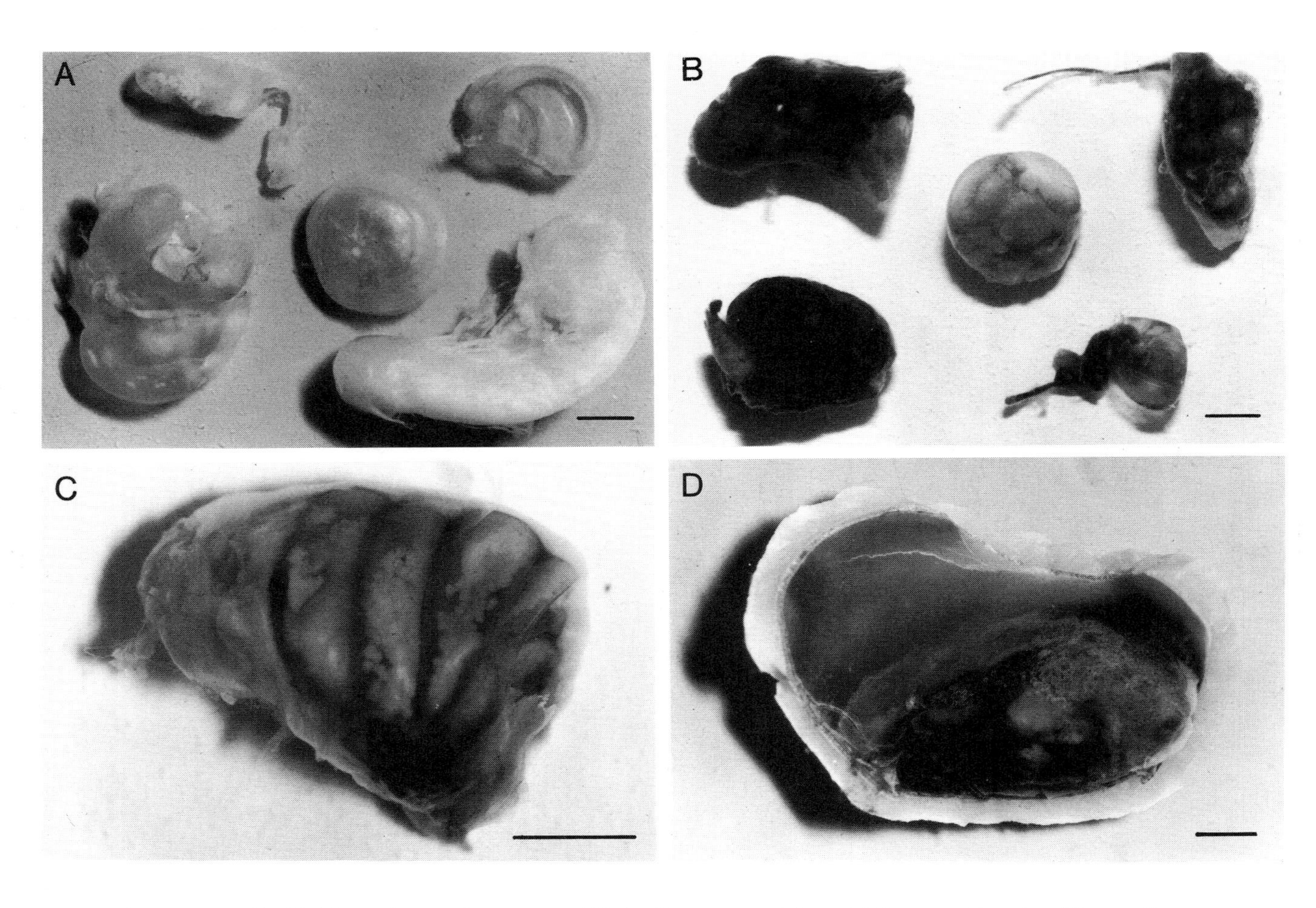

Figure 13.5 (A–C) Aborted capsule masses of *N. lapillus* removed from VDS stage 6 females (Plymouth Sound populations). (D) Interior of shell with fused capsule mass. Scale lines: 2 mm.

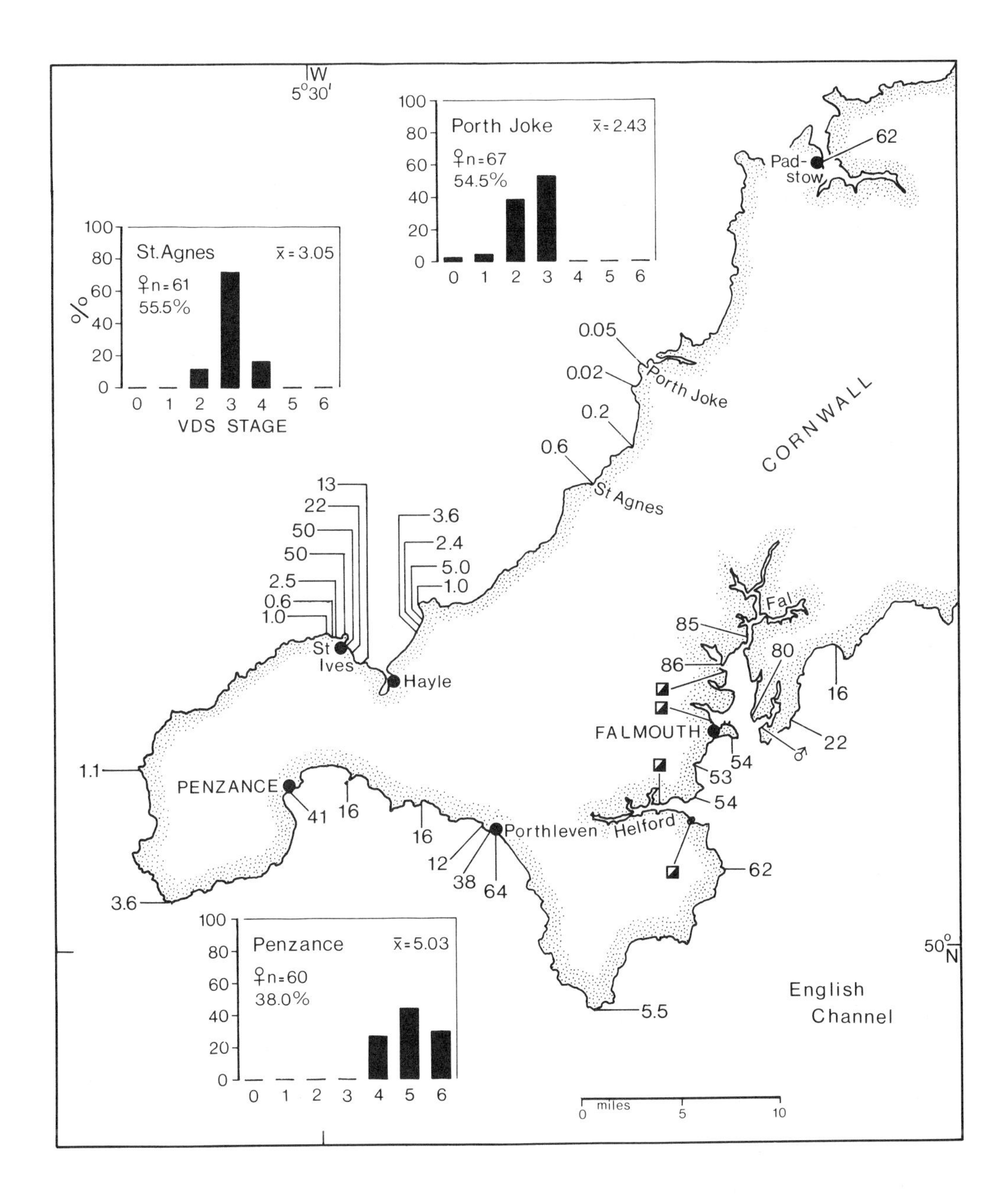

Figure 13.6 Status of imposex in *N. lapillus* populations around west Cornwall, 1984–1986. Values are relative penis size (RPS) indices. One site is shown where males only were found (♂) and four sites where the species has disappeared although formerly common (◪). Histograms of vas deferens sequence (VDS) stages are illustrated for the populations at Porth Joke, St Agnes, and Penzance; number of females examined (♀n), percentage of females in population (%), and VDS index ($\bar{x}$) are given.

to 40–50% and higher close to boating areas (St Ives, Penzance, and Porthleven). Similarly, of the VDS stages found in populations far removed from harbors (e.g. Porth Joke and St Agnes), stage 3 predominates with no female beyond stage 4: as evidenced by the presence of abundant egg capsules and juveniles, the breeding capability of these populations is unimpaired. But close to a fishing harbor such as Penzance, most females are sterile (74% at stages 5 and 6, VDS index = 5.03); and the continuance of the population depends on the relatively few females remaining at stage 4. A significant trend in the population associated with advanced imposex (i.e. stages 5 and 6) is a reduction in the relative proportion of females. In remote areas, females compose >50% of the population; but near harbors this drops to <40% (e.g. 38% at Penzance); and where the number of individuals has declined to the point of scarcity, females are few or absent in samples, as in those from the Fal Estuary. A higher rate of mortality of females over males is most likely caused by the effects of advanced imposex, especially rupture of the capsule gland (Fig. 13.5). The demise of the formerly abundant populations within the Fal Estuary is now virtually complete; the same applies to the Helford Estuary and many other estuaries along the Channel coast of southern England (Bryan *et al.*, 1986).

Bailey and Davies (1991) continued the earlier surveys investigating the degree of TBT pollution arising primarily from two substantial salmon farms (Davies *et al.*, 1987) undertaken in Loch Laxford, Sutherland, on the west coast of Scotland. In the subsequent study it was determined that low residual concentrations of TBT were still present in the loch during 1988 and 1989, even though the concentrations were lower than in the previous years. A mathematical model of the data found that there was a smaller degree of imposex development in the subadult and juvenile dogwhelks in 1988 than in 1987. This suggests that the 1987 UK legislation curtailing the use of TBT in mariculture is beginning to be reflected in lower imposex conditions in dogwhelk populations.

13.3.2. FIELD EXPERIMENTS

The results of field surveys, such as that of west Cornwall (Fig. 13.6), provide strong evidence of the link between the extent of imposex development and proximity to centers of boating activity. Further evidence of this link was obtained by three transplantations of marked adult *N. lapillus* from sites on the north Cornish coast (St Agnes, Widemouth, and Bude) to harbors at Plymouth (native *N. lapillus* extinct) and Dartmouth (sparse population, all females sterile) (Fig. 13.7). In each case some 300–600 animals were deposited and subsequently sampled over 2–3 y. Initially, the mean lengths of female penes in all three transplants measured 0.8–0.9 mm; 12 months after transplantation, mean values >3 mm were recorded in all three experiments and thereafter remained fairly constant. Corresponding values for females at the sites of origin of the transplants showed no similar increase. An increase in the VDS index following transplantation was also noted. For example in transplant 2, animals originally at stage 3 had advanced to late stage 4 (n = 13) and stage 5 (n = 2) 18 months after transplantation (Bryan *et al.*, 1987).

TBT concentrations in the clean water off north Cornwall (low tide subsurface samples) are below the level of detection (<0.5 ng l^{-1} Sn), whereas at the transplant deposition sites they are high: at Sutton Harbour, monthly means (fortnightly samples, high and low water) show the winter minimum and summer maximum TBT levels ranged between 5–15 and 36–46 ng l^{-1} Sn over the years 1986–88, while at the Dart Estuary site, TBT concentrations varied between 9 and 19 ng l^{-1} Sn (mean ± SD = 13.3 ± 3.3, n = 9) between June 1985 and October 1987, falling

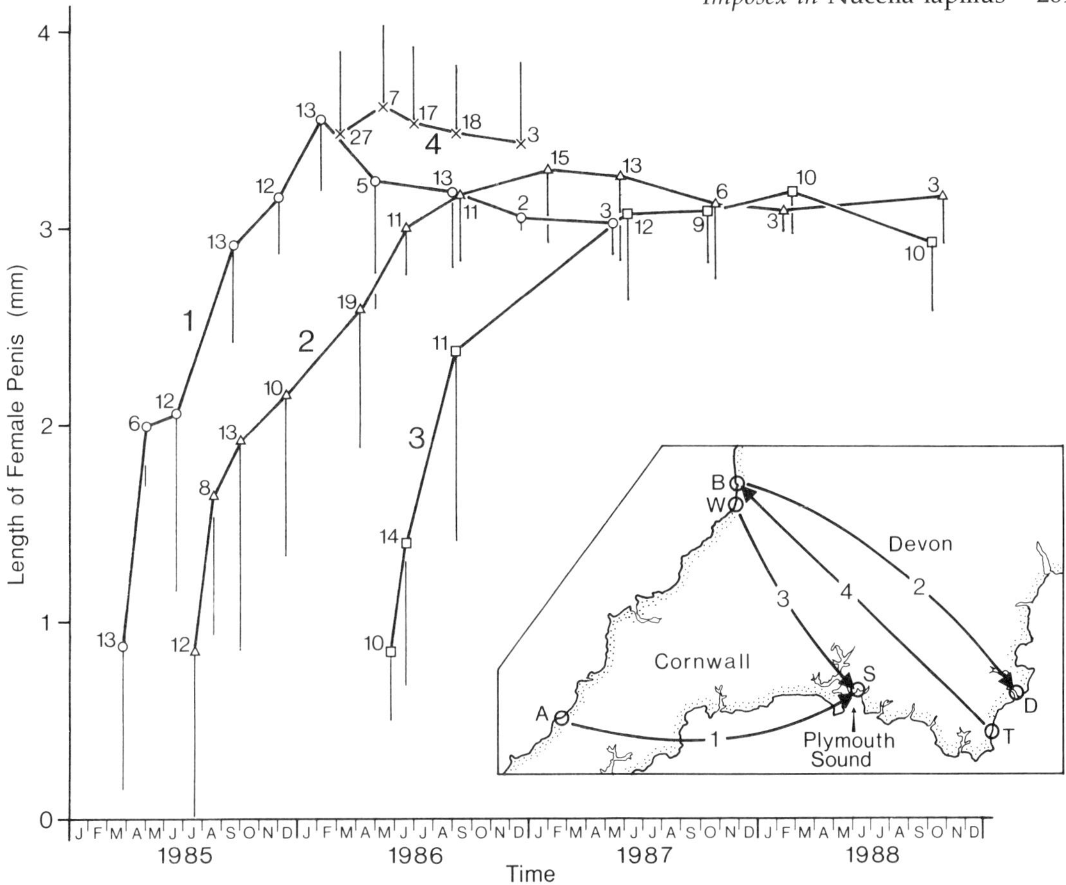

Figure 13.7 Effect of transplantation on female penis length in *N. lapillus*. Results of four transplants (1–4) are shown: transplant 1 transferred adults from St Agnes (A) to Sutton Harbour (S); 2, from Bude (B) to the Dart Estuary (D); 3, from Widemouth Bay (W) to Sutton Harbour (S); and 4, from Torcross (T) to Bude (B). Numbers recovered at each sampling are shown; points are means, bars are standard deviations.

to ~5 ng l^{-1} Sn in 1988 (n = 2). These data support the view that transfer from waters with a low TBT concentration to waters of high TBT content results in the promotion of imposex development (see also Bailey and Davies, 1989).

The question as to whether or not penis resorption would occur following transplantation in the reverse direction to transplants 1, 2, and 3 obviously required investigation. The population at Torcross (South Devon) was chosen because the females all had well-developed imposex (RPS index = 44%). These were transferred to Bude (Fig. 13.7; transplant 4) in March 1986. Over a period of 9 months no appreciable change in the initial mean penis length of 3.5 mm was observed. This result suggests that imposex is irreversible.

Direct evidence for the effect of the leachate of antifouling paints on imposex in *N. lapillus* was gained by a simple field experiment involving the application to the shell spires of either a TBT-based or a copper-based antifouling paint (Blakes 'Tiger' brand, Blakes Paints Ltd, Gosport, Hants, England,

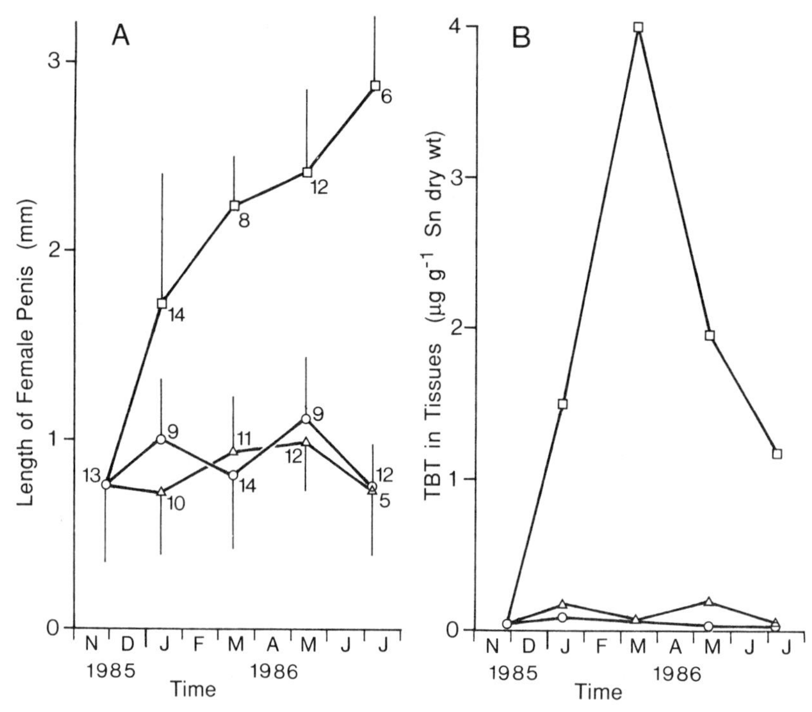

Figure 13.8 Effect on (A) penis length and (B) tin tissue concentrations of female *N. lapillus* (Widemouth population) of coating shell spires with TBT-containing and copper-containing antifouling paints. Symbols: □, TBT paint; △, copper paint; ○, untreated controls. In (A) numbers recovered at each sampling are shown; points are means, bars are standard deviations. In (B) concentrations are based on samples pooled from six females.

'copolymer' and 'tin-free', respectively, both red); unpainted individuals were used as controls. This treatment was carried out on the Widemouth population in November 1985, and bimonthly samples were taken until July 1986 (221 d later). Little change was observed in the mean penis lengths of those females with copper-painted shells and of those females left untreated (Fig. 13.8). But an increase was observed in the TBT-painted females, the mean penis length rising from 0.76 mm to 2.88 mm within 8 months; this was associated with an increase in the tissue TBT concentration.

13.3.3. LABORATORY EXPERIMENTS

The results of the field experiments suggested that the leachate of TBT-containing paints emanating from boating centers could be a causative agent responsible for imposex, and the fact that imposex can be detected in populations far removed from these centers further suggested that, if TBT is responsible, initiation must be effected at very low concentrations. Initial laboratory experiments (see section 13.2, Methods) were designed to cover a wide range of TBT concentrations [i.e. means ± SD = 1.35 ± 0.7 (n = 27), 18.7 ± 5.1 (n = 78), and 107 ± 30 (n = 58) ng l^{-1} Sn] and to maintain exposures for 1 y or more. It soon became obvious that imposex was promoted in the adults used (from Bude) at well below 20 ng l^{-1} Sn, and thus a further TBT concentration of 3.4 ± 0.8 (n = 32) ng l^{-1} Sn was introduced into the experiment. The results are summarized in Fig. 13.9; increases in the penis sizes of females were observed at

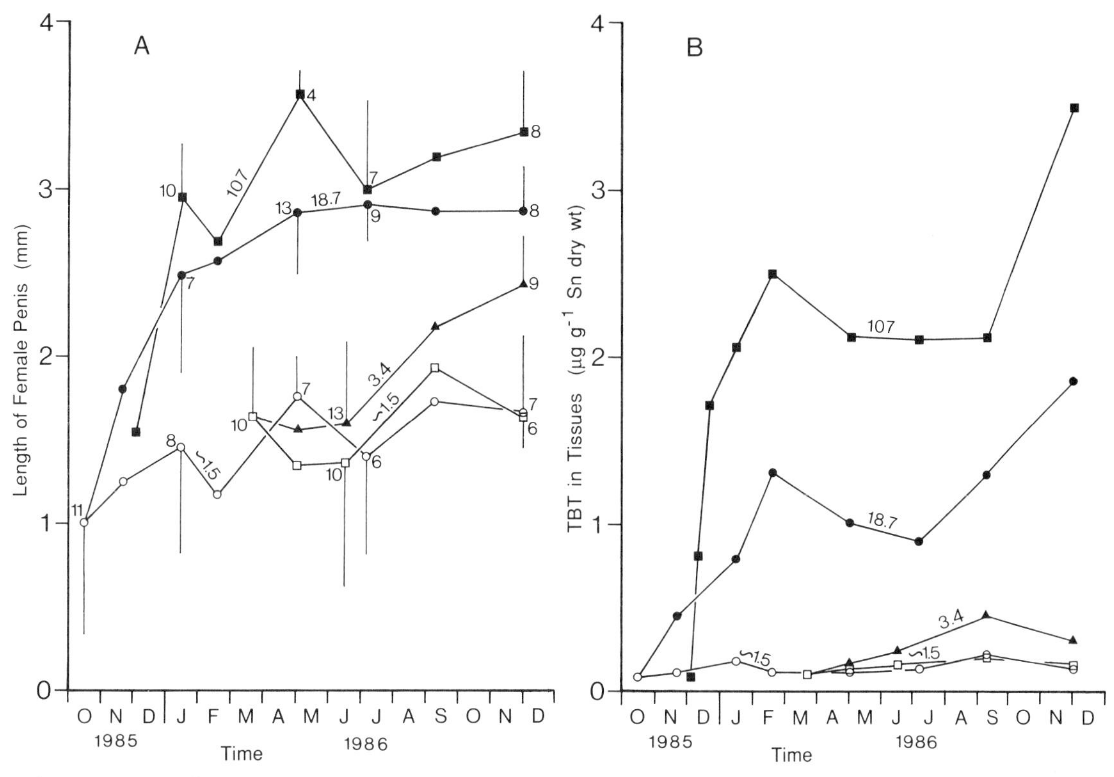

Figure 13.9 Effect of exposing female *N. lapillus* to different concentrations of leachate of TBT-containing antifouling paint on (A) penis length, and (B) TBT tissue concentrations. Plymouth Sound water ($\simeq$1.5 ng l^{-1} Sn) supplied untreated to two 'controls' and experimental TBT concentrations enhanced to mean levels of 3.4, 18.7, and 107 ng l^{-1} Sn. In (A) examples of numbers in samples are shown; points are means, bars are standard deviations. In (B) concentrations are based on samples pooled from 5–6 females.

all concentrations, significantly so even at the mean level of 3.4 ng l^{-1} Sn ($p < 0.01$), and the initial rates of uptake of tin as TBT were approximately proportional to the concentrations in the water. An increase in VDS indices was also noted (see Bryan *et al.*, 1987).

In the environment, ambient TBT levels often fluctuate widely, largely on a seasonal basis (see Waldock *et al.*, 1987); short-term (1–2 weeks) exposure of *N. lapillus* to high concentrations, particularly during the spring and summer months, would be a frequent occurrence in many sheltered localities. The effect of such short-term pulses was investigated by exposing two samples (n = 200 each) of adult *N. lapillus* (from Bude) to a TBT concentration of 107 ng l^{-1} Sn in a tidal laboratory tank for 7 d and 14 d (see Bryan *et al.*, 1987). A third (control) group was kept in laboratory water ($\simeq$1.5 ng l^{-1} Sn) for 14 d. All three samples were then returned to the shore close to the site of origin. Appreciable penis development was noted in the experimental females (but not in the males), and after a period of 8 months the initial RPS index of ~5% remained in the control group but had increased to 30% and 45% for the 7-d and 14-d pulse-exposed groups, respectively. At the start of the experiment, females exhibited predominantly VDS stage 3; at the end, control females were in the same condition, but all the experimental females had

Table 13.1 Effect of different TBT concentrations on the degree of imposex and tissue TBT–DBT levels in 24-month-old populations of *N. lapillus* exposed from hatching under experimental and field conditions

TBT in water ($ng\ l^{-1}$ Sn) ($\bar{x} \pm SD$)	*Sex[a] and no. in sample*	*Penis length (mm) ($\bar{x} \pm SD$)*	*Relative penis size index*	*Vas deferens sequence index*	*Tin concentration ($\mu g\ g^{-1}$, dry tissue)*		
					TBT fraction	*DBT fraction*	*Total*
			Experimental				
1.63 ± 0.75	M 15	3.71 ± 0.22	48.4		0.238	0.085	0.323
(*n* = 48)	F 18	2.91 ± 0.25		4.4	0.239	0.051	0.290
3.83 ± 1.59	M 10	3.44 ± 0.33	96.6		0.602	0.230	0.832
(*n* = 54)	F 10	3.40 ± 0.41		5.1	0.569	0.098	0.666
20.4 ± 6.2	M 9	3.56 ± 0.26	109.0		1.464	1.484	2.948
(*n* = 97)	F 15	3.66 ± 0.23		5.0	1.696	0.544	2.240
100 ± 29	M 29	3.43 ± 0.33	90.4		2.520	2.054	4.304
(*n* = 84)	F 15	3.32 ± 0.39		5.0	3.164	2.193	5.357
			Field				
<0.5	M 20	3.44 ± 0.47	3.7		0.092	0.099	0.191
(Bude)	F 23	1.15 ± 0.62		3.2	0.039	0.041	0.080

[a]M, male; F, female.

advanced to VDS stage 4. The uptake of tin in the TBT fraction was proportional to exposure time (i.e. the 14-d exposure group contained about twice that of the 7-d exposure group). Subsequent losses of excess TBT had half-times of ~48 and 58 d, respectively, for the two groups (data given in Bryan *et al.*, 1987). Significantly, loss of TBT from tissue did not result in any resorption of penial tissue formed before and subsequent to the increased TBT exposure.

All of the above experiments, both laboratory and field, demonstrating the promotion of imposex with increasing TBT concentrations in water, were carried out on fully grown individuals [i.e. those with shells having a thickened, rounded aperture lip, often with internal teeth (see Crothers, 1985)]. Field samples of such animals would be composed of various age groups ranging from 2–3 y to 7–10 y or even older, and the level of imposex exhibited would thus be a response to ambient TBT levels over several to many years. Interestingly, the extent of imposex exhibited by immature, or subadult, individuals in a population is often similar to that shown by adults. This suggests that the imposex level may be largely determined in the first year or two of life, although development can be advanced subsequently with increased TBT exposure in the adult stage, as demonstrated experimentally.

The extent to which the level of imposex in the adult is influenced by TBT exposure during the juvenile and immature phases of life was investigated by rearing animals from the time of hatching from the capsule to the stage of maturity at ~2 y of age. For this experiment the same tidal tanks as used for the experiments on adults were used. The data relating to the TBT water concentrations, development of imposex, and tissue tin concentrations (TBT and DBT fractions) at an age of 24 months are summarized in Table 13.1 (see also Gibbs *et al.*, 1988). Imposex was well developed in all four experimental populations; at TBT levels ⩾3 ng l^{-1} Sn not only did the RPS index exceed 90, but virtually every female was sterile (i.e. at VDS stage 5 or 6). The only females capable of seemingly normal breeding activity and producing capsules in some quantity were those reared at the lowest TBT concentration of 1–2 ng l^{-1} Sn, although even in this population about one-third of the females were sterile (VDS stages 5 or 6). Individuals of similar age (based on shell characteristics) taken from the original stock population at Bude (TBT concentration in water <0.5 ng l^{-1} Sn) gave an RPS index of 3.7%, a value comparable to that of the adult population. These individuals lacked any sign of sterility, conforming to VDS stages 2–4.

Clearly, imposex can be developed to an advanced stage before a female reaches maturity. In the above experiment 40% of females (n = 10) had reached VDS stage 5 at an age of 12 months when reared at a TBT concentration of only 1–2 ng l^{-1} Sn. Thus initiation of imposex probably occurs at TBT concentrations in water of <1 ng l^{-1} Sn, and this would be one explanation for the widespread occurrence of the syndrome, even on shorelines far removed from any boating centers.

13.3.4. CHEMICAL SPECIFICITY

The effectiveness of tri-n-butyltin chloride and five other organotin compounds in promoting the development of imposex in *N. lapillus* has been compared both by water exposure and by injection (Bryan *et al.*, 1988). No detectable effect was induced by the chlorides of dibutyltin (DBT), monobutyltin (MBT), or triphenyltin (TPhT). Tetrabutyltin (TTBT) was found to produce a significant response, partly because of TBT contamination, but debutylation to TBT in the tissues may have been another cause. Imposex was also increased significantly by tri-n-propyltin chloride (TPrT); but since this compound is not extensively employed commercially this effect is considered of little consequence. By far the most effective of the organotins tested for imposex promotion was tri-n-butyltin chloride, either injected or dissolved in seawater. Other tributyl compounds have now been tested. Adult *N. lapillus* were injected with 1 μl of TBTCl (equivalent to 1 μg of Sn or 2.44 μg of TBT in ethanol) and equimolar ethanol solutions of tributylamine, tributylsilane, and tributylphosphine oxide. Compared with non-injected and ethanol-injected controls, only tributyltin chloride produced any statistically significant promotion of penis development after 45 d (t = 12.8, df = 20, p <0.001).

13.4. TBT AND GAMETOGENESIS

Observations of the condition of the gonads of those females that were reared at different TBT concentrations for 24 months (see section 13.3.3, Laboratory experiments) provided important insights into the effect of TBT on gametogenesis (see Gibbs *et al.*, 1988). Histological sections of the gonads of females reared at the two lowest TBT concentrations (1–5 ng l^{-1} Sn) all appeared to show oogenesis progressing normally with pre-vitellogenic, vitellogenic, and post-vitellogenic oocytes (see Feare, 1970b) all being present. However, in those females reared in water containing 20 and 100 ng l^{-1} Sn as TBT, two features of the gonad were notable: (1) oogenesis appeared suppressed with pre-vitellogenic oocytes being either rare or seemingly absent, and those vitelline oocytes present were degenerating (probably undergoing resorption); and (2) in all examples (n = 30), seminiferous tubules had developed – in some the tubules were sparse, in others more numerous – and for the most part only the early stages of spermatogenesis were evident, although later stages including spermatozoa (described in Walker and MacGregor, 1968) were identified in some females. These features are illustrated in Fig. 13.10.

Such a marked modification of the process of gametogenesis not only occurs in experimental populations but is manifest also in field populations, such as that in the Dart Estuary, a popular boating center (Fig. 13.7, location D on inset). At the study site, TBT concentrations in water samples taken 1985–

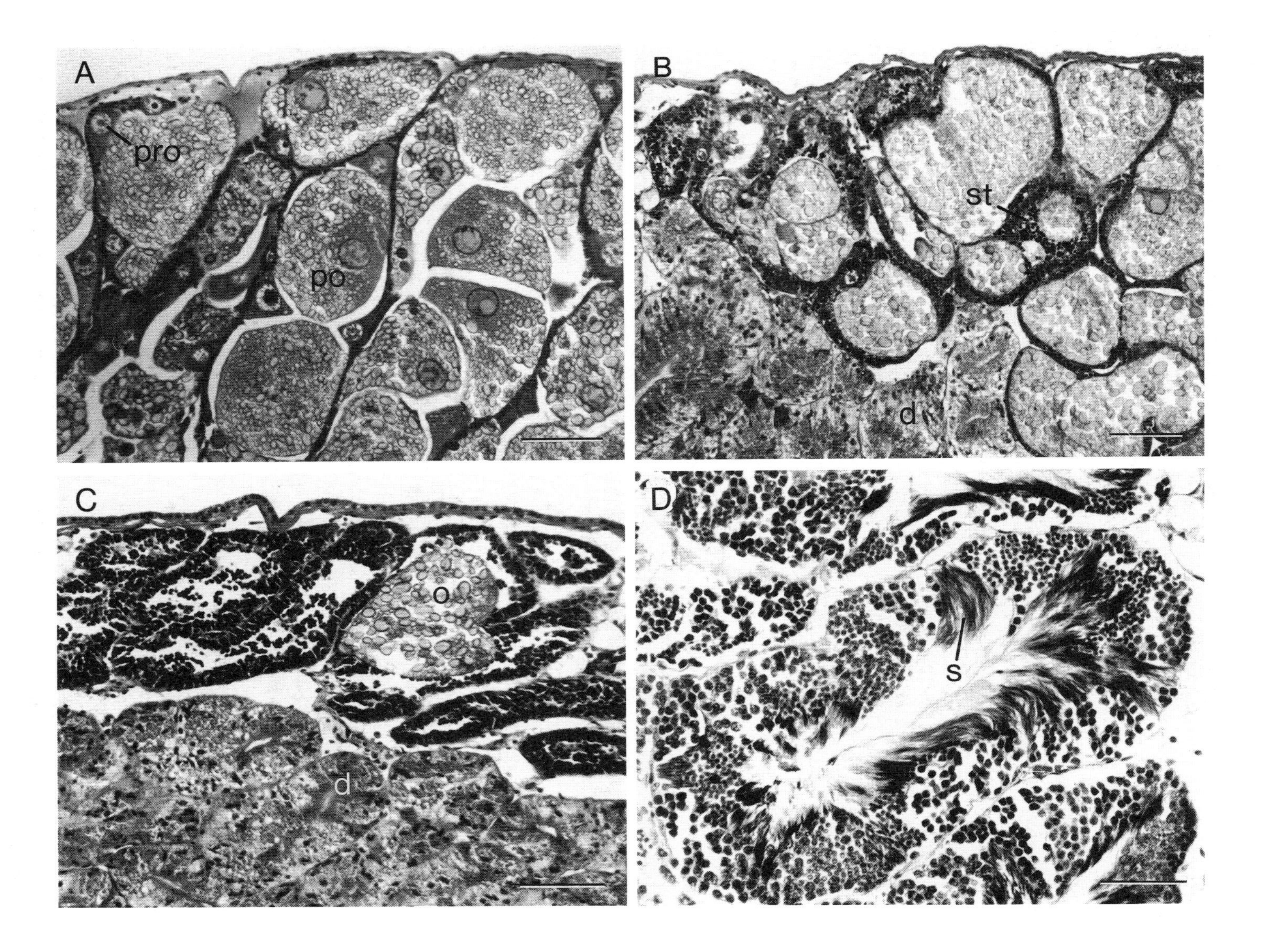
A
pro
po
B
st
d
C
o
d
D
s

1987 were between 9 and 19 ng l^{-1} Sn (see section 13.3.2, Field experiments). At such concentrations, the native *N. lapillus* survived but in low numbers, and the few remaining females were all sterile (VDS stages 5 and 6), their 'ovaries' dominated by seminiferous tubules. The adults of transplant 2 (Fig. 13.7) produced offspring in 1986, the females of which when examined in 1988 were all sterile (VDS stage 5). Significantly the ovaries of these young females (like the native adults at the same site) exhibited the same transition to testes as was noted in the young females reared at a TBT concentration $\geq$20 ng l^{-1} Sn. This field evidence suggests that the TBT concentration necessary to modify gametogenesis may lie around the 10 ng l^{-1} Sn mark.

13.5. DISCUSSION

In general, it would appear that the intensity of imposex displayed by the adult female *N. lapillus* is largely determined by the level of TBT exposure and hence level of imposex development during its early life. The results of the rearing experiments (see Gibbs *et al.*, 1988) suggest that the most sensitive period is when the reproductive system is developing during adolescence; in populations studied this occurs during the period between 9 and 18 months after hatching. Structural aberrations of the anterior pallial oviduct, such as replacement of the bursa copulatrix with a prostate gland and inhibition of genital papilla formation, have been noted in subadults (~18 months old) that have been reared at TBT concentrations $\geq$3 ng l^{-1} Sn. Such abnormalities do not appear to be rectified with age, indicating that sterilization is irreversible and permanent. Modifications of the fully developed, as opposed to the developing, oviduct can be induced, but the process takes a relatively long time. When mature females capable of breeding (i.e. at VDS stage 4 or below) are transferred to a TBT level of 10–20 ng l^{-1} Sn, the penis enlarges but vas deferens and prostate development proceeds slowly so that blockage of the oviduct takes several years of elevated exposure. Thus the VDS index for any population is interpreted as largely reflecting the ambient TBT pollution levels during the adolescent and subadult stage of its component cohorts. Because imposex is irreversible (see also Bailey and Davies, 1989), amelioration of TBT pollution will be signaled by a lowered imposex level in young individuals.

The relationship among the indices used as comparative measures of imposex development (RPS and VDS) and TBT concentrations in both ambient seawater and in female tissues is shown in Fig. 13.11. These correlations show that penis and vas deferens formation increase rapidly within very low ranges of TBT concentrations in water and in tissues. In the laboratory, exposure to ambient concentrations of TBT in water as low as 1–2 ng l^{-1} Sn over 24 months are sufficient to cause sterilization (Table 13.1): and with any increase above this level, the reproductive capacity of a population is rapidly reduced to zero. Only in those populations comprising females containing

Figure 13.10 Effect of TBT exposure on gametogenesis in *N. lapillus* females. (A) 1–2 ng l^{-1} Sn: maturing ovary with normal oogenesis, pre-vitellogenic to post-vitellogenic oocytes (laboratory-reared female at 24 months old); (B) 10+ ng l^{-1} Sn: early stage of seminiferous tubule development with spermatogonia and spermatocytes, pre-vitellogenic oocytes scarce or absent (Dart Estuary juvenile of transplant experiment, 18–24 months old); (C) 20 ng l^{-1} Sn: gonad virtually testes with seminiferous tubules dominating, few residual oocytes remaining (laboratory-reared, 24 months old); (D) 20 ng l^{-1} Sn: spermatogenesis complete with all stages to spermatozoa, oogenesis apparently entirely suppressed (laboratory-reared, 27 months old). Scale lines: A–C 100 μm; D, 50 μm. Abbreviations: d, digestive gland; o, residual oocyte; pro, pre-vitellogenic oocyte; po, post-vitellogenic oocyte; s, spermatozoa; st, seminiferous tubule.

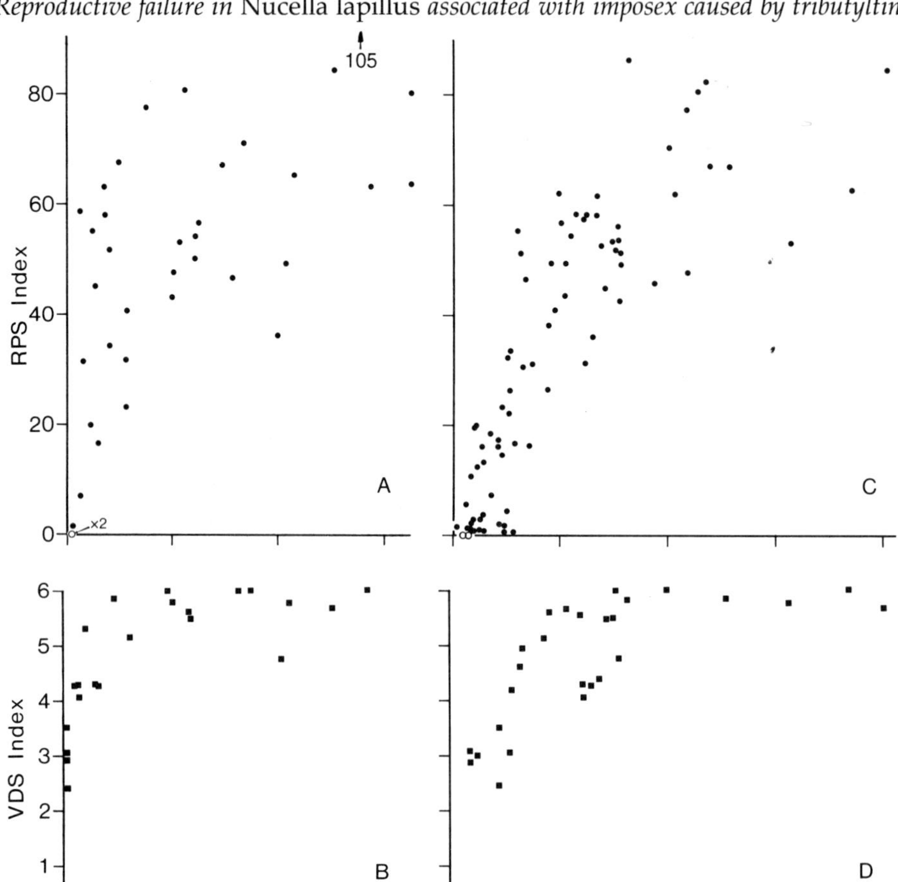

Figure 13.11 Relationship among tin as TBT in ambient seawater (A,B), body tissues of *N. lapillus* females (C,D) and relative penis size (RPS), vas deferens sequence (VDS) indices. Water and animal samples taken at same time at sites around southwest England, 1984–1986. Values for two clean water site samples (Isle of Mull, Scotland: open symbols) are shown for comparison.

TBT at <0.1 μg g^{-1} Sn in their tissues does breeding appear to proceed normally. That TBT disrupts the endocrine system of stenoglossans seems fairly certain, but as yet there is little information on this aspect. Other compounds may equally disturb their steroid metabolism. Spooner *et al.* (1991) have detected a significant increase in testosterone levels in female *N. lapillus* exposed to TBT for 28 d with accumulation of TBT in the tissues and an increase in penis length compared with control animals. This suggests an association between change in testosterone levels, TBT exposure, and imposex in the dog-whelk. TBT has been shown to activate the neuroendocrine mechanism responsible for penis differentiation in *Ocenebra erinacea* (L.) (Féral and Le Gall, 1982, 1983), but attempts using histochemical methods to demonstrate neurosecretory activity in the ganglia of the central nervous system of *N. lapillus* after TBT exposure have been unsuccessful. Overall, the response of *N. lapillus* after TBT exposure appears to be a graded one. The separate

Table 13.2 Female *N. lapillus* reproductive system: summary of effects of TBT exposure during first 2 y of life, based on field and laboratory data

TBT in water ($ng\ l^{-1}$ Sn)	*Effect on reproductive system*
<0.5	RPS index <5%; VDS index <4. Breeding normal. Development of penis and vas deferens
1–2	RPS index 40+%; VDS index 4–6. Breeding capacity retained by some females; others sterilized by oviduct blockage. Aborted capsules in capsule gland
3–4	Virtually all females sterilized. Oogenesis apparently normal
10+	Oogenesis suppressed. Oocytes resorbed. Spermatogenesis initiated
20	Testis developed to variable extent. Vesicula seminalis with ripe sperm in most-affected animals
100	Sperm-ingesting gland undeveloped in some 'females.'

effects of imposex and of gonadal change (Table 13.2) suggest that more than one endocrine system may be involved. While the data demonstrate that females in populations exposed to TBT concentrations <1 ng l^{-1} Sn can be expected to be fecund, despite the inconvenience of bearing male sex organs, the central problem of determining the exposure level that initiates imposex still remains.

Gibbs *et al.* (1991) have investigated the uniformity of the response and effects in adult dogwhelks taken from four populations geographically separated by over 10° of latitude. They found similar levels of accumulation of TBT and increases in female penis lengths. This suggests that the level of sensitivity and development of imposex as a response to TBT is uniform in *N. lapillus* throughout its distributional range. In addition, in surveys in southwest Brittany they found similar patterns to those they found in earlier surveys of the coast of southern England.

The sensitivity of dogwhelks to TBT has also contributed to resolving the controversy as to whether there would be an impact of TBT in an area subjected to only large vessels. Surveys of imposex in dogwhelk populations around oil terminals in Sullom Voe (Shetland) and Scapa Flow (Orkney) in Scotland confirmed that the effects of TBT released from only large vessels can be determined. The 1990 survey of Sullom Voe (a deepwater port specialized for oil and gas tankers) by Davies and Bailey (1991) reconfirmed earlier patterns of imposex development observed in 1987 (Bailey and Davies, 1988b). In the follow-up survey they also found that a large proportion of the female dogwhelks were unable to reproduce. Outside of the Voe, imposex levels declined rapidly. There is no significant pleasure boat or fish farming activity in the area of the Voe, therefore observed effects are directly related to terminal operations and large vessels (Bailey and Davies, 1988b).

Imposex has been reported for ~40 gastropod species. In some, the superimposition of the male gonoduct is such that the functioning of the female gonoduct is apparently unaffected, and reproductive activity continues unhindered, even in high TBT situations. This is the case in *Ilyanassa obsoleta* (see section 13.1, Introduction) and *Hinia* (=*Nassarius*) *reticulatus* (L.). Other species, like *N. lapillus*, suffer malformations of the oviduct; examples include *Ocenebra erinacea* (L.) and *Thais haemastoma* (L.) (Spence *et al.*, 1990), and these malformations are suspected in *Nucella lima* (Short *et al.*, 1989). No doubt future studies will uncover further subtle variations among species. However, because of its high sensitivity, it is improbable that any other species will surpass *N. lapillus* as a TBT indicator. In this context the widespread geographic range of *N. lapillus* is significant in that, throughout its range, the species is known to exhibit imposex, including the northeastern United States (Miller and Pondick, 1984), Norway (Bergen), UK waters from Shetland (Bailey and Davies, 1988b) to the Scilly Isles (Gibbs *et al.*, 1987), Holland, Brittany, France (Féral, 1980), and the Atlantic coast of the Iberian peninsula (Peña *et al.*, 1988).

13.6. SUMMARY

Imposex in the gonochoristic gastropod *Nucella lapillus* is a recent, now widespread, phenomenon that is most intensely developed in populations close to sources of TBT pollution, such as harbors, marinas, and fish farms. Experimentally, its development can be promoted in the field by transplantation of females from 'clean' water sites to areas of high TBT contamination and in the laboratory by exposure to leachates of TBT-containing antifouling paints. Imposex is a graded response, its degree of development being related to ambient TBT concentrations; it appears to be irreversible. The syndrome is initiated on exposure to a TBT concentration of <1 ng l^{-1} Sn; above 2 ng l^{-1} Sn the male gonoduct develops to the extent that the oviduct is blocked, preventing release of egg capsules, thus inhibiting breeding and causing populations to decline. In *N. lapillus* the high sensitivity of the response to TBT permits the use of imposex development as a reliable bioindicator of TBT pollution.

ACKNOWLEDGMENTS

The collaborative contributions of G.R. Burt, L.G. Hummerstone, P.L. Pascoe, and S.K. Spence (all of the Plymouth Marine Laboratory) are gratefully acknowledged. This work was partly supported by the UK Department of the Environment under Contracts PECD 7/7/226 and PECD 7/8/103. The authors are grateful to the proprietors of the *Journal of the Marine Biological Association of the United Kingdom* for permission to reproduce the photographs used in Figs 13.2 and 13.5.

REFERENCES

Bailey, S.K. and I.M. Davies. 1988a. Tributyltin contamination in the Firth of Forth. *Sci. Total Environ.*, **76**, 185–192.

Bailey, S.K., and I.M. Davies. 1988b. Tributyltin contamination around an oil terminal in Sullom Voe (Shetland). *Environ. Pollut.*, **55**, 161–172.

Bailey, S.K. and I.M. Davies. 1989. The effects of tributyltin on dogwhelks (*Nucella lapillus*) from Scottish coastal waters. *J. Mar. Biol. Ass. UK*, **69**, 335–354.

Bailey, S.K. and I.M. Davies. 1991. Continuing impact of TBT, previously used in mariculture, on dogwhelk (*Nucella lapillus* L.) populations in a Scottish sea loch. *Mar. Environ. Res.*, **32**, 187–199.

Blaber, S.J.M. 1970. The occurrence of a penis-like outgrowth behind the right tentacle in spent females of *Nucella lapillus* (L.). *Proc. Malacolog. Soc. London*, **39**, 231–233.

Bryan, G.W., P.E. Gibbs, L.G. Hummerstone and G.R. Burt. 1986. The decline of the gastropod *Nucella lapillus* around south-west England: evidence for the effect of tributyltin from antifouling paints. *J. Mar. Biol. Ass. UK*, **66**, 611–640.

Bryan, G.W., P.E. Gibbs, G.R. Burt and L.G. Hummerstone. 1987. The effects of tributyltin (TBT) accumulation on adult dog-whelks, *Nucella lapillus*: long-term field and laboratory experiments. *J. Mar. Biol. Ass. UK*, **67**, 525–544.

Bryan, G.W., P.E. Gibbs and G.R. Burt. 1988. A comparison of the effectiveness of tri-n-butyltin chloride and five other organotin compounds in promoting the development of imposex in the dog-whelk, *Nucella lapillus*. *J. Mar. Biol. Ass. UK*, **68**, 733–744.

Crothers, J.H. 1985. Dog-whelks: an introduction to the biology of *Nucella lapillus* (L). *Field Studies*, **6**, 291–360.

Davies, I.M. and S.K. Bailey. 1991. The impact of tributyltin from large vessels on dogwhelk (*Nucella lapillus*) populations around Scottish oil ports. *Mar. Environ. Res.*, **32**, 201–211.

Davies, I.M., S.K. Bailey and D.C. Moore. 1987. Tributyltin in Scottish sea lochs, as indicated by degree of imposex in the dogwhelk, *Nucella lapillus* (L). *Mar. Pollut. Bull.*, **18**, 400–404.

Feare, C.J. 1970a. Aspects of the ecology of an exposed shore population of dog whelks *Nucella lapillus* (L). *Oecologia* (Berlin), **5**, 1–18.

Feare, C.J. 1970b. The reproductive cycle of the dog whelk (*Nucella lapillus*). *Proc. Malacolog. Soc. London*, **39**, 125–137.

Féral, C. 1980. Variations dans l'évolution du tractus génital mâle externe des femelles de trois gastéropodes prosobranches gonochoriques de stations Atlantiques. *Cahiers de biologie marine*, **21**, 479–491.

Féral, C. and S. Le Gall. 1982. Induction expérimentale par un polluant marin (le tributylétain), de l'activité neuroendocrine contrôlant la morphogenèse du pénis chez les femelles d'*Ocenebra erinacea* (Mollusque Prosobranche gonochorique). *Compte rendu hebdomadaire des séances de l'Académie des sciences*, **295**, 627–630.

Féral, C. and S. Le Gall. 1983. The influence of a pollutant factor (tributyltin) on the neuroendocrine mechanism responsible for the occurrence of a penis in the females of *Ocenebra erinacea*. In: *Molluscan Neuro-endocrinology*. Proceedings of the International Minisymposium on Molluscan Endocrinology, 1982, J. Lever and H.H. Boer (Eds). North Holland Publishing Company, Amsterdam, pp. 173–175.

Fretter, V. 1941. The genital ducts of some British stenoglossan prosobranchs. *J. Mar. Biol. Ass. UK*, **25**, 173–211.

Gibbs, P.E. and G.W. Bryan. 1986. Reproductive failure in populations of the dog-whelk, *Nucella lapillus*, caused by imposex induced by tributyltin from antifouling paints. *J. Mar. Biol. Ass. UK*, **66**, 767–777.

Gibbs, P.E. and G.W. Bryan. 1987. TBT paints and the demise of the dog-whelk, *Nucella lapillus* (Gastropoda). In: Proceedings of the Oceans '87 International Organotin Symposium, Vol. 4, Institute of Electrical and Electronics Engineers, Piscataway, New Jersey, 28 September–1 October 1987, pp. 1482–1487.

Gibbs, P.E., G.W. Bryan, P.L. Pascoe and G.R. Burt. 1987. The use of the dog-whelk, *Nucella lapillus*, as an indicator of tributyltin (TBT) contamination. *J. Mar. Biol. Ass. UK*, **67**, 507–523.

Gibbs, P.E., P.L. Pascoe and G.R. Burt. 1988. Sex change in the female dog-whelk, *Nucella lapillus*, induced by tributyltin from antifouling paints. *J. Mar. Biol. Ass. UK*, **68**, 715–731.

Gibbs, P.E., G.W. Bryan and P.L. Pascoe. 1991. TBT-induced imposex in the dogwhelk, *Nucella lapillus*: geographical uniformity of the response and effects. *Mar. Environ. Res.*, **32**, 79–87.

Miller, E.R. and J.S. Pondick. 1984. Heavy metals in *Nucella lapillus* (Gastropoda: Prosobranchia) from sites with normal and penis-bearing females from New England. *Bull Environ. Contam. Toxicol.*, **33**, 612–620.

Moore, H.B. 1938. The biology of *Purpura lapillus*. II. Growth. *J. Mar. Biol. Ass. UK*, **23**, 57–66.

Pelseneer, P. 1926. La proportion rélative des sexes chez les animaux et particulièrement chez les mollusques. *Mémoires de l'Académie royale de Belgique. Classe des Sciences*, **8**, 1–258.

Peña, J., M. Guerra, M.J. Gaudencio and M. Kendall. 1988. The occurrence of imposex in the gastropod *Nucella lapillus* at sites in Spain and Portugal. *Lurralde*, **11**, 445–451.

Short, J.W., S.D. Rice, C.C. Brodersen and W.B. Stickle. 1989. Occurrence of tri-n-butyltin caused imposex in the North Pacific marine snail *Nucella lima* in Auke Bay, Alaska. *Mar. Biol.*, **102**, 291–297.

Smith, B.S. 1971. Sexuality in the American mud-snail *Nassarius obsoletus* Say. *Proc. Malacolog. Soc. London*, **39**, 377–378.

Smith, B.S. 1980. The estuarine mud snail, *Nassarius obsoletus*: abnormalities in the reproductive system. *J Mollusc. Stud.*, **46**, 247–256.

Smith, B.S. 1981a. Reproductive anomalies in stenoglossan snails related to pollution from marinas. *J. Appl. Toxicol.*, **1**, 15–21.

Smith, B.S. 1981b. Male characteristics on female mud snails by antifouling bottom paints. *J. Appl. Toxicol.*, **1**, 22–25.

Smith, B.S. 1981c. Tributyltin compounds induce male characteristics on female mud snails *Nassarius obsoletus* = *Ilyanassa obsoleta*. *J. Appl. Toxicol.*, **1**, 141–144.

Spence, S.K., S.J. Hawkins and R.S. Santos. 1990. The mollusc *Thais haemastoma* – an exhibitor of 'imposex' and potential indicator of tributyltin pollution. *Marine Ecology, Naples*, **11**, 147–156.

Spooner, N., P.E. Gibbs, G.W. Bryan and L.J. Goad. 1991. The effect of tributyltin upon steroid titres in the female dogwhelk, *Nucella lapillus*, and the development of imposex, *Mar. Environ. Res.*, **32**, 37–49.

Waldock, M.J., J.E. Thain and M.E. Waite. 1987. The distribution and potential toxic effects of TBT in UK estuaries during 1986. *Appl. Organometal. Chem.*, **1**, 287–301.

Walker, M. and H.C. MacGregor. 1968. Spermatogenesis and the structure of the mature sperm in *Nucella lapillus* (L.) *J. Cell Sci.*, **3**, 95–104.

Ward, G.S., G.C. Cramm, P.R. Parrish, H. Trachman and A. Slesinger. 1981. Bioaccumulation and chronic toxicity of bis (tributyltin) oxide: tests with a saltwater fish. In: *Aquatic Toxicity and Hazard Assessment*, D.R. Branson and K.L. Dickson (Eds). Associate Committee on Scientific Criteria for Environmental Quality, Philadelphia, pp. 183–200.

FLOWTHROUGH BIOASSAY STUDIES ON THE EFFECTS OF ANTIFOULING TBT LEACHATES

14

R. Scott Henderson[1] *and Sandra M. Salazar*[2]

[1]*Environmental Compliance and Protection Department, Building 216, Marine Corps Base Hawaii, Kaneohe Bay, Hawaii 96863-3002, USA*

[2]*EVS Consultants, West Mercer St., Seattle, Washington 98119, USA*

ABSTRACT

The effects of tributyltin (TBT) exposures ranging from 1.4 to 2500 ng l^{-1} were examined on a broad spectrum of shallow-water organisms in long-term flowthrough bioassay tests performed at Hawaii and San Diego sites. Tributyltin was obtained as leachate from panels coated with various antifouling paints. Mature fouling communities at Hawaii sustained major reductions in species and individual abundances at TBT exposures $\geq$500 ng l^{-1}. Taxa most severely affected were those of low phyletic level and species with a high surface area of soft integument. TBT concentrations $\geq$100 ng l^{-1} caused significant reductions in settling epifauna of low phyletic level. Hawaiian infauna experienced abundance reductions of ~ 50% at 500 ng l^{-1} TBT. Affected groups included nematodes, polychaetes, tanaids, isopods, bivalves, ophiuroids, and sipunculids. Abundances and mortality of benthic algae, macrocrustaceans, and fish were unaffected by TBT concentrations up to 2500 ng l^{-1}. Bioconcentration of TBT was found to be very low in swimming crabs and moderate in anchovy fish. Adult American oysters of the Hawaii experiments

Organotin. Edited by M.A. Champ and P.F. Seligman. Published in 1996 by Chapman & Hall, London. ISBN 0 412 58240 6

were unaffected by TBT exposures up to 1800 ng l^{-1} and showed TBT bioconcentration factor (BCF) values that were inversely dose dependent, ranging from 19 600 to 35 000 at TBT exposure levels of 82 and 2 ng l^{-1}, respectively. Reduced growth occurred in San Diego juvenile bay mussels at 70 ng l^{-1} TBT; and their BCF values were inversely dose dependent, ranging from 23 000 to 66 000 for TBT treatments of 450 to 6 ng l^{-1}, respectively. In spite of possible captivity-induced stress during San Diego tests, juvenile American, European, Olympic, and Pacific oysters showed no growth effects from TBT exposures $\leqslant$200 ng l^{-1}. Growth rates and survival of adult Pacific oysters during Hawaii tests were slightly reduced by TBT concentrations of 13 and 29 ng l^{-1}. However, those oysters may also have been nutritionally stressed as indicated by growth, survival, and condition indices of tank oyster controls, which were substantially lower than those of field controls. Pacific oyster BCF values were inversely dose dependent and ranged from 31 600 to 88 000 for TBT exposures of 29 to 14 ng l^{-1}, respectively.

14.1. INTRODUCTION

Prior to 1980, there were very few measurements of tributyltin (TBT) concentrations in natural waters (Hodge *et al.*, 1979; Maguire, 1987). Hall and Pinkney (1985) summarize the early TBT toxicity studies and indicate that most were conducted with freshwater organisms exposed to very high TBT concentrations (0.01–10.0 mg l^{-1}). The majority of those experiments were designed to determine organotin levels that would control snail and insect populations in fresh water without harming nontarget biota.

During the last 15 years, a major new thrust in organotin research has defined environmental concentrations and biological effects of TBT derived from antifouling paints. As these efforts have progressed, significant negative effects on aquatic biota have been measured at increasingly lower concentrations of TBT (Hall and Pinkney, 1985; Maguire, 1987 and Chapter 26). Additionally, many recent studies have delineated lower threshold levels of TBT that cause toxic effects in sensitive species of organisms of economic importance such as bivalve molluscs.

In this chapter we describe several bioassay experiments on TBT effects performed since 1980 by our laboratory (Naval Ocean Systems Center). Benthic organisms, which play a major role in the biological structure and dynamics of bays and harbors, were major targets for toxic effects and bioaccumulation studies. Experimental designs of these tests followed the temporal trends described above. Early experiments exposed a broad spectrum of organisms to relatively high TBT concentrations in the range of 2500–500 ng l^{-1}. As it became apparent that most taxa of test biota showed a significant toxic response to those concentrations, exposure levels for subsequent experiments were decreased to a range of 200–5 ng l^{-1}.

These experiments were performed in part using microcosms, defined here as contained and simplified aquatic systems that can be experimentally manipulated in various ways. These systems are provided with continuous flowthrough of unfiltered seawater and are maintained in open sunlight. As such, they are linked energetically to the natural world, receiving input of ambient sunlight and nutrients, and are subjected to varying degrees of colonization by larvae entering in the supply water. In microcosms, certain organisms can subsist on available natural foods and can experience near-natural seral development and growth. Chronic studies performed under such conditions provide a more realistic test of overall toxicity and bioaccumulation of pollutants than tests performed in static settings. Furthermore, extreme variability of concentrations of dissolved toxins (such as TBT), commonly experienced in static tests due to in-tank degradation and adsorption of toxin, is largely avoided by continuous addition of toxin to water of flowthrough tests.

Microcosms do not mimic complex natural ecosystems precisely, but are composed of interacting biotic and abiotic components and processes that are representative of subsets of larger systems. The term 'synthetic microcosms' has been suggested as appropriate for many of these captive ecosystems (Taub and Crow, 1980) because their structural and functional characteristics have been modified by factors such as effect of enclosure (which alters water motion and food availability, excludes certain predators, and increases substrate area-to-water volume ratios) and intentional introduction of specific organisms. The bioassay communities used in most of the studies reported below are extremely modified synthetic microcosms dominated by a very small number (usually 1–4) of species of organisms selected for fisheries and economic importance.

14.2. EXPERIMENTAL DESIGNS AND METHODS

Two different types of flowthrough seawater systems were used in these experiments (Fig. 14.1). Dilutions of antifouling paint leachate were obtained by either placing painted panels directly into exposure tanks or by providing water with specific leachate concentrations to exposure tanks from a toxicant diluter and flow control unit that received leachate from a separate leaching tank. Tributyltin content of control and leachate exposure waters was measured 5–47 times in the course of each experiment. Details of the experimental designs and analytical methods are available in Henderson and Smith (1980), Henderson (1986, 1988), Salazar *et al.* (1987), and Table 14.1.

Condition index was used in some experiments as a measure of relative health of adult bivalves and is defined as the bivalve meat wet weight (in grams) divided by the bivalve shell cavity volume (in milliliters) multiplied by 100. Wet weight of meat was used instead of dry weight to avoid possible loss of butyltins from the tissues. The status of developing gametes in individual mussels was defined by gonad index, a ratio of wet weight of mantle tissue to total wet weight of soft tissue.

In some experiments, epifauna typical of the test site were introduced to the experimental tanks as mature fouling communities that had settled on plastic panels suspended in harbor waters for several weeks or were allowed to colonize clean plastic panels located in the experimental tanks. These two panel types are defined as 'pre-fouled' and 'settlement' panels, respectively. Infauna, present in two of the studies, were harbored in sediment-filled trays that covered 75% of each tank bottom. Details of epifauna and infauna exposure and sampling are provided in references cited in Table 14.1.

Biological data were analyzed using balanced or unbalanced analysis of variance and significance tests. Methods used were ANOVA, Student's *t*-test, Duncan's new multiple range test and Bonferroni's multiple range test. Probability values ($p < x.xx$) are noted for significant differences.

14.3. EXPERIMENTAL RESULTS

14.3.1. MOKAPU EXPERIMENT 1

This experiment compared the biological effects of leachates derived from equal surface areas of two different antifouling paints. One paint (hereafter called copper paint) contained only cuprous oxide toxicant; the other paint (hereafter called TBT paint) contained TBT oxide and cuprous thiocyanate toxins. Exposure assessments of TBT paint leachate and copper paint leachate will be referred to as TBT and copper exposure experiments, respectively. Leachate concentrations in tank assays were 5600 ng l^{-1} Cu for the copper paint and 900 ng l^{-1} TBT and 1300 ng l^{-1} Cu for the TBT paint. Leach rates were

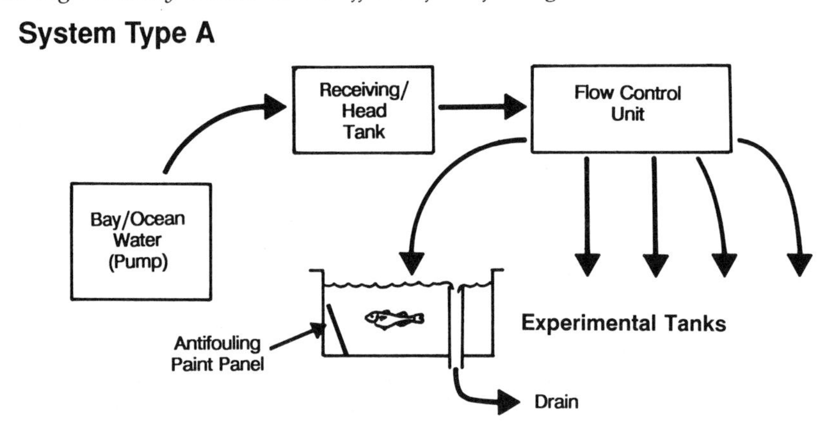

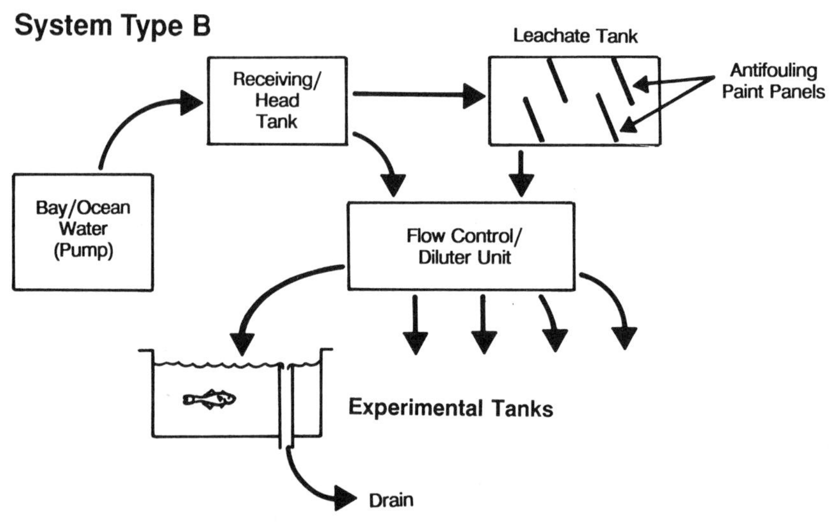

Figure 14.1 Flowthrough bioassay systems schematics for leachate treatment additions (A) using in-tank paint leaching and (B) using external paint leaching tank with diluter unit.

calculated at 0.9 μg $(cm^2)^{-1}$ d^{-1} TBT for the TBT paint and 3.7 μg Cu $(cm^2)^{-1}$ d^{-1} for the copper paint.

At the end of leachate exposure, algal coverage in the TBT-treated tanks was dominated by thick growth of filamentous green alga (*Cladophora socialis*) coated with diatoms. In comparison, algae in the control and copper-treated microcosms were sparse and consisted of diverse assemblages of blue–green, green, brown, and red algae. Benthic diatoms were moderately common in control tanks, and were very abundant in copper-treated tanks where they occurred in uniform, snow-like coatings over attached algae. Macroalgae survived in abundance in the control and copper-treated tanks, but were rare in the TBT-treated tanks. Only a few thalli of a fleshy, brown alga (*Padina japonica*) and a branching brown alga (*Acanthophora spicifera*) were found in TBT exposure experiments.

Approximate abundances of brown anemones (*Aiptasia pulchella*) and sea hares

(*Stylocheilus longicauda*) were monitored as they were usually too numerous and mobile to be counted accurately. At the end of exposure, no live brown anemones were observed in TBT exposure experiments, whereas several hundred were present in control and copper-treated tanks. Less than 10 sea hares were seen in TBT tanks during the exposure, whereas 50–100 were noted in each control and copper tank. High sensitivity of sea hare larvae to TBT leachates may have caused the low abundance of this organism in TBT tanks. Surviving adults appeared healthy and normal.

Two of six sea cucumbers died in the TBT exposure experiments, two developed necrotic lesions, and all lost weight during TBT exposure. All sea cucumbers in control and copper tanks survived and steadily increased in size. Animals that appeared normal in TBT exposure experiments included goby fish, palaemonid shrimp, swimming crabs, zoo-anthid soft corals, and several species of encrusting sponges.

In TBT exposure experiments, all bush coral polyps died after 13 d of exposure, and at 27 d only 11% of wart coral polyps were alive (Fig. 14.2a). Survival of corals exposed to copper paint leachates for 26 d was significantly lower than the survival rates of control animals, but was significantly higher than survival of TBT-exposed corals (Fig. 14.2a). The two coral species showed differing sublethal responses to copper exposure. All surviving bush coral polyps were normally distended and pigmented; whereas most live wart coral polyps were continuously retracted and stressed.

Abundances of all five major infaunal taxa inventoried in TBT exposure experiments were drastically reduced as compared to control abundances (Table 14.2), with the reductions being highly significant ($p < 0.01$) in seven of the nine cases. In copper exposure experiments, significant reductions of infauna occurred in only four of nine cases, and silt-dwelling molluscs actually showed a significant increase in abundance over control values (Table 14.2). Overall reduction of infaunal abundance in TBT tanks was 2.5 times higher than in copper tanks (−84% versus −34%).

Abundances of all naturally recruited epifauna and infauna were similar to those of controls after 104 d of post-exposure recovery.

14.3.2. MOKAPU EXPERIMENT 2

Exposure levels of TBT ranging from 510 to 1700 ng l^{-1} were produced in this experiment by varying the residence time (aging) of paint leachates in supply seawater before it entered the test tanks and by adding particulate organic material (phytoplankton) to tank waters of one exposure experiment (Table 14.1; Henderson, 1988).

After 28 d of leachate exposure, epibiota exposed to 510 ng l^{-1} displayed the lowest overall levels of toxic effects. Several brown anemones, sea hares, and gastropod molluscs were present in 510 ng l^{-1} tanks; whereas all individuals of those species had died in other leachate exposure experiments. Goby fish in all leachate exposure experiments were emaciated in comparison with control goby fish, suggesting that leachate exposure experiments had reduced supply of their preferred food (microcrustaceans) or had caused reductions in their feeding activities. Sea cucumbers ceased normal feeding activity in leachate tanks, and at least one sea cucumber died in each exposure group. The approximate abundances of 12 species of common macroalgae were similar in all control and TBT tanks.

Significant reductions in abundance of infauna were measured in three or four taxa in each leachate exposure experiment following TBT exposure (Table 14.3) and overall reductions were greatest in exposures to 1400 and 1700 ng l^{-1}. Polychaetes and tanaids showed highest sensitivity to TBT, experiencing significant declines in abundance in all TBT exposure experiments. Abundances of

Table 14.1 Experimental system characteristics and experimental design details for flowthrough bioassay studies

Experiment designation, location[a], type system[b], and (references)	*Primary toxin and (leachate source)[c]*	*Number of replicate tanks at measured toxin concentration (ng l^{-1})*	*Tank volume, flow rate, and 90% volume exchange rate*	*Test dates, duration of exposure (E), and recovery (R) phases*	*Test organisms[d] and test methods[e]*
Mokapu Experiment 1, Mokapu, System A, (Henderson and Smith, 1980; Henderson, 1988)	TBTO (1) or Cu_2O (2) (comparison experiment)	3 @ control 3 @ 900 TBT 3 @ 5600 Cu	465 l, 7 l min^{-1}, 2.3 h	10/80–11/81 93 d (E) 104 d (R)	Tank-recruited epibiota – a,h,m,s Tank-recruited infauna – a Goby fish (*Bathygobius fuscus*) [7A] – h,m Sea cucumber (*Holothuria edulis*) [3A] – h,m Wart coral (*Monitpora verrucosa*) [7C] – h,m Bush coral (*Pocillopora damicornis*) [7C] – h,m
Mokapu Experiment 2, Mokapu, System A, (Henderson and Smith, 1980; Henderson, 1988)	TBT methacrylate (3)	3 @ TBT (control) 3 @ 510 TBT 3 @ 1400 TBT 3 @ 1700 TBT	465 l, 7 l min^{-1}, 2.3 h	12/81–2/83 105 d (E) 105 d (R)	Tank-recruited epibiota – a,h,m Tank-recruited infauna – a Goby fish (*B. fuscus*) [7A] – h,m Sea cucumber (*H. edulis*) [3A] – h,m Wart coral (*M. verrucosa*) [7C] – h,m Bush coral (*P. damicornis*) [7C] – h,m
Pearl Harbor Experiment 1, Pearl Harbor System A, (Henderson, 1986)	TBT methacrylate (3)	*TBT* *Cu* 3 @ 10 (control) 1000 3 @ 40 3 @ 100 1200 3 @ 540 1200 3 @ 1800 3100 3 @ 2500 4400 5000	155 l, 4 l min^{-1} 1.4 h	6/84–11/84 60 d (E) 60 d (R)	Field-colonized epibiota – a,s Tank-colonized epibiota – a,s American oyster (*Crassostrea virginica*) [20A] – c,m
San Diego Experiment 1, San Diego Bay, System B, (Salazar *et al.*, 1987)	TBT methacrylate (4)	3 @ 10 TBT (control) 3 @ 70 TBT 3 @ 70 TBT 3 @ 80 TBT 3 @ 200 TBT	340 l, 3 l min^{-1}, 2.2 h	5/86–12/86 196 d (E) 0 d (R)	Bay mussel (*Mytilus edulis*) [16J] – gr

San Diego Experiment 2, San Diego Bay, System B, (Salazar *et al.*, 1987)	TBT methacrylate (4)	3 @ 10 TBT (control) 3 @ 80 TBT 3 @ 90 TBT 3 @ 200 TBT	340 l, 3 l min^{-1}, 2.2 h	5/86–9/86 110 d (E) 0 d (R)	Field-colonized epibiota – a Tank-colonized epibiota – a Bay mussel (*M. edulis*) [83A] – c,b,g,m Bent-nose clam (*Macoma nasuta*) [100A/S] – c,m,b
San Diego Experiment 3, San Diego Bay, System B, (Salazar *et al.*, 1987)	TBT methacrylate (4)	3 @ 10 TBT (control) 3 @ 40 TBT 3 @ 50 TBT 3 @ 160 TBT	340 l, 3 l min^{-1}, 2.2 h	10/86–12/86 56 d (E) 0 d (R)	American oyster (*C. virginica*) [15J] – gr Pacific oyster (*C. gigas*) [18J] – gr European oyster (*Ostrea edulis*) [18J] – gr Olympia oyster (*O. lurida*) [18J] – gr Bay mussel (*Mytilus edulis*) [18J] – gr
Pearl Harbor Experiment 2, Pearl Harbor, System B	TBT methacrylate (4)	3 @ 2 TBT (control) 3 @ 11 TBT 3 @ 32 TBT 3 @ 82 TBT	1500 l, 22 l min^{-1}, 2.5 h	1/88–7/88 191 d (E) 0 d (R)	Anchovy fish (*Stolephorus purpureus*) [25A] – b,m Swimming crabs (*Thalamita admete*) [8A] – b,m American oyster (*C. virginica*) [25A] – b,c,m
Pearl Harbor Experiment 3, Pearl Harbor System B	TBT methacrylate (4)	3 @ 1.4 TBT (control) 3 @ 4.9 TBT 3 @ 13 TBT 3 @ 29 TBT	1500 l, 22 l min^{-1}, 2.5 h	8/88–12/88 148 d (E) 0 d (R)	Pacific oyster (*C. gigas*) [28J] – b,c,gr,m

[a]Location: Mokapu is Mokapu Peninsula (open coast), Kaneohe, Hawaii; Pearl Harbor is Ford Island, Pearl Harbor, Hawaii; and San Diego Bay is Central San Diego Bay, California.
[b]Type system: see Fig. 14.1 for schematic of Type A and Type B flowthrough systems.
[c]Leachate source: antifouling paint (below) applied to plastic panels.

Key no.	*Formulation*	*Manufacturer*	*Toxin content (by weight)*
(1)	SPC-4	International Paint Co.	11.7% bis(tri-n-butyltin)oxide 17.0% cuprous thiocyanate
(2)	F121/63	Military specification	70% cuprous oxide
(3)	OMP-253	US Navy-developed paint	22.6% tributyltin methacrylate
(4)	SPC-9/BFA956	International Paint Co.	9.4% tributyltin methacrylate 0.5% bis(tri-n-butyltin)oxide 44.7% cuprous oxide

[d]Test organism: [ ] = number of organisms per tank; A = adult, S = subadult, J = juvenile, and C = coral colony.
[e]Test methods: a = abundance, b = bioaccumulation, c = condition index, g = gonad index, gr = growth rate, h = health, m = mortality, and s = species diversity.

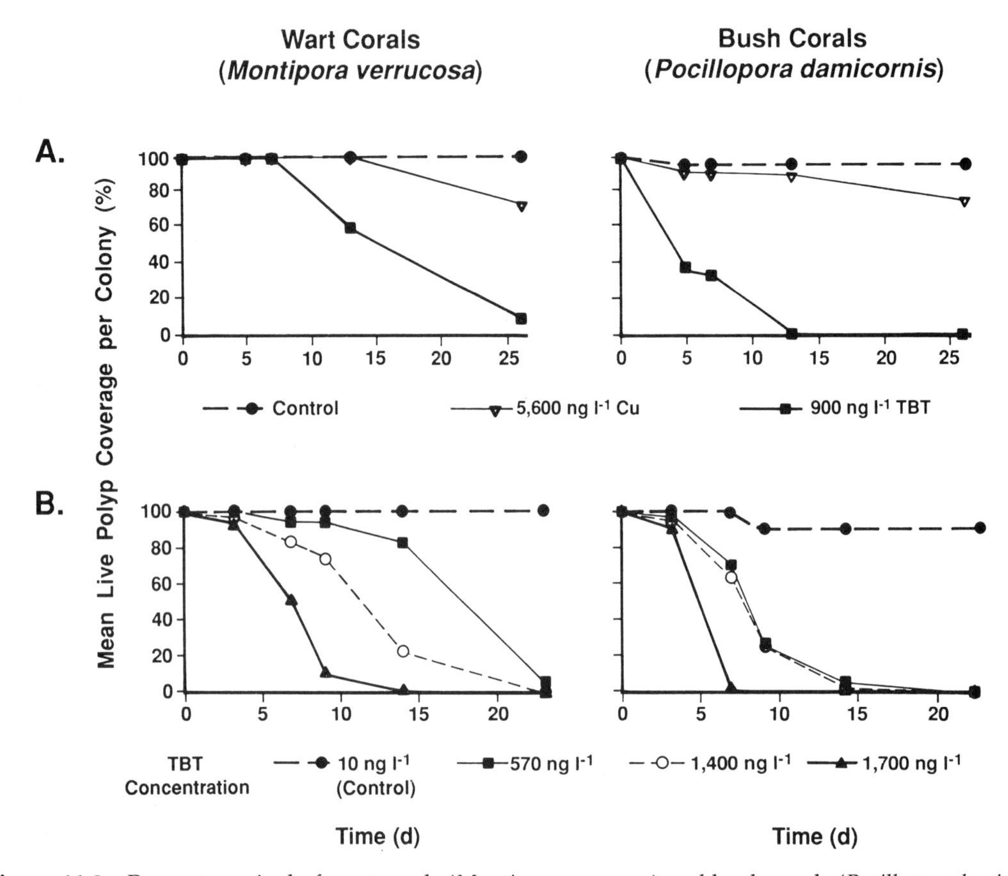

Figure 14.2 Percent survival of wart corals (*Montipora verrucoas*) and bush corals (*Pocillopora damicornis*) during (A) Mokapu experiment 1, and (B) Mokapu experiment 2.

Table 14.2 Abundance (number per m^2, ± 95% confidence limits) of infauna sampled at end of exposure from rubble and silt sediments in the tanks of Mokapu experiment 1

Taxa	*Sediment type*	*Control tanks*[a]	*Copper tank*[a] *(5600 ng l^{-1} Cu)*	*TBT tanks*[a] *(900 ng l^{-1} TBT)*
Polychaetes	Rubble	2 689 ± 2 258	2 318 ± 2 927	167 ± 206[b]
	Silt	13 591 ± 6 331	5 118 ± 3 286[b]	1 836 ± 727[c]
Nematodes	Rubble	575 ± 239	241 ± 250[b]	148 ± 205[c]
	Silt	6 453 ± 2 155	2 800 ± 421[c]	668 ± 461[c]
Tanaids	Rubble	11 238 ± 2 706	11 108 ± 7 132	3 208 ± 2 634[c]
	Silt	11 423 ± 4 917	4 414 ± 1 767[c]	1 372 ± 1 129[c]
Molluscs	Rubble	1 298 ± 864	4 043 ± 3 281	111 ± 148[c]
Ophiuroids	Rubble	195 ± 128	336 ± 315	0 ± 0[c]
	Silt	0 ± 0	0 ± 0	0 ± 0

[a]Number of sediment samples = 6.
[b]Significantly different from control at $p < 0.05$.
[c]Significantly different from control at $p < 0.01$.

Table 14.3 Abundance (number per m^2, ± 95% confidence interval) of infauna sampled at end of the exposure experiment from silt sediments in tanks of Mokapu experiment 2

Taxa	TBT treatment (ng l^{-1})			
	Control[a]	510[b]	1 400[b]	1 700[b]
Nematoda	3 733 ± 2 024	7 705 ± 6 913	1 219 ± 1 343[c]	3 166 ± 171
Polychaeta	25 623 ± 8 486	5 780 ± 5 590[d]	1 779 ± 1 932[d]	7 266 ± 701[d]
Tanaidacea	17 139 ± 9 072	5 790 ± 3 634[c]	1 048 ± 962[d]	2 397 ± 1 711[d]
Isopoda	384 ± 232	60 ± 50[c]	343 ± 543	74 ± 34[c]
Gammaridea	191 ± 141	352 ± 580	250 ± 179	143 ± 155
Copepoda	921 ± 1 729	570 ± 939	454 ± 678	74 ± 98
Gastropoda	1 286 ± 992	576 ± 557	49 ± 72[c]	1 275 ± 1 929
Sipuncula	1 292 ± 2 280	23 ± 29	14 ± 23	19 ± 47

[a]Number of sediment samples = 9.
[b]Number of sediment samples = 6.
[c]Significantly different from control at $p < 0.05$.
[d]Significantly different from control at $p < 0.01$.

all infaunal taxa in TBT tanks recovered to control levels 105 d after termination of exposure.

Because coarser sediments were not present in tanks of Mokapu experiment 2, ophiuroids and bivalves were not as common in those tanks as they were in Mokapu experiment 1. However, at the beginning of the exposure, several dead ophiuroids and bivalves were observed in TBT tanks, and no live individuals of those phyla were seen in any leachate tanks.

Wart corals and bush corals exhibited non-distension of feeding tentacles and pigmentation loss after 2 d of exposure in all concentrations of leachate. By day 23 of exposure, all leachate exposed corals had died, except for wart corals exposed to 510 ng l^{-1} that showed 6% live tissue (Fig. 14.2b). Mortality patterns of both species of corals are seen to be sensitive dose-dependent indicators of TBT concentrations. Tributyltin LD_{50} values of 500 ng l^{-1} occur at 17 and 8 d for wart corals and bush corals, respectively.

14.3.3. PEARL HARBOR EXPERIMENT 1

This experiment was performed in the estuary of Pearl Harbor, Hawaii, where TBT effects on harbor-type organisms could be evaluated in waters containing relatively high levels of dissolved and particulate materials. Tributyltin exposure concentrations ranged from 40 to 2500 ng l^{-1} (Table 14.1).

Total numbers of faunal species and species diversity of fouling communities exposed to 540–2500 ng l^{-1} TBT experienced precipitous declines during leachate exposure (Fig. 14.3) and were significantly different ($p < 0.01$) from controls at the end of the 60-d exposure. After 60 d of recovery, numbers and diversities of faunal species in all exposure experiments were statistically similar to controls. Overall effects on the pre-fouled panel organisms are typified by the mortality data for the most common epifaunal species, which show that four of those six taxa did not survive exposure to 540 ng l^{-1} TBT (Table 14.4).

Twenty taxa of epifauna were present in all tanks on pre-fouled panels and other substrates prior to leachate exposure. At the end of exposure, one of those taxa was absent in the 40 ng l^{-1} exposure experiment and two taxa were absent in the 100 ng l^{-1} exposure experiment. However, 60-d exposure to 540, 1800, and 2500 ng l^{-1} TBT caused common taxa reductions of 55, 60, and 80%, respectively. The only animals observed in the 2500

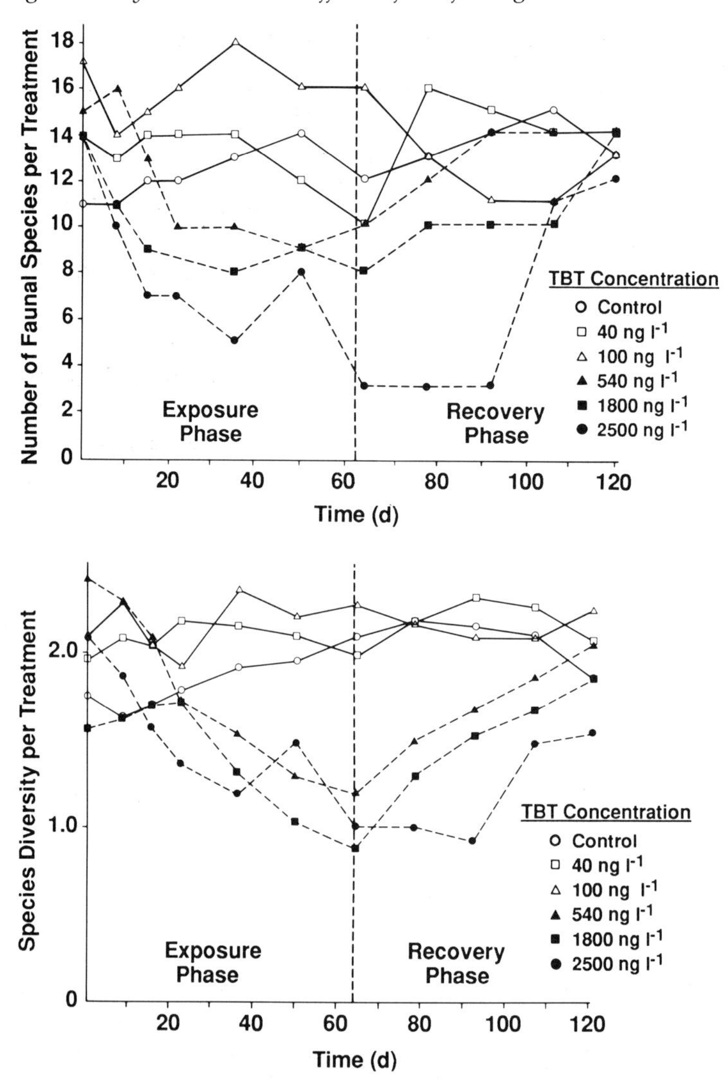

Figure 14.3 Total number of species and species diversity (modified Shannon–Weiner index) of fauna on pre-fouled panels of Pearl Habor experiment 1 plotted versus time.

ng l^{-1} exposure experiment at the end of exposure were the anemone *Haliplanella luciae*, some *Ascidia* spp. solitary tunicates, the tube worms *Hydroides elegans* and *Pileolaria militaris*, and the tube-dwelling gastropod *Vermetus alii*. Species compositions of fouling communities in all tanks were similar at the end of recovery.

The total number of species and species diversity of organisms that colonized settlement panels in 100–2500 ng l^{-1} exposure experiments during exposure were significantly lower ($p < 0.01$) than control and 40 ng l^{-1} exposure tanks. Post-recovery settlement of epifauna in tanks previously exposed to TBT was not significantly different from settlement in control tanks. Presence-or-absence data for the 17 taxa identified on settlement panels indicated that phyla of lower evolutionary level (e.g. Porifera, Cnidaria,

Table 14.4 Mortality summary for the most common species on pre-fouled panels of Pearl Harbor experiment 1

	TBT exposure (ng l^{-1})					
	Control	*40*	*100*	*500*	*1800*	*2500*
Botrylloides spp.[a] (orange colonial tunicate)			100% mortality[b]			
Schizoporella errata[a] (encrusting bryozoan)			49%	100% mortality[b]	84%	100%
Didemnum candidum[c] (white colonial tunicate)				100% mortality[b]		
Anomia nobilis[d] (saddle oyster)				100% mortality[b]		
Hydroides elegans[d] (tube worm)					77%	68%
Ascidia spp.[a] (solitary tunicates)						91%

[a]Significant difference at $p<0.01$.
[b]Any two treatments not underscored by the same line are significantly different; any two exposure concentrations underscored by the same line are not significantly different.
[c]Significant difference at $p<0.05$.
[d]Significant differences not evident in ANOVA analysis of fouling panel data because of sparse control panel populations; indicated mortalities obtained by comparison with control populations on other substrates.

Bryozoa, and Mollusca) were most sensitive to TBT exposure. Organisms least affected were those capable of complete enclosure in hard, tubular shells (e.g. the tube worms, *Hydroides* sp. and *Pileolaria* spp.) or in thick body walls (e.g. solitary tunicates, *Ascidia* spp.). *Hydroides elegans*, *Pileolaria militaris*, and *P. pseudomilitaris* were the only organisms that colonized panels in the 2500 ng l^{-1} exposure.

Survival of oysters (*Crassostrea virginica*) at TBT exposures of ≤1800 ng l^{-1} was similar to that of controls. However, the mortality rate of oysters exposed to 2500 ng l^{-1} TBT was 50 and 82% after 30 and 57 d of exposure, respectively. At the end of the exposure period, condition indices (CI) of oysters in TBT treatments ≥100 ng l^{-1} (Table 14.5) were significantly lower than control values ($p<0.01$) and showed significant dose-dependent responses. After 60 d of recovery, the CI of exposed oysters had returned to near-control levels (Table 14.5).

14.3.4. SAN DIEGO EXPERIMENTS 1, 2, AND 3

These studies were conducted in San Diego Bay and focused primarily on effects of TBT levels of 40–200 ng l^{-1} on bivalve molluscs (Table 14.1). Responses of fouling communities to TBT exposure were also examined in San Diego experiment 2.

Fouling organism abundances on the pre-fouled and settlement panels showed no significant differences among exposure values during exposure. Also, there were no significant among-exposure differences noted in biomass of pre-fouled panel biota. Biomass measurements were not made on the settlement panel organisms because of their very low abundances. Fourteen animal taxa were present on fouling panels exposed to the highest TBT exposure concentration of 200 ng l^{-1} (Table 14.6).

Table 14.5 Mean condition indices of *Crassostrea virginica* measured during Pearl Harbor experiment 1. A, After 57 d of TBT exposure; B, after 59 d of recovery following TBT exposure. The level of significance is given where applicable

(A) after 57 d of TBT exposure

	TBT (ng l⁻¹)					
	Control	*50*	*130*	*300*	*800*	*2000*
Mean condition index	5.0	5.0	3.3	2.9	1.9	1.3
Standard deviation (±)	3.4	2.9	1.2	1.7	0.6	0.5
Number of animals	18	16	18	18	16	17
Control		N[a]	<0.01	<0.01	<0.01	<0.01
50			<0.05	<0.01	<0.01	<0.01
130				N	<0.01	<0.01
300					<0.05	<0.05
800						<0.05

(B) after 59 d of recovery following TBT exposure

	TBT (ng l⁻¹)				
	Control	*50*	*130*	*300*	*800*
Mean condition index	4.9	3.9	3.3	3.6	4.1
Standard deviation (±)	2.4	2.4	1.6	1.4	1.4
Number of animals	19	13	18	22	20
Control		N	<0.05	N	N
50			N	N	N
130				N	N
300					N

[a]N indicates no significant difference between conditions.

Juvenile bay mussels of San Diego experiment 1 showed significant dose-dependent differences in weight increases after 156 d ofexposure (Fig. 14.4). Growth of bay mussels exposed to 70 and 80 ng l^{-1} TBT was statistically similar; but values for both were significantly lower than control values, while growth of bay mussels exposed to 200 ng l^{-1} was significantly lower than growth for all other TBT exposure concentrations and control groups (p <0.01).

Juvenile bay mussels exposed to slightly lower concentrations of TBT (40, 50, and 160 ng l^{-1}) for 56 d during San Diego experiment 3 showed no significant among-exposure differences in weight (Fig. 14.4). However, bay mussels of San Diego experiment 1 did not experience significant differences in weight increases among exposure values until the 60- to 136-d exposure interval when their growth rates increased markedly. Growth rates of bay mussels in both experiments were considerably lower than would be expected in nature as indicated by the high growth of field controls maintained in harbor water adjacent to the site during San Diego experiment 3 (Fig. 14.4).

Condition and gonad indices of adult bay mussels in San Diego experiment 2 steadily decreased in all control and leachate exposure experiments. Significant effects of TBT were noted only in exposure to 200 ng l^{-1} where the CI was less than the control on days 31, 47, and 80; and gonad indices were less than control on days 47 and 95 (p <0.01). Bent nose clams experienced high mortality and highly variable

Table 14.6 Epifauna present on pre-fouled and settlement panels after 68-d exposure to 200 ng l^{-1} TBT during San Diego Bay experiment 2

Pre-fouled panels	*Settlement panels*
Porifera sp. sponge	*Bryozoa* sp. bryozoan
Polychaete spp. polychaetes	*Spirorbis* sp. polychaete
Hydroides pacifica polychaete	*Ericthonius* sp. amphipod
Spirorbis sp. polychaete	Colonial tunicate sp.
Ericthonius sp. amphipod	
Leucothoe sp. amphipod	
Musculista senhousia mussel	
Mytilus edulis mussel	
Ostrea lurida oyster	
Ciona intestinalis tunicate	
Styela montereyensis tunicate	
Styela plicata tunicate	

CI, revealing no significant among-exposure differences in those parameters.

During San Diego experiment 3, the juvenile American, European, Olympia, and Pacific oysters showed no significant differences in growth rates among TBT exposures or controls. All oyster species except Olympic oysters showed growth rates of field control populations (in the harbor) that were significantly higher than corresponding tank populations. Shell thickening was observed, but not quantified, in all tank and field control populations of Pacific oysters. It is not known whether thickening was caused by exposure to TBT present in the ambient and leachate exposure waters or to an undefined factor of water quality.

Bay mussels and bent nose clams exposed to leachate in San Diego experiment 2 accumulated increasing amounts of TBT in their tissues over the first 60 d of exposure (Fig. 14.5). Over the final 50 d of the experiment, TBT body burdens of those bivalves attained approximate steady-state levels that were proportional to exposure concentrations. Steady-state TBT body burdens of mussels were nearly twice as high as those of clams from equivalent exposure experiments. Bioconcentration factor values (BCF, ratio of TBT body burden to TBT water concentration) for mussels and clams ranged from 10 400 to 70 000 (Fig. 14.5). For comparative purposes, TBT uptake data are also shown for bay mussels exposed to 450 ng l^{-1} TBT at Shelter Island marina in San Diego Bay. Mean steady-state BCF of those mussels (23 000) is similar to the BCF observed for the tank bay mussels exposed to 200 ng l^{-1} TBT.

14.3.5. PEARL HARBOR EXPERIMENT 2

This experiment examined the response of three organisms (anchovy fish, swimming crabs, and American oysters) of commercial or potential fisheries interest in Hawaii exposed to TBT concentrations of 11–82 ng l^{-1} (Table 14.1). No significant among-exposure differences in mortalities of those organisms were observed.

The CI values of American oysters exposed to 82 ng l^{-1} TBT for 30 d and to 32 and 82 ng l^{-1} TBT for 57 d were significantly lower than tank control CI values (Fig. 14.6). Through the last 124 d of the experiment, there were no significant differences among CI values of the tank population American oysters. However, the CI values of field control American oysters, maintained in harbor shallows fronting the experiment site, were in all cases significantly higher than those of all tank populations. A few American oysters from all control and exposure tanks were found to contain developing or ripe gametes; however, numbers of those individuals were too sparse for statistical comparisons.

American oysters exhibited TBT body burdens that were dose dependent (Fig. 14.6). After 122 d of TBT exposure, body burdens of populations exposed to TBT had apparently attained steady-state levels as they showed no consistent patterns of TBT accumulation or depuration. Mean steady-state BCF values for the 122- to 181-d exposures ranged from 19 600 to 36 300 (Fig. 14.6).

Body burdens of thoracic tissues from swimming crabs measured at 30, 57 and 91 d of exposure showed very low TBT uptake

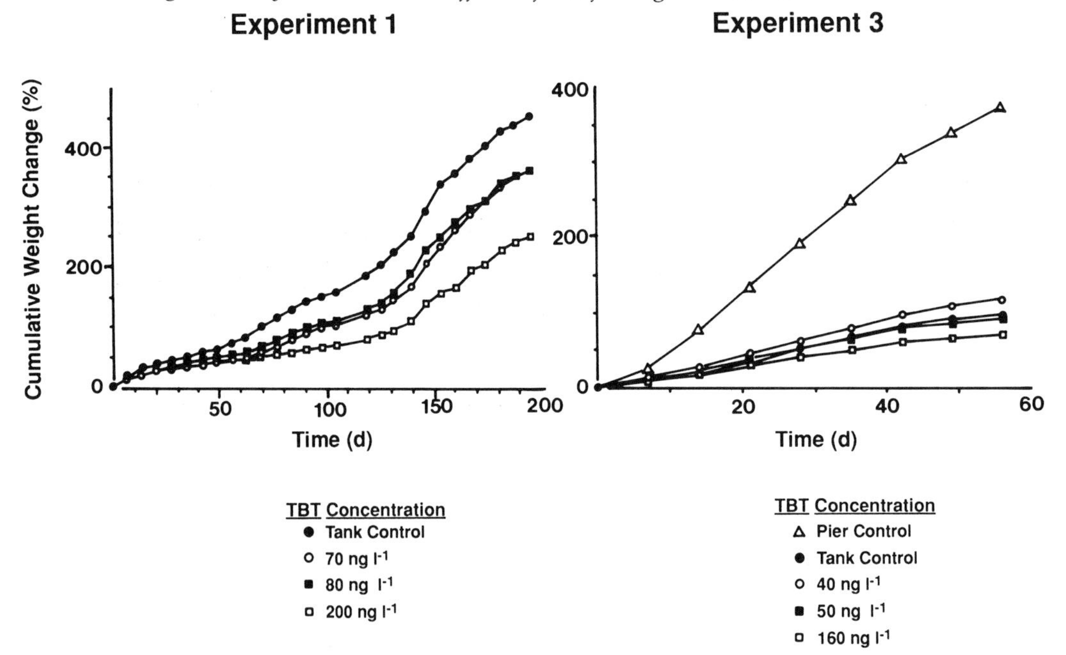

Figure 14.4 Cumulative percent weight change of juvenile bay mussels (*Mytilus edulis*) of San Diego experiments 1 and 3 plotted versus time.

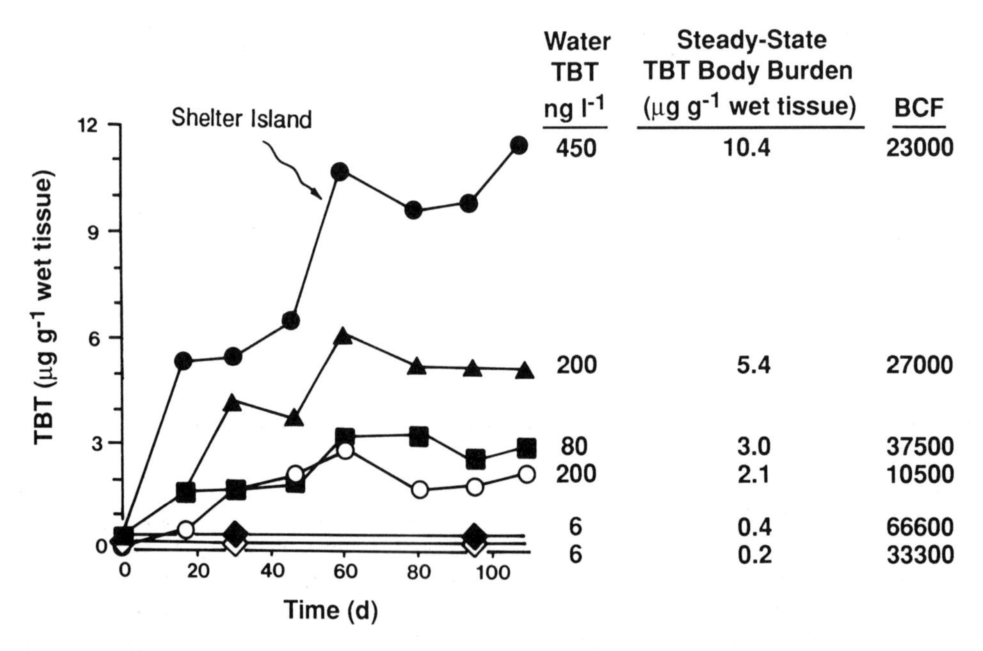

Figure 14.5 Tributyltin bioaccumulation versus water TBT content for adult bay mussels (*Mytilus edulis*) and subadult and adult bent nose clams (*Macoma nasuta*) in tanks and in Shelter Island marina during San Diego experiment 2. Also shown are steady-state TBT body burdens and bioconcentration factors (BCFs). Closed symbols = bay mussels; open symbols = bent nose clams.

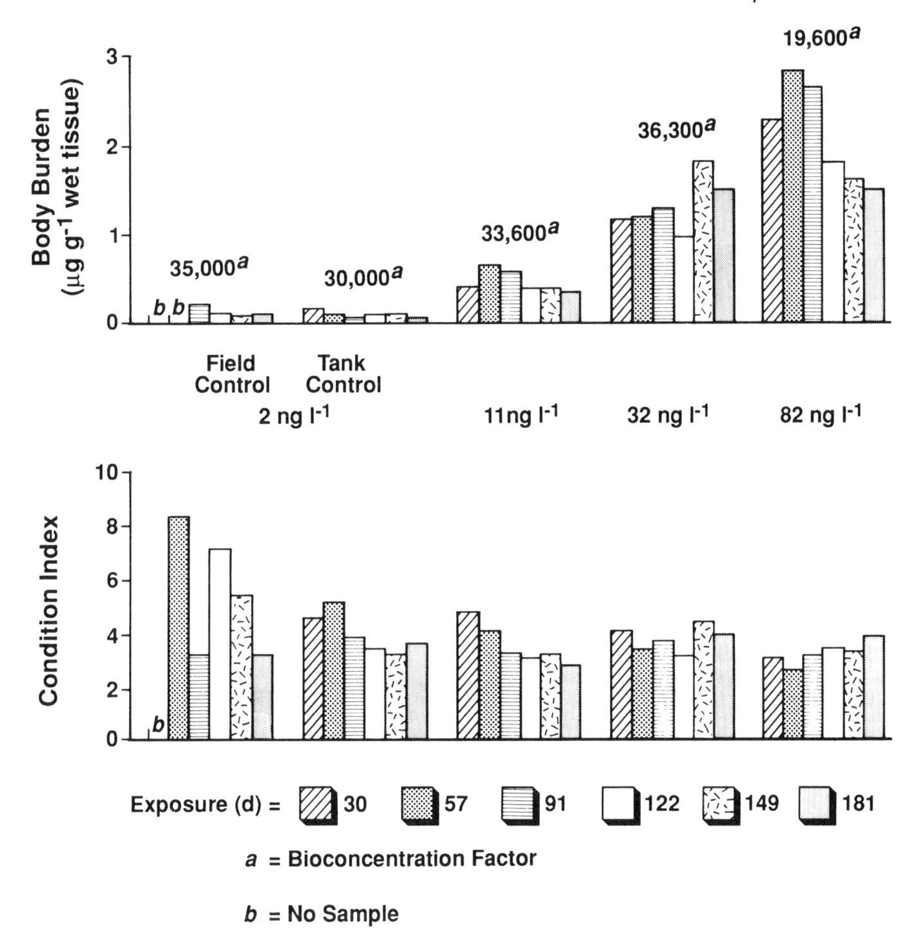

Figure 14.6 Mean TBT body burdens and condition indices of American oysters (*Crassostrea virginica*) of Pearl Harbor experiment 2.

levels and mean BCF values of ~300 (Fig. 14.7). Tributyltin was detected only in the crabs exposed for 30 d to 32 and 82 ng l^{-1} and in crabs exposed for 91 d to 82 ng l^{-1} TBT. Tributyltin may have been rapidly metabolized in crab tissues, as MBT plus DBT accounted for 83–100% of total butyltins present in the tissue samples. Additionally, concentrations of MBT and DBT were relatively low in leachate-treated crabs [averaging 0.18 μg g^{-1} tissue (wet weight)], indicating that those compounds may also be efficiently depurated from crab tissue.

Body burdens of TBT in fish were dose dependent and were relatively constant within exposure groups, showing no clear temporal changes (Fig. 14.7) and suggesting that uptake may have reached steady-state condition soon after commencement of exposure. Fish BCF values ranged from 15 000 to 25 000. No MBT or DBT was detected in control fish, whereas concentra-

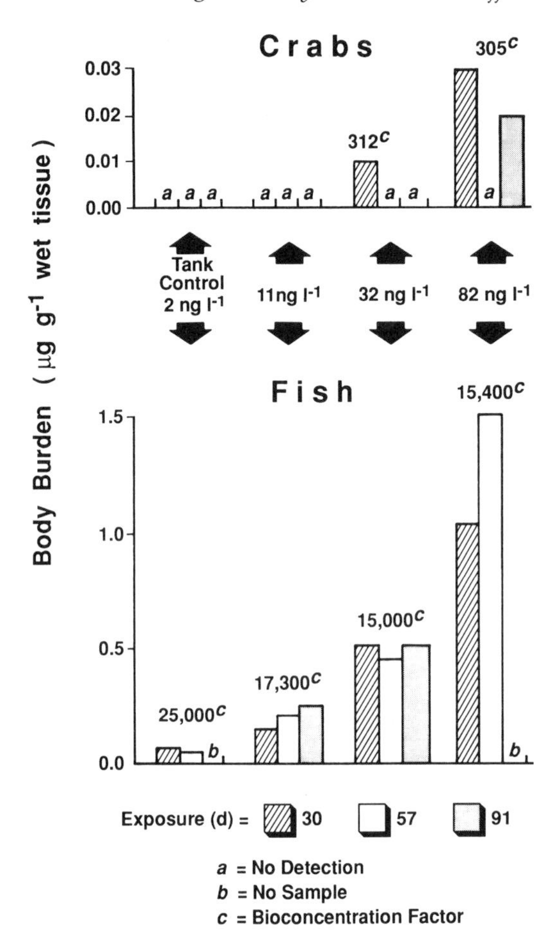

Figure 14.7 Mean TBT body burdens of swimming crabs (*Thalamita admete*) and anchovy fish (*Stolephorus purpureus*) of Pearl Harbor experiment 2.

tions of MBT plus DBT ranged from 16 to 24% of total butyltin in leachate-exposed fish.

14.3.6. PEARL HARBOR EXPERIMENT 3

Pacific oysters were exposed to TBT concentrations of 1.4–29 ng l^{-1} during this experiment (Table 14.1). Condition indices of oysters showed no significant differences between tank control or exposed populations. However, the CI of the field control oyster population was always significantly higher than the CI of tank oyster populations (Table 14.7). Tributyltin concentrations of 13 and 29 ng l^{-1} significantly reduced oyster growth rates below those of control levels, whereas oyster mortality was significantly higher than control level in only those oysters exposed to 29 ng l^{-1} TBT (Table 14.7).

Tributyltin body burdens of Pacific oysters were measured at 87, 120, and 148 d of exposure. Mean body burdens of oysters from the control and exposure tanks were dose dependent and relatively constant (Fig. 14.8). Mean BCF values calculated for those exposure intervals ranged from 31 600 to 88 000 and were inversely related to TBT dosage (Table 14.7 and Fig. 14.8).

14.4. DISCUSSION

Epifauna on the mature fouling panels of these experiments sustained significant reductions in species and individual abundances and species diversity at TBT exposures of ~500 ng l^{-1} and greater. Taxa that were most heavily affected included the saddle oyster (*Anomia nobilis*) and organisms with large surface areas of exposed soft integument (e.g. sponges; the brown anemone *Aiptasia pulchella*; colonial tunicates; and the encrusting bryozoan, *Schizoporella errata*). Significant effects on colonizing epifauna (larval and newly settled life stages) occurred primarily in taxa of low phyletic levels and were noted only at concentrations ≥100 ng l^{-1}.

The two species of coral tested demonstrated mortality rates that were dependent on TBT dosage, and both species sustained 100% mortality within 23 d at TBT concentrations ≥500 ng l^{-1}. No-effect levels of TBT were not determined for corals as they were not exposed to TBT concentrations <500 ng l^{-1}. Other benthic organisms that experienced high mortalities at TBT levels >500 ng l^{-1} were brown anemones (*Aiptasia pulchella*), sea cucumbers (*Holothuria edulis*), and sea hares (*Stylocheilus longicauda*).

We have observed that swimming crabs and palaemonid and alpheid shrimp routinely

Table 14.7 Mean growth rates (percentage of length increases), mean percentage of mortalities, mean condition indices, and mean bioconcentration factors (BCF) for Pacific oysters (*Crassostrea gigas*) of Pearl Harbor experiment 3

Oyster populations	*Mean TBT (ng l^{-1})*	*Mean growth rate (percentage of length increase)*	*Mean mortality (%)*	*Condition index*	*BCF*
Field control	1.4	13.6	4.3	6.3	88 000
Tank control	1.4	6.2	6.3	3.8	48 000
	4.9	5.2	4.8	3.9	41 000
	13	1.4[a]	7.3	3.4	32 100
	29	1.2[a]	24.5[b]	3.3	31 600

[a]Significantly different from tank control and other exposure levels at $p < 0.03$.
[b]Significantly different from tank control and other exposure levels at $p < 0.04$.

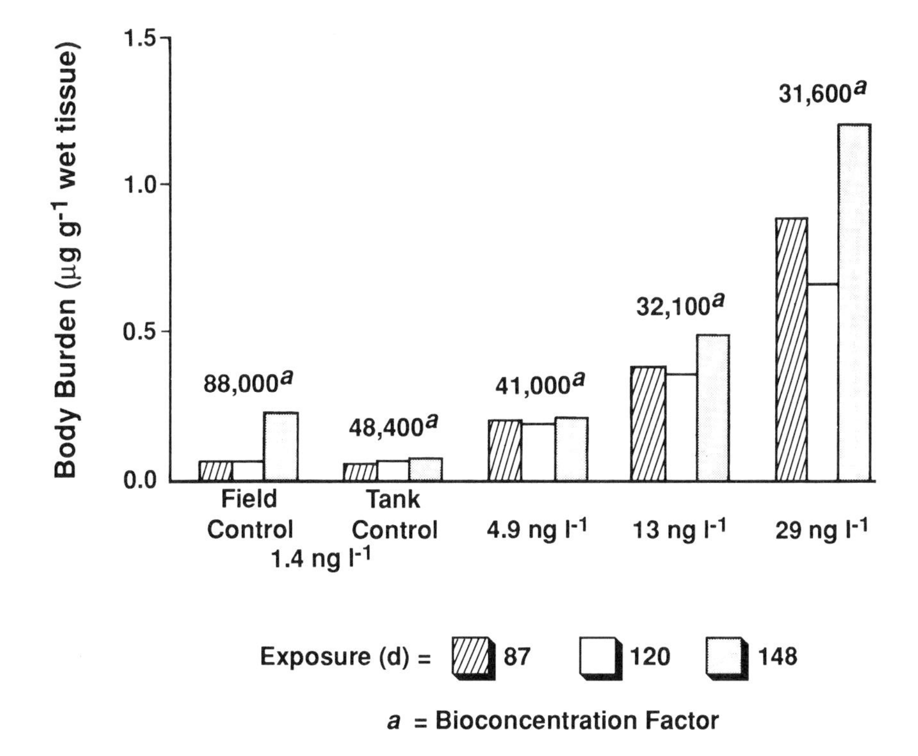

Figure 14.8 Mean TBT body burdens of Pacific oysters (*Crassostrea gigas*) of Pearl Harbor experiment 3.

settled and grew to adult size in all tanks of all of the Hawaii experiments during TBT exposures of 1.4–2500 ng l^{-1}. Females of those species were also commonly observed carrying fertile egg broods. Body burdens of butyltin and BCF values in crabs measured in TBT exposure concentrations up to 82 ng l^{-1} were consistently very low, reflecting the apparent ability of crabs to metabolize TBT efficiently, as also revealed in work by Lee (1986 and Chapter 18). Collectively, these data indicate that macrocrustaceans are extremely tolerant to TBT relative to most other invertebrate taxa.

Certain epifauna living in dense protective integuments or shells such as tube worms (*Hydroides* spp. and *Pileolaria* spp.), vermetid worms (*Vermetus* sp.), and solitary tunicates (*Ascidia* spp.) were observed to have long-term tolerance to TBT levels as high as 2500 ng l^{-1}. As noted by Chesher (1975), these types of organisms are capable of sensing and isolating themselves from pulses of copper-rich water in power plant effluents. It is likely that they may also reduce the long-term effects of TBT exposure using the same intermittent closure strategy. Even when not closed, organisms with thick, impermeable integuments would also be expected to assimilate much less TBT through skin than animals covered only by thin, membranous epidermis. Some organisms capable of closure, such as holothurians and gastropod and bivalve molluscs, are apparently unable to avoid harmful TBT exposure. Those animals may not be able to sense the presence of organotins or cannot tolerate extended periods of inactivity due to specific metabolic or feeding requirements.

Nematodes, polychaetes, tanaids, isopods, gastropods, bivalves, ophiuroids, and sipunculids experienced major infaunal population reductions at TBT exposures in the 500–900 ng l^{-1} range. As infauna were not exposed to TBT concentrations <500 ng l^{-1} in these experiments, it was not possible to define no-effect levels for the more sensitive infauna. However, from these data it would be expected that long-term exposures of typical harbor environs to TBT levels $\geqslant 500$ ng l^{-1} would result in common infaunal species and individual abundance declines of 50% or more.

Macroalgae sustained major reductions in diversity and abundance when exposed to a combined leachate of 900 ng l^{-1} TBT and 1300 ng l^{-1} Cu (as cuprous thiocyanate), but were observed in control abundances in TBT exposures of 1700 ng l^{-1} without copper leachate. Thus, the cuprous thiocyanate (a known algicide) released from the dual-toxin paint was probably responsible for negative effects on algae in those exposure experiments. Beaumont *et al.* (1987) have also documented lack of effects of relatively high TBT levels (1000–3000 ng l^{-1}) on benthic algae in flowthrough microcosms. The combined results of these experiments show that relatively high levels of TBT do not have significant negative effects on benthic algae.

Tank communities of epibiota and infauna that had been significantly impacted by 3-month exposures to TBT levels of 900–1700 ng l^{-1} fully recovered to normal diversity and abundance levels after 3 months of recovery. American oyster CI values that had been reduced by 2-month exposure to 800 ng l^{-1} also rapidly returned to control levels after the exposure experiment. Most benthic organisms, typified by those studied in the present experiments, live in close association with bottom sediments, using the sediment as habitat and, in some cases, as a source of food. However, as we observed no negative post-exposure effects on tank biota, it is apparent that TBT was either rapidly depurated from the tank sediments or was not biologically available. Thus, if the behavior of TBT in field sediments is similar to that experienced in our tank experiments, then harbor benthic communities would also recover rapidly from effects of chronic exposures to relatively high concentrations of TBT.

The overall biological effects of leachate derived from dual-toxin antifouling paint (containing TBT oxide and cuprous thiocyanate) were much greater than those of leachate derived from an equivalent area of cuprous oxide antifouling paint. While 8 of 13 faunal taxa were severely impacted by the dual-toxin paint leachate, only 3 of 13 taxa showed moderate negative response to cuprous oxide paint leachate. Although some of the impact of the dual-toxin paint leachate may have been attributable to a synergistic effect of the TBT and cuprous thiocyanate toxins, we suspect that most of that effect would have been incurred by flora rather than fauna, because of the relatively high susceptibility of algae to copper compounds.

Mortality rates of anchovy fish and goby fish were not affected by maximal TBT exposure concentrations of 82 and 1700 ng l^{-1}, respectively. However, the emaciated state of goby fish in the 1700 ng l^{-1} exposure experiment indicated that their survival over longer exposure periods would likely have been severely affected by reductions in abundance of their food prey by TBT poisoning. This effect of TBT on the food supply of goby fish is an example of a food-chain or 'indirect' effect that a toxin can have on an organism. These effects are usually not observed in acute static bioassay tests where normal food chains and predator–prey interactions are excluded.

The body burdens of anchovy fish reached dose-dependent, steady-state levels within ~1 month of exposure. The overall TBT body burdens and BCF values of anchovy fish, however, showed that their TBT depuration rates were considerably lower than those of swimming crabs of the same experiment.

Long-term survival of American oysters in these experiments was not affected by TBT exposures as high as 1800 ng l^{-1}. However, CI values of individuals exposed to TBT levels ⩾10 ng l^{-1} decreased below control levels for the first 2–3 months of exposure and then recovered to control levels during subsequent exposure. A similar trend was noted in TBT uptake by American oysters, wherein TBT body burdens increased over the first 3 months of exposure and thereafter remained fairly constant or declined.

These results indicate that the American oysters were capable of stabilizing or reducing their chronic uptake of TBT. In a 67-d flowthrough experiment, Valkirs *et al.* (1987) found that survival of subadult American oysters was unaffected by a TBT concentration of 1900 ng l^{-1}, whereas their CI value was decreased by TBT levels ⩾700 ng l^{-1}. Condition indices of oysters in that experiment were not affected by TBT concentrations ⩽300 ng l^{-1}. However, as oyster CI values were measured only at the end of exposure, temporal trends of that measure could not be examined.

Tributyltin BCF values for American oysters were several times higher than those obtained by Waldock *et al.* (1983) and Thain (1986) for Pacific oysters and European oysters exposed for 21–75 d to TBT levels of 200–2600 ng l^{-1}. These differences may be due to differing abilities of the two oyster species to metabolize butyltins or to differing degrees of biological availability of butyltins in the test waters. Reduced nutrition apparently had minimal effect on TBT uptake in tank populations of American oysters as indicated by the fact that mean BCF values of field control and tank control oysters were similar (35 000 and 30 000, respectively).

Juvenile bay mussels of San Diego experiment 1 experienced growth retardation when exposed to TBT concentrations ⩾70 ng l^{-1}. In other chronic flowthrough studies, the lowest TBT exposure of 240 ng l^{-1} used by Thain (1986) caused growth reductions in mussel spat, whereas growth of adult mussels (Valkirs *et al.*, 1987) was reduced by exposure to 310 ng l^{-1} TBT but was unaffected by a concentration of 130 ng l^{-1}. Furthermore, insitu studies at various locations in San Diego Bay (Salazar and Salazar, 1988) indicated that inhibited growth of juvenile mussels occurred in the field only at TBT levels ⩾200 ng l^{-1}. Salazar and Salazar (1988) note that the higher susceptibility of mussels to TBT exposure in San Diego experiment 1 may have been due to nutritional deprivation and temperature stress, possibly causing reduced growth at a TBT level below that which would produce the same effect in otherwise unstressed settings.

BCF values of mussels maintained in tank and field environments in the San Diego experiments, however, were similar ranging from 10 500 to 66 600 and from 5 300 to 44 000, respectively, for exposures to TBT levels of 10–160 ng l^{-1}. Thus, although growth rates and survival of mussels may have been significantly altered by experimental conditions, overall uptake of TBT occurred at about the same rates in tank and field populations.

In spite of possible captivity-induced stress, the four species of juvenile oysters (American, European, Olympia, and Pacific) in San Diego experiments showed no growth or survival abnormalities attributable to TBT exposures as high as 200 ng l^{-1}. In comparison, Thain and Waldock (1986) noted reduced growth of European oyster spat at 60 ng l^{-1} TBT, but found no growth effects on juveniles exposed to 240 ng l^{-1} TBT. They also reported growth reductions in Pacific oyster spat at TBT exposures of 240 ng l^{-1}, but saw no effect on American oyster spat at the same concentration. Among-test differences in responses of the same bivalve species are probably attributable to size-dependent sensitivities of bivalves to TBT and may also be related to differences in levels of captivity stress and amounts of bioavailable TBT in the various experimental environments (Salazar, 1986).

Thickening of Pacific oyster shells, which has been observed to occur at TBT concentrations of 20 ng l^{-1} (Thain *et al.*, 1987), was also noted in field control, tank control, and exposed Pacific oysters of the San Diego experiments. Tributyltin may have been responsible for the shell thickening as the mean ambient concentration of TBT was 10 ng l^{-1}, ranging from non-detectable to 19 ng l^{-1}. However, Key *et al.* (1976) have shown that other factors such as elevated levels of suspended sediments may also cause oyster shell thickening, and this may have been the case with the San Diego test animals.

Adult Pacific oysters of the Hawaii tests showed reduced growth rates at a TBT concentration of 13 ng l^{-1} and no effects at the next lowest exposure concentration of 4.9 ng l^{-1}. In comparison, Thain *et al.* (1987) reported that shell thickening occurred in juveniles of the same species at 20 ng l^{-1} but not at 2 ng l^{-1}. Possible TBT effects on shell thickening could not be effectively examined in the Hawaiian oysters as those individuals all had thickened shells when collected from Kaneohe Bay, where TBT levels would be expected to be extremely low. As early life stages of aquatic organisms are usually found to be more sensitive to toxins than adults of the same species, it is puzzling to note that Pacific oyster spat and juveniles of the San Diego experiments tolerated TBT levels that were 2–6 times higher than levels that caused negative effects on adults in the Pearl Harbor experiments. Again, these incongruities may be due to effects of unmeasured factors such as the stress levels of test organisms and TBT bioavailability that may have varied significantly among experiments.

Condition indices of the Hawaiian Pacific oysters were poor indicators of TBT stress, as they did not show dose-dependent changes as were seen in the growth rates and mortalities of those oysters. The mean CI value of the field control Pacific oysters of the Pearl Harbor experiments, however, was nearly twice as high as the mean tank control CI value. And mean rates of oyster growth and BCF values of field control oysters were nearly twice as high as tank controls. Thus, the healthy, faster-growing field oysters assimilated TBT at a rate that was considerably higher than that of tank oysters maintained at the same TBT exposure. The higher growth rates and TBT uptake of field oysters were very likely due to assimilation of the higher amounts of natural food that were available in the field environment. On the other hand, reduced growth rates and higher mortalities of treated tank oysters may have been caused by combined effects of nutritional stress and toxic response to organotin.

Bioconcentration factors of TBT for the Pearl Harbor Pacific oysters show the same inverse relationship with TBT exposure that has been seen in nearly all other equivalent uptake studies. The authors know of no other studies that have examined the effects of such low TBT concentrations on adult Pacific oysters, although BCF values of 9200 (at 150 ng l^{-1}) and 9900 (at 240 ng l^{-1}) were reported for juveniles exposed in tank experiments [Laughlin (1986) and Thain (1986), respectively]. Additionally, Waldock *et al.* (1987) documented BCF values of 15 000–20 000 for Pacific oysters relaid in harbor environs where TBT levels were 200–1000 ng l^{-1}. Overall, these results show that TBT BCF values of Pacific oysters are inversely related to TBT exposure levels.

14.5. SUMMARY AND CONCLUSIONS

The experiments discussed in this chapter have measured the responses of various taxa of organisms to TBT concentrations of 5–2500 ng l^{-1} in flowthrough bioassay settings. Field-monitoring studies performed in the last few years have shown that TBT concentrations in coastal harbors and embayments of the United States and United Kingdom range from undetectable to ~1500 ng l^{-1} (Waldock *et al.* 1987 and Chapter 11; Seligman *et al.*, 1989). Over 80% of the TBT levels measured in those studies are <50 ng l^{-1}, with nearly all higher concentrations occurring in localized areas of low flushing affected by marina or shipyard activities. Several general predictions of TBT impact on typical harbor macrobiota can be developed by synthesis of the results of the present experiments with updated measurements of concentrations of TBT in inshore waters.

Tributyltin concentrations in the range of 1400–2500 ng l^{-1} would be expected to occur only in localized areas near point sources of organotin. In such areas, direct impact of TBT on adult fish, adult American oyster, and macrocrustacean populations would probably be negligible. However, the health and abundance of some of those fauna may be altered by secondary effects, such as reductions of food prey organisms by TBT poisoning. Fouling populations would be composed largely of taxa that have demonstrated long-term tolerance to high-level TBT exposure such as tube worms, vermetid worms, and solitary tunicates. Benthic algae would flourish on substrates not occupied by dense populations of 'nuisance' fouling organisms. And increased growth of benthic algae would likely be encouraged by decreases in populations of invertebrate grazers, particularly gastropod molluscs, that would be severely impacted by high TBT levels.

The species diversity and abundance of established infaunal and epifaunal communities would be significantly reduced by long-term exposure to TBT concentrations of ~500 ng l^{-1} and higher. Such exposure would be expected to cause mortality ≥50% of typical harbor assemblages of poriferans, polychaetes, tanaids, cnidarians, bryozoans, echinoderms, and molluscs. And benthic species, such as corals, that demonstrated very high mortality during 2–3 weeks of exposure to 500 ng l^{-1} TBT would probably succumb to TBT levels as low as 100 ng l^{-1} over longer time periods. In areas where chronic, high-level exposure to TBT has been terminated, benthic communities of normal population structure would rapidly recolonize both soft and hard substrates.

Although the settlement of larval epifauna of low phyletic levels was reduced by exposure to 100 ng l^{-1} TBT in our tank experiments, that effect would probably not be as great in the field where temporal fluctuations of TBT concentrations above and below a mean value would usually be much greater. Periodic intrusions of low-TBT water masses containing healthy planktonic larvae from less-contaminated areas could probably sustain moderate colonization levels of epifauna that have higher life stages that are TBT tolerant. Settlement of TBT-tolerant organisms would also be enhanced by lowered competition for space and food caused by reductions in populations of TBT-sensitive organisms.

These tests revealed that TBT levels in harbour water ≤200 ng l^{-1} have relatively few significant effects on juvenile and adult life stages of the nine bivalve species of economic importance that were studied. And only Pacific oysters and bay mussels were affected by TBT concentrations <100 ng l^{-1}. Based on these results, mean TBT values should not exceed 10 ng l^{-1} and 50 ng l^{-1} to ensure normal health of Pacific oyster and bay mussel populations, respectively.

At TBT exposures <100 ng l^{-1}, bioaccumulation of TBT varies widely among phyla, being very low for crabs, moderate for fish, and very high for bivalve molluscs. Oysters and mussels can be expected to have BCF values of 20 000–40 000, representing the highest TBT biomagnification levels documented to date.

The testing of captive biota, especially in static environments and to a lesser degree in flowthrough tests, can produce results that may substantially over- or underestimate actual effects of a pollutant in observed field situations (Salazar, 1986; Salazar and Salazar, Chapter 15). Overall examination of the results from the tests described in this chapter and from nearly all other bioassays performed to date indicates that quantities of suspended particulate materials (organic and inorganic) in test waters can significantly affect the health, nutrition, and toxin uptake or bioavailability of nearly all shallow-water aquatic organisms. Therefore, to determine an organism's true response to a toxin, composition and concentration of particulate material in test waters should be similar to those of field waters. Results obtained by testing organisms that are maintained under realistic conditions of water quality and nutrition will be of much greater value for risk assessments than results obtained under altered and possibly stressing conditions.

ACKNOWLEDGEMENTS

We gratefully acknowledge the help of the personnel who assisted in these studies by performing monitoring, sampling, analytical, and editing tasks. These persons include S. Cola, W. Cooke, B. Davidson, C. Dooley, R. Fransham, J. Groves, J. Grovhoug, V. Homer, M. Kram, K. Meyers-Schulte, D. Pawloski, M. Salazar, P. Seligman, M. Stallard, P. Stang, G. Vafa, and A. Valkirs. Funding for this work was provided by the Energy Research and Development Program of the Office of Naval Research. Development of the flowthrough systems was supported by the Naval Facilities Engineering Command. This US government work is not protected by US copyright. Opinions expressed in this chapter are those of the authors and do not necessarily reflect US Government policy.

REFERENCES

Beaumont, A.R., D.K. Mills and P.B. Newman. 1987. Some effects of tin (TBT) on marine algae. In: Proceedings of the Oceans '87 Conference and Exposition, Vol. 4, 28 September–10 October 1987, Halifax, Nova Scotia, pp. 1488–1493.

Chesher, R.H. 1975. Biological impact of a large-scale desalinization plant at Key West, Florida. In: *Tropical Marine Pollution*, E.J.F. Wood and R.E. Johannes (Eds). Elsevier Scientific Publishing Co., New York, pp. 99–153.

Hall, L.W. and A.E. Pinkney. 1985. Acute and sublethal effects of organotin compounds on aquatic biota: an interpretive literature review. *Crit Rev. Toxicol.*, **14**(2), 159–209.

Henderson, R.S. 1986. Effects of organotin antifouling paint leachates on Pearl Harbor organisms: a site specific flowthrough bioassay. In: Proceedings of the Oceans '86 Conference and Exposition, Vol. 4, 23–25 September 1986, Washington, DC, pp. 1226–1233.

Henderson, R.S. 1988. Marine Microcosm Experiments on Effects of Copper and Tributyltin-based Antifouling Leachates. Technical Report 1060, Naval Ocean Systems Center, San Diego, California, 42 pp.

Henderson, R.S. and S.V. Smith. 1980. Semi-tropical marine microcosms: facility design and an elevated-nutrient-effects experiment. In: *Microcosms in Ecological Research*, J.P. Giesy, Jr (Ed.). CONF-781101, National Technical Information Center, Springfield, Virginia, pp. 869–909.

Hodge, V.F., S.L. Seidel and E.D. Goldberg. 1979. Determination of tin (IV) and organotin compounds in natural waters, coastal sediments and macroalgae by atomic absorption spectrometry. *Anal. Chem.*, **51**(8), 1256–1259.

Key, D.R., S. Nunny, P.E. Davidson and M.A. Leonard. 1976. Abnormal shell growth in the Pacific oyster (*Crassostrea gigas*): some preliminary results from experiments undertaken in 1975. *International Council for Exploration of the Sea*, Copenhagen, CM 1976/K:11, pp. 1–12.

Laughlin, R.B, Jr. 1986. Bioaccumulation of tributyltin: the link between environment and organism. In: Proceedings of the Oceans '86 Conference and Exposition, Vol. 4, 23–25 September 1986, Washington, DC, pp. 1206–1209.

Lee, R.F. 1986. Metabolism of bis(tributyltin)oxide by estuarine animals. In: Proceedings of the Oceans '86 Conference and Exposition, Vol. 4, 23–25 September 1986, Washington, DC, pp. 1182–1188.

Maguire, J.R. 1987. Environmental aspects of tributyltin. *Appl. Organometal. Chem.* **1**, 287–301.

Salazar, M.H. 1986. Environmental significance and interpretation of organotin bioassays. In: Proceedings of the Oceans '86 Conference and Exposition, 23–25 September 1986, Washington, DC, Vol. 4, pp. 1240–1245.

Salazar, M.H. and S.M. Salazar. 1988. Tributyltin and mussel growth in San Diego Bay. In: Proceedings of the Oceans '88 Conference and Exposition, Vol. 4, 31 October–2 November 1988, Baltimore, Maryland, pp. 1188–1195.

Salazar, S.M., B.M. Davidson, M.H. Salazar, P.M. Stang and K.J. Meyers-Schulte. 1987. Effects of TBT on marine organisms: field assessment of a new site-specific bioassay system. In: Proceedings of the Oceans '87 Conference and Exposition, Vol. 4, 28 September–1 October 1987, Halifax, Nova Scotia, pp. 1461–1470.

Seligman, P.F., J.G. Grovhoug, A.O. Valkirs, P.M. Stang, R. Fransham, M.O. Stallard, B. Davidson and R.F. Lee. 1989. Distribution and fate of tributyltin in the United States marine environment. *Appl. Organometal. Chem.*, **3**, 31–47.

Taub, F.B. and M.E. Crow. 1980. Synthesizing aquatic microcosms. In: *Microcosms in Ecological Research*, J.P. Giesy, Jr (Ed.). CONF-781101, National Technical Information Center, Springfield, Virginia, pp. 69–104.

Thain, J.E. 1986. Toxicity of TBT to bivalves: effects on reproduction, growth and survival. In: Proceedings of the Oceans '86 Conference and Exposition, Vol. 4, 23–25 September 1987 Washington, DC, pp. 1306–1313.

Thain, J.E. and M.J. Waldock. 1986. The impact of tributyltin (TBT) antifouling paints on molluscan fisheries. *Water Sci. Technol.*, **18**, 193–202.

Thain, J.E., M.J. Waldock and M.E. Waite. 1987. Toxicity and degradation studies of tributyltin (TBT) and dibutyltin (DBT) in the aquatic environment. In: Proceedings of the Oceans '87 Conference and Exposition, Vol. 4, 28 September–2 October 1987, Halifax, Nova Scotia, pp. 1398–1404.

Valkirs, A.O., B.M. Davidson and P.F. Seligman. 1987. Sublethal growth effects and mortality to marine bivalves from long-term exposure to tributyltin. *Chemosphere*, **16**, 201–220.

Waldock, M.J., J. Thain and D. Miller. 1983. The accumulation and depuration of bis(tributyltin) oxide in oysters: a comparison between the Pacific oyster (*Crassostrea gigas*) and the European flat oyster (*Ostrea edulis*). International Council for the Exploration of the Sea, Copenhagen, CM 1983/E:52, pp. 1–9 (mimeo).

Waldock, M.J., J.E. Thain and M.E. Waite. 1987a. The distribution and potential toxic effects of TBT in UK estuaries during 1986. *Appl. Organometal. Chem.*, **1**, 287–301.

Waldock, M.J., M.E. Waite and J.E. Thain. 1987b. Changes in concentrations of organotins in U.K. rivers and estuaries following legislation in 1986. In: Proceedings of the Oceans '87 Conference and Exposition, Vol. 4, 28 September–1 October 1987, Halifax, Nova Scotia, pp. 1352–1356.

MUSSELS AS BIOINDICATORS: EFFECTS OF TBT ON SURVIVAL, BIOACCUMULATION, AND GROWTH UNDER NATURAL CONDITIONS

15

Michael H. Salazar[1] *and Sandra M. Salazar*[2]

[1]*Applied Biomonitoring, 11648 72nd Place NE, Kirkland, Washington 98034, USA*

[2]*EVS Consultants, 200 West Mercer Street, Seattle, Washington 98119, USA*

ABSTRACT

During nine field-transplant experiments (1987–1990), juvenile mussels were exposed to mean tributyltin (TBT) concentrations from 2 to 530 ng l^{-1} for 12 weeks under natural conditions in San Diego Bay. Mussels were used as biological indicators and monitored for survival, bioaccumulation, and growth. Mussel growth was the primary biological response used to quantify TBT effects. Chemical analyses were used to estimate TBT

Organotin. Edited by M.A. Champ and P.F. Seligman. Published in 1996 by Chapman & Hall, London. ISBN 0 412 58240 6

contamination in water and mussel tissues. Integrating intensive measurements of chemical fate and biological effects increased the environmental significance of the data. Multiple growth measurements on individuals increased the statistical power. Size effects were minimized by restricting test animals to 10–12 mm in length, and methods were developed to minimize handling efffects. This monitoring approach also permitted documenting temporal and spatial variability in TBT and its effects that have not been previously reported. Survival, bioaccumulation, and growth were generally higher than predicted from laboratory studies. Survival was not directly affected by seawater or tissue TBT concentrations. Growth was significantly related to both seawater and tissue TBT, with the bioconcentration factor inversely proportional to seawater TBT concentration. Threshold concentrations always causing significant reductions in juvenile mussel growth are estimated at 100 ng l^{-1} TBT for seawater and 1.5 $\mu g\ g^{-1}$ TBT for tissue, but growth could be affected by much lower concentrations of TBT under the most adverse conditions. Temperatures above 20°C were also found to reduce juvenile mussel growth rates.

15.1. INTRODUCTION

Biological indicators provide integrated information about environmental contamination and effects that cannot be defined with chemical analysis of water samples. Nevertheless, chemical analyses quantify contamination in a way that is essential in explaining biological effects. Measuring survival, bioaccumulation, and growth quantifies both contamination and effects. It is important to make the distinction between the use of biological indicators as detectors of environmental contamination by monitoring tissue accumulation with chemical analyses versus their use as indicators of environmental effects by measuring other biological responses (Waldock *et al.*, Chapter 11). Mussels have been used as biological indicators in many field monitoring programs because of their cosmopolitan distribution and ability to concentrate many different contaminants. Their demonstrated utility in transplant experiments and monitoring is another significant advantage. Mussels have been used most extensively to monitor contamination by measuring tissue accumulation with chemical analyses (Phillips, 1980). They have also been used to monitor the biological effects of contamination by measuring biological responses related to growth, physiology, and reproduction (Bayne *et al.*, 1985). It should be recognized that bioaccumulation can be regarded as both a chemical and a biological process; however, biological effects are not necessarily related to tissue accumulation. We suggest that caged mussels should be regarded as a bioindicator system. This indicator system consists of the entire suite of biological responses.

Bioaccumulation is the process through which organisms integrate exposure to environmental concentrations of bioavailable contaminants. The results of these integrated chemical and biological processes can be quantified with chemical analyses. Laboratory and field studies have shown that tributyltin (TBT) is highly toxic to molluscs and that filter-feeding bivalves readily accumulate TBT (Hall and Bushong, Chapter 9; Laughlin *et al.*, Chapter 10; Waldock *et al.*, Chapter 11; Gibbs and Bryan, Chapter 13; Henderson and Salazar, Chapter 14; Laughlin, Chapter 16). Natural mussel populations have been used as indicators of TBT contamination by measuring tissue concentrations of TBT (Wade *et al.*, 1988; Short and Sharp, 1989; Uhler *et al.*, 1989; Roberts *et al.*, Chapter 17). Chemical measurements of TBT concentrations in seawater and tissues of mussels from natural populations (Grovhoug *et al.*, Chapter 25) or transplants (Zuolian and Jensen, 1989) are more informative than measuring seawater or tissue levels alone.

The combination of seawater and tissue measurements of TBT provides a quantitative relationship between the chemistry of the environment and the chemistry of the organism. However, the relative influence of environmental and biological factors on this relationship is highly variable and difficult to predict (Cain and Luoma, 1990).

Survival and growth are biological responses that also integrate exposure to environmental concentrations of bioavailable contaminants. These responses are more directly related to animal health and do not depend on chemical analysis. Survival is the least sensitive to environmental effects because it is an all-or-nothing response. Therefore, survival data are not always informative. Growth is more sensitive because there is a graded response to environmental conditions that can be quantified through repetitive, non-destructive measurements. Reduced growth represents adverse environmental effects and possible effects on the population. Both natural and pollution-related stresses have been shown to reduce mussel growth rates (Bayne *et al.*, 1985). Reduced mussel growth has been associated with TBT in laboratory and field studies (Thain and Waldock, 1985; Stephenson *et al.*, 1986; Salazar and Salazar, 1987; Stromgren and Bongard, 1987; Valkirs *et al.*, 1987; Salazar and Salazar, 1988). Juvenile mussel growth was the most sensitive indicator of TBT measured in San Diego Bay microcosm experiments (Salazar and Salazar, 1987; Salazar *et al.*, 1987; Henderson and Salazar, Chapter 14). Juvenile mussels have advantages over adults as bioindicators: (1) they grow faster and provide a greater range of response; (2) the growth process is not affected by gametogenesis (Rodhouse *et al.*, 1986); (3) bioaccumulation in short-term tests with fast-growing juveniles more accurately reflects recent environmental changes (Fischer, 1983, 1988); and (4) they may be more sensitive to TBT (Hall and Bushong, Chapter 9).

Survival, bioaccumulation, and growth in caged mussels can be routinely measured in field-transplant experiments. Field-transplant experiments provide an opportunity to corroborate and validate mussel performance under laboratory conditions. It is important to measure performance under natural conditions because that is where the biological effects, if any, will occur. The main advantage in using transplanted animals over monitoring naturally settled populations is the experimental control. Experimental control is achieved by using mussels of similar genetic and environmental stocks at all test sites, preselecting test animal size or age group, and monitoring individual animals during the test. Animals can also be transplanted to areas where they might not normally be found. Serial transplants and monitoring facilitate the examination of both short- and long-term trends in contaminant distribution and related effects.

The work reported here had three primary goals: (1) determine the effects of TBT on survival, bioaccumulation, and growth in juvenile mussels under natural conditions; (2) identify long-term trends in the distribution of TBT and its effects; and (3) refine the use of the juvenile mussel bioindicator for environmental assessment.

15.2. METHODS

Nine field-transplant tests were conducted in San Diego Bay between 1987 and 1990 (Salazar and Salazar, 1991c). Juvenile mussels (*Mytilus* sp.) were transplanted from the collection site to test sites and monitored during 12-week exposure periods. Test dates are shown in Fig. 15.1. Serial mussel measurements were made in the field, and water samples collected weekly during tests 1–4 and on alternate weeks (biweekly) during tests 5–9. Growth measurements included whole animal wet weights and lengths. Caged mussels were removed from the water for ~20 min during these measurements, and byssal threads were carefully cut with scissors. Water was measured for chlorophyll-*a* and

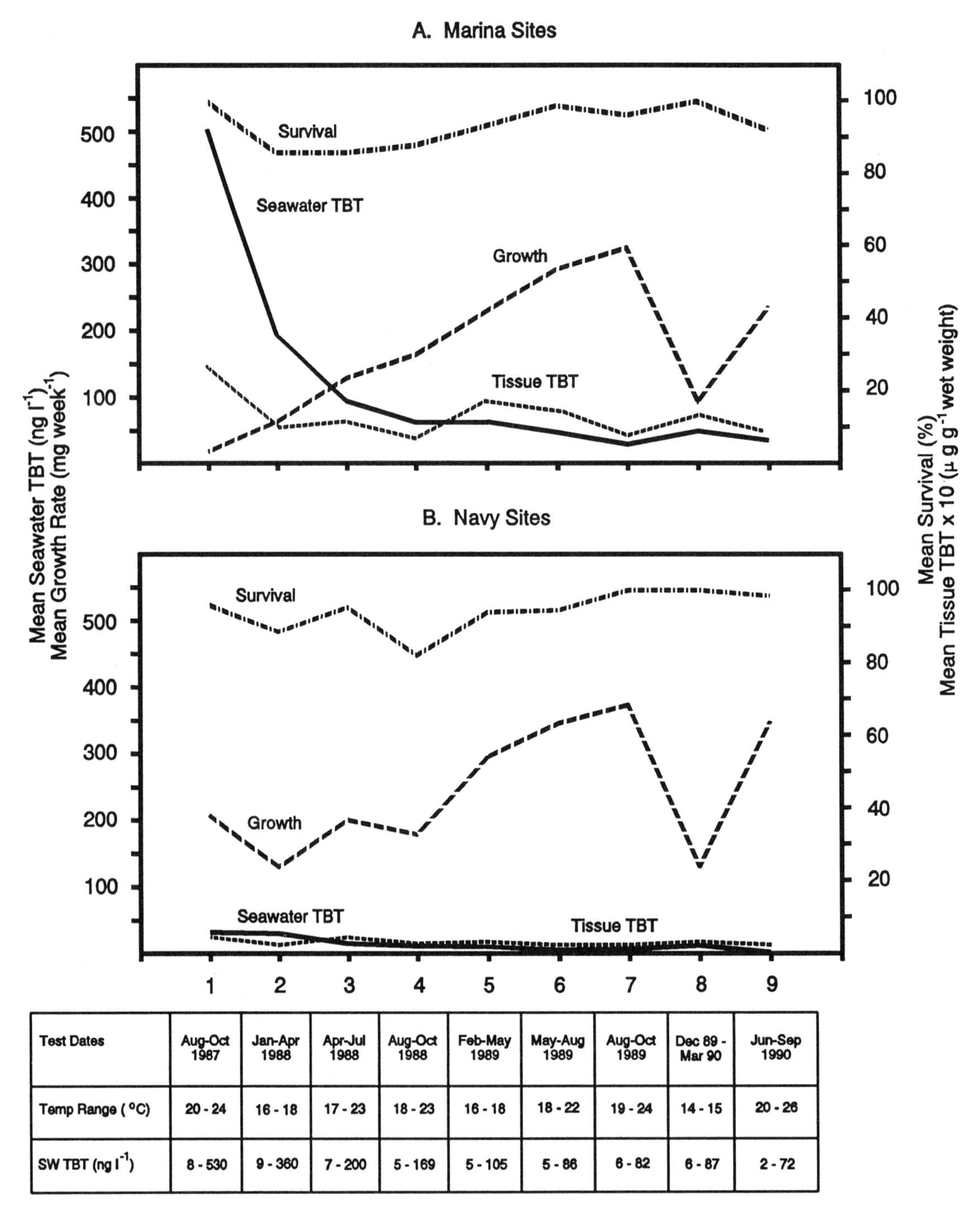

Test Dates	Aug-Oct 1987	Jan-Apr 1988	Apr-Jul 1988	Aug-Oct 1988	Feb-May 1989	May-Aug 1989	Aug-Oct 1989	Dec 89 - Mar 90	Jun-Sep 1990
Temp Range (°C)	20 - 24	16 - 18	17 - 23	18 - 23	16 - 18	18 - 22	19 - 24	14 - 15	20 - 26
SW TBT (ng l^{-1})	8 - 530	9 - 360	7 - 200	5 - 169	5 - 105	5 - 86	6 - 82	6 - 87	2 - 72

Figure 15.1 Regional differences and temporal changes in seawater TBT, tissue TBT, survival, and growth for marina sites (A) and Navy sites (B). Test dates are also given with temperature and seawater TBT ranges (rounded mean values).

TBT concentrations (Venrick and Hayward, 1984; Stallard *et al.*, 1988). Mean chlorophyll-*a* and seawater TBT concentrations for each test were determined from weekly or biweekly measurements. Temperature was measured at half-hour intervals with in-situ monitors. Whole animal wet weights, lengths, shell weights, and tissue wet weights were measured at the end of each study. Tissues for a given site were pooled for TBT analysis on a wet-weight basis (Stallard *et al.*, 1988). Average weekly growth rates based on weight (mg $week^{-1}$) and length (mm $week^{-1}$) were calculated from regression analyses for animals transplanted at each site. Only weight growth rates will be discussed because there is a greater range in response than with length growth rates (25 × compared to 10 ×), and weight measurements are more accurate.

Naturally settled mussels were collected from pilings where mean TBT concentrations in seawater and mussel tissues were low, ~5 ng l^{-1} TBT and ~0.15 μg g^{-1} TBT wet wt, respectively. The collection site, a Navy site, is located near the mouth of San Diego Bay and characterized by high current speeds, nearshore ocean temperatures, few vessels, and low levels of contaminants in mussel tissues. Mussel growth rates in this area are among the highest in San Diego Bay. The collection site was used as a test site for comparative purposes. Animals were sorted by length for convenience. Test mussels were 10–12 mm in length ($\bar{x}$ ~11.0 mm) and initial mean weights ranged from 100 to 250 mg ($\bar{x}$ ~175 mg). Eighteen mussels were caged and transplanted to each site. There were no statistically significant differences in weights or lengths among these mussel groups at the start of any test. Animals were continuously submerged either 1 m below the surface or 1 m above the bottom. The 18 monitoring sites included 11 Navy sites with low seawater TBT concentrations and seven marina sites with high seawater TBT concentrations (Grovhoug *et al.*, 1986). The most significant refinements in methods include the use of (1) field transplants; (2) serial growth measurements on individuals; (3) minimizing the size range and maximum size of test animals; and (4) synoptic measurements of bioaccumulation and growth.

Bioaccumulation in transplanted and natural mussels was compared to determine if the transplanted bioindicator was responding like natural populations of undisturbed mussels. Bioaccumulation in juveniles and adults was compared to determine if there were size or age differences. During tests 4, 6, and 9, juvenile ($\bar{x}$ length <35 mm) and adult mussels ($\bar{x}$ length >50 mm) were collected from natural populations at one marina and one Navy transplant site. These sites were characterized by the highest and lowest concentrations of TBT in seawater, respectively. This Navy site was also the collection site. Tissues from these mussels, along with tissues from juvenile transplants (tests 4, 6, and 9) and adult transplants (test 9), were analyzed for TBT. In a different experiment, handling and measurement effects on growth of juvenile mussels were assessed at five sites, including the most contaminated marina site. Growth rates of animals measured weekly were compared with growth rates from those animals measured only at the beginning and end of the 12-week exposure (untouched). Some comparisons were also made with animals measured on alternate weeks (biweekly).

The data for survival, seawater TBT concentration, tissue TBT concentration, and growth were pooled by test for both marina and Navy sites to calculate regional means for comparison with other San Diego Bay monitoring studies (Grovhoug *et al.*, Chapter 25). Regional means for seawater TBT and growth rates were compared with a one-way analysis of variance on a per-test basis as well as for data pooled across tests. Regional means for tissue TBT and individual site data were pooled across tests and compared using a one-way analysis of variance and multiple linear regression analyses. Although these regional divisions are somewhat arbitrary, they can be used to illustrate the effects of low and high exposure to TBT in seawater and convey important general information about the physical–chemical characteristics of the sites in each region. Details regarding temporal and spatial variability, site locations, and site-specific responses are presented elsewhere (Salazar and Salazar, 1991a,b,c).

Graphical methods are used to display the general relationships among environmental levels of TBT and mussel survival, bioaccumulation, and growth. Each data point represents a 12-week exposure and the response of approximately 18 animals. The significance of each relationship was determined by linear regression analyses. The significance of the regression (p); the correlation coefficient, which provides a measure of association intensity between the two variables (r); and the total variation in the dependent parameter that is explained by the fitted regression (r^2) were used to compare regressions. Stronger relationships are indicated by higher p, r, and r^2 values. Lines and slopes can also be compared on a relative basis. The r^2 statistic estimates the predictive strength of each relationship and possible environmental significance. Statistical significance was determined at the 95% confidence level.

15.3. MUSSELS AS BIOINDICATORS

In recognition of the limitations of laboratory bioassays and chemical monitoring, there has been a shift in emphasis toward biological monitoring and field bioassays (Chapman, 1983; Chapman and Long, 1983; Phillips and Segar, 1986; Parrish *et al.*, 1988). Specific problems with the interpretation and environmental significance of TBT studies have been discussed previously (Stebbing, 1985; Salazar, 1986, 1989; Salazar and Champ, 1988). Using mussels as biological indicators of contamination and effects is a potentially powerful tool, but there are many pitfalls regarding interpretation and environmental significance (Phillips, 1980; White, 1984). Many of these pitfalls were avoided in this mussel bioindicator study by integrating chemical and biological measurements and by demonstrating the differences between using mussels as indicators of contamination versus their use as indicators of effects. Most status and trends programs in the United States have used biological indicators as detectors of environmental contamination of TBT by monitoring tissue accumulation with chemical analyses (Wade *et al.*, 1988; Short and Sharp, 1989; Uhler *et al.*, 1989). The predictive value of these studies is extremely limited. Measuring seawater TBT in addition to accumulation in natural populations (Grovhoug *et al.*, Chapter 25) or field transplants (Zuolian and Jensen, 1989) establishes a quantitative relationship between environment and organism that is much more informative. Such studies are using mussels as detectors of environmental contamination. By measuring survival and growth in addition to bioaccumulation, we used mussels as indicators of environmental contamination and effects (Waldock *et al.*, Chapter 11). The approach was further integrated by including chemical measurements of seawater. This type of integration is very important.

15.3.1. GENERAL TRENDS

Mean concentrations of TBT in seawater at the 18 sites ranged from 2 to 530 ng l^{-1} and tissue TBT concentrations from 0.1 to 3.2 $\mu g\ g^{-1}$ wet wt. Mean 12-week temperatures ranged from 14.3 to 25.7°C and chlorophyll-*a* from 0.79 to 6.07 μg^{-1}. Mean temperatures for winter tests ranged from 14.3 to 14.8°C while mean temperatures for summer tests ranged from 20.1 to 25.7°C. Chlorophyll-*a* was lowest in the winter. End-of-test survival ranged from 50 to 100%. Mean growth rates of caged mussels ranged from 17 to 505 mg $week^{-1}$ (0.2 to 2.5 mm $week^{-1}$). The lowest growth rates were measured at the most contaminated marina site. Mussels increased from ~190 to 420 mg (11 to 14 mm in length). This was an increase of only 230 mg in weight (3 mm in length) after 12 weeks. The highest growth rates were measured at the Navy site nearest the mouth of the bay. Mussels increased from ~160 to 6200 mg (11 to 40 mm in length). These mussel growth rates are among the highest reported (Kiorboe *et al.*, 1981). The minimum concentration of TBT in seawater predicted to always reduce juvenile mussel growth was estimated at 100 ng l^{-1} TBT. The minimum concentration of TBT in tissues predicted to always reduce juvenile mussel growth was estimated at 1.5 $\mu g\ g^{-1}$ (wet wt). These concentrations were estimated from statistical analyses of mussel responses under representative conditions in San Diego Bay. Because biological responses are so site specific, site selection influenced the significance of each relationship and comparisons between regions. Therefore, the most meaningful comparisons are those between sites. Both water and tissue TBT concentrations are similar to those predicted to reduce oyster growth (Waldock *et al.*, Chapter 11). A comparable concentration of tissue TBT was reported to adversely affect adult mussel physiology (Page and Widdows, 1990).

Regional differences and temporal changes in survival, seawater TBT, tissue TBT, and growth for the marina and Navy sites are shown in Fig. 15.1. Seawater and tissue TBT concentrations generally decreased while mussel growth rates increased over time. Although the lowest growth rates were associated with the highest concentrations of seawater TBT, extremely low growth rates were also associated with winter seawater temperatures <15°C. Seawater and tissue TBT concentrations were significantly higher and growth rates significantly lower at marina sites than Navy sites. There were no differences in survival. There was a significant decrease in seawater and tissue TBT concentrations at marina sites and Navy sites. This was associated with a general increase in growth. These changes were not consistent at all sites, however, and argue against pooling (Salazar and Salazar, 1991b). There was also a significant decrease in growth rates during test 8 that demonstrates the significance of natural factors in influencing mussel growth rates. The mean concentration of seawater TBT at marina sites declined rapidly from 1987 to 1988 and by October 1990 approached the mean for Navy sites. The differences between marina and Navy sites as well as the decreases in seawater and tissue TBT concentrations were similar to those described for San Diego Bay by Grovhoug *et al.* (Chapter 25) except that we did not find significant decreases in tissue TBT concentrations at all Navy sites. Neither tissue TBT concentrations, survival, nor growth rates consistently followed seawater TBT concentrations. However, the ratio of tissue to seawater TBT, growth rates, and survival generally increased at both marina and Navy sites after test 4.

15.3.2. SURVIVAL

Although survival is not a sensitive indicator of environmental effects, important information was gained for some sites where exceptionally high or low survival was different than expected based on other measurements and indicators of water quality. Survival is not significantly correlated with seawater or tissue TBT concentration. The four lowest survival measurements were between 50 and 65%. They were associated with mean concentrations of TBT in seawater between 34 and 130 ng l^{-1} at one marina and one Navy site. Conversely, the five highest mean concentrations of TBT in seawater were between 169 and 530 ng l^{-1}. They were measured at a different marina site and were associated with mussel survival between 94 and 100%. There was no difference in measured survival between marina and Navy sites even though seawater and tissue TBT concentrations were significantly higher at marina sites (Fig. 15.1). Survival was higher in tests 5–9 (96.8%) than in tests 1–4 (90.2%), but the difference is not statistically significant. One hundred percent survival was measured in 42% of the transplant in tests 1–4 and in 70% of the transplant in tests 5–9. The increase in survival after test 4 was associated with the decline in TBT concentrations in marina seawater following restrictions imposed in January 1988 (State of California, 1988) and reduced handling after test 4 when weekly growth measurements were changed to alternate weeks. Since handling was reduced at the same time seawater TBT concentrations decreased, the relative effects of each on survival are unclear.

15.3.3. BIOACCUMULATION

There is a significant positive linear relationship between TBT accumulation in juvenile mussel tissues and seawater TBT concentration (Fig. 15.2). Based on regression analysis of all data, only 38% of the variance in tissue TBT concentration can be explained by seawater TBT concentration. The regression equations calculated using all data and only data associated with seawater values >105 ng l^{-1} are very similar. Six of the seven data points for seawater TBT concentrations >105 ng l^{-1} are from the most contaminated marina and strongly influence the regression for all data. The data appear to fall into two separate groups. The slope of the regression for tissue TBT values associated with seawater TBT concentrations <105 ng l^{-1} is almost five times higher than the slope of the

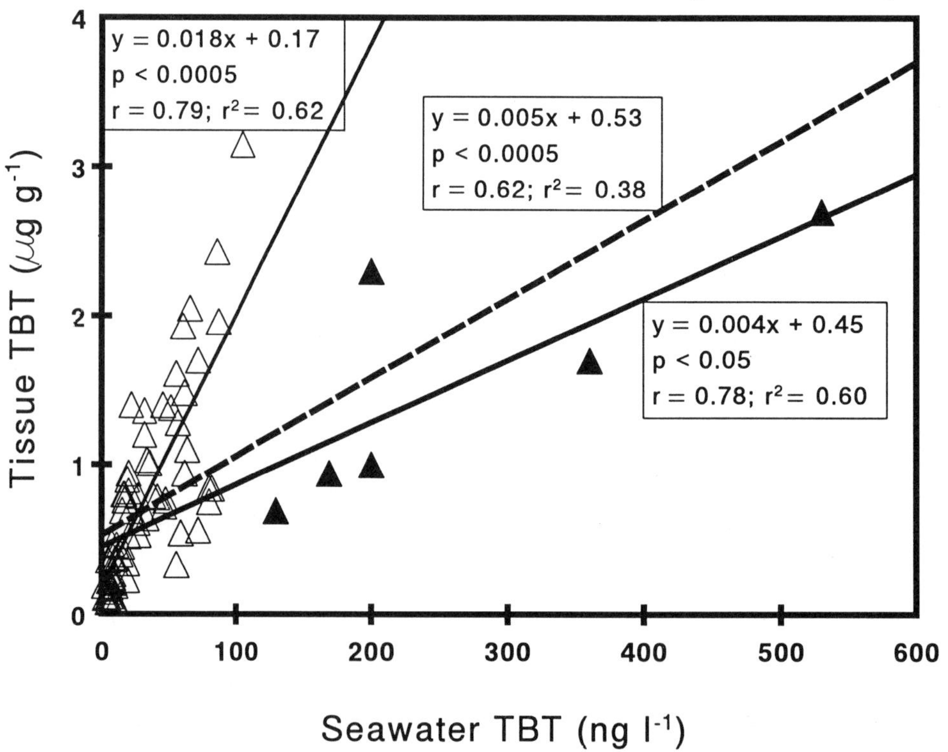

Figure 15.2 Relationship between TBT concentrations in seawater and juvenile mussel tissues. Regression lines show differences and similarities in tissue TBT accumulation between low (≤105 ng l^{-1}) and high (>105 ng l^{-1}) concentrations of TBT in seawater and all data (dashed line). The regression equations and relevant statistics are also given.

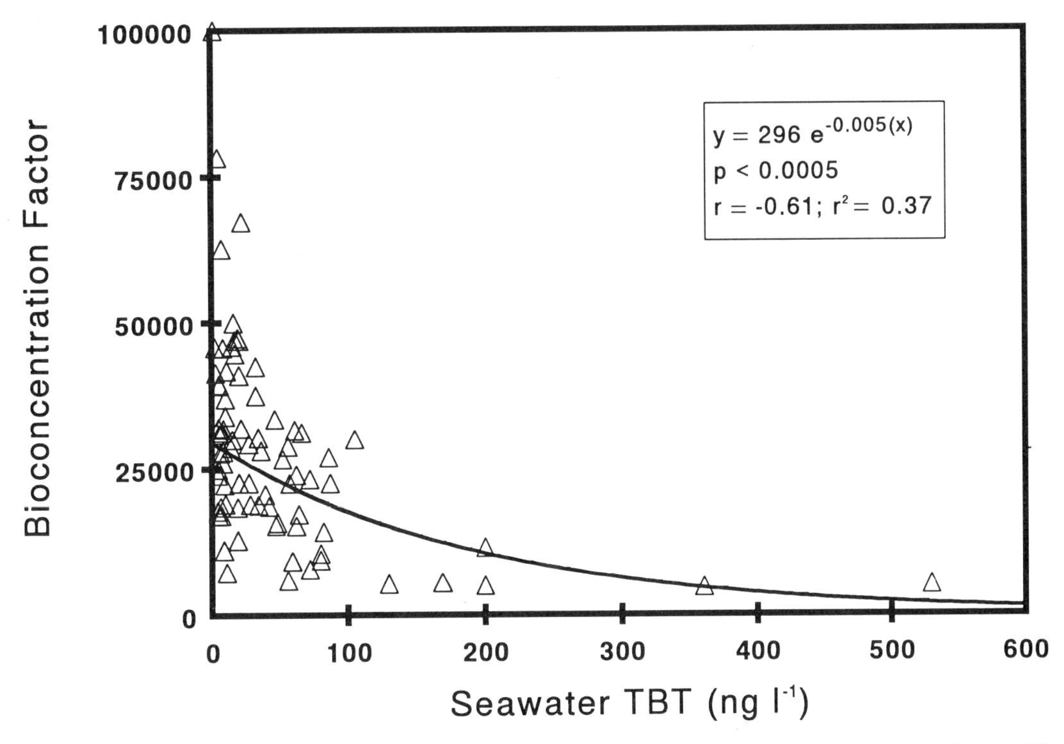

Figure 15.3 Relationship between seawater TBT concentration and bioconcentration factor. The equation of the best-fit line and relevant statistics are also given.

regression for tissue TBT values associated with seawater TBT concentrations >105 ng l^{-1}.

The ratio of tissue TBT to seawater TBT concentration estimates a bioconcentration factor (BCF). BCF values ranged from 4700 to 105 600 with the majority between 20 000 and 40 000 (Fig. 15.3). An inverse exponential function best describes the relationship between BCF and seawater TBT concentration. The higher ratios of tissue TBT to seawater TBT at lower seawater TBT concentrations shown in Fig. 15.3 were suggested by the regressions in Fig. 15.2 and the shift in ratios from the trends in Fig. 15.1. Lower concentrations of TBT in seawater were associated with higher BCF values. At seawater TBT concentrations <105 ng l^{-1}, BCF values range from 5000 to ~100 000. Above seawater TBT concentrations of 105 ng l^{-1}, BCF values were <9000. This inverse exponential relationship between TBT concentrations in seawater and in mussel tissue was found in laboratory studies (Laughlin *et al.*, 1986; Laughlin and French, 1988), microcosm studies (Salazar *et al.*, 1987), other field-transplant studies (Waldock *et al.*, Chapter 11; Zoulian and Jensen, 1989), and natural San Diego Bay populations (Grovhoug *et al.*, Chapter 25).

Using tissue accumulation in field bioindicators to quantify environmental levels of contamination is a potentially powerful tool, but the limitations of this approach must be recognized (Phillips, 1980). Since the relationship between seawater and tissue TBT concentrations (BCF) is not constant, accurate predictions of one using the other are not possible. Even accumulating elevated levels of contaminants does not *a priori* indicate environmental effects on the bioindicator or other species (Peddicord, 1984). Initial reports on mussels as bioaccumulators

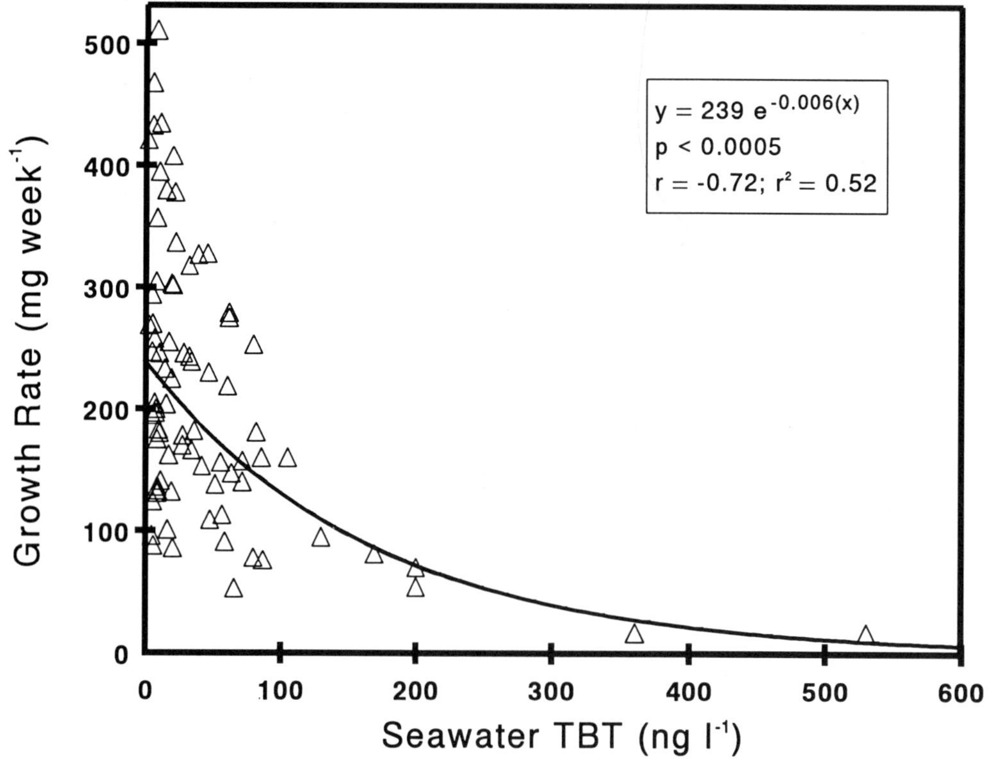

Figure 15.4 Effects of seawater TBT concentration on juvenile mussel growth rate. The equation of the best-fit line and relevant statistics are also given.

emphasized the utility of identifying order-of-magnitude differences in seawater contamination and minimized potential interference from extraneous environmental factors (Goldberg *et al.*, 1978, 1983; Farrington *et al.*, 1983). Other reports have outlined potential problems in using tissue concentrations of contaminants for environmental prediction (Phillips, 1980; Luoma, 1983; White, 1984; Phillips and Segar, 1986; Cain and Luoma, 1990). Bioaccumulation is an important link between the test organism and its environment, but the relationship between concentrations of TBT in seawater and mussel tissue is very complex. Bioavailability of TBT may be affected by a number of natural factors, and the relationship between bioavailability and chemical analysis of TBT is unclear (Laughlin *et al.*, 1986; Salazar, 1986, Laughlin and French, 1988).

15.3.4. GROWTH

Juvenile mussel growth is significantly related to both seawater and tissue TBT concentrations. The statistical relationship is stronger for seawater TBT concentration. The relationship between juvenile mussel growth rate and seawater TBT concentration is negative and exponential (Fig. 15.4). Based on regression analysis, ~52% of the growth variance can be explained by seawater TBT concentration. There is a high degree of variability at the lowest concentrations of seawater TBT. The seven data points for seawater TBT concentration <100 ng l^{-1} strongly influence the significance of the regression for all data. They also demonstrate the influence of the most contaminated marina where six of the seven highest measurements were made. The equations for all data and seawater TBT >100 ng l^{-1} are quite similar. A

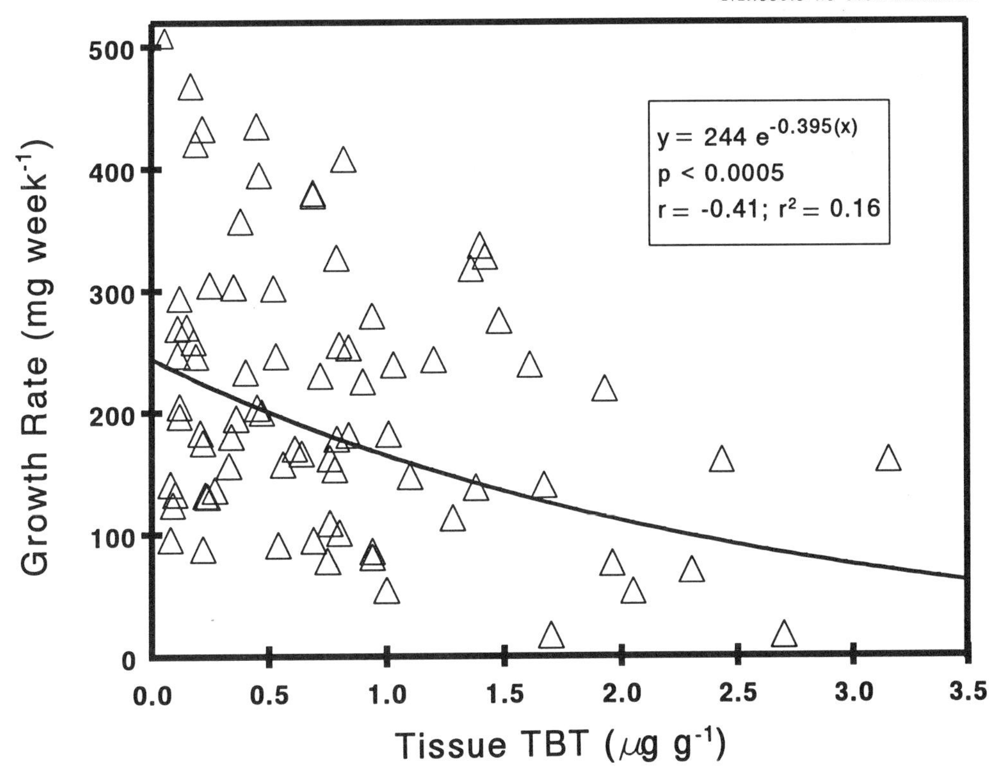

Figure 15.5 Effects of tissue TBT concentration on juvenile mussel growth rate. The equation of the best-fit line and relevant statistics are also given.

power fit best describes the relationship for seawater TBT data >100 ng l^{-1} (p = 0.0001, r = −0.97, r^2 = 0.94). There is also a significant negative exponential relationship when the seawater TBT data <100 ng l^{-1} are analyzed separately (p<0.0025, r = −0.36, r^2 = 0.13), but only 13% of the growth variance can be explained by seawater TBT concentration. Some statistically significant relationships (p = 0.0357) were also found with seawater TBT data <70 ng l^{-1}, but the r^2 value was <0.09 and suggests little environmental significance.

Although it has been suggested that bivalve growth is directly affected by accumulated TBT (Waldock and Thain, 1983), it is not clear from the weak relationships found in these field-transplant studies if accumulated TBT is regulating growth rate or growth rate is regulating the accumulation of TBT. There is a significant negative exponential relationship between juvenile mussel growth and tissue TBT concentration (Fig. 15.5), but variability in growth is high at all tissue TBT concentrations. Based on regression statistics, TBT in seawater has a more direct effect on mussel growth than TBT accumulated in mussel tissues. Only 16% of the variance in growth can be explained by tissue TBT concentration. There are no significant regressions when the analyses are limited to data <1.5 µg g^{-1} TBT in tissue (wet wt). Many high growth rates are associated with elevated concentrations of TBT in tissue even at low concentrations of TBT in seawater.

Growth has a significant effect on the bioaccumulation process as shown by the significant linear relationship between growth rate and BCF. The highest growth rates are associated with the highest BCF values (Fig.

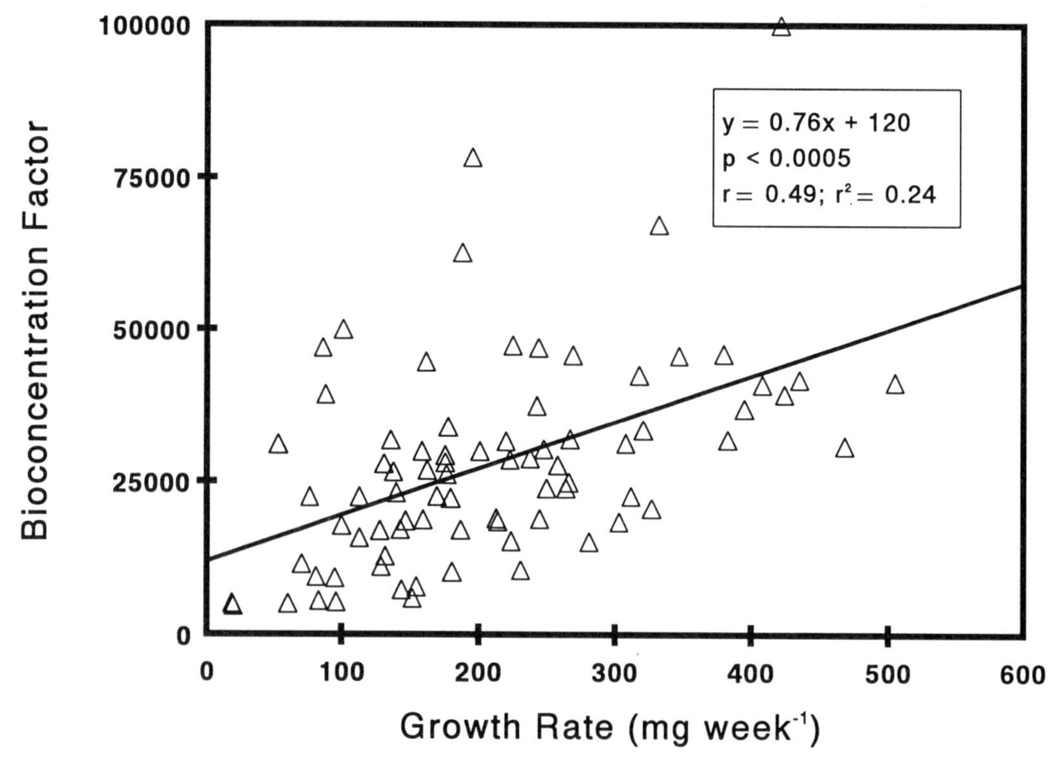

Figure 15.6 The effects of growth rate on bioconcentration factor. The equation of the best-fit line and relevant statistics are also given.

15.6). The highest growth rates are also associated with the lowest concentrations of TBT in seawater and reduced stress from decreased handling. These data suggest that growth rate affects bioaccumulation. At similar seawater TBT concentrations, faster growing mussels may accumulate more TBT. If mussels do not grow, they will not accumulate much TBT. Growth rates were very low in tests 2 and 8 due to winter conditions, but accumulation was similar even though TBT concentrations were significantly lower in test 8.

Using biological responses such as growth in field bioindicators to quantify environmental effects is also a potentially powerful tool, but, as with bioaccumulation, the limitations must be acknowledged (White and Champ, 1983; Cairns and Buikema, 1984; Malins *et al.*, 1984; Moller, 1987; Cairns, 1988). Based on statistical analyses, the data show that TBT accumulated in mussel tissue has less of an effect on mussel growth than TBT in seawater. These relationships demonstrate that other factors have a significant effect on growth rate. Nevertheless, growth rates can provide other information and perhaps be used to calibrate bioaccumulation (Fischer, 1983, 1988). Variable growth rates could explain some of the apparent anomalies in tissue accumulation shown here. For example, mussels severely stressed by TBT or other factors will not grow and will not accumulate much TBT. Without supporting measurements of growth and seawater TBT concentrations, only analyzing tissues could be very misleading. Survival was the least sensitive indicator measured here, but it

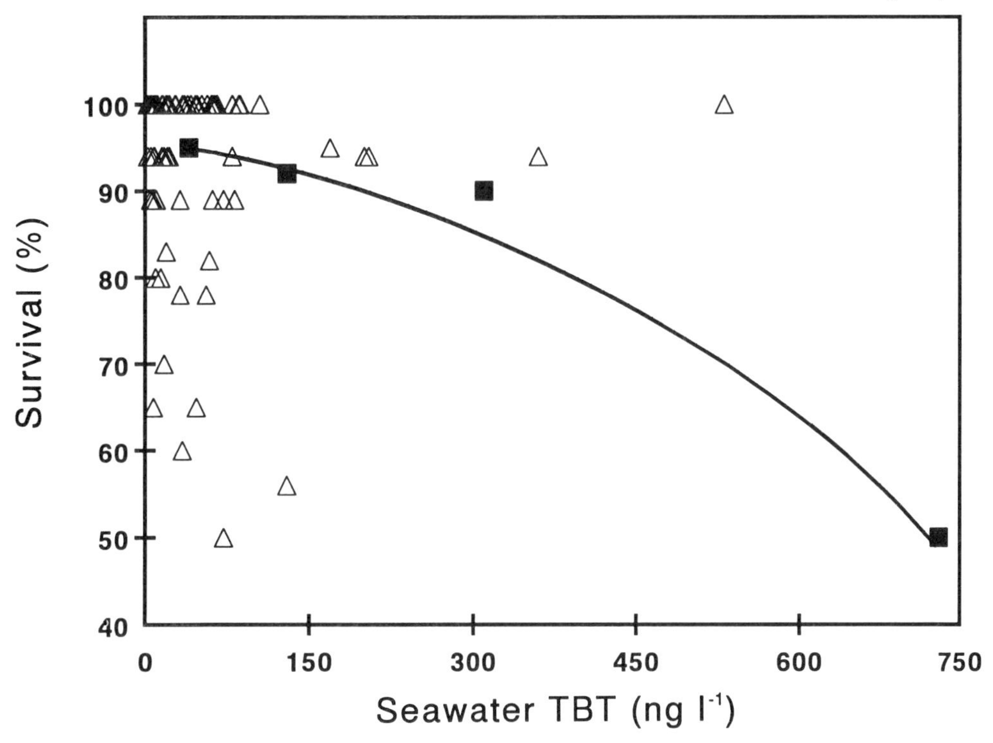

Figure 15.7 Comparison of survival in juvenile mussel field transplants (triangle) with survival in adults from laboratory studies (square) exposed to similar concentrations of TBT in seawater. A best-fit line is estimated.

helped explain the results. Clearly, natural factors can affect survival, bioaccumulation, and growth in juvenile mussels.

15.4. LABORATORY VS FIELD

There is a tremendous gap between correlations and causality when using biological indicators in the field. Biological responses under natural conditions are often very different from biological responses measured under controlled laboratory conditions, and they are affected by many different factors (White and Champ, 1983; Mallet *et al.*, 1987; Salazar, 1989). For example, conditions in laboratory and field tests appear to be as responsible for observed biological responses as the toxicant being tested. Temperature and nutritive stress have been shown to cause adverse physiological changes in mussels under laboratory conditions (Bayne and Thompson, 1970; Bayne, 1973). Mussel survival, bioaccumulation, and growth were generally much higher in these field studies than in comparable laboratory studies. Laboratory-induced stress may have enhanced the effects of TBT and account for the reduced performance of laboratory animals.

Survival of juvenile mussels in the field-transplant experiments was higher and did not demonstrate the dose dependency shown for adult mussels in laboratory toxicity tests (Thain, 1983; Valkirs *et al.*, 1987; Salazar and Salazar, 1989). These differences are best demonstrated by comparing results from a laboratory experiment with conditions most similar to field exposures (Fig. 15.7). In this laboratory experiment adult mussels were exposed to TBT for 66 d in a flowthrough system (Valkirs *et al.*, 1987). Predicted decreases in survival with increasing seawater TBT concentrations found in the laboratory were not observed in the field-transplanted mussels. This apparent dose dependency in the laboratory may be due to a limited number of test concentrations and the use of extremely high seawater TBT concentrations (>500 ng l^{-1}) that strongly influence statistical analyses and yet are environmentally unrealistic. Dose dependency may not be manifested in field studies due to natural factors that modify mussel performance. At similar TBT exposure concentrations, survival was higher in field-exposed animals, even though the exposure period was longer and the test animals were also exposed to high concentrations of other contaminants.

It is extremely difficult to compare tissue TBT concentrations in mussels from these studies with tissue concentrations in mussels from laboratory studies. Analytical methods and units for reporting results are highly variable (Salazar, 1986; Page and Widdows, 1990). Another potential problem is that total contaminant content per individual may be more biologically significant than measured tissue concentrations (Fischer, 1983, 1988; Cain and Luoma, 1990). One apparently common unit for comparison is BCF. This convention is commonly used in the TBT literature. The majority of BCF values calculated for mussels in this and other field-transplant studies, microcosm experiments, and natural populations are ~30 000 (Salazar *et al.*, 1987; Zuolian and Jensen, 1989; Grovhoug *et al.*, Chapter 25). Only one laboratory experiment has provided comparable BCF values, and it was conducted under flowthrough conditions with apparently healthy animals (Waldock *et al.*, Chapter 11). In most other laboratory experiments, BCF values were much lower with a reported maximum of ~7000 (Laughlin *et al.*, 1986; Laughlin and French, 1988). The combined stresses of static-renewal conditions, overcrowding, and poor nutrition probably affected animal health and limited the ability to accumulate TBT. These observations support measuring growth to quantify animal health and calibrate bioaccumulation.

Comparing BCF values from the laboratory and field can be misleading because they do not represent the same conditions. Since mussels may be exposed to TBT through various routes, such as seawater and food, field exposures include TBT from all sources. Technically, the BCF only accounts for the TBT concentration in seawater, which is generally the primary source in laboratory studies. The bioaccumulation factor (BAF) includes TBT from seawater and food. Higher BCF values from these field-transplant studies could be attributable to measuring only the TBT dissolved in seawater. These values could decrease substantially after accounting for TBT associated with the food, if food is a major source of accumulated TBT. Quantifying the food and water components of the BAF is important for understanding the bioaccumulation process and explaining effects attributable to TBT.

Laboratory-held mussels generally grow more slowly than mussels maintained under natural field conditions (Kiorboe *et al.*, 1981). Temperature and nutritive stresses often associated with laboratory experiments may cause physiological changes (Bayne and Thompson, 1970; Bayne, 1973) that preclude optimum growth rates and influence the relative effects of toxicants. In Fig. 15.8 juvenile mussel growth in our field-transplant studies is compared with juvenile mussel growth under laboratory (Thain, 1986) and flow-through microcosm conditions (Salazar and Salazar, 1987; Salazar *et al.*, 1987). The rate of juvenile mussel growth in field-transplant experiments is much higher than in laboratory or microcosm animals. Growth rates under laboratory and microcosm conditions were very similar and suggest a linear relationship between seawater TBT concentration and mussel growth. The field data show an exponential relationship. Growth rates of transplanted mussels approach those of

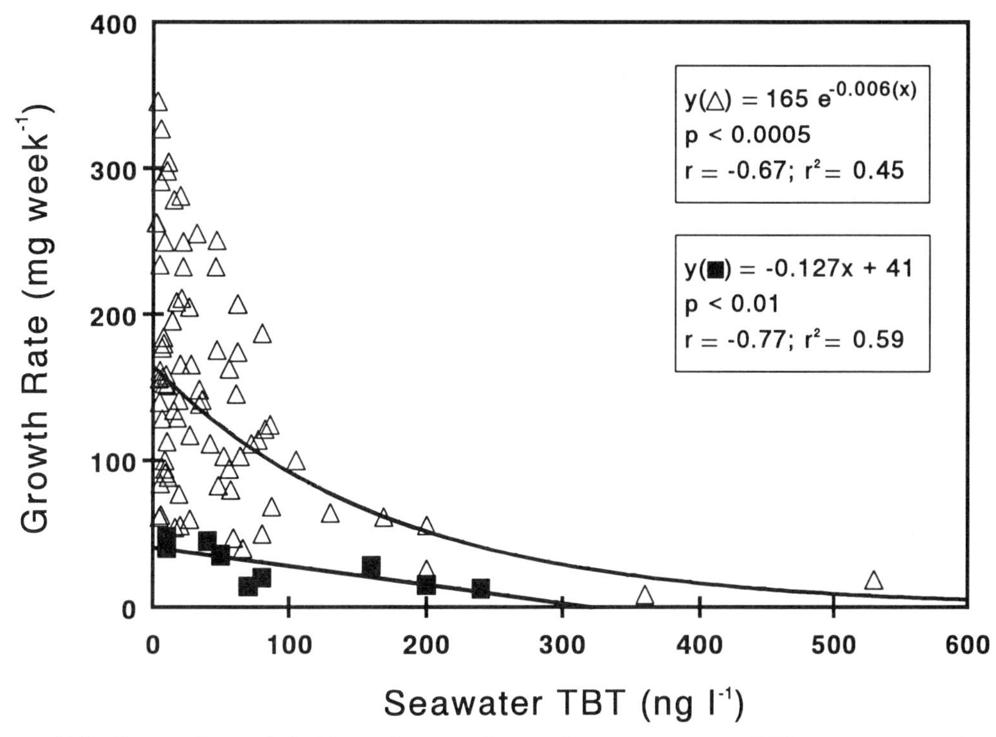

Figure 15.8 Comparison of 49-d juvenile mussel growth rates at various TBT concentrations from this and other studies. The relationship in the field (triangle) is exponential while the relationship in the laboratory and microcosm (square) is linear. The equations for the best-fit lines and relevant statistics are also given.

juvenile mussels under laboratory and microcosm conditions only when exposed to seawater TBT concentrations >100 ng l^{-1}. At seawater TBT concentrations <100 ng l^{-1}, growth rates for transplanted mussels are as much as an order of magnitude greater. Microcosm studies suggest significant reductions in juvenile mussel growth at seawater TBT concentrations between 70 and 80 ng l^{-1} (Salazar and Salazar, 1987; Salazar *et al.*, 1987). However, these mussels were stressed by overcrowding, high temperatures, and inadequate food associated with the test system and experimental procedures. These stressful conditions in laboratory and microcosm tanks may overestimate the effects of TBT in most San Diego Bay environments. They may accurately represent the most stressful San Diego Bay environments.

15.5. INFLUENCING FACTORS

15.5.1. HANDLING EFFECTS

Experimental procedures in the field can also induce stress that affects test results. Transplanting mussels to different environments and removing them from the water for growth measurements both modified performance. This was demonstrated by comparing growth rates of mussels measured weekly to those measured only at the end of the test (untouched) and by comparing bioaccumulation in transplanted and natural populations of mussels. Untouched mussels had significantly higher growth rates than animals measured weekly (Fig. 15.9). The results from five different sites confirm that frequency of measurement affects mussel growth rates. The largest differences were measured at the three highest concentrations of TBT in seawater where growth rates were approximately double those measured weekly. This suggests that stress attributable to handling enhanced the effects of TBT on juvenile mussel growth rates. Rates from biweekly measurements are similar to end-of-test measured rates and suggest they may be a reasonable indicator of growth in the natural population.

Measurements were changed from weekly to biweekly after test 4 to reduce the handling stress affecting growth rates. Increases in survival, growth, and a shift in the relationship between concentrations of TBT in seawater and in tissue after test 4 suggest that handling affected survival, bioaccumulation, and growth (Fig. 15.1). Even though concentrations of TBT in seawater were relatively constant, the largest increase in mean growth rate occurred between tests 4 and 5 when measurements were changed from weekly to biweekly. The dramatic increases in growth and survival after test 4 suggest that switching to biweekly measurements had as much or more of an effect on growth as the concentration of TBT in seawater.

Collectively, these results suggest that transplanting and frequency of measurement both affect bioaccumulation. The effects of handling on bioaccumulation are also shown in the comparison between transplanted mussels and the natural population (Fig. 15.10). At the marina site with the highest concentrations of seawater TBT, transplanted juveniles accumulated much more TBT than natural juveniles in two of the three comparisons. Differences in growth rates associated with lower concentrations of seawater TBT and reduced handling could explain higher

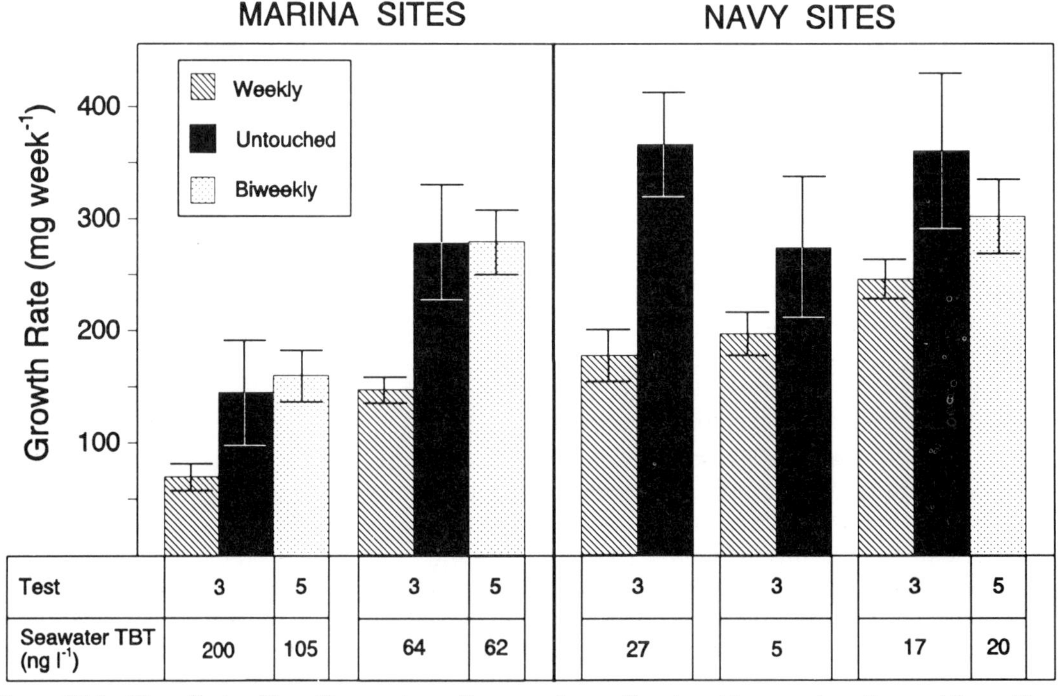

Figure 15.9 The effects of handling on juvenile mussel growth rates at two marina sites and three Navy sites on test 3. Growth rates were compared with animals measured weekly versus animals measured only at the beginning and end of the test. Growth rates for animals measured biweekly in test 5 are shown for comparative purposes. Mean concentrations of TBT in seawater and error bars (± 2 standard errors) are also given.

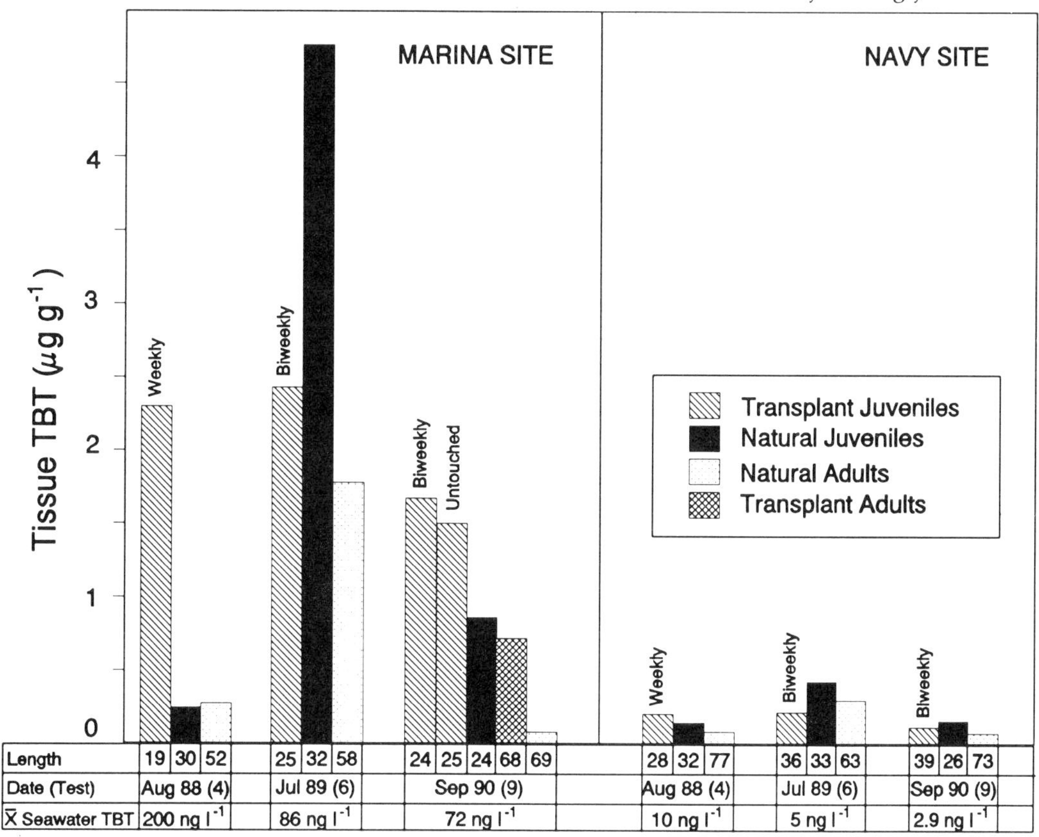

Figure 15.10 Effects of handling on bioaccumulation of TBT at the most TBT-contaminated marina site and the least TBT-contaminated Navy site. Juvenile and adult transplants as well as natural populations of juvenile and adult mussels are compared. Juvenile transplants measured biweekly for growth are also compared with those only measured at the end of the test (untouched).

accumulation than expected at lower seawater TBT concentrations. Although only one comparison was made, transplanted adults accumulated more TBT than natural adults. Juvenile transplants measured for growth on a biweekly basis did not accumulate significantly different amounts of TBT than transplants measured only at the end of the test. Although only one comparison was made, accumulation of TBT in juvenile mussel transplants measured biweekly was slightly higher than in transplants measured only at the end of the test. At the Navy site concentrations of tissue TBT were significantly lower than at the marina site and generally too low for meaningful comparisons.

15.5.2. JUVENILES VS ADULTS

It is often assumed that juveniles of most species are more sensitive to contaminants than adults. Hall and Bushong (Chapter 9) suggest such a difference in sensitivity to TBT based on laboratory studies. In the field, rapid growth and low mortality have been associated with smaller animals while slow growth and high mortality have been associated with larger individuals (Freeman and Dickie, 1979). Juvenile mussels in our field studies exhibited higher survival when exposed to TBT than adults in a laboratory study (Valkirs *et al.*, 1987). These apparent differences between juveniles and adults could be attributable to the differences in laboratory and field experiments. However, test animals in this particular laboratory study were under severe nutritive stress, and mortalities were highest among the largest mussels. Their findings and our results are consistent with relationships established from physiological measurements in the laboratory and field observations. Laboratory experiments demonstrate increasing energy losses from respiration as a function of mussel size (Bayne *et al.*, 1985). These energy losses are enhanced by increasing temperature and decreasing food ration. Additional energy losses occur with gametogenesis. Therefore, under natural conditions the highest mortalities and lowest growth rates would be expected in the largest mussels during gametogenesis when temperatures are highest and chlorophyll-*a* lowest (Freeman and Dickie, 1979; Incze *et al.*, 1980; Bayne *et al.*, 1985; Mallet *et al.*, 1990). The introduction of TBT or any other contaminant could significantly increase the effects of these stressful conditions.

It is well established that juvenile mussels grow faster than adults (Rodhouse *et al.*, 1986), but reported effects of size and age on bioaccumulation are conflicting (Phillips, 1980). Fischer (1983, 1988) suggests that bioaccumulation in juvenile mussels represents a short-term integration while bioaccumulation in adults represents a long-term integration. Many authors have reported that smaller mussels accumulate more heavy metals than larger mussels (Lobel and Wright, 1982; Ritz *et al.*, 1982; Calabrese *et al.*, 1984; Amiard *et al.*, 1986). Juvenile oysters have been shown to accumulate more TBT than adult oysters (Ebdon *et al.*, 1989). In our study juvenile mussels generally accumulated more TBT than adults (Fig. 15.10). Our field-transplanted juveniles also accumulated more TBT than adult mussels in another San Diego Bay monitoring study of natural populations (Grovhoug *et al.*, Chapter 25). The differences in accumulated TBT may be due to the combined effects of size, handling, and natural factors.

15.5.3. NATURAL FACTORS

Even in the absence of contaminants, natural factors affect survival, bioaccumulation, and growth. They could also alter the effects of TBT. Important natural factors include temperature, food, current speed, salinity, suspended sediment, and tidal position (Seed, 1976; Newell, 1979; Kiorboe *et al.*, 1981). In our field-transplant studies there is no statistically significant relationship between chlorophyll-*a* (food) and growth rate, but chlorophyll-*a* measurements may not provide the best estimate of available food. Particulate organic carbon has been related to growth in southern California coastal waters (Page and Hubbard, 1987). Our weekly and biweekly chlorophyll-*a* measurements may have been too infrequent to detect a statistically significant effect. On the other hand, chlorophyll-*a* levels in San Diego Bay may not be a limiting factor. Due to a paucity of freshwater inputs, salinity is not highly variable in most areas of San Diego Bay and is probably not a significant factor in mussel growth rates. Salinity could be an important factor in other more typical estuaries. Current speed appeared to be very important, but measurements were too infrequent to quantify a significant relationship. Suspended sediment was not measured but has been very important in other TBT studies (Waldock and Thain, 1983).

Temperature had more of an effect on growth than any natural factor quantified in our studies. There is a significant linear relationship between temperature and growth rate ($p<0.0005$), but only 11% of the variance in growth can be explained by temperature. The interaction between temperature and seawater TBT on growth of juvenile mussels is shown in Fig. 15.11. The three-dimensional surface plot predicts optimum growth near 20°C and at the lowest concentrations of TBT in seawater. Optimum growth near 20°C has

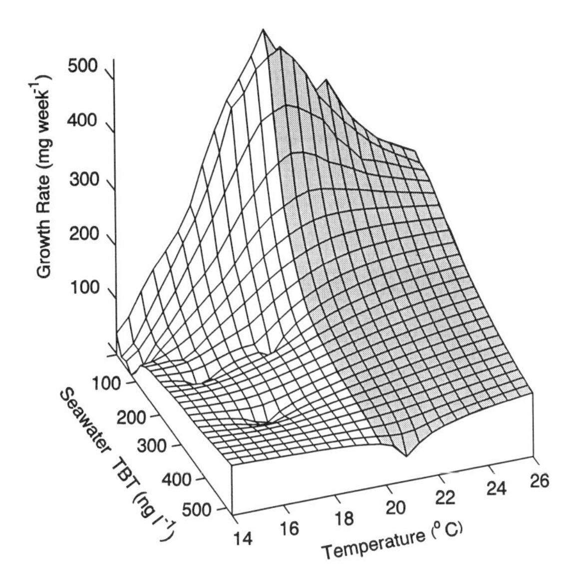

Figure 15.11 Effects of seawater TBT concentration and temperature on growth predicted from three-dimensional surface plots using weighted means. Relative growth is the *z*-axis. Shaded area represents growth rate reductions at temperatures above 20°C. Reduced growth at low temperatures (<16°C) and high concentrations of TBT in seawater (>100 ng l^{-1}) are also apparent.

been reported in several other laboratory and field studies (Incze *et al.*, 1980; Almada-Villela *et al.*, 1982; Bayne *et al.*, 1985). The lowest growth rates are predicted at the highest TBT concentrations (>100 ng l^{-1}) and temperature extremes (>22°C, <16°C). Therefore, the highest and lowest temperatures may impose natural limits on mussel growth rates and survival in San Diego Bay (Wells and Gray, 1960). It is possible that high summer temperatures alone adversely affect survival, bioaccumulation, and growth, and limit natural populations in the southern portion of San Diego Bay.

Comparing the correlation coefficients of growth rate versus temperature and concentrations of TBT in seawater shows how the relationships change (Fig. 15.12). This also establishes a statistical correlation for the graph in Fig. 15.11. There is a significant linear relationship between growth and temperature in the range of 14–20°C ($p<0.0001$, $r = 0.50$, $r^2 = 0.25$). The relationship improves when the seawater TBT data <100 ng l^{-1} are analyzed separately ($p<0.0001$, $r = 0.57$, $r^2 = 0.32$). The combination of TBT masking temperature effects and too few exposures >20°C precluded detecting a statistically significant relationship at higher temperature ranges. Considering all the data, growth rates are correlated with both seawater TBT concentration ($r = -0.72$) and temperature ($r = +0.33$); although TBT concentration is more important. Using only seawater TBT data <100 ng l^{-1}, the importance of seawater TBT concentration is reduced by half. Figure 15.12 also shows the dramatic shift from a positive to a negative correlation at temperatures above 20°C that is consistent with the literature. TBT appears to modify the effects of temperature and temperature appears to modify the effects of TBT on mussel performance.

15.6. MUSSEL BIOINDICATOR MODEL

White (1984) has cautioned against the arbitrary use of mussel monitoring systems without developing a model to be tested. The mussel bioindicator model in Fig. 15.13 emphasizes the importance of natural factors in modifying the environmental effects of TBT in San Diego Bay and depicts the inherent cycles of natural factors, TBT inputs, and mussel biology. It is suggested that natural factors act directly on TBT by altering bioavailability and directly on mussels by altering biochemistry and physiology. Other contaminants are also involved. For example, the marina with the highest seawater TBT concentrations also had the copper concentrations (~10 μg l^{-1}) (Krett Lane, 1980; Johnston, 1989). These copper concentrations were approximately 150 times higher than seawater TBT concentrations and about three times higher than the US Environmental

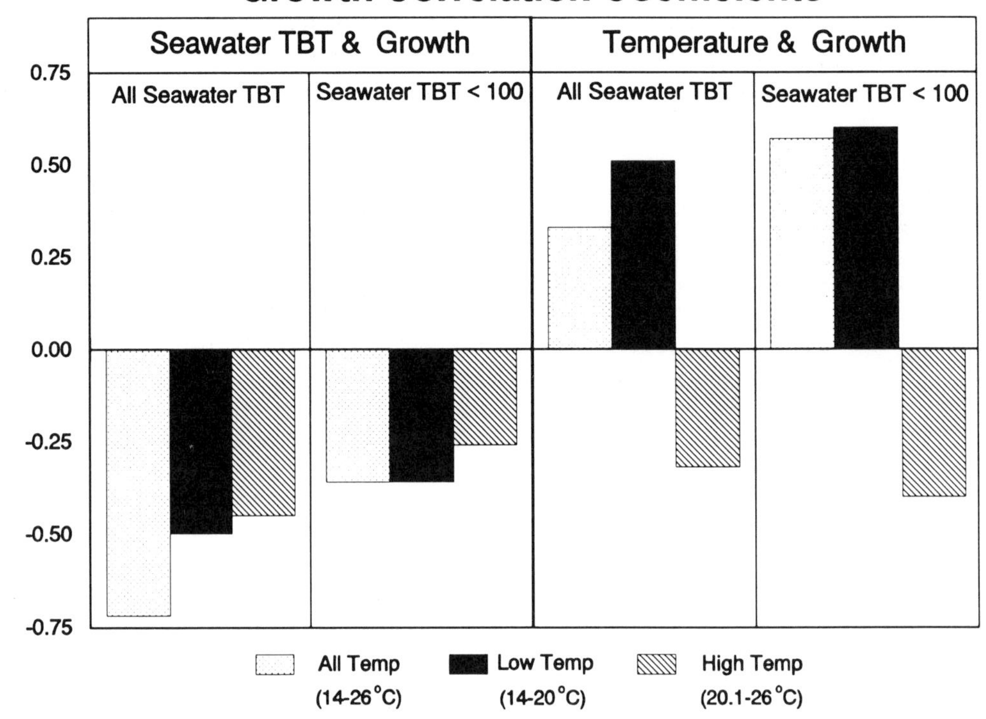

Figure 15.12 Comparison of correlation coefficients of growth rate versus temperature and concentrations of TBT in seawater to show how the relationship with growth changes at different temperatures and TBT exposures.

Protection Agency (US EPA) water quality criterion (US EPA, 1985). Similar copper concentrations reduced mussel growth rates in laboratory studies (Stromgren, 1982; Manley *et al.*, 1984). Even though they are much less toxic than TBT, accumulated petroleum hydrocarbons were shown to have more of an adverse effect on mussel physiology than accumulated TBT (Widdows *et al.*, 1990).

In addition to being modified by natural factors and TBT, mussel biology is also affected by internal biochemical and physiological cycles. These factors act in concert to modify mussel growth, bioaccumulation, and survival. The key to calibrating the mussel bioindicator is separating the effects of natural and biological factors from the effects of TBT. The clear arrows in the model (Fig. 15.13) represent the direct effects of relatively uncontaminated environments on mussel growth and survival. The dark arrows show direct effects from TBT contamination on growth, bioaccumulation, and survival.

This mussel bioindicator model has been developed based on field studies in San Diego Bay and existing knowledge of mussels and TBT. It demonstrates the difficulties in quantifying bioindicator responses to TBT by illustrating the complex and dynamic interactions among various factors that affect mussel growth, bioaccumulation, and survival. All these factors must be measured to test and verify the model and calibrate the bioindicator. The model presented here could be used for other contaminants and other bioindicators; however, given the unique and variable mussel responses to TBT exposures in San Diego Bay, extreme caution should be

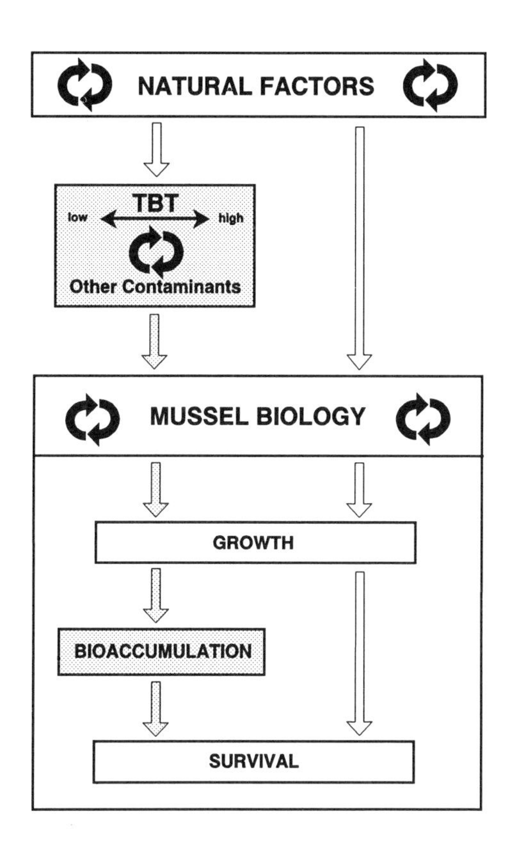

Figure 15.13 Conceptual mussel bioindicator model developed from the results of these studies. The model emphasizes the importance of natural factors in regulating mussel biology and modifying the environmental effects of TBT.

used in extrapolating specific results to other environments, contaminants, or bioindicators.

15.7. CONCLUSIONS

This TBT study in San Diego Bay demonstrates the utility of juvenile mussels as biological indicators. The most significant refinement in approach was the integration of chemical and biological measurements. Chemical measurements of TBT in seawater and mussel tissues were combined with biological response measurements of survival and growth of transplanted mussels. Mussels were used as indicators of contamination and effects. By using this approach and transplanting mussels under natural test conditions, we were able to quantify the distribution and effects of TBT, identify short- and long-term trends, and refine the use of mussels as biological indicators. The frequency of water sampling combined with multiple measurements on the same individual mussels of a very restricted size range facilitated defining statistically significant relationships. The lack of replication for tissue residue may have precluded defining a better relationship between tissue TBT concentration and juvenile mussel growth rates.

It should be emphasized that a statistically significant relationship does not prove an environmentally significant relationship. Conversely, not finding a statistically significant relationship could be attributed to the influence of various factors identified here. Zar (1974) states:

> Although in many cases there is a mathematical dependence of *Y* on *X*, it cannot automatically be assumed that there is a biological cause-and-effect relationship. Causal relationships are concluded only with some insight into the natural phenomenon being investigated and may not be concluded by statistical testing alone. It must also be remembered that a regression function is mathematically nothing more than a line forced to fit between a set of data points, and may not at all describe a natural phenomenon. Although an empirically derived regression function often provides a satisfactory and satisfying description of a natural system, sometimes it does not.

Identifying some of the natural and pollution-related factors affecting mussel performance when exposed to TBT under natural conditions is only the first step in calibrating the field bioindicator and verifying the model. Crucial questions regarding environmental fate and effects of TBT remain unanswered. Factors identified in this study as affecting mussel performance need further investigation. Other contaminants and natural factors require additional study. Differences in results between the laboratory and the field require meaningful explanations. All test conditions modifying results must be identified and their relative influence quantified.

ACKNOWLEDGMENTS

We thank Team Hydride (S. Frank, J. Guerrero, M. Kram, M. Stallard, P. Stang, A. Valkirs, and J. Vaughn) for measuring TBT in >700 water samples. We also thank S. Cola and L. Kear for measuring TBT in >100 tissue samples, B. Davidson for assistance and the Office of Chief of Naval Research through Naval Sea Systems Command.

EPILOGUE

While it is beyond the scope of this chapter to provide all the significant findings and paradigms established since this study was conducted relative to TBT and mussels, we would like to provide an updated perspective. This new perspective will emphasize new results and new ideas concerning the three primary goals of our original work: (1) effects of TBT on juvenile mussels, (2) trends in the distribution of TBT and its effects, and (3) refinements in the mussel bioindicator system.

Most importantly, threshold concentrations predicted in this chapter to always reduce growth in juvenile mussels have not changed. Threshold concentrations for no effects remain speculative based on our data, but are probably lower than originally predicted. Although the data provided in this chapter regarding the size effects on bioaccumulation of TBT were only preliminary, additional San Diego Bay studies conducted in 1990–1991 and 1993, and a Puget Sound study, have confirmed that smaller mussels consistently accumulate higher concentrations of TBT in their tissues and consistently have higher rates of survival (Salazar and Salazar, 1995; Salazar *et al.*, 1995). The interaction between size effects, concentration effects, and natural factors in determining the time required for TBT to reach equilibrium in mussel tissues remains unclear. Although the 1990–1991 study showed that small, fast-growing mussels could reach equilibrium in less than 3 weeks at low TBT concentrations, we still recommend an exposure period of 60–90 d based on the time to reach equilibrium for larger mussels at high TBT concentrations (Salazar and Salazar, 1995). The trends in distribution of TBT and its effects have not changed. The 1993 San Diego Bay study shows that mussel growth rates were higher and tissue TBT concentrations lower than in 1990. This suggests that mussel growth rates have increased as a result of the decreases in TBT.

The mussel monitoring system described in this chapter is operational. A successful test could be conducted using the protocols outlined in Salazar and Salazar (1995). We continue to add improvements in logistics, handling, and interpretation from insight gained in subsequent tests. From 1990 to 1995, we have conducted transplants using six different bivalve species and over 15 000 individuals. The major conceptual refinements in experimental design that we originally introduced in this chapter remain unchanged. Since a 2 mm size range is not always achievable, 10 mm is recommended as a target size range. Absolute size is less important than minimizing the range. Other logistic refinements include using disposable mesh bags with individual compartments instead of rigid plastic cages, and interfacing digital calipers and balances with laptop computers to facilitate data entry in the field.

Although our standard protocol includes measuring whole-animal wet-weights and lengths, shell weights, and tissue weights, we only presented the whole-animal wet-weight data in this chapter. Several subsequent studies have shown that under certain conditions, estimates of growth based on tissue weights are more informative than whole-animal growth. We are in the process of reanalyzing the tissue weight data from these nine transplant experiments to see if these data can provide any further clarification regarding TBT effects on mussels. Our most recent statistical analysis of copper and zinc tissue data collected during tests 5, 6, 7, 9 have shown significant relationships with juvenile mussel growth rates (Salazar and Chadwick, 1991; Salazar and Salazar, 1995). The data further indicate that, in addition to TBT, smaller mussels accumulated higher concentrations of several metals than larger mussels (Salazar and Salazar, 1995; Salazar *et al.*, 1995). Some of the relationships were more informative by analyzing the data on a per animal basis (content) rather than on a concentration basis.

In the final analysis, bivalves in cages could be viewed as an exposure system to make any clinical measurements. This versatility is another advantage of using bivalve transplants. Further, the discriminating power of bivalve monitoring is well beyond the order-of-magnitude differences initially suggested by Goldberg *et al.* (1978) as a limit for detecting significant differences in concentrations of tissue contaminants. A number of studies, including our own, have often found statistically significant differences in both bioaccumulation and growth that we believe are environmentally significant. Some of these site-specific data differ by a factor of 2 or less. In-situ, studies with caged bivalves facilitate measurements that make this field bioassay similar to laboratory bioassays in terms of experimental control and predictive power, and retain the environmental realism of traditional field monitoring. Approaches that combine the integrating ability of filter feeding bivalves, the versatility of in-situ caging, and the relationships between bioaccumulation and bioeffects as demonstrated here, have a number of applications for environmental monitoring, risk assessment, and establishing regulatory criteria.

REFERENCES

Almada-Villela, P.C., J. Davenport and L.D. Gruffydd. 1982. The effects of temperature on the shell growth of young *Mytilus edulis* L. *J. Exp. Mar. Biol. Ecol.*, **59**, 275–288.

Amiard, J.C., D. Amiard-Triquet, B. Berthet and C. Metayer. 1986. Contribution to the ecotoxicological study of cadmium, lead, copper and zinc in the mussel *Mytilus edulis*. I. Field study. *Mar. Biol.*, **90**, 425–431.

Bayne, B.L. 1973. Physiological changes in *Mytilus edulis* L. induced by temperature and nutritive stress. *J. Mar. Biol. Ass. UK*, **53**, 39–58.

Bayne, B.L. and J. Thompson. 1970. Some physiological consequences of keeping *Mytilus edulis* in the laboratory. *Helgo. Meeres.*, **20**, 526–552.

Bayne, B.L., D.A. Brown, K. Burns, D.R. Dixon, A. Ivanovici, D.R. Livingstone, D.M. Lowe, N.M. Moore, A.R.D. Stebbing and J. Widdows. 1985. *The Effects of Stress and Pollution on Marine Animals*. Praeger Special Studies, Praeger Scientific, New York, 384 pp.

Cain, D.J. and S.N. Luoma 1990. Influence of seasonal growth, age and environmental exposure on Cu and Ag in a bivalve indicator, *Macoma balthica*, in San Francisco Bay. *Mar. Ecol. Prog. Ser.*, **60**, 45–55.

Cairns, J., Jr. 1988. What constitutes field validation of predictions based on laboratory evidence. In: *Aquatic Toxicology and Hazard Assessment*, Vol. 10, W.J. Adams, G.A. Chapman and W.G. Landis (Eds). American Society for Testing and Materials, Philadelphia, pp. 361–368.

Cairns, J., Jr. and A.L. Buikema, Jr. 1984. Verifying predictions of environmental safety and harm. In: *Concepts in Marine Pollution Measurements*, H.H. White (Ed.). Maryland Sea Grant Publication, University of Maryland, College Park, pp. 81–111.

Calabrese, A., J.R. MacInnes, D.A. Nelson, R.A. Greig and P.P. Yevich. 1984. Effects of long-term exposure to silver or copper on growth, bioaccumulation and histopathology in the blue mussel, *Mytilus edulis*. *Mar. Environ. Res.*, **11**, 253–275.

Chapman, G.A. 1983. Do organisms in laboratory toxicity tests respond like organisms in nature? In: *Aquatic Toxicology and Hazard Assessment*,

Sixth Symposium, W.E. Bishop, R.D. Cardwell and B.B. Heidolph (Eds). American Society for Testing and Materials, Philadelphia, pp. 315–327.

Chapman, P.M. and E.R. Long. 1983. The use of bioassays as part of a comprehensive approach to marine pollution assessment. *Mar. Poll. Bull.*, **14**, 81–84.

Ebdon, L., K. Evans and S. Hill. 1989. The accumulation of organotins in adult and seed oysters from selected estuaries prior to the introduction of U.K. regulations governing the use of tributyltin-based antifouling paints. *Sci. Total Environ.*, **83**, 63–84.

Farrington, J.W., E.D. Goldberg, R.W. Risebrough, J.H. Martin and V.T. Bowen. 1983. 'Mussel Watch' 1976–78, an overview of the trace-metal, DDE, PCB, hydrocarbon, and artificial radionuclide data. *Environ. Sci. Tech.*, **17**, 490–496.

Fischer, H. 1983. Shell weight as an independent variable in relation to cadmium content of molluscs. *Mar. Ecol. Prog. Ser.*, **12**, 59–75.

Fischer, H. 1988. *Mytilus edulis* as a quantitative indicator of dissolved cadmium. Final study and synthesis. *Mar. Ecol. Prog. Ser.*, **48**(2), 163–174.

Freeman, K.R. and L.M. Dickie. 1979. Growth and mortality of the blue mussel (*Mytilus edulis*) in relation to environmental indexing. *J. Fish Res. Bd Can.*, **36**, 1238–1249.

Goldberg, E.D., V.T. Bowen, J.W. Farrington, G. Harvey, J.H. Martin, P.L. Parker, R.W. Risebrough, W. Robertson, E. Schneider and E. Gamble. 1978. The mussel watch. *Environ. Conserv.*, **5**(2), 101–125.

Goldberg, E.D., M. Koide, V. Hodge, A.R. Flegal and J. Martin. 1983. U.S. Mussel Watch, 1977–1978 results on trace metals and radionuclides. *Estuar. Coast. Shelf Sci.*, **16**, 69–93.

Grovhoug, J.G., P.F. Seligman, G. Vafa and R.L. Fransham. 1986. Baseline measurements of butyltin in U.S. harbors and estuaries. In: Proceedings of the Oceans '86 Organotin Symposium, Vol. 4, Marine Technology Society, Washington, DC, 23–25 September 1986, pp. 1283–1288.

Incze, L.S., R.A. Lutz and L. Watling. 1980. Relationships between effects of environmental temperature and seston on growth and mortality of *Mytilus edulis* in a temperate northern estuary. *Mar. Biol.*, **57**, 147–156.

Johnston, R.K. 1989. The Response of Marine Fouling Communities to a Pollution Gradient in San Diego Bay, CA. Master's Thesis, San Diego State University, San Diego, California, 188 pp.

Kiorboe, T., F. Mohlenberg and O. Nohr. 1981. Effect of suspended bottom material on growth and energetics in *Mytilus edulis*. *Mar. Biol.*, **61**, 283–288.

Krett Lane, S.M. 1980. Productivity and diversity of phytoplankton in relation to copper levels in San Diego Bay. Naval Ocean Systems Center Technical Report #533. Report prepared by San Diego State University Foundation, 68 pp.

Laughlin, R.B., Jr and W. French. 1988. Concentration dependence of bis(tributyl)tin accumulation in the marine mussel *Mytilus edulis*. *Environ. Toxicol. Chem.*, **7**, 1021–1026.

Laughlin, R.B., Jr, W. French and H.E. Guard. 1986. Accumulation of bis(tributyltin) oxide by the marine mussel *Mytilus edulis*. *Environ. Sci. Technol.*, **20**, 884–890.

Lobel, P.B. and D.A. Wright. 1982. Relationship between body zinc concentration and allometric growth measurements in the mussel *Mytilus edulis*. *Mar. Biol.*, **66**, 145–150.

Luoma, S.N. 1983. Bioavailability of trace metals to aquatic organisms – a review. *Sci. Total Environ.*, **28**, 1–22.

Malins, D.C., H.O. Hodgins, U. Varanasi, W.D. MacLeod, Jr., B.B. McCain and S.-L. Chan. 1984. Chemical measurements of organic pollutants and effects criteria. In: *Concepts in Marine Pollution Measurements*, H.H. White (Ed.). Maryland Sea Grant Publication, University of Maryland, College Park, pp. 405–426.

Mallet, A.L., C.E.A. Carver, S.S. Coffen and K.R. Freeman. 1987. Mortality variations in natural populations of the blue mussel, *Mytilus edulis*. *Can J. Fish. Aquat. Sci.*, **44**, 1589–1594.

Mallet, A.L., C.E.A. Carver and K.R. Freeman. 1990. Summer mortality of the blue mussel in eastern Canada: spatial, temporal, stock and age variation. *Mar. Ecol. Prog. Ser.*, **67**, 35–41.

Manley, A.R., L.D. Gruffydd and P.C. Almada-Villela. 1984. The effect of copper and zinc on the shell growth of *Mytilus edulis* measured by a laser diffraction technique. *J. Mar. Biol. Ass. UK*, **64**, 417–427.

Moller, H. 1987. The marine ecologist – scientist or advocate of nature? *Mar. Pollut. Bull.*, **18**, 267–270.

Newell, R.C. 1979. *Biology of Intertidal Animals*, 3rd edn, Marine Ecological Surveys Ltd, Faversham, Kent, 781 pp.

Page, H.M. and D.M. Hubbard. 1987. Temporal and spatial patterns of growth in mussels *Mytilus edulis* on an offshore platform, relationships to water temperature and food availability. *J. Exp. Mar. Biol. Ecol.*, **111**, 159–179.

Page, D.S. and J. Widdows. 1990. Temporal and spatial variation in levels of alkyltins in mussel tissues, a toxicological interpretation of field data. In: Proceedings of the 3rd International Organotin Symposium, Monaco, 17–20 April 1990, pp. 127–131.

Parrish, P.R., K.L. Dickson, J.L. Hamelink, R.A. Kimerle, K.J. Macek, F.L. Mayer, Jr and D.I. Mount. 1988. Aquatic toxicology, ten years in review and a look at the future. In: *Aquatic Toxicology and Hazard Assessment*, Vol. 10, W.J. Adams, G.A. Chapman and W.G. Landis (Eds). American Society for Testing and Materials, Philadelphia, pp. 7–25.

Peddicord, R.K. 1984. What is the meaning of bioaccumulation as a measure of marine pollution effects? In: *Concepts in Marine Pollution Measurements*, H.H. White (Ed.). Maryland Sea Grant Publication, University of Maryland, College Park, pp. 249–259.

Phillips, D.J.H. 1980. *Qualitative Aquatic Biological Indicators*. Applied Science Publishers, London, 488 pp.

Phillips, D.J.H. and D.A. Segar. 1986. Use of bioindicators in monitoring conservative contaminants, programme design imperatives. *Mar. Pollut. Bull.*, **17**, 10–17.

Ritz, D.A., R. Swain and N.G. Elliot. 1982. Use of the mussel *Mytilus edulis* planulatus (Lamarck) in monitoring heavy metal levels in seawater. *Aust. J. Mar. Freshwat. Res.*, **33**, 491–506.

Rodhouse, P.G., J.H. McDonald, R.I.E. Newell and R.K. Koehn. 1986. Gamete production, somatic growth and multiple-locus enzyme heterozygosity in *Mytilus edulis*. *Mar. Biol.*, **90**, 209–214.

Salazar, M.H. 1986. Environmental significance and interpretation of organotin bioassays. In: Proceedings of the Oceans '86 Organotin Symposium, Vol. 4, Marine Technology Society, Washington, DC, 23–25 September 1986, pp. 1240–1245.

Salazar, M.H. 1989. Mortality, growth and bioaccumulation in mussels exposed to TBT: differences between the laboratory and the field. In: Proceedings of the Oceans '89 Organotin Symposium, Vol. 2, Seattle, Washington, 18–21 September 1989, pp. 530–536.

Salazar, M.H. and D.B. Chadwick. 1991. Using real-time physical/chemical sensors and in-situ biological indicators to monitor water pollution. In: L.C. Wrobel and C.A. Brebbia (Eds), *Water Pollution: Modeling, Measuring and Prediction*. First International Conference on Water Pollution Modeling, Measuring and Prediction, 1991. Elsevier Applied Science, London, pp. 463–480.

Salazar, M.H. and M.A. Champ. 1988. Tributyltin and water quality, A question of environmental significance. In: Proceedings of the Oceans '88 Organotin Symposium, Vol. 4, Baltimore, Maryland, 31 October–2 November 1988, pp. 1497–1506.

Salazar, M.H. and S.M. Salazar. 1987. TBT effects on juvenile mussel growth. In: Proceedings of the Oceans '87 Organotin Symposium, Vol. 4, Halifax, Nova Scotia, 28 September–1 October 1987, pp. 1504–1510.

Salazar, M.H. and S.M. Salazar. 1988. Tributyltin and mussel growth in San Diego Bay. Proceedings of the Oceans '88 Organotin Symposium, Vol. 4, Baltimore, Maryland, 31 October–2 November 1988, pp. 1188–1195.

Salazar, M.H. and S.M. Salazar. 1989. Acute effects of (bis)tributyltin oxide on marine organisms, summary of work performed 1981–1983. Naval Ocean Systems Technical Report #1299, 78 pp.

Salazar, M.H. and S.M. Salazar. 1991a. Mussels as bioindicators: a case study of tributyltin effects in San Diego Bay. In: Proceedings of the 17th Annual Aquatic Toxicity Workship, Vancouver, Canada, 5–7 November 1990, P.M. Chapman, F.S. Bishay, E.A. Power, K. Hall, L. Harding, D. McLeay and M. Nassichuk (Eds). *Can. Tech. Report. Fish. Aquat. Sci.*, **1774**, 47–75.

Salazar, M.H. and S.M. Salazar. 1991b. Utility of mussel growth in assessing the environmental effects of tributyltin. In: Proceedings of the 3rd International Organotin Symposium, Monaco, 17–20 April 1990, pp. 132–137.

Salazar, M.H. and S.M. Salazar. 1991c. Assessing site-specific effects of TBT contamination with mussel growth rates. *Mar. Environ. Res.*, **32**, 131–150.

Salazar, M.H. and S.M. Salazar, 1995. *In situ* bioassays using transplanted mussels: I. Estimating chemical exposure and bioeffects with bioaccumulation and growth. In: J.S. Hughes, G.R. Biddinger, and E. Mones (Eds), Third Symposium on Environmental Toxicology and Risk Assessment, STP 1218. American Society for Testing and Materials, Philadelphia, 1994, pp. 216–241.

Salazar, S.M., B.M. Davidson, M.H. Salazar, P.M. Stang and K. Meyers-Shulte. 1987. Field assessment of a new site-specific bioassay system. In: Proceedings of the Oceans '87 Organotin Symposium, Vol. 4, Halifax, Nova Scotia, 28 September–1 October 1987, pp. 1461–1470.

Salazar, M.H., P.B. Duncan, S.M. Salazar and K.A. Rose. 1995. *In situ* bioassays using transplanted mussels: II. Assessing contaminated sediment at a Superfund site in Puget Sound. In: J.S. Hughes, G.R. Biddinger, and E. Mones (Eds), Third Symposium on Environmental Toxicology and Risk Assessment, STP 1218. American Society for Testing and Materials, Philadelphia, 1994, pp. 242–263.

Seed, R. 1976. Ecology. In: *Marine Mussels, Their Ecology and Physiology*, B.L. Bayne (Ed.), Cambridge University Press, Cambridge, pp. 13–65.

Short, J.W. and J.L. Sharp. 1989. Tributyltin in bay mussels (*Mytilus edulis*) of the Pacific coast of the United States. *Environ. Sci. Technol.*, **23**, 740–743.

Stallard, M.O., S.Y. Cola and C.A. Dooley. 1988. Optimization of butyltin measurements for seawater, tissue and marine sediment samples. *Appl. Organometal. Chem.*, **3**, 105–114.

State of California. 1988. Title 3, California Code of Regulations. Section 6488–6489.

Stebbing, A.R.D. 1985. Organotins and water quality – some lessons to be learned. *Mar. Pollut. Bull.*, **16**, 383–390.

Stephenson, M.D., D.R. Smith, J. Goetzl, G. Ichikawa and M. Martin. 1986. Growth abnormalities in mussels and oysters from areas with high levels of tributyltin in San Diego Bay. In: Proceedings of the Oceans '86 Organotin Symposium, Vol. 4, Marine Technology Society, Washington, DC, 23–25 September 1986, pp. 1246–1251.

Stromgren, T. 1982. Effect of heavy metals (Zn, Hg, Cu, Cd, Pb, Ni) on the length growth of *Mytilus edulis*. *Mar. Biol.*, **72**, 69–72.

Stromgren, T. and T. Bongard. 1987. The effect of tributyltin oxide on growth of *Mytilus edulis*. *Mar. Pollut. Bull.*, **18**, 30–31.

Thain, J.E. 1983. The acute toxicity of bis(tributyl tin) oxide to the adults and larvae of some marine organisms. ICES Paper CM 1983/E,13, International Council for the Exploration of the Sea, Copenhagen, 4 pp.

Thain, J.E. 1986. Toxicity of TBT to bivalves, Effects on reproduction, growth and survival. In: Proceedings of the Oceans '86 Organotin Symposium, Vol. 4, Washington, DC, 23–25 September 1986, pp. 1306–1313.

Thain, J.E. and M.J. Waldock 1985. The growth of bivalve spat exposed to organotin leachates from antifouling paints. ICES Paper CM 1985/E,28, International Council for the Exploration of the Sea, Copenhagen, 6 pp.

US Environmental Protection Agency. 1985. Ambient water quality criteria for copper – 1984. EPA publication 440/5-84-031. Office of Water Regulations and Standards, Criteria and Standards Division, Washington, DC, pp. 142–150.

Uhler, A. D., T.H. Coogan, K.S. Davis, G.S. Durell, W.G. Steinhauer, S.Y. Freitas and P.D. Boehm. 1989. Findings of tributyltin, dibutyltin and monobutyltin in bivalves from selected U.S. coastal waters. *Environ. Toxicol. Chem.*, **8**, 971–979.

Valkirs, A.O., B.M. Davidson and P.F. Seligman. 1987. Sublethal growth effects and mortality to marine bivalves from long-term exposure to tributyltin. *Chemosphere*, **16**, 201–220.

Venrick, E.L. and T.L. Hayward. 1984. Determining Chlorophyll on the 1984 CALCOFI Surveys. California Cooperative Oceanic Fisheries Investigations Report, Vol. 25, pp. 74–79.

Wade, T.L., B. Garcia-Romero and J.M. Brooks. 1988. Tributyltin contamination in bivalves from United States coastal estuaries. *Environ. Sci. Technol.*, **22**, 1488–1493.

Waldock, M.J. and J.E. Thain. 1983. Shell thickening in *Crassostrea gigas*, organotin antifouling or sediment induced? *Mar. Pollut. Bull.*, **14**, 411–415.

Wells, H.W. and I.E. Gray. 1960. The seasonal occurrence of *Mytilus edulis* on the Carolina coast as a result of transport around Cape Hatteras. *Biol. Bull.*, **119**, 550–559.

White, H.H. 1984. Mussel madness, use and misuse of biological monitors of marine pollution. In: *Concepts in Marine Pollution Measurements*, H.H. White (Ed.). Maryland Sea Grant Publication, University of Maryland, College Park, pp. 325–337.

White, H.H. and M.A. Champ. 1983. The great bioassay hoax, and alternatives. *In: Hazardous and Industrial Solid Waste Testing*, Second Symposium, R.A. Conway and W.P. Gulledge (Eds), American Society for Testing and Materials, Philadelphia, pp. 299–312.

Widdows, J., K.A. Burns, N.R. Menon, D.S. Page and S. Soria. 1990. Measurement of physiological energetics (scope for growth) and chemical contaminants in mussels (*Arca zebra*) transplanted along a pollution gradient in Bermuda. *J. Exp. Mar. Biol. Ecol.*, **138**(1,2), 99–117.

Zar, J.H. 1974. *Biostatistical Analysis*. Prentice-Hall, Englewood Cliffs, New Jersey, 620 pp.

Zoulian, C. and A. Jensen. 1989. Accumulation of organic and inorganic tin in blue mussel, *Mytilus edulis*, under natural conditions. *Mar. Pollut. Bull.*, **20**, 281–286.

BIOACCUMULATION OF TBT BY AQUATIC ORGANISMS 16

Roy B. Laughlin, Jr

Azurea, Inc., PO Box 561178, Rockledge, Florida 32956, USA

ABSTRACT

Tributyltin (TBT) is accumulated by all taxa that have been examined. Typical tissue burdens range from undetectable levels to as high as 7 $\mu g\ g^{-1}$ (wet weight). Molluscs, as a group, exhibit the highest tissue burdens and also the highest bioaccumulation factors

Organotin. Edited by M.A. Champ and P.F. Seligman. Published in 1996 by Chapman & Hall, London. ISBN 0 412 58240 6

(BAF). Some microorganisms are also notable accumulators, but in this case adsorption to the biofilm (predominantly polysaccharides) rather than sequestration within cells is an important mechanism of uptake. Crustaceans and fish generally accumulate lower burdens of TBT, probably because they possess enzymatic capability to degrade and excrete TBT. Bioaccumulation factors (BAF) as high as 50 000 have been reported, but the route of uptake (e.g. from water or food) was not necessarily differentiated during the experiment. Experimental studies have specifically examined uptake from water, where calculation of a bioconcentration factor (BCF) is appropriate. BCF values range up to 10 000. Both BAF and BCF values are higher than would be caused by partitioning processes alone. Binding is thus proposed as a mechanism that explains enhanced accumulation. A scheme to explain bioaccumulation is presented that proposes a role for (a) chemical speciation of butyltin with major anions in natural waters, (b) routes of exposure, (c) role for partitioning and binding to control both the kinetics of uptake and burdens in tissues, and (d) contribution of depuration mechanisms to control tissue burdens. Findings presented in this review support the use of biomonitors for TBT contamination, but the tissue burdens of TBT in biomonitors likely reflect a steady-state distribution of TBT in all compartments of the ecosystem rather than concentration in a single compartment (e.g. water). While the topical issue of tributyltin use diminished due to restrictions on its use, this compound provides several important opportunities as a model compound to increase understanding of bioaccumulation processes and environmental compartmentalization, in general. Some further studies may thus be warranted.

16.1. INTRODUCTION

Bioaccumulation follows toxicity as the second most intensively examined aspect of the environmental behavior of TBT. Investigators have studied bioaccumulation for several reasons. These include:

1. use of biomonitors for surveillance of aquatic environments for TBT contamination;
2. analysis of seafood to prevent excessive exposure of humans to TBT;
3. characterization of the rates and extent of partitioning of TBT among environmental compartments (water, organisms, and sediment); and
4. collection of information about fundamental mechanisms of bioaccumulation of xenobiotics by aquatic organisms.

Bioaccumulation studies are more limited in number than toxicity studies because chemical analysis of tissues, in particular, has only recently been refined to a level where measurement of trace quantities of TBT was sufficiently reliable for routine analysis of the numerous samples required for bioaccumulation studies.

Bioaccumulation is a complex process of transfer of chemicals between the environment and an organism. The final result, tissue burdens, is the topic of more general interest. Laboratory studies have shown that many factors influence bioaccumulation of TBT, including exposure concentration, route of exposure (from water, food, or sediment), physicochemical form of TBT (charged, neutral, or complexed), and the ability of the organism to modify and excrete the chemical. The dominance of any one of these processes or the proportional contribution of these processes acting in concert (as occurs in natural systems) has yet to be completely elucidated. Tributyltin is a notable compound for model studies since it possesses characteristics of both organic and inorganic compounds, exhibiting the traits of the organic ligands attached to the tin atom and the chemical speciation reactions typical of tin. Chemical speciation occurs within the range of chemical activities of hydrogen ion (pH) and dominant inorganic anions (Cl^{-1} and CO_3^{2-}) typical of seawater (Laughlin *et al.*, 1986b).

A historical account of the study of bioaccumulation of TBT does not begin with strong assertions of uptake because chemical analysis techniques were not sensitive enough to measure TBT in water or tissues at nanogram-per-liter levels. Until 1985, species such as the oyster (*Crassostrea gigas*) were far more sensitive than chemical analysis techniques to demonstrate even qualitatively the presence of TBT. Use of isotopically labeled (^{14}C) tributyltin provided some of the first detailed demonstrations of significant accumulation of TBT by marine organisms (Evans and Laughlin, 1984; Laughlin *et al.*, 1986a). Subsequent refinement of chemical analysis protocols vastly improved the ability of researchers to conduct sensitive, routine analyses of TBT in water and tissue samples (Jackson *et al.*, 1982; Maguire and Tkacz, 1983; Valkirs *et al.*, 1985, 1986; Matthias *et al.*, 1986; Unger *et al.*, 1986; Sasaki *et al.*, 1988b; Weber *et al.*, Chapter 4; Matthias, Chapter 5; Unger *et al.*, Chapter 6; Huggett *et al.* Chapter 7).

16.2. EXISTENCE AND EXTENT OF UPTAKE

Laboratory and field studies of TBT demonstrate that a variety of organisms, including microorganisms, algae, invertebrates, and fish, accumulate TBT (Table 16.1). Molluscs have been the subject of the majority of field studies. They are the group demonstrating the highest burdens. Highest tissue burdens may exceed 5 $\mu g\ g^{-1}$ but are usually much lower. This is not a particularly high value in absolute terms; its importance derives from findings of dissolved TBT concentrations of 10–100 ng l^{-1} in the habitats from which samples were taken. By comparison with many other chemicals, the ambient concentrations in water are fairly low to produce such levels in tissue.

Physicochemical properties of TBT exert the predominant influence on its bioaccumulation, but environmental factors such as salinity and dissolved solids are demonstrably important because of their influence on chemical speciation of TBT and on physiological processes in bioaccumulators. Important biological factors, including taxonomic category and ecological niche, are traits that correlate with characteristics of TBT bioaccumulation. In this section, characterizations and roles of these factors in bioaccumulation of TBT will be briefly analyzed.

Uptake from water is rapid and quantitative in laboratory systems (Ward *et al.*, 1981; Laughlin *et al.*, 1986a). For example, mussels (*Mytilus edulis*) exposed to dissolved TBT reduce concentrations in water to undetectable levels in <3 h, with concomitant increasing tissue concentrations (Fig. 16.1a).

In model laboratory studies, uptake from food is also a significant route that in the field may be more significant than uptake from water (Evans and Laughlin, 1984; Laughlin *et al.*, 1986a). Using mussels (*M. edulis*) as examples, it is observed that accumulation of TBT associated with food (microalgae) is more rapid and extensive than accumulation from water (Fig. 16.1b).

The relationship between TBT concentration in water or food and tissue burdens in bivalve molluscs is particularly enigmatic since microorganisms (bacteria and microalgae) that comprise their food are, on a mass basis, overwhelmingly dominant in estuarine ecosystems. Thus, it is a reasonable assumption that microorganisms in the water column and sediments would be a first and most extensive depot of TBT released into the water column.

Ability to degrade and excrete accumulated TBT is a taxonomic trait that causes a great deal of uncertainty. Bivalves are again a good illustration. Lee (1985) published accounts of abilities of fish, crabs, and oysters to metabolize TBT. The oyster (*Crassostrea virginica*) was shown to metabolize only 10% of TBT, and

Table 16.1 Accumulation of tributyltin by marine organisms

Species or biota	*Tissue burdens*[a]	*Exposure concentration*	*Accumulation factor*[b]	*Comments*[c]	*Reference*[d]
Ankistrodesmus falcatus	155 μg g^{-1} (dry wt)	20 μg l^{-1}	10^4	Single pulsed exposure; highest cell burdens after 1 week. Substantial dealkylation	Maguire *et al.*, 1984
Legionella pneumophila	1×10^4–6×10^7 (molecules per cell)	0.017–1.12 mg l^{-1}	ND	Proportional uptake inversely proportional to concentration	Soracco and Pope, 1983
Microbial biofilm	22 and 59 mg l^{-1} 1.0 and 1.25 mg l^{-1}	ND	ND	Microbes and their biofilm collected from painted panels. Numbers are for two different types of coatings of different TBT content	Guard and Cobet, personal communication
Pseudomonas 244	21 mg g^{-1} (dry wt)	10 mg l^{-1}	487	Uptake at equilibrium in less than 4 min of exposure	Blair *et al.*, 1982
Gram-negative bacteria (eight species)	3.0–7.7 mg g^{-1} (dry wt; no glucose added)	10 mg l^{-1}	356–855	Bioaccumulation attributed to adsorption to cell envelope; non-energy dependent	Blair *et al.*, 1982
	3.2–9.2 mg g^{-1} (dry wt; glucose added)		381–1039	Bacterial isolates were tin resistant. Obtained from Chesapeake Bay	
Zosteria marina	140–800 ng g^{-1}	30 ng l^{-1}	250 100	Significant dealkylation noted	Francois *et al.*, 1989
Crassostrea gigas	1.38 μg g^{-1} 2.17 μg g^{-1}	0.15–1.25 μg l^{-1}	6 000 2 000	Laboratory study; 15-d exposure; apparent steady state	Waldock *et al.*, 1983
Crassostrea gigas	50–189 ng g^{-1}	7–14 ng l^{-1}	7 000 135 000	Pooled tissue; field study	Wolniakowski *et al.*, 1987
Crassostrea virginica	9–834 ng g^{-1}	~30 ng l^{-1}	27 800	Field study; BCF calculated for high value only (Sarah Creek)	Rice *et al.*, 1987
Macoma nasuta	0.22–2.13 μg g^{-1}	6 ng l^{-1} 204 ng l^{-1}	10 400 36 700	Microcosm with flowing seawater; antifouling leachate source of TBT; 90-d exposure	Salazar *et al.*, 1987

Mytilus edulis	gill: 2.5 μg g^{-1} visc: 2.0 μg g^{-1} mant: 2.0 μg g^{-1} musc: 0.5 μg g^{-1}	0.5 μg l^{-1}	5 000 4 000 1 500 1 000	Laboratory study; 47 d duration. No steady state observed	Laughlin *et al.*, 1986a
Mytilus edulis	gill: 2.0 μg g^{-1} visc: 4.0 μg g^{-1} mant: 2.0 μg g^{-1} musc: 1.0 μg g^{-1}	Same mass of TBT as above, associated with diatoms fed to mussels	NA	Laboratory exposure, 29 d. No steady state observed	Laughlin *et al.*, 1986a
Mytilus edulis	4–6 μg g^{-1} (dry wt)	6–53 ng l^{-1}	50 000–500 000	Field study. BCF based on dry weight calculations. On wet weight, values likely one-third of stated values	Zuolian and Jensen, 1989
Pecten maximus	1.86 μg g^{-1}	ND	ND	Juvenile scallops exposed for 16 weeks to TBT leached from pearl net enclosures	Davies *et al.*, 1986
Pecten maximus	gill: 560 ng g^{-1} visc: 510 ng g^{-1} gona: 440 ng g^{-1} musc: 430 ng g^{-1}	ND	ND	Adult scallops exposed to TBT from pearl netting enclosures. Highest burdens in 4 weeks, declining in the following 6 weeks	Davies *et al.*, 1986
Mercenaria mercenaria	visc: 220 ng g^{-1} rems: 175 ng g^{-1} musc: 115 ng g^{-1}	37 ng l^{-1}	5 950 4 750 3 100	Laboratory exposure lasting 56 d. BCF determined by analysis. No steady state; BCF determined by kinetic analysis larger by a factor of ~4	Laughlin, unpublished data
	visc: 369 ng g^{-1} rems: 310 ng g^{-1} musc: 220 ng g^{-1}	71 ng l^{-1}	5 200 4 500 3 100		
	visc: 629 ng g^{-1} rems: 581 ng g^{-1} musc: 424 ng g^{-1}	121 ng l^{-1}	5 200 6 100 3 400		
	visc: 1 122 ng g^{-1} rems: 1 342 ng g^{-1} musc: 748 ng g^{-1}	220 ng l^{-1}	5 100 6 100 3 400		
	visc: 3 146 ng g^{-1} rems: 2 928 ng g^{-1} musc: 2 081 ng g^{-1}	484 ng l^{-1}	6 500 6 050 4 200		

Table 16.1 Continued

Species or biota	*Tissue burdens*[a]	*Exposure concentration*	*Accumulation factor*[b]	*Comments*[c]	*Reference*[d]
Nassarius obsoletus	>20 ng g^{-1} 620 ng g^{-1} (dry wt)	8–16 ng l^{-1}	45 000	First value is for York River, uncontaminated. Second is for Sarah Creek, contaminated	Bryan *et al.*, 1989
Nucella lapillus	14.7–17.2 µg g^{-1} (dry wt)	ND	ND	Field monitoring study. Highest tissue burdens occur in spring; males have slightly higher mean burdens	Bryan *et al.*, 1987
Nucella lapillus	8 µg g^{-1} (dry wt)	760 ng l^{-1}	10 500	Exposure lasted 60 d. No steady state	Bryan *et al.*, 1987
Nucella lapillus	0.1–0.2 µg g^{-1} 2 µg g^{-1} (dry wt)	1.5–18.7 ng l^{-1} 107 ng l^{-1}	100 000 10 000	Tissue burdens varied with season	Bryan *et al.*, 1987
Cyprinus carpio	musc: 1.25 µg g^{-1} kidn: 7 µg g^{-1} livr: 2.0 µg g^{-1} gall: 2.5 µg g^{-1}	2 mg l^{-1}	250 10 000 400 1 000	14-d exposure; di- and monobutyltin produced	Tsuda *et al.*, 1988a
Oncorhynchus tshawytscha	musc: 0.20 µg g^{-1} 0.90 µg g^{-1} 0.08 µg g^{-1} 0.20 µg g^{-1}	ND	ND	First range for fish exposed in TBT fish pens for 3–18 months. Second for aquaculture products purchased in markets	Short and Thrower, 1986

Salmo gairdneri	musc: 0.31 μg g^{-1}	0.42 μg l^{-1}	406	Tissue burdens after 15-d exposure followed by 15 d in clean water. Flowing exposure system. Uptake rate constant = 5.11 h^{-1}; Depuration rate constant = 0.110 h^{-1}. Kinetic model contains three compartments	Martin *et al.*, 1990
	kidn: 1.50 μg g^{-1}				
	livr: 1.32 μg g^{-1}				
	gall: 0.48 μg g^{-1}				
	carc: 2.25 μg g^{-1}				
	brai: 1.80 μg g^{-1}				
	gona: 1.80 μg g^{-1}				
	gill: 0.54 μg g^{-1}				
Cyprinidon variegatus	4.19 μg g^{-1}	1.61 μg l^{-1}	2 600	58-d exposure	Ward *et al.*, 1981
Cyprinidon variegatus	musc: 170 ng g^{-1}	0.18 μg l^{-1}	940	Exposure lasted 167 d. Flowthrough system. Based on analysis of total tin. Approximately 50–75% of all tin present as TBT	Ward *et al.*, 1981
	238 ng g^{-1}	0.32 μg l^{-1}	740		
	788 ng g^{-1}	0.48 μg l^{-1}	1 600		
	1210 ng g^{-1}	1.0 μg l^{-1}	1 200		
	visc: 708 ng g^{-1}	0.18 μg l^{-1}	3 900		
	675 ng g^{-1}	0.32 μg l^{-1}	2 100		
	1390 ng g^{-1}	0.48 μg l^{-1}	2 900		
	1480 ng g^{-1}	1.0 μg l^{-1}	2 500		
	livr: 6650 ng g^{-1}	0.18 μg l^{-1}	37 000		
	6420 ng g^{-1}	0.32 μg l^{-1}	20 000		
	25 000 ng g^{-1}	0.48 μg l^{-1}	52 000		
	40 800 ng g^{-1}	1.0 μg l^{-1}	41 000		
	rem: 750 ng g^{-1}	1.0 μg l^{-1}	750		
Carassius carassius grandoculis	ND	ND	~1 000	Tissue burdens not stated; apparent steady state observed after 15 d	Tsuda *et al.*, 1986
Cyprinus carpio	musc: 250 ng g^{-1}	1.8 mg l^{-1}	ND	Exposures lasted 3 d, flowthrough system	Tsuda *et al.*, 1987
	livr: 400 ng g^{-1}				
	kidn: 1000 ng g^{-1}				
	gall: 1250 ng g^{-1}				
Carassius auratus	25 μg g^{-1}	2 mg l^{-1}	1 230	Exposure lasted 14 d	Tsuda *et al.*, 1988b

[a]Tissue burdens are based on wet weight unless specifically noted otherwise. If a range of water and tissue burdens was provided, ranges were estimated by dividing highest tissue burdens by highest water concentrations; lowest values were calculated similarly. Abbreviations: man: mantle tissue; musc: muscle tissue; visc: visceral tissue; gall: gall bladder; livr: liver; kidn: kidney; carc: fish carcass excluding muscle and organs; brai: brain; gona: gonads; ND: not determined; NA: not applicable.
[b]BCF values used are those given by authors or calculated by present reviewers.
[c]In comments, conversions to wet weight equivalents are made assuming that dry weight is 0.3 of wet weight.
[d]If authors reported more than one experiment, separate results may be shown.

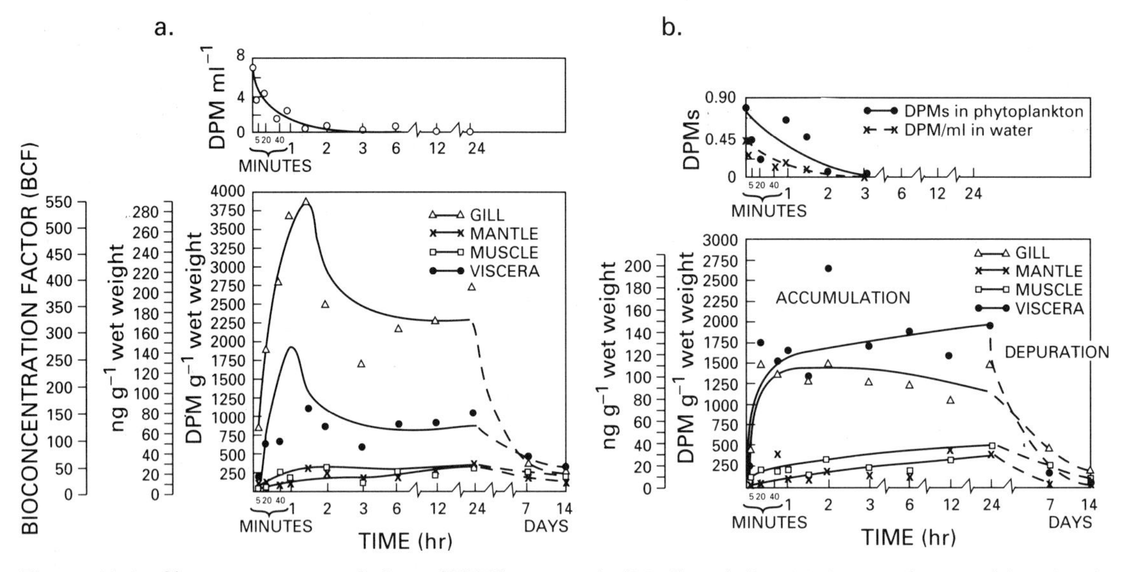

Figure 16.1 Short-term accumulation of TBT by mussels (*Mytilus edulis*). (a) Accumulation of dissolved TBT when the concentration was 0.5 μg l^{-1}. (b) Accumulation from microalgae (*Isochrysis galbana*) when the mass of TBT was identical to that producing a concentration of 0.5 μg l^{-1} in (a) (Laughlin *et al.*, 1986a).

all the product was dibutyltin (DBT). A value of 10% over several days is not extensive, and the finding of only DBT is not strong evidence for enzymatic mechanisms that should also produce monobutyltin (MBT), as is the case in fish (Sasaki *et al.*, 1988a). Yet, many investigators have cited this study to explain the highly variable occurrence of MBT and DBT in bivalve mollusc samples collected from the field (Wade *et al.*, 1988). In my view, a conversion rate of only 10% is too small to explain results found in field studies and is hardly compelling evidence for any significant enzymatic mechanisms capable of degrading TBT in molluscs. It is more likely that these dealkylated homologues reflect uptake of degradation products during environmental exposure rather than significant production of metabolites.

The relationship between duration of exposure and occurrence of a steady state is an important issue in developing a predictive relationship between exposure and tissue burdens. In many studies, exposures lasting weeks to TBT concentrations <500 ng l^{-1}, but sometimes higher, fail to demonstrate any steady state. Studies with fish (Ward *et al.*, 1981) and bivalve molluscs (Laughlin *et al.*, 1986a, 1990; Laughlin and French, 1988) have shown this pattern. (More examples and exceptions will be discussed in section 16.3, Bioaccumulation by taxon) The existence of a steady state, a characteristic of substances accumulated by hydrophobic partitioning from water into nonpolar domains in tissues, sets a limit on the amount of substance an organism may accumulate. Partitioning models based on physicochemical properties have been used to identify substances likely to accumulate to unacceptable levels in living organisms (Clark *et al.*, 1988). Although hydrophobic characteristics correlate with toxicity (Laughlin *et al.*, 1984, 1985), they are not as useful in predicting bioaccumulation processes in a way that allows estimation of exposure from measurements of tissue burdens.

Although investigators have ample evidence of accumulation of TBT from water or food during laboratory studies, interactions of factors that control bioaccumulation in the field are not characterized adequately. Thus, precise correlations between water or food concentrations must be made with caution in an absence of knowledge of additional factors that might influence the process.

16.3. BIOACCUMULATION BY TAXON

At the phylum level, bioaccumulation patterns are characteristic. Presumably, the similarity of bioaccumulation of TBT by species within a phylum (e.g. molluscs) arises from shared physiological and biochemical characteristics. Within major taxonomic groups, significant variability is evident due to natural variation and differences in ecological niche.

16.3.1. MICROORGANISMS

Some marine bacteria display a remarkable ability to accumulate TBT. Cobet and Guard (reported in Laughlin *et al.*, 1987) found that biofilms (consisting of marine bacteria and their exudates) from panels coated with TBT-containing coatings accumulated TBT released from antifouling paints to levels in excess of 50 mg l^{-1}. These levels probably equal or exceed the water solubility of TBT. TBT accumulation by several taxa of TBT-resistant marine bacteria is a non-energy-dependent process. Nine gram-negative species were exposed to 10 mg l^{-1} TBT and accumulated TBT to a range between 3.0 and 7.2 mg g^{-1} (dry wt; Blair *et al.*, 1982). This corresponds to a bioaccumulation factor of 356–1039. These researchers suggested that binding could be a mechanism by which accumulation occurred because concentrations were far in excess of those that could be explained by partitioning alone. Laughlin *et al.* (1987) continued a study of one of the microorganisms used by Blair, a *Pseudomonas* species. They found, using equilibrium dialysis, that even the media from which bacterial cells had been removed by centrifugation accumulated TBT against a concentration gradient. The authors attributed this observation to binding of TBT by exudates released by *Pseudomonas*. Uptake was rapid (~2 h), but release was slow, requiring several days for 90% resolubilization. These kinetics are consistent with, but not proof of, binding.

Soracco and Pope (1983) studied effects of TBT on the pathogenic bacterium *Legionella pneumophila*, responsible for 'legionnaire's disease.' They exposed bacteria to TBT concentrations from 0.017 to 11.2 mg l^{-1} and found that the cells accumulated 1×10^4 to 6×10^7 molecules per cell. The amount of TBT bound was inversely proportional to exposure concentration.

The ability of microorganisms to bind TBT is not necessarily generally shared. Some taxa are clearly sensitive to organotin compounds and thus do not survive to bioaccumulate significant quantities (Laughlin *et al.*, 1987, Cooney and Wuertz, 1989). Biofilm production can be a defensive response to a variety of noxious environmental insults because it sequesters toxic agents outside the cells, preventing them from interfering with metabolism (Gristina *et al.*, 1985). Thus, a paradoxical relationship may exist because bacteria that apparently accumulate highest burdens are not harmed by TBT since they prevent its entry into the cell.

Much of the binding behavior attributed to natural sediments may in fact be due to microorganisms associated with sediment. A frequent lack of correlation between total organic carbon and sorption of chemicals belies the biological basis of the process that is specific to microorganisms present.

16.3.2. MICROALGAE

Studies of bioaccumulation of TBT by marine microalgae, like studies of other micro-

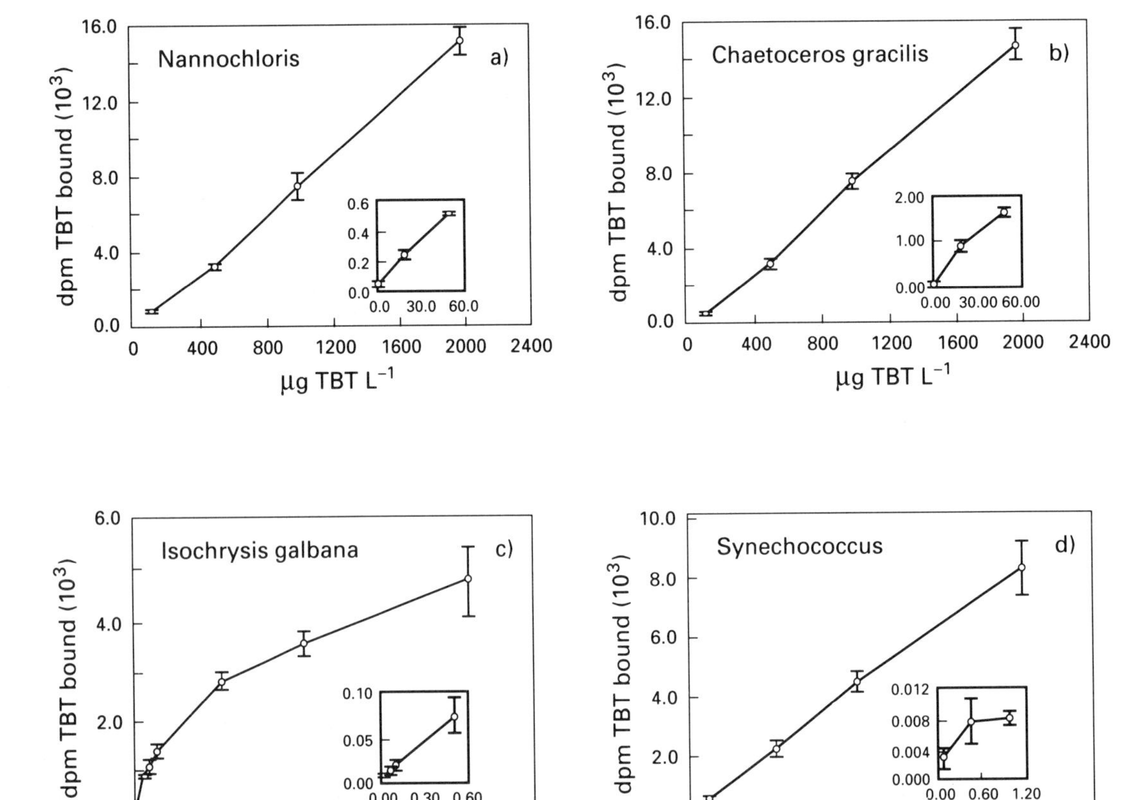

Figure 16.2 Patterns of bioaccumulation by marine microalgae and a cyanobacterium. The density of cells was constant in each experiment. Species used are (a) *Nannochloris* sp., (b) *Chaetoceros gracilis*, (c) *Isochrysis galbana*, and (d) *Synechococcus* sp. Only *I. galbana* demonstrates a nonlinear relationship between exposure concentration and cell burdens characteristic of binding (Chiles *et al.*, 1989). Insets show an exploded view of points near the origin.

organisms, have not been as extensive as one might expect on the basis of their abundance and role in ecosystems. Maguire *et al.* (1984) examined bioaccumulation by the diatom *Ankistrodesmus falcatus* and found a bioconcentration factor of 30 000 when exposure concentration was 20 $\mu g\ l^{-1}$ TBT. Again, this bioaccumulation factor exceeds one that can be explained by partitioning processes alone.

Chiles *et al.* (1989) examined the pattern of TBT accumulation by several species of marine microalgae, *Nannochloris* sp., *Chaetoceros gracilis, Isochrysis galbana,* and the cyanobacterium *Synechococcus* sp., as a function of exposure concentration. Only *I. galbana* demonstrated a nonlinear relationship between exposure concentration and burden of TBT per cell, indicating saturable binding by TBT (Fig. 16.2). Competitive binding using ^{12}C-TBT and ^{14}C-TBT showed that radioactive label was displaced by increasing concentration of cold TBT (Fig. 16.3). An important finding of this study was that the total binding capacity of *I. galbana* was not high in comparison with the other microalgal species tested and was not saturated until the external

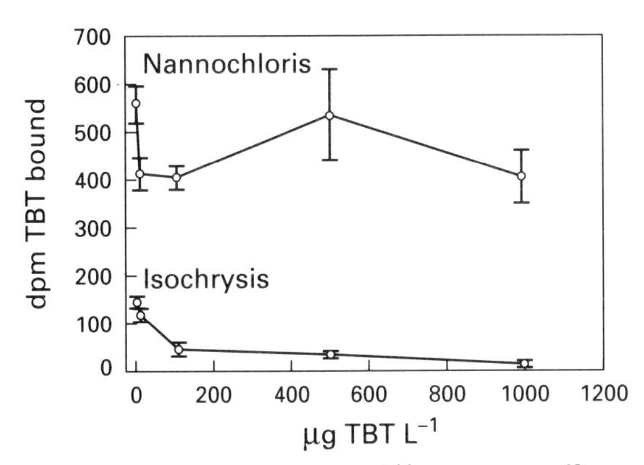

Figure 16.3 Displacement of ^{14}C by excess ^{12}C in a competitive binding experiment. The two species tested were the same as in Fig. 16.2, *Nannochloris* sp. and *Isochrysis galbana*. Only *I. galbana* demonstrates saturated binding characteristics (Chiles *et al.*, 1989).

TBT concentration reached 100 μg l^{-1}, a concentration greater than reported in the environment. The study by Chiles *et al.* is similar to those with bacteria in showing that generalizations are difficult to make, and even dominant patterns may not be as significant as the exceptions. *Isochrysis galbana* is a species with high food value to marine invertebrate larvae, so its bioaccumulation characteristics may be more important in some instances than the behavior of most other microalgal species. Again, the characteristics and extent of bioaccumulation by microalgae are topics that deserves more careful study.

16.3.3. MACROALGAE

The marine macrophyte *Zostera marina* was found to accumulate significant quantities of TBT (Francois *et al.*, 1989). When exposed to 1500 ng l^{-1} TBT in a single pulse exposure, TBT concentrations in the plant reached 140 ng g^{-1} TBT after 7 d before declining as degradation of TBT commenced. A burden of 140 ng g^{-1} is not high in comparison with those frequently encountered in animal tissues following similar exposures to TBT, perhaps because the high proportion of structural materials (cellulose) in plants plays no role in bioaccumulation. A notable finding of this study was that *Z. marina* rapidly dealkylates tributyltin.

In another study, Weber and his colleagues (Donard *et al.*, 1987) studied the accumulation of inorganic tin and production of methyltin by the green alga *Enteromorpha*. They found that mono-, di-, and trimethyltin were formed from inorganic tin. The algae they used were collected from free-living populations in New Hampshire coastal waters and were found to contain n-butyltin compounds as well as methyltin compounds. Thus, these studies show that the organotin fraction found in marine macrophytes may have both endogenous and exogenous sources.

16.3.4. INVERTEBRATES

Studies of tributyltin accumulation in aquatic invertebrates have been confined primarily to two phyla: molluscs (bivalves) and crustaceans (decapods). These groups are of interest because they are important seafood resources supporting extensive fisheries, are ecologically dominant in many habitats and are of sufficient size to facilitate experimentation, and the research community has established a tradition of their use. Because bivalves are known to accumulate significant amounts of many xenobiotics, they have been used as sentinel species for environmental contamination.

Since the first reports of the toxicological effects of TBT concerned the oyster *Crassostrea gigas* (Alzieu *et al.*, 1982; Waldock and Thain, 1983), it should not be surprising that oysters and mussels have been the focus of many experimental studies. These studies have consistently demonstrated that marine bivalves accumulate significant quantities of TBT (sometimes to >5 μg g^{-1}). Experimentally determined bioaccumulation factors may approach 10^3–10^4 (Waldock *et al.*, 1983;

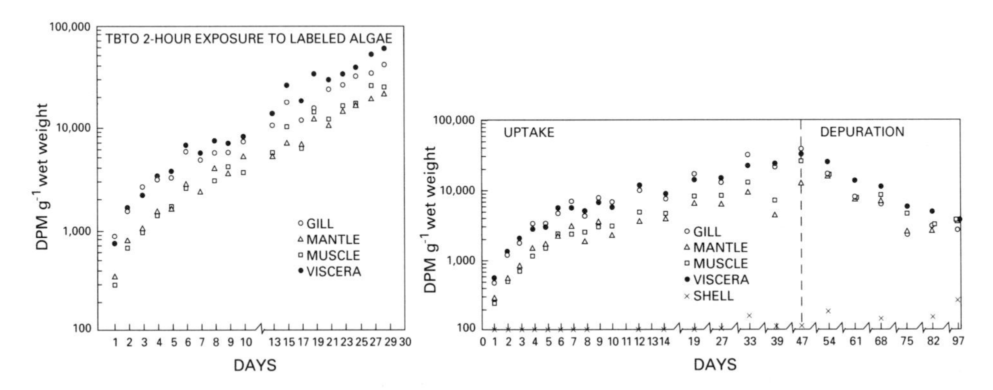

Figure 16.4 Long-term accumulation of TBT in *Mytilus edulis* from food (left) or water (right). Accumulation during 28-d exposure. Accumulation from food is more rapid and at least as extensive as from water. (From Laughlin *et al.*, 1986a.)

Laughlin *et al.*, 1986a) while those inferred from field monitoring are up to two orders of magnitude greater. Bioaccumulation factors of this magnitude are important because they firmly established a link between low nanogram-per-liter exposures typically measured in environmental samples and a dose sufficient to cause chronic and acute effects observed in laboratory and field studies. These studies also lent partial support to a dose-dependent concept of species sensitivity. For example, the oyster *Ostrea edulis* appears to be far less sensitive to effects of TBT than the Japanese oyster *C. gigas*. Under identical exposure conditions, *O. edulis* had bioconcentration factors of 1000 and 1500 compared with values of 2000 and 6000 for *C. gigas*. (Bioconcentration factors were determined at two different exposure concentrations.) Although genetic factors play a role in shell chambering that has become accepted as a diagnostic response to TBT exposure (Okoshi *et al.*, 1987), a better understanding of factors that control dose of TBT among different species is more important to understanding why some species are particularly sensitive.

When studies showing significant TBT accumulation by microorganisms were first published, with the suggestion that microorganisms might play an important role in environmental partitioning of TBT released from antifouling coatings (Laughlin *et al.*, 1987), accumulation through the food chain seemed a potentially important route for filter feeders such as marine bivalves. Mussels (*Mytilus edulis*) accumulate TBT from food rapidly (Fig. 16.4) so that tissue burdens at 28 d equaled those accumulated from water in 56 d (Laughlin *et al.*, 1986a).

Antifouling paints from ships are a prime source of TBT found in field monitoring studies, but antifouling coatings used on enclosures for aquaculture were a significant source of TBT in cultured species before this practice was discontinued several years ago. Accumulation of TBT from treated pearl nets by cultured scallops (*Pecten maximus*) reached 1.86 $\mu g\ g^{-1}$ in juvenile scallops exposed for 16 weeks. Adult scallops accumulated somewhat less TBT, up to 560 $ng\ g^{-1}$ after 4 weeks' exposure. After 4 weeks, tissue burdens fell, presumably because the rate of TBT release from the netting declined (Davies *et al.*, 1986).

It is not clear if differences among tissue burdens of juveniles and adults reflect life history stage or exposure concentration since exposure concentration was not measured.

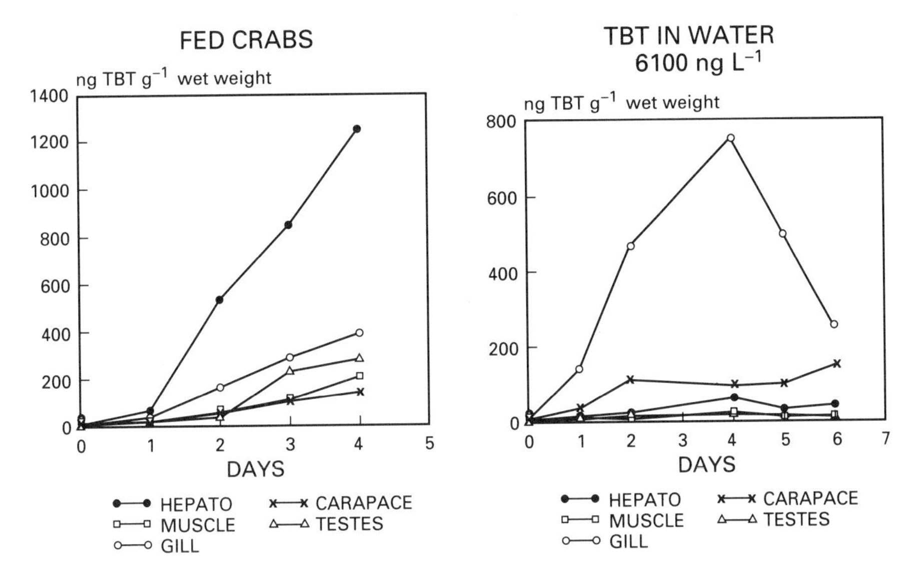

Figure 16.5 Accumulation of TBT from water and food by adult male crabs (*Rhithropanopeus harrisii*). (Drawn from Table 1, Evans and Laughlin, 1984.)

The notable difference between scallops (*P. maximus*) and oysters (*Crassostrea gigas*) is that while oysters could depurate 90% of accumulated TBT within a designated depuration period of several weeks, scallops retained 60–80% in spite of some evidence for metabolic degradation of TBT in the adductor muscle. Also notable was that while TBT caused significant mortality and reductions of growth in oysters, these effects were not observed in cultured scallops (Davies *et al.*, 1986). Investigations conducted since restrictions on use of TBT for antifouling on small ships have clearly demonstrated that a reduction of TBT addition to estuaries closely correlates with a decrease in tissue burdens of TBT and a concomitant improvement of the condition index of *Crassostrea gigas* in France (Alzieu *et al.*, 1989).

Oysters were the first molluscs known to exhibit serious sublethal toxicity from TBT; however, the dogwhelk (*Nucella lapillus*), a gastropod, is among the marine organisms most sensitive to TBT (Bryan *et al.*, 1986). Concentrations in tissues of dogwhelks as high as 8 $\mu g\ g^{-1}$ (dry wt) have been reported (Bryan *et al.*, 1987). These concentrations appear to be similar to and perhaps a bit higher than those reported for bivalve molluscs, but direct comparison is difficult since Bryan *et al.* report their concentrations based on dry weight. (Additional details of findings about *N. lapillus* appear in Gibbs and Bryan, Chapter 13.)

Decapod crustaceans have been the other invertebrate phylum to serve as experimental subjects for TBT bioaccumulation. Evans and Laughlin (1984) studied accumulation of TBT from food by the crab *Rhithropanopeus harrisii*. They found that TBT was taken up from either water or tainted brine shrimp nauplii (*Artemia*). The extent of uptake from food was greater in crabs fed tainted brine shrimp compared with crabs exposed to TBT in the water (Fig. 16.5). Concentration factors, based on the ratio of TBT concentration in the water and in tissues of crabs, varied from >4000 for hepatopancreas to 1200 for the testes. Tissue burdens of both molluscs and crabs exposed experimentally are similar if the route of exposure to TBT is from food, but very much less in crabs if the exposure is from TBT dissolved in water.

16.3.5. FISH

Like bivalve molluscs, fish have been extensively studied since the earliest interest in TBT. In the case of fish, however, interest has focused nearly exclusively on the potential for fish to act as a source of TBT to humans consuming seafood. Investigations show fish accumulate TBT from environmental exposure.

As was the case with oysters, studies of bioaccumulation received major impetus from reports of toxicity of TBT to an aquaculture species: salmon made ill or killed by TBT used to treat nets in which they were raised (Short and Thrower, 1987). Investigation showed that fish from pens contained 0.28–0.90 $\mu g\ g^{-1}$ TBT in muscle. Fish for sale (not from groups previously tested by the author) in markets had TBT concentrations of 0.08–0.20 $\mu g\ g^{-1}$ in muscle tissue. In a study of yellowtails (*Seriola quinqueradiata*) purchased from markets, Sasaki *et al.* (1988a) found TBT concentrations ranging from undetectable to 1730 ng g^{-1} in different fish. Fish from the wild had average TBT concentrations of 126 and 56 ng g^{-1} for two samples of different ages while a sample of older fish thought to come from aquaculture stocks had a mean value of 516 ng g^{-1} (range of this sample: 78–1730 ng g^{-1}). Di- and monobutyltin compounds were also detected in significant quantities in these fish, reflecting active detoxification mechanisms and perhaps explaining why tissue levels are lower than found in bivalve molluscs. Analysis of several species of wild and cultured fish collected primarily from Puget Sound (USA; Sullivan *et al.*, 1988) and sea bass, (*Leteolabrax japonicus*) and gray mullet (*Mugil cephalus*) collected from the Japanese Inland Sea (Takami *et al.*, 1987) indicate that fish in the wild or in aquaculture systems may contain TBT in their tissues, but that levels are fairly low (0.09–0.19 $\mu g\ g^{-1}$). Ambient TBT concentrations were not reported for the studies above, but wild populations are likely exposed to TBT levels <25 ng l^{-1}, based on other studies from open water bodies. Except for the case of fish poisoned by TBT used on aquaculture pens, no adverse effects of TBT have been specifically attributed to environmental exposure to TBT.

Experimental studies of bioaccumulation of TBT by fish comprise a significant portion of the studies on fish. Ward *et al.* (1981) exposed sheepshead minnows (*Cyprinidon variegatus*) to an average concentration of 1.6 $\mu g\ l^{-1}$ TBT for 56 d. At the end of the uptake phase, TBT burdens were 4.19 $\mu g\ g^{-1}$, corresponding to a BCF of 2600. Depuration during a 28-d post-exposure period led to 74, 80, and 46% loss of TBT or its residues in muscle, head, and viscera, respectively. In a second experiment, Ward *et al.* exposed *C. variegatus* to five TBT concentrations from 0.125 to 2.0 $\mu g\ l^{-1}$ for 177 d in a flowing seawater exposure system. Final tissue burdens were proportional to exposure concentration. For fish exposed to 1 ng l^{-1} TBT, highest burdens were in the liver (40 800 ng g^{-1}; BCF = 41 000) followed by viscera (2480 ng g^{-1}; BCF = 2500) then muscle (1210 ng g^{-1}; BCF = 1200). The reported values are based on total tin analyses expressed as TBT equivalents.

Tsuda *et al.* (1986, 1988a) exposed carp (*Cyprinus carpio*) to 2.1 $\pm$ 0.2 mg l^{-1} TBT for 14 d. At the end of the exposure period, tissue concentrations were as follows: muscle, 1.25 $\mu g\ g^{-1}$; liver, 2.0 $\mu g\ g^{-1}$; kidney, 7 $\mu g\ g^{-1}$; and gall bladder, 2.5 $\mu g\ g^{-1}$. High concentrations of TBT in kidney and gall bladder suggest active elimination processes. Dealkylation of TBT to di- and monobutyltin was also evident in carp.

In another set of experiments with round crucian carp (*Carassius carassius grandoculis*), Tsuda *et al.* (1986) showed that the gegen ion associated with TBT did not greatly influence the bioconcentration factor. The BCF values for TBTCl and TBTO were 363 and 589, respectively. Following a 15-d exposure to 5 mg l^{-1} TBT, carp appeared to approach a steady state with a BCF of ~1000. This steady state may, to a large part, be due to active excretory processes in carp (Tsuda *et al.*, 1987).

Goldfish (*Caracius auratus*) also accumulate TBT to levels similar to carp. Following 14-d exposure to 2 mg l^{-1} in a flowing system, tissue burdens were ~25 $\mu g\ g^{-1}$; the BCF was ~1000.

Martin *et al.* (1990) exposed freshwater trout (*Salmo gairdneri*) to 0.52 $\mu g\ l^{-1}$ TBT in a flowing water exposure protocol for 15 d, followed by 15-d depuration. (Concentrations given in Martin *et al.* are assumed to be expressed as tin equivalents. To express values as TBT, values given by Martin *et al.* have been multiplied by 2.48.) At intervals of a couple of days, they sampled fish for TBT analysis. At the end of the depuration period (30 d after the start of the experiment), TBT concentrations were highest in the liver, kidney, brain, gonads, and carcass (Table 16.1). Measured concentrations were liver, 3.27 $\mu g\ g^{-1}$; gall bladder, 0.48 $\mu g\ g^{-1}$; kidney, 3.72 $\mu g\ g^{-1}$; carcass, 5.58 $\mu g\ g^{-1}$; muscle, 0.77 $\mu g\ g^{-1}$; brain and gonads, both 4.5 $\mu g\ g^{-1}$. During the uptake phase, peritoneal fat comprised an important depot (13.8 $\mu g\ g^{-1}$ after 15-d exposure), but, following a move to clean water, TBT concentrations in peritoneal fat dropped markedly (1.86 $\mu g\ g^{-1}$). These authors also suggested that high TBT concentrations in the kidney are due to an affinity of this tissue for cations, and excretion through the kidney does not represent a significant depuration pathway. The primary loss of TBT from trout occurs through the liver–gall bladder. This study indicated extensive metabolic debutylation of TBT in rainbow trout.

Martin *et al.* also determined BCF values of TBT for trout, based either upon analysis or as a ratio of uptake to depuration rate constants. When trout were exposed to 0.89 $\mu g\ l^{-1}$ TBT for 64 d, the BCF determined by analysis was 406. A BCF estimated as the ratio of rate constants was 465. The model used to estimate kinetic rate constants indicated that peritoneal fat was a third compartment during uptake, but during the depuration phase, only two compartments (all tissues and water) were apparent.

16.3.6. BIRDS AND MAMMALS

A review of bioaccumulation of TBT by birds and mammals is outside the intended scope of this review since only food serves as a significant source of TBT; birds and mammals do not absorb significant quantities directly from water. Furthermore, few studies of birds and mammals have been published. Those species that lack efficient abilities to metabolize and excrete TBT will likely accumulate large quantities from tainted food and may be at risk of poisoning. An excellent description of the factors that would influence accumulation of TBT is given in Clarke *et al.* (1988).

16.3.7. SUMMARY OF BIOACCUMULATION BY TAXON

Available data suggest some limited taxonomic distinctions may be applied to describe accumulation of TBT. Molluscs, including both bivalves and gastropods, attain high tissue burdens of TBT, ranging to at least 5–10 $\mu g\ g^{-1}$ (wet wt). These high values are found both in laboratory exposures and in field monitoring. The biological effects of TBT on molluscs, however, are at best only partially explainable by characteristics of their accumulation of TBT. Even though the oyster (*Crassostrea gigas*) and the dogwhelk (*Nucella lapillus*) are among the most effective accumulators, shell thickening and imposex have thresholds of 60 ng l^{-1} and 2 ng l^{-1} TBT, respectively (Laughlin *et al.*, Chapter 10; Bryan *et al.*, 1986). Since these levels are well within ranges of TBT accumulated by most

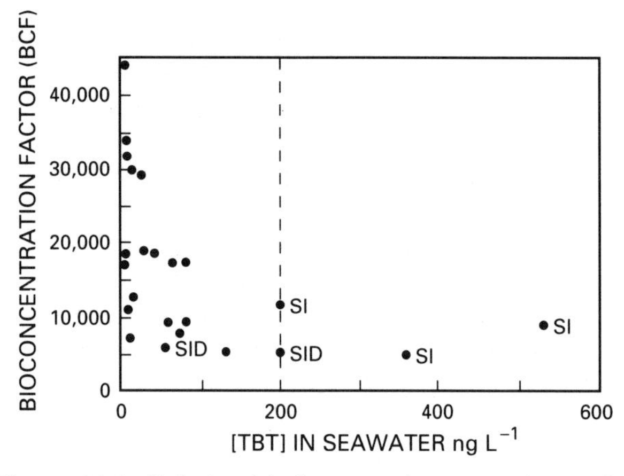

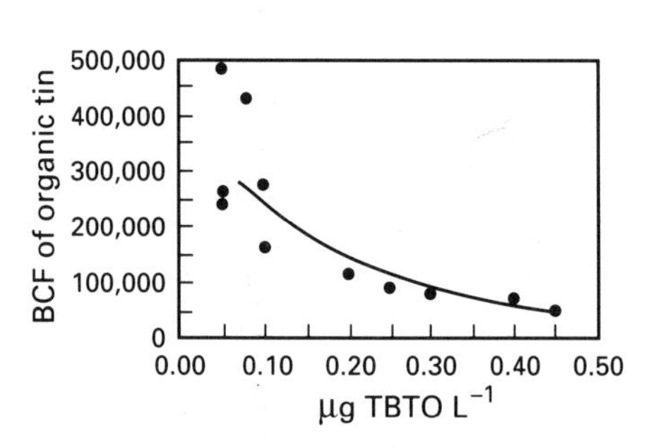

Figure 16.6 Relationship between bioaccumulation factors and exposure concentrations for marine mussels (*Mytilus edulis*) in San Diego Bay, California (left) (Salazar and Salazar, 1988) and Denmark (right) (Zuolian and Jensen, 1989). (Both illustrations used with permission.)

species tested, no evidence exists suggesting that sublethal effects in sensitive species occur because they accumulate more TBT than non-sensitive taxa. Crustaceans and fish fall into a second category of taxa that is characterized by much lower accumulation of TBT. The major factor for their placement is their possession of efficient enzymatic mechanisms that degrade TBT (Lee, 1985, 1986, and Chapter 18). Results of field monitoring studies clearly show that fish have far lower tissue burdens of TBT as a result of environmental exposure than do molluscs even when exposures were likely to have been high, such as in aquaculture pens (Table 16.1). Marine plants and microorganisms fall into a third group that is difficult to characterize succinctly. Some marine bacteria (*Pseudomonas*) 'accumulate' very high levels of TBT, apparently by binding it to biofilm outside the cell. Microalgae also show diverse accumulation patterns with occasional examples of those that accumulate high burdens. Representative algae tested (*Zostera marina*) do not bioaccumulate TBT extensively on a weight basis. As a concluding remark, even in the most notable accumulators of TBT, the actual tissue burdens are not, in comparison with many other anthropogenic contaminants, high enough to raise a warning. Other compounds such as polychlorinated biphenyls may easily occur at higher levels in organisms from contaminated habitats.

16.4. BIOACCUMULATION FACTORS

High TBT bioaccumulation factors from environmental contamination are a primary cause of concern about environmental contamination by TBT. It was entirely unexpected that these relatively water-soluble compounds present at concentrations usually well below 100 ng l^{-1} would accumulate in marine organisms to any extent, and particularly unexpected for one that caused significant chronic toxicity. Bioaccumulation factors for TBT range from 200 in some fish tissues to perhaps as high as 100 000 in American oysters (*Crassostrea americana*). Values from 5000 to 50 000 are not uncommon (Table 16.1).

16.4.1. RELATIONSHIP BETWEEN BCF AND EXPOSURE CONCENTRATION

Bioaccumulation factors show an inverse relationship to concentration. Lower exposures to TBT lead to higher bioaccumulation factors, at least in bivalves (Fig. 16.6). That is, molluscs may accumulate several micrograms of TBT per gram of tissue if exposed to a relatively wide range of TBT concentrations. However, a point of diminishing returns is reached such that increasing TBT exposures does not proportionally increase TBT burdens. Thus, some bioaccumulation factors may appear to be very large because of accumulation from low concentrations rather than because of high burdens in tissue. This phenomenon is shown in two experimental field studies, the first by Salazar and Salazar (1988) and the second by Zuolian and Jensen (1989). Attempts to corroborate these observations in laboratory experiments have met with limited success. Laughlin and French (1988) exposed mussels (*Mytilus edulis*) to a series of five dissolved TBT concentrations from 23 to 670 ng l^{-1} for periods up to 52 d. A steady state was not consistently observed, and no clear relationship between BCF and exposure concentration was apparent. In a similar experiment with clams (*Mercenaria mercenaria*) exposed to dissolved TBT concentrations between 37 and 484 ng l^{-1}, Laughlin (unpublished data) found only a weak indication of an inverse relationship between BCF and exposure concentration. In this experiment, steady state was not observed at any exposure concentration. The lack of a clear agreement between laboratory and field experiments may arise from one of several factors. The range of laboratory exposure levels in experiments by Laughlin and coworkers is narrower than the range of levels in field studies in which the phenomenon was observed. Even in field studies, burdens in tissues of bivalves exposed to TBT concentrations between 10 and 100 $\mu g\ l^{-1}$ did not show a clear correlation, the study by Salazar and Salazar (1988) being the best example. In field studies, the route of exposure was possibly by food as well as water. In laboratory studies, food played little role since microalgae were not added to aquaria. Finally, the lack of steady-state characteristics of most laboratory experiments exerts a problematic qualification on interpretation. At this time, it is possible to state that only a weak correlation exists between tissue burdens and water concentrations of TBT. Because the relationship between dissolved TBT concentration and BCF is inversely proportional, and usually only weakly so at water concentrations below 100 ng l^{-1}, BCF should be used cautiously for prediction. A better characterization of the relationship between exposure and tissue burdens deserves more effort because many compounds appear to follow this pattern. A clear explanation for TBT could have broad application to many other examples.

16.4.2. THE WIDE RANGE OF BIOACCUMULATION FACTOR VALUES

The wide variation of bioaccumulation factors makes use of predictive models unreliable. Partitioning behavior, governed by thermodynamic stability factors, should be constant for a given set of physical factors such as temperature, salinity, and chemical concentrations (Hansch *et al.*, 1968; Hansch and Leo, 1979). Prediction of partitioning is based on behavior of a chemical in an octanol–water model system following widely used test protocols. The resulting coefficient, K_{ow}, allows estimation of a BCF for many chemicals using established relationships (Mackay, 1982). This approach has been applied to TBT with equivocal results. A wide range of values has been reported for the K_{ow} value for TBT (Table 16.2), ranging from several hundred to several thousands. Laughlin *et al.* (1986b) suggested that small quantities of mono- or dibutyltin contamination might be responsible for the wide range of values reported, but other factors may also influence the value measured such as chemical speciation of TBT cation with OH^{1-}, Cl^{1-}, or CO_3^{2-} (Laughlin *et al.*, 1986b) or measurement technique. Tsuda *et al.* (1988a) found that

Table 16.2 Synopsis of octanol–water partition values, K_{ow}, for tributyltin compounds[a]

Compound	K_{ow}	*Comments*	*Reference*
TBTO	200	Distilled water (assumed). Measured total tin; graphite furnace atomic absorption spectrometry using L'vov's platform. Calculated from K_{ow} of 2.3 in their Fig. 5, p. 487	Wong *et al.*, 1981
TBTF	1 400	Distilled water. Oxidation of organotins with acid then colorometric determination of total tin	Slesinger and Dressler, 1978
TBTO	2 185	4 ml octanol, 240 ml water	Wulf and Byington, 1975
TBTO	1 550	Distilled water, pH 6.0; purified TBTO. Derivatization with n-pentylmagnesium bromide, gas chromatography with modified flame photometric detector.	Maguire *et al.*, 1983
TBTC1	1 300	Total tin in octanol phase measured, aqueous phase by difference	Maguire *et al.*, 1983
TBTO	5 500	25‰ S	Laughlin *et al.*, 1986b
	7 000	Distilled water	
		^{14}C-TBT used. 3–5 ml octanol to 5 ml seawater. Range of values shows K_{ow} for seawater between 0 and 45‰ S.	
TBTO	158	Octanol:water ratio = 1:1	Tsuda *et al.*, 1988a
	5 011	Octanol:water ratio = 1:10	
		Tap water from tanks containing fish. Analysis by solvent extraction, hybridization followed by electron capture gas chromatography	

[a]Abbreviations: TBTO, tributyltin oxide; TBTF, tributyltin fluoride; TBTCl, tributyltin chloride; ‰ S, parts-per-thousand salinity.

simply varying the ratio of water to octanol from 1:1 to 1:10 produced K_{ow} values of 158 or 5011, respectively. The K_{ow} should not be significantly influenced by different volume ratios as long as the chemical of interest is not saturated in either phase. In the case of TBT, perhaps the volume of water influences chemical speciation and thus indirectly modifies measured K_{ow} values. Regardless of which value is selected for the true K_{ow} value for TBT, using Mackay's equation, BCF values should be <500. Only values reported for fish and macroalgae (Table 16.1) frequently fall into this range. Values for most other organisms, notably bacteria, microalgae, and molluscs, are significantly higher. Thus, partitioning models are not always the most accurate quantitative models of TBT accumulation.

16.4.3. DOES BINDING INFLUENCE TBT ACCUMULATION?

Since TBT displays accumulation in microorganisms and tissues in excess of that predicted by partitioning, other mechanisms likely play a role. Binding, either electrostatic or covalent, is an important mechanism for metals that form ions in water. Precise characterization of binding by TBT is impeded by a lack of complete understanding of the chemical form of TBT in fresh and marine waters. Solubility varies markedly with pH (Maguire *et al.*, 1983), indicating that the electron structure of tin is amenable to influence by the chemical environment in aqueous solution. A charged (TBT cation) as well as an uncharged (nonpolar) species in water occurs. In this respect, during bioaccumulation TBT may possess the reactive properties of a metal cation as well as the hydrophobic character of an organic compound.

Binding sites and products remain to be identified and characterized. Carboxyl groups in components of the cell walls have been shown to be very important in metal binding in algae (Gardea-Torresdey *et al.*, 1990; Majidi *et al.*, 1990) and probably in bacteria as well (Blair *et al.*, 1982). Covalent bonding with saccharides or sulfur-containing ligands may also play an important role in excretion processes, particularly in the liver–bile pathway. Such binding may be the explanation for the consistently high TBT burdens in digestive glands (liver or hepatopancreas) of nearly all species examined. A better understanding of binding processes is important for an effective ability to predict bioaccumulation. Binding processes, particularly to mucus or other complex mucopolysaccharides, warrant further study.

16.5. THE HYDROPHOBIC MODEL OF BIOACCUMULATION

Much of the experimental work on bioaccumulation of TBT is based on a model or assumption of hydrophobic partitioning as an explanation of TBT accumulation. This 'model' has developed from very simple, non-quantitative generalizations such as 'like dissolves like' and progressed heuristically during the past 15 y. The fundamental concept on which hydrophobic partitioning models exist is that dissolved substances will, at equilibrium, attain a predictable, consistent ratio between water and nonpolar media that are in contact. This idealized concept is most closely accurate for systems of low-molecular-weight organic solutes lacking large, strongly polarized or charged functional groups, and using hydrocarbons (hexane) as the nonpolar phase. Some chemical classes exhibit remarkably predictable behavior, but expanding experience has demonstrated that a general predictive model becomes less accurate as the range of size, charge, functional groups, and water–solvent system is expanded to test the model. Hydrophobic qualities are not characterized by presence or absence, but rather by the degree that atoms and their bonding configurations influence the solubility interactions with water. Organotin compounds are one class of compounds that possess physicochemical properties that influence water solubility and hydrophobic behavior. The following paragraphs analyze details of the application of the hydrophobic models to behavior of TBT.

16.5.1. CHEMICAL FORMS OF TBT IN WATER

Size and charge are the two most important factors in determining whether a molecule will be taken up by living tissue. Uptake of TBT is not significantly limited by size, but charge characteristics may be important. The chemical form of organotins in aqueous solution, as noted above, is the product of a dynamic equilibrium controlled by reversible chemical speciation. Maguire and coworkers refer to the tributyltin cation as the major species in fresh water. Their studies (Maguire *et al.*, 1983) of the effect of pH on solubility of TBT are good evidence that the concentration of cation is important. Laughlin *et al.* (1986b) suggested that, at the pH of seawater (~7–8), TBT may exist as the hydroxide, chloride, or carbonate (Cl^{1-}, OH^{1-}, and CO_3^{2-}). Their methods of analysis did not permit specific identification of the cation in solution, but significant variation of the octanol–water partition coefficient with salinity supported the idea that chemical speciation equilibria are strongly displaced from the TBT cation. The two studies indicate that fresh water with a pH of ~5 favors TBT cation formation while seawater, which is alkaline, with a total anion concentration of ~0.5 M is not favorable to the existence of free cation. In seawater, the equilibrium is apparently easily displaced by relatively minor changes in salinity. This chapter considers both the cationic or uncharged form of TBT in water as the dissolved form.

Binding to dissolved organic matter and exopolysaccharides produced by some micro-organisms is another example of reversible chemical speciation. In this case, the organotin is weakly held by specific functional groups of the organic polymers. The significance of these binding processes is in direct proportion to the abundance of the organic compounds in water or the organism.

16.5.2. ENTRY INTO THE ORGANISM

Uptake of uncharged TBT across epithelia or membranes is the apparent mechanism. A cation would have to be bound and carried across; no evidence for active or passive transport of bound TBT into organisms exists. Most evidence, in fact, suggests that binding, when it occurs, as in the case of the biofilm, acts to reduce the availability of TBT to the cell interior.

16.5.3. RATES OF BIOACCUMULATION

Rates of bioaccumulation are closely correlated with mechanisms of uptake. Partitioning, dependent upon linear free energy changes as a solute is transferred from water to a hydrophobic phase (Hansch *et al.*, 1968; Hansch and Leo, 1979), is clearly the most rapid. Partitioning into single cells or cell components is virtually instantaneous, being limited only by diffusion rates (Wieth and Tosteson, 1979). In macroorganisms, the bioaccumulation is controlled by the kinetics of diffusion between two or more compartments (Mackay and Hughes, 1984; Martin *et al.*, 1990). Bioaccumulation is consequently slower and tends to be mediated by physiological factors such as blood flow and gill irrigation rate that assist diffusion-mediated functions. For most pollutants, a steady state usually occurs in several days to weeks. Interestingly, only a minority of investigators have observed a steady state while exposing aquatic organisms to TBT, and then only with exposure concentrations $>1\ \mu g\ l^{-1}$ or for periods of perhaps 60 d or longer (Table 16.1). Occurrence of this prolonged period in the absence of any reduced transfer into the organism is the strongest circumstantial evidence for a role for TBT binding in tissues.

When feeding is an important uptake mechanism, rates are controlled by bulk transfer during ingestion. Final burdens do not depend strictly on a physicochemical equilibrium between the dissolved concentration in two phases. The bioaccumulation rate is dependent only on the feeding rate if the excretion rate is small or insignificant in relation to feeding.

16.5.4. COMPARTMENTALIZATION IN TISSUES

Some of the studies cited in Table 16.1 show that TBT burdens vary as a function of the tissue. When TBT was dissolved in water, gills of bivalves usually had highest burdens. Gonads, generally a fatty tissue during part of the reproductive cycle, were also characterized by relatively high burdens. Muscle burden was generally lower. Similar patterns were also evident in fish. For example, Martin *et al.* (1990) found that peritoneal fat in trout was a major component of a kinetic model describing uptake of TBT from water.

An explanation of this pattern that relies on chemical dynamics invokes the occurrence of a moiety, either lipid or one that binds TBT, and the proportion that substance comprises of the entire organ (Mackay, 1982; Mackay and Hughes, 1984). The physiological function of the tissue is also important since neither gills nor, in some species, liver (digestive gland) tissue is a fatty tissue. Comparisons of relative tissue burdens are most problematical in a discussion of the mechanisms of uptake. Lipid-rich tissues consistently show higher TBT burdens, but excess accumulation is true even for fish muscle, thus suggesting that TBT bioaccumulation is more than a simple partitioning process.

16.5.5. ROLE OF DEPURATION

Depuration of TBT can be influenced by both physiological and enzymatic mechanisms. Physiological processes such as gill irrigation rates, circulatory rates, or filtration rates in kidneys or other excretory organs will increase diffusion-mediated processes. (No active transport mechanism for tin or organotins is known.) Circumstantial evidence for the existence of inducible physiological mechanisms influencing depuration following TBT exposures >50 ng l^{-1} was noted for mussels in laboratory experiments (Laughlin and French, 1988) and during field studies (Zuolian and Jensen, 1989). Suggestions of a role for inducible physiological responses in controlling TBT depuration rates were viewed skeptically when initially proposed, even though other investigators have shown similar behavior for other compounds, naphthalene, for example (Widdows *et al.*, 1983). The role of diffusion-mediated release of accumulated compounds can easily be demonstrated by comparing effects of temperature on loss rates of TBT from any organism lacking enzymatic systems that structurally modify TBT. Although more limited in capability than enzymatic mechanisms, inducible physiological processes appear capable of playing a significant role when environmental exposures to TBT are very low (several manograms per liter), concentrations generally characteristic outside of marinas or locally heavy traffic.

Enzymatic mechanisms of modification and elimination of TBT, the topic of another chapter in this book (Lee, Chapter 18), are the most effective means for reducing the burdens of TBT in tissues. Molluscs, in particular, lack well-developed enzymatic capability, accounting, at least in part, for their high tissue burdens. Degradation pathways have received the most attention, but conjugation reactions that increase the solubility of hydrophobic compounds may also be important for organotins and should be further characterized.

16.5.6. SUMMARY OF THE BIOACCUMULATION SCHEME

Tributyltin is rapidly and nearly quantitatively transferred from water or food to tissues of exposed animals. Few environmental factors, such as presence of dissolved organic matter, appear to significantly reduce tissue burdens. Release rates are slow so that TBT accumulated over several days will still be detectable in tissues after several weeks or perhaps months. Different tissues within the same animals may have markedly different tissue burdens, reflecting both their biochemical composition and physiological role. Organisms without effective enzymatic pathways depend upon diffusion to rid themselves of accumulated TBT.

16.6. BIOMONITORS FOR TBT

Because of concerns about consumption of tainted seafood, several species, primarily bivalve molluscs, have been used as biomonitors. Explicit in the majority of papers included in this review is the assumption that tissue burdens reflect water concentrations of TBT. Less clearly evident in some cases was an assumption that tissue burdens were consistently proportional to the level of TBT inputs at all locations of large sampling programs. Interpretation of a program using biomonitors for TBT will be most effective if it assumes that TBT burdens in tissues represent a steady state among environmental compartments, including water, microorganisms, sediments, and the biomonitor species (Clark *et al.*, 1988). Since microorganisms quantitatively remove TBT from water rapidly and their biomass dominates ecosystems, they are the compartment that controls flux of TBT, and they do so fairly quickly. Except in marinas where TBT inputs are high and continuous, it is unlikely that TBT persists in water for any length of time. If one considers environmental monitoring in this way, the real question becomes, 'How consistent is environmental compartmentalization in different habitats of the marine environment and across geographical areas?' The answer is unknown. Since TBT is a compound with a nearly unique use confined to aquatic systems, it is a prime model compound for a simplified model for biomonitoring.

16.7. CONCLUSIONS

Tributyltin is taken up from low ambient levels in the environment and accumulated to relatively high concentrations in many organisms. Bioaccumulation factors appear to be high, but field studies, in particular, have not necessarily carefully characterized the route of uptake (water or food). Values reported as 'bioconcentration factors' of 50 000–100 000 may not be as alarming if the route of exposure were actually from food. (The values in this case would be more accurately characterized as 'bioaccumulation factors'.) Both partitioning and binding appear to play a role in bioaccumulation, but in some, perhaps most, cases binding is quantitatively dominant in influencing final tissue burdens. Tributyltin is taken up rapidly but released slowly from tissues, even in aquatic organisms possessing enzymatic mechanisms that degrade TBT or aid in its elimination. Biomonitors thus may be useful for determining the extent of contamination of the environment by TBT. Legislative and administrative restrictions on use of TBT have decreased, at least somewhat, the environmental risk of this compound, but TBT remains a useful model compound for environmental studies and warrants continued attention.

ACKNOWLEDGMENTS

This research was supported by the Office of the Chief of Naval Research. I thank the staff of the Center for Academic Publication, Florida Institute of Technology, for their help. I also thank H.E. Guard for many helpful discussions during manuscript preparation.

REFERENCES

Alzieu, C., M. Heral, Y. Thibaud, M.J. Dardignac and M. Feuillet. 1982. Influence des peintures antisalissures à base d'organostanniques sur la calcification de la coquille de l'huître *Crassostrea gigas*. *Rev. Trav. Inst. Pêches Marit.*, **45**, 100–116.

Alzieu, C.L., J. Sanjuan, J.P. Deltriel and M. Borel. 1986. Tin contamination in Arcachon Bay: effects on oyster shell anomalies. *Mar. Pollut. Bull.*, **17**, 494–498.

Alzieu, C., J. Sanjuan, P. Michel, M. Borel and J.P. Dréno. 1989. Monitoring and assessment of butyltins in Atlantic coastal waters. *Mar. Pollut. Bull.*, **20**, 22–26.

Blair, W.R., G.J. Olson, F.E. Brinckman and W.P. Iverson. 1982. Accumulation and fate of tri-n-butyltin cation in estuarine bacteria. *Microbiol. Ecol.*, **8**, 241–251.

Bryan, G.W., P.E. Gibbs, G.L. Hummerstone and G.R. Burt. 1986. The decline of the gastropod *Nucella lapillus* around south-west England: evidence for the effect of tributyltin from antifouling paints. *J. Mar. Biol. Ass. UK*, **66**, 611–640.

Bryan, G.W., P.E. Gibbs, G.R. Burt and L.G. Hummerstone. 1987. The effects of tributyltin (TBT) accumulation on adult dog-whelks, *Nucella lapillus*: long-term field and laboratory experiments. *J. Mar. Biol. Ass. UK*, **67**, 525–544.

Bryan, G.W., P.E. Gibbs, R.J. Huggett, L.A. Curtis, D.S. Bailey and D.M. Dauer. 1989. Effects of tributyltin pollution on the mud snail, *Ilyanassa obsoleta*, from the York River and Sarah Creek, Chesapeake Bay. *Mar. Pollut. Bull.*, **20**, 458–462.

Chiles, T.C., P.E. Pendoley and R.B. Laughlin, Jr. 1989. Mechanisms of tri-n-butyltin bioaccumulation by marine phytoplankton. *Can. J. Fish. Aquat. Sci.*, **46**, 859–862.

Clark, T., K. Clark, S. Peterson, D. Mackay and R.J. Norstrom. 1988. Wildlife monitoring, modeling and fugacity. *Environ. Sci. Technol.*, **22**, 120–127.

Cooney, J.J. and S. Wuertz. 1989. Toxic effects of tin compounds on microorganisms. *J. Indust. Microbiol.*, **4**, 375–402.

Davies, I.M., J.C. McKie and J.D. Paul. 1986. Accumulation of tin and tributyltin from antifouling paint by cultivated scallops (*Pecten maximus*) and Pacific oysters (*Crassostrea gigas*). *Aquaculture*, **55**, 103–114.

Donard, O.F.X., F.T. Short and J.W. Weber. 1987. Regulation of tin and methyltin compounds by the green alga *Enteromorpha* under simulated estuarine conditions. *Can. J. Fish. Aquat. Sci.*, **44**, 140–145.

Francois, R., F.T. Short and J.H. Weber. 1989. Accumulation and persistence of tributyltin in eelgrass (*Zostera marina* L.) tissue. *Environ. Sci. Technol.*, **23**, 191–196.

Evans, D.W. and R.B. Laughlin, Jr. 1984. Accumulation of bis(tributyltin) oxide by the mud crab, *Rhithropanopeus harrisii*. *Chemosphere*, **13**, 213–219.

Gardea-Torresdey, J., M.K. Becker-Hapak, J.M. Hoses and D.W. Darnall. 1990. Effect of chemical modification of algal carboxyl groups on metal ion binding. *Environ. Sci. Technol.*, **24**, 1372–1378.

Gristina, A.G., M. Oga, L.X. Webb and C.D. Hobgood. 1985. Adherent bacterial colonization in the pathogenesis of osteomyelitis. *Science*, **228**, 990–993.

Hansch, C., J.E. Quinlan and G.L. Lawrence. 1968. The linear free-energy relationship between partition coefficients and the aqueous solubility of organic liquids. *J. Org. Chem.*, **33**, 347–350.

Hansch, C. and A.J. Leo. 1979. *Substituent Constants for Correlation Analysis in Chemistry and Biology*. John Wiley & Sons, New York.

Jackson, J.A., W.R. Blair, F.E. Brinckman and W.P. Iverson. 1982. Gas chromatographic speciation of methylstannanes in the Chesapeake Bay using purge and trap sampling with a tin selective detector. *Environ. Sci. Technol.*, **16**, 110–119.

Langston, W.J. and G.R. Burt. 1991. Bioavailability and effects of sediment-bound TBT in deposit-feeding clams, *Scrobicularia plana*. *Mar. Environ. Res.*, **32**, 61–77.

Laughlin, R.B., Jr, W. French, R.B. Johannesen, H.E. Guard and F.E. Brinckman. 1984. Predictive structure–activity relationships for triorganotin compounds using computed topological values. *Chemosphere*, **13**, 575–584.

Laughlin, R.B., Jr, R.B. Johannesen, W. French, H. Guard and F.E. Brinckman. 1985. Structure–activity relationships for organotin compounds. *Environ. Toxicol. Chem.*, **4**, 343–351.

Laughlin, R.B., Jr, W. French and H.E. Guard. 1986a. Accumulation of bis(tributyltin) oxide by the marine mussel *Mytilus edulis*. *Environ. Sci. Technol.*, **20**, 884–890.

Laughlin, R.B., Jr, H.E. Guard and W.M. Coleman, III. 1986b. Tributyltin in seawater: speciation and octanol–water partition coefficient. *Environ. Sci. Technol.*, **20**, 201–204.

Laughlin, R.B., Jr, A.B. Cobet and H.E. Guard. 1987. Mechanisms of triorganotin toxicity to macroinvertebrates and the role of the biofilm as a controlled release medium. In: *Marine Biodeterioration. Advanced Techniques Applicable to the Indian Ocean*, M.-F. Thompson, R. Sarojini and R. Nagabhushanam (Eds). Oxford & IBH Co, New Delhi, pp. 757–767.

Laughlin, R.B., Jr and W. French. 1988. Concentration dependence of bis(tributyltin) accumulation in the marine mussel, *Mytilus edulis*. *Environ. Toxicol. Chem.*, **7**, 1021–1026.

Laughlin, R.B., Jr, P.D. Pendoley and H.E. Guard. 1990. Concentration dependence of tributyltin bioaccumulation by the hard shell clam, *Mercenaria mercenaria*. NOSC Progress Report, Naval Ocean Systems Command, San Diego.

Lee, R.F. 1985. Metabolism of tributyltin oxide by crabs, oysters and fish. *Mar. Environ. Res.*, **7**, 145–149.

Lee, R.F. 1986. Metabolism of bis(tributyltin) oxide by estuarine animals. In: Oceans '86 Conference Record, Vol. 4, IEEE Service Center, 445 Hoes Lane, Piscataway, New Jersey, pp. 1182–1188.

Mackay, D. 1982. Correlation of bioconcentration factors. *Environ. Sci. Technol.*, **16**, 274–278.

Mackay, D. and A.I. Hughes. 1984. Three-parameter equation describing the uptake of organic compounds by fish. *Environ. Sci. Technol.*, **18**, 439–444.

Maguire, R.J. and R.J. Tkacz. 1983. Analysis of butyltin compounds by gas chromatography. *J. Chromatog.*, **268**, 99–101.

Maguire, R.J., J.H. Carey and E.J. Hale. 1983. Degradation of tri-n-butyltin species in water. *J. Agr. Food Chem.*, **31**, 1060–1065.

Maguire, R.J., P.T.S. Wong and J.S. Rhamey. 1984. Accumulation and metabolism of tri-n-butyltin cation by a green alga, *Ankistrodesmus falcatus*. *Can. J. Fish. Aquat. Sci.*, **41**, 337–540.

Majidi, V., D.A. Laude, Jr and J.A. Holcombe. 1990. Investigation of the metal–algae binding site with ^{113}Cd nuclear magnetic resonance. *Environ. Sci. Technol.*, **24**, 1309–1312.

Martin, R.C., D.G. Dixon, R.J. Maguire, P.V. Hodson and R.J. Tkacz. 1990. Acute toxicity, uptake, depuration and tissue distribution of tri-n-butyltin in rainbow trout, *Salmo gairdneri*. *Aquat. Toxicol.*, **15**, 37–52.

Matthias, C., J.M. Bellama, G.J. Olson and F.E. Brinckman. 1986. A comprehensive method for the determination of aquatic butyltin and butylmethyltin species at ultra-trace levels using simultaneous hybridization/extraction with GC–FPD. *Environ. Sci. Technol.*, **20**, 609–615.

Okoshi, K., K. Mori and T. Nomura. 1987. Characteristics of shell chamber formation between the two local races in the Japanese oyster, *Crassostrea gigas*. *Aquaculture*, **67**, 313–320.

Rice, C.D., F.A. Espourteille and R.J. Huggett. 1987. Analysis of tributyltin in estuarine sediments and oyster tissue, *Crassostrea virginica*. *Appl. Organometal. Chem.*, **1**, 541–544.

Salazar, S.M., B.M. Davidson, M.H. Salazar, P.M. Stang and K.J. Meyers-Schulte. 1987. Effects of TBT on marine organisms: field assessment of a new site-specific bioassay system. In: Oceans '87 Conference Record, Vol. 4, IEEE Service Center, 445 Hoes Lane, Piscataway, New Jersey, pp. 1461–1470.

Salazar, M.H. and S.H. Salazar. 1988. Tributyltin and mussel growth in San Diego Bay. In: Ocean '88 Conference Record, Vol. 4, IEEE Service Center, 445 Hoes Lane, Piscataway, New Jersey, pp. 1188–1195.

Sasaki, K., T. Suzuke and Y. Saito. 1988a. Determination of tri-n-butyltin and di-n-butyltin compounds in yellowtails. *Bull. Environ. Contam. Toxicol.*, **41**, 888–895.

Sasaki, K., T. Ishizaka, T. Suzuke and Y. Saito. 1988b. Determination of tri-n-butyltin and di-n-butyltin compounds in fish by gas chromatography with flame photometric detection. *J. Assoc. Off. Anal. Chem.*, **71**, 360–365.

Short, J.W. and F.P. Thrower. 1986. Accumulation of butyltins in muscle tissue of Chinook salmon reared in sea pens treated with tri-n-butyltin. *Mar. Pollut. Bull.*, **17**, 542–545.

Short, J.W. and F.P. Thrower. 1987. Tri-n-butyltin caused mortality of Chinook salmon, *Oncorhynchus tshawytscha*, on transfer to a TBT-treated marine net pen. In: Oceans '86 Conference Record, Vol. 4, IEEE Service Center, 445 Hoes Lane, Piscataway, New Jersey, pp.193–197.

Slesinger, A.E. and I. Dressler. 1978. Degradation of TBT by bacteria. In: Report of the Organotin Workshop, M.L. Good (Ed.), New Orleans, Louisiana.

Soracco, R.J. and D.H. Pope. 1983. Bacteriostatic and bactericidal modes of action of bis(tributyltin) oxide on *Legionella pneumonophila*. *Appl. Environ. Microbiol.*, **45**, 48–57.

Sullivan, J.J., J.D. Torkelson, M.M. Wekell, T.A. Hollingworth, W.L. Saxton and G.A. Miller. 1988. Determination of tri-n-butyltin and di-n-butyltin in fish as hydride derivatives by reaction gas chromatography. *Anal. Chem.*, **60**, 626–631.

Takami, K., H. Yamamoto, T. Okumura, A. Sugimae and M. Nakamoto. 1987. Application of 'clean-up' cartridge for gas chromatographic determination of di- and tri-n-butyltin in fish. *Anal. Sci.*, **3**, 63–70.

Tsuda, T., H. Nakanishi, S. Aoki and J. Takebayashi. 1986. Bioconcentration of butyltin compounds by round crucian carp. *Toxicol. Environ. Chem.*, **12**, 137–138.

Tsuda, T., M. Wada, S. Aoki and Y. Matsui. 1987. Excretion of bis(tri-n-butyltin) oxide and triphenyltin chloride from carp. *Toxicol. Environ. Chem.*, **16**, 17–25.

Tsuda, T., H. Nakanishi, S. Aoki and J. Takebayashi. 1988a. Bioconcentration and metabolism of butyltin compounds in carp. *Water Res.*, **22**, 647–655.

Tsuda, T., M. Wada, S. Aoki and Y. Matsui. 1988b. Bioconcentration, excretion and metabolism of bis (tri-n-butyltin) oxide and triphenyltin chloride by gold fish. *Toxicol. Environ. Chem.*, **18**, 11–20.

Unger, M.A., W.G. MacIntyre, J. Greaves and R.J. Huggett. 1986. GC determination of butyltins in natural waters by flame photometric detection of hexyl derivatives with mass spectrometric confirmation. *Chemosphere (UK)*, **15**, 461–468.

Valkirs, A.O., P.F. Seligman, G. Vafa, P.M. Stang, V. Homer and S.H. Lieberman. 1985. Speciation of Butyltins and Methylbutyltins in Seawater and Marine Sediments by Hydride Derivatization and Atomic Absorption Detection. Technical Report 1087, Naval Oceans System Center, San Diego, California, 25 pp.

Valkirs, A.O., P.F. Seligman, P.M. Stang, V. Homer, S.H. Lieberman, G. Vafa and C.A. Dooley. 1986. Measurement of butyltin compounds in San Diego Bay. *Mar. Pollut. Bull.*, **17**, 319–324.

Wade, T.L., B. Garcia-Romero and J.M. Brooks. 1988. Tributyltin contamination in bivalves from

United States Coastal Estuaries. *Environ. Sci. Technol.*, **22**, 1488–1493.

Waldock, M.J. and J.E. Thain. 1983. Shell thickening in *Crassostrea gigas*: organotin antifouling or sediment induced? *Mar. Pollut. Bull.*, **14**, 411–415.

Waldock, M.J., J. Thain and D. Miller. 1983. The accumulation and depuration of bis (tributyltin) oxide in oysters: a comparison between the Pacific oyster (*Crassostrea gigas*) and the European flat oyster (*Ostrea edulis*). International Council for the Exploration of the Sea CM1983/E:52.

Ward, G.S., G.C. Cramm, P.R. Parrish, H. Trachman and A. Slesinger. 1981. Bioaccumulation and chronic toxicity of bis(tributyltin) oxide (TBT): tests with a saltwater fish. In: Aquatic Toxicology and Hazard Assessment: Fourth Conference, D.R. Branson and K.L. Dickson, (Eds). American Society for Testing and Materials, Philadelphia, ASTM STP 737, pp. 183–200.

Wieth, J.O. and T. Tosteson. (1979). Organotin-mediated exchange diffusion of anions in human red cells. *J. Gen. Physiol.*, **73**, 765–788.

Widdows, J, S.L. Moore, K.R. Clarke and P. Donkin. 1983. Uptake, tissue distribution and elimination of [^{14}-C]naphthalene in the mussel, *Mytilus edulis*. *Mar. Biol.*, **76**, 109–114.

Wolniakowski, K.U., M.D. Stephenson and G.S. Ichikawa. 1987. Tributyltin concentrations and Pacific oyster deformation in Coos Bay, Oregon. In: Oceans '87 Conference Record, Vol. 4, IEEE Service Center, 445 Hoes Lane, Piscataway, New Jersey, pp. 1438–1442.

Wong, P.T.S., Y.K. Chau, O. Kramar and G.A. Bengert. 1981. Structure–toxicity relationship of tin compounds on algae. *Can. J. Fish. Aquat. Sci.*, **39**, 483–488.

Wulf, R.G. and K.H. Byington. 1975. On the structure–activity relationships and mechanisms of organotin induced, non energy dependent swelling of liver mitochondria. *Arch. Biochem. Biophys.*, **167**, 176–185.

Zuolian, C. and A. Jensen. 1989. Accumulation of organic and inorganic tin in blue mussel, *Mytilus edulis*, under natural conditions. *Mar. Pollut. Bull.*, **20**, 281–286.

TRIBUTYLTIN BIOCONCENTRATION FROM SOLUTION AND SUSPENDED SEDIMENTS BY OYSTERS, WITH A COMPARISON WITH UPTAKE IN A FIELD EXPERIMENT

17

M.H. Roberts, Jr[1], *R.J. Huggett*[1], *M.E. Bender*[1,2], *H. Slone*[1] *and P.F. De Lisle*[3]

[1]*Virginia Institute of Marine Science, School of Marine Science, College of William and Mary, Gloucester Point, Virginia 23062, USA*
[2]*Deceased April 11, 1993*
[3]*Coastal Bioanalysts, Inc., PO Box 626, Gloucester Point, Virginia 23062, USA*

ABSTRACT

The bioconcentration of tributyltin (TBT) by oysters was examined in laboratory and field experiments. The bioconcentration factor (BCF) in laboratory experiments ranged from 5940 to 15 460 for uptake from solution (TBT concentration: 283 ng l^{-1} in experiment 1; 415 ng l^{-1} in experiment 2), while the BCF for uptake from suspended sediment (TBT concentration in water + sediment: 259 ng l^{-1}) was 9179. The BCF for the field experiment was 21 971, which is of a similar order of magnitude to values derived from laboratory

Organotin. Edited by M.A. Champ and P.F. Seligman. Published in 1996 by Chapman & Hall, London. ISBN 0 412 58240 6

experiments. Time to 90% of equilibrium was estimated to be 42–49 d in the laboratory compared with an estimated 56 d to equilibrium in the field.

17.1. INTRODUCTION

It is well accepted that bivalve molluscs accumulate various substances to a high degree. Since many bivalves are also an important food source for humans, consumption of bivalves that have accumulated toxic or carcinogenic materials may represent a significant source of a pollutant to humans. Further, the sedentary nature of bivalves coupled with their tendency to accumulate many classes of compounds make these organisms important as indicators of pollutants in natural environments (Anon, 1980; Langston and Burt, 1991; Lee, 1984; Martin and Severeid, 1984; Peddicord, 1984; White, 1984; Bender and Roberts, 1986; Abel, Chapter 2; Hall and Bushong, Chapter 9).

The American oyster (*Crassostrea virginica*) is harvested for human consumption from many locations throughout the Chesapeake Bay and its tributaries including areas near marinas and major naval and commercial ship moorings. This species is also routinely used to monitor the presence of various xenobiotics in estuarine environments. Laboratory evaluation of uptake provides an important basis for interpretation of field-collected data.

Tributyltin (TBT) is present in various locations within the Chesapeake Bay estuary, principally because of its application as a major biocide in a variety of antifouling paints. Recreational vessels, many of which may still be coated with TBT-containing paints, are moored in restricted portions of the estuary where there is poor tidal or wind mixing; this often results in high concentrations of TBT in the vicinity of marinas (Huggett *et al.*, 1986; Hall *et al.*, 1987, 1988). Tributyltin-containing paints are also applied to commercial vessels and to a limited number of naval vessels, leading to potential contamination of other areas as well, though the application of TBT is now prohibited in Virginia and some other states.

Tributyltin bioconcentration factors have been reported in the literature for only a few molluscs: the Pacific oyster, *Crassostrea gigas* (BCF = 2300–3100; Waldock and Thain, 1983); the European flat oyster, *Ostrea edulis* (BCF = 1000–1500; Waldock *et al.*, 1983); and the blue mussel, *Mytilus edulis* (BCF = 5000; Laughlin *et al.*, 1986b). The BCF values are realistic for a chemical like TBT with a very low octanol–water partitioning coefficient (Mackay, 1982; Laughlin *et al.* 1986a).

The present study consisted of three experiments. First, the bioconcentration factor for TBT in the American oyster, *C. virginica*, was determined. Second, the effect of suspended sediment on bioconcentration by *C. virginica* was examined (see Harris *et al.*, Chapter 22; Unger: *et al.*, Chapter 23). Finally, uptake of TBT was monitored in oysters transplanted into a location known to be contaminated with TBT.

17.2. MATERIALS AND METHODS

17.2.1. LABORATORY BIOCONCENTRATION STUDIES

The initial experiment examining uptake and clearance of TBT by oysters was conducted in 1987; the effects of sediment on uptake and clearance were examined during 1988. In both cases, the basic experimental design involved exposing oysters to TBT introduced to estuarine water either dissolved in triethylene glycol or associated with sediments. Simultaneously, an equal number of oysters were exposed to dilution water to provide control animals for analysis. At specified intervals, oysters were sacrificed and analyzed for TBT and DBT.

Oysters (6–9 cm in length) were collected from the Rappahannock River, cleaned of fouling organisms, and acclimated to the test temperature of 25°C for 4–6 weeks prior to use in an experiment. During the acclimation period, the oysters were fed cultured unicellular algae (*Isochrysis galbana* and *Pavlova lutheri*). A subsample of 10 oysters from each test population was determined to be free of significant infection by *Perkinsus marinus*, a common disease-causing organism in oysters, using the Ray test (Ray, 1954, 1966). Oysters were numbered for later random assignment to treatments and sampling.

For the first experiment, oysters were exposed in two hexagonal six-compartmented plastic trays. Each compartment is an equilateral triangle 33 cm on a side and 9 cm deep (4.2-l capacity). A distribution box was located above each hexagonal tray with an outlet to each chamber delivering 50 ml min^{-1} of solution to the apex of the compartment. Flow was controlled by calibrated pipet tips. Test medium exited each compartment by overflowing weirs cut in the base of the triangle. The turnover time was ~1.4 h.

Estuarine water was delivered to the distribution box from a head box at 300 ml min^{-1}. Cultured unicellular algae (*Isochrysis qalbana* and *Pavlova lutheri*) were delivered to the estuarine water headbox prior to delivery to the distribution box to yield a final algal concentration of ~10^{5} cells l^{-1} (Epifanio and Ewart, 1977). The toxicant was introduced to the estuarine water just prior to entry into the distribution box.

For the second experiment, oysters were exposed in four 74-l aquaria. The four treatments were TBT-contaminated sediment, uncontaminated sediment control, TBT in solution, and solvent control.

Filtered (1-μm pore size) estuarine water was introduced at a rate of 300 ml min^{-1} centrally into each aquarium from a head tank. The flow of estuarine water was controlled by means of calibrated pipet tips. Each tank was equipped with two external standpipes, one located at each end. Toxicant was pumped by a peristaltic pump (Harvard Apparatus Co., Cambridge, Massachusetts, USA) into the estuarine water stream through a glass T-connector just before entry into the aquaria. Sediment slurries were kept in suspension in stock jars by mechanical stirrers. Cultured unicellular algae (*I. galbana* and *P. lutheri*) were delivered to the head tank to provide a final concentration of ~10^{5} cells ml^{-1}. A separate head tank with pipet tips calibrated to deliver 100 ml min^{-1} provided water to 37-l depuration tanks.

Toxicant stocks were renewed daily. Sediment stocks were prepared by addition of previously prepared 23-ml sediment aliquots (418 mg ml^{-1} dry sediment) to a glass carboy containing 17 l of filtered (1-μm pore size) estuarine water. Sediment was collected at a 10-m depth in the York River, Virginia, ~400 m from the laboratory and sieved to a particle size of <500 μm. Sediment was spiked with TBT at 25 μg g^{-1} sediment and mixed for 4 h prior to dispensing to aliquots. Aliquots were stored in the dark at 4°C. The stock slurry was pumped into the exposure tank at a rate of 10 ml min^{-1} to yield a final concentration of 20 mg l^{-1} sediment (equivalent to 500 ng l^{-1} TBT). Control sediment stocks were similarly prepared using clean sediment aliquots. Dissolved phase stock solutions consisted of reagent grade triethylene glycol containing 1.5 mg l^{-1} TBT or clean triethylene glycol (control). Triethylene glycol stock solutions were pumped to exposure chambers at a rate of 0.1 ml min^{-1} to yield a nominal final concentration in the experimental chamber of 500 ng l^{-1} TBT.

Initial shell length was measured on each oyster prior to random assignment to aquaria. For experiment 1, each treatment included 60 oysters, 10 per compartment. For experiment 2, each aquarium contained 42 oysters. Oysters were exposed to the dissolved and suspended sediment phases of TBT for 28 d (uptake phase); the remaining oysters were then removed, scrubbed clean, and exposed to clean flowing water for a period of 28 d (clearance phase). During experiment 1, oysters (three per treatment) were removed on days 0, 3.5, 7, 14, and 28 of the uptake phase and days 3.5, 7, 14, and 28 of the clearance phase. During experiment 2, oysters were removed on days 0, 1.5, 3, 7, 14, 21, and 28 of the uptake phase and days 1.5, 3, 7, 14, 21, and 28 of the clearance phase. All oysters were transferred to depuration tanks for 24 h prior to sacrifice to allow them to clear their guts of any contaminated food or sediment. Final shell height, wet meat weight, and dry shell weight were measured on each animal. Meats were wrapped individually in aluminum foil and stored at about −20°C until chemical analysis for TBT and DBT using the method of Rice *et al.* (1987).

Water samples were collected on days 0, 8, 15, 21, and 22 during the first experiment and on the same days as biological samples during the first 14 d of the second experiment. No water samples were taken on days 21 and 28 because of the tank cleaning necessary at these times. Water samples were analyzed for TBT and DBT using the method of Unger *et al.* (1986). Dissolved oxygen concentration, salinity, pH, and temperature were measured daily in each exposure tank. Oxygen was measured with a YSI Model 57 oxygen meter (Yellow Springs Instrument Co., Yellow Springs, Ohio). Salinity was measured with an AO Spenser Refractometer. A Fisher Acumet pH Meter was used to measure pH. Temperature was measured with a stem thermometer.

Data for TBT body burdens were analyzed statistically using the BIOFAC model (Blau and Agin, 1978). This is a computerized implementation of a first-order pharmacokinetic model used to describe biological uptake of nonpolar organic compounds that reach a dynamic equilibrium with the environment in which the exposed organism lives. Since pharmacokinetic parameters for aquatic organisms are by convention calculated on a wet-weight basis, the tissue concentration data were converted to a wet-weight basis using percent wet-weight data collected in experiment 2. All other data were analyzed using subroutines of the SPSSx statistical package.

17.2.2. FIELD BIOCONCENTRATION STUDY

Oysters were collected from a relatively unpolluted region of the Rappahannock River, Virginia, on 3 February 1988. Some oysters were set aside for tissue analysis for TBT; the remainder were planted in Sarah Creek. The location selected for planting was in the northwest branch of the creek, downstream from a marina and upstream from a condominium. At the marina, boats are hauled for scraping and painting as well as moored. The condominium also has mooring slips, though many are not occupied.

The oysters were placed in plastic-coated wire mesh trays supported off the bottom on metal legs. They were placed so that the trays were submerged during all phases of the tide.

Oysters were sampled from the trays weekly on flood tide stage. Little settlement of silt or debris was observed at any sampling time. Whenever oysters were collected for analysis, temperature and salinity were measured, and a water sample was collected for butyltin analysis. Sampling continued until 20 July 1988.

Surviving oysters were brought to the laboratory and suspended in trays from Virginia Institute of Marine Science's oyster pier to allow for depuration of any TBT that had been accumulated. This group of oysters was sampled on 27 July, 8 August, and 29 August. On 17 August, a group of feral oysters was removed from the pier pilings for analysis.

Water samples for TBT analysis were acidified upon collection and stored at 4°C until they could be analyzed. All water samples were analyzed by the method of Unger *et al.* (1986). For oyster tissue analysis, two composites of oyster tissue were prepared. For each composite, 5–10 oysters (~10 g total wet meat weight) were cleaned and shucked; the meats were composited, homogenized, and desiccated. The dry weight was then determined, and the tissue samples were analyzed for butyltins using the method of Rice *et al.* (1987). On day 84 of the uptake phase, eight oysters were analyzed individually to obtain an estimate of the variance masked by the compositing procedure.

The bioconcentration factors (BCF) for oyster composites were calculated as the body burden based on wet weight divided by the average TBT concentration as measured at the beginning of the preceding week and the time of collection. The resultant BCF values were examined and, once reasonably stable with time, considered to be at equilibrium. The subsequent BCF values were averaged to obtain a mean equilibrium BCF with standard deviation and coefficient of variation.

17.3. RESULTS AND DISCUSSION

17.3.1. EXPERIMENT 1 – UPTAKE FROM SOLUTION

Throughout the uptake and clearance phases of the experiment, the temperature was 24.4 ± 1.0°C. Salinity gradually increased from 16‰ at the start of the 2-month experiment to around 20‰ at the end with a mean of 18 ± 1‰ . Dissolved oxygen averaged 87 ± 7% of saturation in all treatments. All treatments were aerated to increase oxygen concentrations, which initially were below 60% of saturation during the pretest acclimation period.

The measured concentration of TBT in the nominal 500 ng l^{-1} treatments was consistently below the desired concentration. The exposure concentration was 283 ± 71 ng l^{-1} (Table 17.1). The discrepancy between the nominal and measured concentrations may reflect an effect of aeration, but this possibility was not experimentally evaluated. The discrepancy may also reflect significant sorption to the acrylic exposure chambers. Measured TBT concentrations were used to calculate BCF values.

Table 17.1 Measured tributyltin concentrations (ng l^{-1}) in control and test chambers during laboratory exposure of oysters (experiment 1)

	Chambers	
Days	*Control (ng l^{-1})*	*Exposed (ng l^{-1})*
0	ND[a]	266
8	11	269
15	9	552
21	ND	174
22	ND	158
Means	4.0[b]	283.8
Std dev.	2.5	70.8
Coeff. var.	62	25

[a]ND = non-detectable.
[b]To calculate the mean, samples with ND concentrations were treated as a zero. The detection limit was 1 ng l^{-1}.

Oysters used for the experiments had an initial average shell length of 95 ± 9 mm. There was no difference in initial size of control and exposed oysters. No significant growth in shell length was observed. The lack of measurable growth in shell length does not necessarily reflect a lack of growth in biomass, but the latter cannot be measured non-sacrificially.

The ratio of wet meat weight to dry shell weight times 100 was the only usable condition index (Mann, 1978). The condition indices (CI) were initially high (CI = 15.18). They decreased slightly during the 2-month experiment, but the coefficient of determination was low (R^2 <0.20) indicating that there was no reliable relationship between animal condition and time.

The TBT body burdens for oysters in the first experiment seem nearly constant in a plot of log body burden versus exposure time (Fig. 17.1). Since control oysters contained no measurable TBT, there was clearly accumulation by oysters exposed in this experiment. Contrary to this subjective evaluation, TBT uptake by oysters, as modeled by the one-compartment model implemented by BIOFAC, yielded predictions that are reasonable. The model predicted that oysters

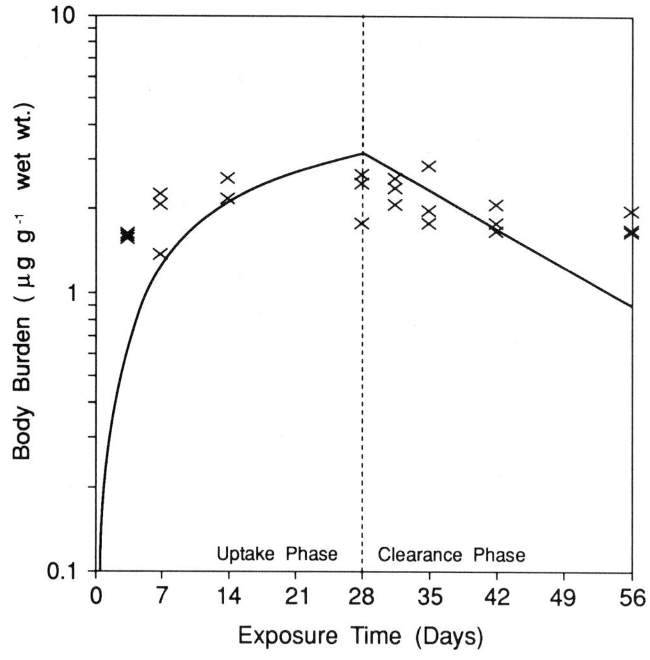

Figure 17.1 BIOFAC plots of tributyltin accumulation by oysters exposed to TBT in solution, experiment 1.

Table 17.2 TBT uptake–clearance parameters for oysters exposed to tributyltin in solution (experiment 1)

Parameter	*Mean ± standard deviation*
Uptake rate, k_1	722 ± 242 d^{-1}
Clearance, k_2	0.047 ± 0.023 d^{-1}
BCF	15460 ± 9124
Time to 90% of BCF	49 ± 24 d
$t_{1/2}$ for clearance	15 ± 7 d

require 49 d to reach 90% of an equilibrium body burden (Table 17.2), within the 30–60 d range observed in other studies. The estimated clearance rate is 0.047 ± 0.023 d^{-1}. From the plot of the data (Fig. 17.1), the clearance rate was apparently not linear, but rather changed markedly during the exposure period. The BCF calculated by the BIOFAC model was 15 460 ± 9124, whereas a direct calculation (mean body burden at 28 d) divided by (mean exposure concentration) yields a BCF of 8215.

17.3.2. EXPERIMENT 2 – EFFECTS OF SEDIMENT ON BIOCONCENTRATION

Treatments receiving triethylene glycol developed a bacterial slime on surfaces from day 15 of the uptake phase to the end; oxygen depression (minimum ~65% of saturation) due to the slime was countered by additional aeration of all tanks. Continued slime production in tanks receiving triethylene glycol required that they be cleaned every 48–72 h; tanks receiving either sediment or triethylene glycol were cleaned at these times. Salinity ranged from 17 to 20% and pH from 7.50 to 7.91 over the 56-d study. Temperature was maintained within 2°C of the desired temperature (25°C) except for a brief period during day 53 on which an electrical failure resulted in loss of temperature control. Oxygen concentrations were >90% of saturation.

Table 17.3 Measured tributyltin concentrations (ng l^{-1}) in control and test treatments receiving TBT in solution or associated with sediment (experiment 2)

	Control		*Exposed*	
Days	*Solution*	*Sediment*	*Solution*	*Sediment*
0	65.0[a]	1.5	268	254
1	ND[b]	5.0	501	333
3	9.0	7.5	405	268
7	ND	ND	475	179
14	ND	ND	425	NS[c]
Mean	14.8[d]	2.8	414.8	258.5
Std dev.	12.7	1.5	40.5	31.6
Coeff var.	85.0	53.0	10.0	12.0

[a]Unexplained peak.
[b]ND = non-detectable.
[c]NS = sample lost.
[d]To calculate the mean, samples with non-detectable concentrations were treated as a zero. The detection limit was 1 ng l^{-1}.

The measured concentrations of TBT were 415 ng l^{-1} in the solution exposure, and 259 ng l^{-1} in the sediment exposure (Table 17.3). Although both treatments had a nominal concentration of 500 ng l^{-1}, sorption to glass in the system and settlement of sediment in the delivery tubing (for the sediment treatment) are believed to explain the deviations from nominal concentrations. The mean measured concentrations are considered to represent the actual exposure concentrations and were used in all data analyses.

Condition indices of oysters [100 × (wet meat weight) (dry shell weight)$^{-1}$] ranged from 18.3 to 22.7 and did not appear to be affected by exposure to TBT in solution or on sediments. Shell growth [(initial shell length) – (final shell length)] was also not affected by exposure to TBT. Oysters exposed to TBT in solution (triethylene glycol treatment) did not reach an apparent steady-state concentration by day 28 (Fig. 17.2). Based on the day-28 body burden, a non-equilibrium BCF [body burden × (water concentration)$^{-1}$] of ~3410 can be calculated. This result does not agree with that obtained during experiment 1 in which apparent equilibrium was achieved in 7 d with a crude BCF of 8215. However, when the BIOFAC model was applied to data from experiment 2, the BCF calculated from the new data was 5940 with an uptake rate of 329 d^{-1} and a clearance rate of 0.055 d^{-1} (Table 17.4). The time to 90% of the BCF was 42 d.

Oysters exposed to TBT associated with sediments also did not reach an apparent steady-state condition by day 28 (Fig. 17.3). At that time the body burden was 2.2 μg g^{-1} TBT tissue (wet wt), yielding a non-equilibrium BCF of ~4400. The BCF calculated from these data using BIOFAC is 9180 with an uptake rate of 431 d^{-1} and a clearance rate of 0.047 d^{-1} (Table 17.4). The time to 90% of the BCF was 49 d.

The apparent steady state in the oysters exposed to TBT in solution may reflect significant bias stemming from the bacterial bloom growing on surfaces. One might hypothesize that the bacteria sorb significant amounts of TBT from water. Since the bacteria exist as a film on surfaces, they and the sorbed TBT would be unavailable to the oysters. The oysters, exposed to an available concentration of TBT lower than planned, would reach an equilibrium at a lower body burden than if they had been exposed to the full amount of TBT. The coincidence of the appearance of the bacterial bloom and the onset of the apparent steady state on or about day 14 seems to lend credence to this interpretation. The BIOFAC analysis yields a BCF of 5940, which is about half that for oysters exposed to TBT with sediment.

The BCF values derived by application of the BIOFAC model for presentation in solution or with sediment are not significantly different (Student's t = 0.201). While it has been shown that TBT does sorb to sediments, TBT also desorbs rapidly, reaching 80% of equilibrium in ~15 min and equilibrium in 2–3 h (Unger, 1988). It is reasonable to assume therefore that the aqueous concentration of TBT was essentially the same in both treatments of our study, although this cannot be

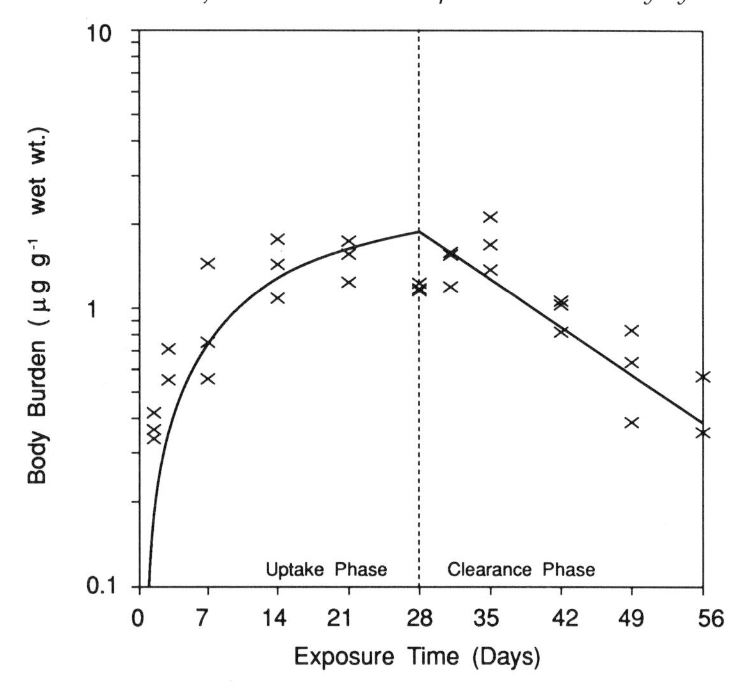

Figure 17.2 BIOFAC plots of tributyltin accumulation by oysters exposed to TBT in solution, experiment 2.

Table 17.4 Comparison of TBT uptake–clearance parameters for oysters exposed to tributyltin in solution or in association with sediments

	Exposure to TBT[a]	
Parameters	*From solution*	*From sediment*
Uptake rate, k_1	329 ± 46 d^{-1}	431 ± 28 d^{-1}
Clearance, k_2	0.055 ± 0.008 d^{-1}	0.047 ± 0.004 d^{-1}
BCF	5940 ± 1219	9179 ± 923
Time to 90% of BCF	42 ± 6 d	49 ± 4 d
$t_{1/2}$ for clearance	13 ± 2 d	15 ± 1 d

[a]Mean ± standard deviation from BIOFAC model calculation.

confirmed by the analytical data available, which was compromised by the bacterial slime during much of the experiment. Laughlin *et al.* (1986b) in short-term accumulation tests with blue mussels (*M. edulis*) also found no difference in uptake from solution with or without kaolin particles (a clay mineral common to coastal estuaries) present. In both the treatments, TBT was present at a concentration of 500 ng l^{-1}.

17.3.3. FIELD BIOCONCENTRATION STUDY

During the field study, the temperature increased from ~4°C in early February to 30.5°C at the end of the uptake phase (Table

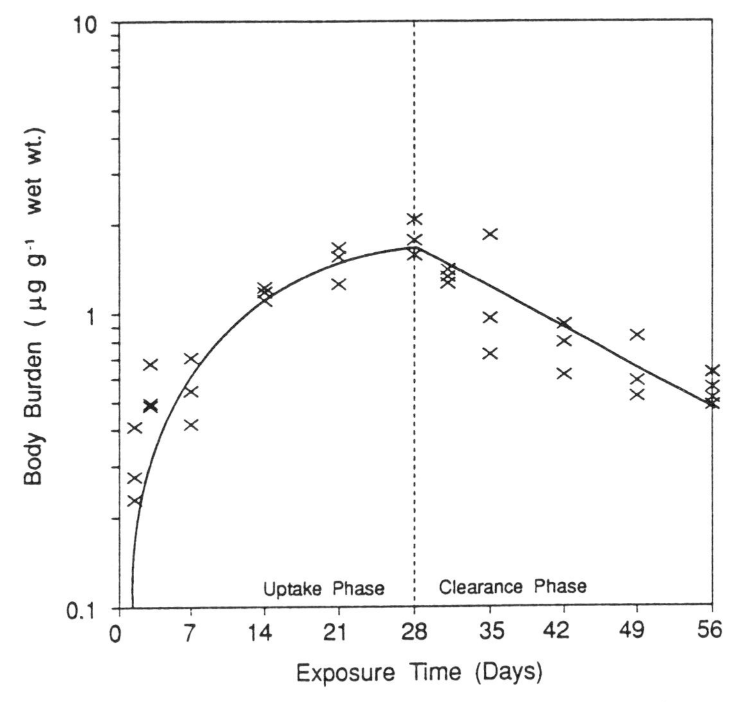

Figure 17.3 BIOFAC plots of tributyltin accumulation by oysters exposed to TBT associated with suspended sediments, experiment 2.

17.5). Salinity at the experimental site in Sarah Creek ranged from 15.5 to 20‰. Water quality conditions were typical of a tidal estuary during this period of the year.

The concentration of TBT based on weekly samples was quite uniform with a mean of 16.8 ± 4.7 ng l^{-1} [coefficient of variation (CV) = 27.8%], ranging from 10.1 to 25.7 ng l^{-1}. The highest concentrations were observed during mid- to late February, mid- to late March, and the last week in April. Dibutyltin concentrations gave an impression of less uniformity, increasing from 6 to 8 ng l^{-1} in February and March to 10 to 19 ng l^{-1} during May and June. The mean concentration was 10.3 ± 3.7 ng l^{-1} (CV = 36.3%). Monobutyltin was uniformly low, ranging from 1 to 3.2 ng l^{-1}, which approximates the detection limit.

The TBT body burden of the oysters when removed from the Rappahannock River was 0.0074 µg g^{-1}. After 3 d in Sarah Creek, the body burden had increased to 0.0207 µg g^{-1}. By day 56 the body burden had increased to 0.211 µg g^{-1}; thereafter the body burden ranged between 0.211 and 0.629 µg g^{-1} with a mean of 0.327 ± 0.120 µg g^{-1} (CV = 36.7%).

From an examination of the BCF valves calculated for each sampling interval, it is estimated that equilibrium was reached in ~56 d. The mean BCF for the period during which oysters were at apparent equilibrium with their environment (56–163 d) was 21 971 ± 7938 (CV = 36%). In tissue from deposit-feeding clams (*Scrobicularia plana*), the concentration factor in sediment was 9.0 ± 1.7 and in water 96 + 12 × 10^3 (Langston and Burt, 1991).

The estimated time to equilibrium for the field-exposed animals (56 d) is only slightly longer than the time to 90% of equilibrium estimated from the laboratory experiments (42–49 d); these estimates are quite consistent with each other. Further, the estimates of BCF for the field and laboratory situations were reasonably consistent with each other, though the field estimate is twice the largest

Table 17.5 Field bioconcentration observations with oysters

Date	Day	Water quality: Temperature	Water quality: Salinity	TBT (ng l^{-1})	DBT (ng l^{-1})	MBT (ng l^{-1})	Body burden of TBT: Wet (μg g^{-1})	Body burden of TBT: Dry (μg g^{-1})	BCF (wet wt)
Uptake phase									
03 Feb	0	ND[a]	9.0	19.4	7.3	2.0	0.0074	0.055	[b]
06 Feb	3	4.0	18.0	ND	ND	ND	0.0207	0.133	1067
09 Feb	6	6.0	18.0	12.5	6.8	1.4	0.0213	0.171	
16 Feb	13	6.0	16.0	22.8	7.2	1.7	0.0312	0.227	1768
23 Feb	20	10.5	18.0	25.7	5.8	1.6	0.0279	0.200	1151
08 Mar	34	11.0	18.0	16.7	6.9	1.7	0.113	0.787	5307
11 Mar	37	ND	ND	12.5	7.2	2.2	ND	ND	
16 Mar	42	7.0	18.0	20.2	7.3	1.0	0.120	0.871	7309
22 Mar	48	9.5	18.0	20.2	9.5	1.3	0.165	1.24	8168
30 Mar	56	15.5	16.0	21.7	10.7	3.2	0.211	1.61	10072
05 Apr	62	19.0	16.0	20.4	10.3	1.5	0.330	2.38	15672
13 Apr	70	13.0	18.0	13.8	8.4	2.8	0.263	2.89	15380
19 Apr	76	14.0	18.0	12.2	7.2	1.4	0.224	1.86	17231
27 Apr	84	17.0	18.0	23.4	13.3	1.6	0.324	1.65	18223
04 May	91	17.5	18.0	13.7	12.6	2.1	0.444	2.61	23935
11 May	98	20.5	18.0	15.8	12.0	1.7	0.629	3.53	42610
23 May	110	27.0	15.0	10.1	17.0	2.9	0.324	1.78	24981
06 Jun	124	24.0	15.5	11.2	10.7	1.7	0.289	1.60	27089
28 Jun	146	25.5	18.0	14.0	15.9	2.7	0.230	1.54	18214
15 Jul	163	30.5	20.0	11.9	19.0	3.0	0.212	1.40	16371
Mean				16.8	10.3	2.0	0.327	2.12	21971
Std dev.				4.7	3.7	0.6	0.120	0.67	7938
Coeff. var.				27.8	36.3	32.1	36.7	31.4	36
Minimum				10.1	5.8	1.0	0.212	1.34	15380
Maximum				25.7	19.0	3.2	0.629	3.53	42610
Clearance phase									
27 Jul	7	28.0	21.5	3.6	4.0		0.162	1.03	45506
09 Aug	20	29.0[c]	22.0[c]	3.5	3.2		0.170	1.17	47881
29 Aug	40	27.5[c]	22.0	4.5	2.0	0.7	0.063	0.42	15711
Mean				3.9					
Std dev.				0.5					
Coeff. var.				11.7					
Feral oysters									
17 Aug		30.0[c]	22.0[c]				0.039	0.265	11088

[a]ND indicates not detectable.
[b]Blank indicates no data.
[c]Data from seawater system drawing water on pier where clearance phase was conducted.

laboratory estimate, and both are higher than the theoretical estimate based on partitioning coefficients (Laughlin, 1986b; Langston and Burt, 1991).

There is no evidence from the present study that temperature affected the uptake of TBT. Despite a 15°C increase from the time at which apparent equilibrium was reached to the end of the exposure period, there was no apparent increase in body burden. This is consistent with prior laboratory studies examining the effect of temperature on uptake of various polycyclic aromatic hydrocarbons (Neff, 1979).

During the field clearance phase conducted in the York River near the mouth of Sarah Creek, water temperatures remained high, and salinity was similar to that at the exposure site. The TBT concentration was 3.86 ± 0.45 ng l^{-1} and approximately equal to the DBT concentration. The body burdens declined from 0.212 to 0.063 $\mu g\ g^{-1}$ during the 40-d clearance phase. At the end of the clearance phase, however, the mean concentration in the experimental oysters was still nine times that in the oysters originally collected from the Rappahannock River, and 1.5 times that in a sample of feral oysters collected at the site of the clearance phase.

During the clearance phase, only three samples were taken, which limits the interpretation of these results. It would appear that the concentration declined ~25% during the first 7 d; this is consistent with results of the laboratory experiments. During the next 13 d, the body burden remained essentially constant, and then declined 63% during the next 20 d; this latter decline is consistent with the laboratory results. However, these field clearance data do not represent a typical clearance curve. These data could be explained by a spike of TBT passing through the site of the clearance phase during late July and early August without being detected by the infrequent water analyses. Regardless of the true explanation, no completely reliable estimate of clearance rate can be derived from the field data.

17.4. CONCLUSION

The BCF for TBT calculated from field experimental data is higher in absolute value than that observed in laboratory experiments, but of a similar order of magnitude and not statistically different. The BCF values derived from laboratory experiments for TBT presented in solution were 15 460 and 5940. Bioconcentration from TBT associated with suspended sediments was similar in magnitude to that from TBT in solution.

The large variances associated with both laboratory and field estimates allow the actual values to differ by a factor of 2.6 even when not significantly different (based on a Student's t-test, $p > 0.05$).

ACKNOWLEDGEMENTS

These studies were supported in part by funding from the Virginia Water Control Board under cooperative state agreements.

We appreciate the efforts of R. Williams, D. Sved, and P. Balcom who assisted with the conduct of the laboratory exposure studies. J. Greene assisted with the field transplantation study. D. Westbrooke and E. Travelstead assisted with the chemical analyses for field and laboratory samples.

REFERENCES

Anonymous. 1980. The International Mussel Watch. Report of a Workshop sponsored by the Environmental Studies Board, Commission on Natural Resources, National Research Council, National Academy of Sciences, Washington, DC, 248 pp.

Bender, M.E. and M.H. Roberts, Jr 1986. Bioconcentration of PAH from Contaminated Sediments by Bivalve Mollusks. Final Report by the Virginia Institute of Marine Science, Gloucester Point, Virginia, to the Virginia Water Control Board, Richmond, Virginia, 8 pp.

Blau, G.E. and G.L. Agin. 1978. A User's Manual for BIOFAC: a computer program for characterizing the rates of uptake and clearance of chemicals in aquatic organisms. Central Research/Physical

Research/Math Applications, The Dow Chemical Co., Midland, Michigan (unpaginated).

Epifanio, C.E. and J. Ewart. 1977. Maximum ration of four algal diets for the oyster *Crassostrea virginica* Gmelin. *Aquaculture*, 11, 13–29.

Hall, L.W., Jr, M.J. Lenkevich, W.S. Hall, A.E. Pinkney and S.J. Bushong. 1987. Evaluation of butyltin compounds in Maryland waters of Chesapeake Bay. *Mar. Pollut. Bull.*, **18**, 78–83.

Hall, L.W., Jr, S.J. Bushong, W.E. Johnson and W.S. Hall. 1988. Spatial and temporal distribution of butyltin compounds in a northern Chesapeake Bay marina and river system. *Environ. Monit. Assess.*, **10**, 229–244.

Huggett, R.J., M.A. Unger and D.J. Westbrook. 1986. Organotin concentrations in southern Chesapeake Bay. In: Proceedings of the Oceans '86 Organotin Symposium, Vol. 4, Marine Technology Society, Washington, DC, 23–25 September 1986, pp. 1262–1265.

Langston, W.J. and G.R. Burt. 1991. Bioavailability and effects of sediment-bound TBT in deposit feeding clams, *Scrobicularia plana*. *Mar. Environ. Res.*, **32**, 61–77.

Laughlin, R.B., Jr, H.E. Guard and W.M. Coleman, III. 1986a. Tributyltin in seawater: speciation and octanol–water partition coefficient. *Environ. Sci. Technol.*, **20**, 201–204.

Laughlin, R.B., Jr, W. French and H.E. Guard. 1986b. Accumulation of bis(tributyltin) oxide by the marine mussel *Mytilus edulis*. *Environ. Sci. Technol.*, **20**, 884–890.

Lee, R.F. 1984 Factors affecting bioaccumulation of organic pollutants by marine animals. In: *Concepts of Marine Pollution Measurements*, H.H. White (Ed.). A Maryland Sea Grant Publication, University of Maryland, College Park, pp. 339–354.

Mackay, D. 1982. Correlation of bioconcentration factors. *Environ. Sci. Technol.*, **16**, 274–278.

Mann, R. 1978. A comparison of morphometric, biochemical, and physiological indexes of condition in bivalve mollusks. In: *Energy and Environmental Stresses in Aquatic Systems*, J.H. Thorpe and I.W. Gibbons (Eds). Department of Energy Symposium Series 48, Washington, DC, pp. 484–497.

Martin, M. and R. Severeid. 1984. Mussel watch monitoring for the assessment of trace toxic constituents in California marine waters. In: *Concepts of Marine Pollution Measurements*, H.H. White (Ed.). A Maryland Sea Grant Publication, University of Maryland, College Park, pp. 291–324.

Neff, J.M. 1979. *Polycyclic Aromatic Hydrocarbons in the Aquatic Environment: Sources, Fates, and Biological Effects*. Applied Science Publishers, London, 262 pp.

Peddicord, R.K. 1984. What is the meaning of bioaccumulation as a measure of marine pollution effects? In: *Concepts of Marine Pollution Measurements*, H.H. White (Ed.). A Maryland Sea Grant Publication, University of Maryland, College Park, pp. 249–260.

Ray, S.M. 1954. Biological Studies of *Dermocystidium marinum*. The Rice University Pamphlet, Monograph in Biology, Special Issue, 114 pp.

Ray, S.M. 1966. A review of the culture method for detecting *Dermocystidium marinum*, with suggested modifications and precautions. *Proc. Nat. Shellfish. Assoc.*, **54**, 55–69.

Rice, C.D., F.A. Espourteille and R.J. Huggett. 1987. Analysis of tributyltin in estuarine sediments and oyster tissue, *Crassostrea virginica*. *Appl. Organometal. Chem.*, **1**, 541–544.

Unger, M.A. 1988. Investigation of Tributyltin Water/Sediment Interactions. PhD Dissertation, College of William and Mary, Williamsburg, Virginia, 80 pp.

Unger, M.A., W.G. MacIntyre, J. Greaves and R.J. Huggett. 1986. Determination of butyltins in natural waters by flame photometric detection of hexyl derivatives with mass spectrometric detection. *Chemosphere*, **15**, 461–470.

Waldock, M.J. and J. Thain. 1983. Shell thickening in *Crassostrea gigas*: organotin antifouling or sediment induced? *Mar. Pollut. Bull.*, **14**, 411–415.

Waldock, M.J., J. Thain and D. Miller. 1983. The accumulation and depuration of bis (tributyltin)oxide in oysters: a comparison between the Pacific oyster (*Crassostrea gigas*) and the European flat oyster (*Ostrea edulis*). International Council for Exploration of the Sea, CM 1983/E:52, 9 pp.

White, H.H. 1984. Mussel madness: use and misuse of biological monitors of marine pollution. In: *Concepts of Marine Pollution Measurements*, H.H. White (Ed.). A Maryland Sea Grant Publication, University of Maryland, College Park, pp. 325–338.

METABOLISM OF TRIBUTYLTIN BY AQUATIC ORGANISMS 18

Richard F. Lee

Skidaway Institute of Oceanography, 10 Ocean Science Circle, Savannah, Georgia 31411, USA

ABSTRACT

This chapter summarizes work on the metabolism of tributyltin (TBT) by marine fauna and flora and how this metabolism may relate to some observed effects. Fish and crustaceans have an active cytochrome P-450 dependent monooxygenase system that oxidizes TBT to a series of hydroxylated derivatives. These hydroxylated derivatives are conjugated to sulfate or carbohydrate by phase-two enzyme systems, which facilitates the elimination of TBT. Molluscs have low cytochrome P-450 content and mixed function oxygenase activity, which result in TBT accumulation and slow depuration due to the low rate of TBT metabolism. Some of the TBT effects observed for molluscs include imposex in stenoglossan gastropods; shell thickening in oysters; reduced growth rates in mussels; and breakdown of sexual differentiation, oogenesis, and egg production in *Ostrea edulis*. I suggest that these effects are related to the slow metabolism of TBT, binding of TBT metabolites to cellular proteins, and inhibition of detoxifying enzyme systems (e.g. cytochrome P-450 systems and glutathione S-transferases) by TBT. TBT and hormones share common metabolic pathways, and many effects of TBT are hormonally regulated.

18.1. INTRODUCTION

Toxic pollutants enter aquatic organisms from the water, food or sediment. After toxicants enter they can be stored or eliminated. The kinetics of both storage and

Organotin. Edited by M.A. Champ and P.F. Seligman. Published in 1996 by Chapman & Hall, London. ISBN 0 412 58240 6

elimination of a toxic compound are affected by its metabolism, or lack of it. Many toxic compounds are highly hydrophobic, leading to accumulation in lipid-rich parts of tissues and cells. The elimination of such hydrophobic compounds is facilitated by their biotransformation to water-soluble polar compounds. Thus, metabolism of a compound generally reduces persistence, increases elimination, and reduces toxicity. However, for some compounds the metabolites are more toxic than the parent compounds. For example, the binding of certain reactive benzo(*a*)-pyrene metabolites (i.e. arene oxides) to DNA initiates carcinogenesis (Miller and Miller, 1971; Ames *et al.*, 1972). Toxic pollutants and their metabolites can bind to various cellular macromolecules (e.g. protein, DNA, and RNA); and for some compounds cellular damage has been found only in organs where there was covalent binding of metabolites to macromolecules (Bartolone *et al.*, 1989). Tributyltin (TBT), a polar compound, enters organisms through lipid membranes and is metabolized in two phases. The phase-one reactions involve the cytochrome P-450 dependent monooxygenase system (MFO), which hydroxylates TBT to alpha-, beta-, gamma-, and delta-hydroxydibutyltin derivatives (Fish *et al.*, 1976). The phase-two reactions conjugate sugars or sulfate to hydroxybutyldibutyltin, and these highly polar conjugates are rapidly eliminated from the animal. The MFO system of vertebrates and invertebrates is associated with the endoplasmic reticulum of the cell and is a multicomponent enzyme system composed of phospholipid, cytochrome P-450, and NADPH cytochrome P-450 reductase (Lu, 1976; Lee, 1981; Stegeman, 1981). The MFO system oxidizes foreign compounds by hydroxylation, dealkylation, or epoxidation. In mammals, exposure to certain inducers results in elevated MFO activity and P-450 content and often in the production of new forms of cytochrome P-450. Phase-two reactions result in the formation of water-soluble conjugates with such compounds as glucose, glutathione, or sulfate.

The pathways used by mammals in metabolizing TBT have been the subject of several investigations (Fish *et al.*, 1976; Kimmel *et al.*, 1977; Iwai *et al.*, 1981; Wada *et al.*, 1982; Fish, 1984; Krigman and Silverman, 1984). TBT binds to the microsomal cytochrome P-450 from rat liver, producing a type I binding spectrum (Rosenberg and Drummond, 1983). The microsomal MFO system of rat liver metabolizes tributyltin acetate to alpha-, beta-, gamma-, and delta-hydroxybutyldibutyltin derivatives, with a predominance of the beta-carbon hydroxylated (Fish *et al.*, 1976). The hydroxy derivatives are dealkylated to form dibutyltin (DBT). Also, both the alpha- and beta-hydroxylation products are unstable and spontaneously go to dibutyltin. Thus, in-vivo studies show dibutyltin and monobutyltin (MBT) as the major TBT metabolites (Kimmel *et al.*, 1977; Iwai *et al.*, 1981; Wada *et al.*, 1982). Presumably, phase-two enzyme systems conjugate sugars for sulfate to the hydroxylated derivatives in mammals. These highly polar conjugates should then be rapidly eliminated from the animal.

This chapter summarizes recent work on the metabolism of TBT by enzyme systems in aquatic plants and animals and the entrance of TBT and metabolites into certain digestive cells. Information is provided that shows the binding of TBT metabolites to protein but not to nucleic acids in crab hepatopancreas cells. I suggest that the metabolism and metabolites of TBT may be related to some of the observed effects.

18.2. DISCUSSION

18.2.1. METABOLISM OF TBT BY FISH

Fish exposed to TBT in the water had tributyltin, dibutyltin, and monobutyltin in their tissues when exposure times varied from 2 to 15 d (Table 18.1; Ward *et al.*, 1981; Lee, 1986; Tsuda *et al.*, 1988; Martin *et al.*,

Table 18.1 TBT metabolites produced in fish after TBT exposure via the water

Species	*Tissue*	*Initial TBT conc. (μg l⁻¹)*	*Exposure time (d)*	*Depuration time (d)*	*Tissue conc. (μg g⁻¹)* TBT	DBT	MBT	*Reference*
Rainbow trout (*Salmo gairdneri*)	Liver	0.42	15	—[a]	1.21	1.04	0.42	Martin *et al.*, 1989
	Liver	0.42	15	15	1.32	2.17	1.27	Martin *et al.*, 1989
	Gall bladder	0.42	15	—	0.34	0.90	1.30	Martin *et al.*, 1989
	Gall bladder	0.42	15	15	0.48	1.37	1.93	Martin *et al.*, 1989
	Kidney	0.42	15	—	2.30	0.96	0.25	Martin *et al.*, 1989
	Kidney	0.42	15	15	1.50	0.53	0.75	Martin *et al.*, 1989
	Muscle	0.42	15	—	0.32	0.11	0.02	Martin *et al.*, 1989
	Muscle	0.31	15	15	0.32	0.10	0.09	Martin *et al.*, 1989
Carp (*Cyprinus carpio*)	Liver	2.1	7	—	2.0	0.25	0.50	Tsuda *et al.*, 1988
	Liver	2.1	15	—	1.2	0.60	3.5	Tsuda *et al.*, 1988
	Gall bladder	2.1	7	—	2.0	0.5	1.0	Tsuda *et al.*, 1988
	Gall bladder	2.1	15	—	2.0	2.5	14	Tsuda *et al.*, 1988
	Kidney	2.1	7	—	0.8	0.5	0.5	Tsuda *et al.*, 1988
	Kidney	2.1	15	—	6.2	0.8	3.0	Tsuda *et al.*, 1988
Spot (*Leiostomus canthurus*)	Liver	2.0	3	—	0.08	0.03	0.02	Lee, 1986
	Liver	2.0	2	—	0.12	0.02	0.01	Lee, 1986

[a]Dashes indicate no data.

1989). Both dibutyltin and monobutyltin increased with time in both liver and gall bladder (Table 18.1). Spot (*Leiostomus canthurus*) exposed to ^{14}C-TBT in the water accumulated radioactivity in the gill and liver. In the gills 90% of the radioactivity was associated with TBT, while in the liver dibutyltin was the major radioactive compound (Fig. 18.1). The lack of TBT metabolites in the gill suggests that gills were not important in TBT metabolism. In contrast, the high level of MFO activity and presence of TBT metabolites in fish liver was indicative of the importance of this tissue in TBT metabolism. The primary butyltin species in the liver of fish collected from sites where the sediment was contaminated with TBT was dibutyltin (Krone *et al.*, 1989). Spot exposed to radiolabeled TBT via food accumulated radioactivity in the intestine followed by an increase in the radioactivity of liver and gall bladder (Lee, 1986). After 3 d, most of the radioactivity found in the fish was associated with dibutyltin, monobutyltin, and various conjugates (Fig. 18.1). Small amounts of hydroxylated derivatives were detected. These hydroxylated derivatives are likely to be conjugated rapidly to sulfate or glucuronic acid moieties by phase-two enzyme systems. It is likely that part of the TBT in the food was metabolized in the intestine, since fish intestinal MFO can metabolize many dietary toxicants (Van Veld *et al.*, 1988a,b; Van Veld, 1990).

The bioconcentration factors reported for fish that take TBT up from water vary from 400 to 500 (Ward *et al.*, 1981; Martin *et al.*, 1989). Sheephead minnow (*Cyprinodon variegatus*) exposed to ^{14}C-TBT in the water and transferred to TBT-free water had a TBT depuration half-life of 7 d (Ward *et al.*, 1981). As noted above, TBT taken up from the food

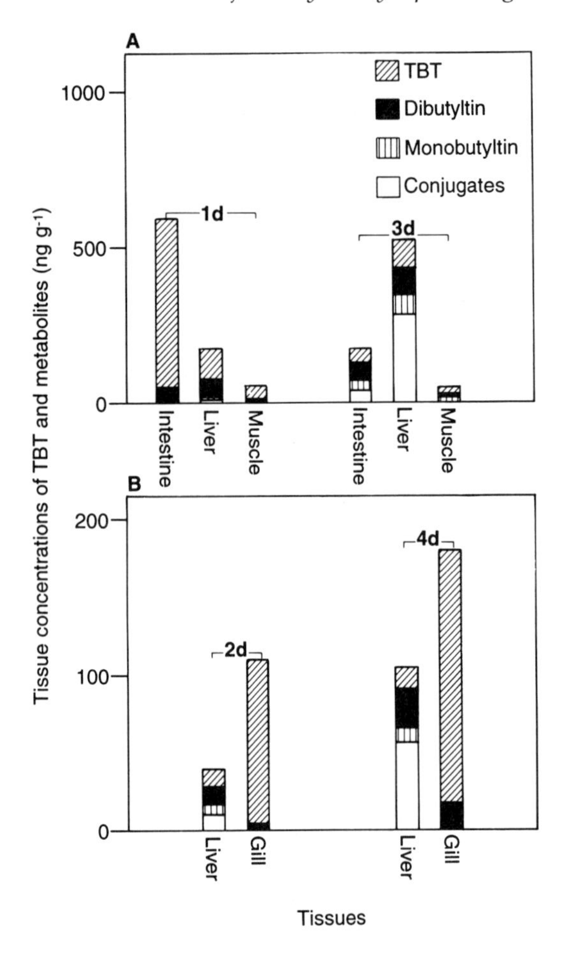

Figure 18.1 Spot (*Leiostoma canthurus*) were exposed to ^{14}C-TBT via food (group A) or water (group B) and then analyzed for TBT and its metabolites. The A group of spot (*Leiostoma canthurus*) was exposed to ^{14}C-TBTO via their food (12 μg g^{-1} food – fed 2 g). After 1 and 3 d, fish were dissected and TBT and metabolites were extracted and analyzed. For exposure via water (group B) individual fish were placed in glass aquaria containing ^{14}C-TBT (1 μg l^{-1} – total volume of 4 l). Water was changed daily with new ^{14}C-TBT added each day. For each time period three fish were removed followed by tissue dissection, extraction, and analysis. Each time period bar is the mean for tissues from three fish, analyzed separately.

by *Leiostomus canthurus* was largely metabolized after 3 d (Fig. 18.1). Thus, active metabolism of TBT by fish is reflected in the relatively short residence time of the compound in fish.

Table 18.2 Metabolism of ^{14}C-TBTO by fish liver microsomes[a]

Compound	Relative activity (%) With NADPH	Relative activity (%) Without NADPH
TBTO	72	91
Beta-hydroxybutyldibutyltin	4	1
Gamma-hydroxybutyldibutyltin	2	1
Delta-hydroxybutyldibutyltin	1	0.2
Dibutyltin	8	2
Monobutyltin	7	3

[a]Each incubation flask contained 1 mg of microsomal protein from spot (*Leiostomus canthurus*) liver, 0.025 μmol of ^{14}C-TBT and 0.05 M TRIS buffer at pH 7.4 in a total volume of 1.0 ml. Where NADPH was added, its final concentration was 1 mM. After incubation for 30 min at 30°C, TBT and its metabolites were extracted with ethyl acetate and metabolites identified and quantified. Data from Lee (1986).

The pathway of TBT into the liver includes metabolism and transfer of metabolites to the bile (gall bladder); this pathway is used for eliminating many toxicants by fish (Lee *et al.*, 1972a; Statham *et al.*, 1976; Melancon and Lech, 1980). TBT metabolites from the gall bladder were not extracted with ethyl acetate, but were extracted by ethyl acetate after acid hydrolysis. After acid hydrolysis of bile conjugates, the hydroxyl derivatives of dibutyltin, which can be solvent extracted, were formed.

Microsomal preparations of *Leiostomus canthurus* liver were able in the presence of NADPH to metabolize TBT (Table 18.2). Some of the TBT metabolites that were found in in-vitro and in-vivo studies with aquatic animals are shown in Fig. 18.2. The postulated reactions involved in the hydroxylation of TBT by the MFO system of aquatic animals are shown in Fig 18.3. Similar reactions have been shown to occur during the metabolism of various toxicants by mammalian microsomal systems. In summary, the TBT binds to

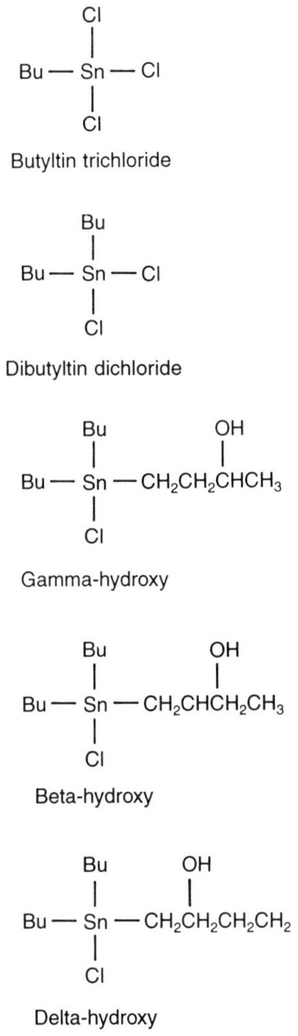

Cl
|
Bu — Sn — Cl
|
Cl

Butyltin trichloride

Bu
|
Bu — Sn — Cl
|
Cl

Dibutyltin dichloride

Bu OH
| |
Bu — Sn — $CH_2CH_2CHCH_3$
|
Cl

Gamma-hydroxy

Bu OH
| |
Bu — Sn — $CH_2CHCH_2CH_3$
|
Cl

Beta-hydroxy

Bu OH
| |
Bu — Sn — $CH_2CH_2CH_2CH_2$
|
Cl

Delta-hydroxy

Figure 18.2 Metabolites of TBT produced by aquatic animals and plants.

the oxidized cytochrome P-450 (Fe^{+3}), and the complex undergoes reduction to cytochrome P-450 (Fe^{+2}), which then interacts with oxygen. A hydroxylated substrate (e.g. beta-hydroxybutyldibutyltin) and a molecule of water leave the now reoxidized cytochrome P-450. The substrate-oxidized cytochrome P-450 complex is reduced by two electrons from NADPH carried by NADPH cytochrome P-450 reductase. Superoxide anion (O_2^-) formed during the reaction may participate in the hydroxylation of the substrate. The evidence suggesting that such reactions occur in fish microsomes is that TBT forms a type I binding spectrum with fish microsomes; NADPH and oxygen are required for TBT oxidation by fish microsomes; and carbon monoxide inhibits TBT oxidation (Lee, 1986).

18.2.2. TBT METABOLISM BY AQUATIC INVERTEBRATES

Marine species in the Annelida, Arthropoda, and Mollusca phyla were able to metabolize TBT. The marine polychaete *Nereis virens* and freshwater oligochaetes showed significant concentrations of dibutyltin and monobutyltin after exposure to TBT-contaminated sediment, suggesting that these annelids were able to metabolize TBT (Maguire and Tkacz, 1985; Thain *et al.*, 1990).

The decapod crustaceans have been the most studied group of invertebrates with respect to TBT metabolism. After exposure of crabs [blue crabs (*Callinectes sapidus*) and spider crabs (*Libinia emarginata*)] and shrimp (*Penaeus aztecus*) to TBT through either the food or water, various metabolites, primarily dibutyltin, were found in the hepatopancreas (Lee, 1985, 1986; Rice *et al.*, 1989). Mud crabs (*Rhithropanopeus harrisii*) exposed to ^{14}C-TBT via the water showed a bioconcentration factor of 4400 in the hepatopancreas (Evans and Laughlin, 1984). This bioconcentration included both TBT and its metabolites.

Exposure of blue crabs to ^{14}C-TBT in the water (at microgram per liter levels) showed the crabs initially accumulated radioactivity in the gill with >90% of the radioactivity associated with TBT (Fig. 18.4). After 6 d, the gill continued to have a high concentration of TBT, but most of the radioactivity in the hepatopancreas was associated with TBT metabolites. Blue crabs fed food containing TBT (12 $\mu g\ g^{-1}$ food) had high radioactivity in the stomach and hepatopancreas with most

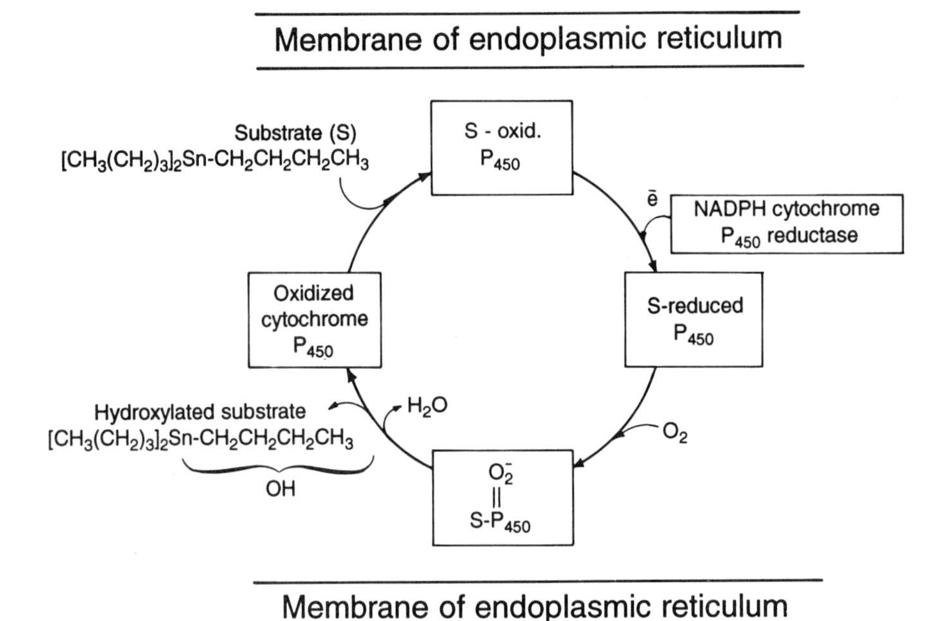

Figure 18.3 Postulated microsomal electron transport reactions involved in the metabolism of TBT by marine animals.

radioactivity associated with metabolites after 3 d (Fig. 18.4). Crab stomach microsomes in the presence of NADPH carried out the oxidation of TBT to various hydroxylated derivatives as well as to dibutyltin and monobutyltin (Table 18.3).

The hepatopancreas of aquatic anthropods plays an important role in the accumulation and metabolism of many toxicants (Lee *et al.*, 1976; James *et al.*, 1979; Lee, 1981; James and Little, 1984). Different cell types found in the hepatopancreas tubules include E-, F-, R-, and B-cells. The F-, R-, and B-cells are derived from embryonic or E-cells. The R-cells are storage cells with large amounts of lipid, while the F- and B-cells are thought to be important in protein synthesis (Johnson, 1980). The F-cells have an extensive rough endoplasmic reticulum and Golgi network. The components of the mixed function oxygenase system (cytochrome P-450 and cytochrome P-450 reductase) and glutathione S-transferase activity are primarily in the F-cells (Lee, 1986; Keeran and Lee, 1987). When blue crabs were exposed to ^{14}C-TBT, most radioactivity was initially in the lipid of R-cells, but after 1–2 d most of the activity was in the F-cells where TBT was metabolized (Lee, 1989). Within the cells of the hepatopancreas, TBT and its metabolites are distributed between the cytoplasm, outer membrane, and different organelles. The TBT and its metabolites can be 'dissolved' in lipid droplets or lipid. TBT and its metabolites in the membranes or organelles can be bound (covalently or noncovalently) to such macromolecules as DNA, RNA, or proteins. Further work indicated that this bound radioactivity was associated with proteins and not nucleic acids. The data shown in Fig. 18.5 indicate that 1 d after feeding food with ^{14}C-TBT to crabs the cytosol, nuclear fraction, and microsomes had the highest activity. The compounds in the cytosol were primarily

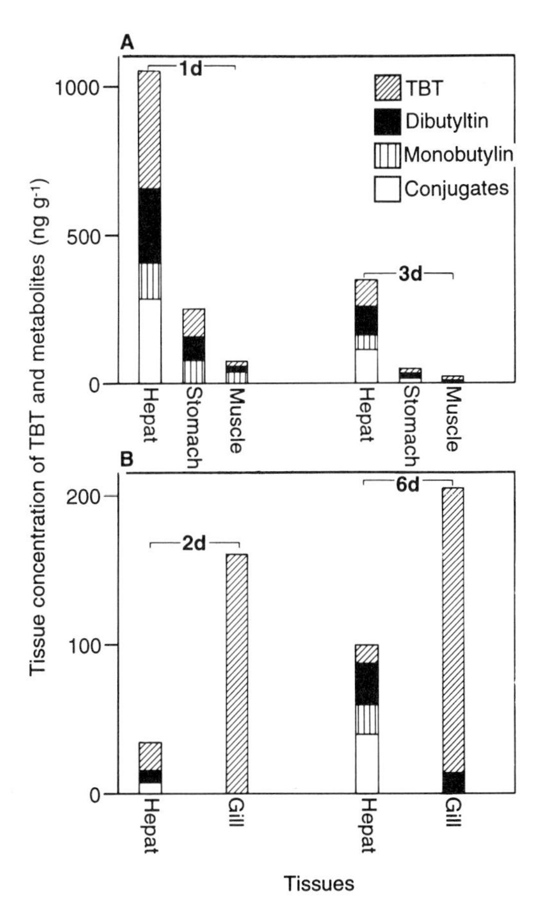

Figure 18.4 TBT and its metabolites in the tissues of blue crab (*Callinectes sapidus*) exposed to ^{14}C-TBT via food or water. One group (Group A) of blue crabs was exposed to ^{14}C-TBT via food (12 μg g^{-1} food – fed 2 g). After 1 and 3 d TBT and its metabolites extracted from the various tissues were analyzed. For exposure via water (Group B) individual crabs were placed in glass aquaria containing ^{14}C-TBT (1 μg l^{-1}). Water was changed daily with new ^{14}C-TBT added each day. For each time period three crabs were removed for tissue dissection, extration, and analysis. Hepat stands for hepatopancreas.

conjugates of hydroxylated derivatives of TBT.

Molluscs, in contrast to crustaceans, show somewhat limited ability to metabolize TBT. For example, when oysters were exposed

Table 18.3 Metabolism of ^{14}C-TBT by blue crab stomach microsomes[a]

Compound	*Relative activity (%)* With NADPH	*Relative activity (%)* Without NADPH
TBT	85	96
Beta-hydroxybutyldibutyltin	5	1
Gamma-hydroxybutyldibutyltin	1	0
Delta-hydroxybutyldibutyltin	1	1
Dibutyltin	7	2
Monobutyltin	1	0

[a]Blue crab (*Callinectes sapidus*) stomach microsomes were added to incubation flasks that contained 1 mg of microsomal protein, 0.025 μmol of ^{14}C-TBT, and 0.05 M TRIS buffer at pH 7.4 in a total volume of 1.0 ml. Where NADPH was added its final concentration was 1 mM. After incubation for 30 min at 30°C, TBT and its metabolites were extracted with ethyl acetate and metabolites separated by thin-layer chromatography. Data from Lee (1986).

to ^{14}C-TBT in water for up to 6 d, only a small amount of the TBT was metabolized, primarily to dibutyltin (Fig. 18.6). Clams (*Mya arenaria*) exposed to water containing TBT showed very little DBT after exposure for 1 month, suggesting that this species has a very limited ability to metabolize TBT (Langston, 1990). For deposit-feeding clams (*Scrobicularia plana*), exposed to sediments under laboratory conditions, equilibrium concentrations in tissues were approached after 40 d (Langston and Burt, 1991). Mussels (*Mytilus edulis*) and oysters (*Crassostrea virginica*) accumulated primarily TBT in the digestive gland and gills after exposure to TBT in food or water (Laughlin *et al.*, 1986; Lee, 1986). Oysters and other bivalves have very low mixed-function oxygenase activity, which correlates with the slow rate of toxicant oxidation in bivalves (Lee *et al.*, 1972b; Lee, 1981; Anderson, 1985; Livingstone and Farrar, 1985).

Bivalve molluscs bioconcentrate TBT to a much greater extent than other aquatic

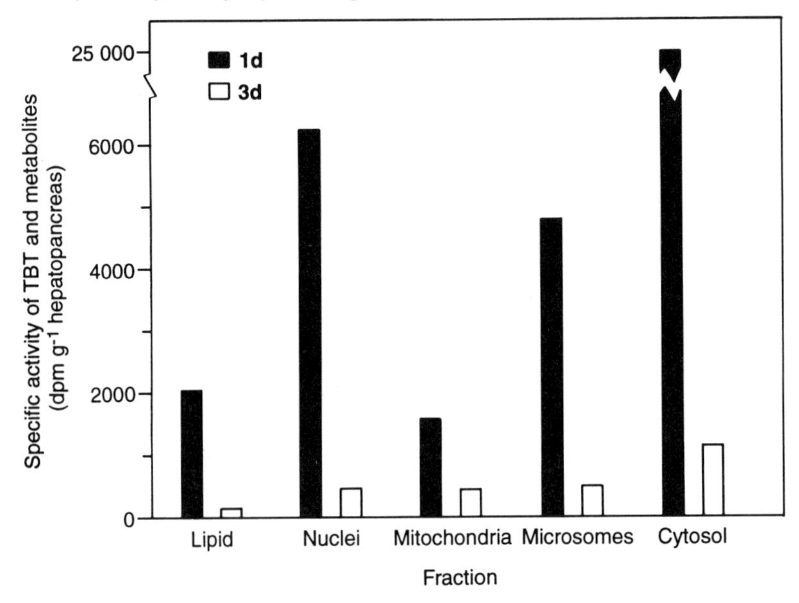

Figure 18.5 Distribution of TBT and metabolites within blue crab (*Callinectes sapidus*) hepatopancreas cells. Male blue crabs were fed food containing ^{14}C-TBT (8×10^5 dpm). The hepatopancreas was removed and homogenized and the homogenate was subjected to differential centrifugation to collect a floating lipid layer, nuclear and cellular debris fraction (640 *g*), mitochondrial (13 000 *g*), microsomal (100 000 *g*), and cytosol (supernatant from 100 000 *g* spin) fraction. Radioactvitiy (dpm) was determined in each fraction. Data taken from Lee (1990).

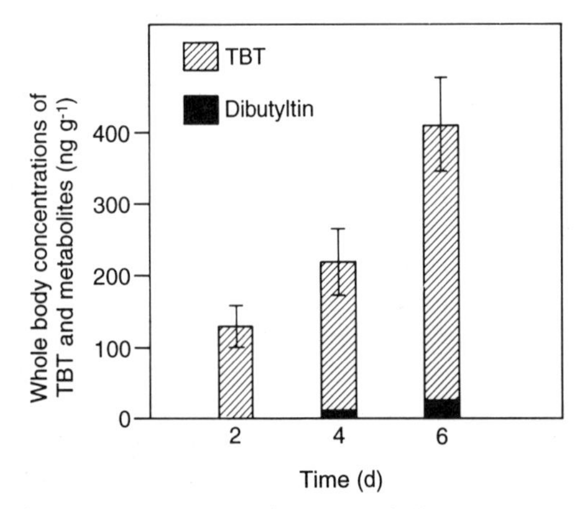

Figure 18.6 TBT and its metabolites in oysters (*Crassostrea virginica*) exposed to ^{14}C-TBT in the water. Oysters were exposed to ^{14}C-TBT (1 μg l^{-1} – total volume of 6 l) in glass aquaria. Filtered water was changed daily with new ^{14}C-TBT added each day. Each aquarium had only one oyster. After 2, 4 and 6 d three oysters were removed. Tissues were removed and analyzed for TBT and metabolites. Each time period bar is the mean for three oysters, which were separately analyzed; standard errors are indicated.

animal groups that have been studied. The bioconcentration factors for the Pacific oyster (*Crassostrea gigas*) were 2000–6000 (Waldock *et al.*, 1983), and 1000–1500 for the European oyster (*Ostrea edulis*) Waldock and Thain, 1983, and 5000–70 000 for mussels (*Mytilus edulis*; Laughlin *et al.*, 1986; Salazar and Salazar, 1990, Chapter 15; Laughlin, Chapter 16). When bivalves that have bioaccumulated TBT are transferred to TBT-free water, the depuration half-lives vary from 14 to 56 d for *Mytilus edulis* (Laughlin *et al.*, 1986; Bryan *et al.*, 1989) and 23 d for *Crassostrea gigas* (Waldock *et al.*, 1983). Presumably, the relatively slow depuration rates were due to the very slow rate of TBT metabolism, since TBT metabolites are eliminated more rapidly than TBT. Based on its octanol–water partition coefficient, the bioconcentration factor should be 160 (Laughlin *et al.*, 1986). Laughlin *et al.* (1986) suggest that the high bioconcentration factor determined for TBT accumulation by bivalves was due to irreversible binding of TBT by cellular macromolecules. Such binding was noted for crabs as described above. Marine gastropods are able to metabolize TBT, but the rate is still slow compared with crustaceans. The gastropods *Nucella lima* and *Nucella lapillus*, which show development of imposex after exposure to TBT, metabolize TBT to DBT and MBT (Bryan *et al.*, 1987, 1989; Short *et al.*, 1990). After a 1-month exposure to a TBT concentration of 0.914 $\mu g\ l^{-1}$, *N. lima* contained 0.70 $\mu g\ g\ l^{-1}$ TBT, 0.21 $\mu g\ g^{-1}$ DBT, and 0.03 $\mu g\ g^{-1}$ MBT (Short *et al.*, 1990).

18.2.3. METABOLISM OF TBT BY AQUATIC PLANTS

TBT degradation in natural waters was faster under sunlight relative to dark degradation (Olson and Brinckman, 1986; Lee *et al.*, 1987, 1989). The lack of TBT degradation when microorganisms were removed from sunlit natural waters indicated that photolysis was not contributing to this observed degradation. Good evidence is available that algae are important degraders of TBT in sunlit waters, since several species of algae can degrade TBT (Maguire *et al.*, 1984; Lee *et al.*, 1987, 1989; Table 18.4). Metabolites produced include hydroxybutyldibutyltins, dibutyltin, butyltin, and inorganic tin. The hydroxylation of polycyclic aromatic hydrocarbons by algae appears to be due to a dioxygenase system (Warshawsky *et al.*, 1988). It is likely that this algal dioxygenase system oxidizes TBT. Eel grass (*Zostera marina*) found in intertidal areas can degrade TBT to DBT and MBT (Francois *et al.*, 1989).

18.3. LINKAGE OF TBT METABOLISM TO OBSERVED EFFECTS

Metabolites produced during the biotransformation of certain foreign compounds by the mixed-function oxygenase system can be highly reactive electrophilic compounds. These reactive metabolites can be destroyed by certain detoxification systems (e.g. epoxide hydrase) or bind covalently to cellular proteins and nucleic acids. The cytotoxic effects of some toxicants are due to covalent binding of their reactive metabolites to cellular proteins (Mitchell *et al.*, 1973; Gillette *et al.*, 1974; Jollow *et al.*, 1974).

In crabs exposed to ^{14}C-TBT in the food, up to 35% of the TBT metabolites were covalently bound to cellular proteins (Lee, 1989, 1990). Very little of the TBT or its metabolites was bound to the cellular nucleic acids. Similar results were observed when the snail *Nasarrius obsoletus* was exposed to ^{14}C-TBT. The effects of TBT on aquatic animals include production of reproduction anomalies in stenoglossan gastropods (imposex), inhibition of glutathione S-transferase and cytochrome P-450, thickening of oyster shells, reduced growth rates for mussels, and breakdown of sexual differentiation, oogenesis and egg production in *Ostrea edulis* (Smith, 1981; Waldock and Thain, 1983; Alzieu *et al.*, 1986; Gibbs and Bryan, 1986 and Chapter 13; Thain, 1986; Bryan *et al.*, 1987; Davies *et al.*, 1987; Salazar and Salazar, 1988). The biochemical basis for these observed effects is not understood, but I hypothesize that many of the effects are linked to enzymes involved in TBT metabolism. As noted above, the observed effects of several toxicants are due to covalent binding of metabolites to cellular proteins. TBT metabolites covalently bind to the cellular proteins of snails and crabs. The binding of TBT to glutathione S-transferase and cytochrome P-450 results in the inhibition of these two detoxifying enzyme systems (Henry and Byington, 1976; Rosenberg and Drummond, 1983; Fent and Stegeman, 1993). The presence of a penis and vas deferens in female stenoglossan snails, referred to as imposex, has been linked to the presence of TBT in the surrounding water (Smith, 1981; Bryan *et al.*, 1987). As noted earlier, TBT metabolites bind to cellular proteins of

Table 18.4 Degradation of tributyltin (TBT) by marine algae[a]

Algal species	*TBT (μg l⁻¹)*	*Dibutyltin (μg l⁻¹)*	*Hydroxylated products (μg l⁻¹)*	*Polar products (μg l⁻¹)*
Diatoms – Bacillariophyta				
Skeletonema costatum	0.1	0.2	0.1	0.04
Chaetoceros curvisetus	0.20	0.11	0.05	ND[b]
Dinoflagellates – Pyrrhophyta				
Procentrum triestinum	0.30	0.05	0.01	ND
Golden algae – Chrysophyta				
Isochrysis glabana	0.40	ND	ND	ND
Cricosphaera ricoco	0.35	0.02	ND	0.01
Green algae – Chlorophyta				
Dunaliella tertiolecta	0.40	0.05	ND	ND

[a] ^{14}C-TBT was added to cultured algae to give an initial concentration of 0.4 μg l^{-1}. After 2 d the water and algae were extracted with chloroform and analyzed for TBT and metabolites. Polar products are the compounds that did not move off the origin in the thin-layer system used. Data from Lee *et al.* (1989).
[b] ND stands for not determined (<0.01 μg l^{-1}).

Nassarius obsoletus, a snail that shows imposex after TBT exposure (Smith, 1981). There has been speculation that TBT results in a hormonal imbalance in these snails. Recent work by Spooner and Goad (1990) showed an increased concentration of testosterone in the dogwhelk (*Nucella lapillus*) after injection with 1 μg of TBT. Testosterone injections into this snail resulted in imposex. Hormones and toxicants share common metabolic pathways (e.g. cytochrome P-450 systems) that can give rise to important interactions (Conney and Klutch, 1963; Brown, 1975; Gessner, 1976). Cytochrome P-450 systems control the conversion of cholesterol into a variety of hormones (e.g. testosterone). Inhibition or stimulation of cytochrome P-450 systems can result in changes in hormone production or clearance (Levin, *et al.*, 1974; Kupfer and Bugler, 1976). The binding and inhibition of cytochrome P-450 by TBT may result in the observed production of testosterone by female whelks. In verbebrates, a calcium metabolism cytochrome P-450 system is involved in the synthesis of vitamin D, which in turn regulates calcium metabolism. Thus, the abnormal shell growth in oysters after TBT exposure may be related to inactivation of cytochrome P-450.

Among various invertebrate groups, the molluscs appear to be the most affected by TBT. Molluscs are characterized by very low cytochrome P-450 content and mixed-function oxygenase activity in their digestive gland (Livingstone and Farrar, 1985). This low detoxifying activity may explain this group's susceptibility to TBT. Further work is necessary to provide evidence of the linkage of TBT metabolites and TBT-metabolizing enzyme systems to observed effects in molluscs.

18.4. SUMMARY AND CONCLUSIONS

Both aquatic plants and animals have enzyme systems to metabolize TBT to dibutyltin, monobutyltin, and hydroxylated derivatives. Among photosynthetic organisms, intertidal eel grass (*Zostera marina*), diatoms, and dinoflagellates have been shown to metabolize TBT. The diatoms produce a series of hydroxylated derivatives of TBT. It is likely that the hydroxylation of TBT is by an algal dioxygenase system.

Among aquatic animals, the crustaceans, annelids, and fish have enzyme systems that rapidly metabolize TBT. Molluscs show a much more limited ability to metabolize TBT.

The microsomal cytochrome P-450 systems present in hepatic, intestinal, and kidney tissues of aquatic vertebrates and invertebrates carry out the hydroxylation of TBT. TBT in these animals is metabolized in two phases. The phase-one reaction (i.e. cytochrome P-450 system) adds a hydroxyl group to TBT. The hydroxybutyldibutyltins are conjugated to sulfate or carbohydrate by phase-two enzyme systems. The action of these enzymes increases the water solubility of TBT metabolites, thereby facilitating the rapid elimination of TBT from the animal. When blue crabs were fed food containing ^{14}C-TBT, most radioactivity initially accumulated in the lipid-rich R-cells of the hepatopancreas. One day after feeding, most radioactivity was in F-cells where TBT was metabolized via phase-one and phase-two enzyme systems. Up to 35% of the radioactivity in the hepatopancreas was covalently bound to macromolecules in mitochondria and cytosolic proteins.

There are a number of TBT effects, including imposex in certain gastropods and shell thickening in oysters, which may relate to TBT metabolism. The relationship among these effects and the binding of TBT or its metabolites to cellular macromolecules and the inhibition of the steroid-synthesizing cytochrome P-450 system by TBT warrant further investigation.

ACKNOWLEDGMENTS

The work reported was sponsored by the Office of Naval Research and the David Taylor Naval Shop Research and Development Center, Energy Research and Development Program.

REFERENCES

Alzieu, C.L., J. Sanjuan, J.R. Deltriel and M. Borel. 1986. Tin contamination in Arcachon Bay: effects on oyster bay anomalies. *Mar. Pollut. Bull.*, **17**, 494–498.

Ames, B.N., P. Sims and P.L. Grover. 1972. Epoxides of carcinogenic polycyclic aromatic hydrocarbons are frameshift mutagens. *Science*, **176**, 47–49.

Anderson, R.S. 1985. Metabolism of a model environmental carcinogen by bivalve molluscs. *Mar. Environ. Res.*, **17**, 137–140.

Bartolone, J.B., W.P. Beiraschmitt, R.B. Birge, S. Hart, S. Wyand, S.D. Cohen and E.A. Khairallah. 1989. Selective acetaminophen metabolite binding to extrahepatic proteins: an in vivo and in vitro analysis. *Toxicologist*, **9**, 48.

Brown, S.S. 1975. Metabolic interactions involving drugs or foreign compounds. In: *Foreign Compound Metabolism in Mammals*, D.E. Hathway (Ed.). Chemical Society, London, pp. 591–630.

Bryan, G.W., P.E. Gibbs, G.R. Burt and L.G. Hummerstone. 1987. The effects of tributyltin (TBT) accumulation on adult dog-whelks, *Nucella lapillus*: long-term field and laboratory experiments. *J. Mar. Biol. Ass. UK*, **67**, 525–544.

Bryan, G.W., P.E. Gibbs, L.G. Hummerstone and G.R. Burt. 1989. Uptake and transformation of ^{14}C-labeled tributyltin chloride by dog whelk, *Nucella lapillus*: importance of absorption from the diet. *Mar. Environ. Res.*, **28**, 241–245.

Davies, I.M., S.K. Bailey and D.C. Moore. 1987. Tributyltin in Scottish sea lochs, as indicated by degree of imposex in the dog whelk (*Nucella lapillus*) (L.). *Mar. Pollut. Bull.*, **18**, 400–404.

Evans, D.W. and R.B. Laughlin. 1984. Accumulation of bis(tributyltin)oxide by the mud crab, *Rhithropanopeus harrisii*. *Chemosphere*, **13**, 213–219.

Fent, K. and Stegeman, J.J. 1993. Effects of tributyltin in vivo on hepatic cytochrome P450 forms in marine fish. *Aquat. Toxicol.*, **24**, 219–240.

Fish, R.H. 1984. Bioorganotin chemistry: a commentary on the reactions of organotin compounds with a cytochrome P-450 dependent monooxygenase enzyme system. *Neurotoxicology*, **5**, 159–162.

Fish, R.H., E.C. Kimmel and J.E. Casida. 1976. Bioorganotin chemistry: reactions of tributyltin derivatives with a cytochrome P-450 dependent monooxygenase enzyme system. *J. Organometal. Chem.*, **118**, 41–54.

Francois, R., F.T. Short and J.H. Weber. 1989. Accumulation and persistence of tributyltin in eelgrass (*Zostera marina* L.) tissue. *Environ. Sci. Technol.*, **23**, 191–196.

Gibbs, P.E. and G.W. Bryan. 1986. Reproductive failure in populations of the dog-whelk, *Nucella lapillus*, caused by imposex induced by tributyltin from antifouling paints. *J. Mar. Biol. Ass. UK*, **66**, 767–777.

Gillette, J.R., J.R. Mitchell and B.B. Brodie. 1974. Biochemical mechanisms of drug toxicity. *Ann. Rev. Pharmacol.*, **14**, 271–288.

Henry, R.A. and K.H. Byington. 1976. Inhibition of glutathione S-aryltransferase from rat liver by organogermanium, lead and tin compounds. *Biochem. Pharmacol.*, **25**, 2291–2295.

Iwai, H., O. Wada and Y. Arakawa. 1981. Determination of tri-, di- and monobutyltin and inorganic tin in biological materials and some aspects of their metabolism in rats. *J. Anal. Toxicol.*, **5**, 330–306.

James, M.O., M.A.Q. Khan and J.R. Bend. 1979. Hepatic microsomal mixed-function oxidase activities in several marine species common to coastal Florida. *Comp. Biochem. Physiol.*, **62C**, 155–164.

James, M.O. and P.J. Little. 1984. 3-Methylcholanthrene does not induce *in vitro* xenobiotic metabolism in spiny lobster hepatopancreas, or affect *in vivo* disposition of benzo(*a*)pyrene. *Comp. Biochem. Physiol.*, **78C**, 241–245.

Johnson, P.T. 1980. *Histology of the Blue Crab, Callinectes sapidus*. Praeger, New York, 440 pp.

Jollow, D.J., J.R. Mitchell, N. Zampaglion and J.R. Gillette. 1974. Bromobenzene-induced liver necrosis. Protective role of glutathione and evidence for 3,4-bromobenzene oxide as the hepatotoxic metabolite. *Pharmacology*, **11**, 151–169.

Keeran, W.S. and R.F. Lee. 1987. The purification and characterization of glutathione S-transferase from the hepatopancreas of the blue crab, *Callinectes sapidus*. *Arch. Biochem. Biophys.*, **255**, 233–243.

Kimmel, E.C., R.H. Fish and J.E. Casida. 1977. Bioorganotin chemistry: metabolism of organotin compounds in microsomal monooxygenase systems in mammals. *J. Agric. Food Chem.*, **25**, 1–9.

Krigman, M.R. and A.P. Silverman. 1984. General toxicology of tin and its organic compounds. *Neurotoxicology*, **5**, 129–140.

Krone, C.A., D.W. Brown, D.G. Burrows, R.G. Bogar, S.-L. Chan and U. Varanaso. 1989. A method for analysis of butyltin species and measurement of butyltin in sediment and English sole livers from Puget Sound. *Mar. Environ. Res.*, **27**, 1–18.

Kupfer, D. and W.H. Bugler. 1976. Interactions of chlorinated hydrocarbons with steroid hormones. *Fed. Proc.*, **35**, 2603–2608.

Langston, W.J. 1990. Bioavailability and effects of TBT in deposit-feeding clams, *Scrobicularia plana*. In: Proceedings of the 3rd International Organotin Symposium, 17–20 April 1990, Monaco. International Atomic Energy Agency, Vienna, pp. 110–113.

Langston, W.J. and G.R. Burt. 1991. Bioavailability and effects of sediment-bound TBT in deposit feeding clams, *Scrobicularia plana*. *Mar. Environ. Res.*, **32**, 61–77.

Laughlin, R.B., W. French and H.E. Guard. 1986. Accumulation of bis(tributyltin) oxide by the marine mussel *Mytilus edulis*. *Environ. Sci. Technol.*, **20**, 884–890.

Lee, R.F. 1981. Mixed function oxygenase (MFO) in marine invertebrates. *Mar. Biol. Lett.*, **2**, 87–105.

Lee, R.F. 1985. Metabolism of tributyltin oxide by crabs, oysters and fish. *Mar. Environ. Res.*, **17**, 145–148.

Lee, R.F. 1986. Metabolism of bis(tributyltin)oxide by estuarine animals. In: Proceedings of the Oceans '86 Organotin Symposium, Marine Technology Society, Washington, DC, pp. 1182–1188.

Lee, R.F. 1989. Metabolism and accumulation of xenobiotics within hepatopancreas cells of the blue crab, *Callinectes sapidus*. *Mar. Environ. Res.*, **28**, 93–97.

Lee, R.F. 1990. Metabolism of tributyltin by marine animals and possible linkages to toxicity. In: Proceedings of the 3rd International Organotin Symposium, 17–20 April 1990, Monaco. International Atomic Energy Agency, Vienna, pp. 70–76.

Lee, R.F. 1991. Metabolism of tributyltin by marine animals and possible linkages to effects. *Mar. Environ. Res.*, **32**, 29–35.

Lee, R.F., R. Sauerheber and G.H. Dobbs. 1972a. Uptake, metabolism and discharge of polycyclic aromatic hydrocarbons by marine fish. *Mar. Biol.*, **17**, 201–208.

Lee, R.F., R. Sauerheber and A.A. Benson. 1972b. Petroleum hydrocarbons: uptake and discharge by the marine mussels, *Mytilus edulis*. *Science*, **177**, 344–346.

Lee, R.F., C. Ryan and M.L. Neuhauser. 1976. Fate of petroleum hydrocarbons taken up from food and water by the blue crab *Callinectes sapidus*. *Mar. Biol.*, **37**, 363–370.

Lee, R.F., A.O. Valkirs and P.F. Seligman. 1987. Fate of tributyltin in estuarine areas. In: Proceedings of the Oceans '87 Organotin Symposium, 29 September–1 October 1987, Halifax, Nova Scotia, Marine Technology Society, Washington, DC, pp. 1411–1415.

Lee, R.F., A.O. Valkirs and P.F. Seligman. 1989. Importance of microalgae in the biodegradation of tributyltin in estuarine waters. *Environ. Sci. Technol.*, **23**, 1515–1518.

Levin, W., R.M. Welch and A.H. Conney. 1974. Increased liver microsomal androgen metabolism by phenobarbital: correlation with decreased androgen action on the seminal vesicles of the rat. *J. Pharmacol. Exp. Therap.*, **188**, 287–292.

Livingstone, D.R. and S.V. Farrar. 1985. Responses of the mixed function oxidase system of some bivalve and gastropod molluscs to exposure to polynuclear aromatic and other hydrocarbons. *Mar. Environ. Res.*, **17**, 101–105.

Lu, A.Y.H. 1976. Liver microsomal drug-metabolizing enzymes: functional components and their properties. *Fed. Proc.*, **35**, 2460–2463.

Maguire, R.J., P.T.S. Wong and J.S. Rhamey. 1984. Accumulation and metabolism of tri-*n*-butyltin cation by a green alga, *Ankistrodesmus falcatus*. *Can. J. Fish. Aquat. Sci.*, **41**, 537–540.

Maguire, R.J. and R.J. Tkacz. 1985. Degradation of the tri-n-butyltin species in water and sediment from Toronto Harbor. ~~*Environ. Sci. Technol.*~~, **33**, 947–953. Journal of Ag + Food Chemistry

Martin, R.C., D.G. Dixon, R.J. Maguire, P.V. Hodson and R.J. Tkacz. 1989. Acute toxicity, uptake, depuration and tissue distribution of tri-n-butyltin in rainbow trout, *Salmo gairdneri*. *Aquat. Toxicol.*, **15**, 37–51.

Melancon, M.J. and J.J. Lech. 1980. Uptake, metabolism and elimination of ^{14}C-labeled 1,2,4-trichlorobenzene in rainbow trout and carp. *J. Toxicol. Environ. Health*, **6**, 645–658.

Miller, J.A. and E.C. Miller. 1971. Chemical carcinogenesis: mechanisms and approaches to its control. *J. Natl Cancer Inst.*, **47**, 5–14.

Mitchell, J.R., D.J. Jollow, W.Z. Potter, D.C. Davis, J.R. Gillette and B.B. Brodie. 1973. Acetaminophen-induced hepatic necrosis. I. Role of drug metabolism. *J. Pharmacol. Exp. Ther.*, **187**, 195–202.

Olson, G.J. and F.E. Brinckman. 1986. Biodegradation of tributyltin by Chesapeake Bay microorganisms. *In*: Proceedings of the Oceans '86 Organotin Symposium, 23–25 September 1986, Marine Technology Society, Washington, DC, pp. 1196–1201.

Rice, S.D., J.W. Short and W.B. Stickle. 1989. Uptake and catabolism of tributyltin by blue crabs fed TBT-contaminated prey. *Mar. Environ. Res.*, **27**, 137–146.

Rosenberg, D.W. and G.S. Drummond. 1983. Direct in vitro effects of bis(tri-n-butyltin)oxide on hepatic cytochrome P-450. *Biochem. Pharmacol.*, **32**, 3823–3829.

Short, J.W., J.L. Sharp-Dahl, W.B. Stickle and S.D. Rice. 1990. Comparison of imposex in *Nucella lima* induced by tributyltin ingested with food and absorbed from seawater. In: Proceedings of the 3rd International Organotin Symposium, 17–20 April 1990, Monaco. International Atomic Energy Agency, Vienna, pp. 105–109.

Salazar, M.H. and S.M. Salazar. 1988. Tributyltin and mussel growth in San Diego Bay. In: Proceedings of the Oceans '88 Organotin Symposium, 31 October–2 November, Baltimore, Maryland, Marine Technology Society, Washington, DC, pp. 1188–1195.

Salazar, S.M. and M.H. Salazar. 1990. Bioaccumulation of tributyltin in mussels. In: Proceedings of the 3rd International Organotin Symposium, 17–20 April 1990, Monaco. International Atomic Energy Agency, Vienna, pp. 79–83.

Smith, B.S. 1981. Male characteristics on female mud snails caused by antifouling bottom paints. *J. Appl. Toxicol.*, **1**, 22–25.

Spooner, N. and L.J. Goad. 1990. Steroids and imposex in the dog whelk *Nucella lapillus*. In: Proceedings of the 3rd International Organotin Symposium, 17–20 April 1990, Monaco. International Atomic Energy Agency, Monaco, pp. 88–92.

Statham, C.N., M.J. Melancon and J.J. Lech. 1976. Bioconcentration of xenobiotics in trout bile: a proposed monitoring aid for some waterborne chemicals. *Science*, **193**, 680–681.

Stegeman, J.J. 1981. Polynuclear aromatic hydrocarbons and their metabolism in the marine environment. In: *Polycyclic Hydrocarbons and Cancer*, Vol. 3, H.V. Gelboin and P.O.P. Ts'o (Eds). Academic Press, New York, pp. 1–60.

Thain, J. 1986. Toxicity of TBT to bivalves: effects on reproduction, growth and survival. In: Proceedings of the Oceans '86 Organotin Symposium, Vol. 4, 23–25 September 1986, Marine Technology Society, Washington, DC, pp. 1306–1313.

Thain, J.E., M.J. Waldock, J. Finch and S. Bifield. 1990. Bioavailability of the residues in sediments to animals. In: Proceedings of the 3rd International Organotin Symposium, 17–20 April 1990, Monaco. International Atomic Energy Agency, Vienna, pp. 84–86.

Tsuda, T., H. Nakanishi, S. Aoki and J. Takebayashi. 1988. Bioconcentration and metabolism of butyltin compounds in carp. *Water Res.*, **22**, 647–651.

Van Veld, P.A., J.S. Patton and R.F. Lee. 1988a. Effect of pre-exposure to dietary benzo(*a*)pyrene (BP) on first pass metabolism of BP by the intestine of toadfish (*Opsanus tau*): *in vivo* studies using portal-vein-catheterized fish. *Toxicol. Appl. Pharmacol.*, **92**, 255–265.

Van Veld, P.A., J.J. Stegeman, B.A. Woodin, J.S. Patton and R.F. Lee. 1988b. Induction of monooxygenase activity in the intestine of spot (*Leiostomus xanthurus*), a marine teleost, by dietary aromatic hydrocarbons. *Drug. Metab. Dispos.*, **16**, 659–665.

Van Veld, P.A. 1990. Absorption and metabolism of dietary xenobiotics by the intestine of fish. *Rev. Aquat. Sci.*, **2**, 185–203.

Wada, O., S. Manabe, H. Iwai and Y. Arakawa. 1982. Recent progress in the study of analytical methods, toxicity, metabolism and health effects of organotin compounds. *Jap. J. Ind. Health*, **24**, 24–54.

Waldock, M.J. and J.E. Thain. 1983. Shell thickening in *Crassostrea gigas*: organotin antifouling or sediment induced? *Mar. Pollut. Bull.*, **14**, 411–415.

Waldock, M.J., J. Thain and D. Miller. 1983. The accumulation and depuration of bis(tributyltin) oxide in oysters: a comparison between Pacific oyster (*Crassostrea virginica*) and the European flat oyster (*Ostrea edulis*). International Council for the Exploration of the Sea, Copenhagen. CM 1983/E:52, pp. 1–9.

Ward, G.S., G.C. Cramm, P.R. Parrish, H. Trachman and A. Slesinger. 1981. Bioaccumulation and chronic toxicity of bis(tributyltin)oxide (TBTO): tests with a saltwater fish. In: *Aquatic Toxicology and Hazard Assessment*, D.R. Branson and K.L. Dickson (Eds). American Society for Testing and Materials, Philadelphia, pp. 183–200.

Warshawsky, E., M. Radike, K. Jayasimhulu and T. Cody. 1988. Metabolism of benzo(*a*)pyrene by a dioxygenase enzyme system of the freshwater green alga *Selenastrum capricornutum*. *Biochem. Biophys. Res. Commun.*, **152**, 540–544.

MEASUREMENT AND SIGNIFICANCE OF THE RELEASE RATE FOR TRIBUTYLTIN

19

Paul Schatzberg
746 Warren Drive, Annapolis, Maryland 21403-2817, USA

ABSTRACT

Reliable measurement of the rate of release of tributyltin (TBT) from antifouling paints is important for a number of reasons, principally for minimizing the potential environmental impact on nontarget organisms. When the TBT is chemically bound to a key constituent of the paint, its release rate is controlled by the rate of hydrolytic cleavage of the chemical bond; therefore, the release rate becomes an intrinsic property of the paint. When the TBT is present as a mixed component of the paint, its rate of release is not controlled and often is much higher than necessary for antifouling performance.

Laboratory measurements of the TBT release rate have shown that it can be measured reliably. These measurements provided the following information: (1) release rates of different paints range over several orders of magnitude; (2) freshly painted surfaces show an initial rate of release very much higher than the subsequent 'steady-state' rate; (3) temperature dependence of release rates demonstrated that the release mechanism involves a chemical reaction (hydrolysis) in those formulations where the TBT is chemically bound; (4) reduction in release rate when the pH is lowered also indicates that the primary mechanism is hydrolytic cleavage of the carboxyl–TBT bond, which releases the biocide allowing it to diffuse out; and (5) for the chemically bound TBT paint formulations, reducing their TBT content reduces the

Organotin. Edited by M.A. Champ and P.F. Seligman. Published in 1996 by Chapman & Hall, London. ISBN 0 412 58240 6

release rate as well as the surface softening and resulting hull smoothing. The rotating cylinder method for measuring the release rate in the laboratory has been applied by the US Environmental Protection Agency (US EPA) to regulate the use of TBT paints. This action achieved the following results: (1) paint suppliers became aware that the biocide release rate is an intrinsic property of the coating and has a direct bearing on evaluating potential environmental consequences from that product; (2) many TBT paints were found to have release rates unnecessarily high for antifouling protection and therefore placed an unnecessary burden on the environment; and (3) the US Congress, in enacting regulatory legislation, and the US EPA, in implementing such legislation, incorporated limits of TBT release rates as one key element in the regulatory action, removing thereby a large number of unacceptable products from the market and making a total ban unnecessary. Nevertheless, the rotating cylinder method appears to be flawed in that it simulates more closely the condition of a ship under way than a ship at rest in a harbor; consequently, very much higher biocide release rates are measured by this method than is actually the case. The rotating cylinder method, as it is currently used, can be improved by increasing the number of measurements and eliminating the current practice of regressing a single data point through the origin of the time–concentration plot to determine the rate of release.

Release of TBT in the environment is complicated by the presence of a biological film on the painted surfaces. This biofilm or slime layer consists of a complex community of microorganisms that presents an attractive environment for large settling organisms such as barnacles or tubeworms. The microorganisms in the biofilm become acclimated to the TBT biocide. As TBT is released from the paint surface, it concentrates in the biofilm due to exudates from the microorganisms, creating a hostile chemical environment that prevents permanent attachment of fouling organisms.

A novel, simple apparatus, easy to use in the laboratory or the field, has been developed to determine the minimum effective biocide release rate needed to prevent the attachment of settling organisms. The apparatus simulates different rates of release of biocides, and can be positioned in the marine environment and exposed to settling organisms. Data collected in this way will establish the optimum rate of release for a given biocide; that is, the minimum rate needed to prevent attachment by fouling organisms and at the same time to make a negligible impact on the environment. This information can then be used by paint chemists to formulate an antifouling paint with a high level of confidence that the new paint will perform well and meet regulatory requirements.

19.1. INTRODUCTION

Antifouling hull paints that contain tributyltin (TBT) compounds as the biocide were introduced approximately 20 y ago. The TBT compounds were mixed into the paint matrix in the same way as the traditional biocide, cuprous oxide. Those paints usually released the biocide at a rate one to two orders of magnitude greater than the paints developed later in which the TBT is chemically bound to an acrylic polymer by means of a carboxyl group. The carboxyl–TBT linkage is hydrolytically unstable under slightly alkaline conditions, which is usually the case in marine waters, resulting in a slow, controlled release of the biocide. By chemically modifying this rate of hydrolysis, different rates of release of TBT, or other organotin compounds, could be obtained. The release mechanism of these copolymer paints has been described (Anderson and Dalley, 1986; Champ and Seligman, Chapter 1). A summary of this description follows, including Fig. 19.1, which illustrates how this reaction proceeds at the molecular level. The copolymer of TBT methacrylate/

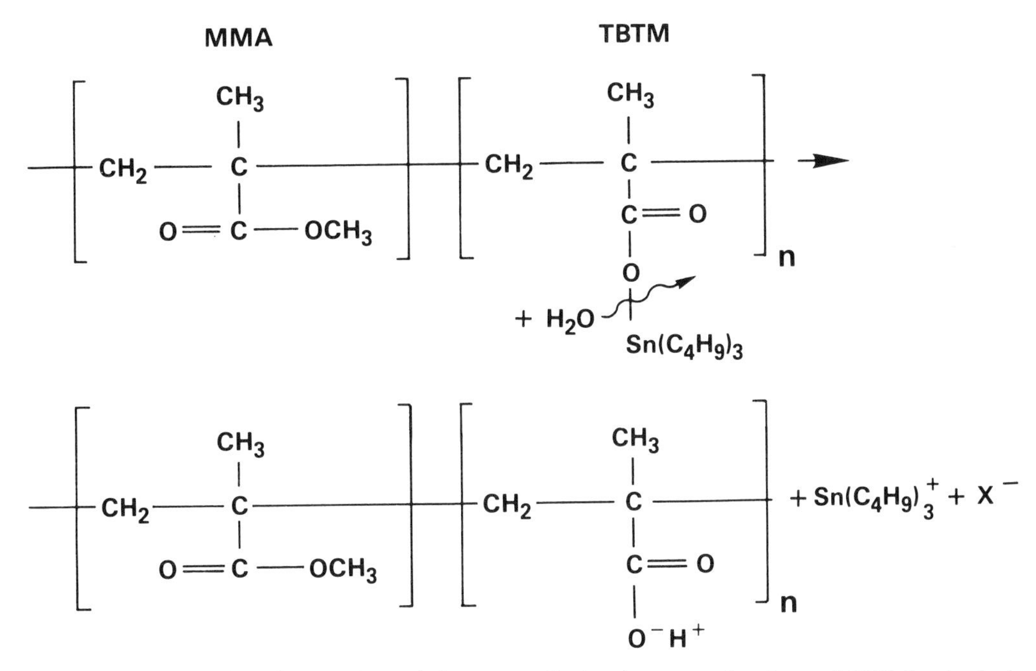

Figure 19.1 A molecular description of the controlled release mechanism of TBT by hydrolysis (Anderson and Dalley, 1986). MMA = methylmethacrylate; TBTM = tributyltinmethacrylate.

methylmethacrylate in the paint is hydrophobic, which prevents seawater from penetrating the paint film. At the surface, however, the seawater reacts with the film, resulting in an alkaline hydrolysis that cleaves the TBT from the copolymer, releasing it, as shown in Fig. 19.1. As discussed in section 19.5, Role of the biofilm, the released TBT concentrates in the biofilm covering the paint (see Champ and Seligman, Chapter 29). This concentration of the biocide creates a chemically hostile environment that prevents permanent attachment of settling organisms. It was also discovered that in some formulations of these new controlled-release paints, the paint surface softened slightly as part of the hydrolyzing process. This phenomenon turned out to be coextensive with the release rate of TBT and had the serendipitous effect of retaining hull smoothness by the preferential removal of paint from rough spots once the vessel was under way. This effect essentially eliminated the heretofore accepted increase in hull roughness over time and the associated fuel penalty. Consequently, the new coatings were effective not only in preventing the attachment and growth of calcareous organisms, largely by the controlled release of TBT, but also in providing hull smoothing. The effectiveness of these paints resulted in a growing market not only in commercial ocean-going shipping and military ships, but also for coastal fishing vessels and recreational craft. The importance of the TBT release rate was essentially forgotten until the potential environmental consequences of this biocide for nontarget organisms began to cause concern.

Quantitative measurement of the TBT release rate from antifouling paints is important because such information helps in (1) selecting more effective materials, (2) providing product quality assurance, (3) determining the minimum rate needed to provide antifouling performance, (4) predicting the distribution and fate of the biocide in harbors and estuaries with numerical models when the hull areas of painted ships are known, and (5) regulating the use of antifouling paints by eliminating those with rates above an acceptable upper limit. It is for this last reason particularly that a reliable, repeatable method for determining release rates is required.

19.2. EARLY METHOD DEVELOPMENT

The goal at the Naval Surface Warfare Center (NSWC) was to develop a laboratory procedure for measuring the biocide release rate under reproducible conditions (Schatzberg *et al.*, 1983). The released TBT had to be at a low but measurable concentration in the parts-per-billion range. Inhibition of release rate will occur at elevated TBT levels in a closed system. Accordingly, design of the system required that TBT concentrations be maintained well below such levels. The controlled laboratory conditions did not simulate the cyclic environmental conditions nor the biological activity occurring at surfaces in a marine environment, and therefore would not evaluate the effect of those conditions on the release rate of TBT. The measurement apparatus that was developed is shown schematically in Fig. 19.2. Steel panels, 32 mm thick with area dimensions 25.4 × 30.5 cm, were coated on both sides, resulting in a painted area of 1549.4 cm^2; the coating consisted of an epoxy anticorrosion paint with a dry film thickness of 200 μm over which was applied an organotin paint with a dry film thickness of 100 μm. Aluminum tanks with volume dimensions of 5.1 × 30.5 × 40.6 cm were used. Aluminum was selected due to its corrosion resistance and ease of fabrication. The tank dimensions were intended to maximize the sensitivity of the system to changes in release rate by maximizing the panel surface area-to-volume ratio. Since these panels were used routinely for conducting evaluations of antifouling coatings, the release rate could be measured after extensive field exposures. The tank and system were initially exposed to TBT by filling with synthetic seawater spiked with the biocide and allowing equilibrium to be established. A peristaltic pump forced the seawater at a measured rate through a column containing activated carbon. The carbon removed organotin so that the concentration in the tank could be controlled at a low value of ~40 $\mu g\ l^{-1}$ Sn. The water in the tank was circulated at 9–10 $l\ min^{-1}$ using a standard aquarium filter pump. The tanks were placed in a water bath to maintain a temperature constant within 0.5°C. Most measurements were conducted at 25°C, some

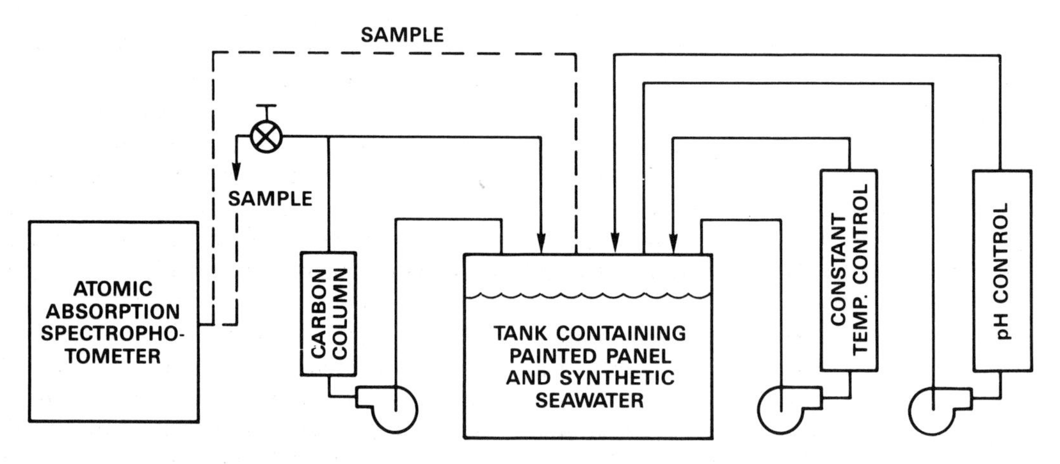

Figure 19.2 Biocide release rate measurement apparatus.

Table 19.1 Release rates of organotin coatings exposed to the environment

Coating	*Site*[a]	*Months exposed*	*Release rate at 25°C ($\mu g\ Sn\ cm^{-2}\ d^{-1}$) Mean ± standard deviation*	*No. of measurements*
Experimental A	M	27	2.5 ± 0.6	9
	H	27	1.3 ± 0.2	10
	SR	42	1.9 ± 0.2	7
Experimental B	M	27	1.3 ± 0.3	10
	H	27	1.2 ± 0.5	10
	SR	42	1.6 ± 0.3	22
Commercial C	H	36	0.6 ± 0.2	8
	LAB	12	0.9 ± 0.3	8

[a]M, Miami; H, Hawaii; SR, Severn River, Annapolis.

at 10°C. The pH was monitored daily, and adjusted if necessary.

A painted test panel was placed in the tank after equilibrium had been established between the TBT in solution and on the surfaces of the apparatus. The pump circulating some of the seawater through the carbon column was turned on and the pumping rate adjusted to maintain a concentration of ~40 $\mu g\ l^{-1}$ Sn in the seawater. The seawater in the tank and effluent from the carbon column were analyzed for inorganic Sn three times weekly. The release rate from each coating was monitored for 3–4 weeks. The panel was then removed and replaced with a new panel. The pumping rate of seawater through the carbon column was adjusted when a new panel was placed in the tank. Activated carbon was replaced when Sn began to appear in the column effluent.

Analysis for organotin was conducted by extracting Sn from the seawater with methylisobutyl ketone (MIBK) and measuring for total Sn in the MIBK with a Perkin-Elmer Model 603 graphite furnace atomic absorption spectrophotometer (AAS, made by the Perkin-Elmer Corporation of Norwalk, Connecticut). Replicate injection into the graphite furnace of the AAS was by an automatic injection system. Standards were prepared from a stock solution of TBT oxide in synthetic seawater. This analysis does not distinguish among different species of organotin. Any organotin homologs of TBT oxide would be extracted and measured as total Sn. The release rate is computed as follows:

$$\text{Release rate} = \{(T_c - E_c)\ (F)\ (1440)\}/A$$

where T_c = concentration of Sn in the tank ($\mu g\ l^{-1}$); E_c = concentration of Sn in carbon column effluent ($\mu g\ l^{-1}$); F = pumping rate through the carbon column ($l\ min^{-1}$); 1440 = $min\ d^{-1}$; and A = painted area of the test panel (cm^2). The release rate and concentrations are expressed in micrograms of Sn. It is assumed, however, that the biocide measured is the TBT cation.

Organotin release rates were determined for several commercially available paints and two experimental paints. Table 19.1 presents release rates of organotin coatings that had been exposed in the environment or in the laboratory system for a number of months. All coatings exposed to the environment were completely free of calcareous fouling organisms. The painted panels were shipped to the laboratory in wet condition and returned to the environment after measurements were completed. Table 19.2 shows the results of measurements on painted panels that had not been exposed to the marine environment, and Table 19.3 shows the effect of temperature on release rate. On some

Table 19.2 Release rates of coatings not exposed to the environment (25°C)

Coating	*Release rate of Sn* ($\mu g\ cm^{-2}\ d^{-1}$)	*No. of measurements*	*Period of measurement (d)*
Experimental A	15.4–4.4	7	21
Experimental B	10.0–3.3	8	21
Commercial C	0.4	14	100
Commercial D	<0.4	8	19
Commercial E	<0.4	8	19
Commercial F	<0.4	8	19

Table 19.3 Effect of temperature on release rate after 42 months exposure in the Severn River, Annapolis

Experimental coating	*Temperature (°C)*	*Release rate of Sn* ($\mu g\ cm^{-2}\ d^{-1}$) *Mean ± standard deviation*	*No. of measurements*
B	25	1.6 ± 0.30	22
	10	0.36 ± 0.16	11
A	25	1.9 ± 0.23	7
	10	0.54 ± 0.15	9

Table 19.4 Effect of biocide concentration on coating release rate (25°C)

			Release rate ($\mu g\ Sn\ cm^{-2}\ d^{-1}$)	
Coating	*Sn concentration* ($\mu g\ l^{-1}$)	*Time without Sn removal (d)*	*With carbon column*	*Without carbon column*
Experimental A	550	6	1.9	0.4
	675	2	2.5	1.3
	325	3	1.3	0.4
Experimental B	490	3	1.6	0.8
	410	2	1.3	0.8
	280	2	1.4	0.5
Commercial C	175	75	0.4	0.004
	90	6	0.4	0.06

occasions, circulation of seawater through the carbon column was stopped and the concentration of TBT allowed to rise for a number of days. The average release rate of biocide during that period was calculated from the tin concentration at the end of that period. In all cases, the release rate was substantially reduced as the tin concentration increased, as shown in Table 19.4.

Table 19.5 shows results on a series of three commercially available, controlled-release paints, of similar formulation, for which the TBT content is known. This provided the opportunity to evaluate the effect of TBT content, temperature, and pH on release rate.

Results from the early method demonstrated that organotin release rates can be

Table 19.5 Effect of temperature, pH, and tributyltin (TBT) concentration (μg TBT cm^{-2} d^{-1}) on release rate

	25°C, pH 8		*10°C, pH 8*		*25°C, pH 7.5*	
Tributyltin (% wt/wt)	*Mean ± SD*	*No. of measurements*	*Mean ± SD*	*No. of measurements*	*Mean ± SD*	*No. of measurements*
10.6	2.4 ± 0.79	14	1.3 ± 0.32	10	1.6 ± 0.12	18
8.9	1.4 ± 0.29	15	0.48 ± 0.11	10	0.80 ± 0.05	18
6.8	0.52 ± 0.05	34	0.12 ± 0.005	22	0.19 ± 0.002	30

reliably measured using steel panels freshly painted or aged in the field. It also provided the following information:

1. The organotin release rates of different paints, all providing adequate fouling protection, range over more than one order of magnitude. Later measurements extended this range to three orders of magnitude (Table 19.6).
2. Freshly coated panels show an initial rate of release very much higher than the subsequent 'steady-state' rate, occurring after several weeks of exposure.
3. The temperature dependence of organotin release rates demonstrated that the release mechanism involves a chemical reaction (hydrolysis) in those formulations where the organotin is chemically bound. Table 19.3 shows a reduction in release rate of a factor of approximately 4 between 25°C and 10°C. Chemical reaction rates are generally affected by a factor of 2–3 for each 10°C temperature change. The temperature effect shown in Table 19.3 is generally confirmed by the results in Table 19.5, where the same temperature change (15°C) influences the release rate by a factor of 1.8–4.3.
4. Table 19.5 also shows the effect of pH on release rate for these formulations. The reduction in release rate for all three paints when the pH was lowered from 8.0 to 7.5 again indicates that the primary mechanism is a hydrolytic cleavage of the carboxyl–TBT bond, which releases the biocide allowing it to diffuse. The consistent reduction in release rate at reduced alkalinity may be one reason why paint manufacturers add chemically unbound TBT compounds to controlled release paints; namely, to assure efficacy in waters of near-neutral pH.
5. Table 19.5 further shows that reducing the TBT content in the paint formulation reduced the release rate and confirms the theoretical description of the release mechanism presented in section 19.1, Introduction, and Fig. 19.1. This demonstrated that paint formulations can be modified to reduce release of TBT to the environment, while maintaining antifouling performance. However, since the controlled-release effect is coextensive with softening of the paint surface (and therefore hull smoothing as mentioned in the Introduction), there will be a lower limit of chemically bound TBT below which the serendipitous softening phenomenon may disappear. This phenomenon is understood better by looking at Fig. 19.1. The number of methylmethacrylate polymer units that have a chemically bound TBT is proportional to the amount of TBT that can be released and therefore proportional to the release rate; in turn, the number of hydrolyzed sites that have released TBT have also become water soluble, and therefore that number is proportional to the softening capacity of the paint film. These phenomena are illustrated further in Fig. 19.3, which shows the data of Table 19.5 in graphic form. While the correlation

Table 19.6 Tributyltin release rates of EPA-registered paints

Paint no.	Release rate for first 14 d ($\mu g\ cm^{-2}$)	Average release rate for weeks 3–5 ($\mu g\ cm^{-2}\ d^{-1}$)	TBT species[a] TBTM (%)	TBTO (%)	TBTF (%)	TBTX (%)	Total TBT (X+M) (%)
1	1	0.02		1.3		1.3	1.3
2	102	1.54		2.1		2.1	2.1
3	113	4.82		3.5		3.5	3.5
4	1	0.24		2.4		2.4	2.4
5	1	0.06		2		2	2
6	2	0.07		2		2	2.0
7	99	3.84		4.8		4.8	4.8
8	177	1.73		3.6		3.6	3.6
9	2	0.10		1.3		1.3	1.3
12	9	0.02		1.3		1.3	1.3
16	1	0.06		1		1	1.0
18	157	6.82	14.5			0	14.5
19	97	5.86	12.7			0	12.7
20	65	2.98	12.1			0	12.1
21	77	6.72	13.8			0	13.8
22	30	1.38		1		1	1.0
23	495	9.41			13.8	13.8	13.8
24	104	7.92	13.4	0.2		0.2	13.6
25	61	4.16	7.9	0.5		0.5	8.4
26	56	3.28	9.5	0.5		0.5	10.0
27	90	5.00	9.4	0.5		0.5	9.9
28	217	21.53	24.5			0	24.5
29	220	8.82	11.8		4.1	4.1	15.9
30	116	5.40	9.9			0	9.9
31	37	4.47	12.7			0	12.7
35	22	4.93			0.51	0.51	0.5
37	24	2.14		0.8		0.8	0.8
39	1128	4.2			6.6	6.6	6.6
43–45	116	3.12	3.71	1.21		3.19	6.9
52	29	2.69		1.9		1.9	1.9
54	55	3.91		2.2		2.2	2.2
55	35	7.07			3	3	3.0
57	105	7.43	15.73			0	15.7

[a]TBTM = TBT methacrylate; TBTO = TBT oxide; TBTF = TBT fluoride; TBTX = total of TBTO, TBTF, and TPTF; total % TBT (X+M) = total of TBTX and TBTM.

of the three curves is good (R^2 >90%), the statistical significance is marginal; since the results in terms of TBT content (the independent variable) are similar, it seemed reasonable to combine the data and fit it to a common slope (but with different intercepts). This yielded a good correlation and statistical significance that a relationship exists (R^2 = 95%; p <0.001). The three curves are seen to converge as the TBT content decreases; that is, the effect of pH and temperature on release rate decreased as the TBT content of the paint decreased. The box in Fig. 19.3 tabulates this observation.

6. A rise in organotin concentration in the measurement tank as shown in Table 19.4 resulted in a reduced release rate, indicat-

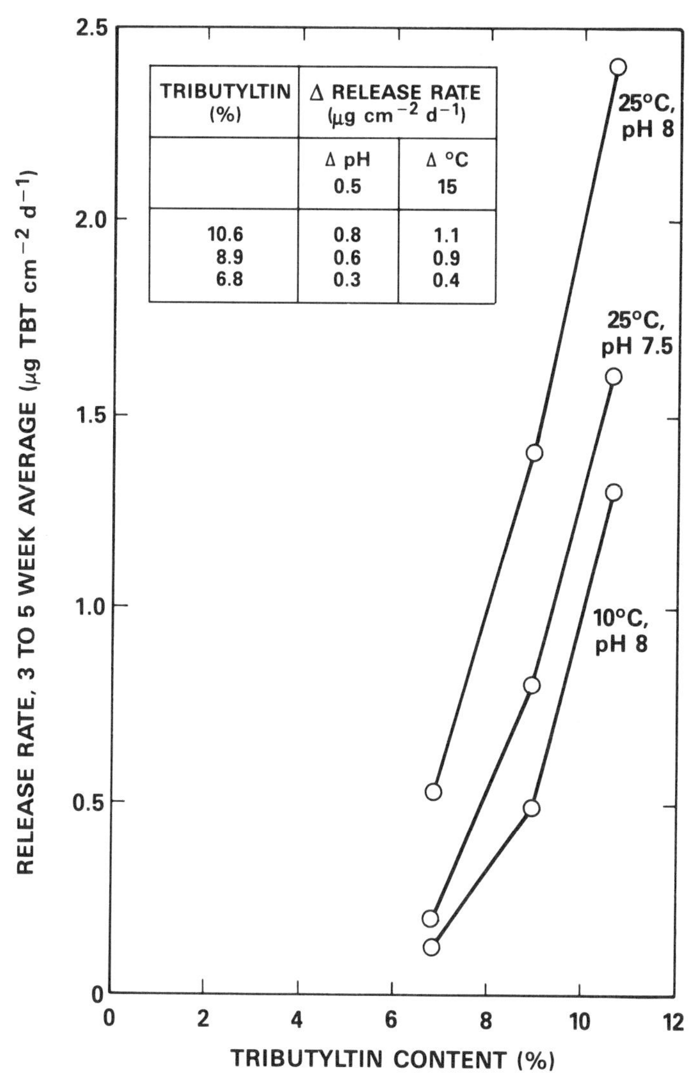

Figure 19.3 Effect of tributyltin content, pH, and temperature on release rate.

ing that the release mechanism is partially dependent on a surface diffusion process that is affected by local high concentrations.

19.3. LATER METHOD DEVELOPMENT

In response to growing concern over the potential environmental consequences resulting from the increasing use of TBT antifouling paints, particularly by recreation craft, the US Environmental Protection Agency (US EPA) announced a Special Review of all pesticide products containing TBT as paint antifoulants, under authority of the US Federal Insecticide, Fungicide and Rodenticide Act (US EPA, 1986a; Champ and Wade, Chapter 3). The US EPA then issued a Data Call-in Notice to all registered vendors of TBT antifouling paints, requiring them to provide extensive data on their products (US EPA, 1986b). This notice included a requirement to provide information on TBT release rates for all registered paints. It was apparent from the data shown in Tables 19.1 and 19.2 that the TBT release rate is an intrinsic property of antifouling paint that could be used effectively to regulate its use by minimizing the quantity of biocide released to the environment.

In preparing to respond to the EPA's Data Call-in Notice requiring the measurement of TBT release rates, the paint industry, with >300 antifouling paint products to be measured, found the procedure developed by the NSWC too costly and time consuming. Accordingly, the American Society for Testing and Materials (ASTM) was asked to develop a less expensive procedure. ASTM Subcommittee D01.45, Marine Coatings, formed a special task group consisting of representatives from industry, the US Navy, the Canadian Navy, the US EPA, and the National Institute of Standards and Technology. After a number of meetings and extensive discussions, a proposed method was devised. In this method a 10-cm band of paint is applied to the exterior curved surface of a polycarbonate cylinder that has one end capped with a polycarbonate disc. This results in a painted area of 200 cm^2, compared with nearly 1550 cm^2 for the panel used in the earlier method. The measurement container is a 2-l polycarbonate jar ~19 cm high and 13.5 cm in diameter. Three 6.5-mm polycarbonate rods, serving as baffles, are evenly spaced and attached to the inside of the measurement jar, which contains 1.5 l of synthetic seawater. The cylinder rotates at 60 rpm. The release rate is calculated from the slope of a concentration–time plot based on a single water sample taken at the end of the measurement period and regressed through the origin. A detailed description of the method is provided in the EPA Data Call-in Notice (US EPA, 1986b). The method described is not an ASTM procedure. An official interlaboratory calibration exercise resulting in a statement of measurement uncertainty (variance) has not been conducted. It is a draft method, which has since been revised a number of times by the ASTM subcommittee working group. However, the basic approach remains the same and provides a laboratory procedure to measure solvent-extractable organotin release rates which occur during a period of immersion under specified conditions of constant temperature, pH, and salinity. Results obtained do not reflect actual release rates in service, but they provide relative comparisons of the release rate characteristics of different antifouling formulations.

Results from the US EPA's Data Call-in Notice for a number of commercial antifouling formulations have been published (US EPA, 1987), and they are reproduced in part in Table 19.6. It seemed worthwhile to determine what additional inferences can be drawn about the nature of these coatings, especially in view of the relationships demonstrated in Table 19.5 and Fig. 19.3. Figure 19.4, based on Table 19.6, shows data on release rates for paints containing TBT methacrylate (TBTM) only and paints containing TBTM plus 0.5% or less free bis-TBT oxide (TBTO). The data used for release rates are the 3- to 5-week averages after the high release occurring during the first 14 d. A general trend to lower release rates with lower TBTM content is seen, in agreement with the data in Fig. 19.3. In addition, those TBTM paints containing 0.5% or less free TBTO can be included in that general trend. The regression of the data in Fig. 19.4 yields a good correlation (R^2 = 0.91 for the TBTM data; R^2 = 0.82 for all data); however, it is dependent on the single high value shown. Without that data point, the correlation is not statistically valid. Inspection of the data in Table 19.6, for paints containing free organotin compounds only, reveals no clear relationship between release rate and organotin content.

The requirement by the US EPA to measure the organotin released by all EPA-

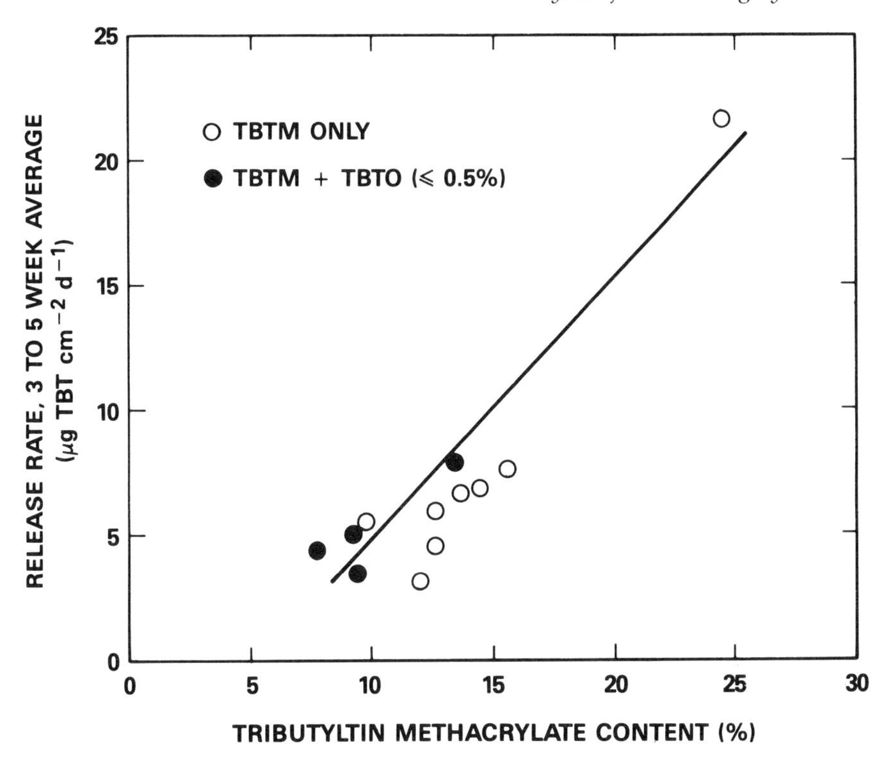

Figure 19.4 Release rates and TBTM content.

registered antifouling paints achieved the following results:

1. Many paint suppliers became aware that the biocide release rate is an intrinsic property of the dry coating and has a direct bearing on evaluating potential environmental consequences from that product. Any new products, as long as they released a toxic substance, would have to be evaluated on the rate of such release.
2. A large number of paints were found to have release rates unnecessarily high for antifouling protection and therefore placed an unnecessary burden on the environment.
3. The US Congress, in enacting regulatory legislation, and the US EPA, in implementing such legislation, incorporated limits on TBT release rates as one key element in the regulatory action. Consequently, a large number of products were removed from the market, and a total ban became unnecessary.

However, data on release rates alone could not be used to determine antifouling performance, because a minimum acceptable rate for reliable protection has not been established. Nevertheless, a sufficient number of high-performance commercial products were found to have acceptable TBT release rates. This established confidence in the validity and fairness of the limits on release rates incorporated in the subsequent regulatory action (US EPA, 1988).

19.4 ANALYSIS OF THE ROTATING CYLINDER METHOD

The primary impact of the regulatory action taken by the US EPA was in establishing limits for release rates for TBT as measured by the rotating cylinder method and a total ban on TBT for all non-aluminum craft up to 25 m in length. The latter action was taken primarily because recreational craft are moored in sheltered harbors where tidal mixing is small, and there exists a relatively large painted hull surface area-to-water volume ratio. Extensive monitoring conducted by the Navy Command, Control and Ocean Surveillance Center (NCCOSC) in 15 US harbors and adjacent estuaries concluded that the highest concentrations of TBT were found in the vicinity of marinas, small boat harbor areas, commercial docks, and repair facilities (Grovhoug *et al.*, 1986). High TBT concentrations in marinas due to antifouling paints from recreational boating activity have been confirmed by a number of independent monitoring efforts, including one study conducted by the US EPA (Batuik, 1987).

As stated previously, the rotating cylinder method used by the US EPA in its Data Call-in Notice to paint manufacturers and distributors was a key factor in an effective regulating action. Many hundreds of measurements were made using this method, and a large body of data was collected. Table 19.6 is a partial presentation of that data. Now that effective regulatory action has been taken and the large effort involved in responding to the requirement for release rate measurement in the Data Call-in Notice has been essentially completed, it is important to examine the rotating cylinder method to identify limitations and means to overcome them.

This method, using a painted cylinder that rotates at 60 rpm, more closely simulates a ship in motion; the procedure developed at the NSWC uses a painted flat plate in a tank with water circulating over the surface; the NCCOSC developed a field method for measuring the release rate directly on the wetted hull of a ship (Seligman *et al.*, Chapter 20). The two US Navy procedures more closely simulate a ship at rest in a harbor, which is an important distinction. The rotating cylinder method is being widely adopted, and despite the disclaimers in the protocol (US EPA, 1986b) that 'results obtained may not necessarily reflect actual TBT release rates which will occur in service', a potential application is to estimate or predict the release into the environment of organotins from antifouling hull paints on ships at berth in a sheltered harbor. Because this method more closely simulates a ship under way, and the two US Navy methods ships at berth, much higher release rates can be expected from this method than would actually occur in a harbor. As measurements by this method became available and were compared to results on commercial ablative organotin paints measured by the US Navy methods, the rotating cylinder method gave results that were substantially higher. An immediate consequence of this was that state and congressional legislation for the regulation of organotin paint, which had been drafted with a limit on release rates of 0.5 $\mu g\ cm^{-2}\ d^{-1}$ TBT based on US Navy data, had to be changed to a value 10 times higher. Table 19.7 shows a comparison of release rates of paints measured by the rotating cylinder method in the laboratory and as measured by NCCOSC directly on the hull of a ship. This demonstrates that the rotating cylinder method yields more than one order of magnitude more TBT than is measured by the in-situ field method, which is more realistic.

Table 19.7 Comparison of rotating cylinder and Navy release rate measurements ($\mu g\ cm^{-2}\ d^{-1}$ TBT)

	Paint identity[a]				
Test method	*43–45*	*55*	*27*	*26*	*54*
Cylinder method	3.12	7.07	5.00	3.28	3.91
Navy method	0.60	0.36	0.44	0.25	0.10
Cylinder/Navy ratio	5.2	19.6	11.4	13.1	39.1

[a]Taken from Table 19.6.

Table 19.8 summarizes the results obtained by six different independent laboratories for the same copolymer standard test paint as

Table 19.8 Summary of results on 'standard' tributyltin test paint for each laboratory

Laboratory	*Average release rate for weeks 3–5 ($\mu g\ cm^{-2}\ d^{-1}$)*
A	4.5
B	1.6
C	1.7
D	4.1
E	3.0
F	5.2
Naval Surface Warfare Center	5.3 ± 1.2

described in the US EPA's Position Document 2/3, October 1987. In addition, Table 19.8 includes results from the NSWC for the same paint. The differences among the six laboratories are surprisingly large. Laboratories B and C appear to be in error. It is not clear to what extent the differences among all laboratories are due to laboratory error, inherent problems in the cylinder method itself, or the nature of the copolymer paint used (identified in Table 19.6 as No. 27). Another concern with the cylinder method lies in the frequency of sampling the test jar in which a painted cylinder is rotating at 60 rpm. Although painted cylinders are used in triplicate, only one aliquot of water is taken from the jar at the end of each measurement period. That single data point is regressed through the origin, and the slope of that line used to calculate the release rate. This assumes that the concentration–time plot routinely intercepts the origin. Experience at the NSWC shows this is not usually the case.

In attempting to improve the rotating cylinder method for screening paints, the NSWC conducted an extensive series of experiments to establish an acceptable balance between precision of measurement of release rates and minimum number of samples needed to determine that measurement. As a result of this effort, the following sampling scheme was adopted for each determination of release rate.

A water blank was taken from the measurement jar prior to the introduction of the cylinder to the jar. Ten minutes after immersion, three samples of 4 ml each were withdrawn from the jar. The cylinder continued to rotate until the concentration of tin (Sn) in the jar reached ~50 $\mu g\ l^{-1}$ Sn, at which time another set of three samples of 4 ml each was taken. A total of seven water samples per cylinder per measurement session were taken. Each sample was analyzed in triplicate by graphite furnace atomic absorption spectrophotometry. Analysis time for this sampling scheme was ~1 h.

This procedure was used to measure the release rates of six different paint formulations, with two cylinder test specimens used for each paint. The paints were controlled release, TBT–acrylic copolymer formulations. The results permitted a comparison between regressing through the origin data from a single sample taken at the end of a test period and the linear regression by the above method. Figure 19.5 shows that regression through the origin introduces a positive bias to the release rate determinations. These data were determined by one experienced operator at one laboratory. It is reasonable to expect larger variability among operators and laboratories. This approach for determining the release rate probably contributed to the large differences among laboratories using the same paint and procedure, as shown in Table 19.8. The probable cause is as follows: when the cylinder is transferred from the holding to the measurement tank, a short period is required during which steady-state conditions are established. The surface of the painted cylinder was exposed to different mixing conditions in the holding tank than in the test tank; in the latter, it is rotating at 60 rpm. Experience has shown that samples taken earlier than 10 min after the cylinder is placed into the measurement tank yield variable results. The TBT released during that period becomes part of the cumulative con-

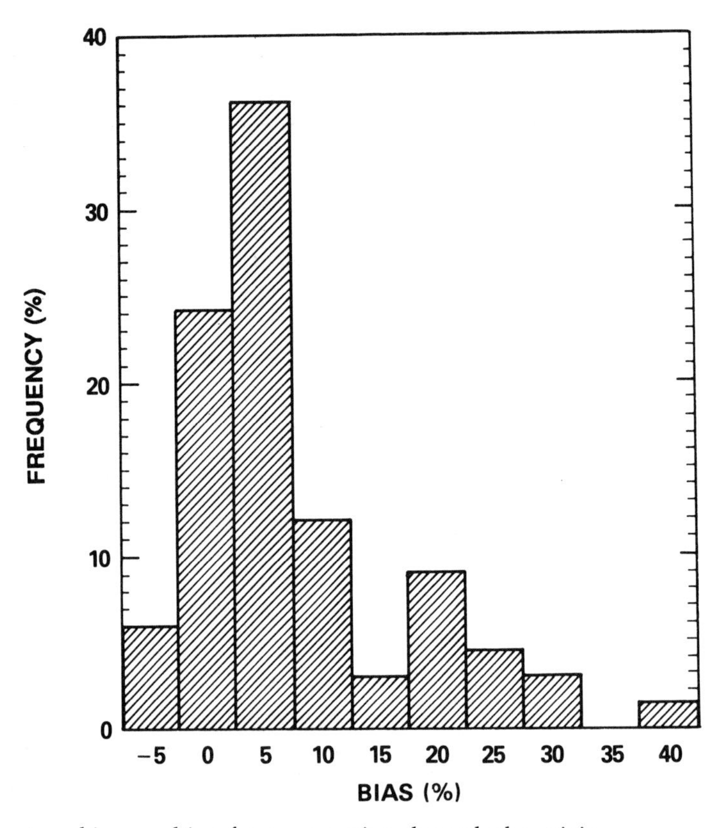

Figure 19.5 Percentage bias resulting from regression through the origin.

centration measured and is apparently sufficient to affect the concentration–time slope, usually by decreasing it. A single data point regressed through the origin ignores this effect, generating an artificially higher slope from which the release rate is determined.

The rotating cylinder method is intended to determine the long-term average release rate for a paint. Shown in Table 19.9 is the average release rate for weeks 3–5, based on three cylinder test specimens, obtained at the NSWC for the same copolymer paint used by the six laboratories. The variability of this result for the 'standard' test paint illustrates the problem with establishing a cutoff point for registered paints. Assuming that the variability shown by the NSWC result in Table 19.8 is typical, a paint could easily meet the criterion and yet have a release rate above the limit. Unless the measurement is made more precise, many paints may be incorrectly rejected or accepted. Assigning a specific long-term release rate as an upper limit should be accompanied by statistically sound sampling and analysis that result in a specific confidence interval. Results of comparisons of methods for determining release rates and measurement of precision of the methods were recently presented (Copeland, 1989). Table 19.9, taken from that presentation, illustrates the precision that is attained when using the procedure described earlier. It is the only statement of variability that has been made on the rotating cylinder method.

Table 19.9 Precision of long-term release rate of cylinders

Paint	*Cylinder*	*Average release rates (confidence intervals)*	*Variability (%)*	
			Cylinders	*Paint*
1	A	3.57 (3.34, 3.81)	6.6	5.6
	B	3.37 (3.03, 3.70)	9.9	5.6
2	A	2.79 (2.45, 3.13)	12.2	7.4
	B	3.03 (2.74, 3.32)	9.6	7.4
3	A	2.58 (2.23, 2.93)	13.6	8.9
	B	2.24 (1.97, 2.51)	12.1	8.9
4	A	8.00 (6.83, 9.16)	14.6	—[a]
	B	5.36 (4.80, 5.92)	10.5	—
5	A	4.27 (3.67, 4.87)	14.1	8.6
	B	3.88 (3.43, 4.32)	11.5	8.6
6	A	3.75 (3.39, 4.12)	9.7	6.6
	B	4.11 (3.73, 4.50)	9.4	6.6

[a]Dash indicates that there was insufficient data to perform the necessary calculation.

19.5. ROLE OF THE BIOFILM

Surfaces exposed to natural waters, even if they are releasing biocides to prevent macrofouling, become coated with a biofilm or slime layer. This layer consists of a complex community of bacteria, protozoa, diatoms, and other microalgae together with their cellular exudates (Daniel *et al.*, 1980). This slime layer can increase ships' frictional resistance to motion by up to 10% (Marriott, 1980; Loeb, 1981). Controlled release, ablative TBT antifouling paints, which do not accumulate biofilms under dynamic flow conditions, become coated with slime under static conditions (Evans, 1981; Ghanem *et al.*, 1981). The role of the biofilm in modifying the release of TBT from painted surfaces was investigated at the Navy Biosciences Laboratory several years ago (Guard *et al.*, 1983). This study showed that microbial biofilms formed on painted panels that release TBT accumulated high concentrations of the biocide. The slime layers were removed from the panels and centrifuged to separate solids from the liquid portion. The TBT content of the centrifugate was measured at >20 mg l^{-1}, which is considered to be above its solubility limit in seawater, while the interstitial water of the biofilm contained 1 mg l^{-1}, considered to be near the solubility limit. In a recent investigation at the NSWC, biofilms were formed in the laboratory with either TBT-resistant bacteria or algae on surfaces painted with a commercially available, controlled-release TBT paint (Mihm and Loeb, 1988). Steel panels with an area of 7.6 × 12.7 cm were coated with a standard anticorrosion paint followed by the TBT paint. The panels were immersed in glass jars containing synthetic seawater held at constant temperature. Mixing was accomplished with a reciprocal shaker. The seawater was replaced regularly to prevent accumulation of TBT and for analysis to determine the release rate. Microorganism growth was prevented with an ultraviolet sterilizer. After a stable release rate was obtained, the jars were inoculated with bacterial and algal cultures originally obtained from the hull of a ship coated with a TBT paint. After a 24-h period for biofilm formation on the panel surfaces, the seawater and jars were replaced. The biofilms remained on the panel surfaces for 38 d or longer, and then were removed. Figure 19.6 is a histogram of the release rate data obtained. An apparent effect on release rate

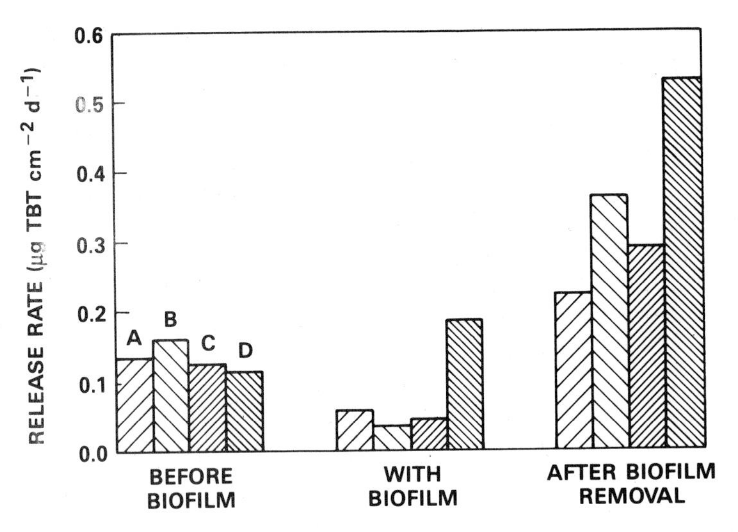

Figure 19.6 Tributyltin release rates before and after biofilm formation and after biofilm removal (Mihm and Loeb, 1988). A, B, C = bacteria; D = algae.

of the biofilms is shown. A very noticeable effect is observed after biofilm removal. Although the removal was gentle, a substantial disturbance of the partially hydrolyzed paint film at the biofilm–paint interface seemed to have occurred, resulting in higher release rates than before. This phenomenon may be an additional clue to why it is inadvisable to determine the release rate in the rotating cylinder method from the concentration–time slope by taking a single data point and regressing through the origin. The soft, hydrolyzed paint film is easily disturbed, temporarily affecting the release rate.

Additional biofilm–TBT studies have been conducted by the National Institute of Standards and Technology (Blair *et al.*, 1988). In these experiments biofilms were formed on glass microscope slides, but in the absence of a paint film. This facilitated analysis of the biofilm for TBT content without the obvious complication that occurs when the biofilm is removed from a paint film. Biofilms were grown from a mixed culture of estuarine microorganisms originally obtained from the surface of a steel panel coated with a commercial, controlled-release TBT paint. Once biofilms had formed on the glass slides, the biofilms were exposed to sterilized water from the Chesapeake Bay that contained TBT. The accumulation of biocide by the biofilm was determined. Some results of these experiments are reproduced in Table 19.10, which shows that a substantial accumulation of TBT from solution has occurred in the biofilms and confirms the prior investigation at the Navy Biosciences Laboratory.

Table 19.10 Accumulation of tributyltin (TBT) by microbial biofilms

Incubation	*TBT solution concentration* ($\mu g\ l^{-1}$)	*TBT biofilm concentration* ($\mu g\ g^{-1}$)	*Bioconcentration factor (dry wt)*
Dark, 97 h	50	353	7060
Light, 168 h	50	373	7460

19.6. AN ANTIFOULING PAINT–BIOFILM MODEL

The experimental evidence thus far has shown the following: TBT accumulates in biofilms; the nature of the biofilm has an apparent effect on release rates; a negative effect on release rate due to localized high TBT concentration in the seawater is observed; and lowered alkalinity has a negative effect on release rate.

Based on this information, the following physical model, which has been proposed previously (Schatzberg, 1987), is described. A biofilm, consisting of microorganisms acclimated to TBT, forms on a stationary surface exposed to marine waters in a harbor. Tributyltin released from the antifouling coating as a result of hydrolytic cleavage is taken up by the biofilm. The TBT is partly bound by the microbial exudates such as polysaccharides and partly held in concentrated solution in the interstitial water of the biofilm. Metabolism in the biofilm lowers the pH at the biofilm–paint interface. The biofilm inhibits rinsing of the stationary painted surface by tidal action in the harbor. The combination of reduced pH at the paint–biofilm interface (Table 19.5), increased TBT concentration within the biofilm (Table 19.10), and the effect of TBT concentration on release rate (Table 19.4), all tend to reduce the release of TBT by the following three processes: by reducing the rate of hydrolysis, which frees the TBT; by reducing the slope of the concentration gradient at the paint surface, reducing thereby the diffusion rate in accordance with Fick's first law of diffusion; and by the chemical binding of TBT by the biofilm exudates. The biofilm under these conditions can be thought of as a complex chemical and physical capacitor that accumulates TBT by various mechanisms. At some point, however, the capacitor leaks, since TBT is clearly released into the environment. Furthermore, when the painted hull surface is in motion, the thickness of the biofilm is reduced and its full restoration under static conditions may require some time. The purpose of this model is heuristic; that is, to illustrate the complex chemical, physical, and biological phenomena taking place at an antifouling paint–biofilm interface under natural conditions and to demonstrate that the microchemistry and microbiology taking place need to be understood in order to design future antifouling coatings that are effective but that have a negligible impact on the environment.

19.7. A NOVEL APPROACH

What is needed now is a relatively simple procedure to determine the minimum rate of a given biocide that will prevent the attachment of settling organisms prior to the expensive process of formulating and testing a paint. Such a procedure has recently been developed at the Naval Surface Warfare Center (Haslbeck *et al.*, 1992).

The apparatus used consists of a circular, inert, permeable, polycarbonate membrane filter, 47 mm in diameter, with a pore size of 0.2 μm, and held in place by a modified filter holder. A controlled release of TBT from the pores of the exposed surface of this membrane is achieved by a metered flow through the membrane of sterilized seawater containing a known concentration of TBT. Various release rates of TBT are obtained by changing the concentration of TBT and the metered flow rate.

Such membrane assemblies or test cells are suspended in a flow of marine water containing a natural population of organisms that includes peritrichs, a group of ciliated protozoans that are an important part of the biofilm in the Chesapeake Bay. Peritrichs attain significant populations on submerged surfaces within a day. After 24-h exposure to estuarine water, the membrane surfaces were appropriately prepared and the number of attached peritrich organisms counted. An example of a result using this arrangement is shown in Fig. 19.7. Another result, Fig. 19.8, illustrates that a TBT release rate of 0.2 μg cm^{-2} d^{-1} is effective in preventing the attachment of barnacle larvae (Haslbeck *et al.*, 1996). Minimum effective release rates of

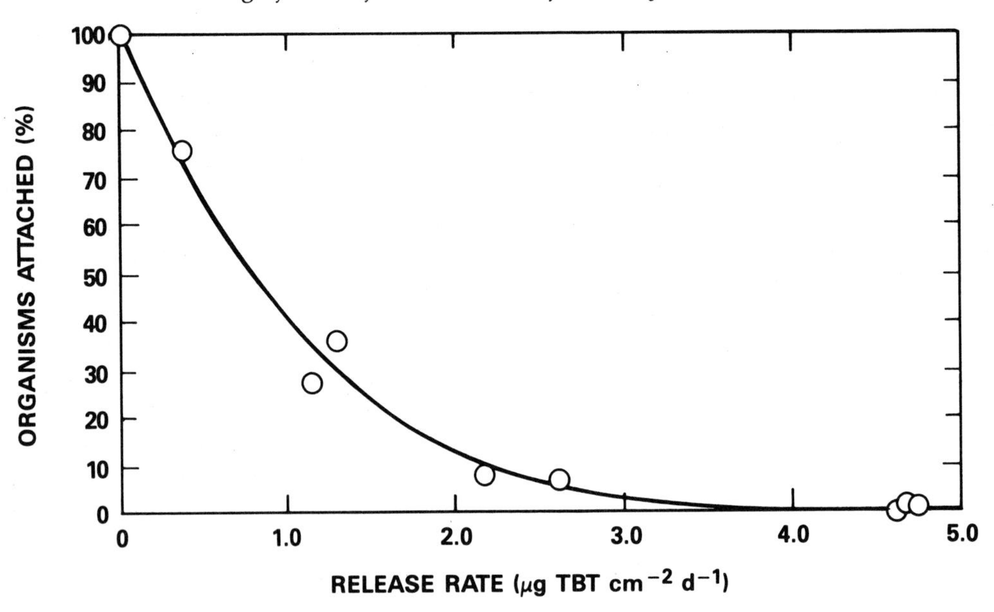

Figure 19.7 Effect of release rate on peritrich attachment Haslbeck *et al.*, 1992).

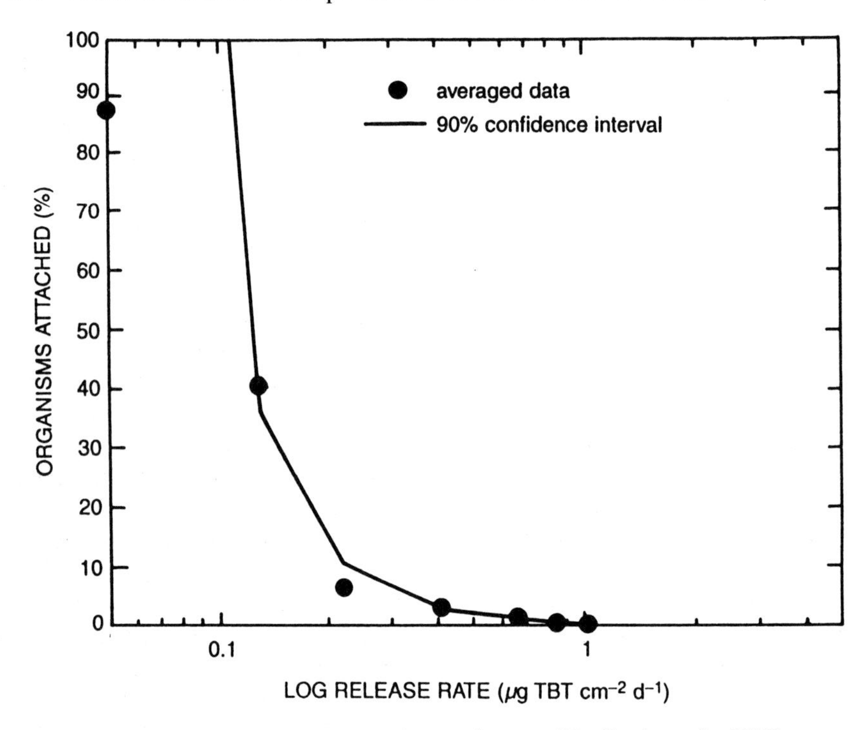

Figure 19.8 Effect of release rate on barnacle attachment (Haslbeck *et al.*, 1996).

biocide can now be determined with this simple apparatus, using the settling larvae of different organisms under controlled as well as under different field conditions. In addition, the synergistic effect of two or more biocides can be determined.

This type of information can provide antifouling paint chemists with reliable data on the optimum (environmentally minimum) biocide release rate needed prior to paint formulation. With such information in hand, paint chemists could then formulate an antifouling paint with adequate confidence that the cost of formulation and testing would result in an effective, environmentally acceptable product.

19.8. SUMMARY

Measurement of the release rate of antifouling paints is important principally for minimizing the potential impact of this biocide on nontarget organisms; additionally, as the results presented in this chapter have shown, such data are useful in enabling an understanding of the mechanism of release and the effect of environmental conditions on the rate. Chemically combining the TBT into a key paint constituent is clearly superior to simple mixing, because this permits a controlled release of the biocide, and, when the minimum release rates needed are known, careful design of the antifouling paint.

Laboratory measurements of the TBT release rate have shown that it can be measured reliably. These measurements provided the following information: (1) release rates of different paints range over several orders of magnitude; (2) freshly painted surfaces show an initial rate of release very much higher than the subsequent 'steady-state' rate; (3) temperature dependence of release rates demonstrated that the release mechanism involves a chemical reaction (hydrolysis) in those formulations where the TBT is chemically bound; (4) reduction in release rate when the pH is lowered also indicates that the primary mechanism is hydrolytic cleavage of the carboxyl–TBT bond, which releases the biocide, allowing it to diffuse out; and (5) for the chemically bound TBT paint formulations, reducing their TBT content reduces the release rate as well as the surface softening and resulting hull smoothing.

The rotating cylinder method for measuring the release rate in the laboratory has been applied by the US EPA to regulate the use of TBT paints. This action achieved the following results: (1) paint suppliers became aware that the biocide release rate is an intrinsic property of the coating and has a direct bearing on evaluating potential environmental consequences from that product; (2) many TBT paints were found to have release rates unnecessarily high for antifouling protection, which therefore caused an unnecessary burden on the environment; and (3) the US Congress, in enacting regulatory legislation, and the US EPA, in implementing such legislation, incorporated limits on TBT release rates as one key element in the regulatory action, removing thereby a large number of unacceptable products from the market and making a total ban unnecessary.

Nevertheless, the rotating cylinder method appears to be flawed in that it simulates more closely the condition of a ship under way than a ship at rest in a harbor; consequently, very much higher biocide release rates are measured by this method than is actually the case. This is dramatically illustrated by Table 19.7, which compares results on five paints by the rotating cylinder method with the US Navy procedure of measurement directly on the hull of the painted ship.

Results show that the biofilm covering the painted surface in the marine environment accumulates and concentrates the TBT biocide released by the paint. The biofilm appears to act as a biocide capacitor, establishing a TBT concentration which presents a chemically hostile environment that prevents permanent attachment. This model suggests that paints that have very low release rates can nevertheless prevent fouling attachment because the biofilm accumulates TBT.

The benefits of measuring the TBT release rate in the laboratory have been demonstrated. Formulating and testing antifouling paints by the paint industry is a time-consuming and expensive process, especially now that these products must meet stringent regulatory requirements. What is needed prior to the paint formulating process is the minimum release rate for a given biocide or combination of biocides to prevent the attachment of fouling organisms. A novel, simple apparatus, easy to use in the laboratory or the field, has been developed to determine such a minimum effective release rate of biocide. The apparatus simulates different rates of release of biocides, can be positioned in the marine environment, and can be exposed to settling organisms. Data collected in this way will establish the optimum rate of release for a given biocide; that is, the minimum rate needed to prevent attachment by fouling organisms and at the same time, a negligible impact on the environment. This information can then be used by paint chemists to formulate an antifouling paint with a high level of confidence that the new paint will perform well and meet regulatory requirements.

ACKNOWLEDGMENTS

During the preparation of this chapter the author was guided by conversations with H.H. Singerman, Head, Chemical and Physical Processes Division; and L.L. Copeland, Environmental Protection Branch, of the Naval Surface Warfare Center. These helpful discussions are gratefully acknowledged. Statistical analyses of release rate data were performed by K.C. Burns of Desmatics, Inc. Preparation of this chapter was supported by the Naval Sea Systems Command, US Department of the Navy, Washington, DC.
 Opinions expressed in this chapter are those of the author and do not necessarily reflect US government policy.

REFERENCES

Anderson, C.D. and R. Dalley. 1986. Use of organotins in antifouling paints. In: Proceedings of the Oceans '86 Organotin Symposium, Vol. 4, 23–25 September 1986, Marine Technology Society, Washington, DC, pp. 1108–1113.

Batuik, R. 1987. *Survey of Tributyltin and Dibutyltin Concentrations at Selected Harbors in Chesapeake Bay – Final Report*. Prepared by US EPA Chesapeake Bay Liaison Office, CBP/TRS 14/87, 98 pp.

Blair, W.R., G.J. Olson, T.K. Trout, K.L. Jewett and F.E. Brinckman. 1988. Accumulation and fate of TBT species in microbial biofilms. In: Proceedings of the Oceans '88 Organotin Symposium, Vol. 4, 31 October–2 November 1988, Marine Technology Society, Baltimore, pp. 1668–1671.

Copeland, L. 1989. Organotin paint release rate methodology. In: Proceedings of the 1989 Marine Offshore Maintenance and Coatings Conference, 28–30 June 1989, Charleston, SC, National Paint and Coatings Association, Washington, DC, pp. 1–12.

Daniel, G.F., A.H.L. Chamberlain and E.B.G. Jones. 1980. Ultrastructural observations on the marine biofouling diatom *Amphora*. *Helgolander Meeresun.*, **34**, 123–140.

Evans, L.V. 1981. Marine algae and fouling: a review, with special reference to ship fouling. *Bot. Mar.*, **24**, 167–171.

Ghanem, N.A., N.N Messiha, M.M. Abd El Malek, N.E. Ikladious and A.F. Shaban. 1981. New terpolymers with pendant organotin moieties as antifouling coatings. *J. Coating Technol.*, **54**, 57–60.

Grovhoug, J.G., P.F. Seligman, G. Vafa and R.L. Fransham. 1986. Baseline measurements of butyltin in US harbors and estuaries. In: Proceedings of the Oceans '86 Organotin Symposium, Vol. 4, 23–25 September 1986, Washington, DC, Marine Technology Society, pp. 1283–1288.

Guard, H.E., A.B. Cobet, R.B. Laughlin, Jr and D.C. Evans. 1983. Interactions between marine biofilms and tributyltin compounds. Paper presented at 4th International Conference on the Organometallic and Coordination Chemistry of Germanium, Tin, and Lead, 8–12 August 1983, Montreal, Canada. National Sciences and Engineering Research Council of Canada, pp. 1–6.

Haslbeck, E.G., W.C. Banta and G.I. Loeb. 1992. Minimum effective release rate of antifoulants: (1) measurement of the effect of TBT on peritrich biofilms. *Biofouling*, **6**, 97–103.

Haslbeck, E.G., C.J. Kavanagh, H. Shin, W.C. Banta, P. Song and G.I. Loeb. 1996. Minimum effective release rate of antifoulants: (2) measurement of the effect of TBT and zosteric acid on biofilms. *Biofouling* (accepted for publication).

Loeb, G.I. 1981. Drag Enhancement of Microbial Slime Films on Rotating Discs. Naval Research Laboratory Memorandum Report 4412, 16 pp.

Marriott, H. 1980. Antifouling developments smooth the way for tomorrow's fleet. *Reed's Special Ships*, **3**, 18–19.

Mihm, J.W. and G.I. Loeb. 1988. The effect of microbial biofilms on organotin release by antifouling paints. In: *Biodeterioration 7*, D.R. Houghton, R.N. Smith and H.O.W. Eggins (Eds). Elsevier Applied Science, London, pp. 309–314.

Schatzberg, P., C.M. Adema and D.F. Jackson. 1983. Biocide release rates of organotin from antifouling paints. Paper presented at 4th International Conference on the Organometallic and Coordination Chemistry of Germanium, Tin, and Lead, 8–12 August 1983, Montreal, Canada, Natural Sciences and Engineering Research Council of Canada, pp. 1–11 numbered separately.

Schatzberg, P. 1987. Organotin paints and the US Navy – a historical perspective. In: Proceedings of the Oceans '87 Conference, Vol. 4, Marine Technology Society, Halifax, Nova Scotia, 19–37 pp. 1324–1333.

US EPA. 1986a. Initiation of a special review of certain pesticide products containing tributyltins used as antifoulants, availability of support document. *Federal Register*, **51**, 778.

US EPA. 1986b. Data call-in notice for data on TBTs used in paint antifoulants, US EPA, Washington, DC, 29 July 1986, 25 pp.

US EPA. 1987. Tributyltin Technical Support Document. Position Document 2/3. Office of Pesticides and Toxic Substances, Office of Pesticides Programs, Washington, DC, October, pp. iv–8.

US EPA. 1988. Tributyltin antifoulants; notice of intent to cancel; denial of applications for registration; partial conclusion of special review. *Federal Register*, **53**, 39022–39041.

ENVIRONMENTAL LOADING OF TRIBUTYLTIN FROM DRYDOCKS AND SHIP HULLS

20

P.F. Seligman,[1] C.M. Adema,[2] J. Grovhoug,[1] R.L. Fransham,[3] A.O. Valkirs,[3] M.O. Stallard[3] and P.M. Stang[4]

[1]*Naval Command, Control and Ocean Surveillance Center RDT&E Division, Environmental Sciences (Code 52), San Diego, California 92152-5000, USA*
[2]*Naval Surface Warfare Center, Carderrock Division, Annapolis Detachment, Code 2184, Maryland 21402, USA*
[3]*Computer Sciences Corporation, 4045 Hancock Street, San Diego, California 92110-5164, USA*
[4]*PRC Environmental, 4065 Hancock Street, San Diego, California 92110, USA*

Organotin. Edited by M.A. Champ and P.F. Seligman. Published in 1996 by Chapman & Hall, London. ISBN 0 412 58240 6

ABSTRACT

Environmental loading factors for tributyltin (TBT) were studied in Pearl Harbor, Hawaii, between 1987 and 1988. During this period three test ships were painted with TBT-containing antifouling paint. The drydocks used to paint the vessels were monitored for overspray and discharges to both the harbor and the sewer system. Use of extensive drydock cleanup procedures, including masking 50% of the drydock floor, resulted in discharges to the harbor of <15 g TBT. Harbor concentrations were largely unaffected by TBT discharges of <3–149 ng l^{-1} from the drydocks. Cleanup procedures were found to be successful in removing at least 99.8% of the residual paint overspray from the drydocks, thus eliminating environmentally significant discharges to the receiving waters.

The TBT flux from underwater hull coatings was evaluated by a device for measuring in-situ release rates, which documented order-of-magnitude differences (0.1–2.8 μg cm^{-2} d^{-1}) in release rates among types of coatings. Tributyltin loading from the test ship hulls, ranging from 2 to 115 g d^{-1}, generally covaried in a linear fashion with regional mean concentrations of TBT in the water column. An empirically derived model was used to predict future water column concentrations of TBT in Pearl Harbor of <10 ng l^{-1} from Navy use, assuming application of low-release-rate (<2.0 μg cm^{-2} d^{-1}) paints and effective drydock cleanup procedures.

20.1. INTRODUCTION

The various sources of tributyltin (TBT) are well understood. However, the mass loading rates of this biocide into the aquatic environment have not been extensively reported. The evaluation of environmental risk from any contaminant is based on the toxicology and ecological effects of the compound, the biogeochemical fate, and the environmental loading or rate of flux of the toxicant to the environment. The ecotoxicity of TBT has been heavily studied (see Chapters 9–16), and the fate of TBT is reasonably well understood (see Chapters 17–20 and 22–23).

Total use of organotins worldwide was ~23–25 × 10^6 kg in 1985 with ~4.5 × 10^6 kg used as biocides (see Champ and Seligman, Chapter 1, for additional information). Use of organotins in the last 20 y has increased by ~10-fold while the biocidal use has grown about twice this fast. The principal source of triorganotin in the aquatic environment is from tributyltin-containing antifouling coatings, which are released (leached) directly from the hull or from residual paint that is released into the environment from painting and removal processes. In the United States, ~2.4 × 10^6 l of TBT-containing antifouling paint was used in 1985 with ~4.5 × 10^5 kg of

TBT used in the formulation [US Environmental Protection Agency (EPA), 1987]. Approximately 60% of the TBT antifouling paint sold is applied to commercial vessels, 33% is applied to recreational vessels, and an additional 7% is sold for other uses.

Because of concerns about the potential effects of TBT on nontarget organisms, several countries have restricted the use of TBT. These concerns arose after environmental measurements were made that documented concentrations in the water column in some bays and estuaries that exceeded the toxic limits for certain sensitive species. These restrictions include limiting the size of vessels eligible for painting, limiting the release rate of the product into water, or otherwise restricting application of the product. These laws and regulations will likely reduce the total production and use of TBT and should significantly reduce both the introduction of the compound into the environment and the measured concentrations.

There are two principal pathways by which butyltin compounds associated with antifouling paints can enter the aquatic environment. The first is the release of butyltin compounds from the underwater hull of the vessel, which represents a continuous source of the biocide when the coated vessels are in a harbor or estuary. The second is the intermittent release of butyltin compounds from boat and shipyard activities associated with hull painting, cleaning, and paint-removal activities.

Field studies have generally found that the highest TBT concentrations have occurred in yacht harbors and marinas (Grovhoug *et al.*, 1986, 1987, 1989, and Chapter 25; Seligman *et al.*, 1986, 1989; Valkirs *et al.*, 1986, 1991; Hall *et al.*, 1987. Alzieu, 1991; Alzieu *et al.*, 1991; Ritsema *et al.*, 1991; Wade *et al.*, 1991; Abel, Chapter 2; Waldock *et al.*, Chapter 11; Salazar and Salazar, Chapter 15; Huggett *et al.*, Chapter 24; Waite *et al.*, Chapter 27). This led to the conclusion that in many harbors and estuaries small craft were the principal source of TBT. This problem was compounded by the fact that small craft tend to be berthed or moored in shallow areas, generally with limited flushing, and in some instances near to shellfish culturing areas. The use of high-release-rate free-association paints and often poor hull maintenance practices exacerbated the problem. In recognition of this, France, the United Kingdom, and the United States have restricted the use of TBT on small craft <25 m in length. (Abel, Chapter 2; Champ and Wade, Chapter 3). During the last several years, a substantial effort has been devoted to measuring the release rates of TBT from various coatings and formulations on test panels (Schatzberg, Chapter 19), primarily from US Navy studies and the data call-in procedure required by the US EPA in its Special Review process (US EPA, 1987, 1988). Our in-situ field studies suggest that measurements of TBT release rates from painted panels made in the laboratory may not yield realistic estimates of TBT releases from hulls under natural conditions with a biological slime layer present. Between 1985 and 1987 high concentrations of TBT were measured periodically in the Elizabeth River, Virginia, which suggested that there were intermittent kilogram-quantity inputs from drydocks in the vicinity (Seligman *et al.*, 1987; Huggett *et al.*, Chapter 24). However, there has been very little work to describe quantitatively the rate or mass of butyltin inputs into adjoining harbor waters during the painting and undocking of vessels. Likewise, there has been very little work that relates measured environmental loading of TBT and actual environmental concentrations to evaluate the risks from use of this antifouling biocide.

In 1984, the US Navy announced its intention to begin implementation of TBT antifouling coatings for its surface fleet (US Naval Sea Systems Command, 1984, 1985). Because of increasing concerns about potential effects on nontarget organisms from the use of TBT antifoulants, it was agreed that only a limited test program would be conducted to further evaluate environmental loading and risks until the US EPA completed its Special Review and finalized its TBT regulation. This chapter reports the results of a 3-y study, conducted between 1987 and 1989, in Pearl Harbor, Hawaii, which compares the measured TBT loading from drydock operations, TBT release rates from painted hulls, and the resultant concentrations in the harbor.

20.2. EXPERIMENTAL DESIGN AND ANALYSIS

20.2.1. BACKGROUND AND OBJECTIVES

The experimental plan was jointly developed by the Naval Command, Control and Ocean Surveillance Center, San Diego, California, and the Naval Surface Warfare Center, Annapolis, Maryland. Pearl Harbor was selected as the case study site because it (1) was primarily used by Navy vessels, making TBT sources unambiguous; (2) had drydock facilities for test painting; and (3) had existing test ships and three ships scheduled for painting to evaluate TBT loading from painted hulls.

The principal objective of the Pearl Harbor case study was to evaluate the environmental risk of US Navy use of TBT antifouling coatings. The specific objectives of the study reported in this chapter were to:

1. measure and evaluate TBT inputs to the marine environment from direct release from hulls and from drydock operations;
2. evaluate the efficacy of drydock TBT containment and cleanup procedures;
3. determine the concentration and distribution of butyltins in the water column in relation to release rates for TBT loading on ship hulls; and
4. evaluate the predictability of TBT concentrations in the harbor from measured loading factors.

Other portions of the case study are reported elsewhere in this volume: degradation and fate (Seligman *et al.*, Chapter 21), biological effects on Pearl Harbor organisms (Henderson and Salazar, Chapter 14), and monitoring studies (Grovhoug *et al.*, Chapter 25).

20.2.2. SITE DESCRIPTION

Pearl Harbor is located midway along the southern side of the island of Oahu in the Hawaiian Islands. The entire harbor is under the jurisdiction of the Navy and contains a naval shipyard, naval station, submarine base, naval supply center, and an inactive ship maintenance facility. Pearl Harbor is divided into three primary regions: East, Middle, and West Lochs (Fig. 20.1). Adjoining East Loch is the smaller Southeast Loch basin, the most heavily used Navy berthing area within the harbor. Civilian vessels visiting the harbor include freighters and tankers to the Naval Supply Center piers (adjacent to Southeast Loch), tuna fishing boats collecting bait fish, and daily commercial harbor tour vessels. Rainbow Marina, a small-boat facility with a capacity of ~70 vessels, is located in the Aiea Bay area in the northeastern portion of East Loch.

Tidal flow and circulation are weak and variable with a maximum ebb flow of ~25 cm s^{-1}. Surfacewater circulation is primarily driven by the predominant northeasterly trade winds. Freshwater inputs are irregular from eight major streams, which drain stormwater runoff into West, Middle, and East Lochs. Measured salinities in the harbor during a previous survey ranged from 14.1 to 37.5‰ with an average of ~32.8‰. The drainage area of the harbor is ~285 km^2. Tidal currents are weak and variable. The mean tidal current velocity at the harbor entrance averages <15 cm s^{-1}, with a maximum velocity of 30 cm s^{-1} (US Department of Commerce, 1986a,b). Tides are mixed with a mean range of 0.4 m.

The Pearl Harbor Naval Shipyard has four graving drydocks (Fig. 20.1). A graving drydock is an excavated portion of shoreline into which a ship can be floated. The harbor end of the drydock is sealed with a caisson and

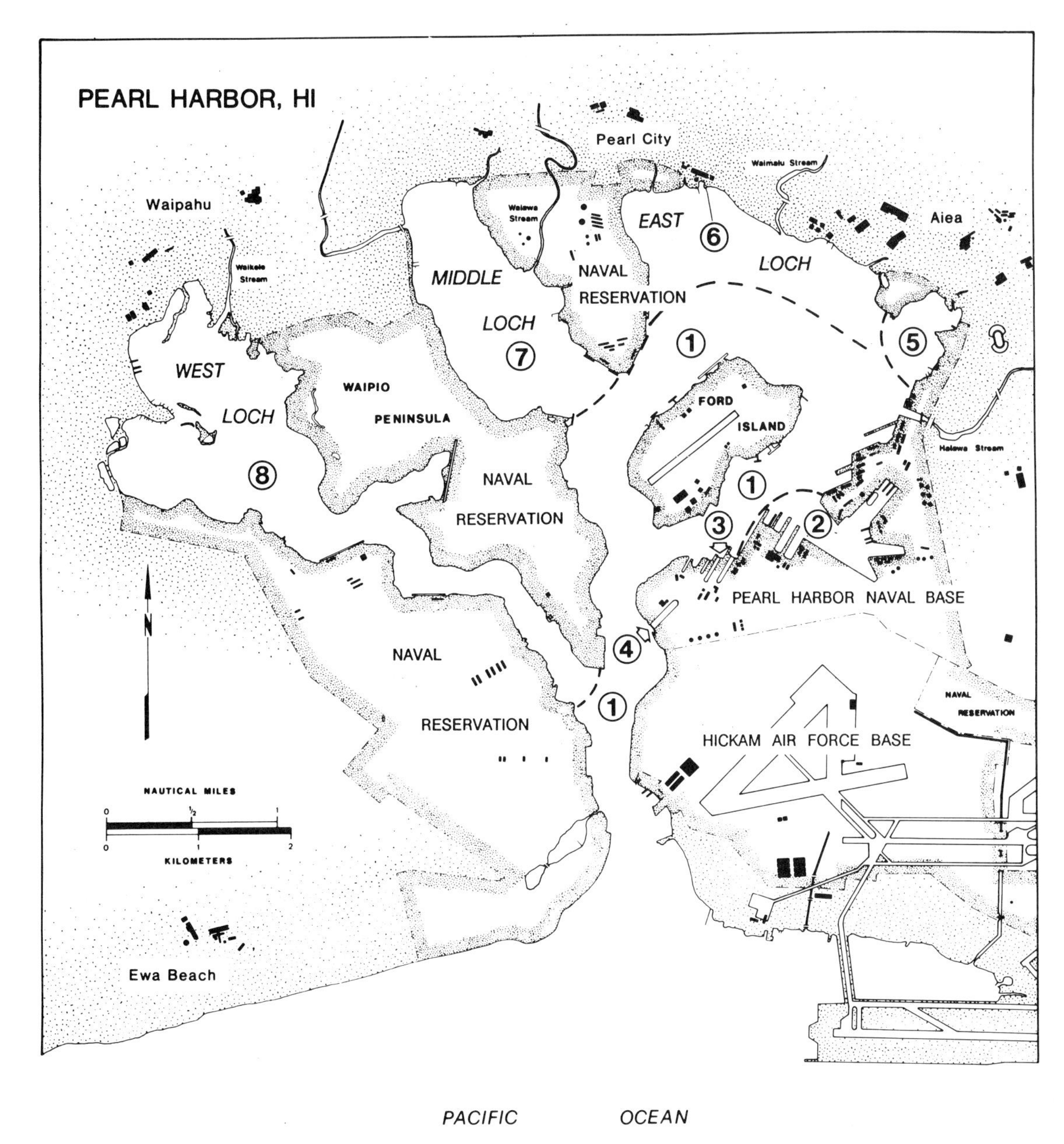

Figure 20.1 Map of Pearl Harbor showing near-field and far-field sampling stations and regions used for averaging TBT concentrations.

water is pumped out, lowering the vessel onto blocks so that hull maintenance can be performed. The process is reversed to undock and release the vessel.

20.2.3. HARBOR MONITORING

A total of eight monitoring surveys were conducted from April 1986 to January 1989.

Prior to 1986, three ships were coated with TBT-containing paint at the Pearl Harbor Naval Shipyard. During 1987, as part of the case study, three vessels were painted with TBT coatings. Extensive monitoring was conducted prior to painting and after the undocking of each of the test vessels. Mean TBT concentrations for various regions (see Grovhoug *et al.*, Chapter 25) are compared with the daily TBT mass from ships loading derived from in-situ hull release measurements. Samples were collected from 0.5 m below the surface and 1.0 m above the bottom using a polycarbonate sampling device, which eliminated adsorption and contamination problems. Three to nine samples were collected from each region and at each depth and averaged for a regional mean value. Pearl Harbor regions are shown in Fig. 20.1. For the purposes of comparing loading with regional means, the three channel areas were combined.

20.2.3a. Undocking phase surveys

Three surveys were conducted during the undocking phase of the study. The first survey was conducted during the undocking of the frigate *USS Badger* in March 1987. Stations were selected for 3 d of water sampling to evaluate the impact of the naval shipyard's clean-up and undocking procedures on harbor TBT concentrations. Most of the stations were located in the immediate area of the drydock region, although outlying areas were also sampled. Six near-field sampling stations were selected within 2 km of the drydocks (Fig. 20.1) and were sampled on the first and second day after the vessel was undocked and moved to Southeast Loch.

The second and third surveys in this series were conducted after undocking of the *USS Brewton* in August 1987 and *USS Davidson* in September 1987. The water column was examined more closely in the outlying areas of the harbor during test ship undocking intervals. Six stations were selected from the outlying regions (Fig. 20.1) and comprised far-field stations. Six near-field stations were also identified (Fig. 20.1). Samples were collected on the first and third days following the undocking of each ship. The stations were sampled in the same order over each of the 72-h periods. Triplicate 1-l water samples were obtained at 0.5 m below the surface and at 1 m above the bottom at each station using a Navy-developed sampler, which collected the water in the polycarbonate bottles in which they were frozen and stored until analysis.

20.2.3b. In-situ ship hull release rate study

The TBT release rates from the test ships were determined by using especially designed, enclosed, recirculating, diver-placed dome systems (Lieberman *et al.*, 1985). A partial vacuum is created in a 30-cm diameter polycarbonate dome fitted with a double knife-edge rubber gasket by placing the dome against the hull and evacuating water from the system via a surface-controlled peristaltic pump (Fig. 20.2). When sufficient vacuum is attained to keep the dome affixed to the hull, water is circulated through the dome and FEP Teflon® (Du Pont, Wilmington, Delaware) tubes via the pump (Fig. 20.2). Six water samples of 20 ml each were collected from the system at 10- to 15-min intervals over 1–1.5 h without compromising the vacuum or allowing ambient harbor water to enter the enclosed system. The samples were placed immediately on ice, frozen at the end of the day, and analyzed by the purge-and-trap hydride derivatization (HD–AAS) method (Stallard *et al.*, 1989; Weber *et al.*, Chapter 4 for determining TBT concentration. Three separate dome systems (near surface, mid-depth, and near bottom) were used to collect replicate release rate data from three regions (fore, mid-, and aft portions of the ship) of the underwater hull. With knowledge of the dome system volume, hull area enclosed,

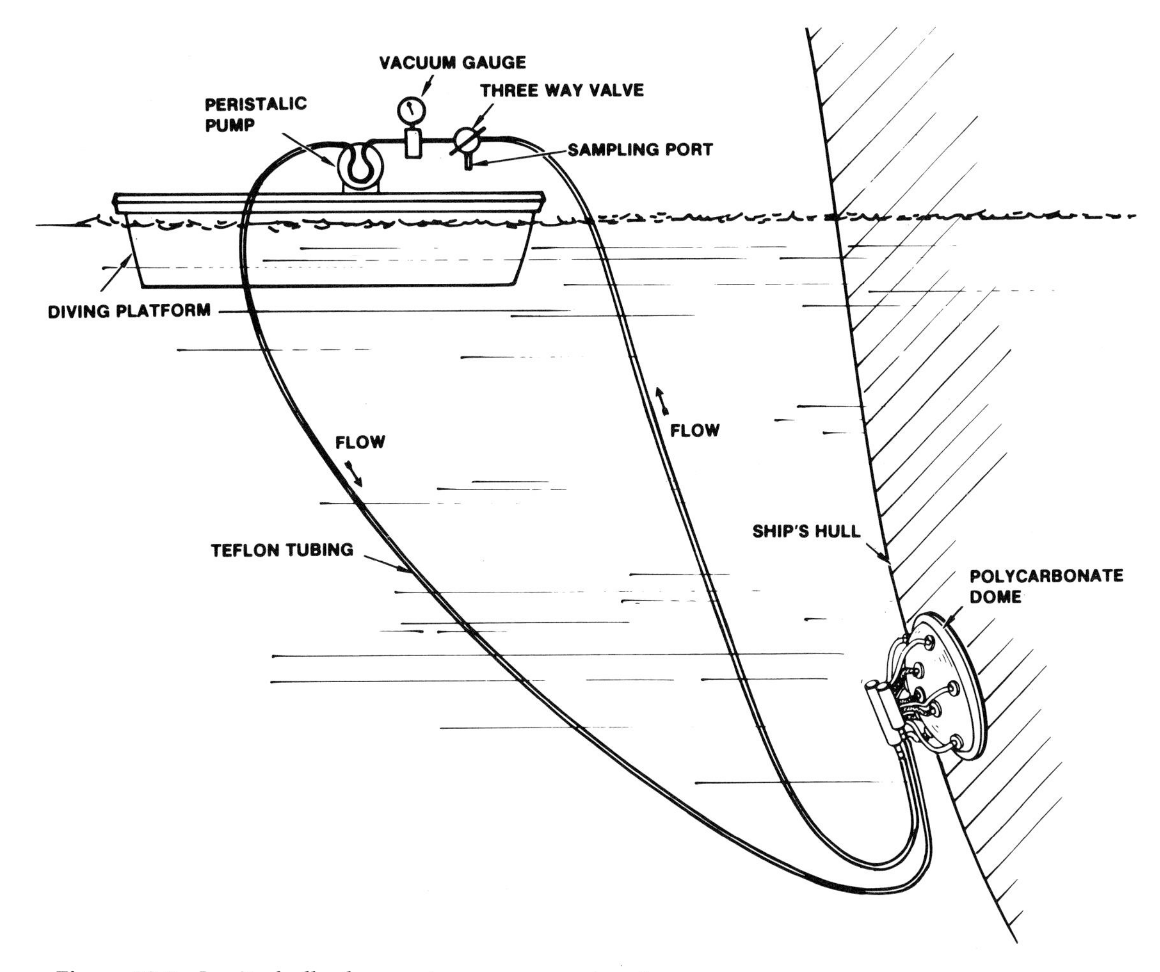

Figure 20.2 In-situ hull release rate measurement system.

duration of sample collection, and sample concentration, a TBT release rate can be calculated in micrograms per square centimeter per day from the slope of the regression curve of TBT concentration in the dome versus time in minutes. Usually three separate measurements were made for a release rate determination and the 95% confidence limits were calculated from the pooled data. This method has been described in greater detail by Lieberman *et al.*, (1985). Because this method measures the release of TBT from the hull of a ship under ambient conditions with the natural slime layer intact, it is more representative of the actual TBT flux from vessels than laboratory release rate measurements of painted panels under the standard procedure developed by the American Society for Testing and Materials (ASTM; US EPA, 1986).

20.2.3c. Near-hull TBT concentration-gradient study

Measurements of the TBT release from the antifouling coating applied to the *USS Badger* were collected on 14 April 1987 while the ship was moored in Southeast Loch. Concurrent with these measurements, seawater samples were collected along five transects extending

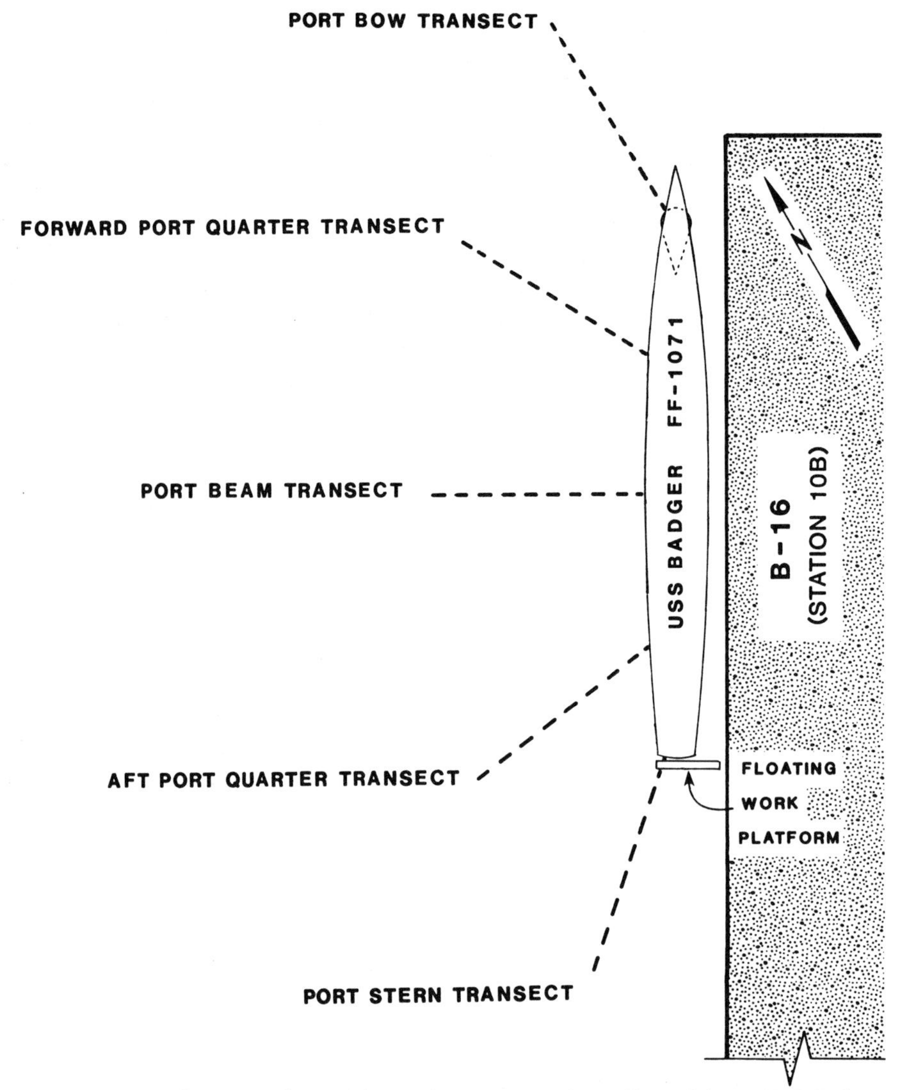

Figure 20.3 Transects of near-surface and near-bottom samples collected from *USS Badger* to evaluate butyltin concentration gradients in the vicinity of the hull.

away from the port side of the hull at distances of 0.5, 2.0, 5.0, 20, and 50 m. Wind and current were nil. Single samples were collected at 0.5 m below the surface and at 1 m above the bottom using the same apparatus and procedures employed during the Pearl Harbor monitoring survey series. Deep-water sample depths ranged from 12 to 13 m. The transects surveyed are depicted in Fig. 20.3. This gradient study was conducted to evaluate the size of the zone and magnitude of elevated TBT concentrations around a freshly painted vessel under low current conditions.

20.2.4. DRYDOCK MONITORING

20.2.4a. Painting operations

The *USS Badger* and *USS Brewton* were painted in the aft section of Drydock No. 2 in

February and August 1987, respectively. This section of the drydock is ~153 m long by 43 m wide, and when flooded the water depth is ~14 m. Longitudinal drains along each side of the drydock transport precipitation and infiltration water to a common pump well from which water is pumped into the south channel of the harbor. The *USS Davidson* was painted in September 1987 in Drydock No. 4, which is located on that part of the shipyard that faces the main channel of the harbor (Fig. 20.1). Drydock No. 4 is ~331 m long by 44 m wide. Longitudinal drains on each side of the drydock empty into individual sumps from which drydock water is discharged into the main channel of the harbor.

All three ships were painted with ABC-2, an organotin antifouling paint manufactured by DeVoe Marine Coatings. The paint contains 0.88% tin (2.2% TBT) and 33.6% copper (as cuprous oxide) by weight. Three coats of ~5 mils (0.13 mm) each were applied for a total of 0.4 mm mean dry film thickness.

Generally, the sequence of events leading to the undocking of the ship was as follows: application of antifouling paint (three coats at one coat per day); final cleanup of the drydock (4–8 d); and undocking of the ship and dewatering of the drydock (~1 d).

To evaluate the efficiency of painting and drydock cleanup techniques and to document the environmental release during the ship undocking, we monitored every aspect of the operation from painting through undocking. This involved measuring the paint overspray, measuring the discharge to the sewer system, documenting the cleanup procedures, and measuring the concentration of butyltins in the drydock water before and after undocking the ship.

Preparation, Drydock No. 2

Both the *USS Badger* and *USS Brewton* are 133.5 m long, have a 14.3 m beam, and have 2104 m^2 of painted underwater surface area. When in drydock, the ships rest on concrete blocks ~3 m above the drydock floor. The waterline of the ships was ~5 m below the top of the drydock wall.

Substantial efforts were made to contain and collect the paint overspray in the drydock. Half of the openings to the longitudinal drains in the drydock were covered completely and the other half were fitted with 6-mm mesh steel screens lined with cheese cloth to collect paint particles. Water from the drains was diverted to the sanitary sewer instead of being pumped to the harbor. The concrete blocks, staging, caisson walkways, painting baskets, ship brows, and ~50% of the drydock floor (that portion directly under the ship) were masked with Herculite,® a phenolformaldehyde-reinforced plastic cloth (Herculite Products, Inc., New York).

Preparation, Drydock No. 4

USS Davidson is 126.2 m long, and has a maximum beam of 13.5 m and an underwater surface area of 1879 m^2. Trenches in Drydock No. 4 were dammed upstream from the pump well, and water was pumped through an activated carbon filtering system and into the sanitary sewer. As in Drydock No. 2, all equipment in the drydock in the vicinity of the ship was masked with Herculite during painting.

Cleanup

Three coats of paint were applied over a 3-d period to each ship. When painting was complete, the Herculite masking was removed and cut into 1.2–1.5 m strips, rolled, and placed in 207-l drums for disposal. Then the entire drydock floor was vacuum-swept using industrial-sized manually operated vacuum sweepers. In only one 6 m × 6 m area during painting of the *USS Brewton* did overspray adhere to the drydock floor such that it required removal with wire brushes and solvents. The entire cleanup operation took 4–8 d and generated ~250 207-l drums of waste material.

The undocking of a ship from a graving drydock is achieved by flooding the drydock with harbor water, removing the caisson, floating the ship out of the drydock, replacing the caisson, and pumping the water out of the drydock. The process normally takes 4–6 h.

20.2.4b. Sampling procedures

Sampling in the drydock was designed to quantify the amount of butyltin from overspray during painting and release of butyltin during undocking of the ship. Samples were collected to quantify the amount of antifouling paint that fell to the drydock floor during the painting process (paint overspray), the amount that was discharged to the sanitary sewer system during painting and clean-up, and the amount released to the harbor while the ship was being undocked. Paint overspray samples were collected from polyethylene-wrapped 20-cm aluminum panels placed on the drydock floor. Drydock discharge samples were collected from the drainage sumps. Water samples were collected at three depths and four or six locations within the flooded drydock. Triplicate water samples were collected at two depths in the harbor at the locations shown in Fig. 20.1. Water samples were collected using a 1.5-l Kemmerer sampler with Teflon®-wetted surfaces. The samples were placed in polycarbonate bottles and frozen at −20°C within 4 h of collection. The samples remained frozen until they were thawed for analysis.

20.2.4c. Analytical procedures

Water samples

Water samples were analyzed using a simultaneous hybridization/extraction procedure followed by gas chromatographic separation with flame photometric detection (GC–FPD) (Matthias *et al.*, 1986). A detection limit for TBT of 2 ng l^{-1} was achieved with this method.

Filters

A number of water samples collected from the drydock during undocking of the ship were filtered, using 0.45-μm glass fiber filters to distinguish between dissolved and particulate forms of the TBT. Analysis of six DBT and TBT standards, three at 100 ng l^{-1} and three at 1000 ng l^{-1} filtered through these filters, showed that <5% of either butyltin species is adsorbed onto the filter. Within 4 h after filtering a sample the filters were frozen at −20°C.

In preparation for analysis the filters were first thawed and then Soxhlet-extracted for 6 h with methanol. Then 10 ml of the methanol was added to 200 ml of water and analyzed in the same manner as the water samples. The addition of the methanol does not affect the analysis at concentrations up to 17% (v/v). The detection limit for tributyltin was 5 ng.

Overspray samples

A 1.25-cm^2 piece of the polyethylene was cut from the panel. This sample was prepared for the analysis by dissolving the paint with dichloromethane. Toluene was added and the dichloromethane was allowed to evaporate. The samples were analyzed for total tin by graphite furnace atomic absorption spectrophotometry.

20.3. RESULTS OF CASE STUDY

20.3.1. SHIP HULL LOADING

20.3.1a. In-situ hull release rates

In-situ release rate measurements of TBT from the test ships in Pearl Harbor were conducted from February 1987 to March 1988. The dates of surveys and release rate data for each vessel are presented in Table 20.1. The

Table 20.1 In-situ TBT release rates from the hulls of US vessels painted with TBT-containing paint

Vessel	*Hull number*	*Paint type*	*Date*	*TBT release rate* ($\mu g\ cm^{-2}\ d^{-1}$)
USS Badger	FF-1071	ABC-2	3 Mar 1987	0.47 ± 0.20
			5 Mar 1987	0.54 ± 0.18
			14 Apr 1987	0.31 ± 0.14
			2 Sept 1987	0.28 ± 0.03
			15 Mar 1988	0.37 ± 0.05
USS Brewton	FF-1086	ABC-2	16 Mar 1988	0.10 ± 0.02
USS Davidson	FF-1045	ABC-2	17 Mar 1988	0.11 ± 0.01
USS Beaufort	ATS-2	IPC-254 Hisol	28 Feb 1987	0.86 ± 0.17
USS Omaha	SEN 692	F-170	6 May 1987	2.77 ± 0.27
USS Leftwich	DD-984	ICP-254 Hisol	—	Not measured (assumed same as ATS-2)

TBT release rates from the hull of the *USS Badger* were conducted five times during the 12-month period immediately after the undocking. A steady-state release rate appears to have been reached by 14 April 1987, 44 d after undocking, as the release rates from this time through 1 y are not significantly different. Consequently, the steady-state release rate was determined to be the mean of the three release rate calculations between 14 April 1987 and 15 March 1988 or 0.32 $\mu g\ cm^{-2}\ d^{-1}$. The *USS Brewton* and *USS Davidson* were surveyed ~6 months after undocking. At this time both of these ships had release rates of ~0.1 $\mu g\ cm^{-2}\ d^{-1}$, which were assumed to be at steady state since sufficient time was allowed for steady-state release rates to be achieved prior to measurement.

The amount of TBT released from the hull per day for each of the test ships is provided in Table 20.2. These values are calculated from the measured release rate and the wetted hull surface area. Two paint systems on the *USS Leftwich* (SPC-254) and *USS Omaha* (SPC-4) were not measured directly; however, previous release rate measurements with the same paint system were used to calculate loading factors. Release rates on the other vessels were measured directly. The F-170 paint system on the *USS Omaha* (SPC-4) is an 'old technology' free-association paint that is heavily loaded with a mixture of TBT fluoride and TBT oxide mixed into the resin. The very high release rate paint on the *USS Omaha* (SPC-4) (2.8 $\mu g\ cm^{-2}\ d^{-1}$) leached TBT at a rate that was at least an order of magnitude greater than the copolymer paints used on the test ships. Tributyltin loading from the *USS Omaha* represented ~60% of the total TBT loading of all six of the test ships (Table 20.2).

20.3.1b. Near-hull TBT gradient study

Water samples were collected at various distances from the hull of the *USS Badger* over a 2.5-h period on 14 April 1987, 44 d after painting. Five transects away from the hull were made as shown in Fig. 20.3 with 10 samples per transect. The analytical results for shallow (0.5 m) and deep (1 m off bottom) water are shown in Fig. 20.4a and b, respectively.

The mean TBT release rate of the antifouling hull paint at this time was 0.31 $\mu g\ cm^{-2}\ d^{-1}$. The relationship of TBT concentration as a function of distance from the hull of the *USS Badger* was demonstrated by least-squares linear regression to be very low for the deepwater samples. The mean of all the deepwater samples was 11.7 $ng\ l^{-1}$.

The surfacewater samples showed a significant correlation ($r^2 = 0.95$) of TBT concentra-

Table 20.2 Tributyltin mass loading from US Navy painted test vessels in Pearl Harbor based on underwater hull areas and measured release rates

Test ship	*Date painted*	*Wetted hull area* (m^2)	*Paint type*	*TBT release rate* ($\mu g\ cm^{-2}\ d^{-1}$)	*Calculated ship TBT load factor* ($g\ d^{-1}$)	*Total loading* (%)
USS Beaufort	Sept 1986	1 337	ICP HiSOL	0.86 ± 0.17	11.5	8.9
USS Leftwich	Dec 1985	3 321	ICP HiSOL	0.86	28.6	22.3
USS Davidson	Sept 1987	1 879	ABC-2	0.11 ± 0.01	2.1	1.6
USS Badger	Mar 1987	2 104	ABC-2	0.32 ± 0.05	6.7	5.6
USS Brewton	Aug 1987	2 104	ABC-2	0.10 ± 0.02	2.1	1.6
USS Omaha	Jan 1982	3 264	SPC-4/F-170	2.62[a]	77.5	60.3
Prediction from Navy use:	100% surface fleet painted					
Worst case (100% in harbor)		55 309		0.10	55	
Average case (67% in harbor)		37 057		0.10	37	37

[a]Composite release rate based on 1643 m^2 SPC-4 AF paint coverage (TBT release rate: 2.5 $\mu g\ cm^{-2}\ d^{-1}$) and 1342 m^2 F-170 AF paint coverage (TBT release rate: 2.77 ± 0.27 $\mu g\ cm^{-2}\ d^{-1}$).

tion with distance from the hull. A concentration gradient of 0.38 ng $l^{-1}\ m^{-1}$ was displayed for the surfacewater samples. Geometric regression analysis of the surfacewater data by the least-squares method resulted in a gradient equation of $f(x) = 33.0\ (x\ (-)\ 0.19)$ ng $l^{-1}\ m^{-1}$, $r^2 = 0.95$, with a standard error of the estimate of 0.086. These results suggest that an envelope of water around the ship with elevated TBT concentrations extended ~50 m from the vessel and that the maximum concentration within a few centimeters of the hull was ~42 ng l^{-1}, and the concentration at 50 m distance was ~15 ng l^{-1}.

20.3.1c. TBT mass loading from test ships

The amount of TBT (g d^{-1}) released from each test ship is given in Table 20.2. Tributyltin mass loading was calculated from the coating release rate and the wetted hull area of the ship. As shown in Table 20.2, there is a very large difference in TBT flux from the lowest release rate coating (i.e. *USS Brewton*, 0.10 μg $cm^{-2}\ d^{-1}$) and the highest release rate coating (i.e. *USS Omaha*, 2.62 $\mu g\ cm^{-2}\ d^{-1}$), ~26-fold. Because of the high release rate of the F-170 paint, related to both the high level of TBT and the nature of free-association paints, the *USS Omaha* represented 60% of the total TBT loading of the six test vessels in Pearl Harbor.

The total daily TBT loading into Southeast Loch during each of the butyltin surveys was calculated from summing the daily mass loading data of TBT for individual ships. The loading calculations were adjusted for the interval of port stay of the test ships within Southeast Loch for a 2-week interval prior to, and the duration of, each survey period. The daily calculated mass loading of TBT into Southeast Loch for each of the surveys and the corresponding mean TBT concentrations in the Southeast Loch water column are summarized in Table 20.3. A range of 2.1 to 115.5 g d^{-1} was calculated from the release rates and ship schedules. Least-square regression curves were plotted for mean regional TBT concentration versus the calculated total ship mass loading for each of the eight regions (Figs 20.5 and 20.6).

Least-squares linear analysis of the Southeast Loch data yielded a correlation coefficient of 0.87 (Fig. 20.5b). Similar testing to determine the relationship between the calculated tributyltin loading and the mean surfacewater TBT concentration data from

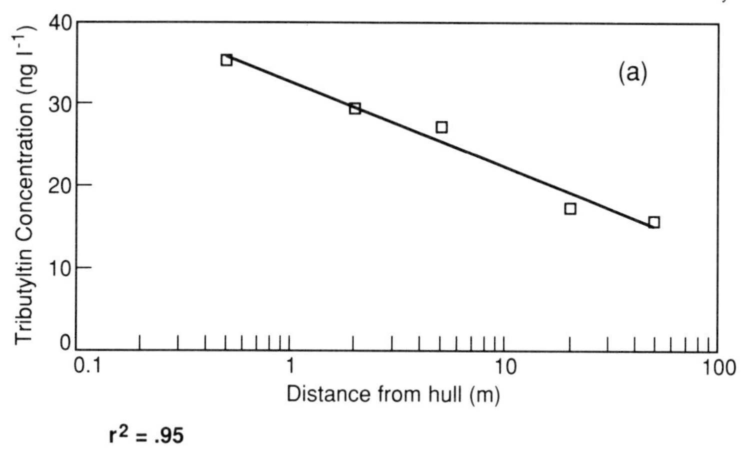

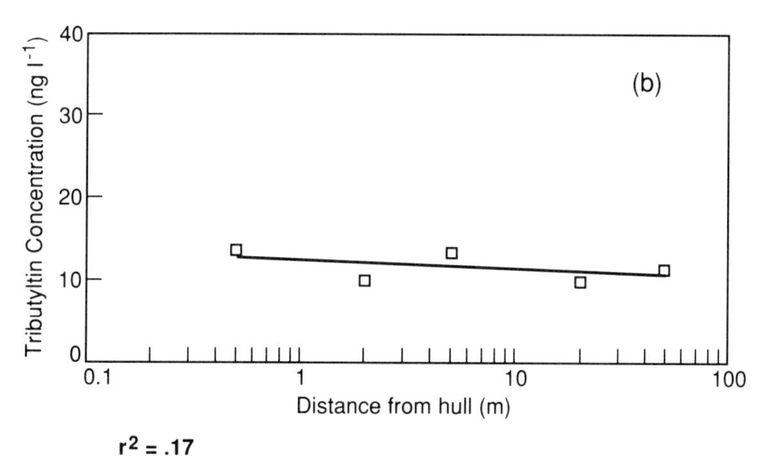

Figure 20.4 Mean concentration of tributyltin at various distances from the hull of the *USS Badger* at (a) 0.5 m below the surface, and (b) at 1 m off the bottom. Least-squares logarithmic regression.

other regions in Pearl Harbor produced correlation coefficient values of 0.83 for the main channels; 0.44, 0.65, and 0.48 for the West Loch, Middle Loch, and East Loch (at Waiau Shoal) areas, respectively; 0.75 for Drydock No. 2; 0.71 for Drydock No. 4; and 0.08 for Rainbow Marina (Figs 20.5 and 20.6). The linear relationships are relatively good in Southeast Loch, where the source ships are docked, and at the nearby channel and drydock areas (Fig. 20.5). Exponential least-squares regression analyses resulted in slightly higher correlation coefficients than linear regression for all but the Drydock No. 4 and Middle Loch regions due to the presence of elevated TBT in surface water samples particularly during the period where test ship loading was lowest (2.4 g d^{-1}). This suggests that other sources of TBT are present in Pearl Harbor that contribute to environmental TBT loading. Possible sources include aluminum-hulled craft, visiting naval vessels from other nations, and harbor-tour boats; and possibly re-release of TBT adsorbed onto the sediments. These TBT sources likely represent only a few grams per day input or generally

Table 20.3 Estimated mass loading of TBT into Southeast Loch from test ships

Survey period	*Test ships present*	*Total TBT mass loading (g d^{-1})[a]*	*Mean surface water concentration (ng l^{-1})*
6 Apr 1986–17 Apr 1986	DD-984, SSN-692	100.1	13.8 ± 10.4
9 Feb 1987–10 Feb 1987	AT-2, DD-984, SSN-692	107.5	14.6 ± 8.3
15 Apr 1987–16 Apr 1987	FF-1071, DD984, SSN-892	111.0	13.9 ± 5.3
12 May 1987–14 May 1987	FF-1071, ATS-2, DD-984, SSN-692	115.5	20.5 ± 11.7
28 Jul 1987	FF-1071, SSN-692	76.6	5.8 ± 2.5
15 Oct 1987–16 Oct 1987	FF-1071, FF-1086, FF-1045, SSN-692	38.1	2.5 ± 1.3
19 Jan 1988–20 Jan 1988	FF-1071, FF-1086, FF-1045, DD-984, SSN-692	2.6	9.2 ± 3.6
12 Oct 1988–13 Oct 1988	FF-1045	2.1	3.4 ± 1.4
16 Jan 1989–17 Jan 1989	FF-1071, FF-1086, SSN-692	76.7	7.0 ± 4.7

[a]Load estimates are adjusted for percentage of duration of test (d) ship assignment at any Southeast Loch berth for 2 weeks prior to, and the duration of, each survey period.

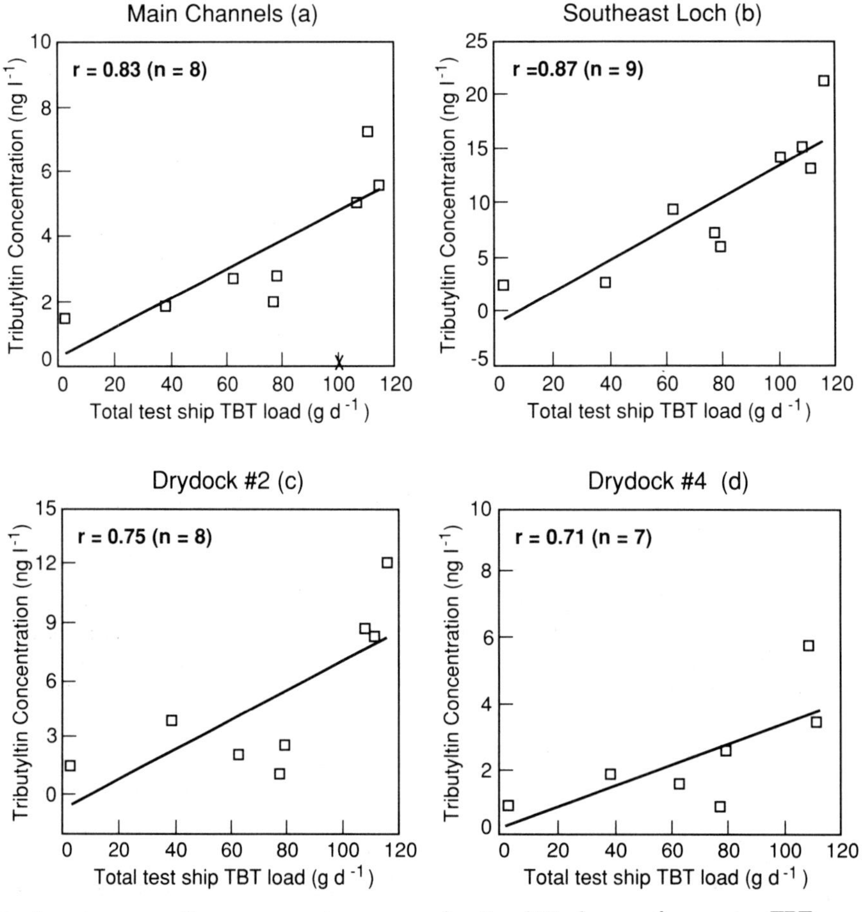

Figure 20.5 Least-squares linear regression curves for Pearl Harbor surface water TBT concentrations versus total TBT loading from test ships.

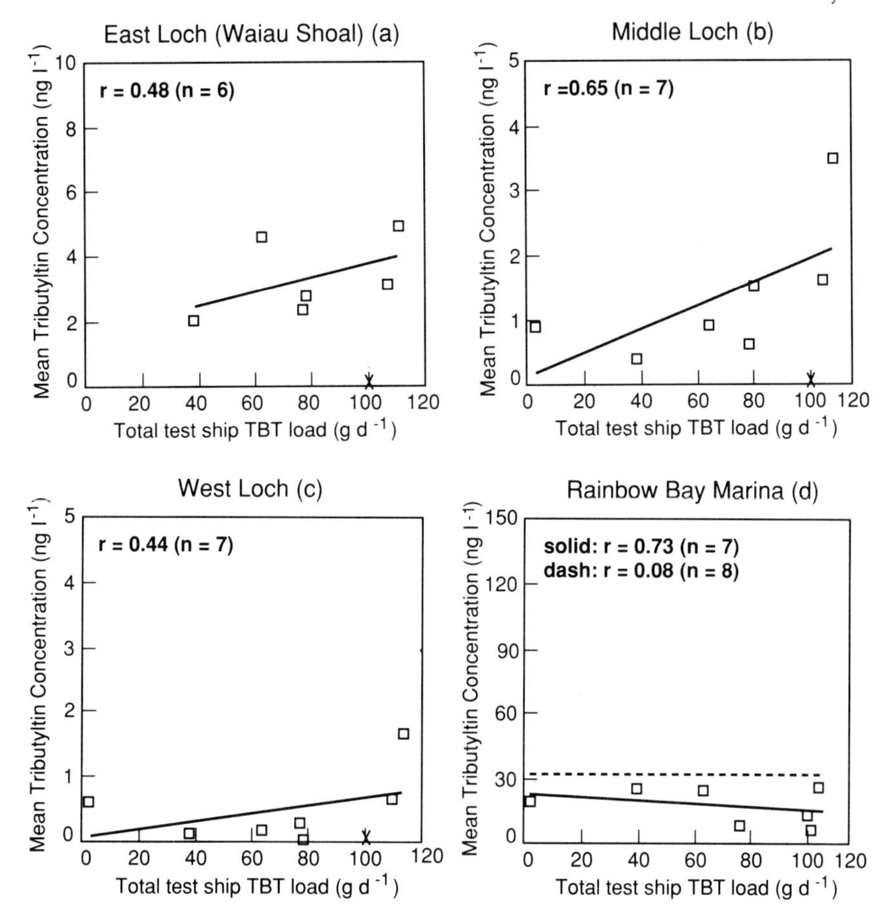

Figure 20.6 Least-squares linear regression curves for Pearl Harbor surface water TBT concentrations versus total TBT loading from test ships. Rainbow Marina only: July 1987 data (solid line = data included in analysis, broken line = excluded data).

<10% of the defined sources. Correlation coefficient values are reduced in the outlying areas of East, Middle, and West Lochs (Fig. 20.6) where circulation is relatively independent of Southeast Loch, and intermittent sources may not have been noted. There was no linear relationship at the Rainbow Bay Marina where local TBT sources from the small yacht harbor masked any input from the TBT test hulls.

Table 20.4 Amount of TBT applied versus amount of overspray

Ship	*TBT in paint applied (kg)*	*TBT in paint overspray (kg)*	*Overspray (%)*
USS Badger	68	5.4	8
USS Brewton	65	3.3	5
USS Davidson	58	1.0	2

20.3.2. DRYDOCK RELEASES

20.3.2a. Paint overspray

The mass of TBT applied to the ship using paint spray systems and the amount of overspray that was deposited on the drydock floor are given in Table 20.4. The percentages of paint overspray from *USS Badger* and *USS*

Brewton are within the 5–9% range previously reported by Adema and Schatzberg (1984). However, the percentage of overspray during the painting of the *USS Davidson* was much lower, ~2%. This low percentage of overspray was observed on a previous occasion when the paint spraying operation was conducted very carefully. The improved paint transfer in this case was probably due to the lower wind velocity during painting and the improved painting techniques gained through experience in working within the constraints of the personnel protection equipment.

The distribution of solvent-extractable organotin on the drydock floor is shown in Fig. 20.7a, b, and c from the three test paintings. This distribution is typical of that seen in other paint applications in which 90–95% of the overspray falls directly beneath the profile of the hull. The patterns seen in these figures also indicate that only very small amounts of tributyltin are transported outside of the drydock.

Analysis of the paint overspray from the *USS Badger* indicated a DBT:TBT ratio of 0.23. The liquid paint applied to the *USS Brewton* and *USS Davidson* revealed a DBT:TBT ratio of 0.74. Both paints contained the same amount of TBT, ~2%; however, the paint used on the *USS Brewton* and *USS Davidson* contained approximately three times as much DBT. The dibutyltin is usually considered a contaminant, but this paint contained nearly as much DBT as TBT.

20.3.2b. Drydock discharge during painting and cleanup

Table 20.5 shows the concentration of butyltin in the drydock discharge to the sanitary sewer before, during, and after painting each ship. The greater discharge of TBT during the painting of the *USS Badger* was probably caused by the moderate to heavy rainfall during the painting and cleanup process. Rain would cause the butyltin to be released from the paint and solubilized into the water that was discharged to the sewage treatment plant.

The drydock discharge water was diluted into the daily sewage treatment plant flow, which varied from 5.6 million gallons (21 200 m^3) per day during painting of *USS Badger*, 5.1 million gallons (19 300 m^3) per day during painting of the *USS Brewton*, and 5.3 million gallons (20 100 m^3) per day during painting of the *USS Davidson*. As a result, the average calculated TBT concentrations entering the sewage treatment plant were ~80, 7, and 1.5 ng l^{-1}, respectively. Previous studies have shown that these concentrations of TBT will not have a measurable effect upon the operation of the sewage treatment plant and that the butyltin compounds are absorbed onto particulate material and very little is discharged in the effluent (Avendt and Avendt, 1982; see Fent, Chapter 28, for further discussion of TBT and sewage treatment discharges).

20.3.2c. Drydock release during ship undocking

Water that is in the drydock when the ship is undocked will contain butyltins from paint overspray remaining from the cleanup and butyltins released from the freshly painted underwater hull of the ship. The contribution of butyltin to the flooded drydock water by the underwater hull of *USS Badger* was estimated by calculating the initial in-situ release rate (0.5 $\mu g\ cm^{-2}\ d^{-1}$) times 0.125 d, the amount of time required for undocking. This estimate yielded ~1.3 g of TBT, resulting in a concentration of 14.4 ng l^{-1} in the 90 000 m^3 of water in the drydock. Table 20.6 shows the initial concentrations of butyltins in the drydock prior to ship undocking. The estimated TBT contribution from the hulls of *USS Brewton* and *USS Davidson* would be about one-third that of the *USS Badger* or ~0.4 g, resulting in ~4.8 ng l^{-1} in the drydocks.

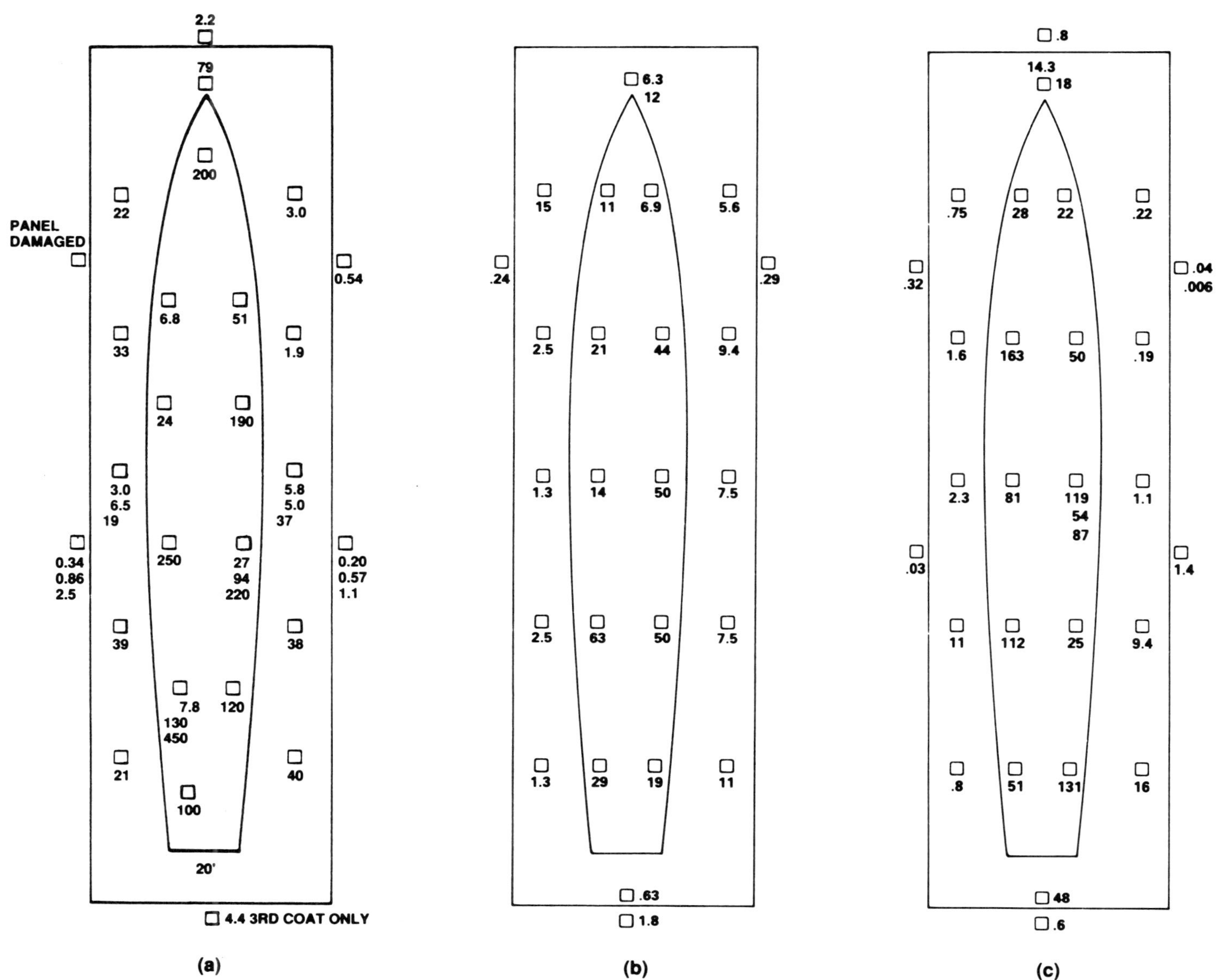

Figure 20.7 Distribution of dichloromethane-extractable organotin (μg cm^{-2} TBT) on collection panels on drydock floor from painting of (a) *USS Badger*, (b) *USS Brewton*, and (c) *USS Davidson*.

Table 20.5 Butyltin concentration in drydock water discharged to the sewage treatment plant (ng l^{-1})

	USS Badger		USS Brewton		USS Davidson	
Drydock operation	*DBT*	*TBT*	*DBT*	*TBT*	*DBT*	*TBT*
Prior to painting	12	17	97	9	2	2
Painting	5 200	7 800	625	380	2 100	1 850
Cleanup	3 300	9 900	600	540	9 300	2 380
After cleanup	2 100	500	—[a]	—	—	—
Volume discharged (l d^{-1})	2.26×10^5		2.26×10^5		1.13×10^4	
Total mass TBT discharged to the sewer (g)	20.5		0.9		0.2	

[a]Dash indicates not determined.

Table 20.6 Butyltin concentration of drydock water before the caisson was removed

	Butyltin concentration (ng l^{-1})	
Ship	*DBT*	*TBT*
USS Badger	390	180
USS Brewton	148	5
USS Davidson	47	3

The in-situ release rate measurement on the hull of a vessel under ambient conditions tends to yield values substantially lower than the EPA/ASTM laboratory method (Schatzberg, Chapter 19). It is our judgement that the in-situ method represents a more realistic estimate of the actual flux of TBT into the aquatic environment than does the laboratory method. The lower in-situ measurements are consistent with the low TBT concentration measured in the flooded drydock prior to releasing the test ship.

Samples were taken at three depths and four or six locations within the drydock. These samples, as well as concurrent fluorescent dye studies, showed that the butyltin was well mixed in Drydock No. 2; however, Drydock No. 4 (*USS Davidson*) showed some degree of stratification from top to bottom. The stratification was apparently due to a difference in the way the water enters the drydock during flooding.

Water in the drydock during undocking of the *USS Badger* contained an average of 180 ng l^{-1} of TBT. Of the 180 ng l^{-1}, 17 ng l^{-1} was contributed as background by the harbor water, and ~14 ng l^{-1} was contributed by the initial release of TBT from the ship hull. Therefore, ~149 ng l^{-1} or 13.4 g of TBT (0.2%) remained from the paint overspray, which represented a cleanup efficiency of 99.8%.

Those few drydock water samples that contained paint particles were not included in the calculation of average concentrations. Water in the drydock during undocking of the *USS Brewton* and *USS Davidson* contained very low concentrations of butyltins, about the same as the calculated contribution from the underwater hull of the ships, suggesting nearly 100% cleanup efficiency.

The low paint release rate coupled with a more efficient cleanup of the drydock, assisted in part by lack of rain, resulted in a very low release of TBT from the drydock into the adjacent harbor. A total of ~15 g of TBT was released from the drydock when the *USS Badger* was undocked.

In all three tests, water samples taken from the flooded dry docks prior to ship discharge contained high concentrations of DBT relative to TBT. This is probably because DBT is not chemically bound to a polymer as is TBT, is more water soluble, and, therefore, is more quickly released from the coating. The correspondingly higher DBT:TBT ratio in the samples from *USS Brewton* and *USS Davidson* as compared with the samples from the *USS Badger* is consistent with this assumption.

Table 20.7 Butyltin concentration in filtered drydock samples from *USS Badger* (ng l^{-1})

	Filter		*Water*		*Total*	
	DBT	*TBT*	*DBT*	*TBT*	*DBT*	*TBT*
Prior to ship undocking	46	77	243	87	315	153
After ship undocking	109	97	22	31	131	128

Table 20.7 shows a comparison of filtered and unfiltered samples of drydock water from *USS Badger*. Filtered samples from before undocking the ship show that ~16% of the DBT and 47% of the TBT is associated with suspended solids, even when the water samples did not contain paint particles. Filtered samples collected 1 d after the *USS Badger* was removed from drydock showed that 83% of the DBT and 76% of the TBT was associated with particulate material in the water samples.

20.3.2d. Harbor TBT concentrations during undocking of test ships

Table 20.8 provides the mean regional TBT concentrations in the water column from both the near-field and far-field sampling efforts as well as additional monitoring for some of the pre-undocking data. The post-undocking column represents the average of samples collected 1, 2, and 3 d after the undocking event.

In a few cases, samples collected at the drydock showed elevated TBT levels during the period directly following the undocking process. Other regions in the harbor showed little, if any, increases in concentration. This finding is consistent with the low concentration of TBT in the drydock water.

20.4. DISCUSSION

The study conducted in Pearl Harbor has found that with a combination of drydock cleanup measures, including covering portions of the drydock floor with plastic, masking various exposed areas, and careful vacuuming procedures, TBT discharges to the environment can be limited to small (<15-g) quantities. Previous monitoring studies have suggested that drydock discharges of TBT can result in substantial increases in concentration of the biocide in surrounding waters. For example, in the Elizabeth River, Virginia, average TBT concentrations as high as 87 ng l^{-1} over an 8-km section of the river were measured, suggesting periodic kilogram releases of TBT from commercial drydocks in the vicinity (Seligman *et al.*, 1987; Huggett *et al.*, Chapter 24). High TBT concentrations in the water column in Honolulu Harbor near a commercial drydock (>100 ng l^{-1}) also suggest painting activities as a significant source (Grovhoug *et al.*, Chapter 25). Table 20.9 summarizes TBT levels in the drydocks and discharges during and after the test painting process. The largest residual TBT discharge observed, after the painting of the *USS Badger*, was 149 μg l^{-1} (a total mass of 13.4 g), which did not result in measurable increases in TBT concentrations in the harbor except in the close proximity of the drydock. In all cases, the 2–8% overspray (1–5.4 kg) was cleaned up with an efficiency of 98.8% or greater. Discharges from the drydock after the painting of the *USS Brewton* and *USS Davidson* were approximately equal to the TBT estimated to have leached from the hull (~2 g d^{-1}), suggesting nearly 100% removal of the residue.

It is apparent from this study that leaching from the underwater hulls as summarized in Table 20.2 was by far the largest TBT loading factor during the Pearl Harbor case study. The release of TBT varied by as much as a factor of 26 between the newer developed ablative coating (ABC-2) and the older technology free-association paint, which was a simple mixture of high-percentage tributyltin fluoride and tributyltin oxide mixed into the formulation (F-170). One vessel (the *USS Omaha*) that was painted in 1982 with a

Table 20.8 Regional mean water column tributyltin concentrations during test-ship undocking-phase surveys[a]

Test ships sample region	*Layer*	USS Badger		USS Brewton		USS Davidson	
		Pre-[b]	*Post-*	*Pre-*	*Post-*	*Pre-*	*Post-*
Drydock[c]	S[d]	11.0 ± 4.9	34.0 ± 29.0	2.6 ± 0.2	6.0	2.6 ± 0.2	73
	D	6.5 ± 2.1	6.2 ± 1.8	3.9 ± 1.3	<2	3.9 ± 1.3	5.0
Southeast Loch	S	22.0 ± 23.0	13.0 ± 5.9	5.8 ± 2.4	−0.0	5.8 ± 2.4	—[e]
	D	7.5 ± 6.4	8.5 ± 7.2	2.4 ± 0.6	4.0	2.4 ± 0.6	—
South Channel	S	6.3 ± 5.1	6.9 ± 4.5	<2	<2	<2	4.0
	D	13.0 ± 8.5	2.9 ± 1.2	3.5 ± 0.7	6.7 ± 4.2	4.0	—
North Channel	S	4.0 ± 2.8	2.4	5.0 ± 2.5	1.3 ± 0.6	5.0 ± 2.5	0.4 ± 0.3
	D	17.0 ± 9.9	1.1	2.0 ± 0.8	1.2 ± 0.8	2.0 ± 0.8	1.2 ± 1.2
Entrance Channel	S	9.0 ± 8.5	3.9 ± 2.1	2.7 ± 1.0	0.9 ± 0.9	2.7 ± 1.0	0.8 ± 1.7
	D	6.5 ± 2.1	1.7 ± 0.8	0.8 ± 0.2	1.0 ± 0.6	0.8 ± 0.2	0.5 ± 0.5
West Loch	S	—	2.1 ± 1.9	0.0 ± 0.0	0.3 ± 0.9	<0.5	<0.5
	D	—	0.8 ± 0.2	0.0 ± 0.0	0.7 ± 0.4	<0.5	0.2 ± 0.2
Middle Loch	S	—	2.2	1.5 ± 0.2	1.4 ± 1.3	1.5 ± 0.2	0.2 ± 0.3
	D	—	3.2	0.9 ± 0.6	0.7 ± 0.5	0.9 ± 0.6	0.4 ± 0.4
Waiau Shoal	S	—	—	2.7	1.3 ± 1.1	2.7	—
	D	—	—	2.7	1.3 ± 0.9	2.7	—

[a]Water column levels in ng l^{-1} TBTCl (mean ± SD).
[b]Pre-/post-undocking.
[c]Drydock No. 2, except for *USS Davidson* in Drydock No. 4.
[d]S = surface (0.5 m); D = depth (1 m off bottom).
[e]Dash indicates no data.

Table 20.9 Pearl Harbor case study – summary of drydock TBT monitoring and discharge levels

Ship	USS Badger (*FF1071*)	USS Brewton (*FF1086*)	USS Davidson (*FF1045*)
Painted	Feb 1987	Aug 1987	Sept 1987
Total TBT applied	68 kg	65 kg	58 kg
Overspray	5.4 kg (8%)	3.3 kg (5%)	1.05 kg (2%)
Discharged to sewage treatment plant during painting	20.5 g	0.9 g	0.2 g
TBT concentration in drydock effluent from overspray	149 ng l^{-1}	<5 ng l^{-1}	<3 ng l^{-1}
Discharged to harbor	11.7 g	<5 g	<5 g
Cleanup efficiency	99.8%	>99.8%	>99.8%

combination of F-170 and an early copolymer paint accounted for >60% of the TBT loading from the six test ships. This demonstrates the very large differences in release rate among paint types and provides support for the legislative and regulatory approach in the United States of only allowing use of TBT paints with demonstrated low release rate characteristics.

One of the objectives of this study was to predict loading and harbor concentrations from full TBT-paint utilization by the US Navy fleet in Pearl Harbor. If it is assumed that the paint with the lowest release rate yet measured (0.10 μg cm^{-2} d^{-1} TBT), the 5.5 × 10^3 m^3 surface area of the ships home-ported in Pearl Harbor would have a mass loading factor of ~55 g d^{-1} TBT or about half the flux that was calculated from the six test ships described in this study (Table 20.2). On average, at least one-third of the ships are absent from the harbor, yielding a mean predicted TBT flux loading factor of 36 g d^{-1} from full fleet use. This demonstrates the significance of the lower paint release rate, which is likely the most important factor in controlling the environmental loading of TBT and therefore the principal factor in determining the steady-state TBT concentration in many harbors and estuaries.

The nearly linear covariance between TBT loading and mean regional concentrations over the nine harbor survey periods (Figs 20.5 and 20.6) provides an empirical model from which future TBT concentrations with defined loading parameters can be predicted. By assuming that the entire Pearl Harbor surface fleet was painted with antifouling coatings that released TBT at 0.1 $\mu g\ cm^{-2}\ d^{-1}$, the average steady-state TBT concentration (calculated from the regression curves) in Southeast Loch would be ~4 ng l^{-1}, the main channels and East Loch ~2 ng l^{-1}, and other regions 1 ng l^{-1} or less given an average loading rate of 37 g d^{-1} (Table 20.2).

From the perspective of assessing environmental risks from the use of TBT coatings by naval vessels in Pearl Harbor, it is probable that the steady-state concentrations, including the intermittent inputs from periodic drydock discharges, would remain substantially below the 10 ng l^{-1} water quality criteria proposed by the US EPA as a safeguard to marine organisms, similar to the results obtained by Harris *et al.* (1991) for the Tamar Estuary. This conclusion is based on the assumption that the lowest available release rate coatings (0.2 $\mu g\ cm^{-2}\ d^{-1}$) would be used and that environmental management practices for drydocks would be implemented that would achieve cleanup efficiencies approximately equivalent to those measured during the test program.

It should be noted that the US Navy has recently decided not to use TBT antifouling coatings on its ships. This chapter and the following conclusions and recommendations should continue to be a useful guide for evaluating and reducing TBT loading from continued commercial use of TBT coatings.

20.5. SUMMARY, CONCLUSIONS, AND RECOMMENDATIONS

20.5.1. SUMMARY

The environmental loading characteristics from use of TBT paints on ships were investigated in Pearl Harbor during 1987 and 1988. Tributyltin contributions to Pearl Harbor from drydock discharges during and after painting and direct release from painted hulls were measured. Drydock cleanup procedures including masking portions of the drydock floor were successful in removing at least 99.8% of the residual paint overspray so that TBT discharges to the harbor remained <15 g and ranged from <3 to 149 ng l^{-1} in the effluent.

Tributyltin loading from underwater hull coatings was measured from six ships using a NOSC-designed device for measuring in-situ release rates. Release rates varied from 0.1 to 2.8 $\mu g\ cm^{-2}\ d^{-1}$ TBT, documenting the large paint-dependent differences in TBT loading. Actual loading from the test ships varied from ~2 to 115 g d^{-1} and generally covaried linearly with regional mean TBT concentrations in the water column. Empirical predictions were made that suggested that potential future use of TBT paint by the US Navy would result in water column levels below the 10 ng l^{-1} existing regulatory limits.

20.5.2. CONCLUSIONS

1. Tributyltin discharges to receiving waters from painting operations in a graving drydock are controllable to low levels by using effective cleanup measures.
2. Release rates of various TBT paints measured in situ on ship hulls are highly variable (greater than a factor of 20)

suggesting the efficacy of using leach rate limitations as a means of regulatorycontrol of TBT loading into receiving waters.

3. TBT concentrations in the water column can be predicted from hull TBT loading factors using an empirically derived model.
4. Use of low-release-rate TBT paints (<0.2 $\mu g\ cm^{-2}\ d^{-1}$) by the US Navy fleet in Pearl Harbor and effective drydock cleanup procedures were predicted to result in TBT levels <10 ng l^{-1} near berthing areas and <5 ng l^{-1} in outlying areas.

20.5.3. RECOMMENDATIONS

1. Investigate the effects of hull sandblasting and other maintenance procedures on TBT release to receiving waters.
2. Continue in-situ hull release studies to evaluate TBT loading characteristics on new coating systems.
3. Define lowest effective release rates to minimize environmental loading of TBT.
4. Optimize numerical hydrodynamic and fate models to use in prediction of TBT environmental levels given various loading scenarios.
5. Optimize drydock cleanup, water treatment, and disposal procedures within funding limitations and water quality requirements.

ACKNOWLEDGMENTS

The authors extend their appreciation of the many contributions to this study including field sampling, sample preparation and chemical analysis, and data reduction and analysis made by W. Thomas, G. Smith, and S. Mangum of Naval Surface Warfare Center, Carderrock Division; S. Lieberman, S. Clavell, and R. Henderson of Naval Command, Control and Ocean Surveillance Center, RDT&E Division; and B. Davidson, D. Bower, S. Frank, and J. Groves of Computer Sciences Corporation. We thank Linnell Kunavich for typing the manuscript and her constant untiring support. We also thank the various members of the Pearl Harbor Naval Shipyard, Naval Station, Rainbow Bay Marina, and the officers and crew of *USS Badger*, *USS Beaufort*, *USS Brewton*, *USS Davidson*, *USS Leftwich*, and *USS Omaha* for their assistance.

This work was co-sponsored by the Office of Chief of Naval Research under the Energy Research and Development Program and the US Naval Sea Systems Command, Shipboard Pollution Abatement Program.

Opinions expressed in this chapter are those of the authors and do not necessarily reflect US government policy.

REFERENCES

Adema, C.M. and P. Schatzberg. 1984. Organotin antifouling paints and the environment – drydock phase. *Naval Eng. J.*, **96**, 209–217.

Alzieu, C. 1991. Environmental problems caused by TBT in France: assessment, regulations, prospects. *Mar. Environ. Res.*, **32**, 7–17.

Alzieu, C., P. Michel, I. Tolosa, E. Bacci, L.D. Mee and J.W. Readman. 1991. Organotin compounds in the Mediterranean: a continuing cause for concern. *Mar. Environ. Res.*, **32**, 261–270.

Avendt, R.J. and J.B. Avendt. 1982. Performance and Stability of Municipal Activated Sludge Facilities Treating Organotin Contaminated Wastewater. David Taylor Naval Ship Research and Development Center. Report SME-CR-29-82, David Taylor Research Center, Annapolis, Maryland, 105 pp.

Grovhoug, J.G., P.F. Seligman, G. Vafa and R.L. Fransham. 1986. Baseline measurements of butyltin in U.S. harbors and estuaries. In: Proceedings of the Oceans '86 Organotin Symposium, Vol. 4, Washington DC, 23–25 September 1986, IEEE, New York, pp. 1283–1288.

Grovhoug, J.G., R.L. Fransham and P.F. Seligman. 1987. Butyltin Concentrations in Selected U.S. Harbor Systems. Naval Ocean Systems Center Technical Report No. 1155, Naval Ocean

Systems Center, San Diego, CA 92152–5000, 66 pp.

Grovhoug, J.G., P.F. Seligman, R.L. Fransham, S.Y. Cola, M.O. Stallard, P.M. Stang and A.O. Valkirs. 1989. Measurement of Butyltin Concentrations in Pearl Harbor, Hawaii, April 1986–January 1988. Pearl Harbor Case Study. Naval Ocean Systems Center Technical Report No. 1293, Naval Ocean Systems Center, San Diego, CA 92152–5000, 40 pp.

Hall, L.W., Jr, M.J. Lenkevich, W.S. Hall, A.E. Pinkney and S.J. Bushong. 1987. Evaluation of butyltin compounds in Maryland waters of Chesapeake Bay. *Mar. Pollut. Bull.*, **18**, 78–83.

Harris, J.R.W., C.C. Hamlin and A.R.D. Stebbing. 1991. A simulation study of the effectiveness of legislation and improved dockyard practice in reducing TBT concentrations in the Tamar Estuary. *Mar. Environ. Res.*, **32**, 279–292.

Lieberman, S.H., V. Homer and P.F. Seligman. 1985. In-situ Determination of Butyltin Release Rates from Antifouling Coatings on Navy Test Ships. Naval Ocean Systems Center Technical Report No. 1027, Naval Ocean Systems Center, San Diego, CA 92152–5000, 17 pp.

Matthias, C.L., J.M. Bellama, G.J. Olson and F.E. Brinckman. 1986. Comprehensive method for determination of aquatic butyltin and butylmethyltin species at ultratrace levels using simultaneous hydridization/extraction with gas chromatography–flame photometric detection. *Environ. Sci. Technol.*, **20**, 609–615.

Ritsema, R., R.W.P.M. Laane and O.F.X. Donard. 1991. Butyltins in marine waters of the Netherlands in 1988 and 1989. *Mar. Environ. Res.*, **32**, 243–260.

Seligman, P.F., J.G. Grovhoug and K.E. Richter. 1986. Measurement of butyltins in San Diego Bay, CA: a monitoring Strategy. In: Proceedings of the Oceans '86 Organotin Symposium, Vol. 4, Washington, DC, 23–25 September 1986, IEEE, New York, pp. 1289–1296.

Seligman, P.F., C.M. Adema, P.M. Stang, A.O. Valkirs and J.G. Grovhoug. 1987. Monitoring and prediction of tributyltin in the Elizabeth River and Hampton Roads, Virginia. In: Proceedings of the Oceans '87 Organotin Symposium, Vol. 3, Halifax, Nova Scotia, 28 September–1 October 1987, IEEE New York, pp. 1357–1363.

Seligman, P.F., J.G. Grovhoug, A.O. Valkirs, P.M. Stang, R. Fransham, M.O. Stallard, B. Davidson and R.F. Lee. 1989. Distribution and fate of tributyltin in the United States marine environment. *Appl. Organometal. Chem.* **3**, 31–47.

Stallard, M.O., S.Y. Cola and C.A. Dooley. 1989. Optimization of butyltin measurements for seawater, tissue, and marine sediment samples. *Appl. Organometal. Chem.*, **3**, 105–114.

US Department of Commerce. 1986a. *United States Coast Pilot: California, Oregon, Washington, and Hawaii*, 22nd edn. National Oceanic and Atmospheric Administration, National Ocean Service, Washington, DC, 421 pp.

US Department of Commerce. 1986b. *Tidal Current Tables 1987 – Pacific Coast of North America and Asia*. National Oceanic and Atmospheric Administration, National Ocean Service, Washington, DC, September 1986, 279 pp.

US Environmental Protection Agency (EPA). 1986. Data Call-in Notice for Data on Tributyltins Used in Paint Antifoulants. Office of Pesticides and Toxic Substances, US EPA, Washington DC, July 1986, 25 pp.

US EPA. 1987. Tributyltin Technical Support Document, Position Document 2/3. Office of Pesticides Programs, Office of Pesticides and Toxic Substances, US EPA, Washington, DC, October 1987, 123 pp.

US EPA. 1988. Tributyltin antifoulants; notice of intent to cancel; denial of applications for registration; partial conclusion of special review. *Fed. Reg.*, **53** (192), 39022–39041.

US Naval Sea Systems Command. 1984. Environmental assessment for fleetwide use of organotin antifouling paint. US Navy, Naval Sea Systems Command, Code 56YP, Washington, DC, 128 pp.

US Naval Sea Systems Command. 1985. Fleetwide implementation of organotin antifouling hull paints: interim finding of no significant impact. *Fed. Reg.* **50** (120), 25748–25750.

Valkirs, A.O., P.F. Seligman, P.M. Stang, V. Homer, S.H. Lieberman, G. Vafa and C.A. Dooley. 1986. Measurement of butyltin compounds in San Diego Bay. *Mar. Pollut. Bull.*, **17**, 319–324.

Valkirs, A.O., B. Davidson, L.L. Kear, R.L. Fransham, J.G. Grovhoug and P.F. Seligman. 1991. Long-term monitoring of tributyltin in San Diego Bay, California. *Mar. Environ. Res.*, **32**, 151–168.

Wade, T.L., B. Garcia-Romero and J.M. Brooks. 1991. Oysters as biomonitors of butyltins in the Gulf of Mexico. *Mar. Environ. Res.*, **32**, 233–242.

PERSISTENCE AND FATE OF TRIBUTYLTIN IN AQUATIC ECOSYSTEMS

21

P.F. Seligman,[1] R.J. Maguire,[2] R.F. Lee,[3] K.R. Hinga,[4] A.O. Valkirs[5] and P.M. Stang[6]

[1]*Naval Command, Control and Ocean Surveillance Center, RDT&E Division, Environmental Sciences (Code 52), San Diego, California 92152-5000, USA*
[2]*National Water Research Institute, Department of Environment, Canada Centre for Inland Waters, Burlington, Ontario L7R 4A6, Canada*
[3]*Skidaway Institute of Oceanography, PO Box 13687, Savannah, Georgia 31416, USA*
[4]*Graduate School of Oceanography, University of Rhode Island, Narragansett, Rhode Island 02882, USA*
[5]*Computer Sciences Corporation, 4045 Hancock Street, San Diego, California 92110-5164, USA*
[6]*PRC Environmental, 4065 Hancock Street, San Diego, California 92110, USA*

Organotin. Edited by M.A. Champ and P.F. Seligman. Published in 1996 by Chapman & Hall, London. ISBN 0 412 58240 6

ABSTRACT

Studies into the fate and persistence of tributyltin (TBT) in marine, estuarine, and fresh water environments have been conducted in diverse geographic regions and under various experimental conditions to determine rates of degradation and loss. Microbial degradation in water was found to be the most important process limiting the persistence of TBT in aquatic environments. Photolysis and chemical degradation were not significant in the degradation of TBT. Degradation studies were conducted using unfiltered seawater and fresh water incubated under natural conditions with sterilized controls from several geographic regions in the United States and Canada. Degradation half-lives were found to be in the range of 4–19 d in seawater and from a few weeks to several months in Canadian fresh water. The principal degradation product of TBT was dibutyltin (DBT) with lesser amounts of monobutyltin (MBT) formed.

The behavior of TBT and its degradation products was evaluated by introducing radiolabeled TBT into a mesocosm (13-m^3 enclosure) to simulate a marine ecosystem including benthic sediments. Tributyltin disappeared from the water in the mesocosm initially at 0.20 d^{-1} (20%) and slowed to ~0.10 d^{-1} after 15 d. Degradation accounted for the greatest loss of TBT, with transport to the sediment and possibly the atmosphere (volatilization) accounting for the other losses. Degradation proceeded through the process of debutylation to DBT and to MBT with substantial formation of MBT directly.

Degradation of TBT in sediments appears to be associated with two independent processes: (1) a rapid abiotic chemical degradation of dissolved TBT has been identified in fine-grained sediment–water mixtures with a short half-life of 2–4 d; and (2) biological degradation as measured in marine and freshwater sediment, which is much longer, with a half-life of several months.

Environmental butyltin measurements provide additional evidence of the rapid degradation of TBT. Concentrations of MBT and DBT in the water column tend to co-vary with TBT levels, and higher percentages of butyltin degradation products (85–90%) were found in regions with long water residence times than in source regions with shorter residence times (40–45%). In addition, sediment profiles and near-bottom concentrations of butyltins are consistent with the rapid formation of MBT at the sediment–water interface. Simulated environmental degradation studies have found that tributyltin is not a highly persistent compound in freshwater and marine ecosystems under a wide range of environmental conditions.

21.1. INTRODUCTION

The hazard posed by a toxic substance to an organism in the environment is a function of its toxicity and environmental concentration. But equally important in defining the environmental risk of a compound is its bioavailability, bioconcentration and accumulation potential, pathways of exposure, and its persistence. The persistence of a pesticide in the environment is a critical factor in assessing the long-term environmental risk from use of that pesticide. The persistence of tributyltin (TBT) or any xenobiotic chemical in an aquatic ecosystem is a function of chemical and biological degradation and physical removal mechanisms (e.g. Baughman and Lassiter, 1978; Branson, 1978). Persistence may be substantially affected by ecosystem-specific properties such as the concentration of suspended particulate and dissolved organic material, the nature and concentration of microbial populations, turbidity, temperature, pH, salinity, and flushing rates. Physical mechanisms include volatilization; freezing; adsorption to sediment, suspended solids, or particulate materials; and advection. Chemical removal mechanisms include chemical and photochemical reactions. Biological removal mechanisms include adsorption, uptake, and transformation by biota. Although lipophilic chemicals such as TBT can be accumulated by invertebrates, fish, and other higher organisms, it is customary to view microbial degradation and transformation as the most important biological considerations in evaluating the persistence and degradation processes of a xenobiotic compound. As an example, DDT was banned in the United States in 1972 because of its persistence in the environment and its very high lipophilicity resulting in a high rate of food-chain biomagnification and concomitant declines in reproductive success in several top-level predator bird species. A variety of estuarine organisms, including crabs, fish, shrimp, and algae have been reported to be able to rapidly metabolize TBT by forming a number of more water-soluble hydroxylated metabolites (Lee, 1985, 1986 and Chapter 18; Lee *et al.*, 1987, 1989). Because molluscs have low activities in certain enzyme systems, they debutylate TBT at much slower rates, which may in part explain molluscan sensitivity to this compound (Lee, 1986 and Chapter 18).

In aquatic ecosystems, TBT concentrations have been found near sources such as yacht harbors at levels (20–2000 ng l^{-1}) that equal or exceed reported acute or chronic toxicity limits (Maguire *et al.*, 1982, 1985a, 1986, 1991, and Chapter 26; Alzieu *et al.*, 1986, 1991; Grovhoug *et al.*, 1986, and Chapter 25; Huggett *et al.*, 1986, and Chapter 24; Maguire, 1986, 1987, 1991; Seligman *et al.*, 1986a, 1987, 1989; Valkirs *et al.*, 1986, 1991; Hall *et al.*, 1987; Waldock *et al.*, 1987; Alzieu, 1991; Salazar and Salazar, 1991; Stephenson, 1991; Waite *et al.*, 1991, and Chapter 27). However, in open areas of bays and estuaries near sources, TBT levels are generally <10 ng l^{-1} and frequently <5 ng l^{-1} in the water column (Grovhoug *et al.*, Chapter 25; Huggett, Chapter 24). It appeared that TBT in the environment was not accumulating as fast as might be anticipated given the relatively high input rates from antifouling coatings. It was therefore likely that degradation and other removal mechanisms were acting to reduce environmental levels of TBT. Therefore the question was raised: what are the rates, mechanisms, and pathways of degradation and removal processes? The work that is beginning to answer these questions is the subject of this chapter.

21.1.1. THEORY AND BACKGROUND OF DEGRADATION PROCESSES

Inorganic tin, certainly at environmental levels, is considered non-toxic. The attachment of one or more organic groups (R) to the Sn atom has a substantial effect on the biological activity (Laughlin *et al.*, 1985, and Chapters 10 and 16; Walsh *et al.*, 1985; Maguire, 1991). Maximum biological activity in any alkyltin (R_nSnX_{4-n}) where $n \leq 3$ occurs when $n = 3$ [e.g. for the triorganotin compounds such as tributyltin (Bu_3SnX)]. The R_3Sn species is a cation in solution. The counter ion or anionic radical (X) has an equilibrium distribution that is dependent on ionic strength, CO_2, and pH, and has little effect on the biological activity (Laughlin *et al.*, 1986a). Speciation products in seawater are composed of a mixture of tributyltin chloride, tributyltin hydroxide, the aquo complex ($TBTOH_2^+$), and a carbonato species $TBTOHCO_2^-$ (Laughlin *et al.*, 1986a). The tributyltin cation, Bu_3Sn^+, however, is responsible for the high level of observed toxicity.

In general, the degradation of an organotin compound may be defined as the progressive removal of the organic groups attached to the tin atom:

$$Bu_3SnX \rightarrow Bu_2SnX_2 \rightarrow BuSnX_3 \rightarrow SnX_4$$

The removal of butyl groups from the tin atom is accompanied by a progressive reduction in biological activity. Dibutyltin (DBT) has been found to be less toxic than TBT by at least two orders of magnitude in tests with algae (Wong *et al.*, 1982; Walsh *et al.*, 1985), by a factor of 60 with mud crab larvae (Laughlin *et al.*, 1986b), and by over three orders of magnitude in producing a shell-thickening phenomenon in the Pacific oyster, (*Crassostrea gigas*) (Thain *et al.*, 1987).

Degradation of butyltins involves the breaking of Sn–C bonds, which can occur by a number of processes. These processes include ultraviolet (UV) irradiation: biological, chemical, or thermal cleavage; and gamma irradiation (Blunden and Chapman, 1982). Thermal cleavage and gamma irradiation are not considered to be of environmental significance because the Sn–C bond is stable at temperatures up to 200°C, and gamma irradiation is negligible at the Earth's surface.

Light emitted by the sun consists mostly of wavelengths >290 nm at the Earth's surface. Butyltins weakly absorb sunlight in the 300-nm region, and at wavelengths >350 nm very little absorption occurs. Thus, only the near UV component of sunlight could cause direct photodegradation of butyltins in surface waters. Exposure to natural sunlight of TBT dissolved in fresh water demonstrated slow (>89 d half-life) photolytic degradation; DBT, monobutyltin (MBT), and inorganic tin appeared as degradation products (Maguire *et al.*, 1983; Maguire, 1991). Because of the attenuation of sunlight with depth in natural waters, photolysis is negligible at depths greater than a few centimeters in estuarine waters or tens of centimeters in more transparent water or fresh water. Photolysis may play a significant role, however, in the fate of organotins in the surface microlayer.

Chemical cleavage of the Sn–C bond can occur by either nucleophilic or electrophilic reagents such as mineral acids, alkali, and carboxylic acid (Blunden and Chapman, 1986). The aqueous stability of TBT is high. Very little debutylation was found in 1- to 2-month experiments in the dark in fresh water (Slesinger and Dressler, 1978; Maguire *et al.*, 1983). Similarly, we have observed little if any degradation in sterile seawater. Biological cleavage of the C–Sn bond, or biodegradation, appears to be the most significant route of TBT detoxification under aqueous conditions.

21.1.2. REVIEW OF EXISTING WORK ON TRIBUTYLTIN DEGRADATION

Early work using ^{14}C-labeled bis(tri-n-butyltin) oxide (TBTO) demonstrated aerobic degradation in soil to DBT and $^{14}CO_2$ with a 15- to 20-week half-life (Barug and Vonk, 1980). Rapid degradation (3–4 d half-life) to MBT and DBT was found in pure cultures of certain bacteria and fungi (Barug, 1981). Aquatic microorganisms degraded TBT to DBT and MBT (~6-d half-life) at 20°C in estuarine water from Baltimore Harbor and in fresh water from the United Kingdom; however, little degradation was observed at 5°C (Olson and Brinckman, 1986; Thain *et al.*, 1987). In Japan, half-lives of 5–11 d were measured in both fresh water and seawater (Hattori *et al.*, 1987); however, some of the experiments yielded little evidence of degradation products. Tributyltin uptake by certain estuarine bacteria has been measured with little apparent degradation (Blair *et al.*, 1982). Half-lives for TBT in the range of several days to several months have been reported in water and sediment and are the principal topic of this chapter (Maguire *et al.*, 1983; Maguire and Tkacz, 1985; Seligman *et al.*, 1986b, 1986c, 1988, 1989; Hinga *et al.*, 1987; Adelman *et al.*, 1990). TBT metabolism has been identified by both freshwater (Maguire *et al.*, 1984) and marine (Lee *et al.*, 1987, 1989, and Chapter 18) phytoplankton species.

Summaries of research on the persistence of TBT in various environments have been reviewed by Maguire (1987, 1991), Seligman *et al.* (1989), and Huggett *et al.* (1992) In evaluating the aquatic persistence of TBT, there are three important considerations:

1. The persistence of TBT should not be defined in terms of the time required for anion exchange. This gives a false impression of degradation kinetics when the loss of butyl groups is really the significant factor. The toxicity of TBT is largely independent of the nature of the counter ion (Polster and Halacka, 1971; Davies and Smith, 1980), but the toxicity declines greatly with decreasing number of butyl groups (Davies and Smith, 1980; Laughlin *et al.*, 1985, and Chapter 10; Hall and Bushong, Chapter 9).
2. The degradation of TBT in experimental systems is not necessarily equivalent to its disappearance from water. Adsorption to container walls, particulate material, or sediments, or uptake by biota, can represent substantial nondegradative losses from the water. In model ecosystems, transport mechanisms can result in significant removal of the compound from an environmental compartment without any net degradation. This point underscores the importance of determining both the mass balance and formation of degradation products in evaluating the persistence of a compound.
3. Realistic environmental conditions, including light, temperature, and ambient water quality (including its microbial population) should be used to evaluate the fate of TBT or any toxicant of interest. Ideally the best evaluation of fate would be to introduce the contaminant in a controlled fashion to an aquatic ecosystem, thus eliminating laboratory artifacts. This is rarely feasible because environmental contamination for environmental studies is usually not acceptable, and the lack of control of variables makes interpretation difficult. The use of large-scale enclosed self-sustaining ecosystems (mesocosms) is a functional alternative that provides a realistic environment under controllable conditions. However, container and wall effects must be evaluated.

This chapter summarizes recent experimental work by the authors on the fate and degradation of TBT. In most of the studies, we attempted to duplicate natural environmental conditions by using realistic TBT concentrations and ambient light, temperature, and microbial populations. In the mesocosm studies, many of the environmental conditions characteristic of a shallow coastal ecosystem were maintained in the enclosures. By using realistic environmental conditions, we believe that the data describing decay rates and persistence are representative of the decay rates and persistence in the environments discussed.

21.2. METHODS AND MATERIALS

21.2.1. STUDIES OF DEGRADATION IN MARINE AND ESTUARINE SYSTEMS

21.2.1a. Water column: analysis and experimental design

Degradation studies were carried out in four diverse regions of the United States using ambient unfiltered seawater contained in 500-ml to 1-l polycarbonate centrifuge bottles for unlabeled TBT experiments and 2-l Pyrex® glass containers for ^{14}C-TBT radiolabeled studies. Both spiked TBT experiments, with concentrations between 0.5 and 2.0 $\mu g\ l^{-1}$ TBT, and unspiked (die away) experiments with ambient concentrations of 0.2–0.5 $\mu g\ l^{-1}$ were conducted. Containers were held under ambient conditions of light and temperature, with replicate sets of dark bottles and abiotic (poisoned or filtered) controls. Degradation studies were conducted in Pearl and Honolulu harbors in Hawaii; San Diego Bay, California; Elizabeth and James River estuaries, Virginia; and in the Skidaway estuary, Georgia. A total of 46 experiments were conducted during various seasons between 1985 and 1988. Details of the experimental designs are given in Seligman *et al.* (1986b, 1986c, 1988, 1989) and Lee *et al.* (1987).

Labeled ^{14}C-TBTO

Where experiments using radiolabeled TBTO were performed, custom-synthesized bis(tri-n-butyltin) oxide (butyl-1-^{14}C) (^{14}C-TBTO) was used. The ^{14}C-TBTO was added to 2 l of seawater to a final concentration of 2 $\mu g\ l^{-1}$. Tributyltin solutions were extracted with 300 ml of chloroform after appropriate incubation periods. Recovery of spiked ^{14}C-TBTO from test systems was between 97% and 99%. Autoradiography of thin-layer silicic acid plates was used to locate and verify the position of TBT and its degradation products. Unlabeled butyltin standards were visualized with a dithizone or pyrocatechol violet spray (Kimmel *et al.*, 1977). Details of this method have been discussed previously (Seligman *et al.*, 1986c).

Unlabeled TBTCl analysis

Environmental water samples and samples spiked with unlabeled tributyltin chloride (TBTCl) used in degradation studies were analyzed for butyltin species by hydride derivatization and atomic absorption detection (HDAA). The method, which uses a cryogenic trap with thermal separation of the individual species, has been successfully used to measure butyltin in seawater directly without preconcentration (Braman and Tompkins, 1979; Hodge *et al.*, 1979) and has been modified for direct determination of TBT as well as DBT and MBT in seawater (Valkirs *et al.*, 1986; Stallard *et al.*, 1989). Analytical variability with replicated measurements of standards and natural seawater samples has routinely given standard deviations ranging from 5% to 10% of mean values at 0.01–0.05 $\mu g\ l^{-1}$ TBT. Detection limits for MBT, DBT, and TBT are ~0.2 $ng\ l^{-1}$. The details of the method are given by Weber *et al.*, Chapter 4.

21.2.1b. Sediment: analysis and experimental design

The HDAA technique was used to analyze sediment for butyltin compounds (Weber *et al.*, Chapter 4). Details of the procedure are given in Stang and Seligman (1986). Aliquots of 0.1–1 g (wet wt) sediment are added directly to the reaction vessel. The volatile butyltin hydrides measured by this technique represent hydride-reducible organotin. High hydrocarbon or sulfide content may reduce recovery of the butyltins contained in some sediment samples (Valkirs *et al.*, 1985). Replicate aliquots were weighed, dried at 110°C to constant weight, and reweighed to present all butyltin concentrations as mg kg^{-1} (dry wt) of sediment by use of the wet-to-dry weight ratios.

To evaluate the vertical distribution of tributyltin and its degradation products, sediment cores of 5 cm diameter to 35 cm in depth were collected in polycarbonate tubes by scuba divers and immediately frozen. In the laboratory, the frozen cores were cut from 0 to 2.0 cm, and 2.1 to 5.0 cm, and continued in 5-cm increments.

Long-term sediment degradation study

Approximately 6 kg (wet wt) of sediment was collected from a TBT-impacted boat basin in San Diego Harbor. The sediment was placed in a 38-l polycarbonate aquarium, covered with seawater, and aerated for 140 d. The ambient (unspiked) concentrations of butyltin species in the sediment were measured, and the changes in concentration were followed over the duration of the experiment. The study was conducted in the dark at 15°C to approximate the benthic conditions at ~10 m in San Diego Bay.

Short-term comparative sediment degradation studies

Sediments were collected from the top 2–4 mm of the benthic layer in San Diego Bay, the Skidaway River, and Pearl Harbor. All seawater used in the experiments was autoclaved in individual 500- or 1000-ml polycarbonate bottles to ensure that degradation was not the result of metabolism by bacteria or algae but was due to added sediment. Once collected, sediment subsamples were either sterilized, made free of organic material, or left untreated. The untreated sediment was added to the sterilized seawater. Tributyltin chloride was subsequently introduced. Sediments were sterilized by adding sediment to the seawater, and the sediment–water mixture was autoclaved. Sediment was made free of organic material by heating to 550°C in a furnace for 2 h to destroy all organic material (<1% organic fraction remaining), and the dried sediment was then added to the seawater. In all cases, unlabeled TBTCl or ^{14}C-TBTO was added after sediment was added to the seawater. Ratios of 1.0 g sediment to 200 ml seawater were used for the Skidaway River sediment, and 2.0 g to 200 ml for San Diego Bay and Pearl Harbor sediment. A ratio of 5 g sediment to 50 ml sterile seawater was used for ^{14}C-TBTO experiments.

To approximate moderately impacted sediments, between 127 and 136 ng g^{-1} TBTCl was added to the sediment–water mixtures in the unlabeled experiments, and 300 ng g^{-1} was added to ^{14}C-TBTO experiments. The individual bottles were capped and suspended ~1 m below the surface in Pearl Harbor (23–24°C), San Diego Bay, and the Skidaway River (16–19°C). Individual bottles were removed at 1-, 2-, 4- and 7-d intervals and analyzed immediately, either by HDAA analysis or by thin layer chromatography as discussed earlier. If samples could not be analyzed immediately, they were frozen and analyzed within 2 weeks. Controls of autoclaved seawater with TBT were prepared and analyzed to document TBT stability in the absence of sediment. Butyltin recoveries averaged 95% (range of 67–122%) for unlabeled and 92–96% for labeled ^{14}C-TBTO spikes.

21.2.2. STUDIES OF DEGRADATION IN FRESHWATER SYSTEMS IN CANADA

21.2.2a. Water column: analysis and experimental design

In the field, 8-l samples of subsurface water were collected from a depth of 0.5 m in glass bottles. Care was taken to open the bottles below the surface to avoid contamination by the surface microlayer. The contents were acidified to pH 1 and stored in the dark at 4°C until analysis (Maguire *et al.*, 1982). These preservation conditions were effective over a period of at least 3 months. Samples collected to evaluate TBT concentrations in the surface microlayer were either collected with a glass plate sampling technique for 100-ml samples (Harvey and Burzell, 1972) or a rotating drum sampler for 4-l samples (Harvey, 1966) and were preserved in the same way as the subsurface samples.

21.2.2b. Sediment: analysis and experimental design

Sediment samples were usually collected with an Ekman dredge. Prior to analysis, the top 2 cm was scraped off into glass jars and frozen as soon as possible, then freeze-dried, ground, and sieved to pass an 850-μm screen (Maguire, 1984; Maguire *et al.*, 1986). In a more detailed study of four sites in Toronto Harbor, 7-cm diameter sediment cores were taken with a lightweight benthos corer and kept at 4°C overnight (Maguire and Tkacz, 1985). The cores were then extruded in 1-cm slices, and these sediment slices were treated in the same way as the sediment grab samples described earlier. The water above the sediment in these cores was also preserved for analysis.

The tributyltin species and its degradation products dibutyltin, monobutyltin, and inorganic tin were determined by extraction from acidified water samples, or dry sediments, with the complexing agent tropolone dissolved in benzene. Pentylation of the extract was performed to produce the volatile mixed butylpentyltin derivatives ($Bu_nPe_{4-n}Sn$), followed by clean-up by silica gel column chromatography, concentration, and analysis by packed column gas chromatography initially with a modified flame photometric detector (Maguire and Huneault, 1981; Maguire *et al.*, 1982, 1984; Maguire, 1984), and subsequently with a quartz tube furnace atomic absorption spectrophotometric detector (Maguire and Tkacz, 1983, 1985, 1987; Maguire, 1984; Maguire *et al.*, 1985a), which was found to be more reliable. Considering that a fairly specific detector for tin was used in the analyses, identities of the butylpentyltin species were deemed to be confirmed by co-chromatography with authentic standards on two column packing materials of very different polarity.

In the quantitation of the analyses, use was made of appropriate reagent blanks. All results reported were above the limit of quantitation (LOQ), which is defined as the reagent blank value plus ten times its standard deviation (Keith *et al.*, 1983). In practice this was equivalent to stating that a chromatographic peak was not accepted as real unless it was at least 2–3 times as large as any corresponding peak in the reagent blank.

Analytical recoveries of tributyltin, dibutyltin, and monobutyltin from spiked water samples at 1–10 mg l^{-1} varied from 96% ± 4% to 103% ± 8%. Recoveries of Sn(IV) from water at pH 5–8 were poor (35% ± 23%), probably because of the formation of unextractable SnO_2 (Maguire *et al.*, 1983). Recoveries of the three butyltin species and inorganic tin from spiked sediment at 0.01, 0.2, 1, and 100 mg kg^{-1} Sn (dry wt) ranged from 55% ± 26% to 180% ± 100% (Maguire, 1984). Concentrations were not corrected for recovery. Although Sn(IV) was the only inorganic tin species for which recoveries were determined, the tin present in the above samples was reported as total inorganic tin, since any Sn(II) that might have been present would likely have been oxidized to Sn(IV) during extraction and derivatization.

21.2.2c. Laboratory experiments

Experiments were conducted to investigate and define the importance of various environmental pathways and reactions of tributyltin to evaluate its persistence in the environment. Physical, chemical, and biological parameters were investigated, including volatilization, hydrolysis (i.e. debutylation), adsorption to sediment, photolysis, microbial and algal degradation, and uptake and transformation by oligochaetes.

Analyses of water and sediment in the laboratory experiments for tributyltin and its degradation products were done in the same general way as for the field samples. Experiments were designed to distinguish between abiotic and biological degradation of tributyltin in water and water–sediment mixtures, and to test for uptake and degradation of tributyltin by oligochaetes (Maguire and Tkacz, 1985).

The analysis of oligochaetes (primarily *T. tubifex* and *L. hoffmeisteri*) in sediment from Toronto Harbor was accomplished by dispersing 1 g of oligochaetes (~100) in 1–2 ml concentrated HCl for 2 h at room temperature (Maguire and Tkacz, 1985). The resulting mixture was diluted fivefold with water, then extracted in the same way as the water samples described above. Recoveries of the three butyltin species and Sn(IV) from spiked oligochaetes ranged from 65% ± 19% to 77% ± 11% at the 0.1 mg kg^{-1} Sn level.

Analyses of algal and bacterial suspensions for butyltin species and inorganic tin were performed in the same way as analyses of water (Maguire *et al.*, 1984, 1985b). Experiments on the biological degradation of tributyltin under aerobic and anaerobic conditions were performed with cyclone fermenters (Liu *et al.*, 1981).

21.2.3. RADIOLABELED ^{14}C-TBTO DEGRADATION STUDY IN AN ENCLOSED MARINE ECOSYSTEM

A variety of approaches have been used to obtain experimental systems that are more realistic than a beaker but better constrained than a typical coastal ecosystem. One such approach is the use of large microcosms, often called mesocosms (Grice and Reeves, 1982; Odum, 1984). The physical systems are usually large (>10 m^3) enclosures or tanks (Davies and Gamble, 1979; Pilson and Nixon, 1980; Lundgren, 1985). While size is important for a number of reasons, the primary distinguishing feature of most mesocosms is that they contain self-sustaining ecosystems, with water column and benthos, that have primary production, secondary production, and regeneration. This distinguishes them from large aquaria where the organisms must be fed or the water treated.

The behavior of TBT was studied for 278 d in a mesocosm that simulates the conditions in New England coastal waters. The mesocosm at the University of Rhode Island Marine Ecosystem Research Laboratory (MERL) consisted of a 1.8 m diameter × 5.5 m high outdoor tank that, for the TBT experiment, contained sediment and water transferred from adjacent Narragansett Bay. A 37-cm layer of sediment was collected intact as a series of box cores, complete with their active biological communities, and transferred to the mesocosm tank. Water from the bay (~30‰) was added to the mesocosm with a non-disruptive bellows pump. The mesocosm had planktonic primary production and a heterotrophic benthos. The tank was mixed 2 h out of every 6 with a vertical plunger. This mixing scheme homogenized the water column and resuspended sediment material at a rate similar to the rate observed in Narragansett Bay. The mesocosm was maintained at temperatures within 2°C of the adjacent bay. Tank walls were cleaned with powered brushes approximately weekly to prevent extensive growth of fouling organisms.

The fundamental assumption for this type of experiment is that the mesocosm simulates the major biological and chemical processes and rates that are similar to those in the natural coastal ecosystem. If this is so, an introduced compound should behave in the experimental system in a manner similar to the way it would behave in the natural system. A number of studies have documented that the MERL mesocosms do represent closely the conditions in the Narragansett Bay and can therefore effectively support environmental fate studies (e.g. Hunt and Smith, 1982; Santschi, 1982; Donaghay, 1984; Pilson, 1985).

On 13 July 1987, 480 μCi of tri-n [1-^{14}C]butyltin chloride, dissolved in 10 ml of

MeOH, was added to a MERL mesocosm, resulting in an initial tributyltin chloride concentration of 590 ± 20 ng l^{-1}. Detailed sampling and analytical procedures may be found elsewhere (Adelman, 1988; Adelman *et al.*, 1990, 1991). Briefly, the concentrations of TBT, DBT, and MBT were quantified in water samples by organic extraction, separation by thin layer chromatography, and counting by liquid scintillation. The raw counts were corrected for the efficiency of extraction for each compound to determine the concentrations of the butyltins. In addition, the concentrations of radiolabeled butyltins were quantified on particles, in the surface microlayer, benthic sediments, and in the radiolabel found as CO_2. A total of 40 sediment cores were taken over 11 sampling periods during the 278-d experiments.

21.3. RESULTS

21.3.1. DEGRADATION IN MARINE AND ESTUARINE SYSTEMS

21.3.1a. Water column

Numerous TBT degradation experiments using microcosms (polycarbonate and glass bottles) with ambient unfiltered water to evaluate TBT half-lives and formation of degradation products were conducted using waters from four diverse harbors and estuaries (Seligman *et al.*, 1986b,c, 1988, 1989; Lee *et al.*, 1987; Lee, Chapter 18). The experiments were conducted at environmentally realistic concentrations (0.2–2.0 μg l^{-1} TBT) using both ^{14}C-TBTO and unlabeled TBTCl and included several die-away experiments using ambient TBT. These experiments have yielded a range of degradation rates for TBT in the water column of 4–19 d half-lives. Table 21.1 presents a summary of the work showing mean TBT half-lives inclusive of light and dark treatments and labeled and unlabeled TBT spikes. The principal degradation product in both light and dark treatments was DBT with much slower formation of MBT. Figure 21.1b shows the changes in butyltin concentrations over the 12-d period of a typical degradation experiment in which seawater was spiked with TBTCl (1.1 μg l^{-1} compared with a filtered (sterile) control (Fig. 21.1a). Figure 21.2 shows the degradation of ^{14}C-TBTO in a parallel experiment from San Diego Bay. These 4- to 12-d experiments document the essentially linear loss of parent TBT compound with concomitant formation of degradation products (primarily DBT) apparently following zero-order kinetics. The rate of loss of parent compound was determined by regression analysis, and a half-life was calculated. Longer (measured in months) mesocosm degradation experiments described below suggest first-order loss rate kinetics.

Degradation experiments were conducted using water from nine locations in four harbor and estuary regions in both the east and west coasts of the continental United States and Hawaii (Table 21.1). The studies were conducted between 1985 and 1987 during several seasons. Measured degradation rates of the spiked radiolabeled ^{14}C-TBTO and unlabeled TBTCl as well as the unspiked die-away experiments compared closely. Tributyltin half-life values varied from 4 d at the Ala Wai Yacht Harbor in Hawaii to 19 d (dark treatment) in south San Diego Bay. Mean degradation half-lives (averaging all degradation experiments from a given location) varied from 6.3 d in a yacht harbor in Hawaii to 13 d in the James River, Virginia. This suggests that for semi-tropical and temperate marine and estuarine systems, degradation half-lives will generally be between 1 and 2 weeks. Sterile controls exposed to the light showed little or non-measurable degradation demonstrating the biological nature of the degradation process.

In general the dark treatments, simulating near-bottom light conditions, had half-lives that averaged 40% longer than treatments exposed to ambient near-surface light conditions, suggesting that photosynthetic algal

Table 21.1 Tributyltin degradation in natural seawater from four marine and estuarine regions

Location (ambient TBT ng l^{-1})	*Date*	*TBT concentration (μg l^{-1})*	*Temperature (°C)*	*Half-life (d)*[a] *TBTCl*		*^{14}C-TBTO*		*Mean half-life (d ± SD)*
				L	*D*	*L*	*D*	
San Diego Bay, CA	5–85	0.6	13–15	13	19	—[b]	—	12.2 ± 3.2
South Bay	7–85	1–2.0	16–20	9	9	11	12	
(2–10)	4–86	2.0	17–18	10	—	12	15	
Shelter Island	7–85	2.0	16–19	—	—	6	7	
Yacht Harbor	1–86	0.5	14–16	7[c]	—	—	—	6.6 ± 0.5
(200–900)	4–86	1.1	17–18	7	—	—	—	
	4–86	0.2	17–18	6[c]	—	—	—	
Pearl Harbor, HI								
Rainbow Marina	5–87	0.5–0.7	24	7	13	7	14	10.2 ± 3.8
(7–20)								
Southeast Loch	9–87	0.5	29	—	—	6	10	8.0 ± 2.8
(5–25)								
West Loch	9–87	1	29	—	—	6	11	8.5 ± 3.5
(0.5–2)								
Ala Wai Basin	5–87	1	24	—	—	4	9	
(200–550)	9–87	0.4–1		—	—	4	9	6.5 ± 2.9
Norfolk, VA								
James River	6–86	1	26	—	—	12	14	13.0 ± 1.4
(<3.0)								
Elizabeth River	6–86	1	26	—	—	6	9	7.5 ± 2.1
(10–100)								
Skidaway Estuary, GA	12–85	1.5	14	—	—	4	—	
(<5.0)	1/2–86	1.3–1.6	11–12	7	11	8	13	8.3 ± 2.7
	8/9–86	0.5–1.5	28–29	6	9	6	10	
	10–85	1.5	19	—	—	9	—	

[a]L indicates light treatment; D indicates dark treatment.
[b]Dash indicates no data.
[c]'Die-away,' unspiked study.

populations are significant in the degradation process. It is highly unlikely that this increase in degradation rate in the light is due to photolysis or chemical redistribution, since the photolysis half-life was >89 d in fresh water (Maguire *et al.*, 1983), and concentration (744 μg l^{-1}) over a 144-d period, likely due to toxic inhibition of the microbial population. The relationship of photosynthetic organisms to TBT degradation is supported by demonstrated metabolism of TBT in a freshwater green alga *Ankistrodesmus falcatus* (Maguire *et al.*, 1984) and accelerated TBT degradation under light conditions in Baltimore Harbor water (Olson and Brinckman, 1986). More recently, we have found that marine diatoms can rapidly metabolize TBT to hydroxylated derivatives (e.g. beta-, delta-, and gamma-hydroxybutyldibutyltin) in light treatments fortified with added nutrients to stimulate algal growth (Lee *et al.*, 1987, 1989). The formation of hydroxylated derivatives only in the light indicates that bacteria may metabolize TBT to dibutyltin through pathways different from the pathways used by algae. Although bacteria undoubtedly play a major role in the degradation process, eukaryotic organisms such as fungi may be important, particularly in benthic environments.

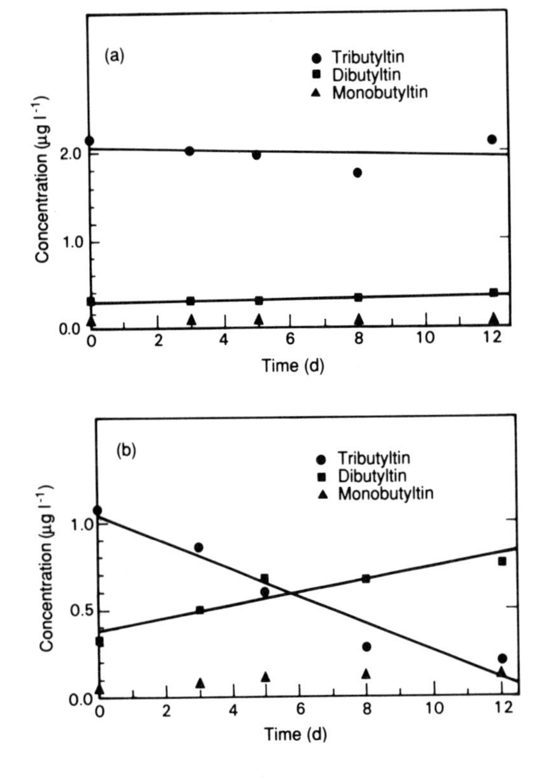

Figure 21.1 Concentration of TBT and degradation products in (a) filtered seawater control, and (b) spiked seawater overtime (modified from Seligman *et al.*, 1986c).

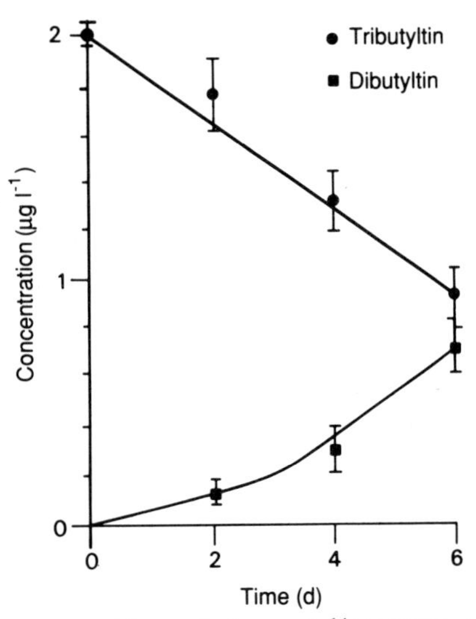

Figure 21.2 Degradation of ^{14}C-TBTO in seawater collected from Shelter Island Yacht Harbor, San Diego, California.

In a number of the ^{14}C-TBTO studies, the formation of $^{14}CO_2$ was used to evaluate the rate of complete mineralization of TBT. The mineralization half-life was calculated between 50 and 75 d in water from south San Diego Bay (Seligman *et al.*, 1986c). The formation of beta-hydroxybutyldibutyltin, which in turn degrades to dibutyltin with the release of butene, has been described in mammals (Kimmel *et al.*, 1977). We have found hydroxylated butyltins in association with light treatments and hypothesize that butene is a by-product. In our system ^{14}C-butene would likely be rapidly released to the air dead space above the water phase in the containers used and would not be collected in the $^{14}CO_2$ trap, potentially resulting in an overestimate of the half-life to complete mineralization.

Within the harbors and estuaries studied, areas of higher TBT concentration, such as marinas or near drydock facilities, typically had faster TBT decay rates by a factor of ~2. This phenomenon was observed in Ala Wai Yacht Basin (6.3-d half-life) versus Pearl Harbor (~9.5-d half-life); Shelter Island Marina (6.6-d half-life) versus south San Diego Bay (12.2-d half-life); and in the Elizabeth River (7.5-d half-life) versus the James River (13-d half-life). The faster decay rates in water from areas of high ambient TBT concentration suggest higher concentrations of TBT-degrading microbes and, in addition, may represent either an adapted or acclimated population or a selection process that favors TBT-metabolizing microbes and algae. Other studies have shown high linear degradation rates when radiolabeled pollutants were added to water that had previous exposure to the same pollutants (PAH) (Lee and Ryan, 1983). Work with phytoplankton, however, did not demonstrate adaptation to TBT after 12 weeks of exposure (Walsh *et al.*, 1985).

Temperature during the seasons evaluated did not seem to cause substantial differences in degradation rates. For example, we measured TBT degradation at water temperatures ranging from 11°C (January) to 29°C (August) in the Skidaway estuary with half-lives of 8 and 6 d, respectively. The differences are less than might be expected on the basis of normal thermal kinetics. At 5°C, however, other investigators have measured half-lives of 60 d or longer, suggesting that very low temperatures substantially reduce the rate of degradation (Olson and Brinckman, 1986; Thain *et al.*, 1987).

We have measured butyltin concentrations in numerous harbors and estuaries (Grovhoug *et al.*, 1986; Seligman *et al.*, 1986c, 1989; Valkirs *et al.*, 1986) and have generally found MBT and DBT in positive correlation with TBT. This association led us to believe that measured levels of MBT and DBT were primarily degradation products from leached TBT. In San Diego Bay we have documented concentrations of degradation products that are generally proportional to the TBT concentrations (Seligman *et al.*, 1986b,c). Relatively strong covariance is found between TBT and DBT in two regions of San Diego Bay (Fig. 21.3a,b), demonstrated by their regression curves. Covariance in the curves provides environmental evidence of a relatively consistent degradation process, while different rates of water flushing (residence times) may explain the variable slopes of the regression curves. Figure 21.4 shows the ratio of tributyltin:total butyltin for four regions within San Diego Bay. Tributyltin, as a percentage of total butyltin, averaged 56% in yacht harbors (yacht), while the South Bay region averaged only 9% TBT or 44% and 91% degradation products (MBT + DBT), respectively. High percentages of TBT in yacht harbors likely resulted from recently released toxicant from hull coatings and relatively rapid flushing (e.g. the tidal exchange coefficient is 0.28 for Shelter Island Marina within San Diego Bay), thereby not

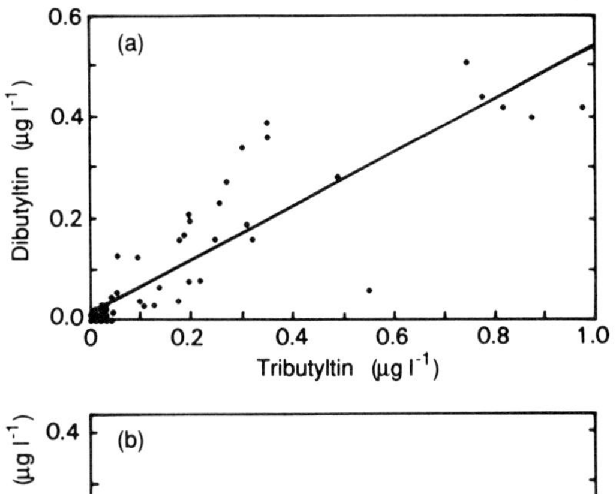

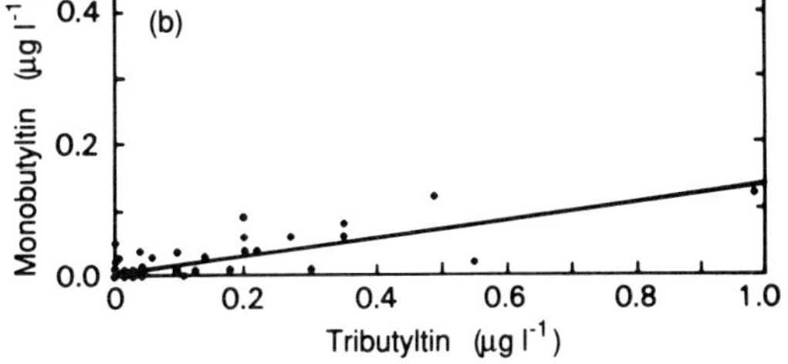

Figure 21.3 Least-squares regression TBT vs DBT and MBT measurements from San Diego Bay, CA water samples. Samples were collected over a 3-y period from various sites: (a) [DBT] = 0.527 [TBT] + 0.013, r^2 = 0.80, n = 78; (b) [MBT] = 0.130 [TBT] + 0.006, r^2 = 0.59, n = 70 (modified from Seligman *et al.*, 1986b).

allowing the buildup of degradation products. Conversely, in the South Bay region (south), which is distant from most TBT sources and characterized by long water residence times, degradation products tend to accumulate at concentrations an order of magnitude higher than the low TBT levels. Intermediate levels, 32% and 24% TBT, were found in the North Bay and Navy regions, respectively (Seligman, 1986b). These observations are consistent with the experimentally derived degradation half-lives of 4–19 d discussed earlier.

As part of a program to predict TBT concentrations in harbors used by the US Navy, a dynamic estuary numerical model of San Diego Bay was developed and tested (Walton *et al.*, 1986). Concentrations of TBT in the central portion of the bay were predicted using measured and estimated

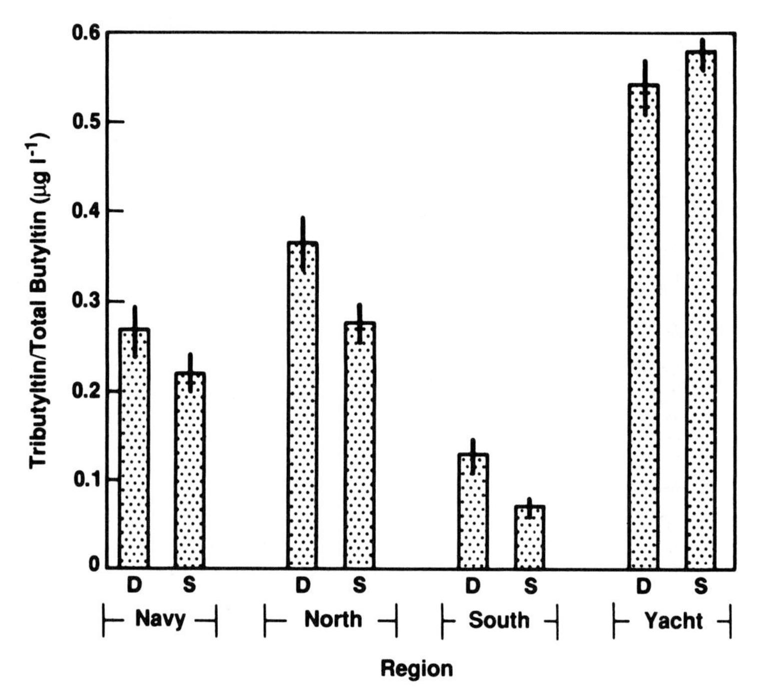

Figure 21.4 Ratios of tributyltin to total butyltin concentrations in the four regions in San Diego Bay, California. Samples were taken 1 m off the bottom (D) and 0.5 m below the surface (S) and averaged over all tides (modified from Seligman *et al.*, 1986c).

inputs of TBT from commercial, private, and military sources. Figure 21.5 shows a comparison of the model simulation with two degradation rates (5 and 10% decay d^{-1}) and the actual survey results for a longitudinal transect of the bay. The modeled concentrations with the 10% (7-d half-life) decay rate show excellent correlation with the measured values in the northern and central portions of the bay (Walton *et al.*, 1986). The decay rate yielding these predicted results again is close to our measured experimental results, which averaged between 6 and 12 d half-life.

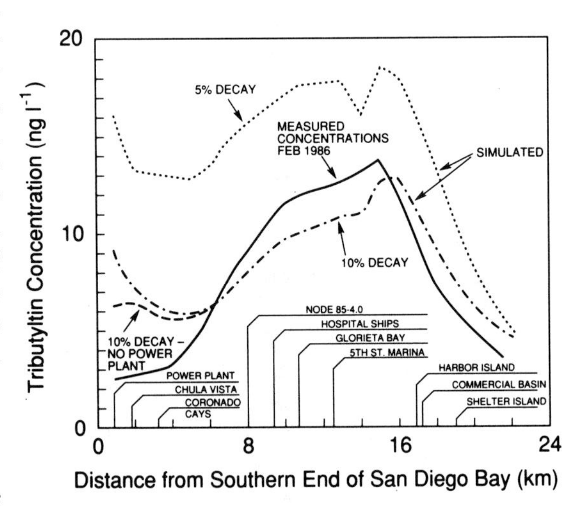

Figure 21.5 Comparison of observed and simulated (5% and 10% decay rates) tributyltin concentrations in San Diego Bay along a transect down the central portion of the bay (modified from Walton *et al.*, 1986).

21.3.1b. Sediments

A preliminary study of the degradation of ambient, unspiked TBT (200 ng g^{-1}) in impacted sediment has indicated a half-life of ~5.5 months (Stang and Seligman, 1986). This is comparable to the several months half-life found in freshwater sediments

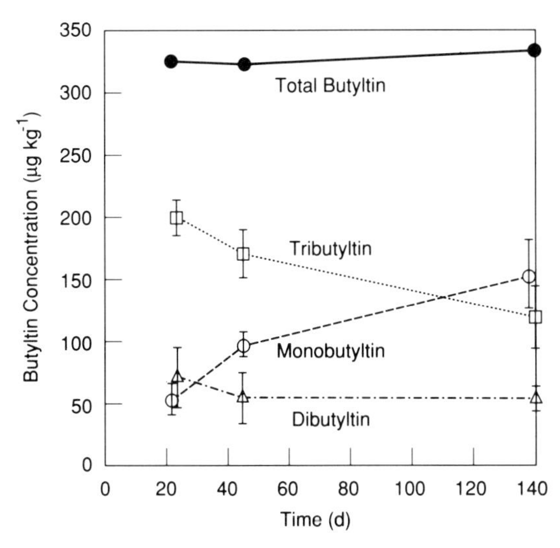

Figure 21.6 Degradation of butyltins in unspiked sediment from San Diego Bay, California (modified from Stang and Seligman, 1986).

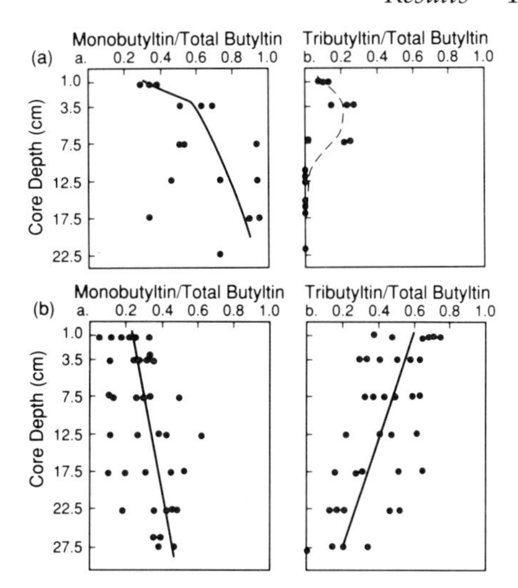

Figure 21.7 Ratios of monobutyltin and tributyltin to total butyltin versus depth in (a) Shelter Island, and (b) Commercial basin, San Diego, California (modified from Stang and Seligman, 1986).

(Maguire and Tkacz, 1985) where DBT was the primary degradation product in Toronto Harbor sediments. Monobutyltin was found to be the primary degradation product in marine sediment from San Diego Bay (Fig. 21.6). The sediment for this experiment was collected near a yacht repair facility and included TBT-containing paint particles. These paint particles were probably more inert and less available to degrading organisms than TBT sorbed to sediment particles. Bacterial degradation was suspected of being the primary pathway in both studies, a conclusion consistent with the work of Barug (1981) who determined that bacteria and fungi can dealkylate TBT. The authors found MBT to be the primary degradation product, but did not exclude the possibility that TBT degraded sequentially to DBT and then to MBT. Additional evidence of the debutylation of TBT to MBT in marine sediments is found by examining the ratio of TBT to MBT in sediment cores from San Diego Bay with respect to depth. The ratio of TBT to MBT decreased with increasing depth in most of the sediment cores (Fig. 21.7a,b), thus indicating the conversion of TBT to MBT over time, as represented by core depth.

Evidence of rapid degradation of TBT under certain conditions by fine-grained sediments is presented in Table 21.2. Sediments that are composed of at least one-third silt and (or) clay can dealkylate dissolved spikes of both radiolabeled (^{14}C-TBTO) and unlabeled TBT (TBTCl) to DBT, MBT, and inorganic tin under environmentally realistic conditions. Monobutyltin represented the largest fraction of the degradation products. Sediments that have well-hydrated particles and low (sub-microgram per liter and microgram per liter) initial TBT spiked concentrations, and are incubated under aerobic submerged conditions in ambient (but sterilized) seawater were found to degrade ~50% of the TBT in 2–4 d (Table 21.2). Sterile seawater controls showed no loss of TBT, and sandy sediments degraded TBT at rates an order of magnitude slower. It appeared that the degradation was physical or chemical in nature as sterilized, fine-grained sediments degraded TBT at nearly the same rate as untreated sediments (Fig. 21.8). Organic-

Table 21.2 Degradation of TBT added to coastal sediments

Location	Organic carbon (%)	Sediment composition Sand (%)	Silt (%)	Clay (%)	Ambient TBT ($ng\ g^{-1}$) sediment	Added TBT ($ng\ g^{-1}$)	Incubation (d)	Degradation ($\%\ d^{-1}$)	TBT half-life (d)
San Diego Bay, California									
North Island	2	95	4	1	<2	127–300	4	3.4	15
Commercial Basin	13	8	53	39	242	300	2	14	4
Shelter Island Yacht Harbor	9	52	28	20	30	300	2	15	4
Honolulu, Hawaii									
Pearl Harbor	22	24	47	29	5	136	2	47	<2
Savannah, Georgia									
Skidaway River	13	61	12	26	<2	300	1–2	29	4

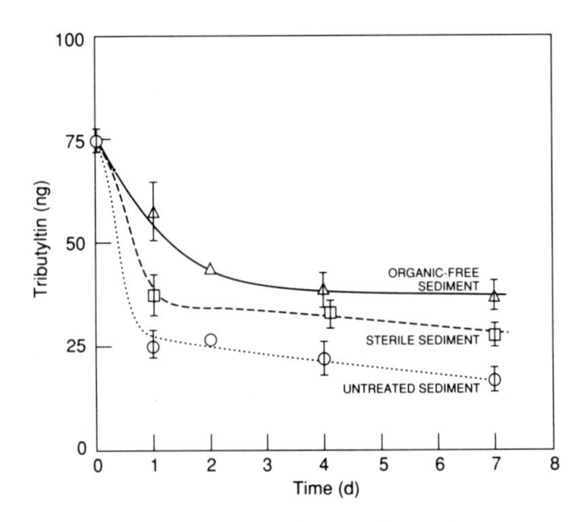

Figure 21.8 Decrease of TBT after addition of 74 ng TBT to 200 ml sterile seawater and 2.0 g of differently treated Skidaway River sediments (modified from Stang *et al.*, 1992).

free, fine-grained sediments, which had been heated to 550°C to destroy the organic material, also degraded TBT but at slower rates. It was not clear whether the loss of organic material or thermal deactivation of the clay mineralogy was responsible for the lowered degradation rates. Sediments from Pearl Harbor, Hawaii; three regions in San Diego Bay, California; and the Skidaway River, Georgia, were studied. The most rapid degradation rates were coincident with both the most fine-grained sediment and the highest MBT:TBT ratios in the field (Pearl Harbor sediment collected from an area far removed from drydock areas). Table 21.2 shows the degradation of TBT in fine-grained sediments from the three areas listed above.

Additional evidence for the rapid dealkylation of TBT to MBT was presented by Stang *et al.* (1989). The ratio of the concentration of TBT to MBT in bottom water was lower than the ratio of TBT:MBT for surface water, indicating potential desorption of the more hydrophilic MBT from the sediment. A basin in San Diego Bay exposed to effluents from yacht repair facilities and containing TBT paint particles in the sediment (Stephenson *et al.*, 1988) did not exhibit this trend, possibly because the paint particles were not subject to rapid degradation. Additionally, the TBT:MBT ratios in the sediment of each basin reflected this trend. The sediment from the yacht basin had a much lower TBT:MBT (2:5) ratio than the sediment from the basin exposed to commercial activity (4:3). This indicates that sorbed TBT is more available

Table 21.3 Concentration of total butyltin and butyltin species in microcosm sediments exposed to tributyltin from an antifouling coating

		Tributyltin		*Dibutyltin*		*Monobutyltin*	
Days of exposure (d)	*Total butyltin* ($\mu g\ kg^{-1}$)	Bu_3SnCl ($\mu g\ kg^{-1}$)	*Percent of total*	Bu_2SnCl^2 ($\mu g\ kg^{-1}$)	*Percent of total*	$BuSnCl^3$ ($\mu g\ kg^{-1}$)	*Percent of total*
14	284	101	35.5	106	37.0	77	27.0
45	339	132	38.9	94	27.7	113	33.3
87	480	181	37.7	123	25.6	176	36.7
Control	<30	<10		<10		<10	

for rapid degradation to MBT and subsequent desorption of the MBT to bottom waters than is TBT in paint particles. The degradation of TBT in sediments where the butyltin is tightly bound or associated with paint particles apparently occurs at slower rates with a half-life of 4–6 months, suggesting moderate persistence in these sediments.

Henderson (1988) reported that fine-grained sediments, which were exposed to dissolved TBT from an antifouling coating leachate in microcosm experiments, apparently limited the quantity of TBT sorbed in the sediment by rapid dealkylation. With constant exposure to $\sim 0.08\ \mu g\ l^{-1}$ water under flowthrough conditions, the Honolulu Harbor sediment consistently averaged 37% (± 2%) TBT and 63% (± 2%) degradation products over the 87-d experiment, approximately evenly distributed between MBT and DBT (Table 21.3). This evidence supports the observations that fine-grained sediments dealkylate TBT to less toxic products in the field as well as experimentally under environmentally realistic conditions. This is also consistent with the rapid formation of monobutyltin in the mesocosm experiments that used spiked ^{14}C-TBTO described below.

21.3.2. FRESHWATER DEGRADATION STUDIES WITH TORONTO HARBOR WATER AND SEDIMENT

Volatilization of tributyltin from natural water–sediment mixtures was negligible over a period of 11 months (Maguire and Tkacz, 1985), probably because of adsorption of tributyltin to the sediment. Such adsorption is fairly strong, and desorption appears to be a slow process (Maguire and Tkacz, 1985); however, even in distilled water there is negligible volatilization of tributyltin over a period of at least 2 months (Maguire *et al.*, 1983). The mesocosm experiments discussed below suggest that under natural estuarine conditions, however, volatilization may be a significant factor in the loss of TBT from aquatic ecosystems.

The chemical degradation of TBT in freshwater ecosystems appears also to be a slow process, with a half-life >11 months in a water–sediment mixture (Maguire *et al.*, 1985b). Sunlight photolysis of TBT is the fastest physical or chemical process that we have identified that causes dissipation from, or degradation in, fresh water. However, even then the half-life is >3 months in the laboratory (Maguire *et al.*, 1983). Because of the attenuation of sunlight with depth in fresh water, photolysis is probably not important at depths greater than a few decimeters to a maximum of 2 m. The rate of photolysis can be accelerated by dissolved organic material such as fulvic acid (Maguire *et al.*, 1983), which indicates the potential influence that an ecosystem-specific property can have on persistence. There is some evidence that the photolysis of TBT in fresh water proceeds through a sequential debutylation pathway (Maguire *et al.*, 1983).

As in seawater, TBT in fresh water appears to be much more susceptible to biological degradation than to chemical or photochemical degradation. There are reports of degradation in fish tissue (Ward *et al.*, 1981) and by microorganisms in natural waters and sediments, as well as in more concentrated cultures of microorganisms (Slesinger and Dressler, 1978; Barug, 1981; Orsler and Holland, 1982; Maguire *et al.*, 1984, 1985b; Maguire and Tkacz, 1985; Olson and Brinckman, 1986). Biological degradation of TBT has also been demonstrated in soil (Kimmel *et al.*, 1977; Sheldon, 1978). Although there are some exceptions, in general it appears that aquatic organisms degrade TBT through a sequential debutylation pathway in a manner analogous to mammalian metabolism (Fish *et al.*, 1976; Kimmel *et al.*, 1977). Metabolic mechanisms can involve hydroxylated butyltin intermediates that are unstable and that lose butene, yielding DBT, MBT, or tin species, for example. This was demonstrated in mammalian microsomal preparations (Fish *et al.*, 1976; Kimmel *et al.*, 1977) and recently in microsomal preparations from crab stomach and fish liver (Lee, 1986, and Chapter 18).

Since biological degradation in water and sediment appears to be the most important factor limiting the persistence of TBT in aquatic ecosystems, it is expected that ecosystem-specific characteristics such as temperature and the kinds and concentrations of TBT-degrading organisms will be important determinants of the persistence of TBT in any particular location. Estimates of the half-life of biological degradation of TBT in fresh water from Toronto Harbor range from a few weeks to 4 months (Maguire and Tkacz, 1985; Maguire, 1986), substantially longer than the 4–19 d reported for seawater (Seligman *et al.*, 1986b,c, 1989; Lee *et al.*, 1987). The difference may be partially explained by the higher TBT concentrations used in the freshwater experiments potentially causing some inhibition of the degrading organisms.

The half-life of TBT from a degradation study with unspiked sediment from Toronto Harbor is similar to the half-life measured in the long-term TBT degradation experiment in San Diego Bay of ~4–5 months (Maguire and Tkacz, 1985; Stang and Seligman, 1986). In short-term (8-d) experiments to evaluate the degradation of spiked (40–45 $\mu g\ l^{-1}$ Sn) TBT in water–sediment suspensions from Toronto Harbor, anaerobic TBT degradation was about twice as fast as aerobic degradation. Half-lives were ~8 and 16 d, respectively. Because of the strong mixing action of the fermentor, the derived rates of degradation are likely upper limits. Thus, in the locations examined in detail to date, TBT exhibits low to moderate persistence in fresh water and water–sediment suspensions and moderate persistence in sediment.

Methylated derivatives of TBT and its degradation products have occasionally been observed in water and sediment in Canada (Maguire, 1984; Maguire *et al.*, 1986). Low concentrations of Bu_3MeSn and Bu_2Me_2Sn have also been detected in laboratory experiments on the biological degradation of TBT (Maguire and Tkacz, 1985). It is likely that these methylated compounds resulted from the methylation of the butyltin species or inorganic tin in water or sediment. However, methyltin species were not detected consistently, and they were usually present in much smaller concentrations than TBT and its degradation products. On the time scale of these experiments, therefore, it appears that methylation of TBT and its degradation products is not a significant pathway of transformation unless there is rapid mineralization of the butylmethyltins. However, the potential importance of methylation on much longer time scales cannot be dismissed.

21.3.3. FATE OF ^{14}C-TBTO IN AN ENCLOSED MARINE ECOSYSTEM

21.3.3a. Water column

The 9-month mesocosm experiment (Hinga *et al.*, 1987; Adelman, 1988; Adelman *et al.*, 1990, 1991) was run with a normal seasonal cycle of temperatures. Water temperatures for the first 40-d of the experiment were 20–23°C. Over the next 60-d the water temperatures in the mesocosm decreased to ~5°C and remained below that temperature for the remainder of the experiment. Hence, the behavior and degradation of TBT in water, which occurred early in the experiment, were under conditions typical for the summer months in New England coastal waters. The longer term behavior of the TBT degradation products was observed under fall and winter conditions.

Tributyltin disappeared from the water column of the mesocosm rapidly (Fig. 21.9a) with the concomitant formation of DBT and MBT (Fig. 21.9b). The decrease in TBT concentration appeared to follow first-order kinetics with an initial rate of 0.20 d^{-1} (half-life = 3.5 d), which slowed to 0.10 d^{-1} (half-life = 6.9 d) at ~15 d. Dibutyltin concentrations increased until about day 16, then decreased (Fig. 21.9b). Monobutyltin concentrations increased to account for ~40% of the original TBT added over the first 40 d. There was no apparent degradation of MBT for the 280-d duration of the experiment.

The overall disappearance rate of TBT was the result of at least three processes: microbial degradation, scavenging to benthic sediments, and, we postulate, removal to the air–water interface followed by volatilization. Adsorption to the walls of the mesocosm tank represented a fourth possible sink; however, sampling of the walls and the overall mass balance of materials in the mesocosm indicated that adsorption to the walls was not important.

The three processes responsible for the overall removal rate of TBT were each determined independently. The rate of net removal of radiolabel to benthic sediments was determined by observing the rate and total amount of radiolabel appearing in sediments. The rate of loss of TBT to the atmosphere was calculated from the overall loss rate of radiolabel from the mesocosm. The rate of TBT degradation was determined, first, by observing the rate of appearance of degradation products, and, second, as the difference between the overall disappearance rate and the sum of losses to sediments and atmosphere. Each process was assumed to follow first-order kinetics and the appropriate

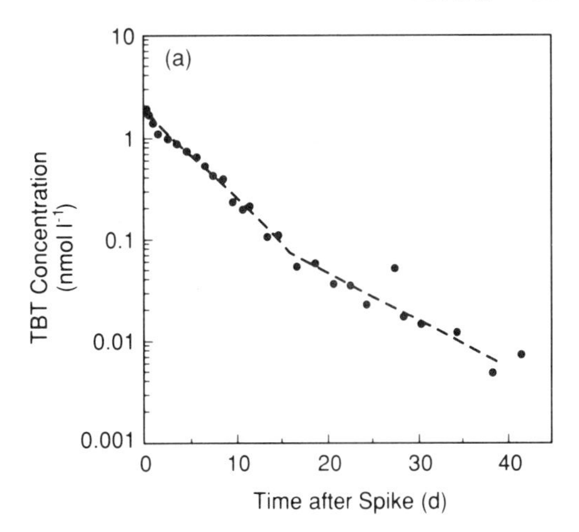

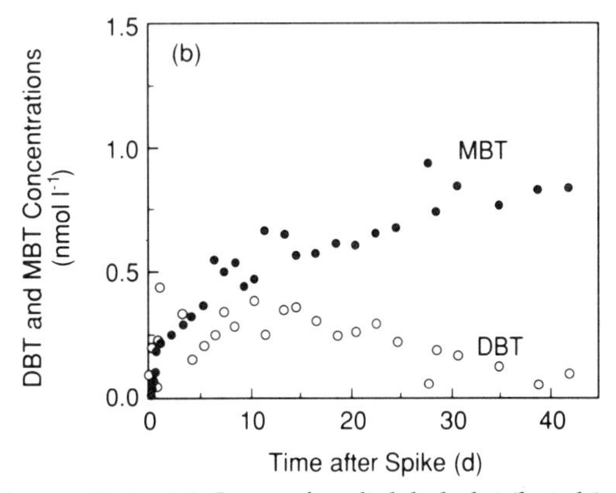

Figure 21.9 (a) Loss of radiolabeled tributyltin from mesocosm, and (b) formation of MBT and DBT degradation products during the initial 40-d fate study (modified from Adelman *et al.*, 1991).

rate constants were determined by fitting first-order curves to the data. Degradation in the water column is likely dominated by microbial processes in the water; however, it also reflects microbial and any abiotic degradation that may take place at the sediment–water interface.

Transport rates to the sediment were calculated assuming that TBT and DBT were removed according to the ratio of their distribution coefficients to suspended particles, which then settle or are harvested by benthic organisms from the water column. Eventually, TBT-contaminated particles mix deeper into the sediment by bioturbation, which removes them from ready exchange with the water column. The calculated removal is the net removal. It is possible in this model that a TBT-carrying particle could make several trips between water and the sediment–water interface before being mixed into deeper sediments. The distribution coefficients for TBT and DBT determined during the experiment were $6 \pm 3 \times 10^4$ and $3 \pm 2 \times 10^4$, respectively. A transport rate for TBT to sediments of 0.045 d^{-1} (half-life = 15.4 d) predicts the total activity found in sediments. This scavenging rate was consistent with previous experiments in MERL with other compounds of similar partitioning behavior (Hinga *et al.*, 1980; Santschi *et al.*, 1980, 1983, 1987; Hinga and Pilson, 1987). This scavenging rate requires a replacement rate for suspended particulate matter of 1.0 d^{-1}. This particle replacement rate was similar to replacement rates previously found in MERL and in Narragansett Bay in the summer (Hinga, 1988).

The rapid loss of total radiolabel (~0.07 μg d^{-1}) from the mesocosm over the first few days was not expected. This loss rate is considerably faster than has been found for volatile organic material or CO_2 in the MERL mesocosms (Bopp *et al.*, 1981; Wakeham *et al.*, 1983, 1985, 1986a,b) and is not consistent with models that accurately predict the volatilization of low-molecular-weight organic material and gases. These models require the chemical to diffuse through a non-mixed diffusive layer at the air–water interface. A different process must be envisioned to account for the rapid loss found in the mesocosm experiment.

For the first few days of the experiment, including a sample taken only 19 h after the spike, surface microlayer samples were clearly enriched in total activity relative to the bulk water column. The surface enrichment in this experiment was no longer found after TBT concentrations in the water decreased to low levels. As has been reported in field studies, whatever mechanisms concentrate chemicals in the surface layer also operate on TBT. Assuming all the activity in the surface layer was TBT, the enrichment factor of 3 ± 0.5 (ratio of microlayer concentration to bulk water concentration) was also similar to enrichments found in the field. It is only the rapidity with which the TBT was brought to the surface layer that is novel.

After rapid transfer to the surface microlayer, TBT was then subject to some removal process that resulted in a significant loss of material from the system, probably due to volatilization. Photolysis is a less likely candidate. Photodegradation rates of TBT are reported to be slow (Maguire *et al.*, 1983; Clark *et al.*, 1988). Since the rates of processes that lead to surface-layer enrichments of chemicals are not known for either the mesocosm or field, it is possible that the rate found in the mesocosm is different from that occurring in the field.

An initial biological degradation rate of 0.08 d^{-1} (half-life of 9 d) may be calculated as the difference between the overall disappearance rate and the two removal rate constants discussed above. This rate is consistent with the rate of appearance of degradation products provided one assumes that the breakdown of the removed butyl group is not instantaneous. A good fit to the concentrations of DBT, MBT, and radiolabeled CO_2 was found assuming first-order kinetics when 33% of the TBT degraded directly to MBT, DBT degraded at a rate of 0.05 d^{-1}, MBT did not further degrade, and the butyl group once removed from the tin degraded to CO_2 at ~0.03 d^{-1}.

21.3.3b. Sediment

A degradation rate for TBT in fine-grained sediments, under primarily fall and winter conditions, may be estimated from the MERL ^{14}C-labeled TBT mesocosm experiment described above. The radiolabel found in the sediment was transferred to the sediment during the first few weeks of the experiment. After that time, there was very little TBT or DBT left in the water column, and MBT did not appear to be subject to removal to sediments. After the first 2 weeks, the radiolabel in the sediments accounted for ~30% of the activity added to the mesocosm. There was considerable variability between duplicate cores, which may obscure some trends, but the activity did not appear to change in the sediment significantly after day 14. Similarly, the radiolabel in water did not appear to change except for the loss of $^{14}CO_2$ to the atmosphere. The sediment extractions and quantification of fractions did not remove enough of the radiolabel [typically only 50%, as determined by combusting the sediments after extraction and collecting $^{14}CO_2$ (Adelman *et al.*, 1990)] to interpret rigorously the concentrations of individual compounds. However, nearly all the radiolabel extracted was TBT, and there did not appear to be any change in the ratio of TBT to the total sediment activity with time. The simplest interpretation is that once TBT was incorporated into the sediments, its degradation was greatly slowed or stopped. This may in part be due to the low temperatures for much of the latter part of the 280-d experiment. Estimating the amount of degradation within the sediment, which would have been observable given the procedures used, the degradation of TBT in the sediments was less than $\sim 1 \times 10^{-4}$ d^{-1}.

21.4. DISCUSSION

Our work has clearly elucidated the labile nature of TBT by documenting debutylation processes in the aquatic environment. A summary of TBT degradation data from our work and that of other authors is provided in Table 21.4. Biological degradation occurs in the water column at rates varying from a few days to a few weeks half-life in marine and estuarine waters (Seligman *et al.*, 1986b,c, 1988; Lee *et al.*, 1987; Lee, Chapter 18). The biological nature of the degradation process was confirmed with a broad range of experiments in which rapid loss of TBT with concomitant formation of debutylated degradation products was observed, generally following zero-order kinetics, while sterile controls (poisoned or filtered) showed very little degradation. In our short-term (<2-week) degradation experiments, it is difficult to determine whether the degradation kinetics are zero order or first order because of the variability in measurements. It is likely in enclosed experiments at low substrate levels that the degradation rate will be proportional to the concentration of declining residual substrate that is typically described by first-order kinetics (Alexander, 1984). Radiolabeled ^{14}C-TBTO under simulated natural conditions in the MERL mesocosm showed loss rates consistent with first-order kinetics over the duration of the experiment. Under natural harbor and estuarine conditions, however, where substrate (TBT) concentrations are low (nanograms per liter) and approach a steady-state level because of relatively constant input from hulls, we would anticipate a relatively constant (zero-order) rate of decay. Therefore, in calculating TBT degradation or loss rates for the water column for use in predictive models, it may be more realistic to define and use a short-term

Table 21.4 Summary of tributyltin degradation from literature

Half-life (d)	*Matrix*[a]	*TBT concentration* ($\mu g\ l^{-1}$)	*Temperature* (°C)	*Comment*	*Author*
Water					
2.5	SW	0.5	Room	Partially filtered	Francois *et al.*, 1989
3.7	SW	0.6	22–24	Loss rate mesocosm	Hinga *et al.*, 1987; Adelman, 1990
9	SW	0.6	22–24	Degradation mesocosm	Adelman 1990; Hinga *et al.*, 1987
3–13	SW	0.5–1.5	11–29	Estuarine	Lee *et al.*, 1987, 1989
4–19	SW	0.5–20	12–20	Biodegradation	Seligman *et al.*, 1986b, 1986c, 1988, 1989
6	FW	0.6	20		Thain *et al.*, 1987
11	FW	5–7	20		Hattori *et al.*, 1987
13	SW	5–7	20		Hattori *et al.*, 1987
~60	SW	0.8	5		Thain *et al.*, 1987
>144	SW	744	12–20		Seligman *et al.*, 1986b,c
150	FW	2750	20		Maguire and Tkacz, 1985
Photosynthetic organisms					
1–2	SW	1.0	2.0	Nitrate supplemented/ diatoms	Lee *et al.*, 1989
6	SW	0.3–1.0	28	Photosynthetic microorganisms	Olson and Brinkman, 1986
9.6	SW	0.5–1.9	Room	Eelgrass *Zostera marina*	Francois *et al.*, 1989
25	FW	20	20	Green alga *Ankistrodesmus falcatus*	Maguire *et al.*, 1984
Sediment					
2–4	Sediment/ SW	127–300 ($ng\ g^{-1}$)	13–29	Fine grained	Stang *et al.*, 1992
15	Sediment/ SW	127–300 ($ng\ g^{-1}$)	16–19	Sandy	Stang *et al.*, 1992
8	Sediment inoculum/ FW	45	20	Sediment microorganisms	Maguire *et al.*, 1985a
16	Sediment inoculum/ FW	45	20	Sediment microorganisms	Maguire *et al.*, 1985a
120	Sediment/ FW	3000	20		Maguire *et al.*, 1985a
160	Sediment/ SW	200 ($ng\ g^{-1}$)	15	Ambient	Stang and Seligman 1986

[a]FW = Fresh water; SW = seawater.

or instantaneous decay rate rather than estimating rates over periods of weeks or months.

Experiments where microcosms were spiked with high concentrations of TBT (744 $\mu g\ l^{-1}$ showed little, if any, degradation, suggesting that toxic inhibition of the degrading microbes can occur. The longer TBT half-lives measured in fresh water from Toronto Harbor fresh water (Maguire and Tkacz, 1985) may have resulted from lower concentrations of degrading microorganisms, toxic inhibition, or the experimental conditions during the long (11-month) incubation. In contrast, a 6-d half-life measured in fresh water from a marina in the United Kingdom under similar conditions (20°C, subdued light, 570 ng l^{-1} TBT) was more representative of marine and estuarine degradation rates (Thain *et al.*, 1987). Cold temperatures (5°C) can significantly lengthen the half-life of TBT, suggesting that under winter conditions in northern waters TBT may be moderately persistent. This should be kept in perspective in that TBT inputs during winter months are substantially reduced because small craft are generally removed from the water. Additionally, release rates of TBT from hulls are significantly reduced under colder conditions (Schatzberg, Chapter 19).

The degradation of ^{14}C-TBTO spiked into a large mesocosm followed similar patterns with a degradation half-life of ~9 d (Adelman *et al.*, 1991). The overall loss rate from the mesocosm following first-order kinetics was initially 0.2 d^{-1}, or a half-life of ~3.5 d. The other losses included scavenging to sediment and volatilization. Volatilization was not observed in Canadian freshwater experiments (Maguire and Tkacz, 1985); however, this may have resulted from a fundamental difference in experimental design (mesocosm versus bottle) rather than from a contradiction in results. The question of volatilization of TBT warrants further investigation to evaluate the process and magnitude of the phenomenon.

Neither chemical degradation nor photolysis seems to be a significant process in debutylation of TBT in the water column. At the sediment–water interface, however, an abiotic degradation process has been identified (Stang *et al.*, 1992) that rapidly degraded TBT to DBT and MBT and inorganic tin in the presence of fine-grained sediments. Monobutyltin was the primary degradation product; because of its hydrophilic nature, MBT may enter the water column shortly after formation. The non-biological degradation of TBT is probably occurring at the surface of particles composed of clay and associated organic material. Other studies have documented that clay minerals were able to catalyze various chemical reactions (Boyd and Mortland, 1985; Laszlo, 1987; Haag and Mill, 1988). The apparent direct degradation of TBT to MBT in the mesocosm system (Adelman *et al.*, 1990, 1991) suggests a similar catalytic mechanism may be occurring at the sediment–water interface with subsequent release of MBT into the water column. It is possible, therefore, that a portion of the MBT measured in the water column originated from both abiotic and biotic degradation processes at the sediment–water interface. In general, MBT formation in water column degradation experiments was slow to negligible, lending further support to the sediment origins of MBT (Seligman *et al.*, 1986b,c). A similar process has been observed in eel grass (*Zostera marina*), which was found to degrade TBT to MBT, which then entered the surrounding seawater (Francois *et al.*, 1989).

Biological degradation in sediments seems to occur at much slower rates, with half-lives of several months (Maguire and Tkacz, 1985; Stang and Seligman, 1986). This slower rate of debutylation is likely due to the non-availability of TBT, which is tightly bound to clay particles or associated with paint particles. It appears, therefore, that degradation of TBT in sediments is associated with two independent processes: a rapid abiotic debutylation forming primarily MBT followed by slower biological degradation. Therefore, sediment can act as a temporary sink for that portion of TBT that is not rapidly degraded and may release TBT and its degradation products slowly back into the water column dependent on solubility and equilibrium conditions.

In general, TBT is not highly persistent in aquatic ecosystems, particularly when compared to the multi-year persistence of well-documented chlorinated hydrocarbons such as DDT, Kepone, and PCB. However, its rate of degradation is dependent on environmental conditions and the physical state of the TBT moiety. Low temperatures can reduce degradation rates, and particulate TBT, either tightly bound to clay or organic material or associated with paint particles, appears to be less available to degrading microbes. The density of degrading microbes is important, and there appears to be an association between prior environmental exposure to elevated TBT concentrations and the rate of debutylation that may be related to an acclimated or selected microbial population.

21.5. CONCLUSIONS AND RECOMMENDATIONS

The bulk of environmental fate studies and research by other investigators has shown that TBT is not highly persistent in aquatic ecosystems. Tributyltin has a half-life in the water column of a few days based on losses from degradation, defined as debutylation to less toxic products, loss to sediments, and possibly volatilization. Degradation half-lives in the water column vary from a few days to a few weeks in seawater and up to several months in fresh water. Degradation in sediments appears to consist of two processes: a rapid abiotic chemical or catalytic debutylation that can result in a very short (2–4 d) half-life of dissolved TBT at the sediment–water interface and a slower biological degradation process yielding a half-life of up to several months. The importance of the sediment–water interface degradation process will be limited by the rate of delivery of TBT to the interface. In addition to bacteria, algae, and fungi, many other aquatic organisms can metabolize TBT (Lee, Chapter 18). The degradation dynamics are important in predicting TBT concentrations from potential future use of the compound and evaluating the environmental loading capacity. It is clear that TBT does not represent the same long-term environmental threat that DDT, PCB, Kepone, and other highly persistent non-metabolizable compounds have because of their tendency to biomagnify in food chains, thus causing toxicity and reproductive failure in mammals and birds or representing cancer risks. Tributyltin is, however, extremely toxic to certain invertebrate species, particularly molluscs. It is apparent therefore, that steady-state concentrations of TBT must be kept below the critical biological action levels that affect population success for sensitive species (see Gibbs and Bryan, Chapter 13; Laughlin, Chapter 10; Salazar and Salazar, Chapter 15; His, Chapter 12). The challenge that regulators face is to restrict the use of this economically important antifouling pesticide within its environmental carrying capacity in order to maintain TBT levels in the environment that do not threaten aquatic species and that can be effectively detoxified through natural degradation and metabolic mechanisms. Future work on the fate of TBT should continue in the following areas:

1. Investigate the environmental significance of sediment abiotic degradation processes, including their kinetics and whether the sites of degradation (catalysis) become saturated.
2. Determine the rate of mobilization, degradation, and bioavailability of TBT paint particles in sediments.
3. Investigate further the biological degradation of TBT in sediments and define the rate of remobilizatiom of butyltin compounds from sediments to the aqueous phase.
4. Conduct more intensive fate studies of TBT in enclosed ecosystems (mesocosms) to define degradation pathways, partitioning, and biological transfer in sediments to better define the loading capacity of the environment.

ACKNOWLEDGMENTS

The authors thank the following people for their valuable assistance in all phases of this work: M. Stallard and M. Kram for their support of the Maine degradation experiments, R.J. Tkacz for his technical assistance in the Canadian work and D. Adelman and

M.E.Q. Pilson for their work and contributions on the MERL mesocosm experiment.

Research funding was provided by the US Navy Office of the Chief of Naval Research, and the Canadian Department of Environment, and through a cooperative agreement between the Centers Program of the US Environmental Protection Agency and the Marine Ecosystems Research Laboratory of the University of Rhode Island.

REFERENCES

Adelman, D.A. 1988. The Geochemistry of Tributyltin in Coastal Waters: An Experiment in a MERL Mesocosm. Thesis for Master of Science, University of Rhode Island, Kingston, 136 pp.

Adelman, D., K.R. Hinga and M.E.Q. Pilson. 1990 Biogeochemistry of butyltins in an enclosed marine ecosystem. *Environ. Sci. Technol.*, **24**, 1027–1032.

Adelman, D.A., K.R. Hinga and M.E.Q. Pilson. 1991 Fractionation of butyltin species during sample extraction and preparation for analysis. *Environ. Monit. Assess.* **16**, 117–125.

Alexander, M. 1984. Biodegradation of organic chemicals. *Environ. Sci. Technol.*, **19**, 106–111.

Alzieu, C. 1991. Environmental problems caused by TBT in France: assessment regulations, prospects. *Mar. Environ. Res.*, **32**, 7–17.

Alzieu, C., J. Sanjuan, J.P. Deltreil and M. Borel. 1986. Tin contamination in Arcachon Bay: effects on oyster shell anomalies. *Mar. Pollut. Bull.*, **17**, 494–498.

Alzieu, C., P. Michel, I. Tolosa, E. Bacci, L.D. Mee and J.W. Readman. 1991. Organotin compounds in the Mediterranean: a continuing cause for concern. *Mar. Environ. Res.*, **32**, 261–270.

Barug, D. 1981. Microbial degradation of bis(tributyltin) oxide. *Chemosphere*, **10**, 1145–1154.

Barug, D. and J.W. Vonk. 1980. Studies on the degradation of bis(tributyltin) oxide in soil. *Pestic. Sci.*, **11**, 77–82.

Baughman, G.L. and R.R. Lassiter. 1978. Prediction of environmental pollutant concentration. In: *Estimating the Hazard of Chemical Substances to Aquatic Life*, J. Cairns, Jr, D.L. Dickson and A.W. Maki (Eds). American Society for Testing and Materials, Philadelphia, pp. 35–54.

Blair, W.R., G.J. Olson, F.E. Brinckman and W.P. Iverson. 1982. Accumulation and fate of tri-n-butyltin cation in estuarine bacteria. *Microb. Ecol.*, **8**, 241–251.

Blunden, S.J. and A. Chapman. 1982. The environmental degradation of organotin compounds. A review. *Environ. Technol. Lett.*, **3**, 267–272.

Blunden, S.J. and A. Chapman. 1986. Organotin compounds in the environment. In: *Organometallic Compounds in the Environment*, P.J. Craig (Ed.). Longman Group, Harlow, pp. 111–159.

Bopp, R.F., P.H. Santschi, Y.-H. Li and B.L. Deck. 1981. Biodegradation and gas exchange of gaseous alkanes in model estuarine ecosystems. *Org. Geochem.*, **3**, 9–14.

Boyd, S.A. and M.M. Mortland. 1985. Dioxin radical formation and polymerization on Cu(II)-smectite. *Nature*, **316**, 532–538.

Braman, R.S. and M.A. Tompkins. 1979. Separation and determination of nanogram amounts of inorganic tin and methyltin compounds in the environment. *Anal. Chem.*, **51**, 12–19.

Branson, D.R. 1978. Predicting the fate of chemicals in the aquatic environment from laboratory data. In: *Estimating the Hazard of Chemical Substances to Aquatic Life*, J. Cairns, Jr, D.L. Dickson and A.W. Maki (Eds). American Society for Testing and Materials, Philadelphia, pp. 55–70.

Clark, E.A., R.M. Sterritt and J.N. Lester. 1988. The fate of tributyltin in the aquatic environment. *Environ. Sci. Technol.*, **22**, 600–604.

Davies, A.G. and J.C. Gamble. 1979. Experiments with large enclosed ecosystems. *Phil. Trans. R. Soc. Lond. B*, **286**, 523–544.

Davies, A.G. and P.J. Smith. 1980. Recent advances in organotin chemistry. *Adv. Inorg. Chem. Radiochem.*, **23**, 1–77.

Donaghay, P.L. 1984. Utility of mesocosms to marine pollution. In: *Concepts in Marine Pollution Measurements*, H.H. White (Ed.). Maryland Sea Grant College, College Park, Maryland, pp. 589–620.

Fish, R.H., E.C. Kimmel and J.E. Casida. 1976. Bioorganotin chemistry: reactions of tributyltin derivatives with a cytochrome P-450 dependent monooxygenase enzyme system. *J. Organometal. Chem.*, **118**, 41–54.

Francois, R., T.T. Short and J.H. Weber. 1989. Accumulation and persistence of tributyltin in eelgrass (*Zostera marina* L.) tissue. *Environ. Sci. Technol.*, **23**, 191–196.

Grice, G.D. and M.R. Reeves (Eds). 1982. *Marine Mesocosms: Biological and Chemical Research in Experimental Ecosystems*. Springer–Verlag, New York, 430 pp.

Grovhoug, J.G., P.F. Seligman, G. Vafa and R.L. Fransham. 1986. Baseline measurements of butyltin in U.S. harbors and estuaries. In: Proceedings of the Oceans '86 Organotin Symposium, Vol. 4, Marine Technology Society, Washington, DC, 23–25 September 1986, Institute of Electrical and Electronics Engineers, New York, pp. 1283–1288.

Haag, W.R. and T. Mill. 1988. Effect of a subsurface sediment on hydrolysis of haloalkanes and epoxides. *Environ. Sci. Technol.*, **22**, 658–663.

Hall, L.W., Jr, M.J. Lenkevich, W.S. Hall, A.E. Pinkney and S.J. Bushong. 1987. Evaluation of butyltin compounds in Maryland waters of Chesapeake Bay. *Mar. Pollut. Bull.*, **18**, 78–83.

Harvey, G.W. 1966. Microlayer collection from the sea surface: a new method and initial results. *Limnol. Oceanogr.*, **11**, 608–613.

Harvey, G.W. and L.A. Burzell. 1972. A simple microlayer method for small samples. *Limnol. Oceanogr.*, **17**, 156–157.

Hattori, J., A. Kosbayashi, K. Nonaka, A. Sugimae and M. Nakamoto. 1987. Degradation of tributyltin and dibutyltin compounds in environmental waters. In: *Specialized Conference on Coastal and Estuarine Pollution*, A. Awaya and T. Kusuda (Eds). Kyushu University, Osaka, 19–21 October 1987, pp. 75–80.

Henderson, R.S. 1988. Marine Microcosm Experiments on Effects of Copper and Tributyltin-based Antifouling Paint Leachates. Naval Ocean Systems Center, Technical Report (NOSC TR) 1060, San Diego, California, 42 pp.

Hinga, K.R. 1988. Seasonal predictions for pollutant scavenging in two coastal environments using a model calibration based upon thorium scavenging. *Mar. Environ. Res.*, **26**, 97–112.

Hinga, K.R. and M.E.Q. Pilson, 1987. Persistence of benz(*a*) anthracene degradation products in an enclosed marine ecosystem. *Environ. Sci. Technol.*, **21**, 648–653.

Hinga, K.R., M.E.Q. Pilson, R.F. Lee, J.W. Farrington, K. Tjessem and A.C. Davis. 1980. Biogeochemistry of benzanthracene in an enclosed marine ecosystem. *Environ. Sci. Technol.*, **14**, 1136–1143.

Hinga, K.R., D. Adelman and M.E.Q. Pilson. 1987. Radiolabeled butyltin studies in the MERL enclosed ecosystems. In: Proceedings of the Oceans '87 International Organotin Symposium, Vol. 4, Halifax, Nova Scotia, 28 September–1 October 1987, Institute of Electrical and Electronics Engineers, New York, pp. 1416–1419.

Hodge, V.F., S.L. Seidel and E.D. Goldberg. 1979. Determination of tin(IV) and organotin compounds in natural waters, coastal sediments and macro algae by atomic absorption spectrometry. *Anal. Chem.*, **51**, 1256–1259.

Huggett, R.J., M.A. Unger and D.A. Westbrook. 1986. Organotin concentrations in the southern Chesapeake Bay. In: Proceedings of the Oceans '86 Organotin Symposium, Vol. 4, Marine Technology Society, Washington, DC, 23–25 September 1986, Institute of Electrical and Electronics Engineers, New York, pp. 1262–1265.

Huggett, R.J., M.A. Unger, P.F. Seligman and A.O. Valkirs. 1992. The marine biocide tributyltin. *Environ. Sci. Technol.*, **26**(2), 232–237.

Hunt, C.D. and D.L. Smith. 1982. Controlled marine ecosystems – a tool for studying stable trace metal cycles: long-term response and variability. In: *Marine Mesocosms: Biological and Chemical Research in Experimental Ecosystems*, G.D. Grice and M.R. Reeve (Eds). Springer-Verlag, New York, pp. 111–122.

Keith, L.H., W. Crummett, J. Deegan, Jr, R.A. Libby, J.K. Taylor and G. Wentler. 1983. Principles of environmental analysis. *Anal. Chem.*, **55**, 2210–2218.

Kimmel, E.C., R.H. Fish and J.E. Casida. 1977. Biorganotin chemistry. Metabolism of organotin compounds in microsomal monooxygenase systems and in mammals. *J. Agr. Food Chem.*, **25**, 1–9.

Laszlo, P. 1987. Chemical reaction on clays. *Science*, **235**, 1473–1477.

Laughlin, R.B., Jr, R.B. Johannesen, W. French, H. Guard and F.E. Brinckman. 1985. Structure–activity relationships for organotin compounds. *Environ. Toxic. Chem.*, **4**, 343–351.

Laughlin, R.B., Jr, H.E. Guard and W.M. Coleman, III. 1986a. Tributyltin in seawater: speciation and octanol–water partition coefficient. *Environ. Sci. Technol.*, **20**, 201–204.

Laughlin, R.B., Jr, W. French and H.E. Guard. 1986b. Accumulation of bis(tributyltin) oxide by the marine mussel *Mytilus edulis*. *Environ. Sci. Technol.*, **20**, 884–890.

Lee, R.F. 1985. Metabolism of tributyltin oxide by crabs, oysters and fish. *Mar. Environ. Res.*, **17**, 145–148.

Lee, R.F. 1986. Metabolism of bis(tributyltin)oxide by estuarine animals. In: Proceedings of the Oceans '86 Organotin Symposium, Vol. 4, Washington, DC, 23–25 September 1986, Institute of Electrical and Electronics Engineers, New York, pp. 1182–1188.

Lee, R.F. and C. Ryan, 1983. Microbial and photochemical degradation of polycyclic aromatic hydrocarbons in estuarine waters and sediments. *J. Fish. Aquat. Sci.*, **40** (Suppl. 2), 86–94.

Lee, R.F., A.O. Valkirs and P.F. Seligman. 1987. Fate of tributyltin in estuarine waters. In: Proceedings of the Oceans '87 International Organotin Symposium, Vol. 4, Halifax, Nova Scotia, 28 September–1 October 1987, Institute of Electrical and Electronics Engineers, New York, pp. 1411–1415.

Lee, R.F., A.O. Valkirs and P.F. Seligman. 1989. Importance of microalgae in the biodegradation of tributyltin in estuarine waters. *Environ. Sci. Technol.*, **23**, 1515–1518.

Liu, D., K. Strachan, W.M.J. Thomson and K. Kwasniewska. 1981. Determination of the biodegradability of organic compounds. *Environ. Sci. Technol.*, **15**, 788–793.

Lundgren, A. 1985. Model ecosystems as a tool in freshwater and marine research. *Arch. Hydrobiol. Suppl.*, **70**(2), 157–196.

Maguire, R.J. 1984. Butyltin compounds and inorganic tin in sediments in Ontario. *Environ. Sci. Technol.*, **18**, 291–294.

Maguire, R.J. 1986. Review of the occurrence, persistence and degradation of tributyltin in fresh water ecosystems in Canada. In: Proceedings of the Oceans '86 Organotin Symposium, Vol. 4, Washington, DC, 23–25 September 1986, Institute of Electrical and Electronics Engineers, New York, pp. 1252–1255.

Maguire, R.J. 1987. Review of environmental aspects of tributyltin. *Appl. Organometal. Chem.*, **1**, 475–498.

Maguire, R.J. 1991. Aquatic environmental aspects of non-pesticidal organotin compounds. *Water Pollut. Res. J. Can.*, **26**(3), 243–360.

Maguire, R.J. and H. Huneault. 1981. Determination of butyltin species in water by gas chromatography with flame photometric detection. *J. Chromatogr.*, **209**, 458–462.

Maguire, R.J. and R.J. Tkacz. 1983. Analysis of butyltin compounds by gas chromatography: comparison of flame photometric and atomic absorption spectrophotometric detectors. *J. Chromatogr.*, **268**, 99–101.

Maguire, R.J. and R.J. Tkacz. 1985. Degradation of the tri-n-butyltin species in water and sediment from Toronto Harbor. *J. Agr. Food Chem.*, **33**, 947–953.

Maguire, R.J. and R.J. Tkacz. 1987. Concentration of tributyltin in the surface microlayer of natural waters. *Water Pollut. Res. J. Can.*, **22**, 227–233.

Maguire, R.J., Y.K. Chau, G.A. Bengert, E.J. Hale, P.T.S. Wong and O. Kramar. 1982. Occurrence of organotin compounds in Ontario lakes and rivers. *Environ. Sci. Technol.*, **16**, 699–702.

Maguire, R.J., J.H. Carey and E.J. Hale. 1983. Degradation of the tri-*n*-butyltin species in water. *J. Agr. Food Chem.*, **31**, 1060–1065.

Maguire, R.J., P.T.S. Wong and J.S. Rhamey. 1984. Accumulation and metabolism of tri-*n*-butyltin cation by a green alga, *Ankistrodesmus falcatus*. *Can. J. Fish. Aquat. Sci.*, **41**, 537–540.

Maguire, R.J., R.J. Tkacz and D.L. Sartor. 1985a. Butyltin species and inorganic tin in water and sediment of the Detroit and St Clair rivers. *J. Great Lakes Res.*, **11**, 320–327.

Maguire, R.J., D.L.S. Liu, K. Thomson and R.J. Tkacz. 1985b. Bacterial Degradation of Tributyltin. National Water Research Institute Report No. 85-82. Department of Environment, Canada Centre for Inland Waters, Burlington, Ontario, 17 pp.

Maguire, R.J., R.J. Tkacz, Y.K. Chau, G.A. Bengert and P.T.S. Wong. 1986. Occurrence of organotin compounds in water and sediment in Canada. *Chemosphere*, **15**, 253–274.

Odum, E.P. 1984. The mesocosm. *Bioscience*, **34**, 558–562.

Olson, G.J. and F.E. Brinckman. 1986. Biodegradation of tributyltin by Chesapeake Bay microorganisms. In: Proceedings of the Oceans '86 Organotin Symposium, Vol. 4, Washington, DC, 23–25 September 1986, Institute of Electrical and Electronics Engineers, New York, pp. 1196–1201.

Orsler, R.J. and G.E. Holland. 1982. Degradation of tributyltin oxide by fungal culture filtrates. *Int. Biodeterioration Bull.*, **18**, 95–98.

Pilson, M.E.Q. 1985. Annual cycles of nutrients and chlorophyll in Narragansett Bay, Rhode Island. *J. Mar. Res.*, **43**, 849–873.

Pilson, M.E.Q. and S.W. Nixon. 1980. Marine microcosms in ecological research. In: *Microcosms in Ecological Research*. Symposium series 52 (CONF-781101), US Department of Energy, pp. 724–741.

Polster, M. and K. Halacka. 1971. Hygienic–toxicological problems of some antimicrobially used organotin compounds. *Ernahrungsforschung*, **16**, 527–535.

Salazar, M.H. and S.M. Salazar. 1991. Assessing site-specific effects of TBT contamination with mussel growth rates. *Mar. Environ. Res.*, **32**, 131–150.

Santschi, P.H. 1982. Applications of enclosures to the study of ocean chemistry. In: *Marine Mesocosms: Biological and Chemical Research in Experimental Ecosystems*, D.G. Grice and M.R. Reeve (Eds). Springer-Verlag, New York, pp. 63–80.

Santschi, P.H., D. Adler, M. Amdurer, Y.-H. Li and J. Bell. 1980. Thorium isotope analogues for 'particle-reactive' pollutants in coastal environments. *Earth Planet. Sci. Lett.*, **47**, 327–335.

Santschi, P.H., D. Adler and M. Amdurer. 1983. The fate of particles and particle-reactive tracer metals in coastal waters: radioisotope studies in microcosms. In: *Trace Metals in Sea Water*, C.G. Wong, E. Boyle, K.W. Bruland, J.D. Burton and E.D. Goldberg (Eds). Plenum Press, New York, pp. 331–349.

Santschi, P.H., D. Adler, P. O'Hara, Y.-H. Li and P. Doering. 1987. Relative mobility of radioactive trace elements across the sediment–water interface MERL model ecosystems of Narragansett Bay. *J. Mar. Res.*, **45**, 1007–1048.

Seligman, P.F., J.G. Grovhoug and K.E. Richter. 1986a. Measurement of butyltins in San Diego Bay, CA: a monitoring strategy. In: Proceedings of the Oceans '86 Organotin Symposium, Vol. 4, Washington, DC, 23–25 September 1986, Institute of Electrical and Electronics Engineers, New York, pp. 1289–1296.

Seligman, P.F., A.O. Valkirs and R.F. Lee. 1986b. Degradation of tributyltin in marine and estuarine waters. In: Proceedings of the Oceans '86 Organotin Symposium, Vol. 4, Washington, DC, 23–25 September 1986, Institute of Electrical and Electronics Engineers, New York, pp. 1189–1195.

Seligman, P.F., A.O. Valkirs and R.F. Lee. 1986c. Degradation of tributyltin in San Diego Bay, California, waters. *Environ. Sci. Technol.*, **20**, 1229–1235.

Seligman, P.F., C.M. Adema, P.M. Stang, A.O. Valkirs and J.G. Grovhoug. 1987. Monitoring and prediction of tributyltin in the Elizabeth River and Hampton Roads, Virginia. In: Proceedings of the Oceans '87 International Organotin Symposium, Vol. 3, Halifax, Nova Scotia, 28 September–1 October 1987, Institute of Electrical and Electronics Engineers, New York, pp. 1357–1363.

Seligman, P.F., A.O. Valkirs, P.M. Stang and R.F. Lee. 1988. Evidence for rapid degradation of tributyltin in a marina. *Mar. Pollut. Bull.*, **19**, 531–534.

Seligman, P.F., J.G. Grovhoug, A.O. Valkirs, P.M. Stang, R. Fransham, M.O. Stallard, B. Davidson and R.F. Lee. 1989. Distribution and fate of tributyltin in the United States marine environment. *Appl. Organometal. Chem.*, **3**, 31–47.

Sheldon, A.W. 1978. Dissipation and detoxification of organotins in the environment. In: Proceedings of the Annual Marine Coatings Conference 18, IX–1–IX–18.

Slesinger, A.E. and I. Dressler. 1978. The environmental chemistry of three organotin chemicals. In: Report of the Organotin Workshop, M. Good (Ed.). University of New Orleans, New Orleans, 17–19 February 1978, pp. 115–162.

Stallard, M.O., S.Y. Cola and C.A. Dooley. 1989. Optimization of butyltin measurements for seawater, tissue, and marine sediment samples. *Appl. Organometal. Chem.*, **3**, 105–114.

Stang, P.M. and P.F. Seligman. 1986. Distribution and fate of butyltin compounds in the sediment of San Diego Bay. In: Proceedings of the Oceans '86 Organotin Symposium, Vol. 4, Washington DC, 23–25 September 1986, Institute of Electrical and Electronics Engineers, New York, pp. 1256–1261.

Stang, P.M., D.R. Bower and P.F. Seligman. 1989. Stratification and tributyltin variability in San Diego Bay. *Appl. Organometal. Chem.*, **3**, 411–416.

Stang, P.M., R.F. Lee and P.F. Seligman. 1992. Evidence for rapid non-biological degradation of tributyltin in fine-grained sediments. *Environ. Sci. Technol.*, **26**, 1382–1387.

Stephenson, M. 1991. A field bioassay approach to determining tributyltin toxicity to oysters in California. *Mar. Environ. Res.*, **32**, 51–59.

Stephenson, M., G. Echikawa, J. Goetzl, L. Aroson, K. Paulson and D. Moore. 1988. Tributyltin Concentrations in Sediments in San Diego Harbor. San Diego Report, August 1988. Report prepared by California Department of

Fish and Game for California State Regional Water Quality Control Board, San Diego, 16 pp.

Thain, J.E., M.J. Waldock and M.E. Waite. 1987. Toxicity and degradation studies of tributyltin (TBT) and dibutyltin (DBT) in the aquatic environment. In: Proceedings of the Oceans '87 International Organotin Symposium, Vol. 4, Halifax, Nova Scotia, 28 September–1 October 1987, Institute of Electrical and Electronics Engineers, New York, pp. 1398–1402.

Valkirs, A.O., P.F. Seligman, G. Vafa, P.M. Stang, V. Homer and S.H. Lieberman. 1985. Speciation of Butyltins and Methyltins in Seawater and Marine Sediments by Hydride Derivatization and Atomic Absorption Detection. Naval Ocean Systems Center Technical Report (TR) 1037, San Diego, California.

Valkirs, A.O., P.F. Seligman, P.M. Stang, V. Homer, S.H. Lieberman, G. Vafa and C.A. Dooley. 1986. Measurement of butyltin compounds in San Diego Bay. *Mar. Pollut. Bull.*, **17**, 319–324.

Valkirs, A.O., B. Davidson, L.L. Kear, R.L. Fransham, J.G. Grovhoug and P.F. Seligman. 1991. Long-term monitoring of tributyltin in San Diego Bay, California, *Mar. Environ. Res.*, **32**, 151–168.

Waite, M.E., M.J. Waldock, J.E. Thain, D.J. Smith and S.M. Milton, 1991. Reductions in TBT concentrations in UK estuaries following legislation in 1986 and 1987. *Mar. Environ. Res.*, **32**, 89–112.

Wakeham, S.G., A.C. Davis and J.A. Karas. 1983. Mesocosm experiments to determine the fate and persistence of volatile organic compounds in coastal seawater. *Environ. Sci. Technol.*, **17**, 611–627.

Wakeham, S.G., E.A. Canuel, P.H. Doering, J.E. Hobbie, J.V. Helfrich and G.R.G. Lough. 1985. The biogeochemistry of toluene in coastal seawater: radiotracer experiments in controlled ecosystems. *Biogeochemistry*, **1**, 307–328.

Wakeham, S.G., E.A. Canuel and P.H. Doering. 1986a. Geochemistry of volatile organic compounds in seawater: mesocosm experiments with ^{14}C-model compounds. *Geochim. Cosmochim. Acta*, **50**, 1163–1172.

Wakeham, S.G., E.A. Canuel and P.H. Doering. 1986b. Behavior of aliphatic hydrocarbons in coastal seawater: mesocosm experiments with ^{14}C-octadecane and ^{14}C-decane. *Environ. Sci. Technol.*, **20**, 574–580.

Waldock, M.J., J.E. Thain and M.E. Waite. 1987. The distribution and potential toxic effects of TBT in United Kingdom estuaries during 1986. *Appl. Organometal. Chem.*, **1**, 287–301.

Walsh, G.E., L.L. McLaughlan, E.M. Lores, M.K. Louie and C.H. Deans. 1985. Effects of organotins on growth and survival of two marine diatoms, *Skeletonema costatum* and *Thalassiosira pseudonana*. *Chemosphere*, **14**, 383–392.

Walton, R., C.M. Adema and P.F. Seligman. 1986. Mathematical modeling of the transport and fate of organotin in harbors. In: Proceedings of the Oceans '86 Organotin Symposium, Vol. 4, Washington, DC, 23–25 September 1986, Institute of Electrical and Electronics Engineers, New York, pp. 1297–1301.

Ward, G.S., G.C. Cramm, P.R. Parrish, H. Trachman and A. Slesinger. 1981. Bioaccumulation and chronic toxicity of bis(tributyltin) oxide (TBTO): tests with a saltwater fish. In *Aquatic Toxicology and Hazard Assessment: Fourth Conference*, D.R. Branson and K.L. Dickson (Eds). American Society for Testing and Materials, Philadelphia, ASTM STP 737, pp. 183–200.

Wong, P.T.S., Y.K. Chau, O. Kramar and G.A. Bengert. 1982. Structure–toxicity relationship of tin compounds on algae. *Can. J. Fish. Aquat. Sci.*, **39**, 483–488.

PARTICLE–WATER PARTITIONING AND THE ROLE OF SEDIMENTS AS A SINK AND SECONDARY SOURCE OF TBT

22

J.R.W. Harris,[1] *J.J. Cleary*[1] *and A.O. Valkirs*[2]

[1]*Plymouth Marine Laboratory, Prospect Place, West Hoe, Plymouth PL1 3DH, UK*

[2]*Computer Sciences Corporation, 4045 Hancock Street, San Diego, California 92110-5164, USA*

ABSTRACT

The sediments are identified as the major environmental sink for tributyltin (TBT) in marine and estuarine systems. Observed distribution coefficients between sediments and the overlying water are in the approximate range from 0.2 to 20 l g^{-1}. Higher values may reflect the presence of antifoulant paint chippings. Equilibrium partition coefficients (K_p) for TBT with natural particulate material from laboratory experiments are reported in the range from 0.1 to 70 l g^{-1}. The partition coefficient appears to decline with increasing particle concentration and to increase in proportion to organic carbon content, consistent with an organic carbon–water partition coefficient (K_{oc}) of ~40 l g^{-1}. Particle–water partitioning of TBT depends on salinity; both increases and decreases of K_p with salinity have been observed. The K_p is also shown to depend markedly on pH, and variations in this or other aspects of the chemical milieu could explain the varied

Organotin. Edited by M.A Champ and P.F. Seligman. Published in 1996 by Chapman & Hall, London. ISBN 0 412 58240 6

relations to salinity. Published data indicate that the partitioning of TBT can show a Freundlich dependence on its own concentration, concentration on particles varying as about the two-thirds power of dissolved concentration in one particular case. A simplified exponential model of the uptake of contaminants into bodies of natural sediment is used to demonstrate the likely importance of the uptake and re-release of TBT in terms of its half-life in the sediments and its maximum rate of release to the water column. With rapid sediment exchange, re-release only exceeds a proportion ζ_μ of the previous input when the dimensionless product of TBT degradation rate and its residence time in the sediments with no degradation exceeds $(1-\zeta_\mu)/\zeta_\mu$. Example plots indicate conditions with maximum re-release >20% of the preceding input and half-life in the sediment exceeding 30 d.

22.1. INTRODUCTION

At concentrations that are insufficient to inhibit its own biological degradation, it appears that the environmental half-life of tributyltin (TBT) is of the order of months (Seligman *et al.*, Chapter 21), not the decades that characterize some other toxicants of environmental concern (e.g. mirex; Halfon, 1984). The biological consequences of earlier release may persist to record its passing once inputs cease, but the TBT itself may be expected to disappear relatively quickly. While environmental inputs continue, however, regions in which TBT will accumulate, and regions that will act as temporary residual sources as inputs decline must be identified. This chapter identifies the sediments as its likely temporary sink, outlines the factors that affect its partitioning onto natural particulates, and discusses the way in which this, in turn, relates to the uptake and release of TBT from sediments in natural systems.

22.2. SEDIMENTS, THE MAJOR SINK

Although the main cause for concern has been the effect of TBT on marine and freshwater life, TBT is relatively insoluble in water [of order 10^{-4} gmol l^{-1} at most (Maguire *et al.*, 1983; Maguire, 1991)] and readily partitions out of it. The high octanol–water partition coefficient [5000–7000, Laughlin *et al.* (1986)] of TBT suggests that it will tend rather to concentrate in hydrophobic, organic phases of an environment, and, accordingly, much higher concentrations are usually to be found in sediments (e.g. Valkirs *et al.*, 1986) and in the organic surface film (Cleary and Stebbing, 1987; Cleary, 1991) than in the water column. Typically concentrations in sediments are around a thousand times those in the water (Table 22.1). Although the apparent enhancement at the surface, sampled using a Garrett screen (Cleary and Stebbing, 1987), is much less, ~2–3 times that in the subsurface water (Fig. 22.1), this probably reflects the thickness of the screen sample [280–300 μm, Cleary and Stebbing (1987)] relative to the thickness of any organic film (<0.01 μm). The results illustrated in Fig. 22.1 would be consistent with partitioning into an organic surface film 0.02 μm thick with a coefficient comparable with the 5500 reported by Laughlin *et al.* (1986) as the octanol–water partition coefficient of TBT in seawater. Although the high surface concentrations of TBT that this implies may be important in themselves [Cleary and Stebbing (1987), see also Connolly and Thomann (1982)], the relatively minute bulk of the surface microlayer compared with that of the underlying water, or sediments, renders the surface microlayer negligible as an environmental sink (Maguire, 1984, 1991). In contrast, particularly in turbid and energetic systems, the uptake potential of sediment in dispersive contact with the water column may much more nearly approach that of the overlying water.

Table 22.1 Reported sediment–water distribution coefficients for TBT

Location	*Ratio of sediment concentration to water concentration ($l\ g^{-1}$)*	*Reference*
Shelter Island, San Diego, California, USA	0.200	Valkirs *et al.*, 1986
	0.223	
	0.257	
Commercial Basin, San Diego, California, USA	0.533	Valkirs *et al.*, 1986
	1.480	
	1.673	
	2.657	
Chesapeake Bay, Virginia, USA	1.0	Unger *et al.*, 1988
	2.5	
	2.6	
	5.5	
	10.0	
Pearl Harbor, Hawaii, USA[a]	14.821	Stang and Seligman, 1987
	24.706	
	55.439	

[a]Determined from desorption experiments.

22.3. PARTICLE–WATER PARTITIONING OF TBT

An understanding of the partitioning of TBT between dissolved and particle-sorbed states provides the initial key to elucidating the occurrence and importance of TBT in sediments. The particle–water partition coefficient of TBT has been determined a number of times under a variety of conditions, both in the laboratory and under natural circumstances. Values obtained are generally of the order of 1 $l\ g^{-1}$ (1000 in the 'dimensionless' terms of $l\ kg^{-1}$), but vary from ~0.1 to ~70 lg^{-1} (Table 22.2). Such wide variation, over nearly three orders of magnitude, is clearly significant in any general calculation of the environmental uptake of TBT on particles. The disparate conditions under which partitions were observed give ample scope for a range of sources for this variation; particle concentration and organic carbon content, salinity, and pH have all been considered (Randall and Weber, 1986; Valkirs *et al.*, 1986; Harris and Cleary, 1987, 1991; Unger *et al.*, 1988; Plymsolve, 1990; Huggett *et al.*, 1992; Unger *et al.*, Chapter 23).

22.3.1. THE EFFECT OF PARTICLE CONCENTRATION

It has been remarked (Valkirs *et al.*, 1986) that the partition coefficient appears to be inversely dependent on particle concentration, and this is apparent in Table 22.2. The relation is similar to the relations demonstrated by O'Connor and Connolly (1980) to be common in interactions with natural sediments (Fig. 22.2). It is not clear whether this appearance indeed reflects an inherent relation or if it is due to some additional factor with which the concentration of particulate material is confounded. Where vessel maintenance occurs, it has been suggested that extremely high partition coefficients observed in the field may simply reflect the presence of particles of antifouling paint within the particulate phase (Valkirs *et al.*, 1987). The presence of paint particles, however, cannot explain the high partition coeffi-

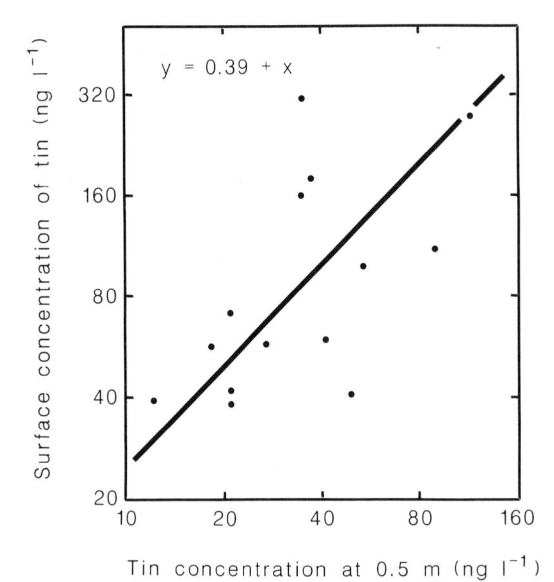

Figure 22.1 Surface–water partitioning of TBT; samples were collected in July 1988 from coastal and estuarine sites in southwest England. They were collected and processed as described in Cleary and Stebbing (1987), surface samples being taken using a Garrett screen. Axes are logarithmic, the line is the least squares fit assuming direct proportionality between concentrations at the surface and at 0.5 m depth.

cients reported in the relatively low turbidity mesocosm studied by Adelman *et al.* (1990). One suggested explanation for partition coefficients that decline as particle concentrations increase has been the presence of dissolved or colloidal organic material in concentrations proportional to those of the particles, which compete with them for the contaminant (Voice *et al.*, 1983; Gschwend and Wu, 1985). Alternatively, or in addition, the high partition coefficient between green algae and water [~30 l^{-1} g, Maguire *et al.*, (1984)] suggests the possibility that the higher partition coefficients associated with low particle concentrations, visible in Table 22.2, may reflect (in part at least) a higher proportion of planktonic algae or other organic material among the particles at lower overall particle concentrations.

22.3.2. THE EFFECT OF ORGANIC CONTENT

It is to be expected that the partitioning of a relatively hydrophobic material such as TBT will depend upon the organic content of the particles concerned (Karickhoff *et al.*, 1979). In general this appears to be so, and this relation may be sufficient to account for the high observed uptake of TBT in the biota (Fig. 22.3). The overall regression for the data plotted in Fig. 22.3, assuming a gradient of unity, implies a general organic carbon–water partition coefficient [K_{oc}; Karickhoff *et al.* (1979), Eq. 2] of ~41 l g^{-1}. This value is somewhat lower than might be expected from the relation to octanol–water partition coefficient (K_{ow}) derived by Karickhoff *et al.* (1979) (Eq. 6; $\log K_{oc} = \log K_{ow} - 0.21$) and the K_{ow} of 5500–7000 reported by Laughlin *et al.* (1986), being consistent with a K_{ow} of ~3900. Be this as it may, considerable variation remains unaccounted for by the organic carbon content of the particles concerned.

22.3.3. THE EFFECT OF SALINITY

The variation unexplained by variation in content of particulate organic carbon or particle concentration might relate in part to the wide range of salinities at which observations were made. Salinity is known to affect particle–water partitioning of both hydrophobic materials and metal ions (e.g. Harris *et al.*, 1984) and has been reported to have a marked effect on the partitioning of TBT (Randall and Weber, 1986; Harris and Cleary, 1987; Unger *et al.*, 1988, and Chapter 23; Harris *et al.*, 1991). Although the relations to salinity reported for TBT partitioning onto natural particulate material are strong, they are inconsistent between the two studies concerned: Harris and Cleary (1987) find the increased particle adsorption with increased salinity typical of hydrophobic materials; Unger *et al.* (1988) show a decline in adsorption similar to that found among metallic cations. TBT is clearly hydrophobic, but in solution appreciable proportions of the free

Table 22.2 Reported particle–water partition coefficients for TBT

K_p (l g^{-1})	*Salinity* (g l^{-1})	*Particle concentration* (mg l^{-1})	*Organic carbon* (%)	*Circumstances*	*Reference*
60 ± 30	Seawater	—[a]	—	Mesocosm	Adelman *et al.*, 1990
71	Seawater	0.68	—	Mesocosm	
39.352	Seawater	1.6	—	In situ	Valkirs *et al.*, 1987
38.919	Seawater	3.0	—	In situ	
4.608	Seawater	8.6	—	In situ	
3.278	Seawater	5.8	—	In situ	Valkirs *et al.*, 1986
3.918	Seawater	6.7	—	In situ	
0.929	Seawater	14	—	In situ	
0.340	Seawater	50	—	In situ	
0.4	0	60	—	Laboratory	Harris and Cleary, 1987
1.4	32	60	—	Laboratory	
0.2	0	60	—	Laboratory	
1.0	32	60	—	Laboratory	
2.18 ± 0.35	41[b]	10 000	2.4	Laboratory	Maguire and Tkacz, 1985
8.0	0	3000–30 000	4.2	Laboratory	Unger *et al.*, 1988
1.3	24	3000–30 000	2.9	Laboratory	
0.6	24	3000–30 000	0.34	Laboratory	
0.11	35	3000–30 000	0.90	Laboratory	

[a]Indicates no information available.
[b]Calculated from reported ionic strength.

cation may occur. The particle–water partition coefficients exhibited by TBT are consistent with its hydrophobic nature, but these coefficients are unfortunately also comparable to those observed for trace metals (e.g. Bale, 1987). It is difficult then to eliminate either electrostatic or hydrophobic interactions *a priori* as the key to the interactions of particles with TBT. To obviate the problems of incomplete recovery that dog studies of this kind, the original experiments of Harris and Cleary (1987) incorporated analysis of both water and particulate phases for tin. As a further check, the studies have since been repeated using ^{14}C-labeled TBT chloride; with >90% recoveries, they gave results completely consistent with those obtained previously (Fig. 22.4). The observations of Unger *et al.* (1987) have similarly been carefully examined for sources of artifactual error (Unger *et al.*, 1988, and Chapter 23).

The wide variation in the general levels of partition coefficient exhibited for the various sediments used in these experiments (Fig. 22.4) are consistent with the variation in organic content among them; from nearly 20% organic carbon in the case of the Belle Glade soil to ~2% for the sediment from the Crouch. This does not, however, account for the differing gradients in Fig. 22.4. In this respect, the consistency within each of the two studies illustrated, despite the disparate sources of the sediments used for each, suggests an explanation within the experimental procedures adopted rather than in the inherent properties of the sediments. The major difference between the two studies is the sediment concentrations used. The declining relation of K_p to salinity was observed with sediment concentrations two to three orders of magnitude higher than concentrations yielding the increasing rela-

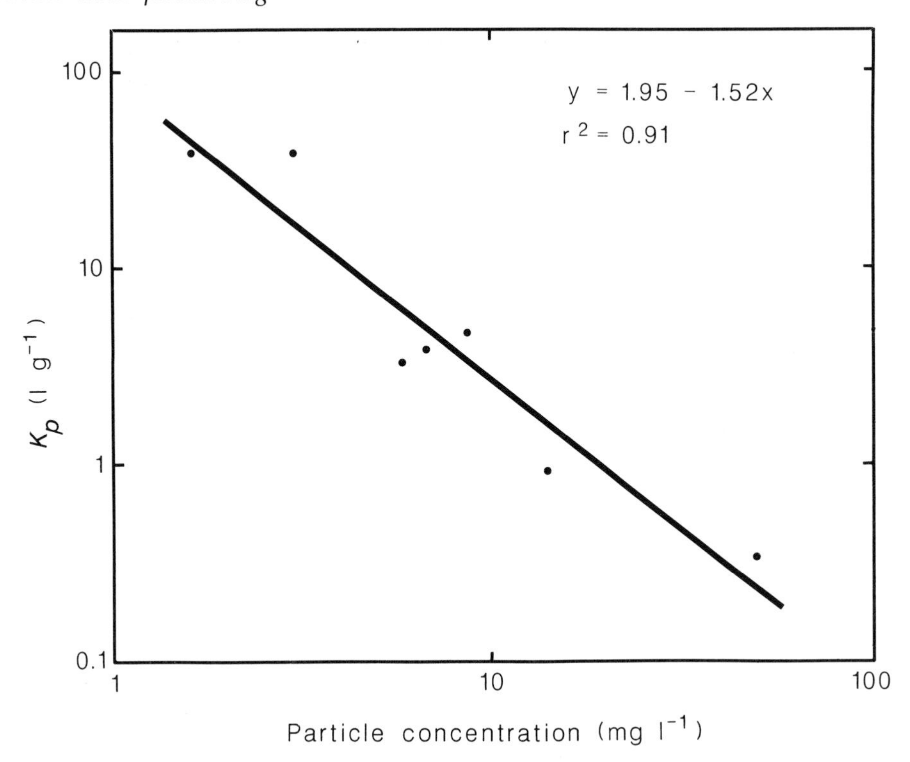

Figure 22.2 The relation of particle–water partitioning of TBT to particle concentration apparent in low-turbidity coastal environments. The line represents the least squares fit to the points. Data from Valkirs *et al.* (1986) and Valkirs *et al.* (1987); see Table 22.2.

tion (Table 22.2). Noting this, Unger *et al.* (1988) suggest that the difference may reflect the limited availability of charged binding sites, which become saturated at the lower particle concentrations. Indeed, although they do not remark it, the results they show [Unger *et al.* (1988), Fig. 1], in which sorbed and dissolved concentrations are perhaps linearly, but not proportionately, related, indicate an increasing K_p at lower TBT concentrations (higher particle concentrations in their experimental context). Replotting the presented results on logarithmic axes (Fig. 22.5) indicates the observations to be well described by the Freundlich relation:

$$c_p = 3.13 c_w^{0.673} \; (r^2 = 0.98, \; n = 10) \qquad 1$$

In this c_p (g g^{-1}) is the concentration of TBT on the particles and c_w (g l^{-1}) is its dissolved concentration. Such a relation is completely consistent with the progressive saturation of heterogeneous binding sites on the particles. Since, however, the experiments reported here and in Harris and Cleary (1987) resulted in concentrations of TBT on particles reaching a maximum of $<10^{-3}$‰ (by wt), this would imply an extremely low availability for charged sites. That much greater uptake of metal ions is found on particles within the Tamar Estuary (Bale, 1987), with no indication of any binding-site limitation, indicates that such a limitation would be unlikely to affect TBT.

22.3.4: THE EFFECT OF pH

Randall and Weber (1986), using a particulate phase of ferric oxide, observed particle concentration to affect partitioning behavior in their experiments. However, in that case, the

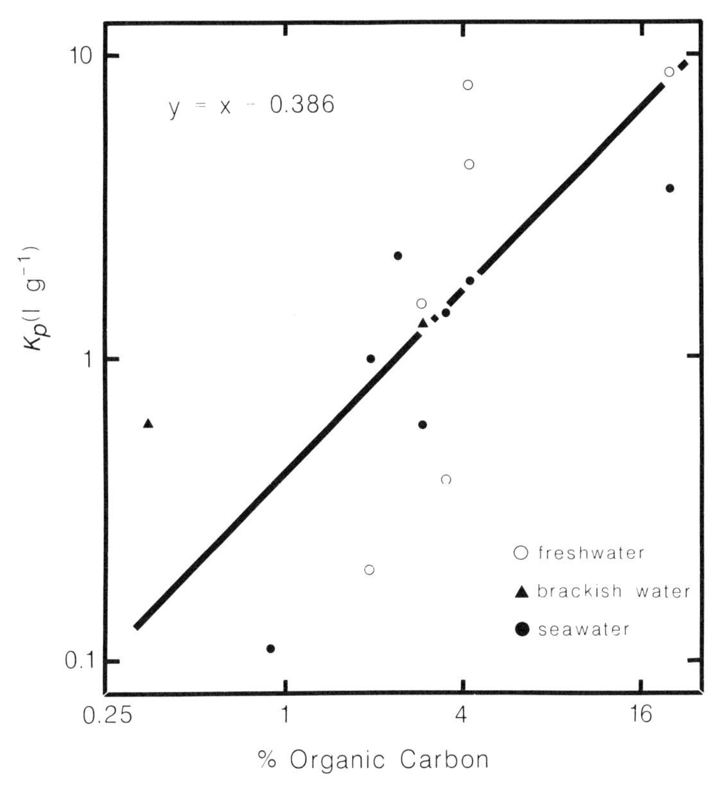

Figure 22.3 The relation of particle–water partitioning of TBT to particle organic carbon content. The line represents the least-squares fit to the points. Data from Maguire and Tkacz (1985), Harris and Cleary (1987) (Tamar sediment organic carbon 3.5%, Crouch 1.92%), and Unger *et al.* (1988), Table 2 and Fig. 4.

partition coefficient declined as particle concentration increased, and their explanation for this, that it reflected a limited supply of an organic surface coating (fulvic acid), does not appear directly to apply here (but see Harris and Cleary, 1987). Randall and Weber also observed, however, that partitioning was sensitive to pH. Seawater is well buffered to a pH of ~8.2. At the particle concentrations used by Harris and Cleary (1987) and in the experiments reported here, this buffering is sufficient to maintain a pH >8 down to salinities <4‰ when diluting with deionized water (the pH of a suspension of sediments from the Tamar Estuary at this concentration in deionized water alone is ~6). Sediments of terrestrial and freshwater origin frequently exhibit a somewhat lower pH than this; and, at the much higher particle concentrations used by Unger *et al.* (1988), the sediments will tend to reduce pH as salinity, and its buffering influence, decline.

The effect of pH on the water solubility of TBT oxide is marked (Maguire *et al.*, 1983), an effect that should be mirrored in its adsorption to particles (Karickhoff *et al.*, 1979). Experiments with Tamar sediments (1 g l^{-1}) in phosphate buffer (ionic strength 0.17–0.25 M), using ^{14}C-labeled TBT chloride (15.5 g l^{-1}) and counting all phases, confirm this (Fig. 22.6), although these experiments do not provide an immediate explanation of the anomalous relations to salinity. It is immediately noticeable that the partition coefficients (0.7–7 l g^{-1}) are rather higher than previously observed with the same sediment (Table 22.2

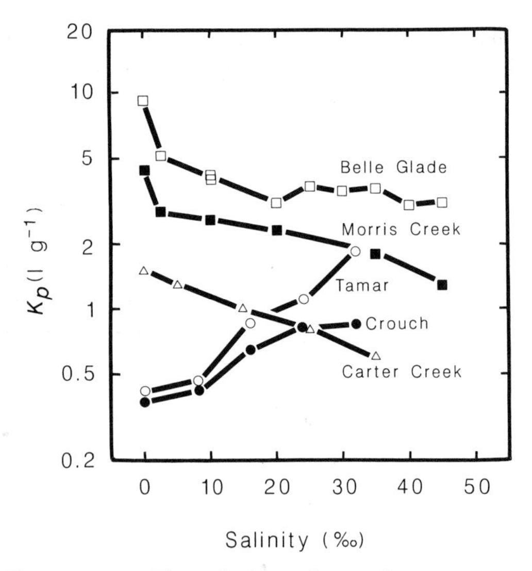

Figure 22.4 The relation of particle–water partitioning of TBT to salinity. The data for Belle Glade (Florida, USA), Morris Creek, and Carter Creek (Chesapeake Bay, USA) are redrawn from Unger *et al.* (1988). Those from the Tamar and the Crouch (England) are new, confirming those of Harris and Cleary (1987).

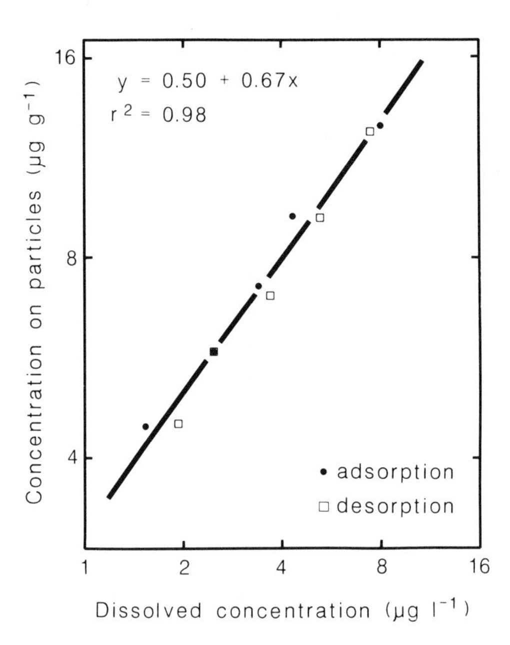

Figure 22.5 The apparent Freundlich sorption isotherm for TBT on sediment from Carter Creek, redrawn from Unger *et al.* (1988). The line is the least-squares fit to the points.

and Fig. 22.4). The increase is about an order of magnitude (~6 l g^{-1} compared with ~0.7 l g^{-1}) when corresponding pH (8.1) and ion strength (0.25 M) are compared. We must presume this to reflect the differing ionic compositions of the aqueous phases, the high phosphate concentration of the buffer yielding a relatively insoluble phosphate of TBT. This reiterates the extent to which alteration of its ionic milieu, including changes in pH, can markedly alter the particle–water partitioning of TBT. At sufficiently high concentrations of natural sediments, such changes might easily accompany changes in particle concentration and be sufficient to cause the observed difference in the salinity dependence.

22.4. CONTAMINANT UPTAKE BY SEDIMENTS

The uptake of contaminants by sediments, particularly in estuarine systems, has been discussed by Harris (1987), who suggested that, neglecting processes of permanent sedimentation and burial, temporal changes in the total sediment contaminant load, Q, under conditions of constant input, may be approximated by an exponential approach toward an asymptotic steady-state load [Harris (1987), Eq. 10].

$$dQ/dt = \omega(Q^{*}-Q)-\eta(t) \qquad 2$$

Here, Q^{*} represents the steady-state load in the absence of the effect of spatial heterogeneity ($\eta(t)$). In addition to the throughput of water and particulate material, this load is largely determined by particle–water partitioning alone, while the uptake constant (ω) depends additionally on the exchange of material between bed and water column. The time-varying factor, $\eta(t)$, which provides a correction for spatial heterogeneity, represents the spatial covariation between the

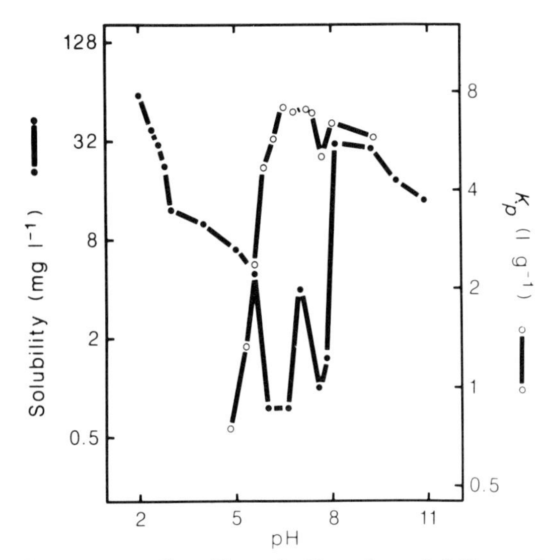

Figure 22.6 The effect of pH on the solubility and the particle–water partitioning of TBT. Solubility data are from Maguire *et al.* (1983). Partitioning data are new, using sediments from the Tamar Estuary, southwest England, and phosphate buffer (see text).

uptake in one region and its influence on uptake elsewhere. It is more convenient to rewrite Eq. 2 in terms of the steady-state load in the presence of spatial heterogeneity (Q_0^*, in which case the equation becomes simply

$$dQ/dt = \omega(Q_0^* - Q) \quad 3$$

where

$$Q_0^* = Q^* - \eta(t)/\omega \quad 4$$

Such an approach is by no means restricted to estuaries in its application, and perhaps provides some basis on which to consider the transfer of TBT into and out of the aquatic sediments. As it stands, however, this representation does not allow for contaminant degradation in the sediments, which is clearly significant in the present instance (Seligman *et al.*, Chapter 21). If degradation occurs at a rate defined by the first-order rate constant μ, Eq. 3 is replaced by

$$dQ/dt = (\omega+\mu)(Q_\mu^* - Q) \quad 5$$

where the steady-state contaminant load in the presence of degradation (Q_μ^* is related to that in its absence (Q_0^*) by

$$Q_\mu^* = Q_0/(\mu/\omega+1) \quad 6$$

The reciprocal of the rate ω is a time that may be considered the sum of two components, the residence time of the contaminant in the sediments in the absence of degradation (τ_0), and the sediment exchange time between bed and water column (θ):

$$1/\omega = \tau_0 + \theta \quad 7$$

The residence time (τ) and the sediment exchange time (θ) are defined by

$$\tau_0 = BK_p/(K_p p + v) \quad 8$$

and

$$\theta = B/r \quad 9$$

In this, K_p is the particle–water partition coefficient (volume of water:mass of particles), p is the total throughput of particles (mass: time), v is the volume throughput of water, B is the total mass of exchangeable bed sediment, and r is the rate at which this sediment exchanges with the water column (mass:time). The partition coefficient is susceptible to experimental determination; and it seems likely that in the case of TBT it will soon be possible to extrapolate its general properties from site to site. Of the remainder, throughputs of water and particles, although specific to the region under consideration, can frequently be estimated with a certain amount of accuracy. The total mass of sediment and its exchange rate, on the other hand, are hard to determine, and typically little known.

22.5. SEDIMENTS AS A SINK AND LONG-TERM SOURCE OF TBT

Ignoring the effects of spatial heterogeneity, which may be important when horizontal mixing is weak, the residence time, τ_0, provides an initial indication of the likely significance of the uptake or release of TBT by sediments. In the absence of degradation, τ_0 effectively indicates the amount of TBT that would be contained by the sediments in the steady state, expressed as its equivalent in days of constant TBT input. To take a concrete example, Bale *et al.* (1987) describe the seasonal dynamics of mobile sediment in the estuary of the Tamar in southwest England. The upper reaches of the estuary contain of the order of 50 kilotonnes (5×10^7 kg) of estuarine mud, which undergoes seasonal migrations within the estuary and is actively exchanged with the water column. About 9×10^8 m^3 y^{-1} of fresh water and ~30 kilotonnes (3×10^7 kg) of particulate material discharge through the estuary in a year. The partitioning of TBT onto this particulate material depends upon salinity (Fig. 22.4), but the partition coefficient will be of the order of 1 m^3 kg^{-1}. Substituting $B = 5 \times 10^7$ kg, $v = 9 \times 10^8$ m^3, $p = 3 \times 10^7$ kg, and $K_p =$ 1 m^3 kg^{-1} in Eq. 8 gives a value of ~20 d for the residence time (τ_0) of TBT in the sediments. This implies that, in the steady state, with a constant input of TBT, the sediments will only contain the equivalent of ~20 d input of TBT, even in the absence of any degradation. In itself, this immediately suggests that any effect of TBT released from the sediments if other inputs cease will be rather small compared with the effect of the original inputs. Either the TBT will be rapidly dissipated or it must be released at rates considerably below the previous inputs. Any processes of degradation would obviously have the effect of reducing the importance of the TBT in the sediments still further.

In summer, flows in the Tamar are greatly reduced, to the extent that it would probably exhibit values of τ_0 in the region of 100 or so days (using figures from Bale *et al.*, 1985); if such summer conditions were permanently maintained, sediment release of TBT would be of greater significance. To explore this it is useful to have some indication of the dynamics of release. This is provided by the sediment exchange time, θ. Determination of θ requires an estimate of the rate of exchange of the sediments with the water column. Numerical modeling of the physical processes involved (Uncles *et al.*, 1985) suggests this to be ~10 kg s^{-1} over the whole estuary. This gives a sediment exchange time of ~58 d. From this we can calculate (still ignoring degradation) that the half-life of TBT in the sediments [$\ln(2)(\tau_0 + \theta)$ or $\ln(2)/\omega$] would be ~110 d, and its maximum natural release rate [$\tau_0/(\tau_0 + \theta)$] would amount to ~63% of the original input. A reduction in θ, reflecting more rapid sediment exchange, would increase the release rate but, by the same token, reduce the half-life of TBT in the sediments.

Degradation of TBT in the sediments has the effect of reducing its residence time. Thus, if τ_μ is the effective residence time with a degradation rate μ, then, in a spatially homogeneous system,

$$\tau_\mu = \tau_0/(\mu(\tau_0 + \theta) + 1) \qquad 10$$

The half-life in the sediments (ϕ_μ) and maximum release rate (ζ_μ) of the TBT would be correspondingly reduced:

$$\phi_\mu = \ln(2)/(\mu + \omega) \qquad 11$$

$$\zeta_\mu = \tau_\mu/(\tau_0 + \theta) \qquad 12$$

It can be seen that a degradation rate of 0.005 d^{-1} (i.e. a degradation half-life of nearly 140 d) would reduce the half-life of TBT in the sediment of the last example from 110 to 61 d, and its maximum release rate from 63% to 35% of the original input. A rate of 0.01 d^{-1} would give, respectively, a 42-d half-life and a 25% maximum release rate.

Conversely, combining Eqs 8 and 10 demonstrates that increasing the particle–water partition coefficient of the TBT increases its residence time in the sediments, whether or not degradation occurs. In the absence of degradation, or if τ_0 remains sufficiently small, this increased residence time (and half-life in the sediments) is associated with a corresponding increase in the maximum release rate of TBT from the sediments. If conditions particularly favored adsorption on particles and the partition coefficient were in the region of $10 \mathrm{~l~g}^{-1}$ (Fig. 22.4) instead of the $1 \mathrm{~l~g}^{-1}$ we have assumed, the effect would be to raise τ_0 by an order of magnitude, under these summer conditions to ~1000 d. Now, in the absence of degradation, the half-life of TBT in the sediments would be ~2 y (733 d), and its maximum release rate, when inputs ceased, would amount to ~95% of the original input. Under these circumstances re-release of TBT from the sediments would have an impact that was much more comparable with that of the original input.

As the timescales for release in this last example are so much longer, introduction of degradation has a marked effect. Degradation at a rate of $0.005 \mathrm{~d}^{-1}$ reduces the half-life to 117 d and the maximum release rate to 15%, while $0.01 \mathrm{~d}^{-1}$ gives a half-life of only 63 d and maximum release that is a mere 8% of the original input. It is noticeable that the release rates for TBT in these last two cases are even lower than they were when τ_0 was an order of magnitude lower; this is because the increased uptake of TBT by the sediments is offset by the degradation that occurs during its longer residence time. The effect of this is that the maximum release rate increases with τ_0 only until the latter reaches a certain value (τ_0^*), after which it declines again. The value of τ_0 at which this maximum release occurs is given by:

$$\tau_0^* = \{\theta(\theta + 1/\mu)\}^{1/2} \qquad 13$$

In the terms used to discuss it here, the maximum rate of release of TBT from sediments depends upon three factors: the exchange time of the sediments (θ), the residence time of TBT in the sediments if there were no degradation (τ), and the degradation rate of TBT in the sediments (μ). If we arbitrarily fix a level below which release from the sediments may be neglected, say when it is <20% of the original input, we can use a diagram such as Fig. 22.7 to indicate

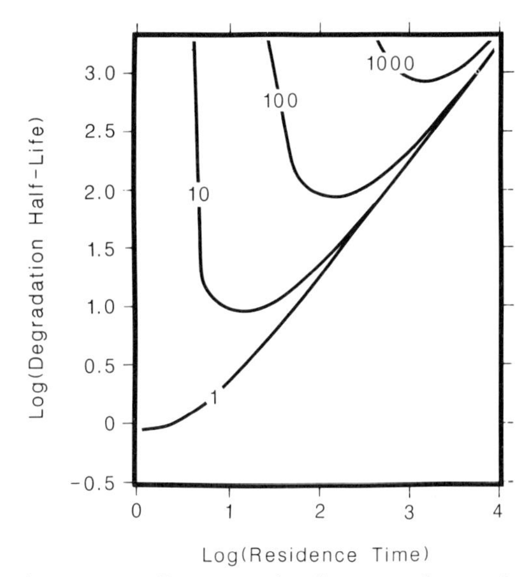

Figure 22.7 Contours of sediment exchange time (θ) in the plane of degradation half-life and residence time in the sediments with no degradation (τ_0) giving a maximum release from the sediments equal to 20% of the original input. All times are in days. Combinations of degradation half-life and residence time which define points above the relevant exchange-time contour will give release rates higher than 20%, those below, lower.

the effects of these three factors. In this, contours of the exchange time corresponding to this release rate have been plotted in the plane of the residence time and the degradation half-life [$\ln(2)/\mu$] of TBT in the sediments. For any particular exchange time, only combinations of degradation half-life and residence time that define points that lie above this contour will give release rates greater than the 20% threshold. In other words, these contours delimit the lower boundaries of conditions in which rates of re-release might give cause for concern. In the limit, with very rapid sediment exchange ($\theta = 0$) the release rate will be less than ζ_μ when the dimensionless product of degradation rate and residence time ($\mu\tau_0$) exceeds $(1-\zeta_\mu)/\zeta_\mu$. The decline in release at any degradation rate, which occurs at high and

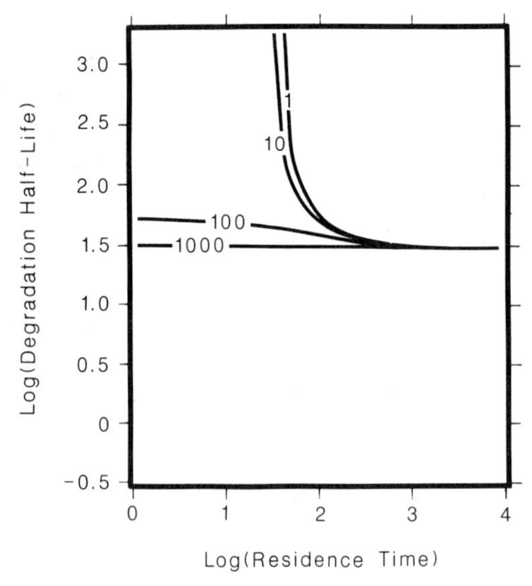

Figure 22.8 Contours of sediment exchange time (θ) in the plane of degradation half-life and residence time in the sediments with no degradation (τ_0) giving a half-life in the sediments of 30 d. All times are in days. Combinations of degradation half-life and residence time defining points which lie above the relevant exchange-time contour will give half-lives exceeding 30 d, those below half-lives of <30 d.

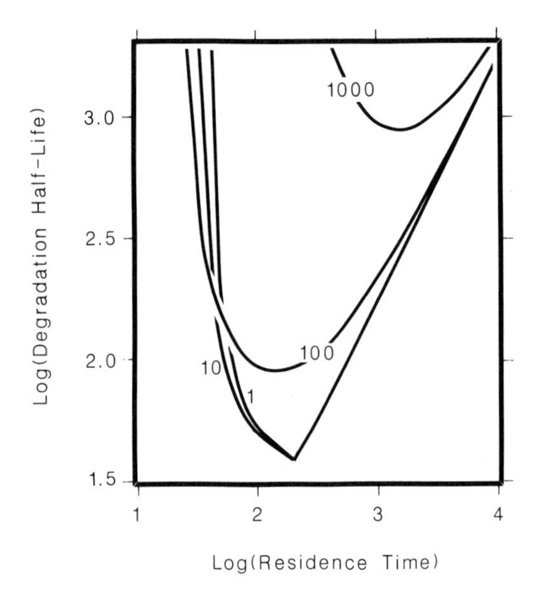

Figure 22.9 Thresholds of sediment exchange time (θ) in the plane of degradation half-life and residence time in the sediments with no degradation (τ_0). All times are in days. Combinations of degradation half-life and residence time that lie above the relevant threshold will combine a maximum release rate that exceeds 20% of the original input with a half-life in the sediments of >30 d.

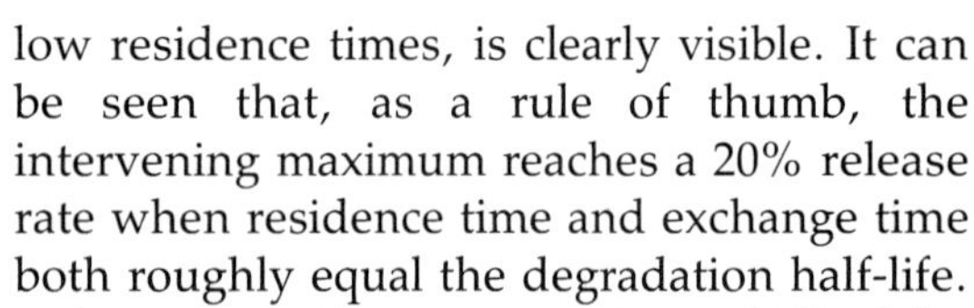

low residence times, is clearly visible. It can be seen that, as a rule of thumb, the intervening maximum reaches a 20% release rate when residence time and exchange time both roughly equal the degradation half-life.

A residual release amounting to 20% of the original input would be of less concern if it persists for only a week or so (the bottom left of Fig. 22.7) than if re-release continues over a much longer period (top right of Fig. 22.7). To make this kind of consideration explicit, we can plot exchange-time contours corresponding to some critical half-life in the same way as for release rate. Figure 22.8 is such a plot for the arbitrary threshold of 30 d. In Fig. 22.8, only combinations of degradation half-life and residence time defining points that lie above the relevant exchange-time contour will lead to a half-life for TBT in the sediments of >30 d.

Figures 22.7 and 22.8 can be combined to provide, for each exchange rate, boundaries for all conditions under which both the threshold release rate and the threshold half-life are exceeded (Fig. 22.9). Outside these boundaries either the maximum re-release of TBT would amount to no more than 20% of the original input, or the half-life of the TBT in the sediments (and hence of its release) would be no more than 30 d. Such diagrams are, at best, a crude representation of reality, with variation in both time and space ignored. With the arbitrary thresholds adopted here replaced by locally acceptable criteria, however, they might provide some means to focus attention on regions of potential concern. The effect of spatial heterogeneity on some aspects of this simplified picture has been discussed in qualitative terms by Harris (1987), but at present it is probably only possible to allow for this, and for the variation through time that characterizes real situations, through dynamic numerical simulation (e.g. Harris and Cleary, 1987); at least one suitable model, designed specifically to simulate the fate of TBT in estuaries, is available for this purpose (Plymsolve, 1990).

22.6. CONCLUSION

The interaction between the hydrophobic organic moiety of TBT and its charge characteristics dictate that its environmental chemistry is highly dependent on its local chemical environment. We have seen that, as a result, its partitioning between dissolved and particle-adsorbed states depends on a wide range of factors; particle concentration and organic carbon content, salinity and pH, as well as the concentration of the TBT itself, have all been demonstrated to have an effect. This relatively complex environmental chemistry has resulted in observed particle–water partition coefficients that vary over nearly three orders of magnitude. The breakdown of TBT to its less-harmful constituents is apparently similarly variable. Dependent, for instance, on the state of sediment anoxia, half-lives that vary from days to months have been reported (Maguire, 1987). This variation, combined with the varied processes of mixing and transport that occur in nature, mean that the significance of the sediments as a sink and, if inputs cease, as a secondary source for TBT will vary considerably from place to place. The examples we have discussed indicate that in relatively well-mixed estuarine systems TBT may be expected to dissipate fairly rapidly from the sediments once other inputs to the water cease, and in the process give water concentrations that are small compared with those resulting directly from the original inputs. If sufficient is known, diagrams such as Fig. 22.9 may be used to assess the potential impact of re-release from the sediments. This will be greatest, and perhaps significant, under conditions that combine a high partition coefficient, say 10 l g^{-1} or more, with low degradation. When degradation is sufficiently slow for potentially significant re-release, the release will be greatest when sediment exchange time and TBT residence time in the sediments are comparable with the degradation half-life.

The conclusion that in many cases the re-release of TBT once inputs have ceased would be slight both in quantity and duration compared with the original inputs does not of course imply that TBT remaining in the sediments would be of no consequence. We have only addressed the situation in which the TBT partitions onto natural particulate material. TBT already incorporated, for example, in particles of antifouling paint, will have quite different properties, and may persist and be released in significant quantities over a much longer period. We have also not considered the effect of TBT in the sediments on the benthic biota. The extent of such an effect would depend upon the bioavailability of the particle-attached TBT, which is not yet well understood and must remain beyond the scope of the present chapter.

ACKNOWLEDGMENTS

We thank R.J. Maguire, P.F. Seligman, K.R. Hinga, and, particularly, M.A. Unger for providing us with information relevant to this chapter. A.R.D. Stebbing provided advice and comments, and M.D. Brinsley drew several of the diagrams. Some of the work reported here is funded by the United Kingdom Department of the Environment.

REFERENCES

Adelman, D., K.R. Hinga and M.E.Q. Pilson. 1990. Biogeochemistry of butyltins in an enclosed marine ecosystem. *Environ. Sci. Technol.* **24**, 1027–1032.

Bale, A.J. 1987. The Characteristics, Behaviour and Heterogeneous Chemical Reactivity of Estuarine Suspended Particles. PhD Thesis, Plymouth Polytechnic, Plymouth, 216 pp.

Bale, A.J., A.W. Morris and R.J.M. Howland. 1985. Seasonal sediment movement in the Tamar Estuary. *Oceanol. Acta*, **8**, 1–6.

Cleary, J.J. 1991. Organotin in the marine surface microlayer and sub-surface waters of south-west England: relation to toxicity thresholds and the UK. Environmental Quality Standard. *Mar. Environ. Res.*, **32**, 212–222.

Cleary, J.J. and A.R.D. Stebbing. 1987. Organotin in the surface microlayer and subsurface waters of southwest England. *Mar. Pollut. Bull.* **18**, 238–246.

Connolly, J.P. and R.V. Thomann. 1982. Calculated contribution of surface microlayer PCB to contamination of Lake Michigan lake trout. *J. Great Lakes Res.* **8**, 367–375.

Gschwend, P.M. and S. Wu. 1985. On the constancy of sediment–water partition coefficients of hydrophobic organic pollutants. *Environ. Sci. Technol.* **19**, 90–96.

Halfon, E. 1984. Error analysis and simulation of mirex behaviour in Lake Ontario. *Ecol. Model.*, **22**, 213–252.

Harris, J.R.W. 1987. Sink or drain: a simulation study of factors affecting the role of an estuary subject to toxic inputs. *Water Res.*, **21**, 975–981.

Harris, J.R.W. and J.J. Cleary. 1987. Particle–water partitioning and organotin dispersal in an estuary. In: Proceedings of the Oceans '87 International Organotin Symposium, Vol. 4, 28 September–1 October 1987. Institute of Electrical and Electronics Engineers, New York, pp. 1370–1374.

Harris, J.R.W., A.J. Bale, B.L. Bayne, R.F.C. Mantoura, A.W. Morris, L.A. Nelson, P.J. Radford, R.J. Uncles, S.A. Weston and J. Widdows. 1984. A preliminary model of the dispersal and biological effect of toxins in the Tamar Estuary, England. *Ecol. Model.*, **22**, 253–284.

Harris, J.R.W., C.C. Hamlin and A.R.D. Stebbing. 1991. A simulation study of the effectiveness of legislation and improved dockyard practice in reducing TBT concentrations in the Tamar Estuary. *Mar. Environ. Res.*, **32**, 279–292.

Hinga, K.R., D. Adelman and M.E.Q. Pilson. 1987. Radiolabelled butyltin studies in MERC enclosed ecosystems. In: Proceedings of the Oceans '87 International Organotin Symposium, 28 September–1 October 1987. Marine Technology Society, Washington, DC, pp. 1416–1419.

Huggett, R.J., M.A. Unger, R.F. Seligman and A.O. Valkirs. 1992. The marine biocide tributyltin. *Environ. Sci. Technol.*, **26**(2), 232–237.

Karickhoff, S.W., D.S. Brown and T.A. Scott. 1979. Sorption of hydrophobic pollutants on natural sediments. *Water Res.*, **13**, 241–248.

Laughlin, R.B., H.E. Guard and W.M. Coleman. 1986. Tributyltin in seawater: speciation and octanol–water partition coefficient. *Environ. Sci. Technol.*, **20**, 201–204.

Maguire, R.J. 1984. Butyltin compounds and inorganic tin in sediments in Ontario. *Environ. Sci. Technol.*, **18**, 291–294.

Maguire, R.J. 1987. Environmental aspects of tributyltin. *Appl. Organometal. Chem.*, **1**, 475–498.

Maguire, R.J. 1991. Aquatic environmental aspects of non-pesticidal organotin compounds. *Water Pollut. Res. J. Can.*, **26**(3), 243–360.

Maguire, R.J., J.H. Carey and E.J. Hale 1983. Degradation of the tri-n-butyltin species in water. *J. Agr. Food Chem.*, **31**, 1060–1065.

Maguire, R.J. and R.J. Tkacz. 1985. Degradation of the tri-n-butyltin species in water and sediment from Toronto Harbor. *J. Agr. Food Chem.*, **33**, 947–953.

Maguire, R.J., P.T.S. Wong and J.S. Rhamey. 1984. Accumulation and metabolism of tri-n-butyltin cation by a green alga, *Ankistrodesmus falcatus*. *Can. J. Fish. Aquat. Sci.* **41**, 537–540.

O'Connor, D.J. and J.P. Connolly. 1980. The effect of concentration of adsorbing solids on the partition coefficient. *Water Res.*, **14**, 1517–1523.

Plymsolve, 1990. TBT simulator, an estuarine dispersal simulator for TBT. Computer software for IBM PC and PS/2 compatible computers, available from Plymsolve, Plymouth Marine Laboratory, Plymouth PL1 3DH, UK.

Randall, L. and J.H. Weber. 1986. Adsorptive behavior of butyltin compounds under simulated estuarine conditions. *Sci. Total Environ.*, **57**, 191–203.

Stang, P.M. and P.F. Seligman. 1987. In situ adsorption and desorption of butyltin compounds from Pearl Harbor, Hawaii, sediment. In: Proceedings of the Oceans '87 International Organotin Symposium, Vol. 4, 28 September–1 October, 1987, Institute of Electrical and Electronics Engineers, New York, pp. 1386–1391.

Uncles, R.J., R.C.A. Elliott and S.A. Weston. 1985. Observed fluxes of water, salt and suspended sediment in a partly mixed estuary. *Estuar. Coast. Shelf Sci.*, **20**, 147–167.

Unger, M.A., W.G. MacIntyre and R.J. Huggett. 1987. Equilibrium sorption of tributyltin chloride by Chesapeake Bay sediments. In: Proceedings of the Oceans '87 International Organotin Symposium, Vol. 4, 28 September–1 October 1987, Institute of Electrical and Electronics Engineers, New York, 1381–1385.

Unger, M.A., W.G. MacIntyre and R.J. Huggett. 1988. Sorption behavior of tributyltin on estuarine and freshwater sediments. *Environ. Toxicol. Chem.*, **7**, 907–915.

Valkirs, A.O., P.F. Seligman and R.F. Lee. 1986. Butyl tin partitioning in marine waters and sediments. In: Proceedings of the Oceans '86 Organotin Symposium, Vol. 4, 23–25 September 1986, Institute of Electrical and Electronics Engineers, New York, pp. 1165–1170.

Valkirs, A.O., M.O. Stallard and P.F. Seligman. 1987. Butyltin partitioning in marine waters. In: Proceedings of the Oceans '87 International Organotin Symposium, Vol. 4, 28 September–1 October 1987, Institute of Electrical and Electronics Engineers, New York, pp. 1375–1380.

Voice, T.C., C.P. Rice and W.J. Weber. 1983. Effect of solids concentration on the sorptive partitioning of hydrophobic pollutants in aquatic systems. *Environ. Sci. Technol.*, **17**, 513–518.

SORPTION BEHAVIOR OF TRIBUTYLTIN 23

M.A. Unger, R.J. Huggett and W.G. MacIntyre

Virginia Institute of Marine Science, School of Marine Science, College of William and Mary, Gloucester Point, Virginia 23062, USA

ABSTRACT

Understanding the sorption behavior of tributyltin (TBT) is important for predicting its fate and effects in the aquatic environment. Equilibrium sorption coefficients ranging from 0.11×10^3 to 350×10^3 l kg^{-1} have been measured in laboratory experiments using a wide variety of sorbent types, but the majority of the sorption coefficients are on the order of 10^3 l kg^{-1}. Apparent sorption coefficients calculated from TBT concentrations measured in environmental sediments and overlying waters generally agree with laboratory measurements and are on the order of 10^3 l kg^{-1}. Higher apparent sorption coefficients near boat maintenance facilities are probably the result of TBT-containing paint chips that are incorporated into sediments. Laboratory studies measuring the rate of TBT sorption or desorption show that the time required to reach equilibrium is relatively fast (hours) and that sorption is reversible. Longer-term (weeks to months) sorption kinetics for TBT have not been reported. For environments where mixing and water mass circulation times are slower than the equilibration process, laboratory-determined sorption coefficients may be reasonable estimates of TBT partitioning in the environment. Changes in salinity over the range encountered in estuaries can alter sorption coefficients, but these reported changes are varied and may depend on the particular system studied. Salinity effects may depend on sorbent-to-solution mass ratios used in equilibration experiments, and could arise from both the ionic and the hydrophobic components of the TBT molecule.

23.1. INTRODUCTION

An understanding of the sorption behavior of tributyltin (TBT) is necessary to predict its

Organotin. Edited by M.A. Champ and P.F. Seligman. Published in 1996 by Chapman & Hall, London. ISBN 0 412 58240 6

Table 23.1 Published equilibrium sorption coefficients

Reference	*Equilibration time (h)*	*Sorption coefficients* ($l\ kg^{-1}$)	*Sorbent*
Maguire and Tkacz, 1985	72	2.18×10^3	Sediment
Randall and Weber, 1986	12	1.5×10^3 to 350×10^3	Hydrous iron oxide + fulvic acid
Harris and Cleary, 1987			
Harris *et al.*, Chapter 22	24	0.2×10^3 to 2.19×10^3	Estuarine sediments
Unger *et al.*, 1988	24	0.11×10^3 to 8.2×10^3	Estuarine and freshwater sediments

fate and effects in the aquatic environment. Reuber *et al.* (1987) have noted that there is often little information on the equilibrium or kinetic behavior of environmental contaminants undergoing sediment–water transfer processes. Karickhoff and Morris (1987) have shown that sorption is an important factor contributing to bioavailability of a pollutant, and Salazar (1986) and Salazar and Salazar (1991, and Chapter 15) have stressed that sorption is important in determining the availability of TBT to filter-feeding organisms.

Laboratory experiments to establish sorption coefficients for TBT and environmental monitoring to compare these measurements with partitioning occurring in the environment are essential. Laboratory experiments should be conducted under a variety of conditions to simulate the different dynamic estuarine environments that are often exposed to TBT inputs.

The works discussed in this chapter have described the equilibrium and kinetic sorption behavior of TBT under a variety of laboratory and environmental conditions. There is generally good agreement among the results from different studies. It is, however, obvious that additional work needs to be done to establish a better understanding of the mechanisms that drive the sorption and desorption of TBT in complex environments (Evans and Huggett, 1991; Huggett *et al.*, 1992, and Chapter 24; Harris *et al.*, Chapter 22; Grovhoug *et al.*, Chapter 25; Waite *et al.*, Chapter 27).

23.2. LABORATORY-DERIVED EQUILIBRIUM SORPTION COEFFICIENTS FOR TBT

Researchers have conducted laboratory experiments to determine equilibrium sorption coefficients for TBT on various types of sorbent (Table 23.1). Equilibrium sorption coefficients derived from laboratory studies have ranged from 0.11×10^3 to 350×10^3 $l\ kg^{-1}$. While the range in these values is large, they represent a great variety of sorbent types, and most sorption coefficients are on the order of $10^3\ l\ kg^{-1}$. In all of these studies, batch techniques were used to determine equilibrium concentrations, and samples were agitated for several hours to insure that equilibrium was reached. Harris and Cleary (1987) and Unger *et al.* (1988) conducted kinetic experiments that showed equilibrium was reached in minutes to hours and that 24 h should be sufficient for equilibration. Maguire and Tkacz (1985) agitated a sample for up to 16 d, and found practically identical concentrations to those at 72 h, indicating that slower long-term sorption processes did not occur over a period of several weeks.

Maguire and Tkacz (1985) used a natural sediment and unfiltered water from Toronto Harbor, Canada, for their sorption experiment. To avoid changing its sorption characteristics the authors did not dry the sediment. The exact sampling location and size characteristics of the sediment were not given, but an organic carbon content of 2.4% was reported. A high sediment-to-water ratio of 10 000 mg l^{-1} was used in this experiment.

In the Randall and Weber (1986) study, the authors assessed the partitioning of mono-, di-, and tributyltin compounds between artificial seawater solution phases and a fulvic acid-coated hydrous iron oxide solid phase. They found that, at the highest particulate matter concentration (1000 mg l^{-1}), the measured sorption coefficients (1.5×10^3 to 4.2×10^3 l kg^{-1}) were on the same order as that found by Maguire and Tkacz (1985). At low and intermediate particulate matter concentrations (10 and 100 mg l^{-1}), Randall and Weber determined that partitioning coefficients were 1–3 orders of magnitude higher. Randall and Weber suggested that this concentration effect may be the result of fulvic acid-associated TBT being preferentially absorbed at the low particulate matter concentrations. These authors also examined the effect of changing pH on the sorption of TBT. They found that an increase in pH caused a reduction in the sorption coefficient. This effect was attributed to reduced adsorption of hydroxide and carbonate TBT complexes at high pH relative to TBT chloride complexes present at lower pH. No direct evidence was given for these mechanistic arguments.

Harris and Cleary (1987) and Harris *et al.* (Chapter 22) conducted a series of sorption experiments to help predict the particle–water partitioning and organotin dispersal in the Tamar and Crouch river (United Kingdom) estuaries. For their sorption experiments, the top 5 cm of estuarine sediment was collected, freeze dried, ground, and sieved (100 mesh) before being used in equilibration experiments with filtered seawater. Like Randall and Weber (1986), the authors found that there was a decrease in sorption coefficient with increasing particle concentration (up to 1000 mg l^{-1}). For River Tamar sediment, they found that the proportion of TBT attached to particles increased with increasing turbidity along a hyperbola consistent with a sorption coefficient of 2.19×10^3 l kg^{-1}. A similar result was noted for River Crouch estuarine sediments, but the asymptote was somewhat lower, 0.92×10^3 l kg^{-1}. The authors suggested that the lower affinity seen for River Crouch sediments may be due to the lower organic content of these sediments (percentage of carbon was not reported). The authors noted that, without further experimentation, detailed mechanistic explanations of these trends would be only speculative.

To examine the equilibrium sorption behavior of TBT, Unger *et al.*, (1988) equilibrated tributyltin chloride with estuarine and freshwater sediments in artificial seawater. Sediments varied in clay (17.9–58.4%) and organic carbon (0.34–19.8%) content and produced sorption coefficients that ranged from 0.11×10^3 to 8.2×10^3 l kg^{-1}. These authors also calculated sorption coefficients normalized to organic carbon content, but cautioned that organic carbon content alone was not a good predictor of TBT sorption. A simple measurement of the percent carbon content of a sediment does not necessarily give a measure of the qualities responsible for the TBT sorption characteristics of the sediment. The chemical characteristics of the organic matter, mineralogy, and other sediment properties can be equally important. The work of Dooley and Homer (1983) showed that there was an increase in organotin sorption with increasing clay content. Until the complex mechanisms involved in TBT sorption on natural sediments are better understood, simple sediment characteristics will probably not be good predictors of TBT sorption. For more precise estimates of TBT partitioning in natural systems, equilibrium sorption experiments should be conducted with sediments from areas of interest.

Unger *et al.* (1988) used sorption and desorption data to construct isotherms for their experiments. They found that desorption data were identical to sorption data and produced isotherms that were linear over the range of sediment-to-water ratios (up to

33 000 mg l^{-1}) used in their experiments. These desorption experiments showed that TBT sorption was reversible, indicating that TBT-contaminated sediments can act as sources for dissolved TBT. These results are in agreement with the earlier work of Dooley and Homer (1983), who showed that tributyltin was released from sediments when the sediments were equilibrated with clean, filtered seawater.

23.3. APPARENT SORPTION COEFFICIENTS FOR TBT CALCULATED FROM ENVIRONMENTAL DISTRIBUTIONS

Apparent sorption coefficients can be calculated from measured TBT concentrations found in environmental sediments and the overlying water. The apparent sorption coefficient K_{app} is defined by:

$$\frac{[\text{TBT in sediment}]}{[\text{TBT in water}]} = K_{app}$$

The apparent sorption coefficient is equal to the equilibrium sorption coefficient only in environmental situations where there is sufficient contact time between sediments and overlying waters. Adequate mixing of the water mass, sediment, and interstitial waters should occur locally for near-equilibrium conditions to exist. Replacement time for the overlying water by advection and dispersion must be considerably greater than the time required to reach equilibrium at the sediment–water interface. Published kinetic data for TBT on various types of sorbents (Dooley and Homer, 1983; Harris and Cleary, 1987; Unger *et al.*, 1988) have all shown that equilibrium concentrations are reached in minutes to a few hours, indicating that equilibrium conditions may exist in many environmental situations.

Maguire and Tkacz (1985) measured TBT concentrations in the water and surface sediment of Toronto Harbor. Apparent sorption coefficients calculated from their data range between 3.5×10^3 and 6.4×10^4 l kg^{-1} for three stations. These values are higher than the sorption coefficient (2.18×10^3 l kg^{-1}) measured in the laboratory when they equilibrated TBT between Toronto Harbor water and sediment. The authors did not specify the sampling location for the sediment used in this laboratory experiment.

Apparent sorption coefficients were calculated for TBT on sediments from Shelter Island Harbor and Commercial Basin of San Diego Bay (United States) by Valkirs *et al.* (1986). These authors measured TBT concentrations in water and sediment and found K_{app} values of 2.8×10^3 and 4.5×10^3 l kg^{-1}, respectively, for the two locations. In a later study (Valkirs *et al.*, 1987), the authors reported apparent partitioning coefficients as high as 3.9×10^4 l kg^{-1} for suspended particulate matter filtered from environmental water samples. They suggested that the unusually high values calculated may be the result of suspended paint particles from nearby boatyard activities.

Stang and Seligman (1987) made in-situ measurements of butyltin adsorption and desorption from sediments in Pearl Harbor, Hawaii. They reported apparent partitioning coefficients for TBT that ranged from 6.23×10^3 to 5.5×10^4 l kg^{-1}. They suggested that the high range of calculated sorption coefficients may be the result of paint chip particles incorporated into the sediments or that they may be an artifact of sediment sampling techniques. They noted that there was little adsorption or desorption of TBT within their enclosed dome systems, indicating that sediments and overlying waters are at, or near, equilibrium in Pearl Harbor.

Unger *et al.* (1988) reported apparent sorption coefficients for several areas around southern Chesapeake Bay. They found that for Sarah Creek, a tributary of the York River, apparent sorption coefficients ranged from 2.5×10^3 to 5.5×10^3 l kg^{-1} and were of the same order of magnitude as equilibrium sorption coefficients determined in the laboratory with sediments from a nearby creek. Apparent sorption coefficients were also calculated for 24 locations in the Hampton Roads area of Virginia. Each sampling location was sampled in July and again in September of 1986. Apparent sorption coefficients at sampling locations away from high vessel activity or berthing locations generally ranged from 10^3 to 10^4 l kg^{-1}. Values up to one order of magnitude higher were reported for 12 of 18 samples in areas where vessel activity was known to be high. These exceptional apparent sorption coefficients may be the result of TBT paint chips within the sediments or localized sediments that have a high sorptive capacity for TBT. In boat berthing areas it is not unusual to find high levels of petroleum contamination. Sediments containing petroleum hydrocarbons may demonstrate enhanced partitioning for TBT, which could raise apparent sorption coefficients.

Simultaneous measurements of butyltins in the microlayer, water column, and sediment of a northern Chesapeake Bay marina were made by Matthias *et al.* (1988). In this work, the authors reported enrichment factors (apparent sorption coefficients) that ranged from 10^3 to 10^4 l kg^{-1}. These values are in close agreement with the apparent sorption coefficients calculated by Unger *et al.* (1988) for portions of the lower Chesapeake Bay.

23.4. KINETICS OF TBT SORPTION

The rate at which TBT comes to equilibrium in a water–sediment system is important when estimating if equilibrium conditions are reached in dynamic natural environments such as estuaries. If mixing and water mass circulation times are slow relative to the time required for sorption processes to reach equilibrium, then laboratory-determined sorption coefficients may be reasonable estimates of TBT partitioning in the natural environment.

Dooley and Homer (1983) measured the uptake of organotin leachates by various sediment types and found that there was an initial fast rate with most of the uptake occurring in the first hour. Similar results were reported by Harris and Cleary (1987), and Harris *et al.* (Chapter 22) for uptake of TBT by estuarine sediments from the Tamar and Crouch rivers. These authors found that there was an initial fast rate in the first hour followed by a slower increase in sorption over the next few hours. Unger *et al.* (1988) measured the rate of TBT desorption from natural sediments that were previously equilibrated with TBT–water solutions, and found that water concentrations reached at least 60% of equilibrium values in <30 min. This was followed by a slow, approximately linear, rise toward equilibrium over the next hour. Differences in the desorption kinetics for two sediments were insignificant. Figure 23.1 illustrates the desorption kinetics for TBT from Carter Creek sediment by showing the change in the ratio of the water concentration (C) to the known equilibrium water concentration (C_{eq}) with time. This curve is typical of the kinetics of TBT sorption–desorption and indicates a two-step sorption process.

This type of two-step kinetics for sorption or desorption was observed for hydrophobic organic compounds by Karickhoff and Morris (1985) and by Wu and Geschwend (1986). Chen *et al.* (1973) measured the kinetics of phosphate sorption on aluminum oxide and kaolinite, finding a fast step followed by a slow first-order sorption they attributed to the formation of $AlPO_4$. Benjamin and Leckie (1981) studied sorption kinetics of metal ions of Sn, Zn, Cu, and Pb on iron oxides and found a rapid uptake attributed to adsorption, and a second slow rate attributed to ion diffusion into the solid substrate.

These works have treated uptake of metal ions, anions, and hydrophobic organic compounds whose range of sorption properties should include the possible sorption behavior of organotin species. It is therefore expected

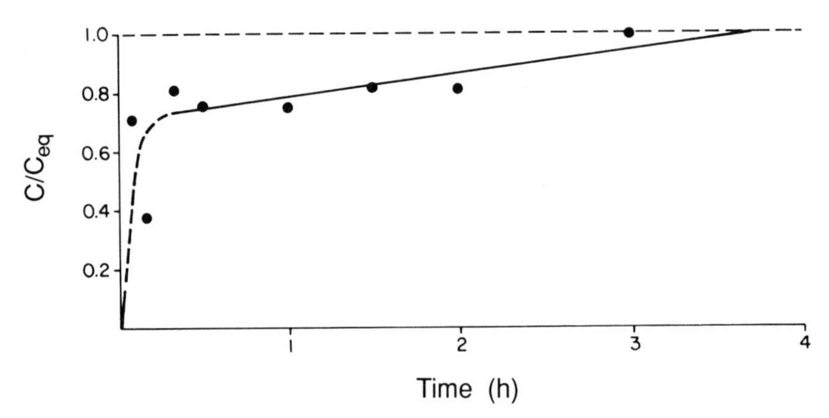

Figure 23.1 TBT desorption kinetics for Carter Creek, Virginia, sediment (adapted from Unger *et al.*, 1988).

that the sorption–desorption kinetics for TBT exhibits similar fast and slow steps. Sorption kinetics cannot establish a unique reaction mechanism without additional information at the molecular scale. The only recourse is to assume the sorption mechanism and produce models that can be fitted to kinetics data. Several authors have made assumptions that can be classified as either chemical multisite sorption models (e.g. Cameron and Klute, 1977) or sorbate transport limited models (e.g. Wu and Geschwend, 1986; Brusseau and Rao, 1989). Such assumptions and models are not mathematically distinguishable from sorption kinetics data alone. Results from sorption kinetics do not give insight into the structure of TBT sorbate species since the kinetic sorption behavior of TBT is similar to that of metals and nonpolar organic compounds. While the mechanism of TBT sorption behavior cannot be determined from these types of experiment, it is interesting that all published data on TBT kinetics are very similar. Variations in organic and clay content of the sorbent have little effect on the overall sorption–desorption rate for TBT. TBT sorption dynamics are so rapid that near-equilibrium conditions are reached in minutes to hours. This supports the hypothesis that localized equilibrium partitioning can occur in natural systems where adequate mixing takes place.

23.5. VARIATION IN TBT SORPTION WITH SALINITY

When TBT is released into the aquatic environment, it comes to equilibrium with anions in solution to produce a variety of dissolved TBT species. Little is known about the nature of these dissolved species or how they change under the varying environmental conditions found in fresh, oceanic, and estuarine waters. Laughlin *et al.* (1986) investigated the speciation of TBT by ^{119}Sn nuclear magnetic resonance (NMR) examination of chloroform extracts of synthetic seawater. While this study showed the qualitative presence of TBT chloride, TBT oxide, and TBT carbonate, it is likely that quantification was inaccurate due to equilibria shifts during the extraction process. These dissolved species probably have considerable ionic properties and are dominated by TBT chloride at high salinities. Variations in salinity will alter the relative abundances of these dissolved TBT species and modify the charged sorbent surfaces they may adsorb to. Since TBT has an ionic character, as well as hydrophobicity due to its butyl substituents, it is not surprising that a variety of results have been reported for changes in TBT sorption with salinity, which may be quite dependent on the experimental conditions.

Unger *et al.* (1988) reported on the variation of TBT sorption coefficients with salinity for three natural sediments. Two of the sediments were from freshwater locations, and one was of estuarine origin. The sediments varied in organic carbon content from 2.9 to 19.8% and were equilibrated at various salinities with synthetic seawater. These authors found a sudden decrease in sorption coefficient between 0 and 5‰ salinity for the freshwater sediments, followed by a linear decrease over the range 5–35‰. The estuarine sediment showed a linear decrease in sorption over the entire range of salinities. They found no evidence of a salting-out effect (increase in sorption with increasing salinity). Their results are similar to those reported for the sorption of metal ions on sediments (Salomons and Forstner, 1984) and suggest that TBT may behave like a simple metal ion in solution.

These results are in apparent disagreement with the experiments describing salinity effects on TBT sorption by Randall and Weber (1986) and by Harris and Cleary (1987). These investigators found an increase in TBT sorption with increasing salinity. Randall and Weber (1986) observed a salting-out effect, but only reported data for the extremes of the salinity range, so changes in salinity effects over the range cannot be inferred. Harris and Cleary (1987) and Harris *et al.* (Chapter 22) observed a sigmoidal increase in sorption of TBT with increasing salinity for sediment from the Tamar and Crouch rivers and noted that their results showed no minimum value at intermediate salinities as did the octanol–water partition coefficients reported by Laughlin *et al.* (1986). If TBT sorption on these sediments was governed by purely hydrophobic interactions, there should be greater agreement with the octanol–water partition coefficients.

Unger *et al.* (1988) suggested that differences in the reported salinity effects may be the result of the different sorbent-to-solution mass ratios used in the experiments. They noted that the salting-out effect observed by Randall and Weber (1986) diminished with increasing particle concentration and may be insignificant at the high (1000 mg l^{-1}) sorbent particle concentration they used. Harris and Cleary (1987) conducted their salinity experiments at 60 mg l^{-1} particle concentrations to best simulate suspended sediment concentrations, while Unger *et al.* (1988) worked at much higher concentrations (up to 33 000 mg l^{-1}) to represent the water–bottom sediment interface. Unger *et al.* (1988) speculated that, at low sediment concentrations, sorbent surface sites may become saturated, and hydrophobic interactions would occur. They did, however, emphasize that mechanistic arguments could not be proven on the basis of the existing data.

23.6. CONCLUSIONS

Equilibrium sorption coefficients for TBT that range from 0.1×10^3 to 350×10^3 l kg^{-1} have been reported in the literature. While there is a large range in these published values, the majority of the sorption coefficients fall in the range of 10^3–10^4 l kg^{-1}. Apparent sorption coefficients calculated from environmental TBT concentrations are in good agreement with laboratory-generated equilibrium values. These relatively high sorption coefficients indicate that significant amounts of TBT may be associated with sediments in aquatic environments. In certain coastal areas where boat maintenance activity is high, TBT concentrations in the sediments greatly exceed those predicted by equilibrium partitioning. This discrepancy may be attributed to TBT-laden paint chips that have been incorporated into the bottom sediments.

The sorption and desorption kinetics for TBT are rapid, indicating that near-equilibrium conditions may exist in many environmental situations. Direct measurement of the TBT flux at the water-sediment interface with in-situ chambers has shown that assumptions of localized equilibrium may be valid. The sorption process is reversible, indicating that TBT-contaminated sediments can act as sources for dissolved TBT at times when inputs from antifoulants are reduced.

Changes in salinity over the range encountered in estuaries can alter sorption coefficients, but these reported changes are varied and may depend on the particular system studied. Salinity effects may depend on sorbent-to-solution mass ratios used in equilibration experiments and could be the result of the ionic and hydrophobic components of the TBT molecule. Additional experiments that include isotherm data for TBT on a wide range of sediment types and sorbent-to-solvent mass ratios may give better insight into the mechanisms governing these salinity effects.

ACKNOWLEDGEMENT

This chapter is VIMS contribution number 1959.

REFERENCES

Benjamin, M.M. and J.O. Leckie. 1981. Multiple-site adsorption of Cd, Cu, Zn, and Pb on amorphous iron oxyhydroxide. *J. Colloid Interf. Sci.*, **79**, 209–221.

Brusseau, M.L. and P.S.C. Rao. 1989. Sorption non-ideality during organic transport in porous media. *CRC Crit. Rev. Environ.*, **19**, 33–99.

Cameron, D.R. and A. Klute. 1977. Convective–dispersive solute transport with a combined equilibrium and kinetic adsorption model. *Water Resour. Res.*, **13**, 183–188.

Chen, Yi Shok, J.N. Butler and W. Stumm. 1973. Kinetic study of phosphate reaction with aluminum oxide and kaolinite. *Environ. Sci. Technol.*, **7**, 327–332.

Dooley, C.A. and V. Homer. 1983. Organotin Compounds in the Marine Environment: Uptake and Sorption Behavior. Naval Ocean Systems Technical Report No. 917. US Navy, San Diego, California, 20, pp.

Evans, D.A. and R.J. Huggett. 1991. Statistical modeling of intensive TBT monitoring data in two tidal creeks of the Chesapeake Bay. *Mar. Environ. Res.*, **32**, 169–186.

Harris, J.R.W. and J.J. Cleary. 1987. Particle–water partitioning and organotin dispersal in an estuary. In: Proceedings of the International Oceans '87 Organotin Symposium, Vol. 4, Institute of Electrical and Electronics Engineers and the Marine Technology Society, Halifax, Nova Scotia, 28 September–1 October, 1987 pp.1386–1391.

Harris, J.R.W., C.C. Hamlin and A.R.D. Stebbing. 1991. A simulation study of the effectiveness of legislation and improved dockyard practice in reducing TBT concentrations on the Tamar Estuary. *Mar. Environ. Res.*, **32**, 279.

Huggett, R.J., M.A. Unger, P.F. Seligman and A.O. Valkirs. 1992. The marine biocide tributyltin: assessing and managing environmental risks. *Environ. Sci. Technol.*, **26**, 232–237.

Karickhoff, S.W. and K.R. Morris. 1985. Sorption dynamics of hydrophobic pollutants in sediment suspensions. *Environ. Toxicol. Chem.*, **4**, 469–479.

Karickhoff, S.W. and K.R. Morris. 1987. Pollutant sorption: relationship to bioavailability. In: *Fate and Effects of Sediment-bound Chemicals in Aquatic Systems*, K.L. Dickson, A.W. Maki and W.A. Brungs (Eds). Pergamon Press, Elmsford, New York, pp. 75–82.

Laughlin, R.B., Jr, H.E. Guard and W.M. Coleman, III. 1986. Tributyltin in seawater: speciation and octanol–water partitioning coefficient. *Environ. Sci. Technol.*, **20**, 201–204.

Maguire, R.J. and J. Tkacz. 1985. Degradation of tri-n-butyltin species in water and sediment from Toronto Harbor. *J. Agr. Food Chem.*, **33**, 947–953.

Matthias, C.L., S.J. Bushong, L.W. Hall, Jr, J.M. Bellama and F.E. Brinckman. 1988. Simultaneous butyltin determinations in the microlayer, water column and sediment of a northern Chesapeake Bay marina and receiving system. *Appl. Organometal. Chem.*, **2**, 547–552.

Randall, L. and J.H. Weber. 1986. Adsorptive behavior of butyltin compounds under simulated estuarine conditions. *Sci. Total Environ.*, **57**, 191–203.

Reuber, R., D. Mackay, S. Paterson and P. Stokes. 1987. A discussion of chemical equilibrium and transport at the sediment–water interface. *Environ. Toxicol. Chem.*, **6**, 731–739.

Salazar, M.H. 1986. Environmental significance and interpretation of organotin bioassays. In: Proceedings of the Oceans '86 Organotin Symposium, Vol. 4, Institute of Electrical and Electronic Engineers and the Marine Technology Society, Washington, DC, 23–25 September 1986, pp.1240–1245.

Salazar, M.H. and S.M. Salazar. 1991. Assessing site-specific effects of TBT contamination with mussel growth rates. *Mar. Environ. Res.*, **32**, 131–150.

Salomons, W. and U. Forstner. 1984. *Metals in the Hydrocycle*. Springer–Verlag, New York. 349 pp.

Stang, P.M. and P.F. Seligman. 1987. In situ adsorption and desorption of butyltin compounds from Pearl Harbor, Hawaii, sediment. In: Proceedings of the Oceans '87 International Organotin Symposium, Vol. 4, Institute of Electrical and Electronics Engineers and the Marine Technology Society, Halifax, Nova Scotia, 28 September–1 October 1987, pp. 1386–1391.

Unger, M.A., W.G. MacIntyre and R.J. Huggett. 1988. Sorption behavior of tributyltin on estuarine and freshwater sediments. *Environ. Toxicol. Chem.*, **7**, 907–915.

Valkirs, A.O., P.F. Seligman and R.F. Lee. 1986. Butyltin partitioning in marine waters and sediments. In: Proceedings of the Oceans '86 Organotin Symposium, Vol. 4, Institute of Electrical and Electronics Engineers and the Marine Technology Society, Washington, DC, 23–25 September pp. 1165–1170.

Valkirs, A.O., M.O. Stallard and P.F. Seligman. 1987. Butyltin partitioning in marine waters. In: Proceedings of the Oceans '87 International Organotin Symposium, Vol. 4, Institute of Electrical and Electronics Engineers and the Marine Technology Society, Halifax, Nova Scotia, 28 September–1 October 1987, pp. 1375–1379.

Wu, S.-C. and P.M. Geschwend. 1986. Sorption kinetics of hydrophobic organic compounds to natural sediments and soils. *Environ. Sci. Technol.*, **20**, 717–725.

TRIBUTYLTIN CONCENTRATION IN WATERS OF THE CHESAPEAKE BAY

24

R.J. Huggett,[1] *D.A. Evan,*[1] *W.G. MacIntyre,*[1] *M.A. Unger,*[1] *P.F. Seligman*[2] *and L.W. Hall, Jr*[3]

[1]*Virginia Institute of Marine Science, School of Marine Science, College of William and Mary, Gloucester Point, Virginia 23062, USA*
[2]*Department of the Navy, Naval Ocean Systems Center, San Diego, California 92152-5000, USA*
[3]*The University of Maryland, Agricultural Experiment Station, Wye Research and Education Center, Box 169, Queenstown, Maryland 21658, USA*

ABSTRACT

Tributyltin (TBT) has been analyzed in unfiltered water samples from the Chesapeake Bay (USA) from 1984 to 1989. Areas with large numbers of recreational vessels and one area with a high density of commercial and naval craft were targeted because of the likely input of TBT from these sources. The latter area also contained several large ship repair facilities that could contribute the antifouling agent. The range of TBT concentrations found at any one station was large with highest levels associated with times of peak usage. Concentrations in excess of 100 ng l^{-1} were common in some areas. The state and federal restrictions on TBT use appear to have partially achieved their desired effect of reducing water concentrations. While only one data set is extensive enough to show

Organotin. Edited by M.A. Champ and P.F. Seligman. Published in 1996 by Chapman & Hall, London. ISBN 0 412 58240 6

statistically significant downward trends, all sets visually indicate a decrease. A model that fits the most extensive data set indicates an 8–30% per year decrease in water TBT concentration for stations near boating activities.

24.1. INTRODUCTION

The major source of tributyltin (TBT) to the Chesapeake Bay is antifouling paints. No industrial inputs are known, and analyses of sewage plant effluents entering directly into the bay reveal no detectable TBT. It is not surprising, therefore, that the highest concentrations of TBT are found in areas with the greatest numbers of water craft.

An evidentiary hearing held by the Commonwealth of Virginia in 1987 reported that ~70% of the TBT entering Virginia's waters came from recreational vessels. Approximately 27% came from large commercial vessels, such as freighters, colliers, and tankers. The remainder came from miscellaneous sources including the military (Huggett *et al.*, 1992). While a number of assumptions went into these estimates (e.g. the amount of time commercial vessels spend at sea relative to that spent in port), the estimates implied that recreational vessels were the major source of contamination. In addition many, if not most, commercial vessels are painted in foreign countries, so it was logical for Virginia to focus on controlling TBT from the recreational sector in order to protect the organisms of the Chesapeake Bay. In 1987 Virginia restricted the use of TBT-containing paints on vessels <25 m in length. Longer vessels and aluminum craft were exempted but could only use paints that release TBT at a rate $<5\ \mu g\ cm^{-2}\ d^{-1}$. Maryland and a number of other states followed Virginia's action with similar legislation. Soon after, federal regulations limiting the use of TBT went into effect.

This chapter describes the efforts to monitor TBT in water in the Chesapeake Bay both before and after implementation of the various regulations. The data set starts in 1984 with intermittent samplings focused on waters near naval facilities. In 1985 programs were expanded to monitor TBT in the vicinity of recreational moorings. The authors believe that the combined TBT monitoring data set for the Chesapeake Bay and its tributaries is the most extensive in existence.

24.2. CHESAPEAKE BAY MONITORING

Monitoring for TBT has been concentrated in regions with intense vessel activity, including restricted waterways, harbors, and marinas (Roberts *et al.*, Chapter 17; Seligman *et al.*, Chapter 20; Unger *et al.*, Chapter 23). Several agencies have monitored TBT at locations identified in Figs. 24.1 and 24.2 and described in Table 24.1. The Naval Ocean Systems Center has sampled in the Elizabeth River, James River, and Hampton Roads, Virginia; and the US Environmental Protection Agency has collected data near Annapolis, Maryland: Oxford, Maryland; and Solomons, Maryland. The Johns Hopkins Applied Physics Laboratory has monitored several Maryland marinas and the Baltimore Harbor. Published results of these programs are presented in Table 24.1.

The Virginia Institute of Marine Science has conducted a program to determine TBT water concentrations and their time variation in and near marinas in Virginia. Collection was initiated in 1986. Individual results from this program are too numerous to include in a summary table, but the density of sampling permits data analyses of long-term series. These are described later in this chapter.

24.2.1. MARYLAND'S WATERS

The northern Chesapeake Bay has areas with large numbers of recreational vessels and docking facilities. Surface water (0.5-m) samples for TBT analyses were collected biweekly at seven stations over a 4-month period (June–September) in 1986 and 1988 in the Back Creek–Severn River area of Mary-

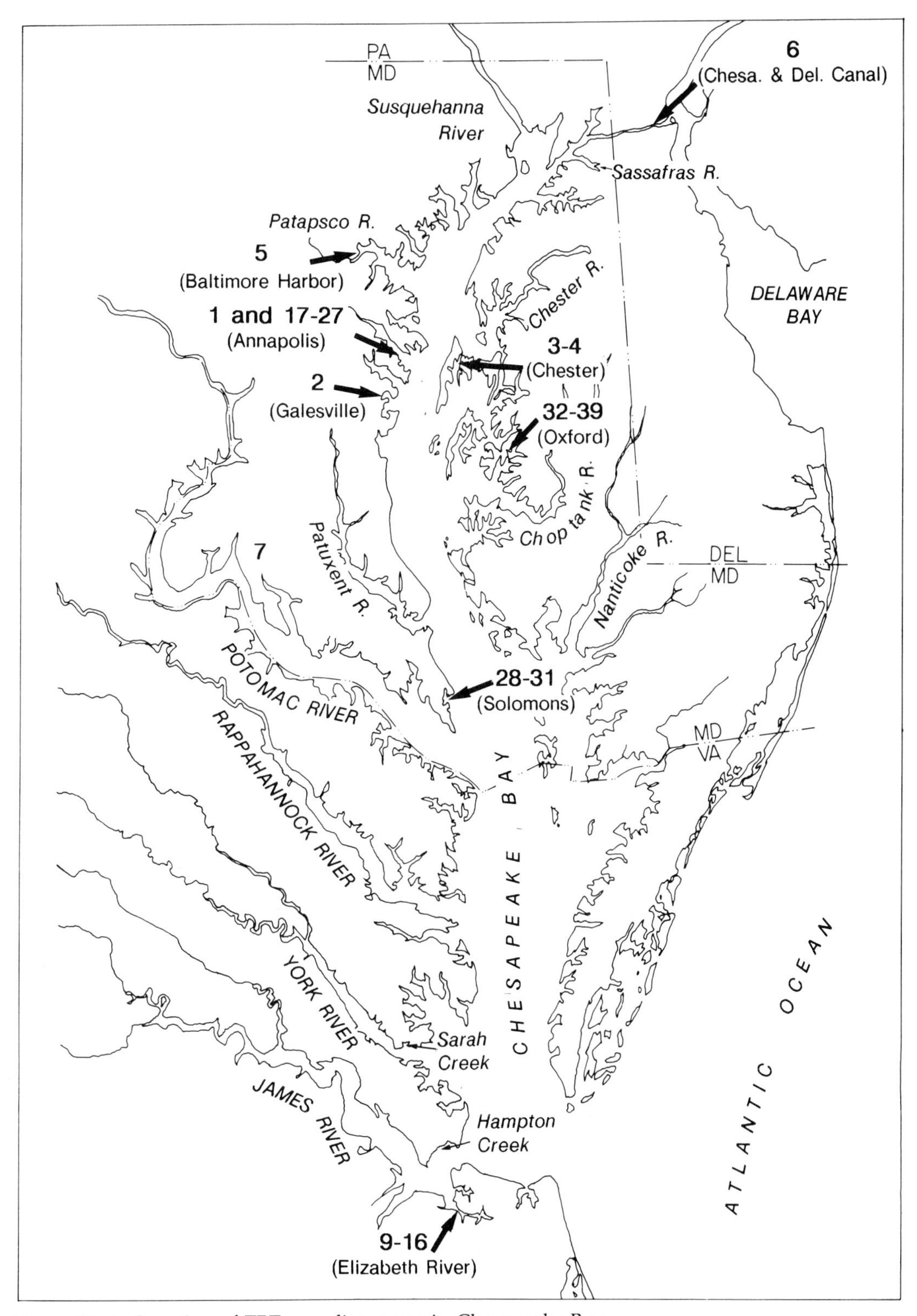

Figure 24.1 Location of TBT sampling areas in Chesapeake Bay.

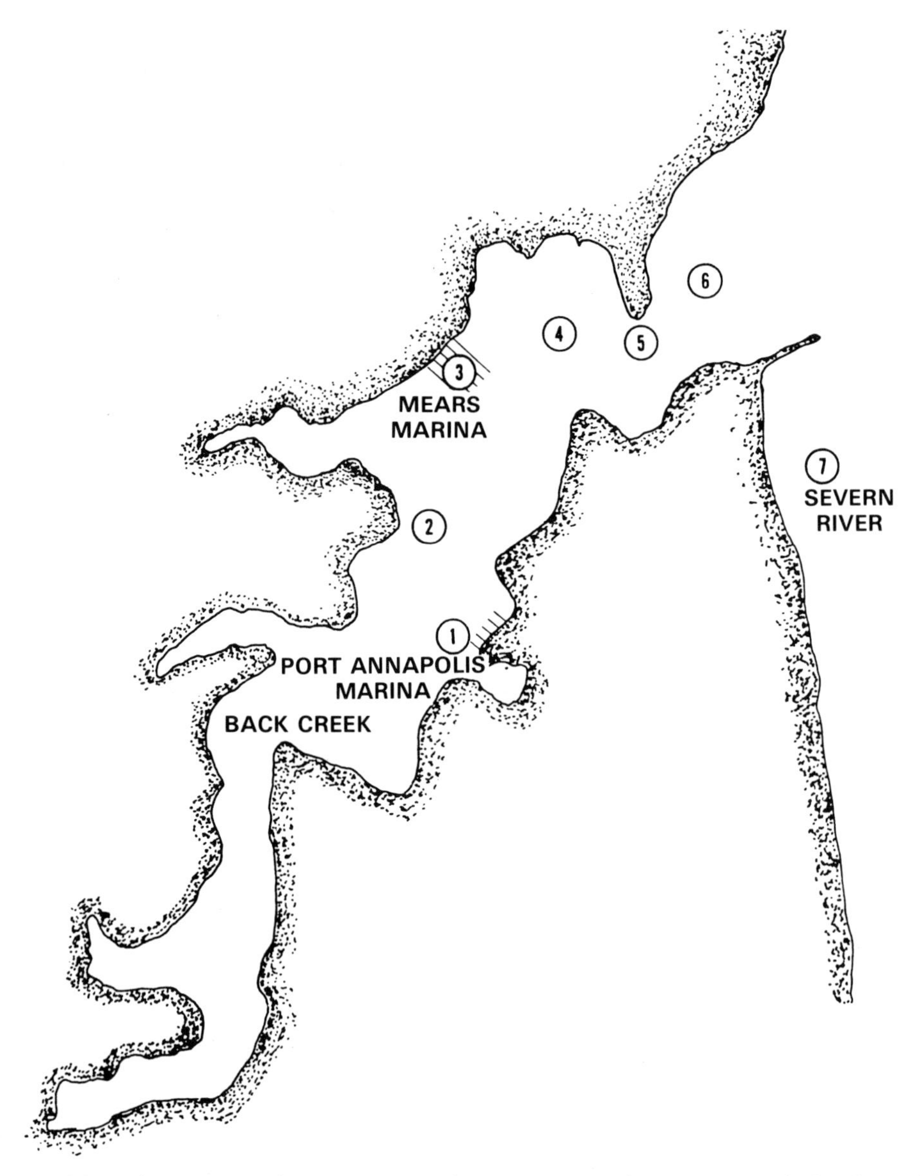

Figure 24.2 Map of seven sampling stations in the Port Annapolis Marina, Back Creek, and Severn River study area.

land (Fig. 24.2). In 1988, studies were conducted during the first year after the passage of legislation in Maryland to prohibit the use of TBT antifouling paints on recreational vessels <25 m in length. A comparison of TBT data from the seven stations in 1986 and 1988 is presented in Table 24.2 (Hall *et al.*, 1989). TBT concentrations were not significantly reduced between 1986 and 1988 when a year-to-year comparison was conducted on the pooled data using a general linear models procedure (Hall *et al.*, 1989). Analysis of the

Table 24.1 Summary of literature values of TBT concentration in waters of the Chesapeake Bay and tributary estuaries (exclusive of VIMS data) (see Fig. 24.1)

			TBT concentration ($ng\ l^{-1}$)				
Location	*No. of samples*	*Date*	*Average*	*Range*	*Station description*	*Ref.*[a]	*Location number*
Annapolis, MD	14	7/86–6/86	408	1056–39	Port Annapolis Marina	1	1
Galesville, MD	14	7/85–6/86	51	210–ND[b]	Hartge Marina	1	2
Stevensville, MD	14	7/85–6/86	123	364–ND	Pier I Marina	1	3
Chester, MD	14	7/85–6/86	119	456–ND	Piney Narrows Marina	1	4
Baltimore Harbour	14	7/85–6/86	44	129–ND		1	5
Chesapeake, DEL	12	7/85–6/86	ND	ND	C and D Canal	1	6
Potomac River	12	7/85–6/86	4	24–ND	Gov. H. Nice Mem. Bridge	1	7
Cambridge, MD	12	7/85–6/86	ND	ND	Choptank River	1	8
Elizabeth River, VA	36	5/84–5/87	4	9–ND	Sewells Point	2	9
Elizabeth River, VA	32	5/84–5/87	2	11–ND	Norfolk Naval Base Piers	2	10
Elizabeth River, VA	39	5/84–5/87	9	25–ND	Craney Island	2	11
Elizabeth River, VA	45	5/84–5/87	19	54–3	Lamberts Point	2	12
Elizabeth River, VA	45	5/84–5/87	33	160–3	Naval Hospital	2	13
Elizabeth River, VA	43	5/84–5/87	51	155–5	Naval Shipyard, St Helena	2	14
Elizabeth River, VA	40	5/84–5/87	42	210–2	Navy Shipyard, North End	2	15
Elizabeth River, VA	41	5/84–5/87	26	120–ND	Navy Shipyard, South End	2	16
Back Creek, MD	18	6/86–9/86	435	1800–110	Port Annapolis Marina	3	17
Back Creek, MD	18	6/86–9/86	222	807–65	Channel between Marinas	3	18
Back Creek, MD	18	6/86–9/86	291	1170–77	Mears Marina	3	19
Back Creek, MD	18	6/86–9/86	154	359–76	Between Marina and Mouth	3	20
Back Creek, MD	18	6/86–9/86	132	368–45	Between Marina and Mouth	3	21
Back Creek, MD	18	6/86–9/86	76	283–12	Back Creek Mouth	3	22
Severn River, MD	18	6/86–9/86	22	48–5	Severn River	3	23
SPA Creek–Annapolis, MD	24	5/86–9/86	22	70–10	Creek Mouth	4	24
SPA Creek–Annapolis, MD	24	5/86–9/86	47	80–30	Naval Academy	4	25
SPA Creek–Annapolis, MD	22	5/86–9/86	121	530–40	Annapolis Yacht Club	4	26
SPA Creek–Annapolis, MD	22	5/86–9/86	99	175–20	Near Acton Cove	4	27
Back Creek–Patuxent River	22	5/86–9/86	19	40–15	Patuxent River	4	28
Back Creek–Patuxent River	24	5/86–9/86	21	38–15	Mouth of Back Creek	4	29
Back Creek–Patuxent River	23	5/86–9/86	47	170–20	James Point	4	30
Back Creek–Patuxent River	23	5/86–9/86	52	120–40	Calverts Marina	4	31
Oxford, MD (Town Creek)	23	5/86–9/86	24	52–13	Tred Avon River	4	32
Oxford, MD (Town Creek)	23	5/86–9/86	23	48–12	Town Creek Mouth	4	33
Oxford, MD (Town Creek)	25	5/86–9/86	30	63–16	Near Bates Marina	4	34
Oxford, MD (Town Creek)	24	5/86–9/86	34	61–17	Near #13/Channel Marker	4	35
Plain Dealing Creek	12	5/86–9/86	16	22–12	Tred Avon River	4	36
Plain Dealing Creek	12	5/86–9/86	28	91–10	Creek Mouth	4	37
Plain Dealing Creek	12	5/86–9/86	19	27–10	0.5 NM up creek	4	38
Plain Dealing Creek	12	5/86–9/86	18	23–10	1.1 NM up creek	4	39

[a]Ref. 1, Hall *et al.*, 1987. Ref. 2, Seligman *et al.*, 1987, 1989. Ref. 3, Hall *et al.*, 1988. Ref. 4, Batiuk, 1987.
[b]ND signifies not detected.

1986 and 1988 TBT data, treating data as a fixed effect with year-to-year comparisons within each station, showed the following: TBT was significantly reduced at station 1 in 1988 (61%); there was no significant difference in TBT between years at stations 2, 3, 4, and 5 (3–45% reductions); and TBT was significantly greater in 1988 than in 1986 at stations 6 and 7 (17–50%; Hall *et al.*, 1989). Due to the limited frequency of sampling (bi-weekly for 4 months) the statistical power to detect changes in TBT concentrations

Table 24.2 Comparisons of mean TBT concentrations in surface water samples by station for 1986 and 1988 in the Back Creek–Severn River area of Maryland waters of Chesapeake Bay (see Fig. 24.2)

	Mean TBT (ng^{-1}) by station[a]						
Year	*1*	*2*	*3*	*4*	*5*	*6*	*7*
1986	435	222	291	154	132	76	22
1988	169	147	160	146	129	89	33
TBT difference	−266	−75	−131	−8	−4	+13	+11
Reduction							
% Reduction	61%	34%	45%	5%	3%	—	—
% Increase	—	—	—	—	—	17%	50%

[a]Samples are two replicates each from nine sampling dates.

between years is reduced. Only large differences in mean concentrations between years were statistically significant.

24.2.2. VIRGINIA'S WATERS

Virginia's portion of the Chesapeake Bay contains numerous recreational vessels but, unlike Maryland's waters, there are a number of large commercial shipyards, a naval shipyard, and a large naval base at Norfolk.

24.2.2a. Regional monitoring

The Elizabeth River and Hampton Roads area in Virginia have substantial commercial and naval shipping and shipyard activity. A survey to evaluate the relative contribution of these activities to harbor butyltin concentrations was undertaken starting in 1985 (Seligman *et al.*, 1987, 1989, and Chapter 20). Surface samples were collected in triplicate at 0.5 m from nine locations on the Elizabeth River (Fig. 24.3) 13 times over 21 months from September 1985 through May 1987. The samples were analyzed for butyltin species by hydride derivatization and atomic absorption detection.

The mean TBT concentration and the total calculated TBT mass for the segment of the river encompassing stations 10, 11, 32, and 19 through time is presented in Fig 24.4. These stations encompass shipyards in the vicinity. Tributyltin mass was estimated by multiplying the volume of the river segment in the vicinity of individual stations and the mean concentration. Large-scale variation in TBT concentration over the study period was evident with the highest concentrations occurring in December 1985 and 1986. High concentrations during December likely resulted from the seasonal painting of cruise ships. The lowest average concentrations, <10 ng l^{-1}, were measured in the spring and summer of 1986 and spring of 1987. If the mass loadings of TBT in the December samplings represent a single drydock release in each case (they may represent more than one), 1–6 kg of dissolved and suspended TBT could be added from a single event. These higher TBT concentrations in the winter are in contrast to the lower values observed during these periods in areas dominated by pleasure craft in the Chesapeake Bay.

The regional monitoring approach described previously (Grovhoug *et al.*, Chapter 25) was applied to five regions starting in October 1988: the James River, Hampton Roads, the Naval Station, the Elizabeth River, and the Hampton River. Station locations within regions are shown in Fig. 24.3. Quarterly samples were taken using methods identical to those described for monitoring

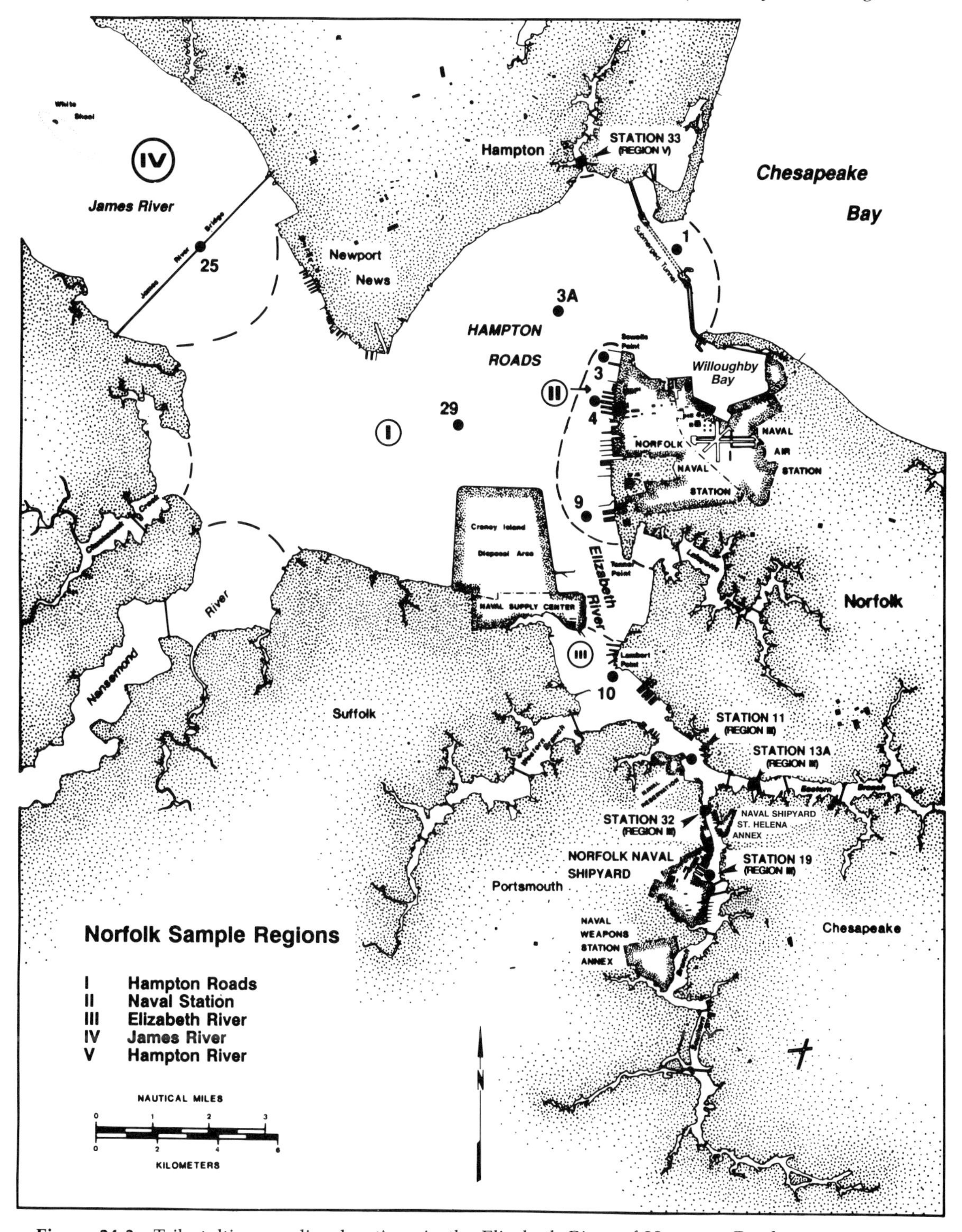

Figure 24.3 Tributyltin sampling locations in the Elizabeth River of Hampton Roads.

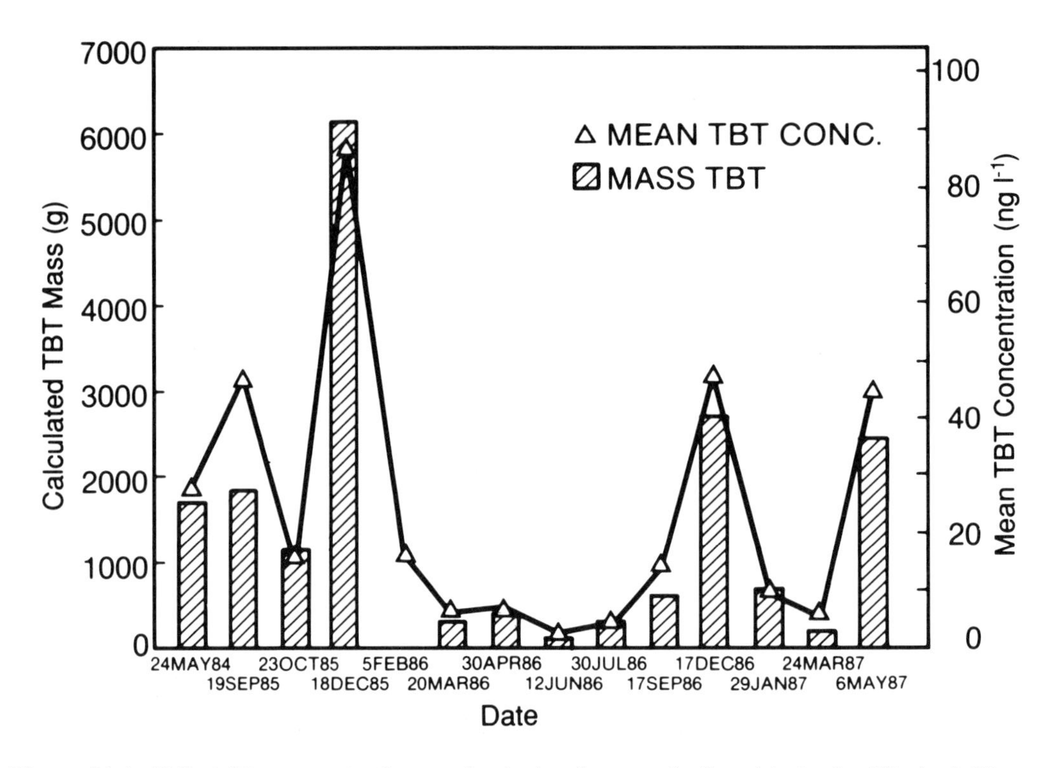

Figure 24.4 Tributyltin concentrations and calculated mass of tributyltin in the Elizabeth River.

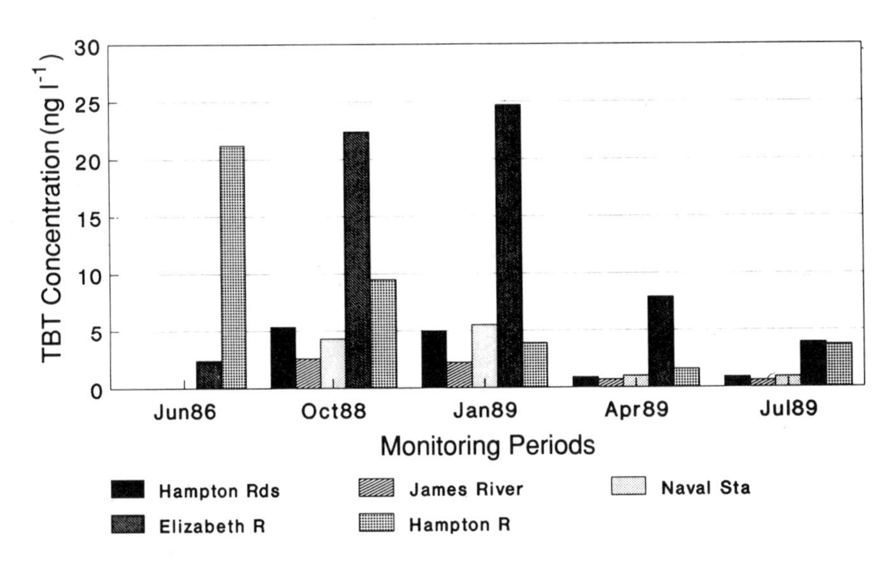

Figure 24.5 Surface water concentrations of tributyltin from the Elizabeth River–Hampton Roads area.

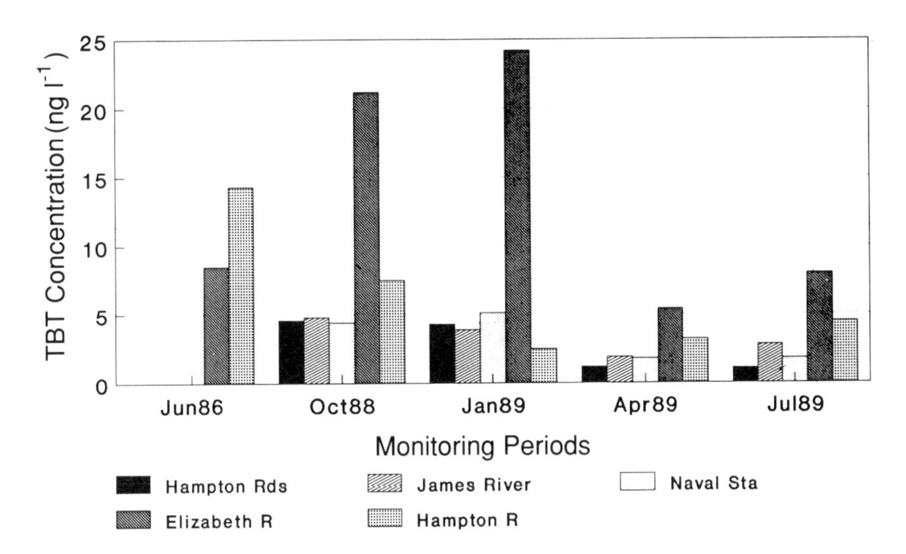

Figure 24.6 Bottomwater concentrations of tributyltin from the Elizabeth River–Hampton Roads area.

studies in Pearl Harbor and San Diego Bay (Grovhoug *et al.*, Chapter 25). Mean TBT values for the five regions are given in Figs 24.5 and 24.6 for the quarterly monitoring periods as well as for the samples collected in June 1986. Significant differences among TBT concentrations in surface- and bottomwater were found in only 5 of 22 comparisons. Definitive long-term changes are difficult to identify; however, the data visually suggest some trends relative to the regulation of TBT, which became effective in 1987. Concentrations of TBT in surface water have decreased or remained very low in the five regions shown in Fig. 24.5 during the last two monitoring periods.

Concentrations of TBT in surface water were statistically similar in the Elizabeth River during the June 1986 period and the April and July 1989 monitoring periods. Significantly higher values were recorded during intermediate periods in October and January 1989. The Hampton River region exhibited significantly higher surface TBT values during the June 1986 and October 1988 periods from the later monitoring intervals. The James River region was characterized by very low values for mean concentrations of TBT in surface water, which were always <3 ng l^{-1} and were not significantly different among monitoring periods.

24.2.2b. Intensive monitoring of Virginia's marinas

To determine TBT levels in waters with a high density of recreational vessels, three locations were selected in Sarah Creek and one in the adjacent York River (Figs 24.1 and 24.7). Four additional stations were occupied in Hampton River (Figs 24.1 and 24.8) and another at the mouth of the Hampton Roads at Old Point Comfort. There are large marinas in Sarah Creek and in the Hampton River. Surfacewater samples were collected for TBT analysis from ~15 cm below the water surface, always on high slack tides.

The Sarah Creek stations were occupied biweekly from mid-January to July 1986 and weekly thereafter through 1989. Hampton

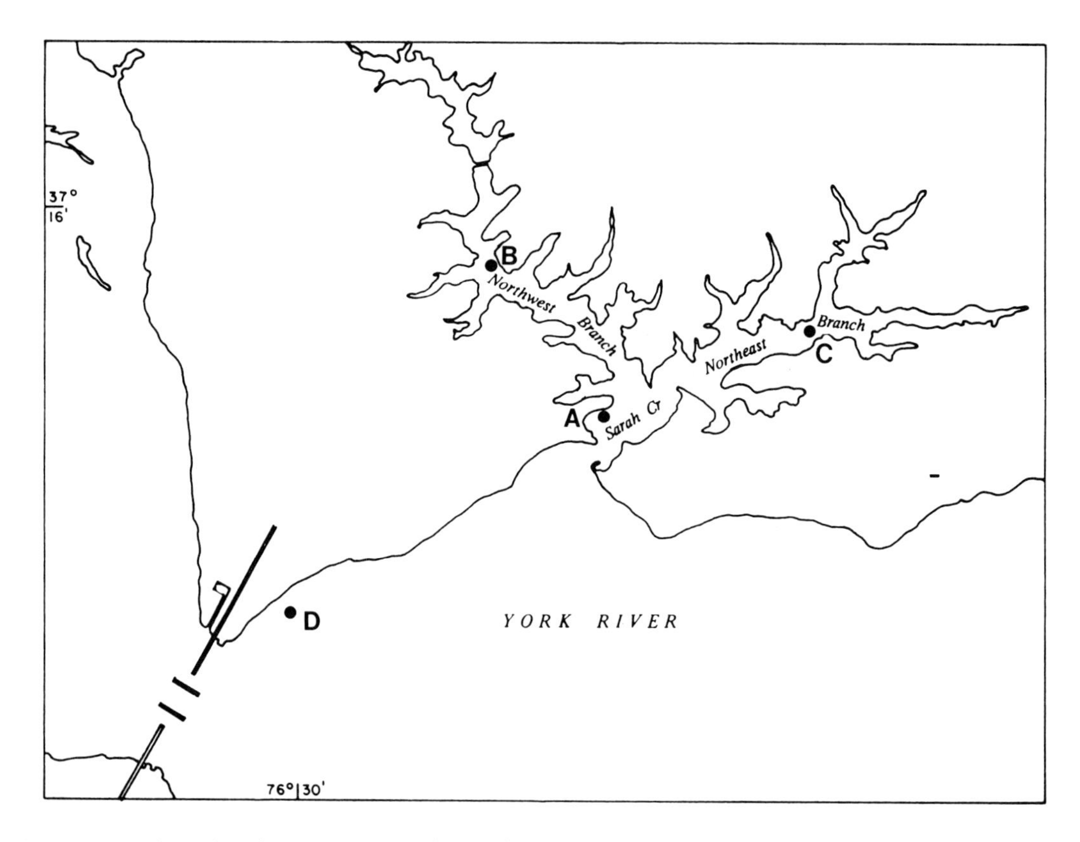

Figure 24.7 Sampling locations in Sarah Creek.

Creek sampling started in July 1986 and has continued on a weekly basis.

The analytical method utilized to determine TBT concentrations has been described by Unger *et al.* (1986). It involves liquid–liquid extraction of the water with an n-hexane–tropolone mixture and conversion to the hexyl derivative via the Grignard reagent, hexylmagnesium bromide. The derivatized TBT solution is cleaned by column chromatography and analyzed by high-resolution gas chromatography with flame photometric detection. Quantification of TBT is accomplished by comparing the derivatized TBT signal with that of a derivatized internal standard, tripentyltin, added as the chloride at the start of the analysis.

Over 1500 samples have been analyzed. The data exhibit considerable variability, which may be attributed to such circumstances as vessel traffic, season of the year, water temperature, and phase of the moon (tidal effects). The entire data set is too extensive to present here, but a mathematical model that 'fits' the data and a statistical evaluation has been developed; see Evans and Huggett (1991) for further discussion of statistical modeling of the tidal creek data.

The data set of more than 3 y of weekly observations of TBT concentration at nine stations is unique and presents a rich source of information. Because of the long sequence and high frequency of observations, the data for each station can be regarded as a time

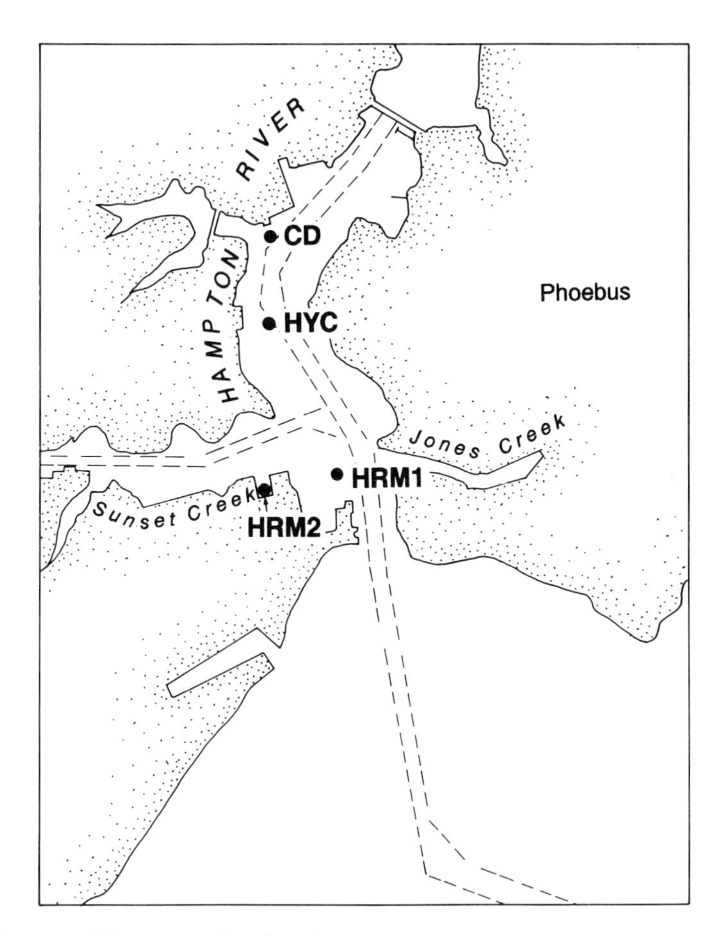

Figure 24.8 Hampton River sampling locations.

series. This analysis adopts this view and is primarily concerned with the structure of temporal variability.

Concentration values are necessarily positive numbers, and preliminary inspection of the data shows variation over several orders of magnitude. Both these considerations indicate that a logarithmic transformation of the dependent variable (concentration) should be made. Subsequent study of the residuals from the model fitting confirm that this transformation is appropriate since residuals are more normally distributed than the residuals from fits to the untransformed concentration data.

The model tested is as follows:

$$\log_{10} [\mathrm{TBT}] = a_0 + a_1\cos(2\pi t) + b_1\sin(2\pi t) + qt \quad (1)$$

where t is time in years expressed in decimal form with $t = 0$ on 1 January 1986. The coefficients a_1 and b_1 describe a sinusoidal variation with a period of 1 y, amplitude $(a^1{}_2 + b^1{}_2)^{1/2}$, and phase angle $\tan^{-1}(b_1/a_1)$. Superimposed upon the annual signal is a constant background a_0 and a linear trend with rate q. Note that a linear decrease (increase) in log[TBT] corresponds to an exponential decrease (increase) in concentration. Equal fractional changes occur in equal time intervals.

Table 24.3 Statistical evaluations of model predictions

Station	a_0	*Amplitude*	*Phase (deg)*	q	p(q)[a]	p(*fit*)[b]	*Root-mean-square residual*
A	1.22	0.171	205	−0.064	0.002	<0.001	0.280
B	1.38	0.102	202	−0.067	<0.001	<0.001	0.154
C	0.92	—[c]	—	−0.037	0.001	0.002	0.154
D	0.35	—	—	—	—	—	—
HRM1	1.51	0.141	153	−0.117	<0.001	<0.001	0.288
HRM2	2.38	0.143	128	−0.107	<0.001	<0.001	0.292
HYC	1.52	0.139	165	−0.095	<0.001	<0.001	0.229
CD	1.42	0.133	128	−0.135	<0.001	<0.001	0.185
OPC	0.47	0.077	56	+0.056	0.002	<0.001	0.184

[a]Probability that $q = 0$.
[b]Probability that all regression parameters are zero (i.e. data do not show any dependence on time).
[c]Dash indicates no trend.

The model (Eq. 1) is linear in the parameters a_0, a_1, b_1, and q, and thus the least-squares fitting process is merely a multiple linear regression procedure. The significance of the overall fit and individual terms can be judged using a standard F-test.

One of the stations (Sarah Creek B) also showed the presence of an additional harmonic variation with a period of ½ year,

$$a_2\cos(4\pi t) + b_2\sin(4\pi t) \qquad 2$$

24.2.2c. Statistical evaluation

The sequence of log[TBT] values for each station was subjected to a least-squares fit to the model described above. Data points where the concentration was undetectably small were not included in the fit since a zero concentration yields negative infinity under the logarithmic transformation. The significance of each fit was evaluated as were the individual significance of the 'q' term and the annual terms. Table 24.3 presents the results of the fits together with the probabilities of type I error for $q = 0$ (i.e. no long-term trend, and for all fit parameters = 0, i.e. no functional dependence on time) and the root-mean-square (r.m.s.) residual for the fit. The observed data and fitted curves for stations Sarah Creek B and HRM1 are shown in Figs 24.9 and 24.10. The other stations show the same general appearance.

Station HYC had five data points that lay more than 2.5 times the r.m.s. residual from the fitted curve. These points were positioned such as to have an undue effect on the fit. The tabulated and plotted results for this station are for calculations that ignore these 'wild' points. The points were not tightly clustered around a particular time, 'ordinary' points lay between them, so it is unlikely that they are produced by an identifiable effect.

The fitted curves for Sarah Creek stations A, B, and C are shown in Fig. 24.11, and those for the Hampton Creek stations are shown in Fig. 24.12.

24.3. DISCUSSION

24.3.1. SARAH CREEK

The parameter a_0 characterizes the overall level of TBT at each station. Station B shows the largest value with 24 ng l^{-1}, A has 17 ng l^{-1}, and C has 10 ng l^{-1}. Within the definition of the model, these numbers represent the average concentration at the beginning of 1986. This is the level about which the annual fluctuation takes place.

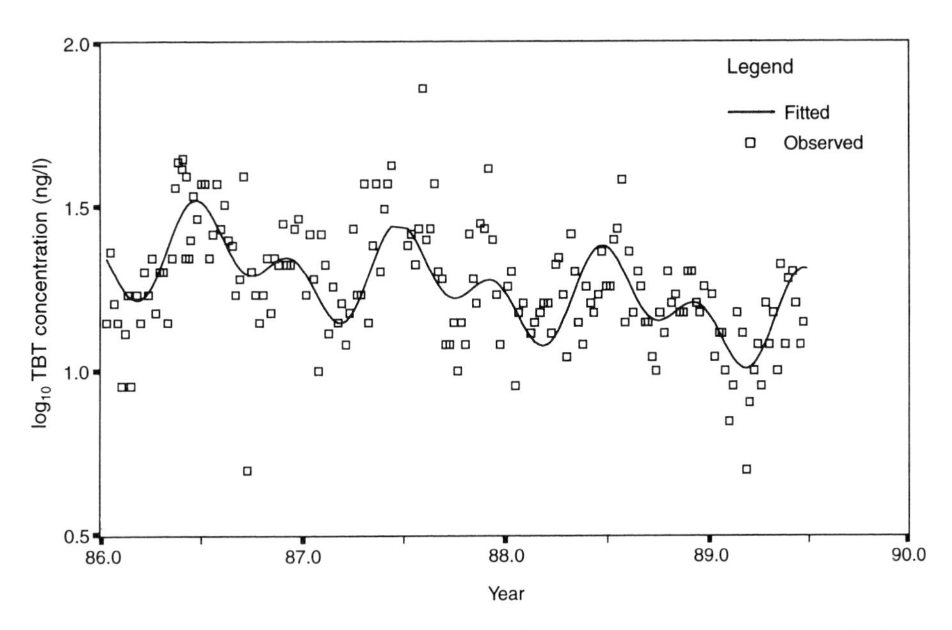

Figure 24.9 Tributyltin concentrations in Sarah Creek (station B) showing observed data and fitted trend.

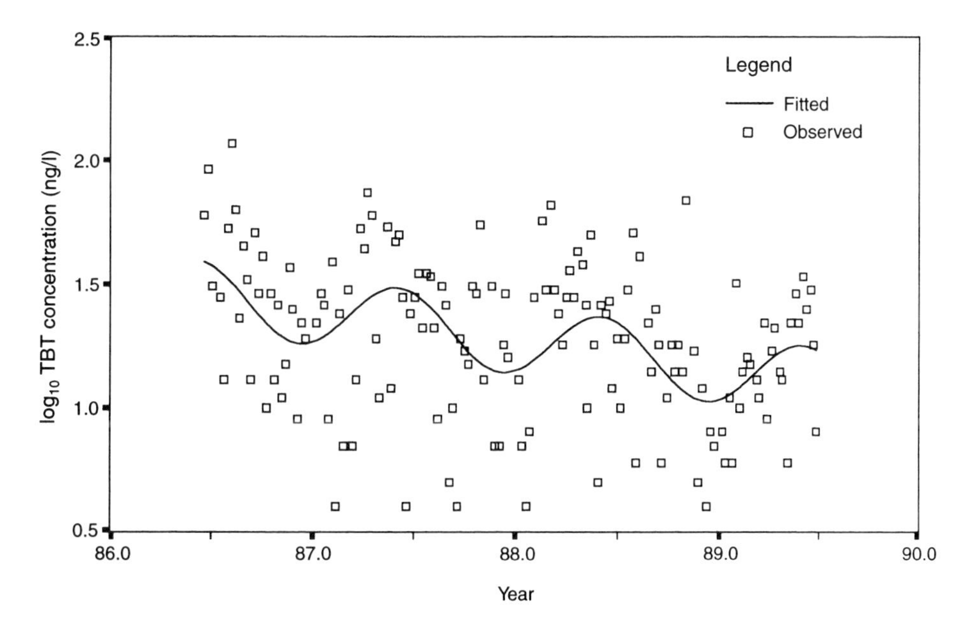

Figure 24.10 Tributyltin concentrations in Hampton River (Hampton Roads marina, station 1) showing observed data and fitted trend.

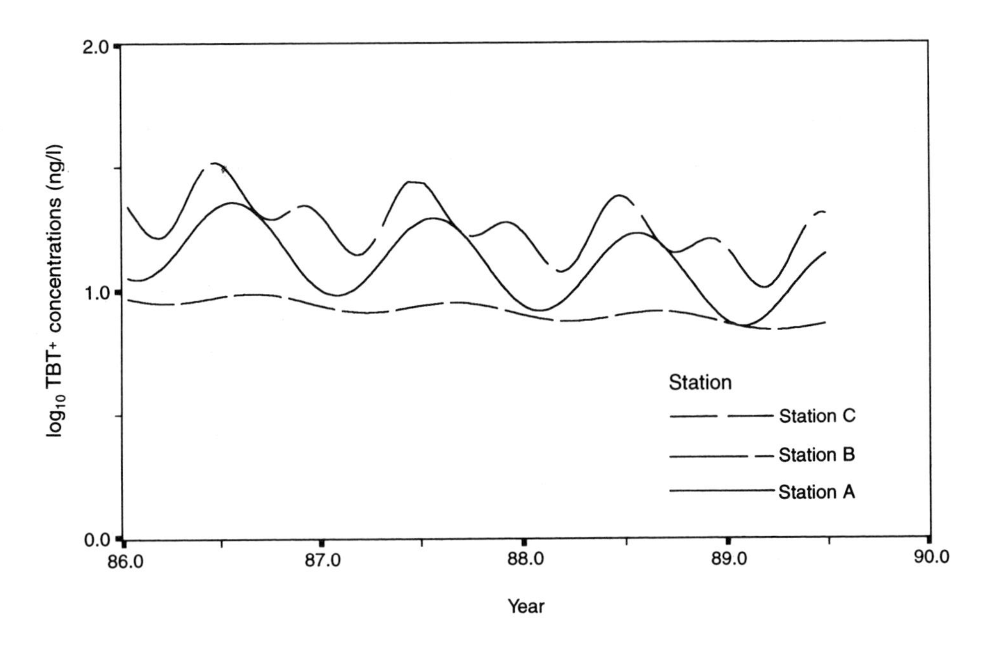

Figure 24.11 Fitted curves for tributyltin in Sarah Creek.

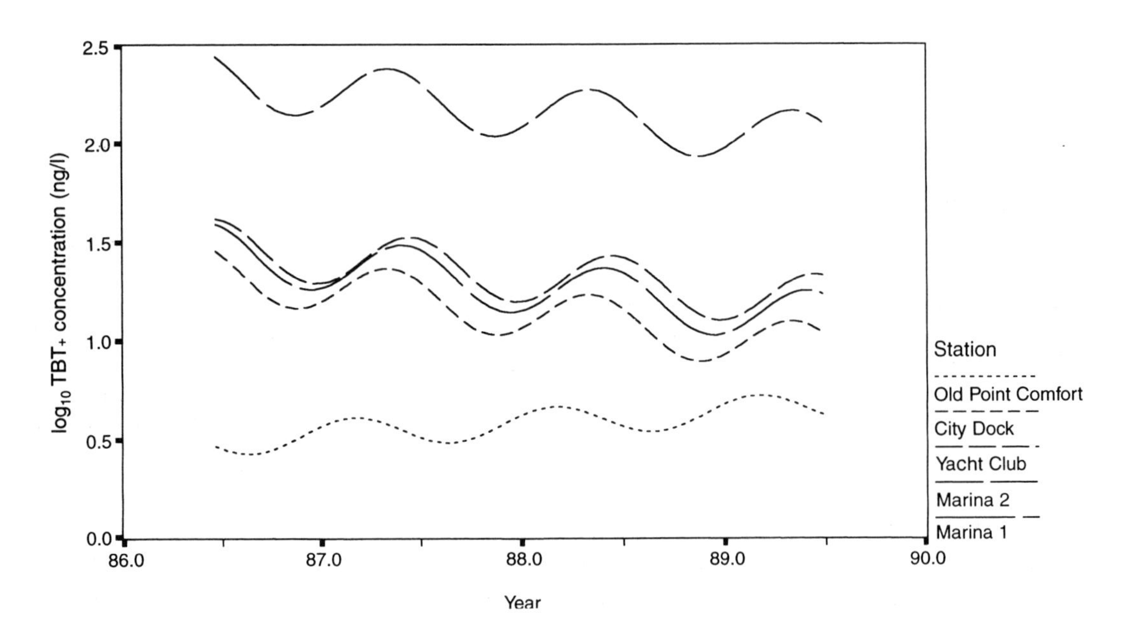

Figure 24.12 Fitted curves for tributyltin for Hampton River stations by year.

Relative to long-term trends, stations A, B, and C, which lie within the creek, all show evidence of a decrease in concentration. Stations A and B give consistent values for q of -0.064 and -0.069, respectively. These values correspond to yearly decreases of 13.7% and 14.7% or, alternatively, half-life residencies of 4.7 y and 4.4 y, respectively. Station C gives a lower value of q of -0.037 corresponding to a yearly decrease of 8%, or half-life residency of 8 y. However, the concentrations at station C are smaller than those at A and B. Station D shows no evidence of a trend.

Only stations A and B show a significant seasonal variation. Station A has an amplitude of 0.171 corresponding to a peak of ~50% above the 'mean' level and a trough of 50% below 'mean' level. Because of the logarithmic transformation, the fitted variation in actual concentrations is not sinusoidal, and the 'mean' value referred to above about which oscillation takes place is the geometric mean.

Station B shows a significant signal for a component with period of half a year. Examination of a plot of the data and the fitted curve indicates a subsidiary maximum in the fall of the year. The station is located close to a marina, and this may correspond to boat-painting activity in the fall.

The other two stations, C and D, do not show any conclusive evidence of a signal with annual periodicity. Station C is on a separate branch of Sarah Creek that has no boatyards, and station D is in the York River at the Virginia Institute of Marine Science pier.

24.3.2. HAMPTON RIVER

Hampton Roads Marina station 2 (HRM2) has the highest 'mean' value of 240 ng l^{-1}; other stations inside the creek, Hampton Roads Marina 1 (HRM1), Hampton Yacht Club (HYC), and City Dock (CD), have values of 32, 33, and 26 ng l^{-1}, respectively. The openwater station Old Point Comfort (OPC) has a much lower value of 3 ng l^{-1}.

All the stations inside the creek show a statistically significant ($p < 0.005$) decreasing trend over the whole time span of the data. Stations HRM1 and HRM2 show yearly decreases of 24% and 22%, respectively, corresponding to half-life residencies of 2.6 and 2.8 y. The HYC and CD stations have decreases of 20% and 27%, respectively, implying half-lives of 3.2 and 2.2 y. Station OPC has much lower concentrations, but there is a significant *increase* of 14% per year.

All stations show the presence of a significant sinusoidal seasonal signal. None requires the addition of a semi-annual component. The amplitude of the components for all stations inside the river are comparable, ranging from 0.133 at CD to 0.143 at HRM2. These values correspond to fluctuation in concentration about the 'mean' of 36% and 39%. The annual signal for OPC has an amplitude of 0.077 (19%). The phase angles for the river stations range from 128° (HRM2, CD) to 165° (HYC) corresponding to a concentration maximum in mid-April and mid-May.

24.4. GENERAL COMMENTS FOR MARINA MONITORING

Because of the apparently natural noise present in the observed data, it may be concluded that short-term periods of monitoring can yield little reliable information (Valkirs *et al.*, 1991; Seligman *et al.*, Chapter 20; Grovhoug *et al.*, Chapter 25; Maguire, Chapter 26; Waite *et al.*, Chapter 27). The high levels of confidence attributed to the fits to the data are due to the length of the data sets and the consistent, regular, and closely spaced measurement intervals.

The inclusion of the annual component into the model is necessary for the correct detection of the long-term trend. Consider, for example, a 1.5-y series with annual signal and no trend. If the series starts at the annual maximum, it will end at the next year's minimum. When this sequence is modeled as only a long-term trend, it will show a spurious decreasing trend, merely caused by the data set not being a whole number of years long.

A general decrease in TBT concentration with time has occurred at the marina sampling locations. This TBT decrease is probably due to diminished use and subsequent regulation of TBT paints in marinas. The rate of decrease may be affected by the illegal use of TBT paints or by leaching from vessels that have not yet been repainted. It is also probable that TBT sorbed on sediment or existing in the sediments as paint chips contributes to the load in the overlying waters. Illegal use, leaching from vessels, and presence in sediment are all conditions that will yield an apparent half-life for TBT in marina waters that is considerably greater than the observed half-life for TBT in laboratory situations that do not account for continual sources such as sediments and paint chips. Long-term monitoring of TBT concentrations in marinas after restriction may lead to a better understanding of the environmental behavior of TBT. The importance of slow TBT desorption and slow release from sedimentary paint chips cannot be proven from the present data, but should be considered as a subject for future research.

ACKNOWLEDGMENTS

The authors gratefully acknowledge the assistance provided by C. Adema, B. Davidson, R. Fransham, J. Greene, J. Grovhoug, H. Slone, M. Stallard, W. Thomas, E. Travelstead, A. Valkirs, and D. Westbrook.

This work was sponsored by the Office of Chief of Naval Research, The Commonwealth of Virginia, and the State of Maryland. VIMS Contribution Number 1704.

REFERENCES

Batiuk, R. 1987. Survey of Tributyltin and Dibutyltin Concentrations at Selected Harbors in Chesapeake Bay. Final Report prepared for US Environmental Protection Agency. Chesapeake Bay Liaison Office, Annapolis, Maryland.

Evans, D.A. and R.J. Huggett. 1991. Statistical modeling of intensive TBT monitoring data in two tidal creeks of the Chesapeake Bay. *Mar. Environ. Res.*, **32**, 169–186.

Hall, L.W., Jr, M.J. Lenkevich, W.S. Hall, A.E. Pinkney and S.J. Bushong. 1987. Evaluation of butyltin compounds in Maryland waters of Chesapeake Bay. *Mar. Pollut. Bull.*, **18**, 78–83.

Hall, L.W., Jr, S.J. Bushong, W.F. Johnson and W.S. Hall. 1988. Spatial and temporal distribution of butyltin compounds in a northern Chesapeake Bay marina and river system. *Environ. Monit. Assess.*, **10**, 229–244.

Hall, L.W., Jr, M.A. Unger, S.J. Bushong and M.C. Ziegenfuss. 1989. Butyltin Monitoring in a Northern Chesapeake Bay Marina and River System in 1988: An Assessment of Tributyltin Legislation. Final Report prepared for Johns Hopkins University, prepared by Applied Physics Laboratory, Shady Side, Maryland.

Huggett, R.J., M.A. Unger, P.F. Seligman and A.O Valkirs. 1992. The marine biocide tributyltin: assessing and managing environmental risks. *Environ. Sci. Technol.*, **26**, 232–237.

Naval Sea System Command. 1986. US Navy Experience on Research and Development of Antifouling Hull Paints Containing Tributyltin Compounds. Presentation to the Department of Agriculture and Consumer Affairs, Commonwealth of Virginia, Richmond, Virginia (5 Nov.).

Seligman, P.F., C.M. Adema, P.M. Stang, A.O. Valkirs and J.G. Grovhoug. 1987. Monitoring and prediction of tributyltin in the Elizabeth River and Hampton Roads, Virginia. In: Proceedings of the Oceans '87 International

Organotin Symposium, Vol. 3, Halifax, Nova Scotia, 28 September–1 October 1987, pp. 1357–1363.

Seligman, P.F., J.G. Grovhoug, A.O. Valkirs, P.M. Stang, R. Fransham, M.O. Stallard, B. Davidson and R.F. Lee. 1989. Distribution and fate of tributyltin in the United States marine environment. *Appl. Organometal. Chem.*, **3**, 31–47.

Stallard, M.O., S.Y. Cola and C.A. Dooley. 1989. Optimization of butyltin measurements for seawater, tissue, and marine sediment samples. *Appl. Organomet. Chem.*, **3**, 105–114.

Unger, M.A., W.G. MacIntyre, J. Greaves and R.J. Huggett. 1986. G.C. determination of butyltins in natural waters by flame photometric detection of hexyl derivatives with mass spectrometric confirmation. *Chemosphere*, **15**, 461–470.

Valkirs, A.O., B. Davidson, L.L. Kear and R.L. Fransham. 1991. Long-term monitoring of tributyltin in San Diego Bay, California. *Mar. Environ. Res.*, **32**, 151–167.

TRIBUTYLTIN CONCENTRATIONS IN WATER, SEDIMENT, AND BIVALVE TISSUES FROM SAN DIEGO BAY AND HAWAIIAN HARBORS

25

Joseph G. Grovhoug,[1] *Roy L. Fransham,*[2] *Aldis O. Valkirs*[2] *and Bradley M. Davidson*[2]

[1]*Naval Command, Control and Ocean Surveillance Center, RDT&E Division, Environmental Sciences (Code 52), San Diego, California 92152–5000, USA*
[2]*Computer Sciences Corporation, 4045 Hancock Street, San Diego, California 92110–5164, USA*

Organotin. Edited by M.A. Champ and P.F. Seligman. Published in 1996 by Chapman & Hall, London. ISBN 0 412 58240 6

ABSTRACT

During the period of 1986 through 1989, tributyltin (TBT) levels in San Diego Bay (California, USA) surface waters averaged 4.7–13 ng l^{-1} in the north bay, 1.3–9.9 ng l^{-1} in the south bay, and 3.5–14 ng l^{-1} in US Navy pier regions. TBT concentrations in the surface water of yacht harbors averaged 19–120 ng l^{-1}, while concentrations in bottom waters ranged from 8.8 to 61 ng l^{-1}. Yacht harbors and the naval pier regions showed a decline in TBT concentrations in water since restrictive paint-use legislation was enacted in January 1988. Average TBT concentrations in sediment in San Diego Bay ranged from 1.7 to 1100 ng g^{-1} (dry weight) with higher values in yacht harbors adjacent to vessel repair and maintenance facilities. TBT concentrations in bottom water <60 ng l^{-1} correlated well with concentrations in sediment. In regions where TBT levels in bottom waters were <30 ng l^{-1}, TBT levels in sediment were highly variable. TBT concentrations in sediment were generally highest in yacht harbors but did not exhibit significant temporal decreases coincident with restrictive legislation on TBT use. TBT burdens in tissues of the bay mussel (*Mytilus edulus*) in San Diego Bay ranged from 32 to 2100 ng g^{-1} (wet wt); with bioaccumulation factors (BAF) ranging from 2.8×10^3 to 6.6×10^4. BAF decreased rapidly as TBT concentrations in surface water approached or exceeded 50 ng l^{-1}. In Pearl Harbor from 1986 through 1989, mean TBT levels in regional surface water ranged from 0.0 to 6.8 ng l^{-1} in the channels; 0.0 to 4.9 ng l^{-1} in the outlying regions; 2.4 to 31 ng l^{-1} in Southeast Loch; and from 6.7 to 130 ng l^{-1} in a small marina. Bottomwater samples throughout Pearl Harbor during this period averaged from 0.0 to 9.7 ng l^{-1}. TBT concentrations in surface water in Southeast Loch were highly correlated with the presence of TBT-coated ships, while no significant relationship was found between TBT concentrations in surface and bottom water in Pearl Harbor. Water samples from Honolulu harbors averaged 4.8–580 ng l^{-1} in TBT concentration at the surface and 2.6–170 ng l^{-1} at the bottom during this same period. Regional mean concentrations of TBT in Pearl Harbor sediments ranged from 10 to 4500 ng g^{-1} (dry wt) and were highest near ship maintenance activities. On an average basis, the sediments from the Honolulu Harbor drydock facility contained three times the amounts of TBT as seen in comparable Pearl Harbor locations. Average TBT concentrations in oyster tissues from Hawaiian waters ranged from 41 to 1000 ng g^{-1} (wet wt), and were directly proportional to ambient surfacewater concentrations. Oysters in Hawaiian waters accumulated TBT at rates of 8.6×10^3–7.0×10^4 times. Bioaccumulation factors in oyster tissue were inversely proportional to the concentration of TBT in ambient surface waters and decreased rapidly as TBT concentrations in surface water rose.

25.1. INTRODUCTION

The Naval Ocean Systems Center (NOSC) has measured butyltin concentrations in US harbors and ports since 1982. The major thrust of these efforts was focused on San Diego Bay, Pearl Harbor, and the Norfolk Harbor/Hampton Roads area confluent with lower Chesapeake Bay (see Huggett *et al.*, Chapter 24) during the period 1985–1989. Butyltin monitoring studies performed in San Diego Bay and several Hawaiian harbors (including Pearl Harbor) are described in this chapter. Other chapters in this volume related to this study include 14, 15, 20, 21, and 24.

Monitoring efforts in San Diego Bay have evaluated major tributyltin (TBT) sources as well as tidal, vertical, and spatial variability (Seligman *et al.*, 1986, 1989; Valkirs *et al.*, 1986a, 1991). The bay consists of a 24-km-long crescent-shaped body of water with one entrance channel. The bay is a well-mixed estuary with depths generally <4.5 m, except for the 7.5–20-m deep dredged channel (Kram *et al.*, 1989a). Currents generally flow along the long axis of the bay with velocities ranging from 0.25 to 1.5 m s^{-1} (mean: 0.75 m s^{-1}). Freshwater input is minimal, resulting in near-oceanic salinities throughout most of the bay. The bay accommodates a commercial port, private shipyards, recreational boating and repair operations, naval station, amphibious training base, commercial and sportfishing berths, and yacht harbors.

Between April 1986 and January 1989, a series of environmental surveys was conducted to evaluate butyltin concentrations in several harbors and basins on the island of Oahu in the Hawaiian Islands. These investigations were performed as part of a series of follow-on studies to a baseline survey conducted during March–April 1984 (Grovhoug *et al.*, 1987, 1989; Seligman *et al.*, 1989). During these surveys, a total of six US Navy vessels were coated with antifouling paints containing TBT to determine both the efficacy and the environmental influence of various TBT antifouling coatings. Pearl Harbor was selected as a test site for these studies because it is a Navy-controlled port with few ambiguous TBT sources. The harbor consists of three main regions: East Loch, Middle Loch, and West Loch. The most heavily used areas within the harbor are the smaller Southeast Loch basin, Naval Shipyard, and Supply Center (Evans, 1974). The tidal currents in Pearl Harbor are weak and variable with maximum ebb flows of ~0.25 m s^{-1}. Surfacewater circulation is primarily driven by northeasterly trade winds. Freshwater inputs are intermittent throughout the year from five major streams that direct runoff from an agricultural and urban drainage area of 285 km^2. Honolulu Harbor and three smaller harbors, Keehi Lagoon, Kewalo Basin, and Ala Wai Boat Harbor were also sampled to provide a comparison with TBT inputs from commercial and recreational vessels. Keehi Lagoon is a small boat harbor adjacent to Honolulu Harbor. Kewalo Basin is primarily a charter and commercial fishing vessel port. The Ala Wai Boat Harbor is the state's largest yacht harbor berthing ~700 vessels.

25.2. APPROACH AND METHODS

25.2.1. MONITORING STRATEGY

The basic approach to designing a harbor monitoring program was guided by several previous years' experience in measuring butyltin at nanogram–per–liter levels during baseline studies (Grovhoug *et al.*, 1987). Detection of changes in butyltin concentrations in water, sediment, and tissue samples over time within and among specific locations was a principal goal. Station selection and coverage was designed to provide sufficient sampling sites to typify various regions within each harbor. Quarterly sampling of the water column was initiated to detect seasonal changes in butyltin concentrations. Because of their integrative nature, sediment and tissue samples were collected from selected stations less frequently, usually quarterly or semi-annually.

Analytical interpretation of the TBT data from San Diego Bay focused on evaluating changes in environmental concentrations and potential effects of restrictive TBT legislation. The effects of vessel distribution and painting activities on TBT concentrations were evaluated in Hawaiian harbors.

25.2.2. REGIONAL APPROACH

Each harbor was divided into study regions that were defined by one or more of the following factors: physiographic features, dredged-channel limits, water motion characteristics, and pattern of vessel use. Four regions were defined in San Diego Bay (Fig. 25.1) primarily based on current velocity patterns and use characteristics (Seligman *et al.*, 1986). Each region contains at least three stations, which provided a minimum of

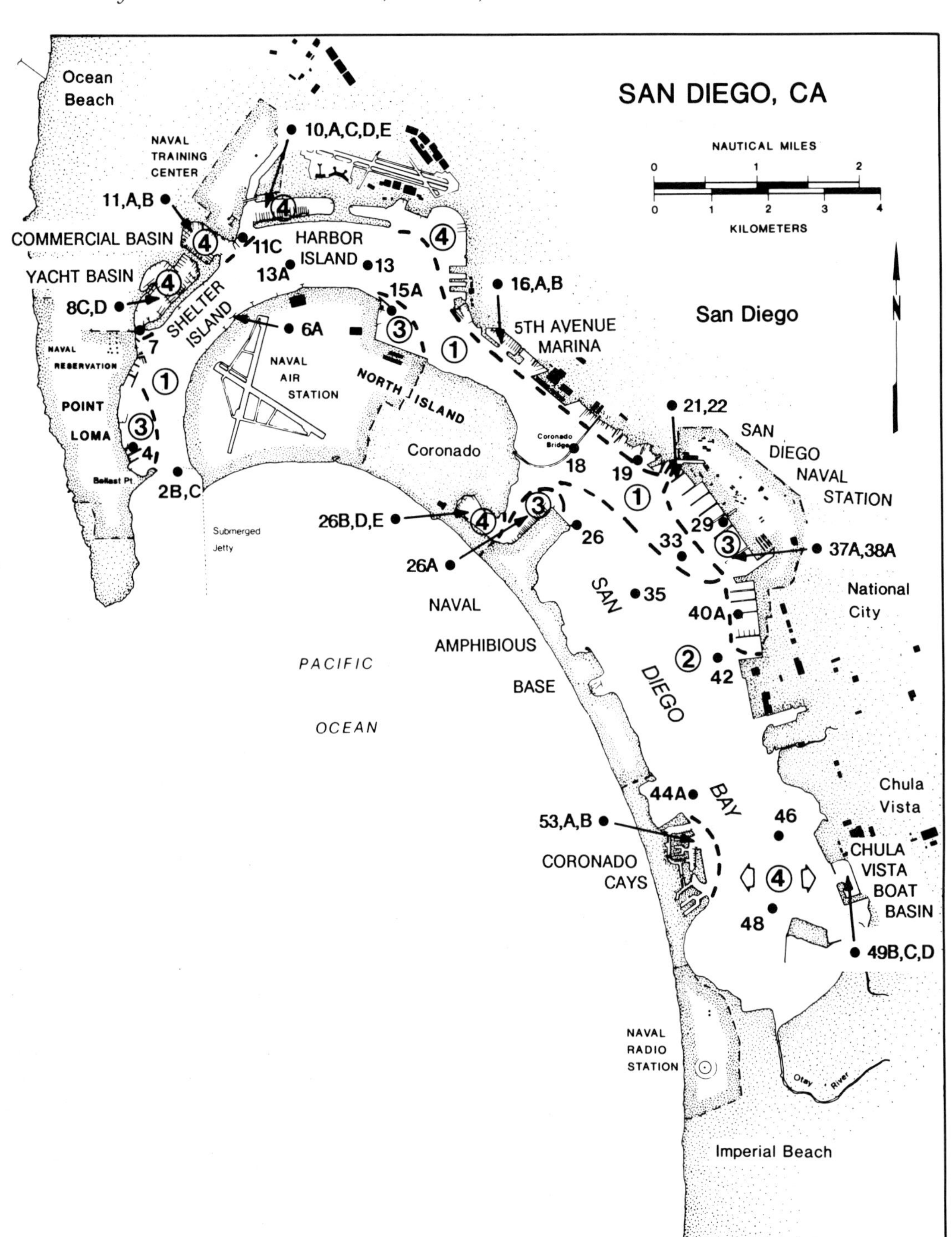

Figure 25.1 San Diego Bay sampling locations. Key to regions (numbers in circles): 1, north bay; 2, south bay; 3, naval piers; 4, yacht basins. Solid circles indicate sample station locations.

three surface and three near-bottom samples during each sampling period. The north bay region is characterized by high current velocities and rapid flushing rates and includes most of the shipping channels and dredged areas of the bay. The south bay region is largely shallow with low-velocity currents and longer water residence times. The south bay area is also important as a nursery ground for numerous fish and invertebrate species. Naval pier areas, including several Navy berthing areas, are moderately well flushed. Yacht basins are generally characterized by moderate to dense aggregations of vessels in enclosed embayments with reduced flushing characteristics (Seligman *et al.*, 1989).

Pearl Harbor was subdivided into eight regions (Fig. 25.2) by geographic and use-pattern factors: main channels (comprised of the entrance channel, south channel, and north channels), Southeast Loch, Rainbow Marina (a small pleasure-boat facility with a capacity of ~70 vessels, located in the northeastern corner of Pearl Harbor), East Loch, Middle Loch, West Loch, drydock #2, and drydock #4. The Honolulu harbors complex was separated into geographic and use-pattern regions along the same lines as Pearl Harbor (Fig. 25.3). Honolulu Harbor itself was divided into entrance channel, main basin, and drydock facility regions. Three other areas, Kewalo Basin, Keehi Lagoon, and the Ala Wai Boat Harbor, are separate from the main harbor basin and were treated independently.

Originally, individual stations within San Diego Bay and Pearl Harbor were compared to characterize TBT concentrations in each harbor, as well as event-specific effects (Grovhoug *et al.*, 1987). The data from these stations were subsequently grouped into several harbor regions for analysis. The regional mean concentrations of TBT were then examined rather than emphasizing a single station where the TBT concentration may vary widely. Large variability in TBT concentrations has been reported by Clavell *et al.* (1986), Huggett *et al.* (1986), Stang *et al.* (1989), Valkirs *et al.* (1991). Sample depth and tidal stage were found to be significant sources of variability in San Diego Bay (Seligman *et al.*, 1986). Combining stations into regions and sampling at specific tidal stages and water depths enabled direct hypothesis testing using a simplified one-way analysis of variance (ANOVA).

The stations within a region were considered as replicates, and all individual samples collected at a particular station and time were pooled. Although resulting in reduced degrees of freedom, this strategy was necessary because among-station variability in a region often proved greater than within-station variability. It was possible to use samples as replicates in some regions of Pearl Harbor since ANOVA results indicated that stations within these regions were statistically similar. In Hawaiian harbors where a single station was located within a given region, samples were used as replicates.

Prior to analysis, data were examined for conformance to the ANOVA assumption of homogeneity of variance by use of the Levene's test. The data showed significant heteroscedasticity and were logarithmically transformed and rechecked prior to hypothesis testing with two one-way ANOVA models. Data from the San Diego area were tested for variability among survey periods within harbor regions. Data from Hawaii were primarily tested for variability among regions within survey periods, but were also tested for variability among survey periods. Multiple range testing (Student–Newman–Keuls test) determined which survey periods or regions had statistically comparable ($p = 0.05$) means. Values presented in figures are standard means of the actual TBT concentrations.

At several stations three collocated samples were collected at each depth to maintain sampling continuity with previous collection periods. During quarterly sampling, three

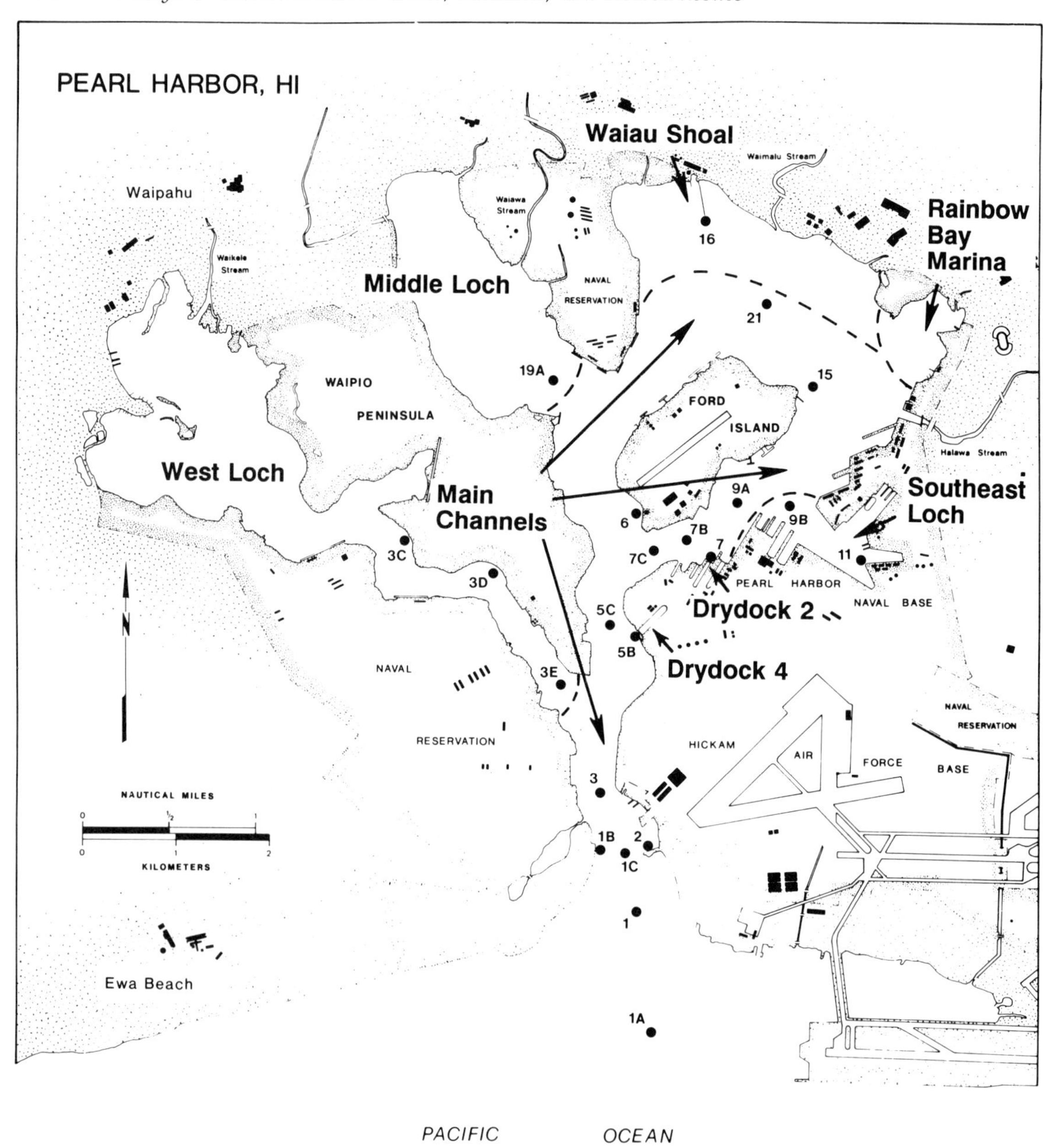

Figure 25.2 Pearl Harbor sampling locations. Key to regions: 1, main channels; 2, Southeast Loch, 3, drydock #2; 4, drydock #4; 5, Rainbow Marina; 6, East Loch; 7, Middle Loch, and 8, West Loch.

samples were taken at each station and depth, which were spread individually to expand spatial coverage. A one-way ANOVA test was used to compare variability between the two sampling procedures. No significant difference ($p = 0.33$) was found between collocated sample collection and collection of spatially separated samples at three stations selected from the north bay region of San Diego Bay. Therefore, collocated and spa-

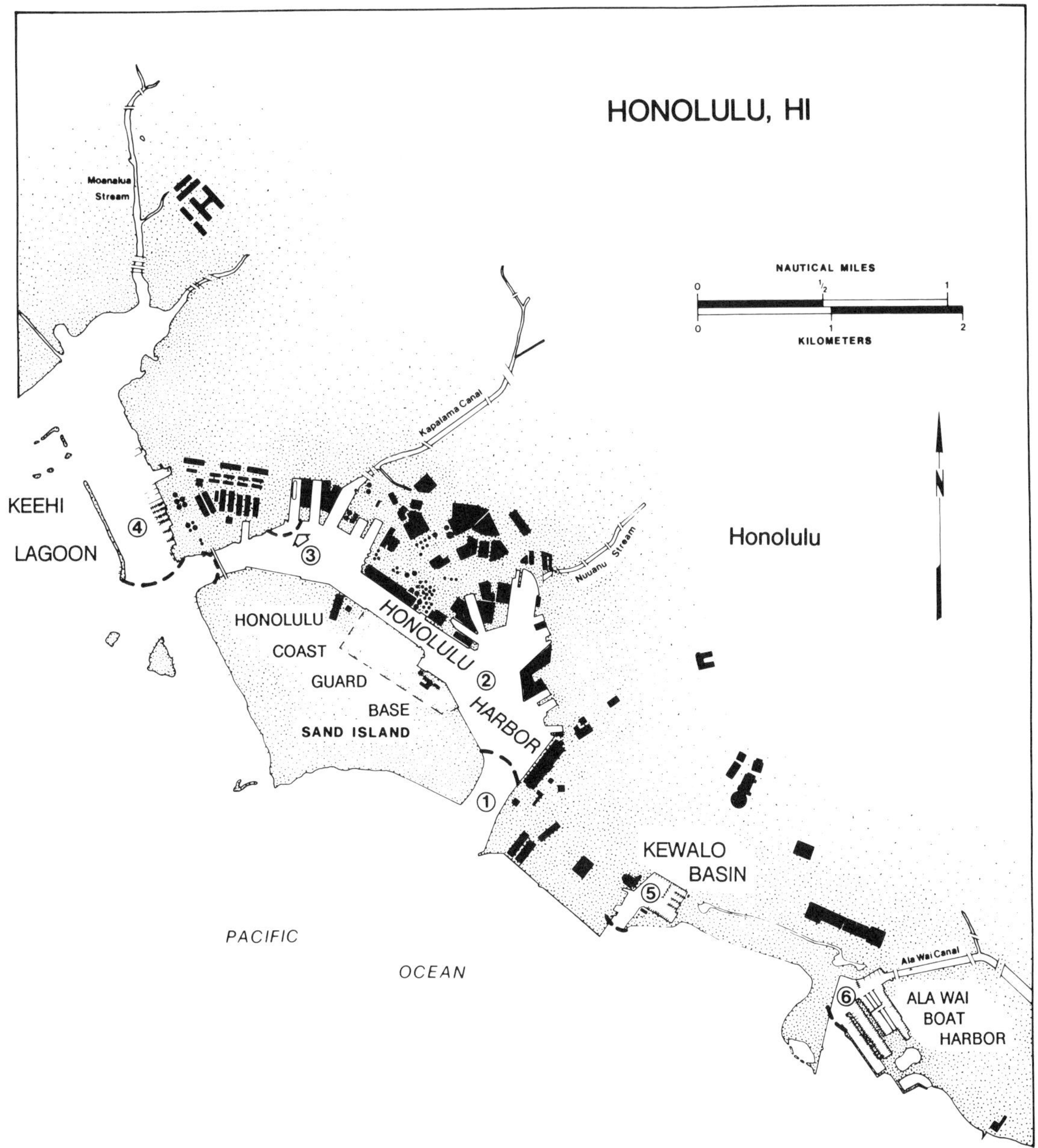

Figure 25.3 Honolulu Harbor sampling locations. Key to regions: 1, entrance channel; 2, main basin; 3, drydock facility; 4, Keehi Lagoon; 5, Kewalo Basin; and 6, Ala Wai Boat Harbor.

tially separated samples were pooled to generate station means that were used to compare regional differences reported in this study.

The regional approach used for data analysis of TBT concentration in water was not considered entirely practical for the analysis of data for TBT concentration in sediment. While water bodies may be typified by circulation and use patterns, marine sediments may be less homogeneous due to origins and activities such as dredging. Kram *et al.*

(1989a) identified differences in sediment grain size, clay mineralogy, and percent organic carbon in sediments from San Diego Bay, Pearl Harbor, Honolulu harbors, and lower Chesapeake Bay. These factors were shown to influence highly the degree of TBT adsorption by sediments (Kram *et al.*, 1989b). A recent study in Puget Sound, Washington, has documented significantly heterogeneous distributions of TBT in sediment from site to site as well as within a single site (Krone *et al.*, 1989). Differences in TBT concentrations in sediment were therefore examined on a station-by-station basis using a one-way analysis of variance where statistical comparisons were made over time using individual samples within a station as replicates. The data were not transformed since sample sizes were constant within and among sampling intervals. Data for TBT concentration in tissue were also compared on a station-by-station basis over time using samples within a station as replicates.

25.2.3. FIELD PROCEDURES

All seawater samples collected during the monitoring surveys were obtained at 0.5 m below the surface and at 1 m above the bottom. Care was taken to avoid contamination from the surface microlayer, which can exhibit high concentrations of butyltins (Maguire and Tkacz, 1987; Maguire, 1991). In Hawaii, water samples were not collected at particular tidal conditions, since tidal effects are insignificant (Grovhoug *et al.*, 1989); however, in San Diego Bay, water samples were collected during low slack tide as tidal effects have been shown to influence butyltin measurements in that embayment (Seligman *et al.*, 1986). Samples were collected and stored in 1-l polycarbonate bottles. All samples were iced immediately and transferred to a laboratory freezer within 8 h of collection.

Sediment samples were collected with a stainless steel grab sampler. About 150 ml (or 100 g) of sediment from the uppermost 2 cm of each grab was carefully removed and placed into 250-ml high-density polyethylene bottles. Three replicate samples were collected from each harbor region and were then treated in the same manner as water samples until analysis.

In Hawaii the eastern oyster (*Crassostrea virginica*) or two species of saddle oysters (*Ostrea sandvicensis* and *O. hanleyana*) were collected from available intertidal substrata at selected regional locations. In San Diego Bay, tissue samples were analyzed from the indigenous bay mussel (*Mytilus edulis*). Oysters do not occur in San Diego Bay, and *M. edulis* is not found in Hawaii. All bivalves were collected from approximately the same tidal level, near mean lower low water (MLLW). Shell lengths of the individual bivalve molluscs were recorded to evaluate TBT bioaccumulation values by size class. After initial freezing, individual soft tissues were excised using stainless steel and Teflon® (E. I. Du Pont de Nemouis, Wilmington, Delaware) implements and were pooled to obtain sufficient mass (3–5 g, wet wt) to provide three replicate samples at each station. The pooled tissues were placed in 85-ml polycarbonate centrifuge tubes and frozen until analyzed.

25.2.4. LABORATORY PROCEDURES

Water samples were collected and stored in polycarbonate plastic 1-l bottles, which have previously been shown to be nonadsorptive to butyltins and have been recommended as a suitable container for water samples containing butyltins (Dooley and Homer, 1983; Valkirs *et al.*, 1986b, Young *et al.*, 1986). Serial time analysis of frozen environmental water samples in polycarbonate bottles for 4.3 months in our laboratory has shown no greater than a 15% loss of the initial TBT concentration. This loss is consistent with the relative standard deviations of methods employing hydride derivatization for analysis of TBT in seawater (Valkirs *et al.*, 1987a).

Whole unfiltered seawater was measured directly by hydride derivatization followed by purging and trapping the evolved hydrides. Tin hydrides were then volatilized and detected by hydrogen flame atomic absorption spectroscopy (AAS) in a quartz burner. Studies by Valkirs *et al.* (1986b, 1987b) and Johnson *et al.* (1987) have shown that unfiltered seawater samples with low particulate concentrations generally have minimal TBT (<5%) associated with the particulate fraction. Thus, measurement of TBT in the dissolved phase and that available to hydride derivatization on particulate material essentially accounts for nearly all TBT present in an unfiltered seawater sample.

Tributyltin detection limits in water were ~5 ng l^{-1} prior to 1987. Recent improvements in the quartz furnace design and silanization of the cryotrap used to collect butyltin hydrides have resulted in detection limits as low as 0.2 ng l^{-1} (Stallard *et al.*, 1989). Data reported prior to 1987 were reviewed and presented if detection limits were below 1 ng l^{-1} A small number of values below detection limits (where no signal was apparent) were reported as zero and not used in summary statistics or for hypothesis testing.

Analysis of extracts of tissue and sediment samples was performed by Grignard derivatization with hexylmagnesium bromide. Butyltin derivatives were separated and detected by gas chromatography and flame photometric detection. Samples were quantified by comparison of analyte signal with the internal standard (tripentylhexyltin) response factors. Detection limits for sediment and tissue analyses ranged from 10 to 20 ng g^{-1}. Concentrations reported below the detection limit were achieved by analyzing large sample aliquots. Complete details of the analytical system and procedure used for water, tissue, and sediment analysis are reported in Stallard *et al.* (1989). All seawater tissue, and sediment values reported are as the respective butyltin chloride species. Water TBT measurements have not been corrected for recovery. In 66 of 76 samples collected from Pearl Harbor, San Diego Bay, and Chesapeake Bay, the percentage recovery of TBT added to samples (ranging from 1 to 6 ng) exceeded 75%.

25.3. RESULTS OF MONITORING STUDIES

25.3.1. SAN DIEGO BAY

A total of 861 water, 234 sediment, and 216 tissue samples were collected and analyzed between February 1986 and August 1989 from San Diego Bay. Surfacewater TBT values measured within five yacht harbors (Shelter Island, Commercial Basin, Harbor Island, Coronado Cays, and Chula Vista Marina) averaged 78 ng l^{-1} and ranged from 2.0 to 450 ng l^{-1} from the winter of 1986 to the summer of 1989. This region exhibited significant (p = 0.001) differences in surfacewater TBT concentration through time. Surface values during February 1986, October 1986, October 1987, and February 1988 were significantly higher than values recorded in April 1989. Mean surface values during October 1988, April 1989, and August 1989 were 26, 19, and 24 ng l^{-1}, respectively, but lower than the other periods, which ranged from 68 to 120 ng l^{-1} (Fig. 25.4). Mean concentrations of TBT in surface water measured during January 1989 were higher than during October 1988, April 1989, and August 1989, and were likely influenced by samples collected in Shelter Island yacht harbor.

The naval pier region had an overall mean surfacewater TBT concentration of 8.0 ng l^{-1} and ranged from 0.9 to 22 ng l^{-1}. A significant reduction was seen in TBT concentration during April 1989 (3.5 ng l^{-1}) and August 1989 (4.7 ng l^{-1}) from levels measured during October 1987 (11 ng l^{-1}) and February 1988 (14 ng l^{-1}). Significant (p = 0.004) statistical trends in TBT concentrations were seen in the south bay region as well. The mean TBT

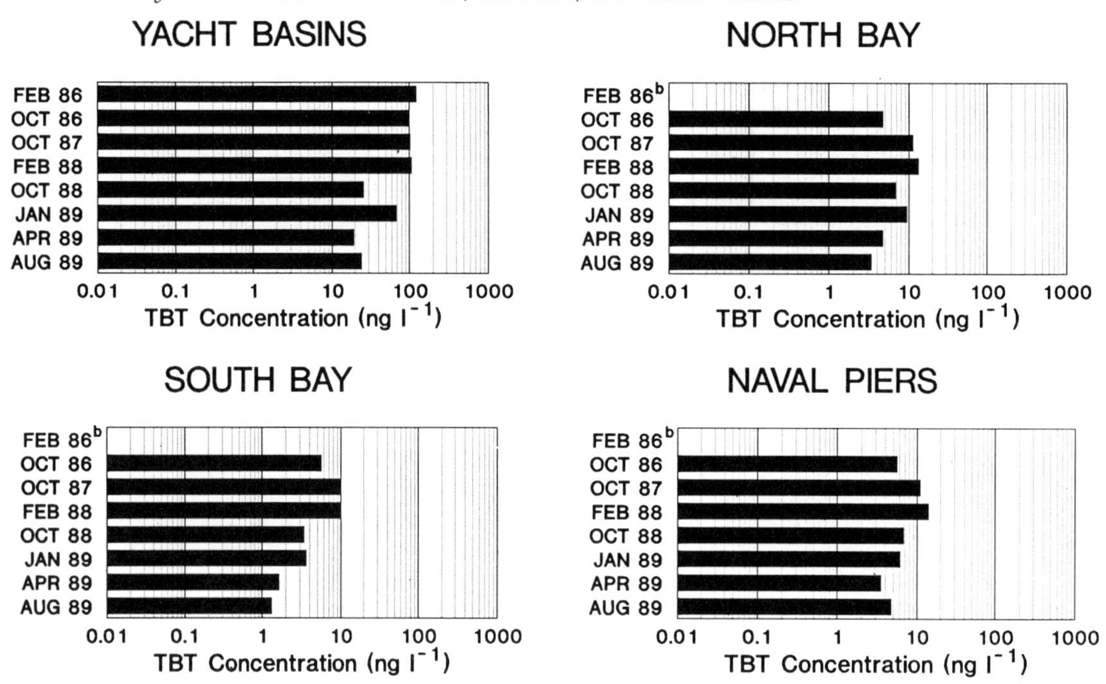

Figure 25.4 Mean concentrations of tributyltin in surface water in San Diego Bay ([a]no data).

surfacewater concentration for this area was 5.8 ng l^{-1} and ranged from 0.0 to 32 ng l^{-1}. Mean TBT values declined significantly from February 1988 values (9.9 ng l^{-1}) during April 1989 and August 1989 (1.6 and 1.3 ng l^{-1}). The north bay region ranged from 1.0 to 27 ng l^{-1}, with an overall mean of 7.9 ng l^{-1} for surface waters. Significantly lower ($p = 0.002$) concentrations were found during August 1989 when compared with October 1987, February 1988, or January 1989 data.

Concentrations of tributyltin measured in bottom water from the yacht basin region were generally low relative to the surface values and consistent with previous data for San Diego Bay reported by Seligman *et al.* (1986). Tributyltin values recorded during April 1989 and August 1989 were significantly lower than those found in other periods (Fig. 25.5). Bottom TBT concentrations were statistically similar in February 1986, October 1986, October 1987, February 1988, and January 1989. Significantly lower TBT concentrations were measured in bottomwater samples from the north bay region during October 1988, April 1989, and August 1989. Bottomwater TBT concentrations were statistically similar during the October 1987, January 1989, and February 1988 sampling intervals with the highest values measured during February 1988. Significant differences ($p = 0.005$) were found in bottomwater TBT concentrations measured in the south bay region among the monitoring periods presented in Fig. 25.5. The highest bottomwater TBT concentrations were measured during February 1988. Tributyltin values in August 1989 were significantly lower than February 1988 values. In the naval pier region a significant difference ($p = 0.0001$) in concentrations of TBT in bottom water was found among monitoring periods. Concentrations of TBT were statistically similar between the April 1989 and July 1989 periods and lower than values measured during February 1988.

Significant changes ($p = 0.05$) in TBT concentrations in sediment over time were recorded in only 6 of 14 stations in San Diego

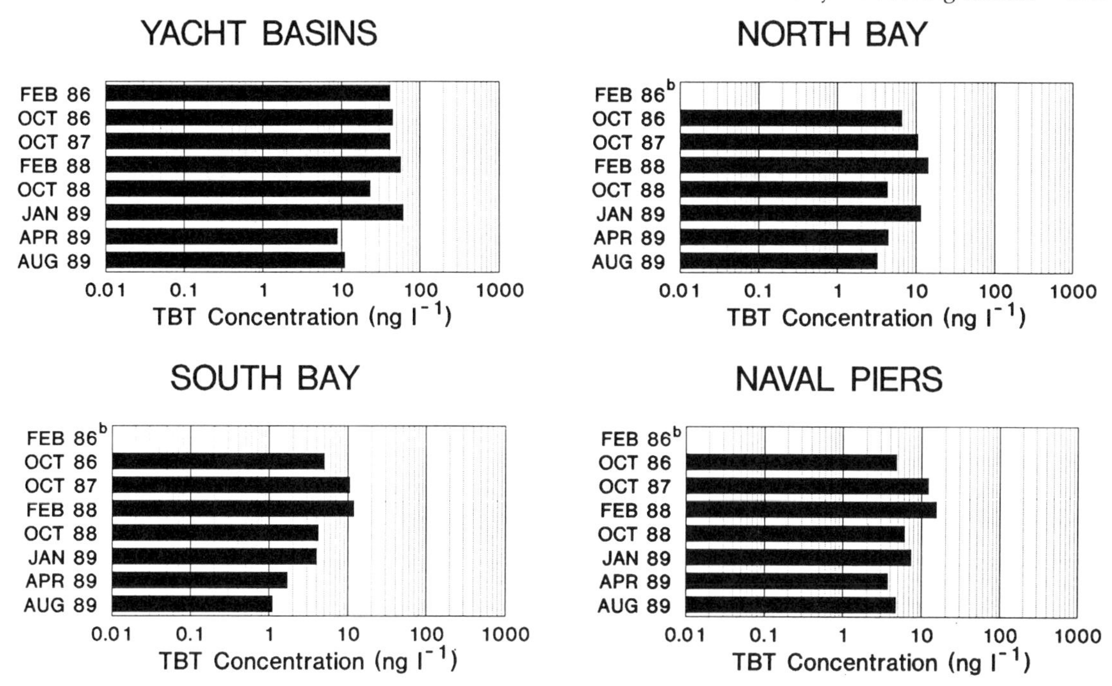

Figure 25.5 Mean concentrations of tributyltin in bottom water in San Diego Bay ([a]no data).

Bay (a summary of TBT concentrations in sediment from San Diego Bay by station is shown in Table 25.1). Analysis of sediment samples collected from San Diego Bay indicated that TBT concentrations were highest in the Island, Commercial Basin, and Harbor Island yacht anchorages. The presence of high TBT concentrations in sediment in these areas is consistent with their reduced circulation, the large number of vessels, and vessel repair activities. Significant decreases in TBT concentrations in sediment have yet to be measured in these areas in contrast to recent decreases in TBT values in water. A significant decrease in TBT concentrations in sediment was recorded at station 22 in the naval pier region where values were previously similar to yacht harbor stations. Concentrations of TBT in sediment were significantly lower ($p = 0.004$) during July 1989 than during January 1989.

Significant differences in TBT concentrations in sediment over time were recorded at stations 35, 46, and 48 in the south bay region, which exhibited an increase in TBT concentration in sediment during April 1989 and July 1989 over the two previous periods, although a significant increase ($p = 0.0049$) was determined only at station 48.

Tributyltin concentrations measured in the bay mussel (*Mytilus edulis*) between October 1986 and July 1989 ranged from 32 to 2100 ng g^{-1} (wet wt). Mean concentrations of TBT in tissues for each station are summarized in Table 25.2. The highest concentrations were generally found in October 1987 and February 1988 and have since been decreasing at all stations with the exception of station 22. A series of one-way ANOVA tests of individual stations shows a significant decrease in tissue concentrations between February 1988 and July 1989.

Stations at yacht basins, generally exhibited the highest mean concentrations of TBT in tissue samples. Shelter Island (station 7) reached a high of 1600 ng g^{-} in October 1987, while the Coronado Cays (stations 53, 53B) reached the maximum level measured, 2100

Table 25.1 Mean concentrations of tributyltin in sediment from San Diego Bay, February 1988–July 1989[a]

		Mean concentrations (ng g^{-1} dry wt)									
		Feb 1988		*Oct 1988*		*Jan 1989*		*Apr 1989*		*Jul 1989*	
Station(s)	*Region*	n[b]	*Mean*	n	*Mean*	n	*Mean*	n	*Mean*	n	*Mean*
02B,C	North bay	3	1.7	3	12	3	70	3	33	3	34
04	Naval piers	3	57								
06A	North bay	3	37	3	16	3	17	3	37	3	21
07	Yacht basins	3	80								
08C,D	Yacht basins	2	96	3	97	3	120	3	90	3	74
10,C,D,E	Yacht basins	6	160	3	56	3	160	3	84	3	80
11,A,B	Yacht basins	3	690	3	1100	3	530	3	430	3	220
13,A	North bay	4	75	4	43	3	39	3	51	3	29
15A	Naval piers	2	91								
16,A,B	Yacht basins	3	110								
18	North bay	3	78	3	53	3	64	3	68	3	28
19	Yacht basins	3	55								
21	Naval piers	3	190								
22	Naval piers	3	56	3	240	3	280	3	160	3	130
26A	Naval piers	3	29								
26C	South bay	3	31								
26B,D,E	Yacht basins	3	56								
29	Naval piers	3	350								
33	North bay	3	38								
35	South bay					3	19	3	25	3	26
38A	Naval piers	3	180	3	37	3	40	3	56	3	43
42	South bay	2	36								
44A	South bay	3	4.7								
46	South bay			3	6.7	3	8.7	3	24	3	22
48	South bay					3	7.7	3	16	3	20
49B,C,D	Yacht basins			3	29	3	29	3	30	3	25
53,A,B	Yacht basins			3	15	3	19	3	45	3	31

[a]Blank indicates no data.
[b]*n* indicates number of samples.

ng g^{-1}, in February 1988. Commercial Basin (station 11C) mussels reached a maximum value of 830 ng g^{-1} in October 1987. In recent surveys, several of the yacht basins have not been sampled because no mussels could be found. The Fifth Avenue Marina (station 16), however, has exhibited TBT concentrations in tissue of 840 and 820 ng g^{-1} during surveys in January 1989 and April 1989, respectively.

In contrast to the stations at yacht basins, station 2B at the mouth of the bay never exceeded a mean concentration of 140 ng g^{-1} TBT in tissue and most recently exhibited a concentration of 53 ng g^{-1}. A TBT concentration of 1700 ng g^{-1} in tissue recorded at station 22 in July 1989 represents an unusually high value for this area. It is possible that the presence of a large oil tanker (the *Exxon Valdez*) awaiting repairs at a nearby shipyard may have influenced uptake of TBT by mussels at station 22.

Individual size and growth rate have been shown to affect the concentration of TBT in tissues (Salazar and Salazar, Chapter 15). An attempt was made to collect mussels of the same general size class; however, the scarcity of individuals at some stations at yacht basins required the collection of variable sizes to

Table 25.2 Mean concentrations of tributyltin in bay mussel (*Mytilus edulis*) tissues from San Diego Bay, October 1986–July 1989

		Mean TBT concentrations (ng g^{-1} wet wt)													
		Oct 1986		Oct 1987		Feb 1988		Oct 1988		Jan 1989		Apr 1989		Jul 1989	
Station(s)	*Region*	n[a]	*Mean*	n	*Mean*	n	*Mean*	n	*Mean*	n	*Mean*	n	*Mean*	n	*Mean*
02B	North bay	3	100	3	120	3	140	3	53	3	32	3	57	3	53
06A	North bay	3	110	3	230	3	220	3	150	3	110	3	110	3	110
07	Yacht basins	3	690	3	1600	3	960	3	1200	3	890				
10,A	Yacht basins	3	670					3	420	3	260	3	290	3	210
11C	Yacht basins			3	830	3	530								
15,A	Naval piers	3	220	3	220	3	490	3	170	3	240	3	210	3	110
16	Yacht basins									3	840	3	820	3	380
18	North bay	3	360	3	330	3	380	3	140	3	320	3	220		
19	Yacht basins					3	500								
22	Naval piers			3	360			3	200	3	390	3	170	3	1700
26C	South bay	3	200	3	190	3	240	3	100	3	160	3	110	3	59
37A	Naval piers					3	290								
38A	Naval piers	3	220							3	190				
40A	Naval piers					3	380								
44A,B	South bay			3	200	3	390			3	95	3	51	3	59
53,B	Yacht basins	3	820	3	1400	3	2100	3	530	3	550	3	320	3	290

[a]*n* indicates number of samples.

collect enough tissue for analysis. Before comparing tissue burdens among time intervals, a one-way ANOVA model was run to test size as a covariate. No statistical differences were found in the individuals, which were typically 30–50 mm long, collected from these locations.

25.3.2. PEARL HARBOR

A total of 741 water, 213 sediment, and 64 tissue samples were collected and analyzed between April 1986 and July 1989 from Pearl Harbor. Comparisons among regions in the Pearl Harbor area were made using a one-way ANOVA test for each time period. During the initial monitoring survey performed in April 1986, detectable levels of TBT were only measured in surfacewater samples obtained in Southeast Loch and Rainbow Marina. Surfacewater TBT concentrations in West Loch were significantly lower than those measured at drydock #2, Southeast Loch, and Rainbow Marina during February 1987. Southeast Loch exhibited a mean concentration of TBT in surface water of 15 ng l^{-1}. This concentration was not, however, significantly different from concentrations measured in regions other than West Loch. All other areas averaged surface TBT concentrations between 3.0 and 9.0 ng l^{-1}. Bottom-water concentrations for all regions averaged $\leq$7.0 ng l^{-1} TBT, and were not significantly different between regions. During April 1987, TBT concentrations in Pearl Harbor surface waters varied from 3.4 ng l^{-1} in West Loch to 27 ng l^{-1} in Rainbow Marina. Differences among the regional surface values were not significant. Bottomwater samples during this period averaged 1.2–25 ng l^{-1} TBT, and significant differences were found among the regional bottomwater values. Multiple range testing indicated that bottom water from drydock #2 exhibited the highest TBT concentration, which was statistically similar to concentrations of TBT in

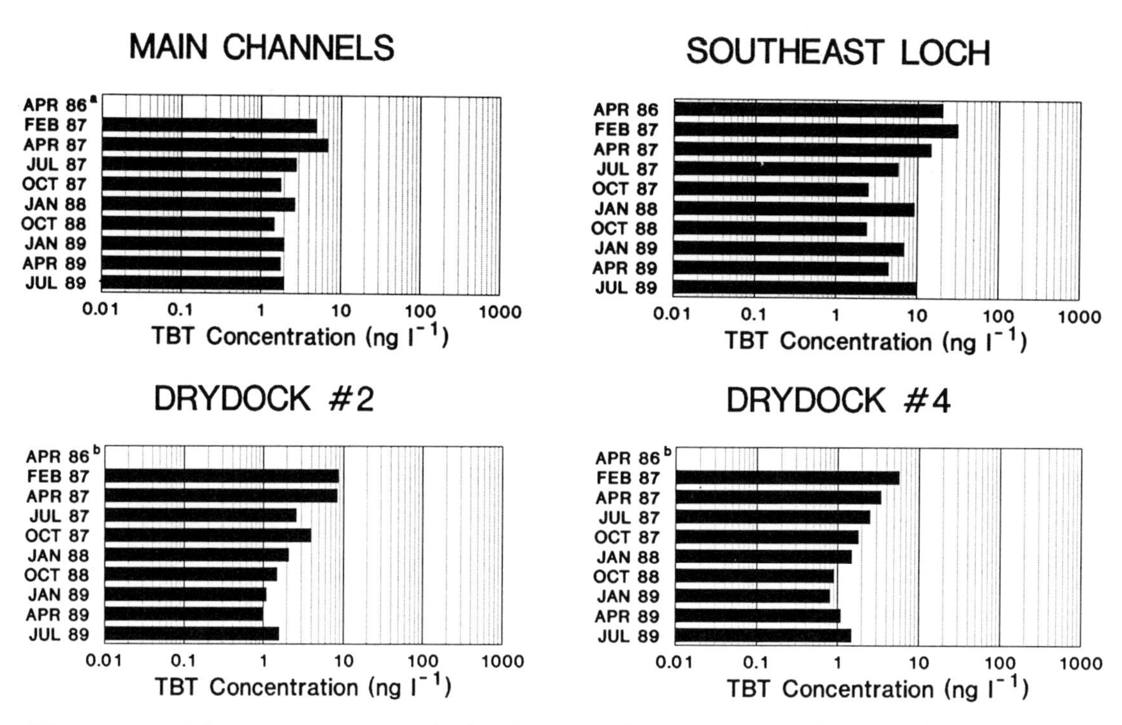

Figure 25.6a Mean concentrations of tributyltin in surface water in Pearl Harbor ([a]no data).

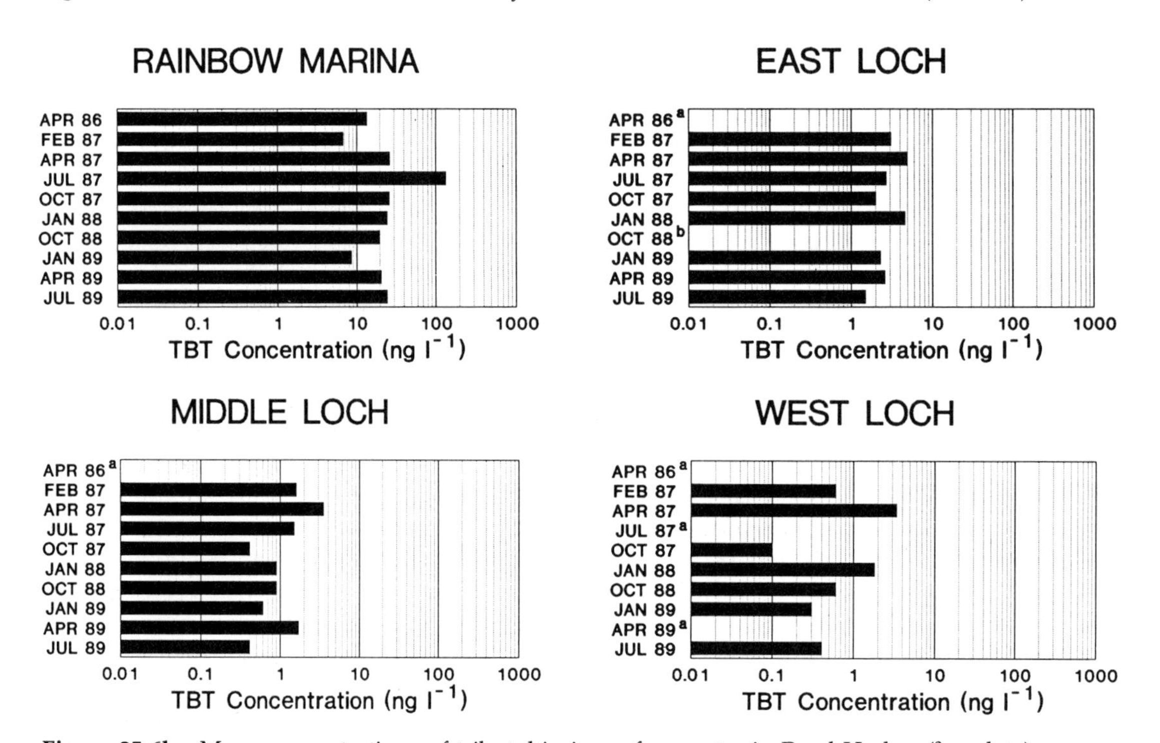

Figure 25.6b Mean concentrations of tributyltin in surface water in Pearl Harbor ([a]no data).

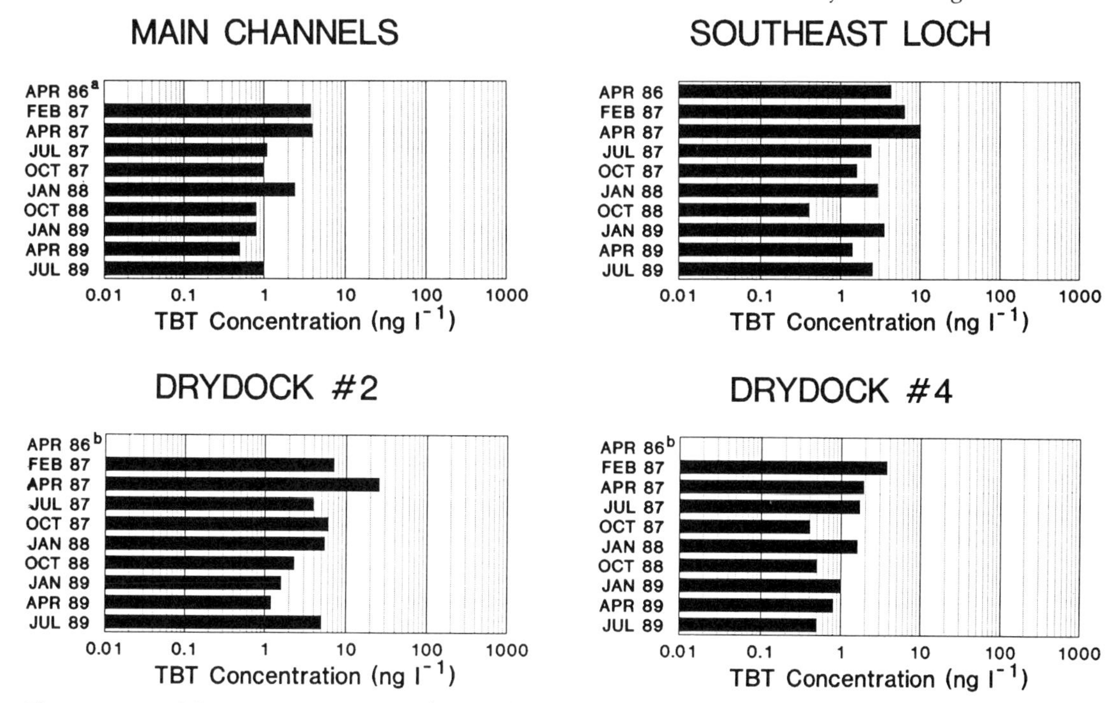

Figure 25.7a Mean concentrations of tributyltin in bottom water in Pearl Harbor ([a]no data).

bottom water from Southeast Loch and Rainbow Marina. Mean regional concentrations of TBT in surface and bottom water from Pearl Harbor during each of the monitoring surveys are illustrated in Figs. 25.6a,b and 25.7a,b.

Except for Rainbow Marina, surfacewater samples collected during July 1987 exhibited lower TBT concentrations throughout all harbor regions than during April. Significantly lower surface TBT concentrations were found at drydock #2. The high levels seen in Rainbow Marina surface waters during July 1987 (130 ng l^{-1}) coincided with the presence of two large non-resident yachts at this time. No other region in the harbor showed similarly elevated water TBT concentrations. During the October 1987 survey, the highest overall levels, of TBT in surface water were again found in Rainbow Marina and were significantly different from the other regions in Pearl Harbor. While lower than levels encountered during the survey of July 1987, the surfacewater TBT concentration at Rainbow Marina remained ~10 times that seen in Southeast Loch. In January and October 1988, water samples from Rainbow Marina continued to exhibit mean concentrations of TBT in surface water several times higher and significantly different from other regions in Pearl Harbor. Both drydock stations showed no significant change in TBT concentrations in the water column from October 1987 levels. During the January 1989 survey, the mean concentration of TBT in surfacewater in Rainbow Marina was significantly higher than the West Loch, Middle Loch, and drydock #2 regions, but similar to the remaining regions. During April 1989, the TBT concentration in surfacewater was highest in Rainbow Marina (20 ng l^{-1}). All regional TBT concentrations were statistically similar during this period with the exception of West Loch, which was significantly lower with no measurable TBT present. In July 1989 the TBT concentration in surface water was again

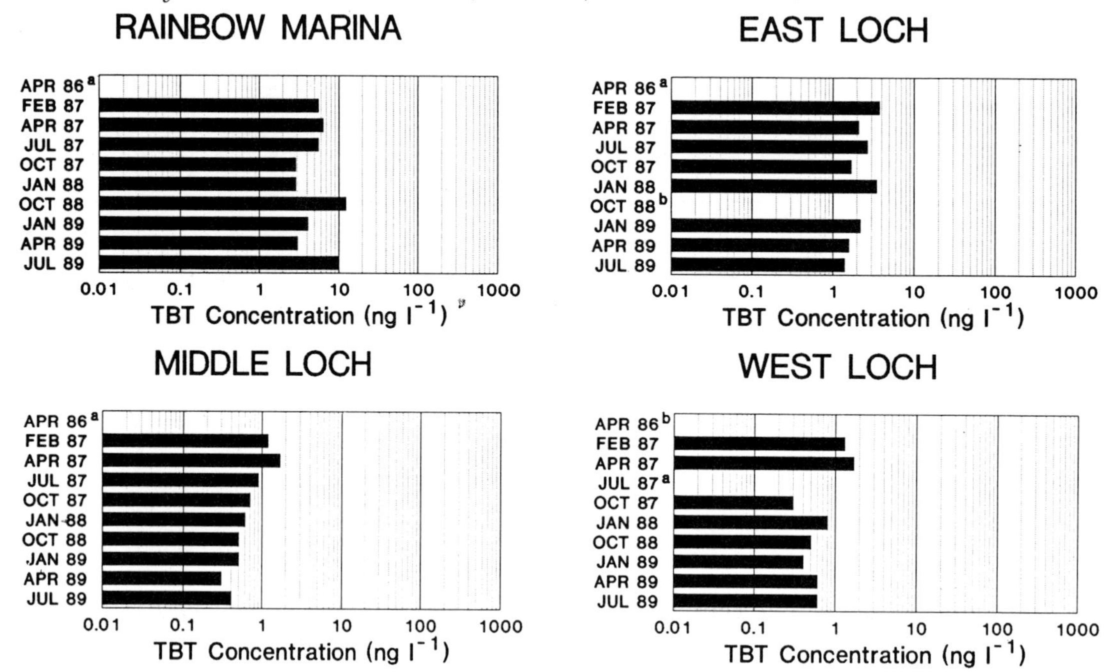

Figure 25.7b Mean concentrations of tributyltin in bottom water in Pearl Harbor ([a]no data).

highest in Rainbow Marina (25 ng l^{-1}) and was significantly higher than the West Loch or Middle Loch regions.

In addition to testing differences among regions during a given monitoring period, regions were also examined over time to determine if significantly different TBT concentrations in water were measurable. Surfacewater samples collected in the East Loch region in July 1989 were significantly lower in TBT concentration than those collected in April 1987 and January 1988 when TBT concentrations were highest (Fig. 25.6b). The concentration of TBT in surface water in drydock #2 was significantly higher during February and April 1987 than during the other sampling intervals (Fig. 25.6a). These periods correspond approximately with the TBT–AF paint application and drydock operations of two of the three US Navy test ships painted under the Pearl Harbor Case Study (Grovhoug *et al.*, 1989).

During October 1987 and October 1988 concentrations of TBT in surface water were lower and significantly different from concentrations measured in several other monitoring periods (April 1986, February 1987, April 1987, January 1988, and July 1989) in the Southeast Loch region. No significant differences were seen in water samples collected from the West Loch region, indicating that this region was unaffected by drydock operations and the level of activity in the main ship mooring areas of the harbor.

Rainbow Marina exhibited statistically similar concentrations of TBT in surface water during all monitoring periods with the exception of July 1987. During the July 1987 interval the TBT concentration was measured at 130 ng l^{-1} as a probable consequence of the presence of pleasure vessels recently painted with TBT antifouling paint. TBT concentration in surface water in the main channels region was significantly higher in April 1987 than in October 1987 and 1988, January 1989, April 1989, and in July 1989.

Mean regional concentrations of TBT in sediment in April 1986 ranged from 16 to 420

Table 25.3 Mean concentrations of tributyltin in sediment from Pearl Harbor, April 1986–January 1989

	Mean TBT concentration (ng g^{-1} dry wt)									
	Apr 1986		*Feb 1987*		*Apr 1987*		*Jan 1988*		*Jan 1989*	
Region	n[a]	*Mean*	n	*Mean*	n	*Mean*	n	*Mean*	n	*Mean*
Main channels	8	70	26	260	27	120	11	400	18	57
Southeast Loch	16	420	15	360	15	420	5	230	3	220
Drydock #2			2	2300	2	4500	3	1900	3	2000
Drydock #4			3	150	3	240	3	350	3	260
Rainbow Marina	2	66	3	33	3	59	1	72	3	406
East Loch	2	28	2	15	3	41	1	32	3	40
Middle Loch	1	48			3	120	2	27	3	87
West Loch	2	16	3	10	3	23	4	24	3	42

[a]*n* indicates number of samples.

ng g^{-1} (dry wt), with the highest levels recorded from the Southeast Loch region and the lowest from West Loch. In February 1987, concentrations of TBT in sediment ranged from 10 to 2400 ng g^{-1}. Drydock #2 had the highest concentration of TBT in sediment (2400 ng g^{-1}). Sediment samples from Pearl Harbor during April 1987 exhibited mean concentrations of TBT from 23 to 4500 ng g^{-1}. The most elevated levels were observed at drydock #2. During January 1988, sediment samples from Pearl Harbor exhibited mean concentrations of TBT ranging from 24 to 1900 ng g^{-1}, the most elevated levels again being observed at drydock #2. Sediments from the north channel region included samples collected adjacent to the US Army 5th Transportation Company Heavy Boat facility at Ford Island, adjacent to the associated small boat maintenance activity. These samples contributed to the north channel's mean regional concentration of 1000 ng g^{-1} TBT. In January 1989 regional mean concentrations of TBT in sediment ranged from 26 to 2000 ng g^{-1}. The highest concentration measured was from samples collected at drydock #2. Concentrations of TBT in sediment in Pearl Harbor are summarized in Table 25.3.

A one-way ANOVA model testing differences in concentrations of TBT in sediment over time at a specific monitoring station indicated that significant ($p = 0.05$) differences were present at only four stations in Pearl Harbor from February 1987 to January 1989. Three of the four stations were distant from regions where painting activity or ship presence were TBT exposure factors. Concentrations of TBT in sediment were significantly higher at one station in Southeast Loch during February 1987 than during January 1988 and January 1989. The higher concentration of TBT in sediment measured in February 1987 corresponded with the presence of three vessels painted with TBT and the highest TBT loading factor calculated during sampling periods (Grovhoug *et al.*, 1989).

Samples of oyster tissue collected from Pearl Harbor in April 1986 exhibited TBT levels ranging from <20 ng g^{-1} to 350 ng g^{-1} (wet wt). The highest TBT levels were observed in samples from Rainbow Marina. Samples of oyster tissue, which exhibited mean TBT concentrations ranging from 64 to 360 ng g^{-1} (wet wt), were collected from six sampling regions in Pearl Harbor during the February 1987 survey. Tissues from Rainbow Marina exhibited the highest mean concentrations of TBT observed during this survey. In July 1987, oyster tissues were collected from West Loch, drydock #2, and from McGrew Point across Aiea Bay from Rainbow

Table 25.4 Mean concentrations of TBT in oyster tissues from Pearl Harbor, April 1986–January 1989

	Mean TBT concentration (ng g^{-1} wet wt)									
	Apr 1986		*Feb 1987*		*Jul 1987*		*Jan 1988*		*Jan 1989*	
Region	n[a]	*Mean*	n	*Mean*	n	*Mean*	n	*Mean*	n	*Mean*
Main channels	2	80	6	86					3	38
Southeast Loch									3	190
Drydock #2			3	240	3	63	3	90	3	65
Rainbow Marina	4	260	3	360	3	60	3	180	3	280
East Loch			3	160			3	140	3	45
West Loch	3	70	3	70	3	20	1	25	3	11

[a] *n* indicates number of samples.

Marina (the oyster population at the regular Rainbow Marina sample station was severely depleted at this time). Tributyltin levels in tissue samples ranged from 20 to 63 ng g^{-1}. By January 1988, the regular Rainbow Marina sample station still did not show any signs of appreciable recovery in oyster population. Oyster samples from McGrew Point exhibited a mean TBT level three times greater than samples collected from the same area six months earlier. Tributyltin levels in tissue samples averaged 180 ng g^{-1}. Oyster populations in the West Loch region were observed to be abundant and apparently healthy.

A series of one-way ANOVAs testing tissue concentrations among surveys at each station individually indicated statistical differences in time at two stations. Concentrations of TBT in oysters from drydock #2 decreased after February 1987. This may reflect the absence of TBT paint application during periods approximating recent surveys. Concentrations of TBT in oysters from East Loch have significantly decreased, with the lowest concentration recorded during January 1989. Concentrations in oyster tissue of TBT in Pearl Harbor are reported in Table 25.4.

25.3.3. HONOLULU HARBORS COMPLEX

A total of 153 water, 19 sediment, and 8 tissue samples were collected and analyzed between April 1986 and July 1989 from the Honolulu Harbors complex. In April 1986, the main basin of Honolulu Harbor exhibited mean concentrations of TBT in water of 68 ng l^{-1} at the surface and 30 ng l^{-1} at depth. The highest values were observed at the Ala Wai Boat Harbor basin and adjacent to a floating drydock, with the lowest levels seen in Kewalo Basin and in Keehi Lagoon Boat Harbor. Sampling at Kewalo Basin coincided with the reconstruction of several of the main piers, which precluded the presence of many tenant vessels. The main basin of Honolulu Harbor during March 1987 exhibited an overall significant decrease in TBT concentration at the surface similar to concentrations seen in Southeast Loch. The greatest reduction was seen at the entrance to the harbor, which showed a decrease of 89%. The stations located near the drydock facility and the containership facility showed minor decreases in surface TBT concentrations (63% and 44%, respectively). Mean surfacewater TBT levels during July 1987 rose again to levels approximating those seen during April 1986. Bottomwater samples throughout this period, however, remained at a near-constant average level. Mean surfacewater TBT levels during July 1989 were generally lower by one-half compared with those measured during January 1988. Except for the Ala Wai Boat Harbor and the Honolulu Harbor entrance channel, however, bottomwater concentrations in most of the regions remained consistent with levels seen during previous surveys. Mean concentrations of TBT in the water column for the Honolulu Harbors complex during each survey are illustrated in Figs 25.8 and 25.9.

Sediment samples were collected from each station in the Honolulu Harbors complex during April 1986 and January 1988. In January 1988, regional TBT concentrations in sediment ranged from a low of 100 ng g^{-1} in the entrance channel to 7000 ng g^{-1} (dry wt) at the drydock facility (Grovhoug *et al.*, 1989).

Oyster samples were obtained from two stations within Honolulu Harbor's main basin in July 1987 and January 1988, concurrent with the collection of water samples. Tributyltin levels in the tissue samples ranged from 440 to 610 ng g^{-1} (wet wt). The mean TBT level in oyster tissues collected from Honolulu Harbor in January 1988 showed an increase of ~45% from July 1987, with individual tissue samples containing from 650 to 1100 ng g^{-1} (Grovhoug *et al.*, 1989).

25.4. DISCUSSION

25.4.1. REGIONAL PATTERNS AND TRENDS

25.4.1a. San Diego Bay

An examination of the data collected before and after restrictive use legislation was passed by the state of California indicates that TBT levels in surface water have decreased in the yacht harbor region. In the south bay, north bay, and naval pier regions, TBT values in surface water have decreased significantly as well (Fig. 25.4). While significant decreases in TBT values in surface water have occurred in all regions recently, the number of occupied boat spaces has increased in San Diego Bay from 6373 in 1986 to 7190 in 1988 and 7425 in 1989 (data compiled by San Diego harbor police). A high degree of variability in TBT measurements in a given area and particularly near yacht harbors has been reported previously (Clavell *et al.*, 1986; Huggett *et al.*, 1986; Stang *et al.*, 1990) and must be anticipated in such areas. However, the substantial decrease in TBT values seen in three of four monitoring periods after restrictive use legislation was enacted in January 1988 strongly suggests that TBT levels have significantly decreased as a consequence. Similar observations were made by Alzieu *et al.* (1986) in Arcachon Bay, France, 3 y after adoption of restrictive legislation governing use of TBT in paint, which was nearly identical to that adopted by California. Tin levels in areas of organotin input to Arcachon Bay were 5–10 times lower in 1985 than those found in 1982.

The significant decreases in TBT concentrations in surface water seen in the yacht basin region have been reflected in the other regions of San Diego Bay as well. Significant reductions in TBT concentrations in surface water were seen during the last two monitoring surveys of 1989. We conclude that pleasure craft activity is a major source of TBT in San Diego Bay, although significant inputs are also provided by large commercial craft, vessel repair, and maintenance activities. The north bay region is exposed directly to the three largest yacht harbors in San Diego Bay, as well as to commercial vessel repair facilities. The presence of these sources and the high current velocities and flushing characteristics undoubtedly contribute to the variable TBT levels seen in the north bay region.

In the south bay region, a significant reduction in the concentration of TBT in surfacewater samples was seen. Although the number of occupied boat spaces in the region increased from 1104 in 1987 to 1670 in 1989, concentrations of TBT in water decreased from October 1988 to July 1989 from previous levels (Fig. 25.4). The addition of more sampling stations in the south bay region will permit testing for differences in TBT concentrations between sampling periods with a greater number of degrees of

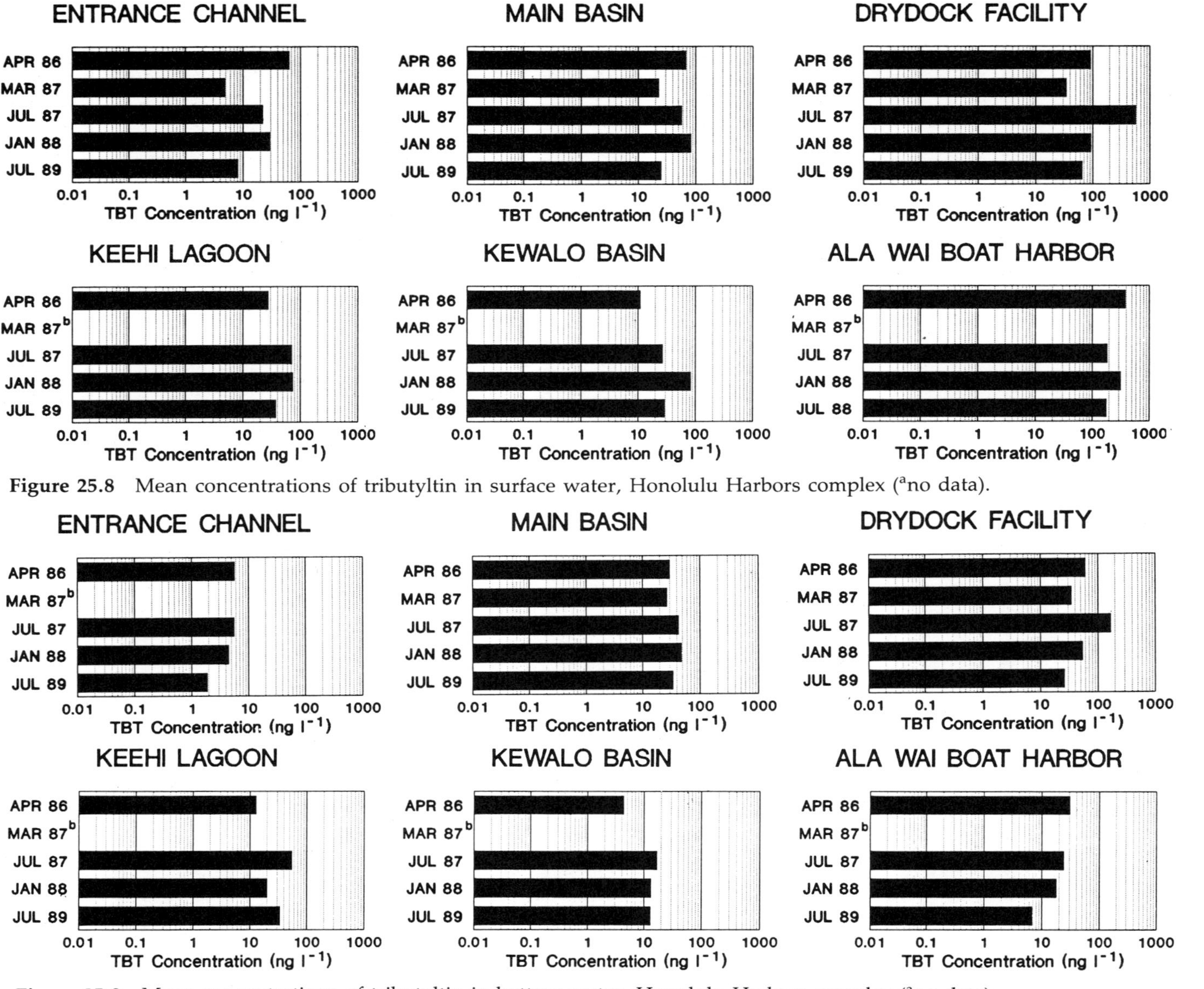

Figure 25.8 Mean concentrations of tributyltin in surface water, Honolulu Harbors complex ([a]no data).

Figure 25.9 Mean concentrations of tributyltin in bottom water, Honolulu Harbors complex ([a]no data).

freedom, and likely confirm future decreases in TBT concentrations in this region.

Significant reductions in TBT concentrations in bottom water were seen in the yacht basin and naval pier regions during the same monitoring periods where significant reductions were recorded in surface waters. The highest TBT concentrations measured in bottomwater samples were recorded during February 1988. The close similarity between decreases in TBT values during given periods and the highest values recorded in surface- and bottomwater samples may be due to increases in occupied boat spaces prior to restrictions on use of TBT paints. The significant decreases in TBT concentrations observed in surface- and bottomwater samples at the end of 1988 and throughout 1989 seem to reflect a decrease in the use of TBT antifouling paints on pleasure craft in accordance with restrictions on use. In the absence of restrictions on use, increases in the number of occupied boat spaces in 1988 and 1989 would likely have led to similar TBT concentrations, such as those presented in Figs 25.4 and 25.5, during February 1988.

TBT concentrations in sediment in San Diego Bay did not reflect recent decreases in water column values and were variable among stations over time. Significant decreases in TBT concentrations in sediment were not recorded in yacht basin stations. It is likely that these sediments would tend to adsorb TBT strongly and release TBT slowly to overlying water since the silt-plus-clay percentages are relatively high (37–92%) in these areas as are the percentages of expandable minerals (32–41%) in the clay fraction (Kram *et al.*, 1989a). A strong linear increase in sediment TBT adsorption capacity with increasing percent clay and percent clay plus silt has been demonstrated by Kram *et al.* (1989b) with San Diego Bay sediments. In most cases, sediments from the yacht harbors surveyed were shown to possess low to moderate ability to desorb TBT in relation to their percent clay–silt fractions and expandable clay mineral index (Kram *et al.*, 1989b).

Stations located in the south bay region have similar physical characteristics to the yacht basin stations and were shown to have a high affinity for TBT adsorption (Kram *et al.*, 1989b). Under these circumstances no significant loss of TBT over time would be expected. No significant decrease in concentrations in TBT in sediments was seen at south bay stations 35, 46, and 48 in agreement with the sediment physical characteristics and overlying TBT concentrations in water measured over time. Increases in the TBT concentration in sediment at stations 48, 46, and 35 during July 1989 may have been influenced by sampling heterogeneity in the absence of an apparent cause for an increase. Since sediments collected in these areas are located in open water bodies at sites not identified clearly by a fixed marker, some variability in site location during a given interval may contribute to variability in TBT concentrations measured.

Changes in TBT concentrations in sediment at stations 22 and 38A were not closely associated with TBT concentrations in overlying bottom water. During February 1988, when concentrations of TBT in bottom water were highest in the naval pier region, concentrations of TBT in sediment were significantly lower at station 22 from those measured during the other monitoring periods. At station 38A, concentrations of TBT in sediment were significantly higher in February 1988 than those recorded during the other monitoring periods. During July 1989 concentrations of TBT in sediment had decreased from the two previous periods at station 22, while an increase was seen at station 38A (Table 25.1). Stations 22 and 38A are located near fixed, well-defined references and are in an area that has similar physical characteristics of sediment with respect to percent organic carbon and grain size content (Kram *et al.*, 1989a). Thus, sampling heterogeneity would not likely contribute to variability in

TBT measurements as much as might be the case in south bay stations. A significantly higher ($p = 0.0003$) concentration of TBT in sediment was recorded at station 13 during February 1988 compared with later periods. Data for the sediment from this station were similar in their distribution over time to station 38A with the exception of the July 1989 survey.

Variability in concentrations of TBT in sediments complicated prediction of trends in TBT concentration in sediment. Tributyltin profiles in sediments in San Diego Bay have not clearly followed water concentrations and likely reflect the complex and variable sediment composition encountered throughout the bay. While some agreement in concentrations of TBT in sediment was found among stations sampled over time in a given area, other data are contradictory. The utility of measuring TBT in San Diego Bay sediments may be limited to reporting the concentration at a given station at a specific time. Little predictive capability is currently possible due to sample heterogeneity and complex physical characteristics. Continued monitoring will contribute additional data, which may identify trends and permit some prediction of future concentrations of TBT in sediment.

Tributyltin concentrations in the tissues of *Mytilus edulis* have generally been declining in San Diego Bay since February 1988. Prior to that survey, concentrations of TBT in tissue gradually increased. This trend follows that of the surfacewater concentrations (Fig. 25.4) and suggests natural tissue burden of TBT in mussel populations as a reasonable indicator of trends in ambient TBT concentration in water when data are available from organisms of comparable age and size class and that have similar histories. Concentrations of TBT in tissue for transplanted mussels in San Diego Bay during this same period show tissue burdens twice as high for areas of similar water concentrations (Salazar and Salazar, 1991, and Chapter 15). This is likely due to the size of mussels measured and their relative growth rates. The Salazar and Salazar study used smaller, juvenile mussels that grew faster, taking up more TBT in the process, while the natural populations sampled here were mature and slower growing. Additionally, the difference in the histories between the natural and transplanted mussels may be important. Tributyltin concentrations measured in tissues of bay mussels were <2000 ng g^{-1} in natural mussel specimens collected in this study with the exception of one sample collected from the Coronado Cays yacht marina. Recent data reported by Salazar and Salazar (1991, and Chapter 15) indicate that concentrations of TBT in tissue above 2000 ng g^{-1} inhibit growth of bay mussels transplanted to natural field locations in San Diego Bay.

25.4.1b Pearl Harbor and Honolulu Harbors complex

Butyltin concentrations in water have been measured from a composite total of 50 Pearl Harbor locations during the period April 1986–July 1989. Increased analytical sensitivities have provided the capability to measure levels down to 0.2 ng l^{-1}; and data are now available from regions of the harbor previously reported as below detection limits. While butyltin levels have appeared to increase from baseline levels measured in 1984 (Grovhoug *et al.*, 1987), part of this apparent increase is due to enhanced analytical sensitivity permitting detection of TBT at concentrations <5 ng l^{-1}.

The movements of individual test ships into various areas of Pearl Harbor have resulted in temporary, highly localized increases in concentrations of TBT in water at specific stations, which soon returned to previous levels after the vessel departed. The relationship between the mean concentration of TBT in surface water in Southeast Loch and the estimated total TBT load factor of the various test ships present demonstrated a

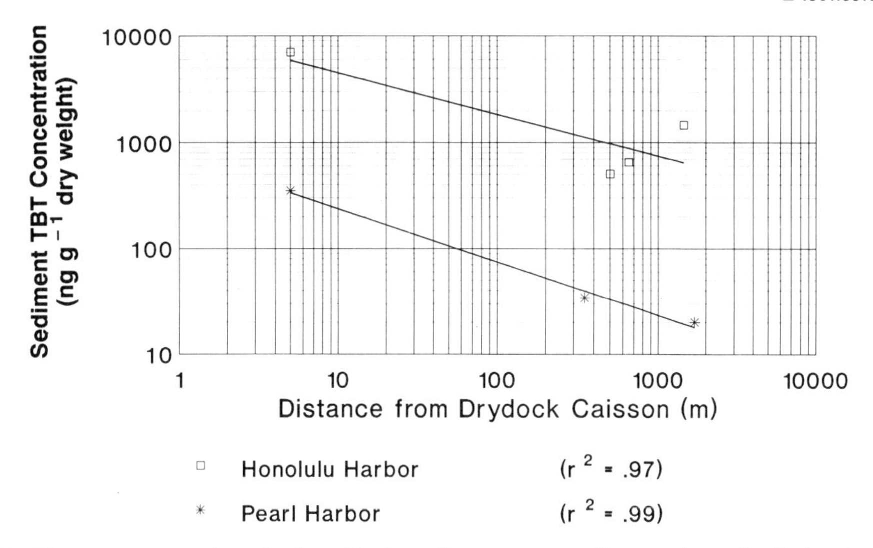

Figure 25.10 Mean concentration of tributyltin in sediment versus distance from drydock caisson, Pearl Harbor drydock #4, and Honolulu Harbor drydock facility.

Pearson correlation coefficient of 0.85 (probability >*F* of 0.02). The association of measured surfacewater concentrations in Pearl Harbor and test ship loading is discussed in Chapter 20. Brief temporal fluctuations in TBT concentrations may indicate that TBT is not highly persistent in the water column at environmental concentrations (Seligman *et al.*, 1989). The consistent low level of tributyltin at depth suggests the possibility that long-term elevated inputs of TBT may have created a reservoir of TBT in the sediments that slowly rediffuses into the deeper waters of the harbor. However, Langston *et al.* (1987) and Langston and Burt (1991) have reported that up to 99% of the TBT present in the water column may be removed into the sediments, with little subsequent desorption back into the water column. Kram *et al.* (1989b) have reported high concentrations of organic carbon in Pearl Harbor sediments relative to those found in Honolulu Harbors, San Diego Bay, or Norfolk Harbor. When the authors tested adsorption and desorption of TBT in Pearl Harbor sediment, poor associations with sediment grain size, clay mineralogy, and percent organic carbon were found.

The elevated sediment loadings seen at drydock facilities appear to be associated with the discharge of particulate material bearing butyltin compounds and paint chips, rather than with butyltins dissolved in the effluent. This particulate material probably settles in the sediment surface layer after remaining temporarily suspended in the water column. At the drydock facility in Honolulu Harbor, TBT loading in sediment during January 1988 was seen to be greater than 10 times that of the rest of the harbor basin. Concentrations of TBT in the water column were nearly identical at both depths.

The extent of migration of the suspended particulate material can be inferred from the sediment samples collected during January 1988 (Fig. 25.10). Drydock #4 in the Pearl Harbor Naval Shipyard opens directly into the northern end of the entrance channel, and sediment samples taken adjacent to the caisson exhibited an average TBT concentration of 350 ng g^{-1} (dry wt). At 350 m to the northwest, sediment samples averaged 34 ng g^{-1} TBT. At 1700 m to the south, a TBT concentration in sediment of 20 ng g^{-1} was measured. Sediment migration appears to be minimal in most areas of Pearl Harbor. Sediment samples collected from other regions did not show a similar gradient in TBT concentrations. Sediment samples collected off the caisson to drydock #2 contained the highest TBT concentrations in the harbor. However, January 1988 values were less than half of those seen in April 1987 samples (1900 and 4500 ng g^{-1}, respectively), suggesting that considerable degradation has occurred at this site.

In Honolulu Harbor, the concentration of TBT in sediment from the immediate vicinity of the drydock facility averaged 7000 ng g^{-1} (dry wt). At 500 m to the south, sediment samples exhibited a mean of 420 ng g^{-1}. At 650 m to the southwest, the mean concentration of TBT in sediment was 690 ng g^{-1}. At 1450 m to the southeast of the drydock facility, the mean concentration of TBT in sediment was 300 ng g^{-1} (Fig. 25.10). Concentration of TBT in surfacewater in these areas averaged between 77 and 95 ng l^{-1}.

In general, concentrations of TBT in the water column in the Honolulu Harbors complex were an order of magnitude greater than those found in Pearl Harbor. An estimate of TBT loading in Honolulu Harbor based on ship data from the Honolulu Harbormaster's Office, the US Coast Guard, and Naval Sea Systems Command revealed that the average daily input into the harbor in 1986 was ~732 g d^{-1}. By comparison, the average daily input of TBT into Pearl Harbor during the course of these surveys was ~88 g d^{-1}. The estimated maximum daily TBT input based on the US Navy's projected Fleet Implementation Plan using low-release-rate paints (0.1 μg cm^{-2} d^{-1}) is 1.1 g d^{-1}. Surfacewater samples from Kewalo Basin contained about one-fourth the amount of TBT measured at Ala Wai, revealing that the total TBT loading from ship hulls docked in Kewalo Basin is notably lower than the total loading from the vessels within the Ala Wai Boat Harbor. The daily average TBT loading in the Ala Wai Boat Harbor in 1986 was estimated to be 103 g d^{-1} based on data provided by the Ala Wai Harbormaster's Office. No data were available, however, to make similar estimates for Kewalo Basin.

The influence of maintenance activities on concentrations of TBT in sediment is also suggested in the data compiled from Kewalo Basin and the Ala Wai Boat Harbor taken in January 1988. Sediment samples collected at the Ala Wai Boat Harbor show mean TBT levels similar to those seen in the main basin of Honolulu Harbor, although the mean concentration of TBT in surface water was nearly four times higher than the mean concentration of TBT in surfacewater samples collected from the Honolulu Harbor basin. The small boat maintenance facility in the area is located near the entrance to the basin adjacent to a major drainage canal that leads directly into the entrance channel of the harbor. No large drainage canal system is associated with Kewalo Basin, as at Ala Wai, and any material emanating from the shipyard within Kewalo Basin would conceivably have added opportunity to settle within the confines of the harbor. Sediment samples collected within the center of Kewalo Basin were seen to contain twice the concentration of TBT of samples from the Ala Wai Boat Harbor.

25.4.2. RELATIONSHIPS BETWEEN CONCENTRATIONS OF TBT IN BOTTOM WATER AND IN SEDIMENT

The data presented in Fig. 25.11 demonstrate relatively constant concentrations of TBT in sediment as concentrations of TBT in bottom water increase beyond 60 ng l^{-1}. In San Diego Bay, a power regression curve ($TBT_{sediment} = 2.3(TBT_{bottom\ water})^{0.6}$) best fit the data with an r^2 value of 0.25. It is clear that little predictive capability is possible if all sediment types are correlated with concentrations of TBT in bottom water. Concentrations of TBT in sediment of 427, 693, and 1057 ng g^{-1} were measured from yacht basin stations located in the commercial basin of San Diego Bay. Yacht

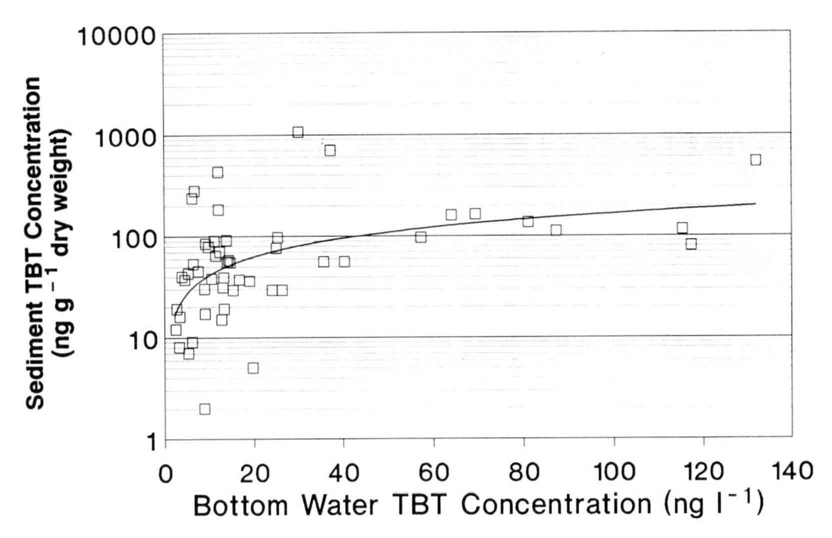

Figure 25.11 Concentration of tributyltin in sediment versus concentration of tributyltin in bottom water in San Diego Bay (r^2 = 0.25, n = 54).

repair and painting activity in this area would contribute TBT paint chips to the nearby water and underlying sediment and potentially influence TBT measurements. It is also probable that the wide range of concentrations of TBT in sediment associated with low (<30 ng l^{-1}) bottomwater concentrations was influenced by differences in sediment type in the regions studied. This variability is reflected by the low correlation with sediment and bottomwater concentrations, although the regression equation was significant (p = 0.0002). Sediment grain size, clay mineralogy, and percent organic carbon content vary a great deal in San Diego Bay and other estuaries and have been shown to greatly affect TBT adsorption (Kram *et al.*, 1989a), partitioning, and bioconcentration (Roberts *et al.*, Chapter 17; Harris *et al.*, Chapter 22; and Unger *et al.*, Chapter 23).

The association of TBT in sediment with the overlying bottom water indicates a pattern related to spatial-use characteristics and possibly water circulation and sediment type. Yacht harbors in San Diego Bay are narrowly confined and have reduced flushing relative to the open bay regions. Additional TBT inputs from releases of antifouling paint from hulls and painting activity would be expected to increase concentrations of TBT in sediment above those in other regions. Additionally, yacht basin sediments from San Diego Bay have been shown to contain moderate to high percentages of organic carbon, silt and clay, and expandable clay minerals (Kram *et al.*, 1989a), which would increase TBT adsorption. Higher TBT concentrations were also measured in the overlying bottom water from samples collected in yacht harbors where concentrations of TBT in sediment were high, suggesting some interaction between the compartment of TBT in sediment and in overlying bottom water. Desorption of TBT from sediment has been shown experimentally (Unger *et al.*, 1987, 1988, and Chapter 23; Kram *et al.*, 1989b). Other studies have, however, shown that TBT desorption occurs slowly if at all (Maguire, 1984; Stang and Seligman, 1986). The degree of sediment desorption of TBT in regions of San Diego Bay is uncertain but would be expected to occur slowly where high percentages of organic carbon, silt, clay, and expandable minerals exist, such as in yacht basins.

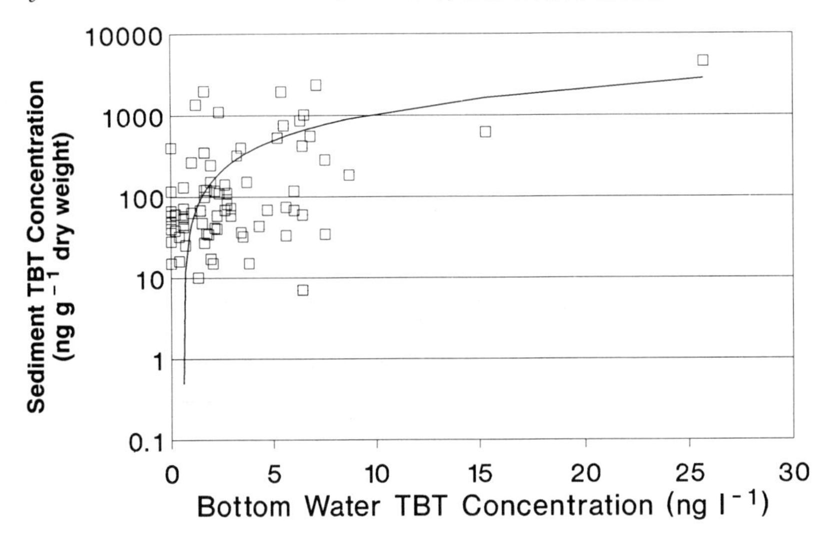

Figure 25.12 Concentration of tributyltin in sediment versus concentration of tributyltin in bottom water in Hawaiian harbors ($r^2 = 0.41$, $n = 82$).

Examination of the correlation between water and sediment samples from Pearl Harbor reveals that no well-defined relationship exists between concentrations of TBT in surface or bottom water and the TBT concentrations found within the sediments at the same station. Elevated concentrations of TBT in bottom water and sediment, such as those seen in San Diego Bay yacht basins, were not measured in Pearl Harbor since only a single small yacht marina exists there. When the Pearl Harbor data were combined with the data from Honolulu Harbors, trend analysis by least-squares linear regression revealed a weak relationship between concentrations of TBT in sediment and the corresponding concentrations of TBT in bottom water ($TBT_{sediment} = -66.6 + 111.9(TBT_{bottom\ water})$, $r^2 = 0.41$), which is displayed in Fig. 25.12.

25.4.3. ENVIRONMENTAL BIOACCUMULATION

Bioaccumulation describes the uptake of a given constituent (in this case, TBT) from water or food by an aquatic organism (Young, 1984). Tissue samples from *Mytilus edulis* collected in San Diego Bay exhibited bioaccumulation factors (BAF) ranging from 2.8×10^3 to 6.6×10^4 (Fig. 25.13). Bioaccumulation factors decreased rapidly as the concentrations of TBT in surface water approached 50 ng l^{-1}. Where concentrations of TBT in surface water ranged from 100 to 300 ng l^{-1}, bioaccumulation factors were 1.0×10^4 or less. Bioaccumulation factors compared using a power regression against concentrations of TBT in surface water fit the regression equations ($BAF = 11.1 (TBT_{water})^{-0.48}$) with an r^2 value of 0.50. Concentrations of TBT in tissue increased with concentrations of TBT in surface water (Fig. 25.13). In 28 of 31 tissue samples collected from areas where concentrations of TBT in water were <25 ng l^{-1}, concentrations of TBT in tissue were <500 ng l^{-1}. As concentrations of TBT in surface water increased, concentrations of TBT in tissue varied considerably. A power regression of concentrations of TBT in tissue versus TBT concentration in water ($TBT_{tissue} = 4.2(TBT_{water})^{0.52}$) best fit the data with an r^2 value of 0.55. Both BAF and tissue

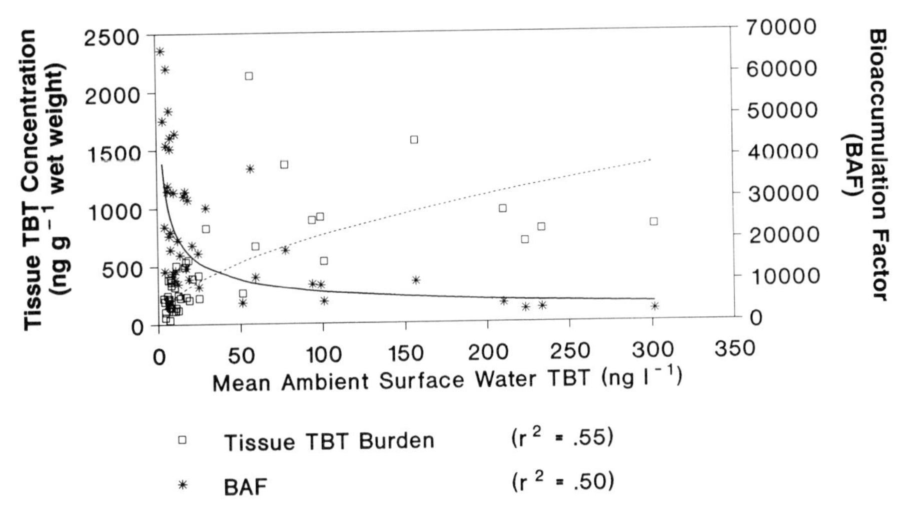

Figure 25.13 Bioaccumulation of tributyltin in the bay mussel (*Mytilus edulis*) in San Diego Bay (broken line: tributyltin concentration curve, solid line: tributyltin bioaccumulation factor curve, $n = 44$).

regressions with concentrations of TBT in water were highly significant ($p = 0.0001$), although low r^2 values indicated little predictive ability.

The data presented in Fig. 25.13 from natural mussel populations closely reflect bioaccumulation data recently reported by Salazar *et al.* (1987), Salazar and Salazar (1988, 1991, and Chapter 15) and Zuolian and Jensen (1989), under natural conditions, in studies where mussels were caged and placed at selected locations in San Diego Bay, or exposed to TBT in in-situ mesocosms. A clear inverse relationship is seen with bioaccumulation factors decreasing with increasing TBT concentrations. Bioaccumulation factors were generally highest when the concentration of TBT in surface water was <100 ng l^{-1}, in agreement with data reported by Salazar and Salazar (1988, 1991, and Chapter 15).

The highest bioaccumulation factors reported by Salazar and Salazar (1988, 1991) were $>2.0 \times 10^4$ and were associated with the highest mussel growth rates measured. At concentrations of TBT in tissue <2000 ng g^{-1}, the authors reported no clear correlation with mussel growth. Only two concentrations of TBT in tissue >1500 ng g^{-1} were measured in mussels collected from natural populations in San Diego Bay in our study.

Tissue samples from Hawaiian waters showed bioaccumulation factors of 8.6×10^3–7.0×10^4 (combined data from Pearl Harbor and Honolulu Harbors surveys). A linear regression analysis of concentrations in oyster tissue shows that the tissue burden of TBT increases with the increase of TBT present in the ambient surface waters according to the function: $TBT_{tissue} = 128 + 8.4(TBT_{water})$, with an r^2 value of 0.84 (Fig. 25.14). The rate of bioaccumulation was seen to initially decline rapidly, then continue decreasing slowly as concentrations of TBT in the surface waters rose. A power regression of bioaccumulation in oyster tissue against concentration of TBT in surface water ($BAF = 55094(TBT_{water})^{-0.39}$) best fit the data with an r^2 value of 0.53. These observations are similar to those reported earlier in this section with *M. edulis* in San Diego Bay under natural and caged conditions.

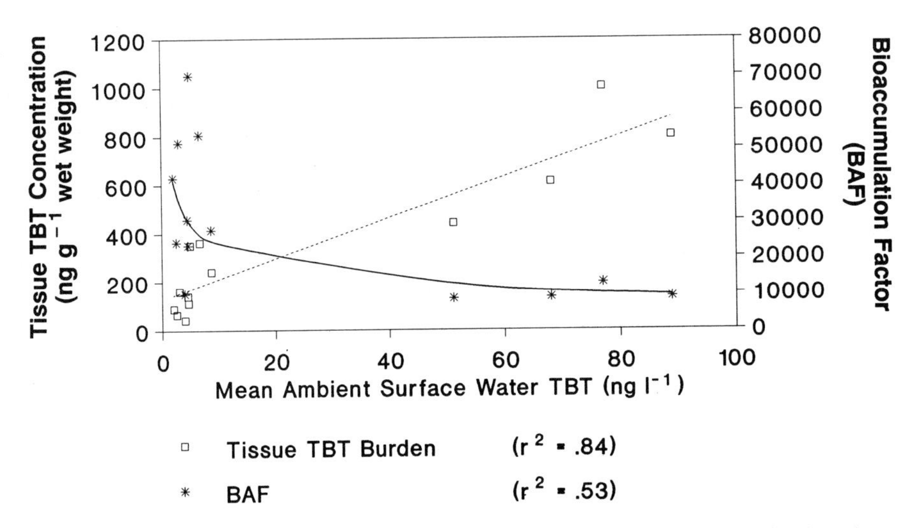

Figure 25.14 Bioaccumulation of tributyltin in oysters in Hawaiian harbors (broken line: tributyltin concentration curve, solid line: tributyltin bioaccumulation factor curve, n = 13).

25.5. CONCLUSIONS

25.5.1. SAN DIEGO BAY

Concentrations of TBT in surface water have significantly decreased in all four regions of San Diego Bay after legislation prohibiting the use of TBT was implemented in January 1988.

Tributyltin concentrations in sediment were not consistent with trends in surface-water data. Concentrations of TBT in sediment were generally highest in yacht basins but did not exhibit significant decreases over time. Variable sediment composition among areas in San Diego Bay appears to be important in determining long-term changes in TBT concentrations, and complicates regional comparative and predictive efforts.

Concentrations of TBT in tissue measured in bay mussels (*M. edulis*) have been decreasing in recent surveys following trends observed in surfacewater data. These data suggest that natural mussel populations are reasonable indicators of large-scale changes in ambient concentrations of TBT in water.

A significant relationship between concentrations of TBT in bottom water and concentrations of TBT in sediment was found when the data were compared using a power regression function. Little predictive capability was possible, however, since the data were collected from various stations having different sediment characteristics, which likely influenced such factors as adsorption and desorption kinetics.

Bioaccumulation factors calculated from concentrations of TBT in surface water and concentrations of TBT in mussel tissue exhibited an inverse relationship, decreasing rapidly as concentrations of TBT in surface water approached 50 ng l^{-1}. This trend is similar to that found with transplanted mussels in San Diego Bay and may be due to reduced uptake at the higher TBT concentrations.

25.5.2. HAWAIIAN HARBORS

Tributyltin concentrations in water generally correlated with calculated TBT loading from ship hull releases documenting that the test ship hulls were the principal source of the compound in Pearl Harbor, except within Rainbow Marina. Rapid changes in harbor concentration based on presence or absence of test ships suggest that a combination of flushing and degradation can effectively remove TBT from the harbor.

When the data were compared using a power regression function, the relationship between TBT concentrations in bottom water and in sediment provided little predictive ability, possibly due to the high percentage of organic material and its effects on the TBT adsorption and desorption rates in the sediments of Pearl Harbor. Concentrations of TBT in sediment were most closely correlated with maintenance activities involving TBT-coated vessels. However, sediment samples collected adjacent to drydocks in Pearl Harbor during three test ship undocking periods showed only moderately elevated levels of TBT.

Tributyltin concentrations in the water column of Honolulu Harbor and adjacent regions were generally an order of magnitude higher than in Pearl Harbor. Burdens of TBT in sediment and tissue were likewise higher, suggesting that overall TBT loading was substantially greater in civilian harbors. The use of paints with higher TBT release rates and less stringent drydock procedures are probable causes.

Concentrations of TBT in oyster tissues correlated with the proximity of TBT sources (test ship hulls, drydocks, and marinas). Bioaccumulation factors for oyster tissues in Hawaiian waters varied inversely with the level of TBT in ambient surface waters.

Restrictive legislation regarding the use of paint that contains TBT has recently been enacted in Hawaii. However, current decreases in concentrations of TBT in tissue at Rainbow Marina in Pearl Harbor may not be representative of a sustained downward trend, but may be due to the demolition of one of the two main piers and subsequent dislocation of the vessels berthed there.

ACKNOWLEDGMENTS

We express our gratitude to the following individuals: D. Bower, S. Cola, S. Frank, J. Groves, M. Guidry, R.S. Henderson, R.K. Johnston, M. Kram, M.O. Stallard, P.M. Stang, and G. Vafa. We also extend our gratitude to the members of Mobile Diving and Salvage Unit One and the MCAS Kaneohe Bay Air–Sea Rescue Team who provided valuable field support in terms of survey craft and personnel. We also thank various members of the Pearl Harbor Naval Station, Pearl Harbor Naval Shipyard, Rainbow Bay Marina, and the officers and crew of *USS Badger*, *USS Beaufort*, *USS Brewton*, *USS Davidson*, *USS Leftwich*, and *USS Omaha* for their assistance and cooperation with these investigations.

REFERENCES

Alzieu, C., J. Sanjuan, J.P. Deltreil and M. Borel. 1986. Tin contamination in Arcachon Bay: effects on oyster shell anomalies. *Mar. Pollut. Bull.*, **17**(11), 494–498.

Clavell, C., P.F. Seligman and P.M. Stang. 1986. Automated analysis of organotin compounds: a method for monitoring butyltins in the marine environment. In: Proceedings of the Oceans '86 Organotin Symposium, Vol. 4, Marine Technology Society, Washington, DC, 23–25 September 1986, pp. 1152–1154.

Dooley, C.A. and V. Homer. 1983. Organotin Compounds in the Marine Environment: Uptake and Sorption Behavior. Technical Report 917, Naval Ocean Systems Center, San Diego, California, 19 pp.

Evans, E.C., III (Ed.). 1974. Pearl Harbor Biological Survey – Final Report. Technical Note 1128. Naval Undersea Center, San Diego, California, 706 pp.

Grovhoug, J.G., R.L. Fransham and P.F. Seligman. 1987. Butyltin Concentrations in Selected US Harbor Systems: A Baseline Assessment. Technical Report 1155. Naval Oceans System Center, San Diego, California, 66 pp.

Grovhoug, J.G., P.F. Seligman, R.L. Fransham, S.Y. Cola, M.O. Stallard, P.M. Stang and A.O. Valkirs. 1989. Measurement of Butyltin Concentrations in Pearl Harbor, Hawaii, April 1986–January 1988: Pearl Harbor Case Study. Technical

Report 1293, Naval Ocean Systems Center, San Diego, California, 39 pp.

Huggett, R.J., M.A. Unger and D.A. Westbrook. 1986. Organotin concentrations in the southern Chesapeake Bay. In: Proceedings of the Oceans '86 Organotin Symposium, Vol. 4, Marine Technology Society, Washington, DC, 23–25 September 1986, pp. 1262–1265.

Johnson, W.E., L.W. Hall, Jr, S.J. Bushong and W.S. Hall. 1987. Organotin concentrations in centrifuged versus uncentrifuged water column samples and in sediment pore waters of a northern Chesapeake Bay tributary. In: Proceedings of the Oceans '87 International Organotin Symposium, Vol. 4, Halifax, Nova Scotia, 28 September–1 October 1987, pp. 1364–1369.

Kram, M.L., P.M. Stang and P.F. Seligman. 1989a. Fate and Distribution of Organotin in Sediments of Four US Harbors. Technical Report 1280, Naval Ocean Systems Center, San Diego, California, 79 pp.

Kram, M.I., P.M. Stang and P.F. Seligman. 1989b. Adsorption and desorption of tributyltin in sediments of San Diego Bay and Pearl Harbor. *Appl. Organometal. Chem.*, **3**, 523–536.

Krone, C.A., D.W. Brown, D.G. Burrows, S.-L. Chan and U. Varanasi. 1989. Butyltins in sediments and waterways in Puget Sound, Washington State, USA. *Mar. Pollut. Bull.*, **20**, 528–531.

Langston, W.J. and G.R. Burt. 1991. Bioavailability and effects of sediment-bound TBT in deposit feeding clams, *Scrobicularia plana*. *Mar. Environ. Res.*, **32**, 61–77.

Langston, W.J., G.R. Burt and Z. Mingjiang. 1987. Tin and organotin in water, sediments and benthic organisms of Poole Harbour. *Mar. Pollut. Bull.*, **18**(12), 634–639.

Maguire, R.J. 1984. Butyltin compounds and inorganic tin in sediments in Ontario. *Environ. Sci. Technol.* **18**(4), 291–294.

Maguire, R.J. 1991. Aquatic environmental aspects of non-pesticidal organotin compounds. *Water Pollut. Res. J. Can.*, **26**(3), 243–360.

Maguire, R.J. and R.J. Tkacz. 1987. Concentration of tributyltin in the surface microlayer of natural waters. *Water Pollut. Res. J. Can.*, **22**(2), 227–233.

Salazar, M.H. and S.M. Salazar. 1988. Tributyltin and mussel growth in San Diego Bay. In: Proceedings of the Oceans '88 International Organotin Symposium, Vol. 4, Baltimore, Maryland, 31 October–2 November 1988, pp. 1188–1195.

Salazar, M.H. and S.M. Salazar. 1991. Assessing site-specific effects of TBT contamination with mussel growth rates. *Mar. Environ. Res.*, **32**, 131–150.

Salazar, S.M., B.M. Davidson, M.H. Salazar, P.M. Stang and K. Meyers-Schulte. 1987. Field assessment of a new site-specific bioassay system. In: Proceedings of the Oceans '87 International Organotin Symposium, Vol. 4, Halifax, Nova Scotia, 28 September–1 October 1987, pp. 1461–1470.

Seligman, P.F., J.G. Grovhoug and K.E. Richter. 1986. Measurement of butyltins in San Diego Bay, CA: a monitoring strategy. In: Proceedings of the Oceans '86 Organotin Symposium, Vol. 4, Washington, DC, Marine Technology Society 23–25 September 1986, pp. 1289–1296.

Seligman, P.F., J.G. Grovhoug, A.O. Valkirs, P.M. Stang, R. Fransham, M.O. Stallard, B. Davidson and R.F. Lee. 1989. Distribution and fate of tributyltin in the United States marine environment. *Appl. Organometal. Chem.*, **3**, 31–47.

Stallard, M.O., S.Y. Cola and C.A. Dooley. 1989. Optimization of butyltin measurements for seawater, tissue, and marine sediment samples. *Appl. Organometal. Chem.*, **3**, 105–114.

Stang, P.M. and P.F. Seligman. 1986. Distribution and fate of butyltin compounds in the sediment of San Diego Bay. In: Proceedings of the Oceans '86 Organotin Symposium, Vol. 4, Washington, DC, 23–25 September 1986, pp. 1256–1261.

Stang, P.M., D.R. Bower and P.F. Seligman. 1989. Stratification and tributyltin variability in San Diego Bay. *Appl. Organometal. Chem.*, **3**, 411–416.

Unger, M.A., W.G. MacIntyre and R.J. Huggett. 1987. Equilibrium sorption of tributyltin chloride by Chesapeake Bay sediments. In: Proceedings of the Oceans '87 International Organotin Symposium, Vol. 4, Halifax, Nova Scotia, 28 September–October 1987, pp. 1381–1385.

Unger, M.A., W.G. MacIntyre and R.J. Huggett. 1988. Sorption behavior of tributyltin on estuarine and freshwater sediments. *Environ. Toxicol. Chem.*, **7**, 907–915.

Valkirs, A.O., P.F. Seligman, P.M. Stang, V. Homer, S.H. Lieberman, G. Vafa and C.A. Dooley. 1986a. Measurement of butyltin compounds in San Diego Bay. *Mar. Pollut. Bull.*, **17**(7), 319–324.

Valkirs, A.O., P.F. Seligman and R.F. Lee. 1986b. Butyltin partitioning in marine waters and sediments. In: Proceedings of the Oceans '86 Organotin Symposium, Vol. 4, Washington, DC, 23–25 September 1986, pp. 1165–1170.

Valkirs, A.O., M.O. Stallard and P.F. Seligman. 1987a. Butyltin partitioning in marine waters. In: Proceedings of the Oceans '87 Organotin Symposium, Vol. 4, Halifax, Nova Scotia, 28 September–1 October 1987, pp. 1375–1380.

Valkirs, A.O., P.F. Seligman, G.J. Olson, F.E. Brinckman, C.L. Matthias and J.M. Bellama. 1987b. Di- and tributyltin species in marine and estuarine waters. Inter-laboratory comparison of two ultratrace analytical methods employing hydride generation and atomic absorption or flame photometric detection. *Analyst*, **112**, 17–20.

Valkirs, A.O., B. Davidson, L.L. Kear, R.L. Fransham, J.G. Grovhoug and P.F. Seligman. 1991. Long-term monitoring of tributyltin in San Diego Bay, California. *Mar. Environ. Res.*, **32**, 151–168.

Young, D.R. 1984. Methods of evaluating pollutant biomagnification in marine ecosystems. In: *Concepts in Marine Pollution Measurements*, H.H. White (Ed.). Maryland Sea Grant Publication, University of Maryland, College Park, Maryland, pp. 261–278.

Young, D.R., P. Schatzberg, F.E. Brinckman, M.A. Champ, S.E. Holm and R.B. Landy. 1986. Summary report – interagency workshop on aquatic sampling and analysis for organotin compounds. In: Proceedings of the Oceans '86 Organotin Symposium, Vol. 4, Marine Technology Society, Washington, DC, 23–25 September 1986, pp. 1135–1140.

Zuolian, C. and A. Jensen. 1989. Accumulation of organic and inorganic tin in blue mussel, *Mytilus edulis*, under natural conditions. *Mar. Pollut. Bull.*, **20**(6), 281–286.

TRIBUTYLTIN IN CANADIAN WATERS 26

R. James Maguire

National Water Research Institute, Department of Environment,
Canada Centre for Inland Waters, Burlington, Ontario L7R 4A6, Canada

ABSTRACT

This chapter is a summary of investigations of the occurrence of tributyltin (TBT) and its less toxic degradation products in water and sediment in Canada in the period 1980–1985. Tributyltin was mainly found in areas of heavy boating or shipping traffic, which was consistent with its use as an antifouling agent. In 8% of the 269 locations across Canada at which samples were collected, TBT was found in water at concentrations that could cause chronic toxicity in a sensitive species, rainbow trout. Tributyltin was occasionally found in the surface microlayer of fresh water at much higher concentrations than in subsurface water. It was also found in 30% of sediment samples collected across Canada. The few fish analyzed that contained TBT were from harbors, a finding consistent with findings in water and sediment.

26.1. INTRODUCTION

In the mid- to late 1970s, concern was raised in Canada about the increasing annual use of organotin compounds, some of which were known to be toxic to aquatic organisms. Organotin compounds as a class were placed on Canada's Environmental Contaminants Act Category III list, which meant that further information was required on their occurrence, persistence, and toxicity in order to make environmental and human-health risk assessments (Environment Canada and Health and Welfare Canada, 1979).

The main organotin compounds that are likely to be released to the environment

Organotin. Edited by M.A. Champ and P.F. Seligman. Published in 1996 by Chapman & Hall, London. ISBN 0 412 58240 6

in Canada are those of triphenyltin (Ph_3Sn^+), tricyclohexyltin (Cy_3Sn^+), di-n-octyltin (Oct_2Sn^{2+}), di-n-butyltin (Bu_2Sn^{2+}), dimethyltin (Me_2Sn^{2+}), and tri-n-butyltin (Bu_3Sn^+) (Thompson *et al.*, 1985). Triphenyltin and tricyclohexyltin are agricultural pesticides. Di-n-octyltin is used as a stabilizer in some food wrappings. Di-n-butyltin is used as a poly (vinyl chloride) stabilizer, as is dimethyltin, and as a catalyst in a number of industrial processes. Tri-n-butyltin was used as an antifouling agent in some paints for boats, ships, and docks prior to its regulation in 1989, as discussed later in this chapter. It is also used as a general lumber preservative and as a slimicide in cooling towers. It is by far the most toxic to aquatic organisms of all organotin compounds used in Canada (Thompson *et al.*, 1985).

Persistence studies on prominent organotin pesticides have indicated that abiotic degradation generally occurs, as does biological degradation, through mechanisms of sequential dealkylation (Maguire *et al.*, 1983) or dearylation (Soderquist and Crosby, 1980). Therefore, the series $Ph_nSn^{(4-n)+}$, $Cy_nSn^{(4-n)+}$, and $Bu_nSn^{(4-n)+}$ (where in each case $n < 4$); $Oct_nSn^{(4-n)+}$ (where $n \leq 2$); and $Me_nSn^{(4-n)+}$ (where $n \leq 4$) may be present in the Canadian environment. The last series includes tri- and tetramethyltin, which are not released as such to the environment, because methylation of tin and methyltin species has been demonstrated in natural water–sediment mixtures (Chau *et al.*, 1981; Guard *et al.*, 1981).

Environmental agencies in many countries have been interested in tributyltin (TBT) because its use as an antifouling agent results in direct contact with aquatic environments, and because of its high toxicity to aquatic organisms (Huggett *et al.*, 1992). Lethal concentrations are in the range 0.04–16 $\mu g\ l^{-1}$ Sn for short-term exposures of copepods (Linden *et al.*, 1979; U'ren, 1983), mussel larvae (Beaumont and Budd, 1984); crab larvae (Laughlin *et al.*, 1984, and Chapter 10), lobster larvae (Laughlin and French, 1980), sheepshead minnow (Ward *et al.*, 1981), bleak (Linden *et al.*, 1979), guppy (Polster and Halacka, 1971), and rainbow trout (Alabaster, 1969; Seinen *et al.*, 1981). See Hall and Bushong (Chapter 9) for a review of the literature on acute effects. Over the past 10 y, research at Environment Canada's National Water Research Institute has been conducted on methods of analysis for butyltin species in water, sediment, and biota (Maguire and Hunealt, 1981; Maguire *et al.*, 1982, 1986; Maguire and Tkacz, 1983; Maguire, 1984, 1991); their occurrence in Canadian waters (Maguire *et al.*, 1982, 1985a, 1986; Maguire, 1984; Maguire and Tkacz, 1985, 1987); the persistence of TBT in fresh water and sediment (Maguire *et al.*, 1983, 1984, 1985b; Maguire and Tkacz, 1985); and the toxicity of TBT to rainbow trout (and its accumulation in various tissues) (Martin *et al.*, 1989). This chapter summarizes the occurrence of TBT in Canada. The occurrence of the other butyltin species and inorganic tin is also described. These other species are far less toxic than is TBT (Davies and Smith, 1980).

In addition to subsurface water, sediment, and fish samples, some samples of the surface microlayer of natural waters have also been analyzed for TBT and its degradation products. The thickness of the surface microlayer is operationally defined by the type of collector used (Garrett, 1965; Harvey, 1966; Baier, 1970; Harvey and Burzell, 1972; Hatcher and Parker, 1974; Larsson *et al.*, 1974; Carlson, 1982), and estimates up to 300 μm are common. There has been interest in the role of the surface microlayer in the dynamics of contaminants in the aquatic environment since it is often enriched in metals, lipophilic contaminants, nutrients, dissolved and particulate organic matter, and microorganisms (e.g. Maguire and Tkacz, 1988). Enrichment of contaminants at the air–water interface will have a negative impact on organisms that spend part or all of their lives at the interface

(Cleary, 1991). Surface microlayer enrichment is also an important phenomenon in the cycling of contaminants between air and water. An enormous quantity of material can be transferred from water to the atmosphere as jet and film drops ejected from air bubbles bursting at the surface of water (MacIntyre, 1974; Blanchard and Szydek, 1975; Liss, 1975; Piotrowicz *et al.*, 1979). In addition, there is evidence that volatilization of lipophilic pesticides from the surface microlayer of natural water after aerial spraying is the main route of dissipation (Maguire *et al.*, 1989).

For brevity, the tributyltin, dibutyltin (DBT), and monobutyltin (MBT) species are referred to in this chapter as though they existed only in cationic form (e.g. Bu_3Sn^+). All TBT compounds dissolved in water appear to yield a species that has a readily exchangeable counter ion depending upon the nature and concentration of other solutes. Support for this contention comes from chromatographic evidence (Fish *et al.*, 1976; Kimmel *et al.*, 1977; Jewett and Brinckman, 1981) and from mammalian (Fish *et al.*, 1976; Davies and Smith, 1980) and fish (Polster and Halacka, 1971) toxicity studies, which have indicated that the toxicity of TBT compounds is independent of the nature of the counter ion.

26.2. EXPERIMENTAL SECTION

The water analyses were done on unfiltered samples. Eight-liter samples of subsurface water were collected from a depth of 0.5 m in glass bottles, and the contents were acidified to pH 1 and stored in the dark at 4°C until analysis (Maguire *et al.*, 1982). Care was taken to open the bottles below the surface to avoid contamination by the surface microlayer. These preservation conditions were effective over a period of at least 3 months. Surface microlayer samples were collected with either a glass-plate sampler for 100-ml samples (Harvey and Burzell, 1972) or a rotating drum sampler for 4-l samples (Harvey, 1966), and were preserved in the same way as the subsurface water samples.

Sediment samples were usually collected with an Ekman dredge. No attempt was made to separate the sediment into different size fractions. The top 2 cm was scraped off into glass jars and frozen as soon as possible, then freeze-dried, ground, and sieved to pass an 850-μm screen (Maguire, 1984; Maguire *et al.*, 1986). In a more detailed study of four sites in Toronto Harbor, 7-cm diameter sediment cores were taken with a lightweight benthos corer and kept at 4°C overnight (Maguire and Tkacz, 1985). The cores were then extruded in 1-cm slices, and these sediment slices were treated in the same way as the sediment grab samples described above. The water above the sediment in these cores was also preserved for analysis.

The TBT species and its degradation products DBT, MBT, and inorganic tin were determined by extraction from acidified water samples, or dry sediments, with the complexing agent tropolone dissolved in benzene; pentylation of the extract to produce the volatile mixed butylpentyltin derivatives, $Bu_nPe_{4-n}Sn$; cleanup by silica gel column chromatography; and concentration and analysis by packed column gas chromatography initially with a modified flame photometric detector (Maguire and Huneault, 1981; Maguire *et al.*, 1982, 1983, 1984; Maguire, 1984), and subsequently with a quartz-tube-furnace atomic absorption spectrophotometric detector (Maguire and Tkacz, 1983, 1985, 1987; Maguire *et al.*, 1985a,b, 1986), which was found to be more reliable. Considering that a fairly specific detector for tin was used in the analyses, identities of the butylpentyltin species were deemed to be confirmed by co-chromatography with authentic standards on two column packing materials of very different polarity.

Several fish netted in Vancouver Harbor (herring, *C. harengus pallasi*) and in harbors and open areas of the Great Lakes (yellow perch, *P. flavescens*; white sucker, *C. commer-*

soni; carp, *C. carpio*; smelt, *O. mordax*; and lake trout, *Salvelinus namaycush*) were also analyzed for butyltin species. The analyses of the fish required some modification of the procedures used for water and sediment. Whole fish (5–200 g) were homogenized in a blender, and the homogenate was dispersed in concentrated HCl, with 10 ml HCl per gram of homogenate (Maguire *et al.*, 1986). Practically complete solution or dispersion was usually effected in <2 h at room temperature with magnetic stirring. Higher temperatures and longer stirring times should be avoided. The resulting mixture was diluted fivefold with water, then extracted in the same way as the water samples described above. Emulsions were broken with large quantities of Na_2SO_4, and lipids were removed by using 3% (w/w) water-deactivated silica gel in the final cleanup, rather than the activated silica gel used in the water and sediment analyses.

Recoveries of TBT, DBT, and MBT from spiked water samples at 1–10 mg l^{-1} varied from 96% ± 4% to 103% ± 8% (Maguire and Huneault, 1981). Recoveries of Sn(IV) from water at pH 5–8 were poor (35% ± 23%), probably because of the formation of unextractable SnO_2 (Maguire *et al.*, 1983). Recoveries of all four species from spiked sediment at 0.01, 0.2, 1, and 100 mg kg^{-1} Sn (dry wt) averaged 90% (range 63–108%), 124% (range 97–180%), 82% (range 55–103%), and 101% (range 96–105%) for TBT, DBT, MBT, and inorganic tin, respectively (Maguire, 1984). Recoveries of these same four species from lake trout (*S. namaycush*) spiked with each species at 0.02–0.10 mg kg^{-1} Sn (wet wt) were 99% (range 94–104%), 75% (range 66–83%), 59% (range 55–63%), and 59% (range 21–97%) (Maguire *et al.*, 1986). Although the method described above was developed for lake trout, it is reasonable to assume that it would be equally effective for other fish. Concentrations reported later were not corrected for recovery. Although Sn(IV) was the only inorganic tin species for which recoveries were determined, the tin present in the above samples was reported as total inorganic tin, since any Sn(II) that might have been present would likely have been oxidized to Sn(IV) during extraction and derivatization.

In the quantitation of the analytes, use was made of appropriate reagent blanks. All results reported were above the limit of quantitation, which is defined as the reagent blank value plus ten times its standard deviation (Keith *et al.*, 1983). In practice this was equivalent to stating that a chromatographic peak was not accepted as real unless it was at least 2–3 times as large as any corresponding peak in the reagent blank.

Three major surveys were made for TBT: around Ontario in 1980–1981 (30 locations) (Maguire *et al.*, 1982), the Detroit River–St Clair River area in 1983 (29 locations) (Maguire *et al.*, 1985a), and across Canada in 1982–1985 (265 locations) (Maguire *et al.*, 1986). Several locations in Ontario were sampled monthly in these surveys, but most locations were sampled only once. Toronto Harbor was also sampled in 1983 at eight locations in a study of the persistence of TBT in water and sediment (Maguire and Tkacz, 1985).

26.3. RESULTS AND DISCUSSION

26.3.1. SUBSURFACE WATER

In the following discussion, TBT concentrations are compared to those which cause acute or chronic toxicity to rainbow trout (*Salmo gairdneri*), a sensitive species for which toxicity data are available. Some reference concentrations are as follows. The 24-h LC_{50} value (the concentration at which 50% of the test organisms died during the specified time period) for rainbow trout is 11.2 μg l^{-1} Sn (Alabaster, 1969). The 12-d LC_{100} value (the concentration at which 100% of the test organisms died during the specified time period) for rainbow trout yolk-sac fry is 1.8

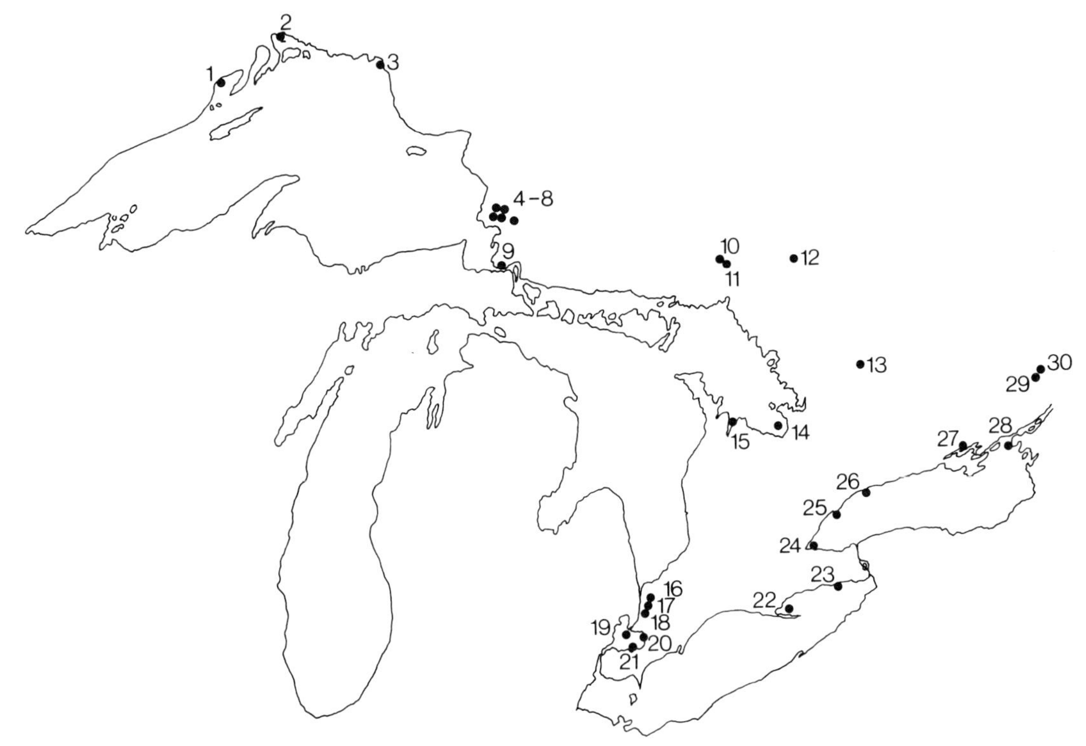

Figure 26.1 Tributyltin sampling locations in 1980–1981 Ontario survey.

μg l^{-1} Sn (Seinen *et al.*, 1981). Concentrations in the range 0.37–1.8 μg l^{-1} Sn have been found, over a period of 110 d, to cause a significant and dose-related growth retardation in rainbow trout yolk-sac fry (Seinen *et al.*, 1981). Concentrations in the range 0.07–0.37 μg l^{-1} Sn have been shown to cause growth retardation in rainbow trout yolk-sac fry over 110 d, but the retardation was significant only during the last few weeks of the exposure period (Seinen *et al.*, 1981).

Tributyltin was first found in Ontario in 1980–1981 in 8 of 30 samples collected on the Canadian side of the Great Lakes (Fig. 26.1) (Maguire *et al.*, 1982). It was found mainly in harbors, marinas, and shipping channels, and not in open areas of the lakes. The presence of TBT was attributed to its use as an antifouling agent. Concentrations for those locations at which it was found are shown in Table 26.1. Concentrations of TBT in a marina in Lake St Clair (site 20), Toronto Harbor (site 25), and Collingwood Harbor adjacent to a shipbuilding facility (site 14) were 15–60% of the 12-d LC_{100} value for rainbow trout yolk-sac fry (Seinen *et al.*, 1981).

Table 26.1 Tributyltin concentrations in subsurface water in 1980–1981 Ontario survey[a]

Sample	*Location*	*[Bu_3Sn^+] (μg l^{-1} Sn)*
3	Lake Superior (Marathon)	0.01
10	Ramsey Lake (Sudbury)	0.31
14	Collingwood Harbor	0.41
20	Lake St Clair (marina)	1.19
24	Hamilton Harbor	0.07
25	Toronto Harbor	0.35
26	Whitby Harbor	0.02
28	Kingston Harbor	0.01

[a]From Maguire *et al.* (1982), with permission of the American Chemical Society. The limit of quantitation was 0.01 μg l^{-1} Sn.

Tributyltin was found in 8 of 23 samples from the Detroit River (Fig. 26.2a–c) in 1983, but not in any of six samples from the St Clair River (Maguire *et al.*, 1985a). Those locations at which it was found are shown in Table 26.2. The highest concentration of TBT was at the mouth of the Ecorse River (site 0311), but it was only 4% of the 12-d LC_{100} value for rainbow trout yolk-sac fry (Seinen *et al.*, 1981). Although most of the locations in the Detroit River were close to ship berths or marinas, it is interesting to note that the second highest TBT concentration was found at site 0330 in the middle of the river north of Fighting Island, possibly due to shipping traffic at the time.

Based on the results of the 1980–1981 Ontario survey, a fairly detailed study was made of the occurrence of TBT in Toronto Harbor in 1983 (Maguire and Tkacz, 1985). Water was sampled at six locations in the harbor (Fig. 26.3), one location close to a drinking water intake outside the harbor (site 7), and one location (site 8, not shown) at the mouth of the Humber River, 5 km west of the harbor. Table 26.3 shows TBT concentrations in the range 0.01–0.20 $\mu g\ l^{-1}$ Sn in the harbor. There was little contamination at the two locations outside the harbor, which supports the contention that the TBT found was due to its use as an antifouling agent. At site 5 a sediment core was taken, and the water overlying the core had a very high TBT concentration of 18.1 $\mu g\ l^{-1}$ Sn. This value is ten times higher than the 12-d LC_{100} value for rainbow trout yolk-sac fry (Seinen *et al.*, 1981) and almost twice the 24-h LC_{50} value for adult rainbow trout (Alabaster, 1969). However, although the sample was filtered through glass wool to remove sediment before extraction, the possibility of contamination by tributyltin-containing fine-grained sediment cannot be eliminated. Site 5 was also sampled once a week for 3 weeks, and concentrations of TBT varied from 0.15 to 0.20 $\mu g\ l^{-1}$ Sn over this period (unpublished observation).

The cross-Canada survey included samples taken from 1982 to 1985 (Maguire *et al.*, 1986). Tributyltin was determined above its limit of quantitation in 42 of 221 samples, and traces (i.e. below its limit of quantitation of 0.01 $\mu g\ l^{-1}$ Sn but above the limit of detection of 0.005 $\mu g\ l^{-1}$ Sn) of it were found in a further 33 samples. Tributyltin occurred most frequently in Ontario, which also had the highest concentrations. In general TBT was found in areas of heavy boating or shipping traffic, which is consistent with its use as an antifouling agent. Table 26.4 summarizes its occurrence in this survey. The highest concentration found was in Port Hope Harbor, Ontario (2.34 $\mu g\ l^{-1}$ Sn). This concentration is about one-quarter of the 24-h LC_{50} value for adult rainbow trout (Alabaster, 1969) and it exceeds the 12-d LC_{100} value for rainbow trout yolk-sac fry (Seinen *et al.*, 1981). At another 7 locations the TBT concentration was $<1.8\ \mu g\ l^{-1}$ Sn but $>0.37\ \mu g\ l^{-1}$ Sn, a concentration that again was found to cause chronic effects in rainbow trout yolk-sac fry (Seinen *et al.*, 1981). At 13 other locations the TBT concentration was $<0.37\ \mu g\ l^{-1}$ Sn but greater than the 0.07 $\mu g\ l^{-1}$ that caused some early growth retardation of rainbow trout yolk-sac fry (Seinen *et al.*, 1981). Therefore, in 21 of these 42 locations at which TBT was determined above its limit of quantitation, there may be cause for concern with regard to chronic toxicity effects in sensitive organisms. These 21 locations represent 10% of all locations at which water samples were taken.

Partitioning of TBT between the (operationally defined) dissolved and particulate phases in water, with possible differences in biological availability, is an important consideration since different organisms may be affected differently, depending upon whether they are exposed to dissolved or particulate-bound TBT, or both. A solids-to-water partition coefficient (K_p) of ~3000 ($\mu g\ kg^{-1}$) ($\mu g\ l^{-1})^{-1}$ has been determined at a suspended solids concentration of ~10 mg l^{-1} (Valkirs *et al.*, 1986), in agreement with

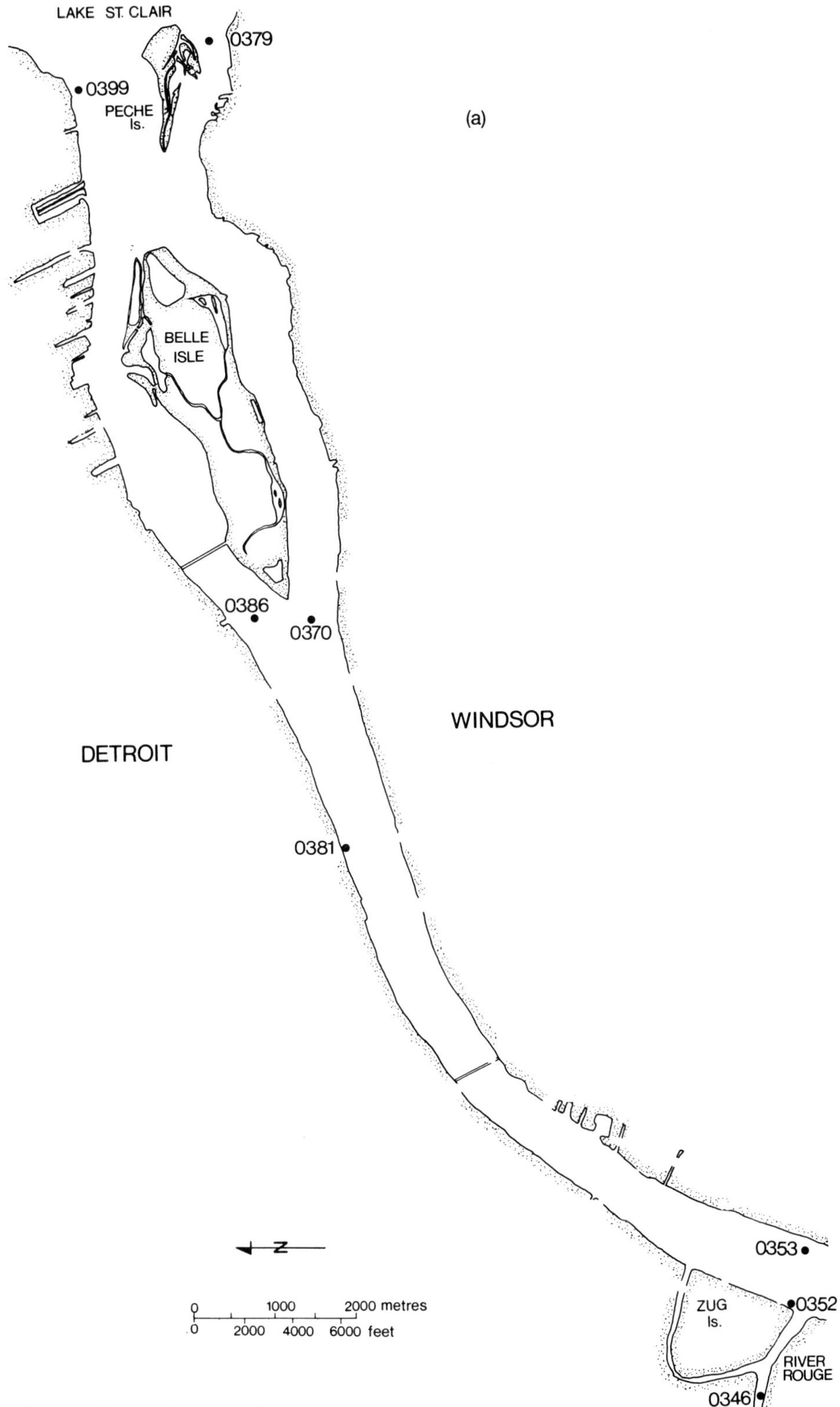

Figure 26.2a–c Tributyltin sampling locations in Detroit River, 1985. From Maguire *et al.* (1985a), with permission of the International Association for Great Lakes Research.

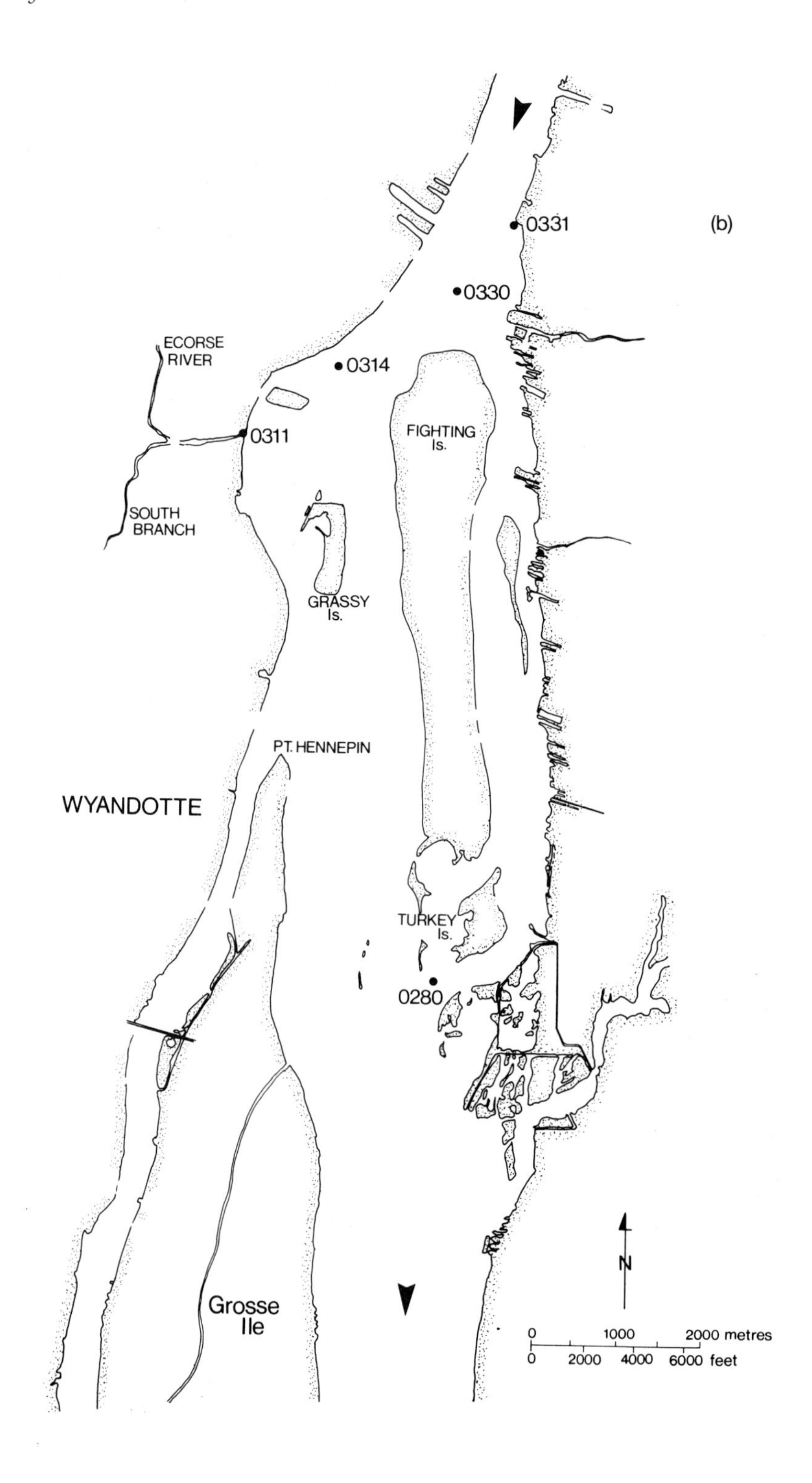
(b)
0331
0330
0314
0311
ECORSE
RIVER
SOUTH
BRANCH
FIGHTING
Is.
GRASSY
Is.
PT. HENNEPIN
WYANDOTTE
TURKEY
Is.
0280
Grosse
Ile
N
0 1000 2000 metres
0 2000 4000 6000 feet

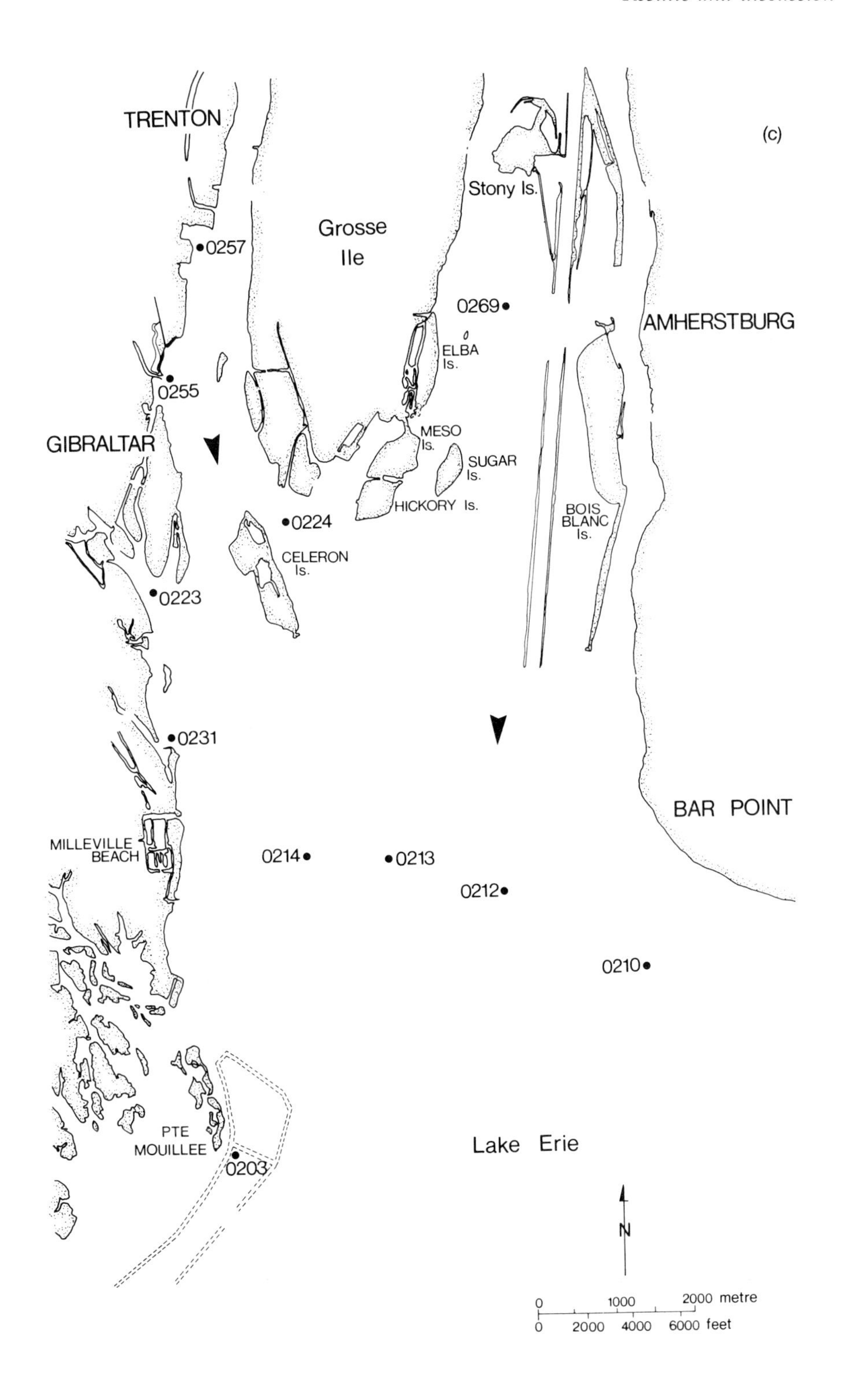
(c)
TRENTON
Grosse
Ile
Stony Is.
0257
0269
AMHERSTBURG
ELBA
Is.
0255
GIBRALTAR
MESO
Is.
SUGAR
Is.
HICKORY Is.
0224
BOIS
BLANC
Is.
CELERON
Is.
0223
0231
BAR POINT
MILLEVILLE
BEACH
0214
0213
0212
0210
PTE
MOUILLEE
0203
Lake Erie
N
0 1000 2000 metre
0 2000 4000 6000 feet

Table 26.2 Tributyltin concentrations in subsurface water in Detroit River, 1983[a]

Location no.	*[Bu_3Sn^+] (μg l^{-1} Sn)*
0399	0.01
0346	0.01
0330	0.05
0314	0.04
0311	0.07
0223	0.04
0231	0.01
0203	0.03

[a]From Maguire *et al.* (1985a), with permission of the International Association for Great Lakes Research. The limit of quantitation was 0.01 μg l^{-1} Sn.

estimates made on the basis of the octanol–water partition coefficient of TBT [log K_{ow} = 3.2 (Maguire *et al.*, 1983)]. (A figure of 10 mg l^{-1} is a typical suspended solids concentration for many fresh waters in Canada.) Such a value for the partition coefficient indicates that most (97%) of the TBT is associated with the aqueous phase of the water column (i.e. that which passes a 0.45-μm filter), and very little is adsorbed to suspended solids. This has been confirmed in samples taken from Toronto Harbor (unpublished observations). In addition, Valkirs *et al.* (1986) found that <5% of butyltin species were associated with the particulate fraction of seawater. The TBT present in the operationally defined dissolved fraction could still, of course, be adsorbed to colloidal material that passes the filter.

In addition to TBT, the much less toxic DBT and MBT species and inorganic tin were found in subsurface water samples in the surveys. Dibutyltin was found in ~10% of all survey samples. It could be introduced to the water directly from DBT compounds used as poly (vinyl chloride) stabilizers, as well as as a degradation product of TBT, which is more likely since TBT was found in the majority of locations at which DBT was found. Monobutyltin was also found in ~10% of the survey samples. Monobutyltin is apparently not used in Canada, so the MBT found in water is probably a degradation product of DBT. The inorganic tin found in water may be present naturally, may be introduced in inorganic form, and/or may be a degradation product of organotin compounds. If it is assumed that the DBT and MBT species are degradation products of TBT, and taking into account the number of times that TBT was found below its limit of quantitation but above its limit of detection, TBT contamination of subsurface water in Canada at the time the surveys were conducted was probably more extensive than the conservative data reported above have indicated. Table 26.5 summarizes the occurrence of DBT, MBT, and inorganic tin, as well as TBT, in subsurface water in Canada.

26.3.2. SURFACE MICROLAYER

Tributyltin was first found in the surface microlayer of natural waters in the 1980–1981 Ontario survey (Maguire *et al.*, 1982), at concentrations up to 10^4 times those of subsurface water. Table 26.6 describes the microlayer results for all surveys. Only the results of the most extensive survey (Maguire and Tkacz, 1987) will be discussed here. Tributyltin was detected in 24 of 74 samples. Concentrations ranged from non-detectable to 473 μg l^{-1} Sn^{-1}. When TBT was found in the microlayer, in general its concentration was much higher than concentrations that have been observed in subsurface water. At all 24 locations mentioned above, the concentration of TBT exceeded the 12-d LC_{100} value for rainbow trout yolk-sac fry. In 6 of these 24 locations the concentration exceeded the 24-h LC_{50} value for adult rainbow trout. The highest concentration, observed in the Moira River at Belleville, Ontario, was 473 μg l^{-1} Sn. These high concentrations have significant implications for organisms that spend part or all of their lives at the air–water interface. It should be borne in mind, however, that the concentration of any toxic

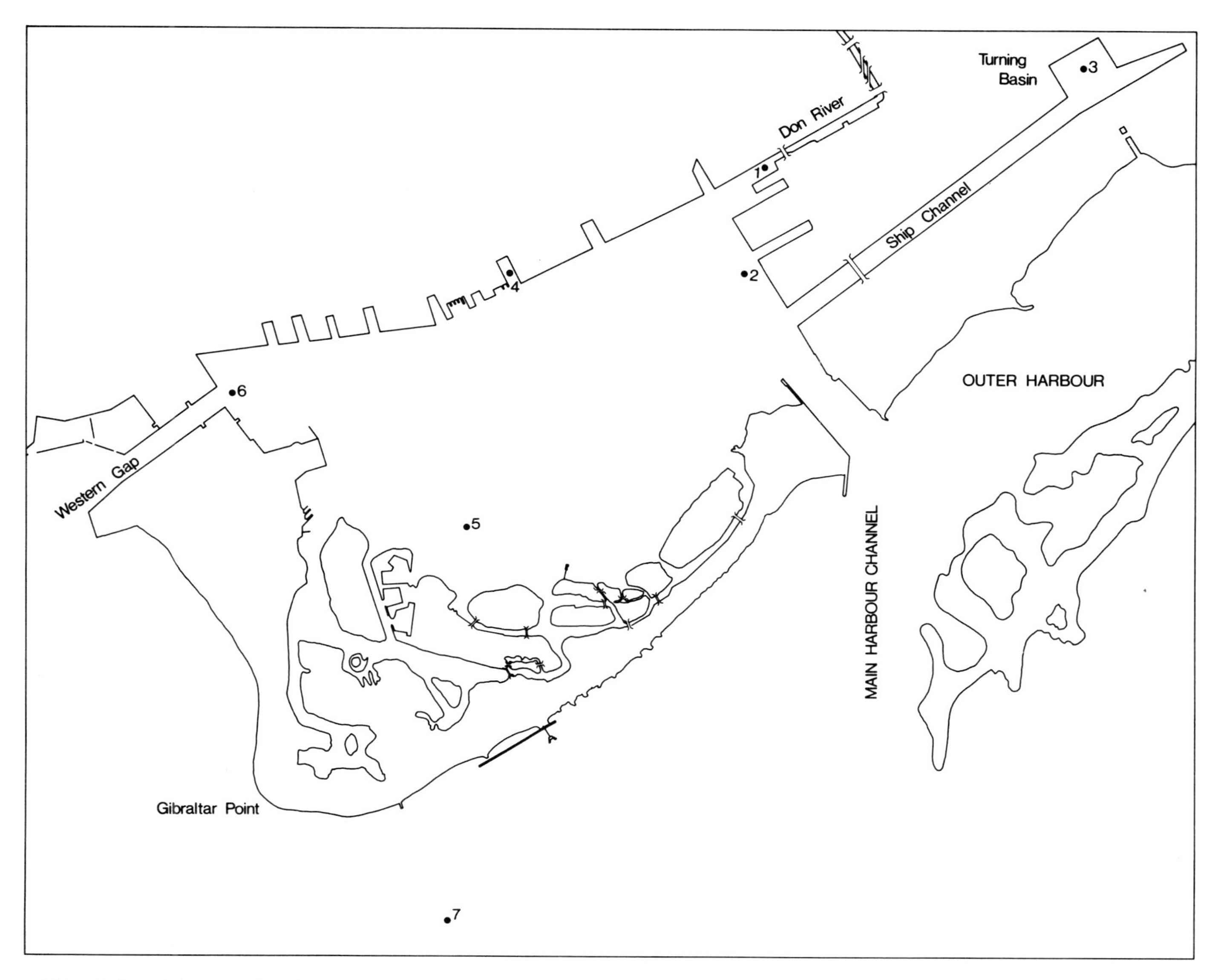

Figure 26.3 Tributyltin sampling locations in Toronto Harbor, 1985. From Maguire and Tkacz (1985), with permission of the American Chemical Society.

Table 26.3 Tributyltin concentrations in subsurface water of Toronto Harbor, 1983[a]

Location no.	*[Bu_3Sn^+] ($\mu g\ l^{-1}$ Sn)*
1	0.02
2	0.02
3	0.01
4	0.11
5	0.20
6	0.04
7	ND
8	0.01

[a]From Maguire and Tkacz (1985), with permission of the American Chemical Society. ND means below limit of quantitation (0.01 $\mu g\ l^{-1}$ Sn).

Table 26.4 Summary of TBT occurrence from national survey, 1982–1985[a]

Province	*Frequency*	*%*	*Range ($\mu g\ l^{-1}$ Sn)*
British Columbia	2/79	3	0.01
Alberta	0/15	0	—
Saskatchewan	0/11	0	—
Manitoba	1/6	17	0.02
Ontario	33/98	34	0.01–2.34
Quebec	5/29	17	0.02–0.36
New Brunswick	1/21	5	0.02
Prince Edward Island	0/1	0	—
Nova Scotia	0/4	0	—
Newfoundland	0/5	0	—

[a]From Maguire *et al.* (1986). Frequency of occurrence is the number of times TBT was determined above its limit of quantitation (0.01 $\mu g\ l^{-1}$ Sn) divided by the total number of samples.

substance in the surface microlayer may vary significantly with time because of turbulence.

Extremely high concentrations of TBT and its less toxic degradation products in surface microlayers relative to concentrations in the subsurface water below the microlayer have led us to estimate the amounts of these chemicals in the microlayer relative to the amounts in the whole depth of subsurface water. With a few spectacular exceptions, the ratios of amounts in the microlayer to amounts in subsurface water are generally negligible, a conclusion reinforced by recent observations with chlorinated hydrocarbons in the Niagara River (Maguire and Tkacz, 1988).

26.3.3. SEDIMENT

Tributyltin was first found in sediment in Ontario in 6 of 20 locations sampled in 1980–1981 (Maguire, 1984). In general the pattern of occurrence of TBT in sediment was similar to its pattern of occurrence in water (i.e. mainly in harbors, marinas, and shipping channels). Those locations at which it was found are shown in Table 26.7 (see also Fig. 26.1). Toronto and Kingston Harbors had the highest TBT concentrations, ~0.2 mg kg^{-1} Sn (dry wt). The toxicological significance of sediment-associated TBT is at present difficult to assess since relatively little work has been done in this area. Tributyltin appears to be adsorbed moderately strongly to sediment. Very little was desorbed in laboratory experiments over a 10-month period (Maguire and Tkacz, 1985). It has been shown that oligochaete worms can (1) accumulate sediment-associated TBT, thus making it potentially available to bottom-feeding fish, and (2) can degrade TBT (Maguire and Tkacz, 1985). Table 26.8 summarizes the results of all sediment surveys for the butyltin species and inorganic tin. The ten highest concentrations of TBT, up to 10.8 mg kg^{-1} Sn (dry wt), were all found in Vancouver Harbor, Canada's busiest harbor. It is possible that the sediment in harbors was contaminated by paint chips <850 μm in size.

26.3.4. FISH

Table 26.9 shows concentrations of TBT and its degradation products in those few fish that were analyzed. Only 18 fish of a variety of species (see section 26.2, Experimental section) were analyzed, so it is impossible to generalize on the results. In addition, the results are not normalized to size or lipid

Table 26.5 Summary of frequency of occurrence and concentrations (μg l^{-1} Sn) of TBT, DBT, MBT, and inorganic tin in subsurface water in Canada, 1980–1985[a]

Location	*[Bu_3Sn^+]*	*[Bu_2Sn^{2+}]*	*[BuSn3+]*	*[Inorganic tin]*
Ontario lakes, rivers, and harbors	8/30	20/30	2/30	23/30
	ND–1.19	ND–3.73	ND–5.73	ND–50.1
Detroit and St Clair Rivers	8/29	13/29	26/29	27/29
	ND–0.07	ND–0.10	ND–0.11	ND–2.95
Toronto Harbor	7/8	5/8	4/8	8/8
	ND–0.20	ND–0.10	ND–0.09	ND–0.88
Bottom water	1/4	1/4	1/4	1/4
	ND–18.10	ND–0.72	ND–0.29	ND–0.11
Canadian lakes, rivers, bays, and harbors	42/221	27/221	33/221	117/221
	ND–2.34	ND–1.36	ND–1.89	ND–37.20

[a]The four studies are reported in Maguire *et al.* (1982, 1985a), Maguire and Tkacz (1985), and Maguire *et al.* (1986), respectively. Frequency of occurrence is the number of times a chemical was determined above its limit of quantitation (0.01 μg l^{-1} Sn) divided by the total number of samples. ND means below limit of quantitation.

Table 26.6 Frequency of occurrence and concentrations (μg l^{-1} Sn) of TBT, DBT, MBT, and inorganic tin in the surface microlayer in Canada[a]

Location	*[Bu_3Sn^+]*	*[Bu_2Sn^{2+}]*	*[$BuSn^{3+}$]*	*[Inorganic tin]*
Ontario lakes, rivers, and harbors	20/28	12/28	0/28	27/28
	ND–24.91	ND–1328	ND	ND–633
St Clair River	2/6	4/6	5/6	6/6
	ND–0.03	ND–0.01	ND	0.06–2.27
Ontario, Quebec, and New York lakes, rivers, and harbors	24/74	19/74	41/74	42/74
	ND–473	ND–365	ND–66.8	ND–14000

[a]The three studies are reported in Maguire *et al.* (1982, 1985a) and Maguire and Tkacz (1987), respectively. Frequency of occurrence is the number of times a chemical was determined above its limit of quantitation (0.01 μg l^{-1} Sn) divided by the total number of samples. ND means below limit of quantitation.

Table 26.7 Tributyltin concentrations in sediment in 1980–1981 Ontario survey[a]

Sample no.	*Location*	*[Bu_3Sn^+] (mg kg^{-1} Sn dry wt)*
20	Lake St Clair (marina)	0.05
22	Port Dover Harbor	0.05
23	Grand River mouth	0.01
25	Toronto Harbor	0.22
26	Whitby Harbor	0.05
28	Kingston Harbor	0.17

[a]From Maguire (1984), with permission of the American Chemical Society. The limit of quantitation was 0.01 mg kg^{-1} Sn (dry wt).

content. Inorganic tin was found frequently in the fish. The only fish that contained TBT were from harbors, which is at least consistent with the findings in water and sediment. It should be noted that the presence of degradation products of TBT in fish may be due to metabolism of TBT or actual uptake of these degradation products by fish.

26.3.5. OTHER ORGANOTIN SPECIES FOUND

The trimethyltin, dimethyltin, and monomethyltin species were also observed in water, sediment, and fish, but much less frequently than the butyltin species (Maguire

Table 26.8 Summary of frequency of occurrence and concentrations [mg kg^{-1} Sn (dry wt)] of TBT, DBT, MBT, and inorganic tin in sediments in Canada, 1980–1985[a]

Location	*[Bu_3Sn^+]*	*[Bu_2Sn^{2+}]*	*[$BuSn^{3+}$]*	*[Inorganic tin]*
Ontario lakes, rivers, and harbors	6/20	6/20	5/20	20/20
	ND–0.54	ND–0.35	ND–0.58	0.08–5.13
Detroit and St Clair Rivers	12/26	10/26	6/26	7/26
	ND–0.07	ND–0.04	ND–0.01	ND–0.39
Toronto Harbor	6/8	5/8	3/8	5/8
	ND–1.28	ND–0.26	ND–0.08	ND–0.62
Sediment core	3/4	3/4	3/4	3/4
	ND–3.52	ND–0.53	ND–0.06	ND–13.80
Canadian lakes, rivers, bays, and harbors	78/235	61/235	47/235	100/235
	ND–10.78	ND–8.51	ND–4.73	ND–15.50

[a]The four studies are reported in Maguire (1984), Maguire *et al.* (1985a), Maguire and Tkacz (1985), and Maguire *et al.* (1986), respectively. Frequency of occurrence is the number of times that a chemical was determined above its limit of quantitation [0.01 mg kg^{-1} Sn (dry wt)] divided by the total number of samples. ND means below limit of quantitation.

Table 26.9 Frequency of occurrence and concentrations (mg kg^{-1} Sn wet weight of whole fish) of TBT, DBT, MBT, and inorganic tin in fish in Canada[a]

Location	*[Bu_3Sn^+]*	*[Bu_2Sn^{2+}]*	*[$BuSn^{3+}$]*	*[Inorganic tin]*
Ontario harbors	2/13	0/13	0/13	12/13
	ND–0.02	ND	ND	ND–0.62
Lakes Ontario, Erie, and Huron	0/4	0/4	0/4	0/4
	ND	ND	ND	ND–0.90
Vancouver Harbor	1/1	1/1	1/1	1/1
	0.24	0.05	0.06	0.04

[a]From Maguire *et al.* (1986). Various freshwater species were analyzed. Herring were the only fish analyzed from Vancouver Harbor. Frequency is the number of times that a chemical was found above its limit of quantitation [0.01 mg kg^{-1} Sn (wet wt)] divided by the number of samples. ND means below limit of quantitation.

et al., 1982, 1986). The presence of these species may be due to the use of dimethyltin compounds as poly (vinyl chloride) stabilizers, or they may result from the methylation of inorganic tin in aquatic environments. The methyltin species are far less toxic to aquatic organisms than is TBT (Davies and Smith, 1980; Laughlin *et al.*, 1984).

The early finding of tributylmethyltin and dibutyldimethyltin in the sediments of four harbors in Ontario at relatively high concentrations with respect to TBT and DBT raised the possibility that methylation of butyltin species in aquatic environments was a significant pathway of transformation (Maguire, 1984). The cross-Canada survey has shown, however, that butylmethyltin compounds were only found infrequently in water and sediment (Maguire *et al.*, 1986). For example, the compound Bu_3MeSn was found in only seven water samples and one sediment sample, compared with 42 water samples and 78 sediment samples for Bu_3Sn^+. Therefore, although it is likely that the butylmethyltin compounds resulted from the methylation of butyltin species in water or sediment, the national survey demonstrated that such methylation is not in general a significant pathway of transformation of butyltin species. Support for such a conclusion is provided by the study of biological degradation of TBT in water and sediment from

Toronto Harbor in which it was demonstrated that butylmethyltin compounds occurred infrequently and in very low concentrations (Maguire and Tkacz, 1985). Tetrabutyltin was found only rarely in water and sediment in the national survey (Maguire *et al.*, 1986). Its presence may have been due to the disproportionation of other butyltin species or, and perhaps more likely, to the contamination of TBT antifouling paint formulations by tetrabutyltin. No other organotin compounds were found in our surveys.

26.3.6. CANADIAN REGULATION OF ANTIFOULING USES OF TBT

In Canada, pesticides are regulated by the Department of Agriculture under the authority of the Pest Control Products Act. In early 1989 the Department of Agriculture announced its regulation of antifouling uses of TBT (Agriculture Canada, 1989). Tributyltin-containing antifouling paint is prohibited on vessels of <25 m in length with the exception of those with aluminum hulls. On larger vessels there is a maximum daily release rate for TBT of 4 $\mu g\ cm^{-2}$. These regulations are consistent with those adopted by the United States of America.

26.4. CONCLUSIONS

In surveys of water and sediment in Canada conducted in the period 1980–1985, TBT was mainly found in harbors, marinas, and shipping channels, a finding consistent with its use as an antifouling agent. Tributyltin was found above its limit of quantitation in 42 of 269 subsurface water samples collected. In 21 of these 42 locations (representing 8% of the total number of samples), TBT was found in water at concentrations that could cause chronic toxicity in a sensitive species, rainbow trout. The frequency of occurrence of TBT in the surveys was probably an underestimate because (1) TBT concentrations above the limit of detection but below the limit of quantitation were not reported, and (2) some degradation products of TBT were found in areas in which TBT was not found, indicating likely antifouling use in those locations. Tributyltin was occasionally found in the surface microlayer of fresh water at much higher concentrations than in subsurface water. It is possible that organisms that spend part or all of their lives may be at risk from such high concentrations of TBT in the surface microlayer. Tributyltin was also found in ~30% of sediment samples collected across Canada, at concentrations up to 11 $mg\ kg^{-1}$ Sn (dry wt); however, the biological availability of sediment-associated TBT has not been established for many benthic organisms. Those few fish analyzed that contained TBT were from harbors, a finding consistent with findings in water and sediment. In addition to TBT, the DBT and MBT species were found in environmental samples, and likely were degradation products of TBT. The trimethyltin, dimethyltin, and monomethyltin species were also found, but much less frequently than the butyltin species. Butylmethyltin species, which may result from environmental methylation of butyltin species, were rarely found.

REFERENCES

Agriculture Canada. 1989. Antifouling paints for ship hulls. Canadian Association of Pesticide Control Officials Note 89–02, 28 February, Department of Agriculture, Pesticides Directorate, Ottawa, Ontario, 3 pp.

Alabaster, J.S. 1969. Survival of fish in 164 herbicides, insecticides, fungicides, wetting agents and miscellaneous substances. *Internat. Pest Control*, **11**, 29–35.

Baier, R.E. 1970. Surface water quality assessment of natural bodies of water. In: Proceedings of the 13th Conference on Great Lakes Research, Buffalo, New York, 1–3 April 1970, Part 1, sponsored by the International Association for Great Lakes Research, pp. 114–127.

Beaumont, A.R. and M.D. Budd. 1984. High mortality of the larvae of the common mussel at low concentrations of tributyltin. *Mar. Pollut. Bull.*, **15**, 402–405.

Blanchard, D.C. and L.D. Szydek. 1975. Electrostatic collection of jet and film drops. *Limnol. Oceanogr.*, **20**, 762–773.

Carlson, D.J. 1982. A field evaluation of plate and screen microlayer sampling techniques. *Mar. Chem.*, **11**, 189–208.

Chau, Y.K., P.T.S. Wong, O. Kramar and G.A. Bengert. 1981. Methylation of tin in the aquatic environment. In: Proceedings of the 3rd International Conference on Heavy Metals in the Environment, 14–18 September 1981, Amsterdam, CEP Consultants Ltd, Edinburgh, pp. 641–644.

Cleary, J.J. 1991. Organotin in the marine surface microlayer and sub-surface waters of south-west England: relation to toxicity thresholds and the UK Environmental Quality Standard. *Mar. Environ. Res.*, **32**, 213–222.

Davies, A.G. and P.J. Smith. 1980. Recent advances in organotin chemistry. *Adv. Inorg. Chem. Radiochem.*, **23**, 1–77.

Environment Canada and Health and Welfare Canada. 1979. Environmental Contaminants Act. Priority Chemicals – 1979. *Canada Gazette*, Part 1, 1 December, pp. 7365–7370.

Fish, R.H., E.C. Kimmel and J.E. Casida. 1976. Bioorganotin chemistry: reactions of tributyltin derivatives with a cytochrome P-450 dependent monooxygenase enzyme system. *J. Organometal. Chem.*, **118**, 41–54.

Garrett, W.D. 1965. Collection of slick-forming materials from the sea surface. *Limnol. Oceanogr.*, **10**, 602–605.

Guard, H.E., A.B. Cobet and W.M. Coleman, III. 1981. Methylation of trimethyltin compounds by estuarine sediments. *Science*, **213**, 770–771.

Harvey, G.W. 1966. Microlayer collection from the sea surface: a new method and initial results. *Limnol. Oceanogr.*, **11**, 608–613.

Harvey, G.W. and L.A. Burzell. 1972. A simple microlayer method for small samples. *Limnol. Oceanogr.*, **17**, 156–157.

Hatcher, R.F. and B.C. Parker. 1974. Laboratory comparisons of four surface microlayer samplers. *Limnol. Oceanogr.*, **19**, 162–165.

Huggett, R.J., M.A. Unger, P.F. Seligman and A.O. Valkirs. 1992. The marine biocide tributyltin: assessing and managing environmental risks. *Environ. Sci. Technol.*, **26**, 232–237.

Jewett, K.L. and F.E. Brinckman. 1981. Speciation of trace di- and triorganotins in water by ion-exchange HPLC–GFAA. *J. Chromatogr. Sci.*, **19**, 583–593.

Keith, L.H., W. Crummett, J. Deegan, Jr, R.A. Libby, J.K. Taylor and G. Wentler. 1983. Principles of environmental analysis. *Anal. Chem.*, **55**, 2210–2218.

Kimmel, E.C., R.H. Fish and J.E. Casida. 1977. Bioorganotin chemistry. Metabolism of organotin compounds in microsomal monooxygenase systems and in mammals. *J. Agr. Food Chem.*, **25**, 1–9.

Larsson, K., G. Odham and A. Sodergren. 1974. On lipid surface films on the sea. I. A simple method for sampling and studies of composition. *Mar. Chem.*, **2**, 49–57.

Laughlin, R.B. and W.J. French. 1980. Comparative study of the acute toxicity of a homologous series of trialkyltins to larval shore crabs, *Hemigrapsus nudus*, and lobster, *Homarus americanus*. *Bull. Environ. Contam. Chem.*, **25**, 802–809.

Laughlin, R.B., Jr, W. French, R.B. Johannesen, H.E. Guard and F.E. Brinckman. 1984. Predicting toxicity using computed molecular topologies: the example of triorganotin compounds. *Chemosphere*, **13**, 575–584.

Linden, E., B.-E. Bengtsson, O. Svanberg and G. Sundstrom. 1979. The acute toxicity of 78 chemicals and pesticide formulations against two brackish water organisms, the bleak (*Alburnus alburnus*) and the harpacticoid *Nitocra spinipes*. *Chemosphere*, **11/12**, 843–851.

Liss, P.S. 1975. Chemistry of the sea surface microlayer. In: *Chemical Oceanography*, J.P. Riley and G. Skirrow (Eds). Academic Press, New York, pp. 193–243.

MacIntyre, F. 1974. The top millimetre of the ocean. *Sci. Am.*, **230**, 62–77.

Maguire, R.J. 1984. Butyltin compounds and inorganic tin in sediments in Ontario. *Environ. Sci. Technol.*, **18**, 291–294.

Maguire, R.J. 1991. Aquatic environmental aspects of non-pesticidal organotin compounds. *Water Pollut. Res. J. Can.*, **26**(3), 243–360.

Maguire, R.J. and H. Huneault. 1981. Determination of butyltin species in water by gas chromatography with flame photometric detection. *J. Chromatogr.*, **209**, 458–462.

Maguire, R.J. and R.J. Tkacz. 1983. Analysis of butyltin compounds by gas chromatography. Comparison of flame photometric and atomic absorption spectrophotometric detectors. *J. Chromatogr.*, **268**, 99–101.

Maguire, R.J. and R.J. Tkacz. 1985. Degradation of the tri-n-butyltin species in water and sediment from Toronto Harbor. *J. Agr. Food Chem.*, **33**, 947–953.

Maguire, R.J. and R.J. Tkacz. 1987. Concentration of tributyltin in the surface microlayer of natural waters. *Water Pollut. Res. J. Can.*, **22**, 227–233.

Maguire, R.J. and R.J. Tkacz. 1988. Chlorinated hydrocarbons in the surface microlayer and subsurface water of the Niagara River, 1985–86. *Water Pollut. Res. J. Can.*, **23**, 292–300.

Maguire, R.J., Y.K. Chau, G.A. Bengert, E.J. Hale, P.T.S. Wong and O. Kramar. 1982. Occurrence of organotin compounds in Ontario lakes and rivers. *Environ. Sci. Techol.*, **16**, 698–702.

Maguire, R.J., J.H. Carey and E.J. Hale. 1983. Degradation of the tri-n-butyltin species in water. *J. Agr. Food Chem.*, **31**, 1060–1065.

Maguire, R.J., P.T.S. Wong and J.S. Rhamey. 1984. Accumulation and metabolism of tri-n-butyltin cation by a green alga, *Ankistrodesmus falcatus*. *Can. J. Fish. Aquat. Sci.*, **41**, 537–540.

Maguire, R.J., R.J. Tkacz and D.L. Sartor. 1985a. Butyltin species and inorganic tin in water and sediment of the Detroit and St. Clair Rivers. *J. Great Lakes Res.*, **11**, 320–327.

Maguire, R.J., D.L.S. Liu, K. Thomson and R.J. Tkacz. 1985b. Bacterial Degradation of Tributyltin. National Water Research Institute Report No. 85–82, Department of Environment, Canada Centre for Inland Waters, Burlington, Ontario, Canada, 17 pp.

Maguire, R.J., R.J. Tkacz, Y.K. Chau, G.A. Bengert and P.T.S. Wong. 1986. Occurrence of organotin compounds in water and sediment in Canada. *Chemosphere*, **15**, 253–274.

Maguire, R.J., J.H. Carey, J.H. Hart, R.J. Tkacz and H.-B. Lee. 1989. Persistence of deltamethrin sprayed on a pond. *J. Agr. Food Chem.*, **37**, 1153–1159.

Martin, R.C., D.G. Dixon, R.J. Maguire, P.V. Hodson and R.J. Tkacz. 1989. Acute toxicity, uptake, depuration and tissue distribution of tri-n-butyltin in rainbow trout, *Salmo gairdneri*. *Aquat. Toxicol.* **15**, 37–52.

Piotrowicz, S.R., R.A. Duce, J.L. Fasching and C.P. Weisel. 1979. Bursting bubbles and their effect on the sea-to-air transport of Fe, Cu and Zn. *Mar. Chem.*, **7**, 307–324.

Polster, M. and K. Halacka. 1971. Hygienic–toxicological problems of some antimicrobially used organotin compounds. *Ernahrungsforschung*, **16**, 527–535.

Seinen, W., T. Helder, H. Vernij, A. Penninks and P. Leeuwangh. 1981. Short term toxicity of tri-n-butyltin chloride in rainbow trout (*Salmo gairdneri* Richardson) yolk sac fry. *Sci. Total Environ.*, **19**, 155–166.

Soderquist, C.J. and D.G. Crosby. 1980. Degradation of triphenyltin hydroxide in water. *J. Agr. Food Chem.* **28**, 111–117.

Thompson, J.A.J., M.G. Sheffer, R.C. Pierce, Y.K. Chau, J.J. Cooney, W.R. Cullen and R.J. Maguire. 1985. *Organotin Compounds in the Aquatic Environment: Scientific Criteria for Assessing their Effects on Environmental Quality*. Publication No. 22494, National Research Council of Canada, Ottawa, 284 pp.

U'ren, S.C. 1983. Acute toxicity of bis(tributyltin) oxide to a marine copepod. *Mar. Pollut. Bull.*, **8**, 303–306.

Valkirs, A.O., P.F. Seligman and R.F. Lee. 1986. Butyltin partitioning in marine waters and sediments. In: Proceedings of the Oceans '86 Organotin Symposium, Vol. 4, 23–25 September 1986, Marine Technology Society, Washington, DC, pp. 1165–1170.

Ward, G.S., G.C. Cramm, P.R. Parrish, H. Trachman and A. Slesinger. 1981. Bioaccumulation and chronic toxicity of bis(tributyltin) oxide (TBTO): tests with a saltwater fish. In: *Aquatic Toxicology and Hazard Assessment: Fourth Conference*, D.R. Branson and K.L. Dickson (Eds). Aquatic Toxicology and Hazard Assessment: Fourth Conference, American Society for Testing and Materials, ASTM STP 737, pp. 183–200.

CHANGES IN CONCENTRATIONS OF ORGANOTINS IN WATER AND SEDIMENT IN ENGLAND AND WALES FOLLOWING LEGISLATION

M.E. Waite,[1] J.E. Thain,[2] M.J. Waldock,[2] J.J. Cleary,[3] A.R.D. Stebbing[3] and R. Abel[4]

[1]*National Rivers Authority, South Western Region, Manley House, Kestrel Way, Exeter, Devon EX2 7LQ, UK*

[2]*Ministry of Agriculture, Fisheries and Food, Fisheries Laboratory, Remembrance Avenue, Burnham-on-Crouch, Essex CM0 8HA, UK*

[3]*Plymouth Marine Laboratory, Prospect Place, West Hoe, Plymouth PL1 3DH, UK*

[4]*Toxic Substances Division, Department of the Environment, Marsham Street, London SW1P 3PY, UK*

Organotin. Edited by M.A. Champ and P.F. Seligman. Published in 1996 by Chapman & Hall, London. ISBN 0 412 58240 6

ABSTRACT

This chapter reports the results of monitoring environmental concentrations of tributyltin (TBT) in the United Kingdom from 1986 to 1988. The UK government implemented legislation in 1986, that limited the amount of tin in copolymer and free-association tributyltin-based antifouling paints. This was followed in 1987 by a ban on the use of TBT-based paints on boats under 25 m in length and on mariculture equipment. Concentrations of TBT have been monitored in seawater and sediments from estuaries that have traditionally supported shellfisheries and are popular centers of boating activity. By 1988, concentrations of TBT in water samples taken close to sites of oyster and mussel cultivation were approximately half of the concentrations recorded in 1986. However, at all of these sites in 1988 the concentrations of TBT in the water exceeded 2 ng l^{-1}, which is the environmental quality standard (EQS) set by the United Kingdom for the protection of marine life. The concentrations in many areas were above the toxicological threshold values of a variety of species. Water samples from marinas have also been analyzed, and in some cases concentrations of TBT in excess of 1000 ng l^{-1} have been recorded; however, by 1988, concentrations of TBT in the water of several marinas had decreased to approximately a quarter of the values recorded in 1986. The concentrations of TBT in sediments were highest in samples taken close to high-density boat moorings or in marinas. Such samples contained concentrations of TBT >1 $\mu g\ g^{-1}$ (dry weight). Water samples have also been taken from commercial harbors, major waterways, and close to anchorages. It is difficult to assess the contribution of shipping to the inputs of TBT to the environment, when ships share the same body of water with small boats and yachts. However, the hosing-down of ships in dry-dock has been identified as a major source of input of TBT to the aquatic environment. Concentrations of TBT in major rivers, lakes, and the Norfolk Broads have been recorded. Levels of TBT in excess of 1000 ng l^{-1} have been measured in samples from some freshwater marinas and boatyards. The spillage of timber-treatment chemicals, containing TBT, from riverside storage facilities, has sometimes been the cause of major inputs of TBT to the freshwater environment. The distribution of organotin (tributyltin and dibutyltin) throughout the water column has been investigated. Maximum concentrations occur in the surface microlayer and minimum values near the bottom. Concentrations in samples from the surface microlayer, taken with a Garrett screen sampler, were as much as 27 times greater than those in subsurface waters. It is probable that such high concentrations in the surface microlayer may have deleterious effects on both the neuston and on organisms of the littoral zone.

27.1. INTRODUCTION

In 1986, the UK government introduced legislation that reduced the amount of tin permitted in antifouling paints available for retail sale. This was followed, in 1987, by further legislation that banned the use of tributyltin-based paints on boats <25 m in length and on mariculture equipment. The UK Department of the Environment (DoE) and the Ministry of Agriculture, Fisheries and Food (MAFF) commissioned a research and monitoring program to evaluate the efficacy of the legislative action (see Abel *et al.*, 1987; Abel, Chapter 2). The aim was to measure the reduction of environmental contamination by TBT in areas used by small boats, to examine the possible recovery of sensitive marine species in affected areas, and to carry out preliminary studies on inputs of TBT from shipping. Investigations into the distribution of organotins throughout the water column, including the surface microlayer, were also undertaken.

In England and Wales, the monitoring program for environmental levels of TBT was carried out by scientists from MAFF at Burnham-on-Crouch and the Plymouth Marine Laboratory (PML). This chapter reports some of the measurements made on estuarine waters and sediments in areas of shellfish cultivation and high pleasure-craft activity, commercial harbors, and drydocks, and the results of similar studies on rivers and lakes. The distribution of organotins throughout the water column and surface microlayer is also discussed.

Water samples taken as part of the aquatic monitoring programs (see section 27.2, Aquatic monitoring), have been analyzed for TBT using methods based on the technique developed by Matthias *et al.* (1986) and Matthias (Chapter 5). Alkyltin compounds were converted to the corresponding hydrides and simultaneously extracted into dichloromethane; the extract was then reduced in volume and injected into a gas chromatograph fitted with a flame photometric detector (Waldock *et al.*, 1989). At the 60 ng l^{-1} level, the mean recovery for TBT (as the cation) is 97% ± 11% ($n = 7$), and for DBT (as the cation) it is 93% ± 15% ($n = 6$). The detection limit for TBT is ~1 ng l^{-1}. Initially, recoveries for $MBTCl_3$ were good (84% ± 5%). After 1986 there was a change in the gas chromatograph used for analysis, which necessitated sample extracts being reduced further in volume before injection, and as a consequence recoveries for MBT (as the cation) were <50%.

Sediment samples were analyzed by a similar technique. Alkyltin compounds were extracted from the sediment by sodium hydroxide and methanol, converted to the hydrides, and partitioned into hexane. The derivatives were then analyzed by gas chromatography (Waldock *et al.*, 1989).

Water samples collected for the investigation into the vertical distribution of organotins through the water column (see section 27.3, Fresh water) were analyzed by atomic absorption spectrophotometry. Organotins were extracted from unfiltered seawater into toluene or hexane before analysis by electrothermal atomic absorption spectrophotometry (EAAS). In the toluene method the water is first acidified with glacial acetic acid and then extracted into toluene (Cleary and Stebbing, 1987a). Alternatively, water is acidified with hydrochloric acid and extracted with hexane (Bryan *et al.*, 1986). These methods extract TBT and some DBT, and therefore results are expressed as organotin rather than TBT. On some occasions the solvent extracts were washed with 1 M sodium hydroxide solution, which back extracted DBT to the aqueous phase, thus enabling the determination of the separate concentrations of DBT and TBT.

27.2. AQUATIC MONITORING

27.2.1. AREAS OF SHELLFISH CULTIVATION AND HIGH PLEASURE-CRAFT ACTIVITY

The MAFF laboratory at Burnham-on-Crouch has been monitoring concentrations of TBT in estuaries and marinas of England and Wales since 1982 (Waldock, 1986; Waldock *et al.*, 1987a, b; Waite *et al.*, 1991). However, after 1986 the monitoring program was greatly

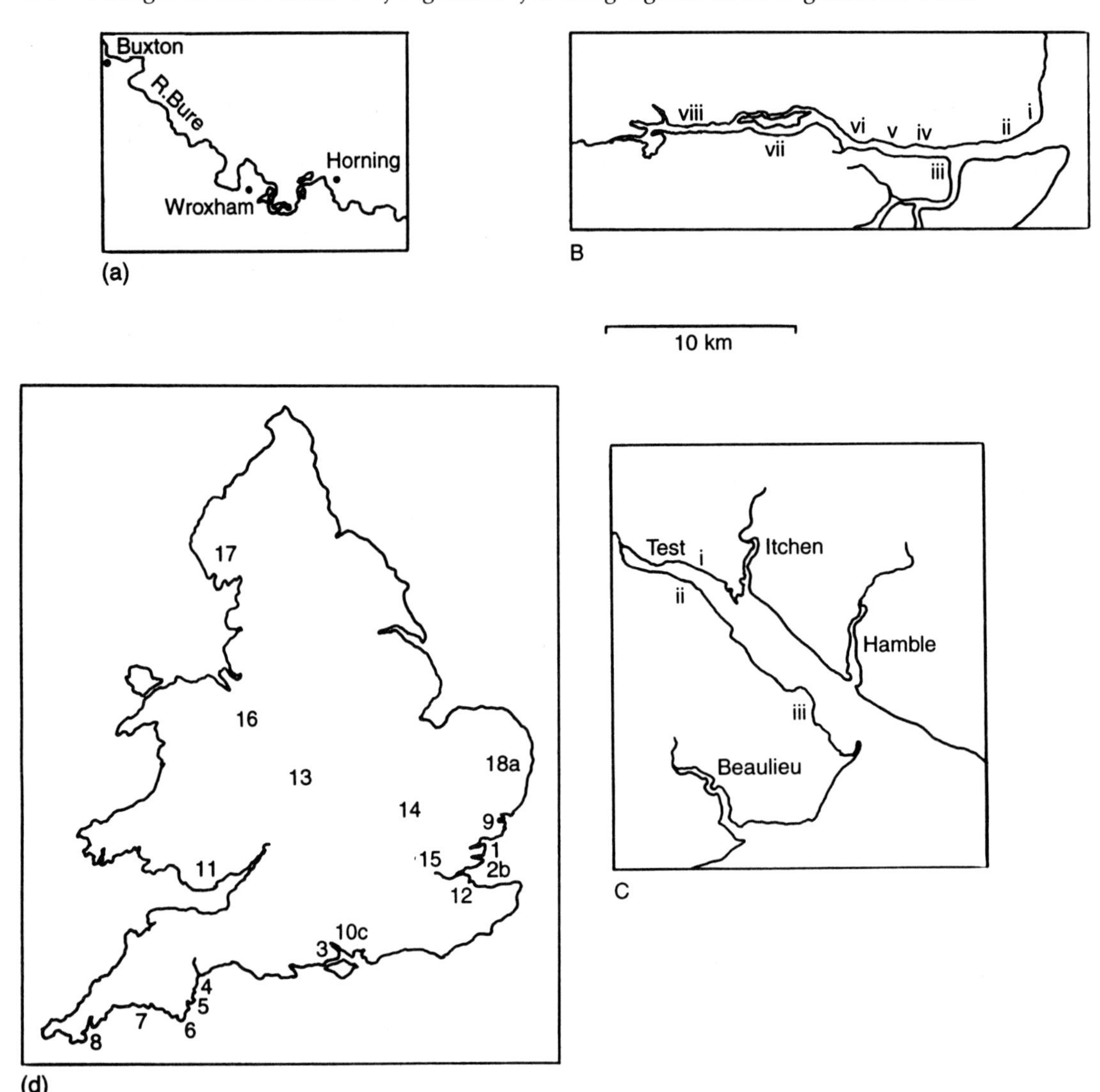

Figure 27.1 MAFF UK monitoring sites for TBT. 1, Blackwater; 2, Crouch (b) (i, Holliwell Point; ii, Holliwell Buoy; iii, Roach Mouth; iv, Bush Shore; v, Burnham; vi, Creeksea; vii, Bridgemarsh Island; viii, Fambridge); 3, Beaulieu; 4, Teign; 5, Dart; 6, Kingsbridge; 7, Plymouth; 8, Fal; 9, Felixstowe Harbour; 10, Southampton Water (c) (i, container port; ii, ship repair yard; iii, Fawley); 11, Ebbw; 12, Medway; 13, Tame; 14, Great Ouse; 15, Thames; 16, Trent Canal; 17, Lake District; 18, Norfolk Broads (a).

expanded (see Abel, Chapter 2, for background discussion). The full range of monitoring sites is shown in Fig. 27.1. Effort has been concentrated on six estuaries (Beaulieu, Blackwater, Crouch, Dart, Kingsbridge and Teign), which have all traditionally supported shellfisheries.

Pacific oyster (*Crassostrea gigas*) and mussel (*Mytilus edulis*) spat have been relaid each year in these estuaries at sites ranging from those remote from boats to one within a marina. A site on the River Teign was chosen as a control but, because of the ubiquitous nature of TBT in English estuaries, this area had to be redesignated as a 'low TBT site' (Waldock *et al*, 1987a).

Oysters and mussels were sampled monthly from March/April to November/ December, growth performance was monitored (increase in whole weight and meat weight), and tissues were analyzed for TBT. In the case of *C. gigas*, shell thickness index was also measured. Some of the results of the shellfish monitoring are reported by Waldock *et al.*, Chapter 11.

Single samples of surface sediment and single water samples, from a depth of 10 cm, were taken monthly from each of the shellfish sites. In addition, water samples were collected at locations up river of the shellfish sites, near the top of the estuaries (to reflect the quality of water flowing into them), and also in harbors or marinas near to the shellfish growing areas. Data obtained for organotin concentrations in water samples collected in 1986 have been published in detail (Waldock *et al.*, 1987a).

Of the six estuaries, the River Crouch was the most extensively studied: bivalves were relaid at eight stations along a 20-km stretch of the estuary from the top to the mouth, and water samples were taken fortnightly (weekly in summer when boating activity was at its height) at Burnham-on-Crouch.

Monthly water samples have also been taken from Sutton Marina, Plymouth, and

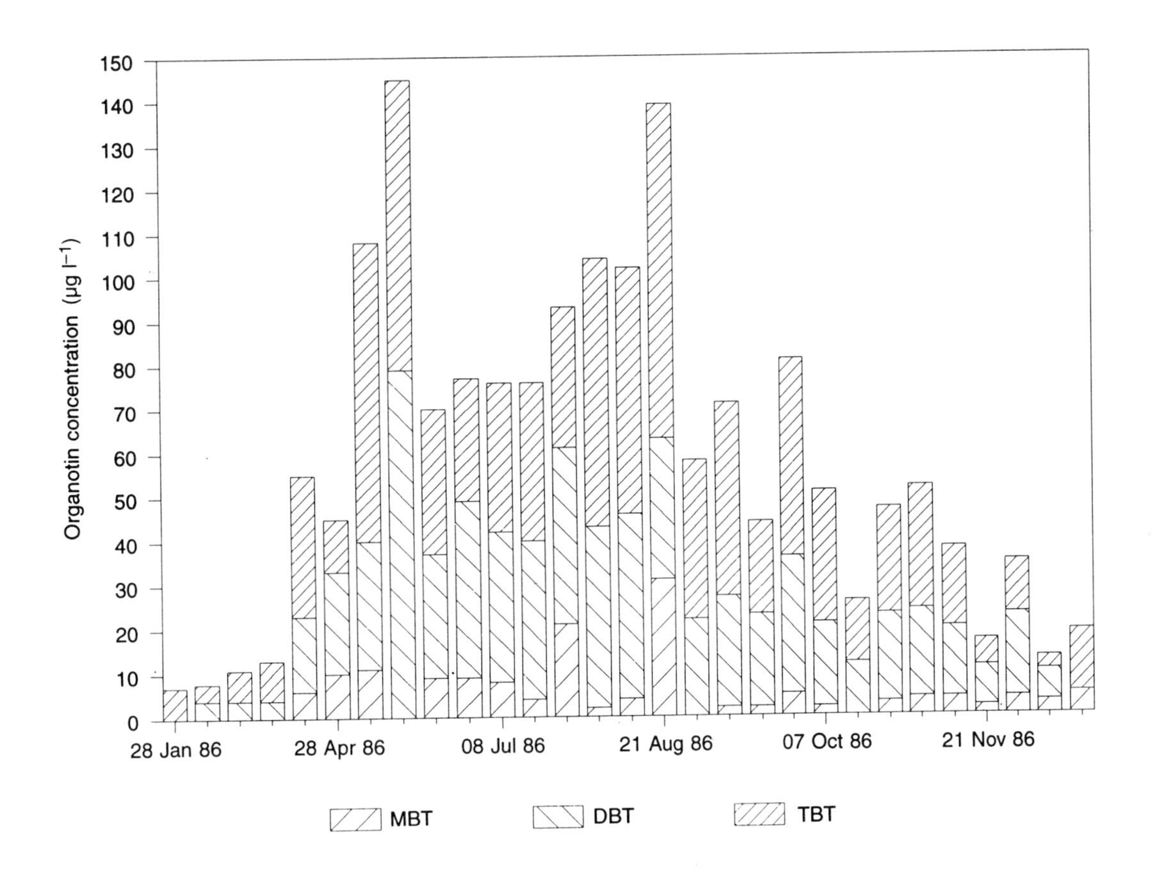

Figure 27.2 Total organotin (MBT + DBT + TBT) concentrations in subsurface (10 cm depth) water at Burnham-on-Crouch during 1986.

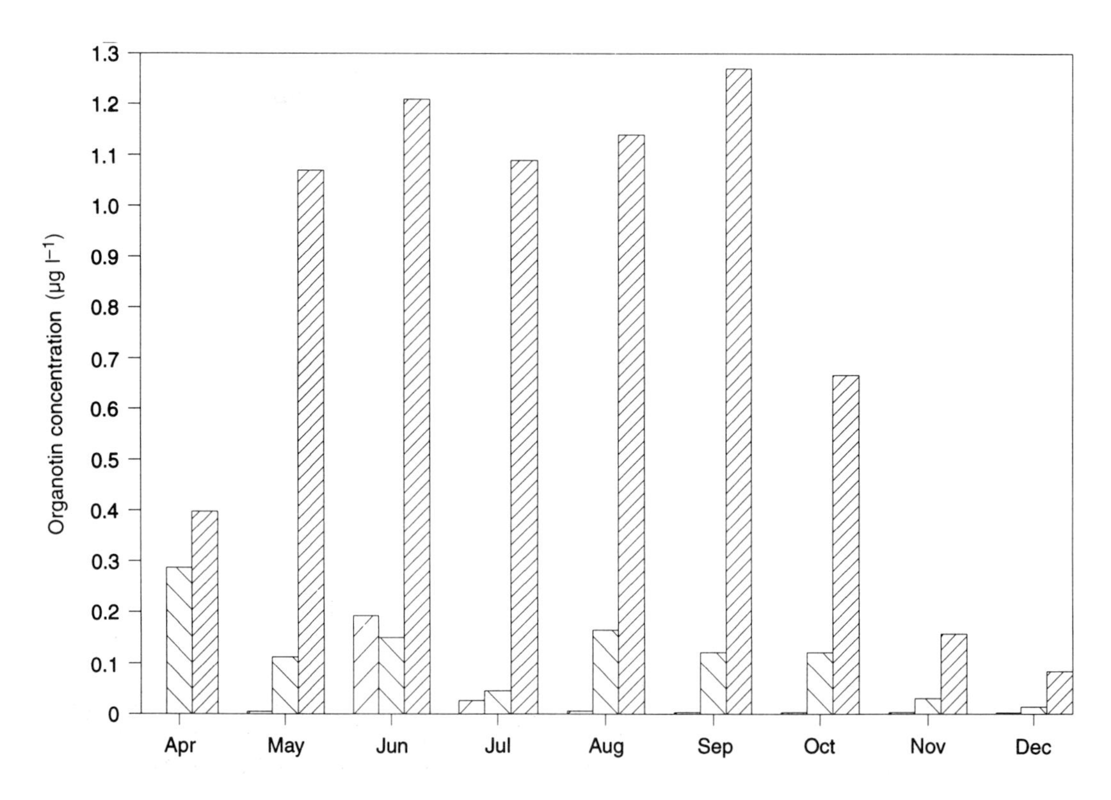

Figure 27.3 Concentrations of MBT, DBT, and TBT in subsurface (10-cm depth) water at Sutton Marina during 1986.

Hythe Marina on Southampton Water (Fig. 27.1). The latter marina has only been in use since 1988, (i.e. after the ban on the use of TBT-based paints for boats <25 m in length; see Abel, Chapter 2).

27.2.1a. Water

In 1986, water samples from the riverine ends of the estuaries contained only low levels of TBT (generally <2 ng l^{-1}), indicating that no significant inputs of TBT were entering the head of the estuaries. When organotins were present in a water sample, TBT was usually the predominant species followed by DBT, with only small amounts of MBT (Fig. 27.2).

However, the ratio of DBT to TBT in water varied from open estuarine sites (i.e. shellfish cultivation sites) to marina sites. For example, the ratio of DBT to TBT in the River Crouch at Burnham was 0.9 ± 0.5 in 1986 and 0.7 ± 0.5 in 1988; the ratio for water from Sutton Marina, where samples were taken close to TBT-painted hulls, was 0.2 ± 0.2 and 0.3 ± 0.1 in 1986 and 1988, respectively (Figs 27.2 and 27.3).

In both the open estuary and marina situations, the seasonal changes in concentrations of TBT were correlated with changes in boating activity. Concentrations of TBT rose rapidly in the spring at the time of boat launching, reduced slightly in mid-summer, and then peaked again in the late summer. Mid-season cleaning or repainting of boats and late season scrubbing off were common practices in all of the estuaries studied, and a

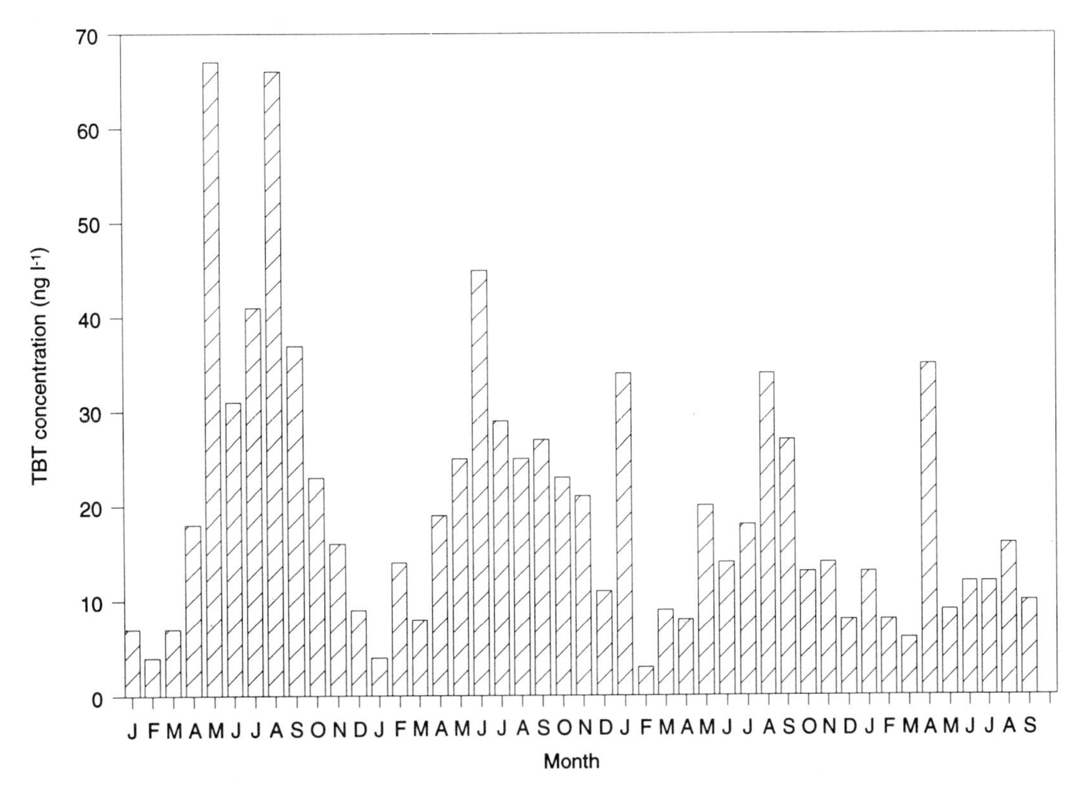

Figure 27.4 Mean monthly concentrations of TBT in subsurface (10-cm depth) water at Burnham-on-Crouch from 1986 to 1989. Mean values are for up to four (weekly) samples. The unusually large TBT value for January 1988 at Burnham was due to a single highly contaminated sample, which probably contained TBT effluent from hosing yachts.

late-season peak in TBT was often observed in water samples collected in 1986 (Figs 27.2 and 27.3). Hall (1988), Seligman *et al.* (Chapter 20), Huggett *et al.* (Chapter 24), Grovhoug *et al.* (Chapter 25), and Maguire (Chapter 26) have reported seasonal changes in the TBT concentration of water which relate to boating activity.

Since 1986 there has been a gradual decrease in TBT water concentrations in the Crouch concurrent with reductions of the amount of TBT allowed in paint formulations early in 1986 and the ban on use of TBT for small boats in 1987. However, the seasonal pattern has remained similar from year to year at Burnham: TBT concentrations in general are low during the winter, rise in the spring, and decline in the autumn (Fig. 27.4).

At other sites, water samples were taken monthly and hence only give an indication of general trends. In 1988, at sites where oysters and mussels were held, mean summer concentrations of TBT had decreased compared with the corresponding 1986 values (Table 27.1). This difference was statistically significant (t-test $p < 0.01$).

At three of the marina and harbor sites the reductions in TBT were more marked, the 1988 TBT water concentrations being approximately a quarter of those recorded in 1986

Table 27.1 Mean summer (May–September) concentrations of tributyltin (ng l^{-1}) in UK estuaries, marinas, and harbors from 1986 to 1988[a]

		TBT concentration (ng l^{-1})		
Estuary/location	*Shellfish site*	*1986*	*1987*	*1988*
Crouch	Fambridge	15 ± 8	33 ± 27	21 ± 8
	Bridgemarsh	22 ± 12	17 ± 12	13
	Creeksea	35 ± 17	17 ± 9	22 ± 14
	Burnham	45 ± 17	31 ± 18	23 ± 18
	Bush Shore	26 ± 9	22 ± 15	13 ± 5
	Roach Mouth	26 ± 12	18 ± 13	15 ± 12
	Holliwell Buoy	11	26 ± 23	10 ± 4
	Holliwell Point	16	6 ± 5	6 ± 5
Blackwater	West Mersea	38 ± 21	36 ± 29	76 ± 43
Dart	Blackness Point	38 ± 33	13 ± 4	13 ± 5
Kingsbridge	Frogmore	15 ± 6	1 ± 6	51 ± 95[b]
Teign	Arch Brook	12 ± 10	7 ± 6	6 ± 2
	Marinas/Harbors			
Plymouth	Sutton Marina	1156 ± 84	882 ± 323	274 ± 79
Dart	Dart Marina	95 ± 66	85 ± 32	21 ± 4
Kingsbridge	Salcombe	117 ± 84	62 ± 71	30 ± 18
Teign	Teignmouth	22 ± 24	23 ± 25	19 ± 12
Beaulieu River	Bucklers Hard[c]	93 ± 45	1087 ± 1845[d]	82 ± 9
Southampton Water	Hythe Marina			1956[e] ± 2472

[a]SD only given where $n \geqslant 4$.
[b]8 ± 2 without August value.
[c]Also a shellfish site.
[d]263 ± 130 without June value.
[e]Mean = 728 without May value.

(Table 27.1). The difference between water concentrations in Sutton Marina in 1986 and 1988 was highly significant (t-test $p < 0.01$); there was also a significant difference between the concentrations of TBT in 1986 and 1988 for Dart Marina and Salcombe Harbour (t-test $p < 0.01$). The concentrations of TBT in these British marinas are in the same range as those reported for marinas in Chesapeake Bay in the United States (see Hall, 1988; Huggett *et al.*, 1992, and Chapter 24).

Bucklers Hard on the River Beaulieu was chosen as a 'worst case' for shellfish cultivation, as the bivalves were grown in cages suspended within the marina. At this site, concentrations of TBT in water were the highest recorded for any of the shellfish sites. Indeed, a concentration of 4400 ng l^{-1} was measured in a water sample taken in June 1987 from Bucklers Hard. Concentrations of TBT in the water were generally much higher in 1987 than those of the other two years. As there is a large boatyard at Bucklers Hard, it was not surprising that from time to time very substantial inputs of TBT occurred, presumably from the hosing-down and repainting of boats.

In 1988, Hythe Marina was included in the list of monitoring sites. It has only been in use since early 1988, and consequently any TBT present in the water must have resulted from the release of TBT from boats alone and not from TBT-contaminated sediments. In fact, water concentrations of TBT were the highest recorded at any of the sites from 1986 to 1988, being in the range 345–5600 ng l^{-1}

Table 27.2 Summer concentrations of tributyltin in sediments from 1986 to 1988

		TBT concentration ($\mu g\ g^{-1}$ dry weight)[a]		
Estuary	*Site*	*1986*	*1987*	*1988*
Crouch	Fambridge	0.08	0.03	0.02
	Bridgemarsh	0.15	0.04	0.08
	Creeksea	0.04	0.11[b]	0.07
	Burnham	0.36	0.15[c]	0.31
	Bush Shore	0.07	0.05	0.27
	Roach Mouth	0.05	0.02	0.02
	Holliwell Buoy	<0.01	0.01	0.02
	Holliwell Point	0.01	0.01	0.05
Blackwater	West Mersea	0.66	0.26	0.15
Dart	Blackness Point	0.08	0.19	0.17
Kingsbridge	Frogmore	0.05[c]	0.03	0.07
Teign	Arch Brook	0.05[c]	0.02	0.03
Beaulieu	Bucklers Hard	4.56	10.83	1.11

[a]All values are for August samples unless otherwise indicated.
[b]July samples.
[c]September samples.

during the summer of 1988. As there are no boatyard facilities at Hythe Marina, the high levels of TBT in the water cannot be attributed to the hosing down of boats coated with old layers of TBT-based paint. The extremely high concentrations of TBT in the water seem most likely to be the result of illegal use of TBT-based paint during 1988.

27.2.1b. Sediments

Concentrations of TBT in surface sediment samples (top 10 mm) collected in the summer from the shellfish cultivation sites are shown in Table 27.2.

The most comprehensive set of sediment data was obtained in 1986 for the River Crouch. Monthly samples were taken from the eight stations along the river. The data highlight the geographical and temporal patchiness of TBT concentrations in the sediment; in some cases consecutive monthly samples varied in concentration by an order of magnitude. The variability showed some seasonal pattern, but observations at the time of sampling suggested that the levels of contamination were also highly influenced by weather conditions. Scouring of some of the sites was noted following windy weather, while deposition of sediment occurred during calm periods. There did not appear to be any decrease in concentrations of TBT in sediment from 1986 to 1988; this is not surprising as laboratory experiments at MAFF indicate that TBT in sediments is not rapidly degraded and has a half-life of more than a year in anaerobic estuarine muds.

In order to provide a working definition of the degree of contamination, sites may be divided into the following four categories (Waite *et al.*, 1991):

1. those lightly contaminated, with TBT concentrations in the range <0.01–0.05 $\mu g\ g^{-1}$ (dry wt), typical of areas within estuaries with large numbers of boats but remote from boat moorings themselves (Holliwell Point, Holliwell Buoy, Roach Mouth on the River Crouch; Arch Brook on the River Teign);

2. those showing medium levels of contamination, in the range 0.06–0.2 $\mu g\ g^{-1}$ (dry wt), closer to moorings (Creeksea, Bridgemarsh, Fambridge on the River Crouch; Blackness on the River Dart; Frogmore on the Kingsbridge estuary);
3. highly contaminated areas, in the range 0.3–1 $\mu g\ g^{-1}$ (dry wt), found within high-density mooring areas or marinas (West Mersea on the River Blackwater; Bush Shore, Burnham on the River Crouch); and
4. those containing paint particles at concentrations >1 $\mu g\ g^{-1}$ (dry wt; Bucklers Hard on the Beaulieu River). This separate category is important because the bioavailability of TBT in such sediments may be different from that adsorbed onto sediment particles.

The TBT concentrations of sediment reported here are similar to those found by other workers for sites influenced by boating/shipping activity. Ashby and Craig (1989) reported TBT concentrations of up to 1.85 $\mu g\ g^{-1}$ in sediment samples collected near to boatyards on the Rivers Dart and Beaulieu. TBT concentrations in the range 0.01–1.4 $\mu g\ g^{-1}$ have been found in sediments from Chesapeake Bay, USA (Rice *et al.*, 1987; Hall, 1988; Matthias *et al.*, 1989), whilst for Lake Zurich sediments Muller (1984) has reported concentrations of up to 0.02 $\mu g\ g^{-1}$; both these areas are used for boating activity. In sediment from the Sado estuarine system, Portugal, TBT concentrations of 0.52 $\mu g\ g^{-1}$ occur as a result of discharges of TBT from shipyards in the area (Quevauviller, 1989). For a review and summary of sediment data, see Harris *et al.* (Chapter 22) and Unger *et al.* (Chapter 23).

27.2.2. SHIPPING

Since 1986, MAFF has monitored the concentration of TBT in water samples from one anchorage and several commercial harbors, and determined inputs of TBT from drydocks (Waldock *et al.*, 1988).

27.2.2a. Anchorages

Commercial vessel activity is strongly influenced by fluctuations in international shipping, and ships may be laid up for a considerable time in one place. Deep and relatively sheltered waters are preferred for such anchorages, which may be some distance away from areas normally used by commercial ships. Two surveys of anchorages were carried out on the Fal estuary (Fig. 27.1) in the southwest of England in September 1986 and October 1987. The results of these surveys are shown in Table 27.3.

At King Harry Ferry on the River Fal, there were usually 5–10 commercial vessels at anchor. In 1986, TBT concentrations in subsurface water at a depth of 10 cm ranged from 6 to 17 $ng\ l^{-1}$ (samples were taken at distances of 20–500 m from the nearest ship). However, in samples taken away from the anchorage upstream toward moored yachts on an ebb tide, the observed concentrations of TBT more than doubled. In 1987, concentrations of TBT in subsurface waters, near the anchorage, ranged from 3 to 35 $ng\ l^{-1}$, and only 8 $ng\ l^{-1}$ was recorded close to the yacht moorings on an incoming tide. Whilst these data suggest that commercial shipping sources contributed to the TBT concentrations in the water, it is clearly difficult to assess the inputs from commercial vessels alone when yachts painted with TBT share the same water.

Both in 1986 and 1987, concentrations of TBT (6–17 $ng\ l^{-1}$ and 3–9 $ng\ l^{-1}$) in the water showed no discernible gradient with distance (2 m to 1.5 km) from the King Harry Ferry anchorage in the Fal. A depth profile study in 1986, at a distance of 200 m from the nearest vessel, indicated that TBT concentrations (6–17 $ng\ l^{-1}$) were fairly uniform throughout the water column, from a depth of 1 cm down to 10 m. On occasions TBT does accumulate

Table 27.3 Concentrations of tributyltin in the River Fal

Site	*Date*	*Distance (m)*[a]	*Depth (m)*[b]	*TBT (ng l^{-1})*
Malpas near yacht moorings	2 Sept 1986	100		44
King Harry Ferry Anchorage	2 Sept 1986	500		8
King Harry Ferry Anchorage	2 Sept 1986	200	0.01	11
King Harry Ferry Anchorage	2 Sept 1986	200		17
King Harry Ferry Anchorage	2 Sept 1986	200	1	15
King Harry Ferry Anchorage	2 Sept 1986	200	5	15
King Harry Ferry Anchorage	2 Sept 1986	200	10	6
King Harry Ferry Anchorage	2 Sept 1986	100		9
King Harry Ferry Anchorage	2 Sept 1986	20		6
Malpas near yacht moorings	20 Oct 1987	100		8
Coombe Creek	20 Oct 1987	100		4
Earl Granville[c]	20 Oct 1987	1		35
Earl Harold[c]	20 Oct 1987	1		10
Earl Godwin[c]	20 Oct 1987	1		9
Earl Godwin[c]	20 Oct 1987	5		5
Castor[c]	20 Oct 1987	1		19
Centaurus[c]	20 Oct 1987	10		15
King Harry Ferry Anchorage	20 Oct 1987	0.1	0.01	137
King Harry Ferry Anchorage	20 Oct 1987	2		6
King Harry Ferry Anchorage	20 Oct 1987	5		7
King Harry Ferry Anchorage	20 Oct 1987	10		3
King Harry Ferry Anchorage	20 Oct 1987	15		9
King Harry Ferry Anchorage	20 Oct 1987	150		8
King Harry Ferry Anchorage	20 Oct 1987	400		7
King Harry Ferry Anchorage	20 Oct 1987	1000		5
King Harry Ferry Anchorage	20 Oct 1987	1500		3

[a]Distance from nearest moored vessel.
[b]Sample taken at depth 0.1 m unless otherwise stated.
[c]Ship.

in the surface water; in 1987 a sample taken 10 cm away from one ship, and at a depth of 1 cm, had a concentration of 137 ng l^{-1}. The high concentrations of TBT in such a sample are partly due to enrichment of the surface microlayer.

27.2.2b. Commercial harbors

The levels of organotins in three commercial harbors have been studied. Water samples were taken from Felixstowe and Harwich harbors, on the east coast, in March and August 1987, and in Southampton Water (Fig. 27.1) on four occasions from 1987 to 1988.

In March 1987, samples were taken from Felixstowe and Harwich harbors at a distance of 2–10 m from the hulls of individual container ships and ferries. Only 3 of the 10 ships in the harbors gave rise to concentrations of TBT in the water >20 ng l^{-1}, and these ranged from 22 to 166 ng l^{-1} (see Tables 27.4 and 27.5). In August of the same year, water samples were collected at a distance of 2 m from the hulls of 17 ships, and all samples except one had concentrations of TBT <20 ng l^{-1}. The low concentrations of TBT found close to commercial vessels are in accordance with the published data for release rates of TBT-based paints intended for use on such ships (see Waldock *et al.*, 1988). Background levels of TBT in Felixstowe

Table 27.4 Concentrations of tributyltin in Felixstowe Harbour

Site	*SRW*	*Distance (m)*[a]	*TBT (ng l^{-1})*
East end of Harbour	10 Mar 1987	35	1
Mid Harbour	10 Mar 1987	100	2
West end of Harbour	10 Mar 1987	100	9
Shotley Spit	10 Mar 1987	1000	<1
Harbour Basin	10 Mar 1987	25	2
Aurora[b]	10 Mar 1987	2	5
Karen Oltmann[b]	10 Mar 1987	3	166
Karen Oltmann[b]	10 Mar 1987	50	2
Ile Maurice[b]	10 Mar 1987	4	3
Dietrich Oldendorff[b]	10 Mar 1987	7	22
Liesel Essberger[b]	10 Mar 1987	5	3
Trabant[b]	10 Mar 1987	10	10
Pegasia[b]	10 Mar 1987	2	5
Merzario Britannia[b]	10 Mar 1987	2	3
Shotley Spit	5 Aug 1987	1000	8
Verrazano Bridge[b]	5 Aug 1987	2	3
Dorado[b]	5 Aug 1987	2	7
Strathbrora[b]	5 Aug 1987	2	3
Doric Ferry[b]	5 Aug 1987	2	4
Doric Ferry[b]	5 Aug 1987	25	3
Pegasus[b]	5 Aug 1987	2	2
Artus[b]	5 Aug 1987	2	24
Nordic Ferry[b]	5 Aug 1987	2	9
Limassol[b]	5 Aug 1987	2	5
Eagle Limassol[b]	5 Aug 1987	2	4
Nordsee Rendsburg[b]	5 Aug 1987	2	<1
Gina S[b]	5 Aug 1987	2	8
Novamarina[b]	5 Aug 1987	2	5

[a]Distance from nearest moored vessel.
[b]Ship.

Table 27.5 Concentrations of tributyltin in Harwich Harbour

Site	*Date*	*Distance (m)*[a]	*TBT (ng l^{-1})*
Clervaux[b]	10 Mar 1987	5	10
Tor Scandinavia[b]	10 Mar 1987	2	78
Laverval[b]	5 Aug 1987	2	12
Bolero[b]	5 Aug 1987	2	19
Dana Anglia[b]	5 Aug 1987	2	4
Hamburg[b]	5 Aug 1987	2	16
Earl William[b]	5 Aug 1987	2	5

[a]Distance from nearest moored vessel.
[b]Ship.

Harbour were no more than a few nanograms per liter.

Southampton Water is fed by three main rivers: the Test, Itchen, and Hamble. It is a busy waterway used by container ships, oil tankers, ferries, and ocean-going liners. In addition to the ships, large numbers of pleasure craft are to be found in marinas in Southampton Water, and moored in the Rivers Itchen and Hamble. Consequently, levels of organotin in Southampton Water are derived from small boat activity as well as from shipping activity.

The highest recorded concentrations of TBT in the water were in June 1987, when 668

Table 27.6 Concentrations of tributyltin (ng l^{-1}) in Southampton Water

	1987		1988	
Site	*March*	*June*	*May*	*November*
Calshot	—[a]	16	11	8
Solent Breezes	—	46	15	5
Fawley	17	78	201	18
Hamble Mouth	—	362	89	12
Mid Hamble	—	—	155	59
Hamble Bursledon	—	668	94	82
Netley Abbey	—	38	21	14
Hythe Pier	14	26	46	18
Itchen Mouth	—	63	14	16
Southampton Ferry Terminal	42	34	7	22
Shipyard	250	—	36	9
Container quay	—	25	10	158

[a]Not sampled.

ng l^{-1} was measured in a sample taken from the top of the Hamble at Bursledon, and 362 ng l^{-1} in a sample from the mouth (see Table 27.6). These values are not surprising, as 5000 pleasure craft are moored in a section of river that is only 5 km long; however, concentrations of TBT in the Hamble have decreased, and at Bursledon in May and November 1988 they were 94 and 82 ng l^{-1}, respectively. Water samples taken at Solent Breezes and Calshot Castle, near the mouth of Southampton Water, had TBT concentrations of 16–46 ng l^{-1} in June 1987, and 11–15 ng l^{-1} and 5–8 ng l^{-1} in May and November 1988, and probably resulted from a TBT-contaminated plume of water coming from the Hamble. The River Itchen has fewer yachts moored in it than the Hamble, but even so in June 1987 the level of TBT at the mouth of the Itchen was in excess of 50 ng l^{-1}.

The other high levels of TBT (>50 ng l^{-1}) recorded in Southampton Water are associated with inputs from container ships, oil tankers, and a shipyard. Water samples collected from the container quay, in the River Test at the top of Southampton Water, had TBT concentrations ranging from 10 to 158 ng l^{-1} in 1987 and 1988. The inputs of TBT from this source are likely to fluctuate as the number of ships moored at the quay is variable.

On the other side of the Test, opposite the container quay, there is a commercial shipyard. In March 1987, a high TBT concentration of 250 ng l^{-1} was recorded for a water sample taken from outside this yard, where a 3000-t ship was being hosed down on the foreshore. The input of TBT from the shipyard is intermittent, however, and in May and November 1988 the concentrations of TBT were only 36 and 9 ng l^{-1}, respectively. Seligman *et al.* (1989) found variable TBT concentrations in the Elizabeth River, Virginia, USA, which they attributed to inputs from commercial drydocks.

Large oil tankers are often moored at the quay close to Fawley oil refinery. A water sample taken in 1987, at a distance of 2 m from an oil tanker, had a TBT concentration of 78 ng l^{-1}, yet 100 m away from the tanker the level of TBT was only 17 ng l^{-1}. A similar pattern was observed in 1988: 200 ng l^{-1} was recorded in a sample taken 1 m away from an oil tanker, whilst a sample at 100 m from the vessel had only 24 ng l^{-1}.

Table 27.7 Inputs of tributyltin from yacht and ship washdown

Formulation	*Vessel*	*Wash water concentration (mg l^{-1})*	*TBT removed from hull (g)*	*Input (mg m^{-2})*
Free association	Yacht	1.4 ± 0.9	0.4	20
Free association	*RV Clione*	0.25 ± 0.09	3.0	10
Copolymer	Yacht	12 ± 6.3	3.6	200
Copolymer	Frigate	2 ± 5	100	60

Our study of Southampton Water has shown that it is difficult to apportion the relative inputs of TBT from yachting activity and commercial shipping, although in 1987 the main source of TBT came from yachts moored in the River Hamble. In major waterways in the United States and Australia, the greatest inputs of TBT are from marinas and shipyards (Batley *et al.*, 1989; Seligman *et al.*, 1989, and Chapter 20; Grovhoug *et al.*, Chapter 25; Maguire, Chapter 26).

27.2.2c. Inputs of TBT to the marine environment from hosing down ships and yachts

To date, two studies of ship drydocking procedure have been carried out by MAFF. The aim of the first study was to determine the amount of TBT generated from high-pressure hosing of a Leander-class naval frigate painted with a TBT copolymer formulation. Environmental loads of TBT from the washing operation were determined by measurements of TBT concentrations in 'hosings' collected from beneath the hull, and, by timing the operation, estimates were made of the total amount of TBT removed from the hull over the hosing period. The second study measured inputs of TBT to the marine environment arising from the washing down of the MAFF research vessel *Clione*, which was coated with a TBT and copper free-association paint.

The concentrations of TBT in 'hosings' were measured using a contact solvent method, and consequently not all of the TBT held within the larger particles of dried paints would have been extracted. The concentrations of TBT have therefore been arbitrarily defined as 'readily extractable residues.'

A comparison of the amounts of TBT generated from the hosing down of ships and yachts coated with either copolymer or free-association paint is given in Table 27.7. The hosing equipment used in each case was slightly different, and the angle and distance of the water jet to the hull are critical in determining the amount of debris and paint blasted away. Consequently, the estimates necessarily incorporate large margins of error. Nevertheless, the data indicate that washing down of both free-association and copolymer-painted yachts gives rise to higher concentrations of TBT in waste water than from the corresponding formulations designed for ships. In the case of the free-association paints, this may be due to the yacht paint being less depleted of TBT at the time of hosing, and in the case of copolymers, the fact that formulations used on yachts may be softer than those employed on larger vessels.

27.2.3. FRESH WATER

Although in fresh water the main fouling organisms are algae and not barnacles as in marine and estuarine waters, boats using the inland waterways are still painted with antifoulant. Consequently, MAFF undertook a series of surveys for inland waterways, including major rivers, canals, the Lake District, and the Norfolk Broads. Accidental spills of wood preservative from timber treatment plants are another source of TBT in rivers that has been investigated by MAFF.

27.2.3a. Major rivers

Water samples submitted by UK Water Authorities from seven major rivers (see Fig. 27.1) have been analyzed for TBT at Burnham. In 1986, samples from the Rivers Ebbw in Wales and Tay in Scotland all contained from <1 to 2 ng l^{-1}. Concentrations of TBT in the Rivers Tame and Aire, during 1987, ranged from 12 to 25 ng l^{-1}. The River Medway was rather more contaminated with TBT, having concentrations of >50 ng l^{-1} in both 1986 and 1987.

More detailed surveys were carried out for the Great Ouse and the Thames. A series of samples was taken from Hartford Marina and St. Ives Marina on the Great Ouse; in June 1986, TBT concentrations of 1340 ng l^{-1}, and 4280 ng l^{-1}, respectively, were recorded. The following year, in June 1987, the concentrations of TBT had sharply decreased at Hartford Marina to 398 ng l^{-1} and at St Ives Marina to 1140 ng l^{-1}, while in an open stretch of the river, between the two marinas, the concentration of TBT was only 15 ng l^{-1}.

Water samples were taken from five marinas on the Thames in February 1987. Concentrations of TBT ranged from 200 ng l^{-1} at Purley Marina to 1290 ng l^{-1} at Shepperton Marina, while a water sample taken at the same time from an open stretch of the river at Teddington contained 24 ng l^{-1}. In September 1987, the concentration of TBT at Teddington was 99 ng l^{-1}, and in September the following year a value of 37 ng l^{-1} was recorded.

The concentrations of TBT recorded in 1987 for UK freshwater marinas, containing several hundred boats, were up to an order of magnitude higher than those reported for Californian river and lake marina waters (<2 to 140 ng l^{-1}) in 1988 (Stang and Goldberg, 1989). The use of TBT-based paints for boats in fresh water does not appear to be so widespread in California as it obviously was in the United Kingdom in 1987. TBT was not detected (i.e. <2 ng l^{-1}) in five Californian freshwater marinas, even though the numbers of boats berthed in the marinas ranged from 200 to 650.

27.2.3b. Canals

Trent Canal in the Midlands region of the United Kingdom (Fig. 27.1) was sampled in November 1987. The level of TBT in a marina was 148 ng l^{-1}, but in the canal itself outside the marina the concentration of TBT was 20 ng l^{-1}. The more open stretches of the canal had TBT concentrations of 3 to 6 ng l^{-1}. As with rivers, the high levels of TBT appear to be localized within marinas.

27.2.3c. The Lake District

The Lake District (Fig. 27.1) is a very popular recreational area of the United Kingdom, and as a result the lakes are well used by pleasure craft. In November 1987, water samples were taken from four lakes: Bassenthwaite, Derwent Water, Ullswater, and Windermere. TBT was only detected in samples from Lake Windermere; the maximum concentration of 277 ng l^{-1} occurred in Windermere Marina, while away from the marina concentrations ranged from 5 to 11 ng l^{-1}.

27.2.3d. The Norfolk Broads

The Norfolk Broads, in the east of England (Fig. 27.1), is a region of shallow lakes (broads) and rivers, which is heavily used by pleasure craft, and is the most popular area in the United Kingdom for freshwater boating. In 1986, a survey of local chandlers suggested that TBT-based paints were commonly used on boats in the area. Consequently, MAFF carried out a detailed TBT-monitoring program in 1986 and 1987 for the River Bure, Wroxham Broad, and the River Yare (Waite *et al.*, 1989).

Table 27.8 Concentrations of tributyltin (ng l^{-1}) in the Rivers Bure and Yare during 1986 and 1987

	1986						*1987*		
Site	*June*	*July*	*Aug*	*Sept*	*Oct*	*Nov*	*Mar*	*June*	*Sept*
River Bure									
Ingworth	—[a]	<1	<1	<1	<1	<1	—	<1	<1
Buxton	—	2	2	<1	<1	1	—	<1	<1
Wroxham	105	112	9	75	7	84	102	121	32
Wroxham Broad	25	898	55	458	17	27	12	44	136
Landamore's Yard	1540	232	83	691	180	346	—	336	385
Horning	50	70	53	213	11	6	20	112	—
River Yare									
Brundall	371	83	7	130	42	9	28	235	47
Brundall Marina	3260	132	46	100	61	22	142	202	162

[a]Dash indicates not sampled.

An aerial survey, in August 1986, showed that there were >1400 boats on the River Bure, the highest densities being around Wroxham and Horning. There were few boats at the upper river site of Buxton, and none at the furthest upstream site of Ingworth. The results of the survey for concentrations of TBT in the water are shown in Table 27.8.

Water samples taken from the sites at Ingworth and Buxton had non-detectable and low concentrations of TBT, respectively. The trace amounts of MBT and DBT that occurred in the water at Ingworth could have come from the degradation of TBT, although diffuse contamination from diorganotin used in catalysts and stabilizers in the plastics industry cannot be ruled out.

At Wroxham, downstream of Buxton, there was a marked increase in the concentration of TBT and breakdown products; the maximum recorded value for TBT was 121 ng l^{-1}, in June 1987. During the summer of 1986, there were ~200 boats moored in the vicinity of the sampling site at Wroxham. It was calculated that the total immersed hull area for these boats was ~1400 m^2. Assuming that all of the boats used TBT-based antifouling paints, with a release rate of ~4 g cm^{-2} d^{-1} (Waldock and Thain, unpublished data; UK DoE, 1986; Gardner, 1987), and a mean volumetric flow for the River Bure of 282 000 m^3 d^{-1} (Anglian Water Authority, personal communication), then the theoretical concentration of TBT at the sampling station of Wroxham would be ~200 ng l^{-1}. The maximum concentration of TBT recorded in the summer of 1986 was 112 ng l^{-1}, showing that the predicted environmental concentration from input estimates is close to the concentrations measured.

Landamore's Yard, in a backwater at Wroxham, is used for boat building and repair and is probably typical of the highest inputs likely to occur in the area. A maximum concentration of 1540 ng l^{-1} was measured in water samples taken close to the boatyard in June 1986.

In the enclosed waters of Wroxham Broad, the potential for build-up of TBT is clearly much greater than in the open river, and this is confirmed by the very high concentrations found close to one yacht club in July (898 ng l^{-1}) and September (458 ng l^{-1}) 1986. Cleary and Stebbing (1987b) found levels of >50 ng l^{-1} Sn organotin (TBT and DBT) in Wroxham Broad in March, when boating activity was reduced. At Horning, another center of boating activity on the River Bure, TBT concentrations exceeded 200 ng l^{-1} during 1986.

The levels of TBT recorded for water samples from the River Yare at Brundall were 375 ng l^{-1} in June 1986 and 235 ng l^{-1} in June 1987, indicating that the River Yare was even more contaminated than open stretches of the River Bure. In the enclosed waters of Brundall Marina, the level of TBT reached the extremely high value of 3260 ng l^{-1}, far greater than the highest value recorded on the Bure of 1540 ng l^{-1} at Landamore's Yard. A winter value for organotin (TBT and DBT) of 180 ng Sn l^{-1} at Brundall Marina was measured by Cleary and Stebbing (1987b).

These levels can be compared with tributyltin values of 10–5700 ng l^{-1}, which have been recorded in freshwater systems in Canada (Maguire *et al.*, 1982, 1985, 1986; Maguire and Tkacz, 1985), and <2–140 ng l^{-1} found in Californian river and lake marina waters (Stang and Goldberg, 1989). The only other documented measurement of TBT in fresh water is from a lake in Switzerland where a relatively low concentration of 13 ng l^{-1} was found (Muller, 1984, 1987). Also, Linden (1987) has reported the presence of TBT in the tissues of freshwater molluscs from Lake Malaren, a popular boating center in Sweden.

27.2.3e. Spills of TBT-based preservatives from timber yards

In the late 1980s, TBT was commonly used in timber preservatives, together with dieldrin, lindane, or pentachlorophenol. Many large timber yards are situated close to rivers for transport reasons, and the timber preservatives are often stored close to the riverside. The MAFF analyzed water samples following three separate spills in 1985 and 1986.

Results for water samples taken in the vicinity of such spills showed variable concentrations depending on the amount of fluid lost and the size of the catchment. However, concentrations of up to 540 μg l^{-1} of TBT were recorded, and storage conditions for timber treatment formulations were identified as a major cause for concern.

27.3. VERTICAL DISTRIBUTION

This section gives an account of the work carried out by the Plymouth Marine Laboratory on the vertical distribution of organotins throughout the water column, with special reference to concentrations in the surface microlayer. Samples of the surface microlayer were generally collected using a Garrett screen (Garrett, 1965), although some samples were also taken with the glass plate sampler (Harvey and Burzell, 1972).

27.3.1. SURFACE MICROLAYER

The naturally occurring surface film at the air– water interface is an organic monomolecular layer with a thickness of ~300 nm, which, because of its nonpolar nature, will readily accumulate lipophilic components (Garrett and Duce, 1980). Substances such as organotins, with octanol–water partition coefficient values of 500–7000 (Laughlin *et al.*, 1986), would be expected to partition into such a monolayer.

The relative enrichment of the surface microlayer is more pronounced in samples taken with the glass plate sampler, which removes a 60–100 μm layer, than in those taken with the Garrett screen, which removes between 280 and 300 μm (Cleary and Stebbing, 1987a).

The sea-surface monolayer will form an integral part of microlayer samples taken with both the Garrett screen and the glass plate samplers. Therefore, the differences in organotin concentrations in samples taken by the two methods may be due, in whole or in part, to dilution of the monolayer in microlayer samples of different thickness.

It is not clear to what extent organotins are distributed throughout the microlayer as distinct from their concentration in the surface monolayer, but it is known that stratification can occur within the microlayer for a variety of components including dissolved substances (Garrett and Duce, 1980). If this process occurs for organotins, it may also account for differences in surface microlayer enhancement in samples of different thickness, such as those collected by the Garrett screen and the glass plate. Since the glass plate sampler acts as a selective adsorber and the Garrett screen is non-selective, the method of sampling may also affect the final concentration.

27.3.1a. Comparison of organotin concentrations in subsurface and surface microlayer water samples, 1986–1988

Water samples were collected from coastal locations in Devon and Cornwall in the summers of 1986, 1987, and 1988 (Fig. 27.5). Maximum organotin concentrations in both surface microlayer and subsurface samples occurred in areas of high boating activity (Fig. 27.6), such as marinas. It would seem that the highest levels existed where tidal exchange and mixing is poor, for example in Torquay Marina and Sutton Marina, Plymouth, whereas lower values occurred at Dartmouth and Plymouth Ocean Court where tidal currents are much stronger and dilution greater. The numbers of boats in the four marinas sampled were similar.

Although annual trends varied from place to place, in 1987 organotin concentrations tended to increase in both the subsurface water and the microlayer at the majority of sites. Trends for the following year were also variable, but there was an indication of a decline in the concentration of organotins in microlayer samples in 1988, although this was not consistently reflected in the subsurface samples. It was clear that the concentration of organotins, in both the surface microlayer and in subsurface waters in 1988, continued to exceed toxicity thresholds for a number of organisms (Cleary, 1991).

Considerable enhancement of organotin concentrations occurred in the surface microlayer compared with subsurface samples at all the sites sampled (Fig. 27.6). As organotin concentrations in the surface microlayer were an order of magnitude greater than those in subsurface waters, organisms exposed to these levels are likely to be affected upon a shorter time of exposure. The organisms most at risk are the neuston (Gucinski, 1986), the community of organisms that for all or part of their life inhabit the surface microlayer (Zaitsev, 1971). These include microorganisms, and eggs and larvae of invertebrates and fish, including some commercial species (Hardy *et al.*, 1985), and therefore larval survival and recruitment could be affected by exposure to toxic concentrations in the microlayer. Furthermore, many other planktonic species that undergo a diurnal vertical migration to surface waters may be exposed to these elevated concentrations of organotin compounds. Toxic components of the surface microlayer may be carried further into the food chain as protozoans graze on bacterioneuston (Zaitsev, 1971), and juvenile fish feed on the neuston (Hempel and Weikert, 1972).

The elevated organotin concentrations in the surface microlayer may also affect organisms resident in the littoral zone, since they will be exposed to the microlayer during tidal changes as will the substrata on which many organisms live and graze. The nature of exposure for littoral organisms is therefore different and potentially much greater than for other communities, such as those that dwell in subsurface waters. Since TBT is readily adsorbed to surfaces, organisms that move over them may be exposed to even higher concentrations than the concentrations in water would suggest. At PML, methods are currently being developed to determine concentrations of contaminants likely to be deposited from the microlayer on to littoral surfaces in this way.

Biological activity in the surface microlayer can transform both natural and anthropogenic materials in the marine environment (Liss, 1975). Chemical and physical processes, such as complexation, photolysis, wind, and wave action, will also affect surface microlayer concentrations and

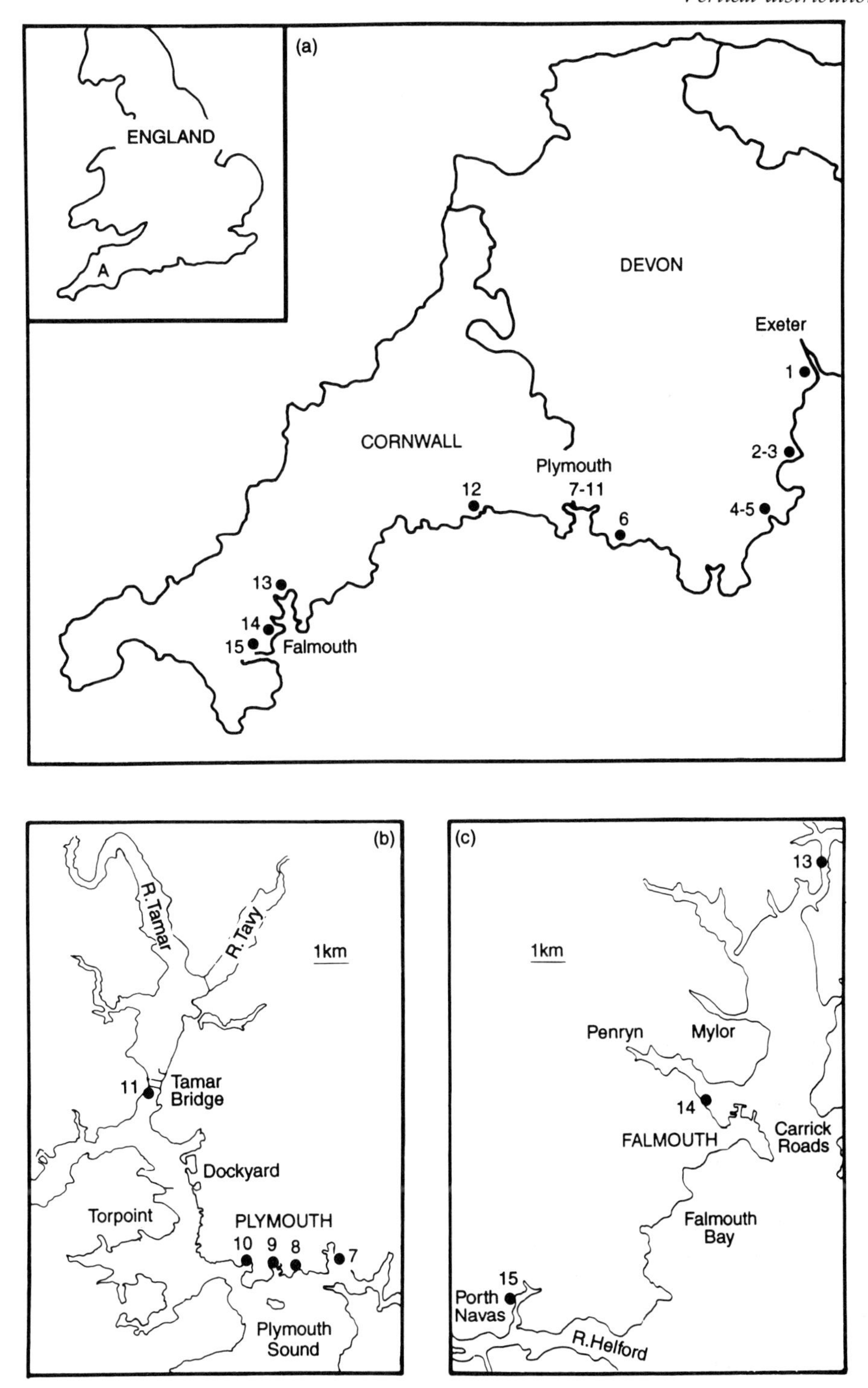

Figure 27.5 Sampling sites (a) in Devon and Cornwall from Starcross in the River Exe (1) to Porth Navas in the Helford River (15); (b) Plymouth area; (c) Falmouth area.

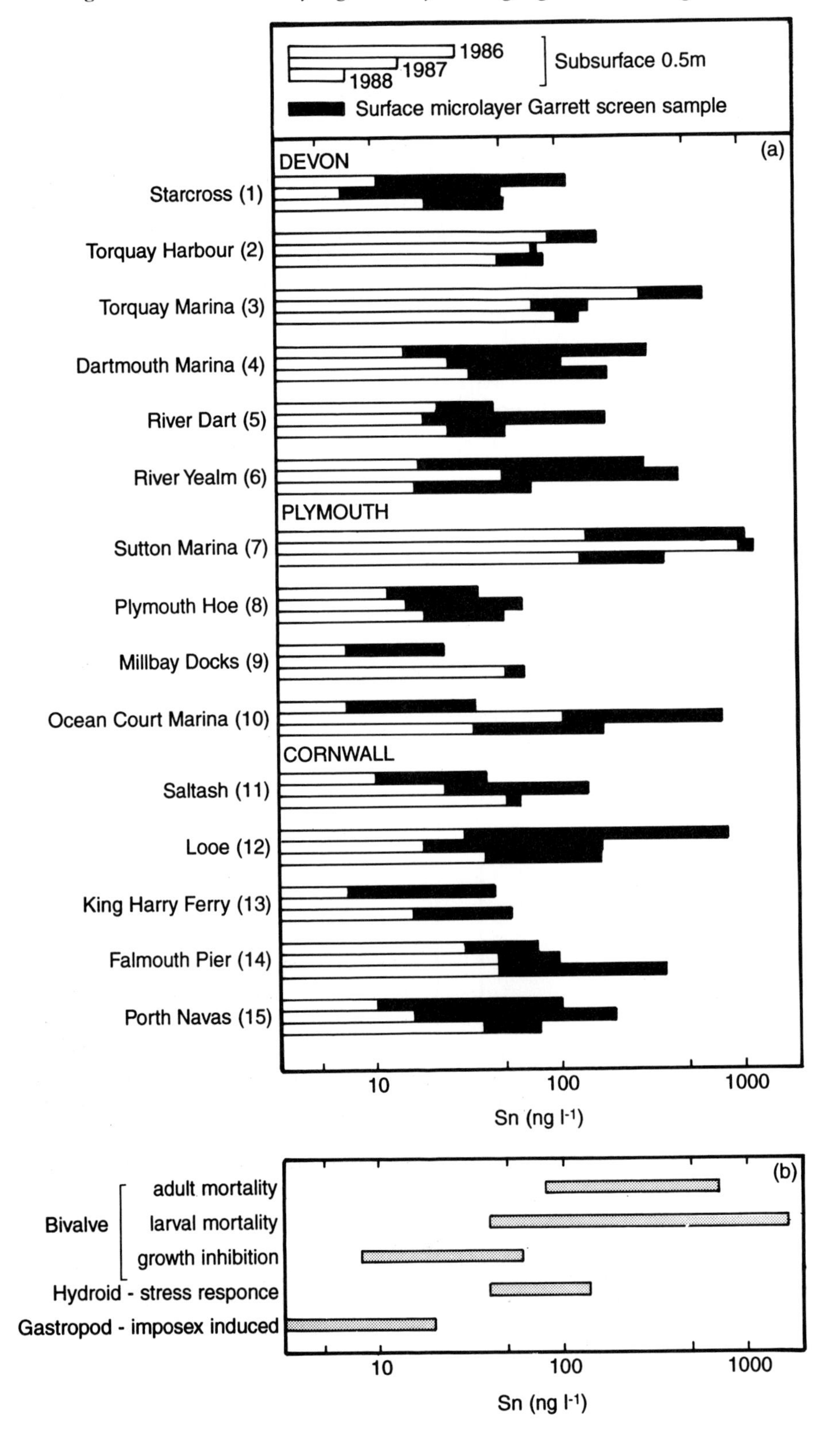
1986
1987
1988
Subsurface 0.5m
Surface microlayer Garrett screen sample
(a)
DEVON
Starcross (1)
Torquay Harbour (2)
Torquay Marina (3)
Dartmouth Marina (4)
River Dart (5)
River Yealm (6)
PLYMOUTH
Sutton Marina (7)
Plymouth Hoe (8)
Millbay Docks (9)
Ocean Court Marina (10)
CORNWALL
Saltash (11)
Looe (12)
King Harry Ferry (13)
Falmouth Pier (14)
Porth Navas (15)
10
100
1000
Sn (ng l-1)
(b)
Bivalve
adult mortality
larval mortality
growth inhibition
Hydroid - stress responce
Gastropod - imposex induced
10
100
1000
Sn (ng l-1)

therefore the toxicological impact of contaminants. These complex processes will have important implications for the bioavailability of organotins in the microlayer (Cardwell and Sheldon, 1986) and their toxic effects on the neuston and other communities (Gucinski, 1986).

Organotin concentrations in surface microlayer samples from the United Kingdom, Canada, and the United States of America show a common pattern of enhancement (Maguire *et al.* 1982, and Chapter 26; Donard *et al.*, 1986; Hall *et al.*, 1986; Valkirs *et al.*, 1986; Cleary and Stebbing, 1987a, 1987b), but direct comparisons are difficult since several types of sampling device are used. Not only do these sample a different thickness of microlayer, but they also operate by different mechanisms.

27.3.1b. TBT and DBT distribution in 1988

Both tributyltin and dibutyltin were detected at all sites visited in 1988 (Fig. 27.7), but the TBT:DBT ratios varied considerably. Enhanced values of both species occurred in the surface microlayer, but those sites most contaminated with TBT were not necessarily high in DBT. For example, samples from Sutton Marina contained high concentrations of both species, but samples from Ocean Court Marina, with a similar amount of TBT, contained much less DBT.

Although the most abundant source of TBT in areas of high boating activity is from antifouling paints, the DBT found may arise from sources other than degradation of TBT. Diorganotins are widely used as stabilizers for polyvinyl chloride (PVC) products and also as industrial catalysts. Therefore, variable TBT:DBT ratios may be due to inputs of DBT from the manufacture or weathering of PVC products, such as water piping, roofing materials, window frames, or from other industrial wastes. High concentrations of DBT, which have been attributed to normal urban activities, have been found in wastewater and sewage sludge (Fent, 1990). High DBT levels in seawater samples from the Menai Strait are also thought to arise from sources other than antifouling paints (Waldock *et al.*, 1987a).

27.3.2. WATER COLUMN PROFILES

27.3.2a. Intercalibration

Water samples taken on vertical profiles at two sites in the River Crouch were analyzed at PML using the toluene extraction/atomic absorption method, and at MAFF, Burnham-on-Crouch, using the dichloromethane/GC–FPD method (Fig. 27.8). The lower values for organotins, obtained by PML, are consistent with the fact that the former method extracts only part of the DBT present in the water sample.

27.3.2b. Vertical distribution of TBT and DBT

Depth profiles taken at locations in Plymouth and Dartmouth showed a similar pattern of distribution with the highest values occurring in the surface microlayer and the lowest in near-bottom samples (Fig. 27.9). TBT concentrations were considerably greater than DBT values, sometimes by as much as an order of magnitude, although concentrations of both organotin species differed markedly in samples from sites only a short distance apart. For example, at Plymouth order-of-magnitude differences occurred between

Figure 27.6 Comparison of (a) organotin (TBT + DBT) levels in surface microlayer and subsurface samples (0.5 m) in southwest England during 1986, 1987, and 1988; (b) toxicity threshold levels (TBT) in seawater (Alzieu *et al.*, 1980; His and Robert, 1980; Stebbing *et al.*, 1983; Waldock and Thain, 1983; Beaumont and Budd, 1984; Thain and Waldock, 1985; Bryan *et al.*, 1986; Gibbs and Bryan, 1987.)

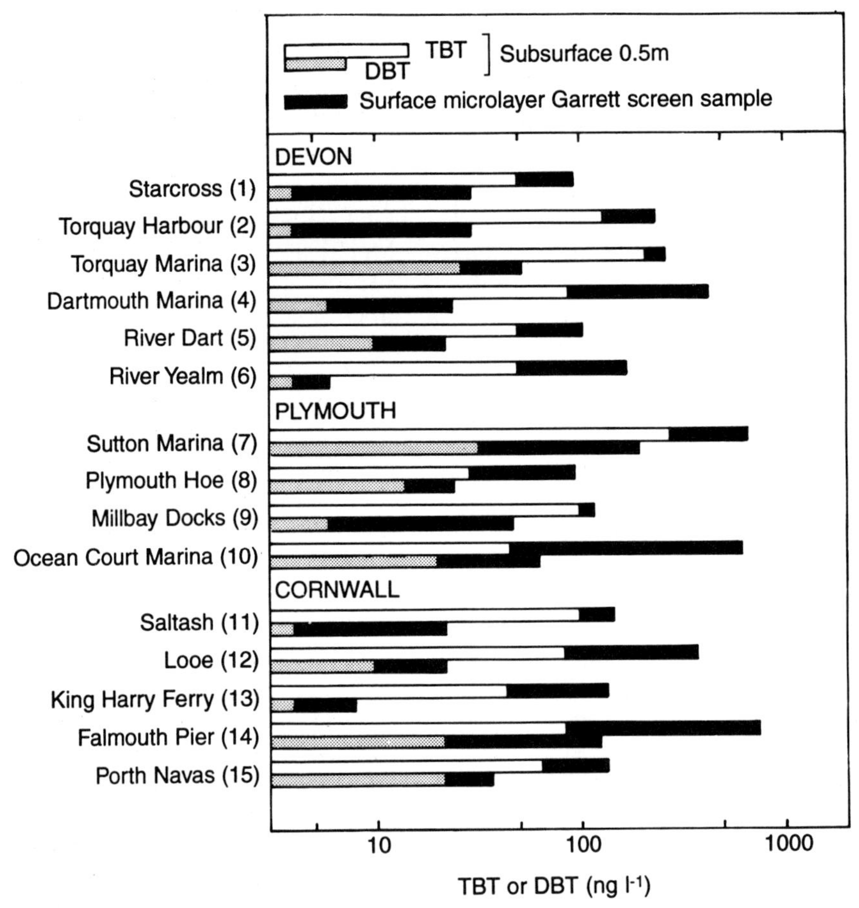

Figure 27.7 Distribution of TBT and DBT in surface microlayer and subsurface waters in southwest England during 1988.

concentrations in samples from Plymouth Hoe and Sutton Marina only 1 km apart; similar differences were seen at Dartmouth between marina samples and river samples.

27.3.2c. Organotin gradients in the water column

It is clear from a number of studies that profile gradients varied from sample to sample (Figs. 27.8–27.10). Detailed examination of profiles (Fig. 27.10), showed that there was virtually no concentration change with depth in the River Crouch at Creeksea, unlike Dartmouth Marina where there was a gradual change in concentration down the profile. Such profiles showed that, independently of microlayer enhancement, concentrations through the water column are not necessarily homogeneous and that samples at any single depth may not be representative of the water column.

There was no evidence of an input of organotins from sediments (Figs. 27.9 and 27.10), since the samples taken very close to the bottom (the maximum depth profile samples) did not show any increase in concentration. In fact, the reverse was true, and bottom samples contained less organotin than the other profile samples, even though

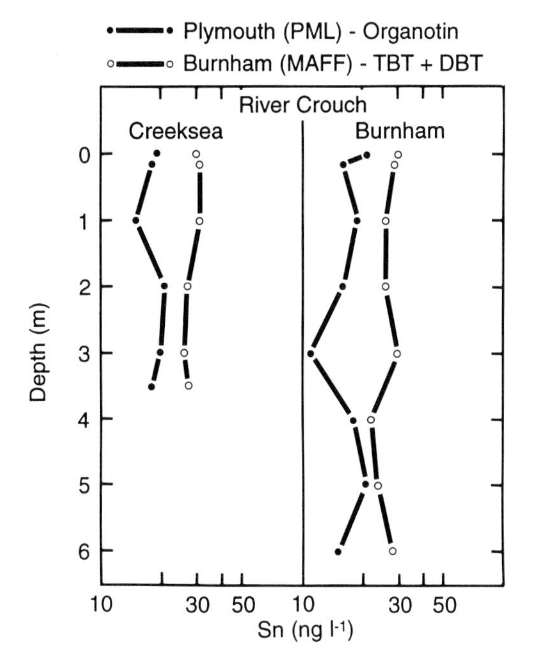

Figure 27.8 Intercalibration profile samples from the River Crouch. A comparison of two analytical techniques.

sediments are considered to be the final sink for organotins.

27.4. DISCUSSION

In the beginning of 1986, the UK government introduced legislation that restricted the amount of TBT in antifouling paints available for retail sale. It was hoped that as a result of the legislation the environmental levels of TBT would be sufficiently reduced to meet the EQT of 20 ng l^{-1} (see Abel, Chapter 2). The water samples collected in 1986 and analyzed for TBT showed that at many estuarine and freshwater sites the Environmental Quality Target (EQS) was greatly exceeded.

Very high concentrations of TBT occurred where there were large numbers of boats moored in areas of enclosed water (e.g. marinas and freshwater marinas on lakes, rivers, and canals). Water samples taken close to boatyards (e.g. on the Beaulieu River and Norfolk Broads) also had extremely high levels of TBT (>1 μg l^{-1}), although such levels tended to fluctuate (presumably relating to the number of boats hosed down and repainted at the yard prior to sampling).

It was difficult to assess the relative inputs of TBT from small boats and ships when they share the same stretch of water (e.g. parts of the River Fal and Southampton Water). On the River Hamble, which feeds into Southampton Water, there are only small boats and no commercial ships, and in 1986 concentrations exceeded 600 ng l^{-1}. In contrast, the background levels of TBT in Felixstowe and Harwich harbors were only a few nanograms per liter. In general, the UK's monitoring program indicated that small boats were the major source of inputs of TBT to the aquatic environment. Consequently, in 1987, the UK government banned the use of TBT-based paints on boats <25 m in length.

Other major sources of input of TBT to the aquatic environment were identified by the monitoring program. Intermittent inputs of TBT came from drydocking practices (e.g. the shipyard on Southampton Water). On rare occasions, spills of wood preservative into rivers caused exceptionally high levels of TBT in the water (>540 μg l^{-1}).

The UK's monitoring program has revealed that concentrations of TBT in water samples can very much depend on the time of year that the samples were collected. In the United Kingdom there is a seasonal pattern to boating activity, and, consequently, water samples taken in the winter contained very much lower concentrations than those taken in the summer. The position at which the sample is collected in the water column is also important, as concentrations can vary widely between the surface microlayer and subsurface and bottom waters. For ease of sampling, subsurface waters are probably the easiest to collect.

The monitoring of sediments has shown that the highest levels of TBT are found close to marinas and boatyards (e.g. Bucklers Hard on the Beaulieu River). However, using

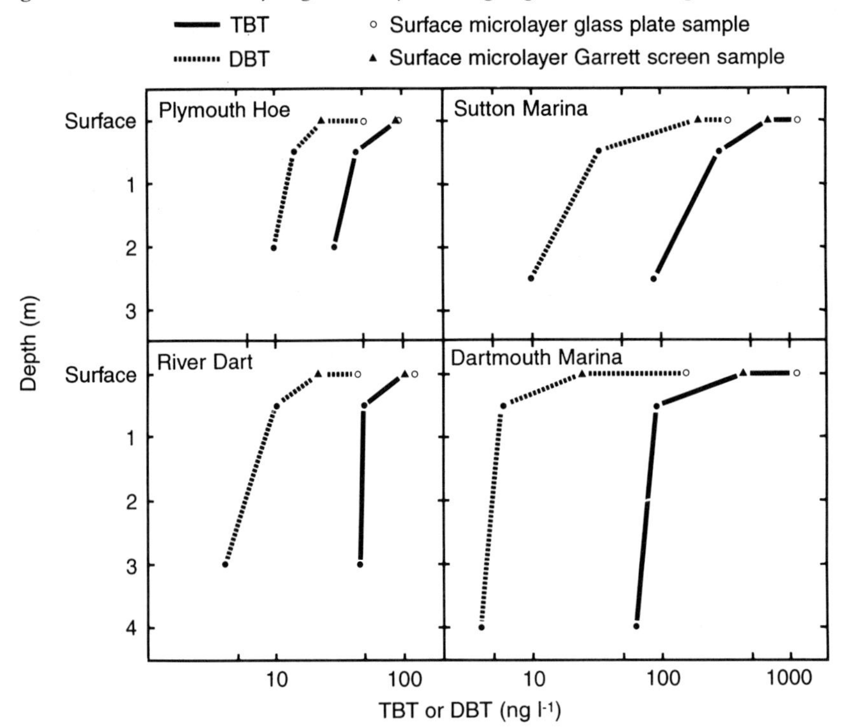

Figure 27.9 Water column profiles for TBT and DBT at four sites in Devon. Comparison of two surface microlayer samplers.

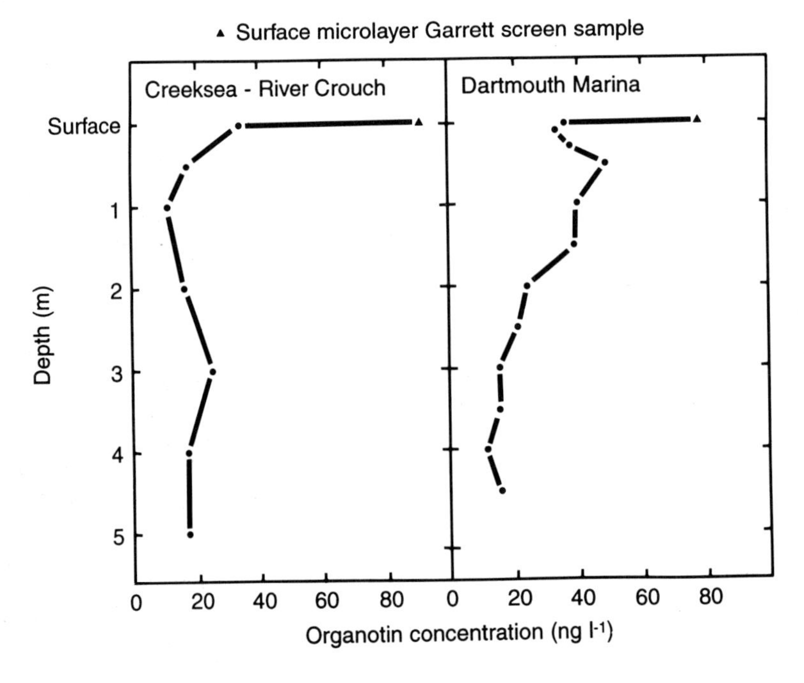

Figure 27.10 Contrasting profile gradients for organotin.

sediments as indicators of contamination may give very variable results. Estuarine sediments are often extremely mobile, so a sample taken from a site one week may have a concentration of TBT very different from that of a sample taken a week later. The movement of sediment means that clean or contaminated sediment may be deposited on the sampling site, or contaminated sediment at the surface may be scoured away. Laboratory studies have indicated that the half-life of TBT in anaerobic sediments may be very long (~2 years). Consequently, the sampling and analysis of sediments are still important in any monitoring program for TBT.

In the United Kingdom, monitoring of TBT in the aquatic environment is continuing, so that the effectiveness of the 1987 ban can be assessed. It is hoped that, as TBT-based paints can no longer be used on small boats, the concentrations of TBT in estuaries, rivers, and lakes will continue to fall and so meet the EQS of 2 ng l^{-1}. Greater effort is now being directed toward monitoring the inputs of TBT from commercial shipping, in harbors, estuaries, and offshore, and from drydocking practices, as the latter is now the major source of input of TBT to the aquatic environment.

27.5 CONCLUSIONS

1. Widespread contamination of UK rivers and estuaries by TBT was evident in 1986.
2. Concentrations of TBT reduced following legislation in 1986 and 1987, but all sites under study were still contaminated by TBT in 1988. Values continued to exceed the UK government's Environmental Quality Standard for TBT of 2 ng $l,^{-1}$ and in many areas were above toxicological threshold values of many taxonomically diverse species. Studies of new marinas suggested that concentrations higher than those expected may have resulted from illegal use of TBT, and the continued presence of TBT may not be a reflection of persistence of the compound alone. Initial results from 1989 showed a further decline in environmental concentrations.
3. Measurements of organotin throughout the water column indicate that maximum concentrations occur in the surface microlayer. Concentrations in the surface microlayer, taken with the Garrett screen sampler, were as much as 27 times greater than those in subsurface waters. The high concentrations found in the surface microlayer may enhance deleterious effects on both the neuston and on organisms in the littoral zone, although this has not yet been demonstrated.
4. By 1988 it was still difficult to gauge the environmental impact of TBT released from large ships due to inputs from yachts; however, drydocks were clearly identifiable as sources of major inputs of TBT.

ACKNOWLEDGMENTS

The authors thank T.G. Wilkinson for his interest and support of the work, D. Smith and K. Evans for help with collection and analysis of samples, and P. Liss for the loan of the Garrett screen sampler. We are also grateful for the assistance of Mrs Brinsley. The work was carried out under Department of the Environment contracts PECD 7/8/74, PECD 7/8/76 and PECD 7/8/103.

REFERENCES

Abel, R., R.A. Hathaway, N.J. King, J.L. Vosser and T.G. Wilkinson. 1987. Assessment and regulatory actions for TBT in the UK. In: Proceedings of the Oceans '87 International Organotin Symposium, Vol. 4, M.A. Champ (Ed.). Institute of Electrical and Electronics Engineers, Piscataway, New Jersey, pp. 1314–1319.

Alzieu, C., Y. Thibaud, M. Heral and B. Boutier. 1980. Evaluation des risques dus à l'emploi des peintures anti-salissures dans les zones conchylicoles. *Rev. Trav. Inst. Scient. Tech. Pêches Marit.*, **44**, 301–348.

Ashby, J.R. and P.J. Craig. 1989. New method for the production of volatile organometallic species for analysis from the environment: some butyl tin levels in UK sediments. *Sci. Total Environ.*, **78**, 219–232.

Batley, G.E., K.J. Mann, C.I. Brockbank and A. Maltz. 1989. Tributyltin in Sydney Harbour and Georges River waters. *Aust. J. Freshwater Res.*, **40**, 39–48.

Beaumont, A.R. and M.D. Budd. 1984. High mortality of the larvae of the common mussel at low concentrations of tributyltin. *Mar. Pollut. Bull.*, **15**, 402–405.

Bryan, G.W., P.E. Gibbs, L.G. Hummerstone and G.R. Burt. 1986. The decline of the gastropod *Nucella lapillus* around southwest England: evidence for the effect of tributyltin from antifouling paints. *J. Mar. Biol. Ass. UK*, **66**, 611–640.

Cardwell, R.D. and A. Sheldon. 1986. A risk assessment concerning the fate and effects of tributyltins in the aquatic environment. In: Proceedings of the Oceans '86 Organotin Symposium, Vol. 4, M.A. Champ (Ed.). Institute of Electrical and Electronics Engineers, Piscataway, New Jersey pp. 1117–1129.

Cleary, J.J. 1991. Organotin in the marine surface microlayer and sub-surface waters of south-west England: relation to toxicity thresholds and the UK Environmental Quality Standard. *Mar. Environ. Res.*, **32**, 213–222.

Cleary, J.J. and A.R.D. Stebbing. 1987a. Organotin in the surface microlayer and subsurface waters of southwest England. *Mar. Pollut. Bull.*, **18**, 238–246.

Cleary, J.J. and A.R.D. Stebbing. 1987b. Organotins in the watercolumn – enhancement in the surface microlayer. In: Proceedings of the Oceans '87 International Organotin Symposium, Vol. 4, M.A. Champ (Ed.), Institute of Electrical and Electronics Engineers, Piscataway, New Jersey, pp. 1405–1410.

Donard, O.F.K., S. Rapsomanikis and J.H. Weber. 1986. Speciation of inorganic tin and alkyltin compounds by atomic absorption spectrometry using electrothermal quartz furnace after hydride generation. *Anal. Chem.*, **58**, 772–777.

Fent, K. 1990. Organotins in municipal wastewater and sewage sludge. In: 3rd International Organotin Symposium Proceedings, Monaco, 17–20 April. IAEA, Monaco, pp. 44–45.

Gardner, M.J. 1987. Examination of Leaching of Organotin Antifoulants (EC 9310 SLD). Water Research Centre, Environment Report to the Department of the Environment, UK, 25 pp.

Garrett, W.D. 1965. Collection of slick-forming materials from the sea surface. *Limnol. Oceanogr.*, **10**, 602–605.

Garrett, W.D. and R.A. Duce. 1980. Surface microlayer samplers. In: *Air–Sea Interaction: Instruments and Methods*, F. Dobson, L. Hasse and R. Davis (Eds). Plenum Press, New York.

Gibbs, P.E. and G.W. Bryan. 1987. TBT paints and the demise of the dog-whelk *Nucella lapillus* (Gastropoda). In: Proceedings of the Oceans '87 International Organotin Symposium, Vol. 4, M.A. Champ (Ed.). Institute of Electrical and Electronics Engineers, Piscataway, New Jersey, pp. 1482–1487.

Gucinski, H. 1986. The effect of sea surface microlayer enrichment on TBT transport. In: Proceedings of the Oceans '86 Organotin Symposium, Vol. 4, M.A. Champ (Ed.). Institute of Electrical and Electronics Engineers, Piscataway, New Jersey, pp. 1266–1274.

Hall, L.W. 1988. Tributyltin environmental studies in Chesapeake Bay. *Mar. Pollut. Bull.*, **19**, 431–438.

Hall, L.W., Jr, M.J. Lenkevich, W.S. Hall, A.E. Pinkney and S.J. Bushong, 1986. Monitoring organotin concentrations in Maryland waters of Chesapeake Bay. In: Proceedings of the Oceans '86 Organotin Symposium, Vol. 4, M.A. Champ (Ed.). Institute of Electrical and Electronics Engineers, Piscataway, New Jersey, pp. 1275–1279.

Hardy, J.T., C.W. Apts, E.A. Crecilius and N.S. Bloom. 1985. Sea-surface microlayer metals enrichment in an urban and rural bay. *Estuar. Coast. Shelf Sci.*, **20**, 29–312.

Harvey, G.W. and L.A. Burzell. 1972. A simple microlayer method for small samplers. *Limnol. Oceanogr.*, **17**, 156–157.

Hempel, G. and H. Weikert. 1972. The neuston of the subtropical and boreal Northeastern Atlantic Ocean. A review. *Mar. Biol.*, **13**, 70–88.

His, E. and R. Robert. 1980. Action d'un sel organo-métallique, l'acetate de tributyle-étain sur les oeufs et les larves d de *Crassostrea gigas* (Thunberg). (mimeo). Comité de la Mariculture, Conseil International pour l'Exploration de la Mer, Copenhagen, Paper CM 1980/F:27, 10 pp.

Huggett, R.J., M.A. Unger, P.F. Seligman and A.O. Valkirs. 1992. The marine biocide tributyltin: assessing and managing environmental risks. *Environ. Sci. Technol.*, **26**, 232–237.

Laughlin, R.B. Jr, H.E. Guard and W.M. Coleman III. 1986. Tributyltin in seawater: speciation and octanol–water partition coefficient. *Environ. Sci. Technol.*, **20**, 201–204.

Linden, O. 1987. The scope of the organotin issue in Scandinavia. In: Proceedings of the Oceans '87 International Organotin Symposium, Vol. 4, M.A. Champ (Ed.). Institute of Electrical and Electronics Engineers, Piscataway, New Jersey, pp. 1320–1323.

Liss, P.S. 1975. Chemistry of the sea-surface microlayer. In: *Chemical Oceanography*, Vol. 2, J.P. Riley and G. Skirrow (Eds). Academic Press, London, pp. 193–243.

Maguire, R.J. and R.J. Tkacz. 1985. Degradation of the tri-n-butyltin species in water and sediment from Toronto Harbor. *J. Agr. Food Chem.*, **33**, 947–953.

Maguire, R.J., Y.K. Chau, G.A. Bengert, E.J. Hale, P.T.S. Wong and O. Kramar. 1982. Occurrence of organotin compounds in Ontario lakes and rivers. *Environ. Sc. Technol.*, **16**, 698–702.

Maguire, R.J., R.J. Tkacz and D.L. Sartor. 1985. Butyltin species and inorganic tin in water and sediment of the Detroit and St. Clair Rivers. *J. Great Lakes Res.*, **11**, 320–327.

Maguire, R.J., R.J. Tkacz, Y.K. Chau, G.A. Bengert and P.T.S. Wong. 1986. Occurrence of organotin compounds in water and sediment in Canada. *Chemosphere*, **15**, 253–274.

Mathias, C.L., J.M. Bellama, G.J. Olson and F.E. Brinckman. 1986. Comprehensive method for determination of aquatic butyltin and butylmethyltin species at ultratrace levels using simultaneous hydridization/extraction with gas chromatography–flame photometric detection. *Environ. Sci. Technol.*, **20**, 609–615.

Matthias, C.L., J.M. Bellama, G.J. Olson and F.E. Brinckman. 1989. Determination of di- and tributyltin in sediment and microbial biofilms using acidified methanol extraction, sodium borohydride derivatization and gas chromatography with flame photometric detection. *Int. J. Environ. Anal. Chem.*, **35**, 61–68.

Muller, M.D. 1984. Tributyltin detection at trace levels in water and sediments using GC with flame-photometric detection and GC–MS. *Fresen. Z. Anal. Chem.*, **317**, 32–36.

Muller, M.D. 1987. Comprehensive trace level determination of organotin compounds in environmental samples using high-resolution gas chromatography with flame photometric detection. *Anal. Chem.*, **59**, 617–623.

Quevauviller, P. (1989). Organo-tins in sediments and mussels from the Sado estuarine system (Portugal). *Environ. Pollut.*, **57**, 149–166.

Rice, C.D., F.A. Espourteille and R.J. Huggett. 1987. Analysis of tributyltin in estuarine sediments and oyster tissue, *Crassostrea virginica*. *Appl. Organometal. Chem.*, **1**, 541–544.

Seligman, P.F., J.G. Grovhoug, A.O. Valkirs, P.M. Stang, R. Fransham, M.O. Stallard, B. Davidson and R.F. Lee. 1989. Distribution and fate of tributyltin in the United States marine environment. *Appl. Organometal. Chem.*, **3**, 31–47.

Stang P.M. and E.D. Goldberg. 1989. Butyltin in California river and lake marina waters. *Appl. Organometal. Chem.*, **3**, 183–187.

Stebbing, A.R.D., J.J. Cleary, M. Brinsley and C. Goodchild. 1983. Responses of a hydroid to surface water samples from the River Tamar and Plymouth Sound in relation to metal concentrations. *J. Mar. Biol. Ass. UK*, **63**, 695–711.

Thain, J.E. and M.J. Waldock. 1985. The growth of bivalve spat exposed to organotin leachates from antifouling paints. Marine Environmental Committee, International Council for the Exploration of the Sea, Copenhagen, Paper CM 1985/E:28. 10 pp. (mimeo).

UK Department of the Environment 1986. Organotin in Antifouling Paints, Environmental Considerations. Pollution Paper No. 25, DoE. London, 82 pp.

Valkirs, A.O., P.F. Seligman, P.M. Stang, V. Homer, S.H. Lieberman, G. Vafa and C.A. Dooley. 1986. Measurement of butyltin compounds in San Diego Bay. *Mar. Pollut. Bull.*, **17**, 319–324.

Waldock, M.J. 1986. TBT in UK estuaries, 1982–1986. Evaluation of the environmental problem. In: Proceedings of the Oceans '86 Organotin Symposium, Vol. 4, M.A. Champ (Ed.). Institute of Electrical and Electronics Engineers, Piscataway, New Jersey, pp. 1324–1330.

Waldock, M.J. and J.E. Thain. 1983. Shell thickening in *Crassostrea gigas*: organotin antifouling or sediment induced. *Mar. Pollut. Bull.*, **14**, 411–415.

Waldock, M.J., J.E. Thain and M.E. Waite. 1987a. The distribution and potential toxic effects of TBT in UK estuaries during 1986. *Appl. Organometal. Chem.*, **1**, 287–301.

Waldock, M.J., M.E. Waite and J. Thain. 1987b. Changes in concentrations of organotins in UK rivers and estuaries following legislation in 1986. In: Proceedings of the Oceans '86 Organotin Symposium, Vol. 4, M.A. Champ (Ed.). Institute of Electrical and Electronics Engineers, Piscataway, New Jersey, pp. 1352–1356.

Waldock, M.J., M.E. Waite and J.E. Thain. 1988. Inputs of TBT to the marine environment from shipping activity in the UK. *Environ. Technol. Lett.*, **9**, 999–1010.

Waldock, M.J., M.E. Waite, D. Miller, D.J. Smith and R.J. Law. 1989. The determination of total tin and organotin compounds in environmental samples. In: *Aquatic Environment Protection: Analytical Methods*, Ministry of Agriculture Fisheries and Food Directorate of Fisheries Research, Lowestoft, No. 4, 25 pp.

Waite, M.E., K.E. Evans, J.E. Thain and M.J. Waldock. 1989. Organotin concentrations in the Rivers Bure and Yare, Norfolk Broads, England. *Appl. Organometal. Chem.*, **3**, 383–391.

Waite, M.E., M.J. Waldock, J.E. Thain, D.J. Smith and S.M. Milton. 1991. Reductions in TBT concentrations in UK estuaries following legislation in 1986 and 1987. *Mar. Environ. Res.*, **32**, 89–111.

Zaitsev, Y.P. 1971. *Marine Neustonology*, K.A. Vinogradov (Ed.). Israel Programme for Scientific Translations, Jerusalem.

ORGANOTINS IN MUNICIPAL WASTEWATER AND SEWAGE SLUDGE

28

Karl Fent

Swiss Federal Institute for Environmental Science and Technology (EAWAG) and Swiss Federal Institute of Technology Zurich (ETH), CH-8600 Dübendorf, Switzerland

ABSTRACT

Direct entry of tributyltin (TBT) into the aquatic environment is primarily due to its use in antifouling paints on boats, which gives rise to contamination of waters and sediments of marinas, lakes, and coastal areas. To date, there is a lack of knowledge on other organotin sources and on the contamination of other environmental compartments. Inputs from wastewater and sewage sludge as well as from landfills are not well known. As the consumption of TBT and other organotins increases, these compounds are of growing importance in wastewater and sewage sludge. This chapter reviews speciation and contamination of wastewater and sludge, and describes the fate of organotins in a treatment plant. Organotins were determined by capillary gas chromatography with

Organotin. Edited by M.A. Champ and P.F. Seligman. Published in 1996 by Chapman & Hall, London. ISBN 0 412 58240 6

flame photometric detection after extraction and derivatization. In untreated wastewater of the city of Zurich, Switzerland, substantial concentrations of butyltins were determined. Phenyltins, dioctyltin, and tricyclohexyltin were not detected. Average values on six sampling days were 245 ng l^{-1} monobutyltin (MBT), 523 ng l^{-1} dibutyltin (DBT), and 157 ng l^{-1} TBT. The mean daily load of organotin was 122 g. Partitioning in wastewater showed that ~90% of the butyltins was associated with suspended solids in the influent, and then the percentage dropped at each successive stage of the treatment process. As a consequence, these compounds are removed from raw wastewater by sedimentation in the primary clarifier. Aerobic digestion did not lead to a significant elimination from wastewater, and anaerobic digestion was not very effective in reducing organotins in sludge; hence, substantial organotin levels were determined in digested sludge. Average concentrations were 0.78 mg kg^{-1} MBT, 0.98 mg kg^{-1} DBT, and 0.99 mg kg^{-1} TBT (dry weight). Organotins are therefore efficiently removed from wastewater (elimination 90%), but they become enriched in sewage sludge (daily load 59 g d^{-1}). Sludge is disposed of in landfills and at sea and is used in agriculture as a soil amendment to a large extent, giving a transfer path into the aquatic and terrestrial environment unrecognized thus far. Mono-, di-, and tributyltin residues have also been determined in the leachate of a landfill where the total butyltin concentration was 373 ng l^{-1}. The environmental input from landfills, however, is not as important as that from untreated wastewater and sewage sludge.

28.1. INTRODUCTION

Organotin compounds are of anthropogenic origin apart from methyltins, which were indicated to be produced by environmental biomethylation as well (Guard *et al.*, 1981). The first commerically registered organotin compound was marketed in 1936 for use as a stabilizer for synthetic polymers. Since the biocidal properties of trisubstituted organotins were recognized in the 1950s, the range of applications and products and the level of consumption have increased greatly. Production of organotins has risen from 5000 metric tonnes (t) in 1955 to at least 35 000 t in 1985 (Zuckerman *et al.*, 1978; Meinema *et al.*, 1982; Thompson *et al.*, 1985; Blunden and Chapman, 1986), which represents ~7% of the total annual world consumption of tin (Blunden and Chapman, 1986). Thompson *et al.* (1985) reported total organotins used in Canada to be >1000 t y^{-1}. In the Federal Republic of Germany total production of bis-tributyltin oxide was reported to be 2000 t y^{-1}. National usage was as follows: 70% antifouling paints, 20% timber protection, 10% textile and leather protection, and small amounts used as a preservative in dispersion paints and as a disinfecting agent [World Health Organization (WHO), 1991]. In The Netherlands some 4100 t passed through various organotin processing industries in 1985, of which two-thirds consisted of trisubstituted organotins. To date, organotins are the most widely used organometallic compounds.

There is a lack of actual estimations of the worldwide annual organotin consumption, partly because consumption varies in different countries; however, a significant increase in use is occurring in the South-East Asian region and in developing countries. Exact worldwide annual consumption and use patterns are unknown; however, principal applications can be addressed. Figure 28.1 shows that of the estimated worldwide annual organotin consumption of 35 000 t in 1985, the biocidal uses of the trisubstituted organotin compounds (~8000 t y^{-1}) were exceeded by the applications of the di- and monoorganotin derivatives (~27 000 t y^{-1}) used for stabilizers and catalysts (Blunden and Chapman, 1986). About 70% of the total annual world production is devoted to the thermal and UV stabilization of rigid and semi-rigid polyvinyl chloride (PVC). Typical

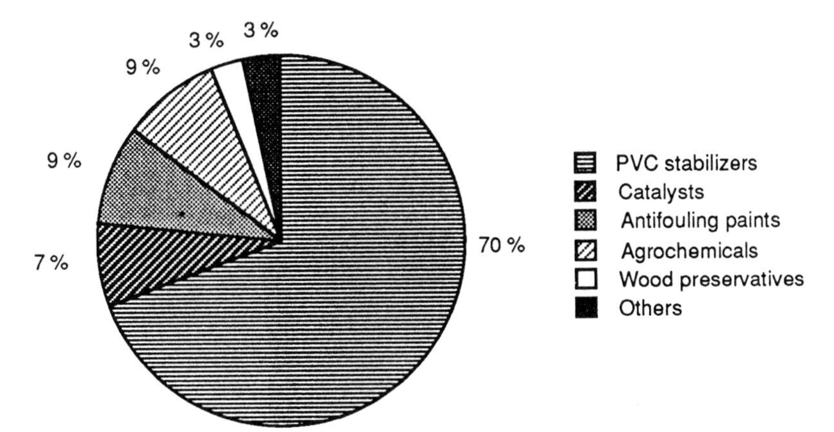

Figure 28.1 Application of organotins. The estimated worldwide annual consumption of organotin compounds in 1985 is ~35 000 t (after Blunden and Chapman, 1986). Data are based on weights of tin + organo groups. Dialkylated and monoalkylated organotins are used as stabilizers for PVC and other synthetic polymers and as catalysts. Trisubstituted compounds, which account for 24%, are used as biocides in agriculture and in the protection of various materials.

stabilizers are dialkyltin compounds [e.g. dibutyltins (DBT)], and the addition of monoalkyltins [e.g. monobutyltin (MBT)] has synergetic effects. Sulfur-containing mercaptide compounds (dibutyltin bis (isooctyl mercaptoacetate)) are used for heat stabilization, whereas non-sulfur DBT compounds impart light stability. The stabilizers are added to the PVC before processing at a level of ~5–20 g kg^{-1} (Evans and Karpel, 1985; Blunden and Chapman, 1986; Lawson, 1986). Organotin-stabilized PVC has many applications, including piping of potable water, wastewater, and drainage water. Monobutyltins and dibutyltins are also used as catalysts in the production of polyurethane foams and of silicones, and in other industrial processes. About 24% of the total worldwide organotin production is used as agrochemicals and as general biocides in a broad spectrum of applications (Fig. 28.1). The agricultural and biocidal use of organotins probably gives rise to the most significant proportion of free organotins in the environment, due to their direct introduction into soil, air, and water. Trialkylated tins have high fungicidal, bactericidal, algicidal, and acaricidal properties. Organic derivatives of tin are far more toxic than inorganic tin; and the toxicity of organotins increases with progressive introduction of organic groups at the tin atom, with maximum toxicity for trialkylated organotin compounds (WHO, 1980, 1991). Of particular importance to the environment is the high toxicity of tributyl-, triphenyl-, trimethyl and tricyclohexyltin derivatives. Growing consumption of tributyltin (TBT) stems from its use as a preservative for wood, textiles, paper, leather, and stonework. A small percentage of this compound is used in dispersion paints and in a variety of other materials (e.g. PVC) as protection against microbial or fungal attack (Evans and Karpel, 1985). Additionally, TBT has widespread applications as an effective biocide in antifouling paint on ships, boats, docks, and nets in fish culture. In the United States the annual consumption of TBT in antifouling paints was reported to be ~450 t in 1987 (Champ and Bleil, 1988).

The use of antifouling paints containing TBT has given rise to concern about the environmental impact of TBT. As TBT leaches directly from paints into the water, high contamination of marinas, harbors, and coastal areas results (Maguire, 1987, and Chapter 26; Fent and Hunn, 1991; Schatzberg, Chapter

19; Seligman *et al.*, Chapters 20 and 21; Huggett *et al.*, Chapter 24; Grovhoug *et al.*, Chapter 25). This route is the major pathway of entry of TBT into the aquatic environment. Because of the hazard it poses to aquatic life (Hall and Pinkney, 1985; Alzieu, 1986, 1991; Bryan *et al.*, 1986; Lawler and Aldrich, 1987; Maguire, 1987; Fent, 1992; Ritsema *et al.*, 1991; Hall and Bushong, Chapter 9; Laughlin *et al.*, Chapter 10; Waldock *et al.*, Chapter 11; His, Chapter 12; Gibbs and Bryan, Chapter 13; Henderson and Salazar, Chapter 14), the use of TBT-containing antifouling paints is now controlled or banned in several countries, resulting in a decrease in TBT contamination of coastal and marina waters (Alzieu *et al.*, 1989, 1991; Alzieu, 1991; Fent and Hunn, 1995). However, after several years of the ban in France, organotin levels are consistently higher than the concentration that causes some effect in the most susceptible species (Alzieu *et al.*, 1989). The question thus arises: are there sources of TBT entering the aquatic environment other than that leaching from antifouling paints?

Figure 28.2 illustrates the routes of entry and disposal of organotin compounds into the environment. While the release of TBT from antifouling paint into the aquatic environment is well documented, there is a lack of knowledge on other sources of TBT release and on organotin levels in other compartments than marina waters and sediments (Champ and Seligman, Chapter 29). It has been estimated that only ~0.5% of total organotin production is lost to the environment at the source of manufacture (Zuckerman *et al.*, 1978). Once applied, organotins can be leached from consumer products into environmental compartments, which may be of growing importance since the variety of materials protected by TBT and the range of TBT in industrial applications is increasing. Leaching and normal weathering lead to inputs in the aquatic and terrestrial environment. Disposal of materials manufactured with organotins in landfills could give rise to leaching into soil and groundwater. Municipal waste incineration gives no significant inputs into the environment, since organotins are decomposed to the relatively low-toxicity tin oxide and to other unspecified combustion products. Emissions into the air during application of agrochemicals or from treated surfaces of preserved materials do not result in substantial inputs, because the

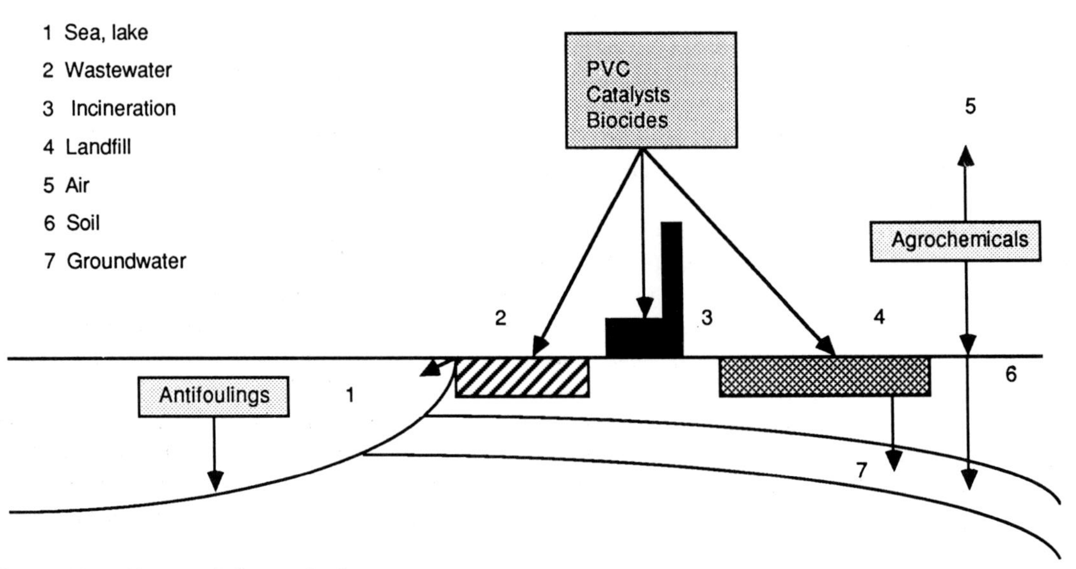

Figure 28.2 Fate and disposal of organotins in the environment.

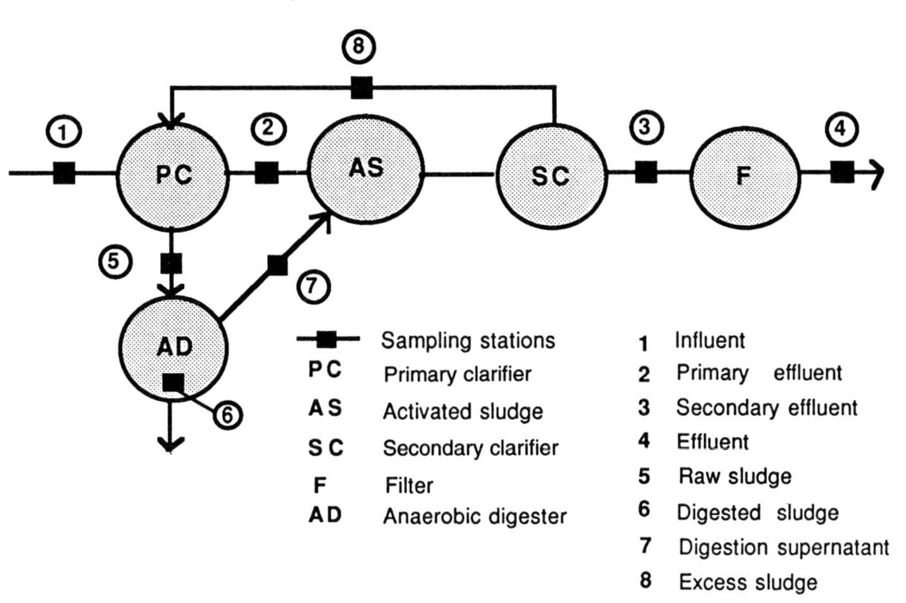

Figure 28.3 Treatment plant of the city of Zurich (350 000 inhabitants), Switzerland, with tanks for wastewater and sludge treatment. Stations where flow-proportional water samples and grab samples of sludge were taken are indicated.

vapor pressure of organotins is low, and photodegradation is reported to be relatively fast (Blunden, 1983).

The growing use of organotin-containing materials may result in contamination of municipal wastewater. This route of entry into the aquatic environment has significance. There is, however, a lack of data on organotin speciation and on the levels in wastewater and sewage sludge. Preliminary data of digested sludge samples indicated a considerable organotin load in Switzerland (Müller, 1987; Fent and Müller, 1991). The first data on organotin levels in municipal wastewater entering a sewage treatment plant and on concentrations in raw and digested sludge were given by Fent (1989), Fent *et al.* (1989), and Fent and Müller (1991), who also have reported on the fate of organotins through various processes of settlement, digestion, and filtration of the wastewater and through the sludge treatment process in a treatment plant. Subsequent studies in Germany (Schebek *et al.*, 1991), Canada (Chau *et al.*, 1992), France (Donard *et al.*, 1993) and Switzerland (Backer-van Slooten *et al.*, 1994; Fent, 1995) confirmed the occurrence of organotin compounds in wastewater and sludge. In this chapter, the existing data on organotin contamination of wastewater and sewage sludge are summarized, and the consequences for the aquatic and terrestrial environment are discussed.

28.2. EXPERIMENTAL METHODS

In the treatment plant of the city of Zurich (350 000 inhabitants), Switzerland, flow-proportional composite water samples were taken automatically on a total of 6 d in February 1988 and January and March 1989. Wastewater and sludge samples were collected at different sites in the plant, in order to elucidate the fate of organotins during wastewater and sludge treatment (Fig. 28.3). The first major stage of wastewater treatment is primary sedimentation in large tanks in which ~70% of the suspended solids, and significant quantities of many contaminants, such as metals, are removed as raw primary sludge. The settled wastewater is then treated by an aerobic biological treatment process (activated sludge process), which removes most of the remaining solids and organic loads as well as further contaminant fractions. At this stage a secondary sludge is produced by sedimentation. Such sludge is mixed with raw primary sludge by recycling to the primary sedimentation tank for resettlement. Raw sludge is then processed by anaerobic digestion for 3–5 weks, where ~40% of the organic matter is removed with the production of methane and carbon dioxide.

Partitioning between the dissolved and particulate-associated organotins is important for the understanding of their fate in the treatment process. Therefore, the water samples were filtered through glassfiber filters (pore size 1.2 μm) prior to organotin trace analysis in order to separate suspended particles and dissolved water phase. Details of the analytical procedure were given by Fent (1989) and Fent and Müller (1991). The method is based on the procedure of Müller (1987), and it speciates butyltins (MBT, DBT, TBT), phenyltins (mono-, di-, triphenyltin), dioctyl-, and tricyclohexyltin at trace levels, but not methyltins. Briefly, organotin compounds were extracted from wastewater by solid-phase extraction. Particulate material and sewage sludge were extracted with ether containing 0.25% tropolone. After extraction, compounds were ethylated by a Grignard reaction to render them volatile. After a cleanup procedure, the organotins were analyzed by high-resolution capillary gas chromatography with flame photometric detection (HRGC–FPD). As organotins with different degrees of alkylation do not behave as chemically identical species in the extraction and derivatization procedure, separate internal standards (mono-, di-, and tripentyltin and tripropyltin) were used for accurate quantification (Fent, 1989). The limit of detection was 1–5 ng l^{-1} for wastewater and 10–50 μg kg^{-1} for sludge (dry wt). All the results refer to the butyltin species as the ion (Bu_1Sn^{3+}, MBT; Bu_2Sn^{2+}, DBT; Bu_3Sn^{+}, TBT) and were corrected for recovery by internal standards.

28.3. RESULTS

28.3.1. ORGANOTINS IN WASTEWATER

28.3.1a. Contamination and fate in the treatment process

Residues of MBT, DBT, and TBT were detected in the incoming wastewater. Phenyltins, dioctyltin, and tricyclohexyltin were not detected and methyltins were not determined (Fig. 28.4a). Concentrations of the different butyltin species and total butyltins in 1988 and January 1989 are given in Fig. 28.5. The concentrations in raw wastewater on six sampling days (February 1988 and January and March 1989) ranged from 136 to 564 ng l^{-1} for MBT, from 127 to 1026 ng l^{-1} for DBT, and from 64 to 217 ng l^{-1} for TBT (total of butyltins 489 to 1484 ng l^{-1}). Average butyltin concentrations were 245 ng l^{-1} MBT, 523 ng l^{-1} DBT, and 157 ng l^{-1} TBT, giving a total mean concentration of butyltins of 925 ng l^{-1}. On a Sunday, concentrations of total butyltins were lower than on the other days that were weekdays. This indicates that additional inputs originate from industrial activities on weekdays. Daily variation was most pronounced for DBT and was lowest for TBT.

The fate of organotins through sewage treatment is described in detail by Fent *et al.* (1989) and Fent and Muller (1991). Figure 28.6a illustrates the behavior of butyltin species on a typical sampling day, and Fig. 28.6b shows partitioning of dissolved and particulate-associated TBT through the treatment process. Briefly, substantial amounts of the incoming butyltins were removed during primary settlement, resulting in a significant reduction to values that were ~70% lower than those in the influent. This significant

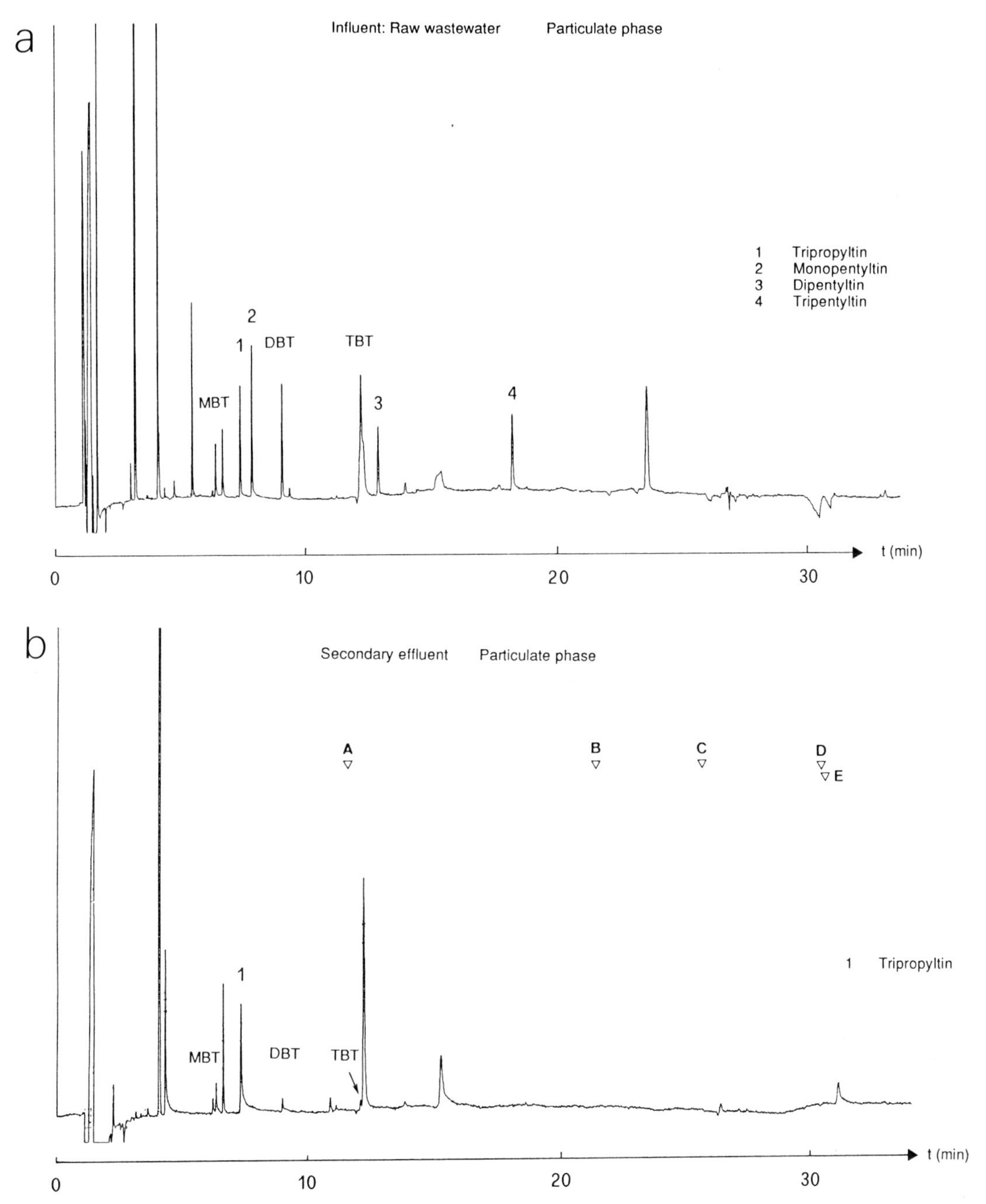

Figure 28.4 HRGC–FPD chromatograms of the particulate phase of wastewater samples (a) in the influent, and (b) after secondary clarification, which represents the outlet of most plants. Four internal standards (mono-, di-, and tripentyltin and tripropyltin) were added to the samples prior to extraction. Monobutyltin (MBT), dibutyltin (DBT), and tributyltin (TBT) were detected. As indicated in the chromatogram (b), no monophenyltin (A), diphenyltin (B), dioctyltin (C), triphenyltin (D), or tricyclohexyltin (E) were detected.

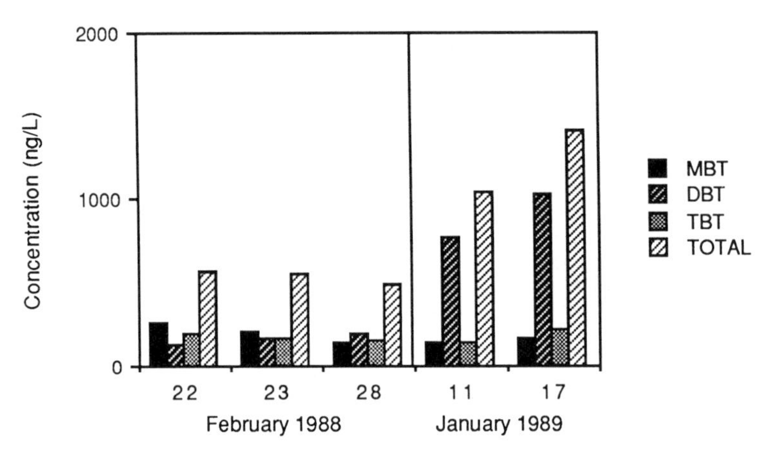

Figure 28.5 Concentrations of monobutyltin (MBT), dibutyltin (DBT), and tributyltin (TBT) in raw wastewater on sampling days in February 1988 and January 1989.

elimination is a consequence of the partitioning of the butyltin compounds between water and particles: in raw wastewater ~90% of the butyltins was associated with suspended solids. Butyltins were further eliminated in the activated sludge treatment, but to a small extent. Figure 28.4a shows that residues of MBT, DBT, and TBT were also present in the secondary effluent. The average concentration of total butyltins was 80 ng l^{-1}. As normal treatment plants have only these steps, this represents the concentration in the outlet of average treatment plants. After passage through a filter, the average concentration of total butyltins was 17 ng l^{-1}, which represents 2% of the incoming level. This shows that organotins are efficiently removed from wastewater in this plant. They are not, however, substantially degraded through aerobic digestion, but are adsorbed into sewage sludge, where they become enriched.

28.3.1b. Partitioning between dissolved and particulate-associated organotins

Low water solubility and its lipophilic character cause TBT to adsorb preferentially onto particles. Adsorption depends on the nature (organic carbon content) and size of particles in suspension, salinity, temperature, and presence of dissolved organic matter (DOM). The octanol–water partition coefficient (K_{ow}) of TBT lies between 7000 for deionized water and 5500 for seawater (Laughlin and Linden, 1985; Laughlin *et al.*, 1986). Unger *et al.* (1988) have reported that TBT adsorbs strongly to sediment particulate matter. Our data show that, in wastewater, MBT, DBT, and TBT exist both in a dissolved form and in association with suspended solids to varying proportions throughout treatment. In untreated wastewater, butyltins are primarily associated with suspended solids: only ~10% of DBT and TBT and 20% of MBT were dissolved. After primary clarification 80%, and after aerobic digestion 53% of total butyltin was associated with suspended solids. Hence there is a change in the partitioning from predominantly particulate-associated to dissolved forms through the wastewater treatment process, because particulate-associated organotins are removed. Partitioning is controlled by the suspended solids concentration and thus by the total surface area onto which organotin can be adsorbed. Characteristics of the particulate material and the composition of the waste-

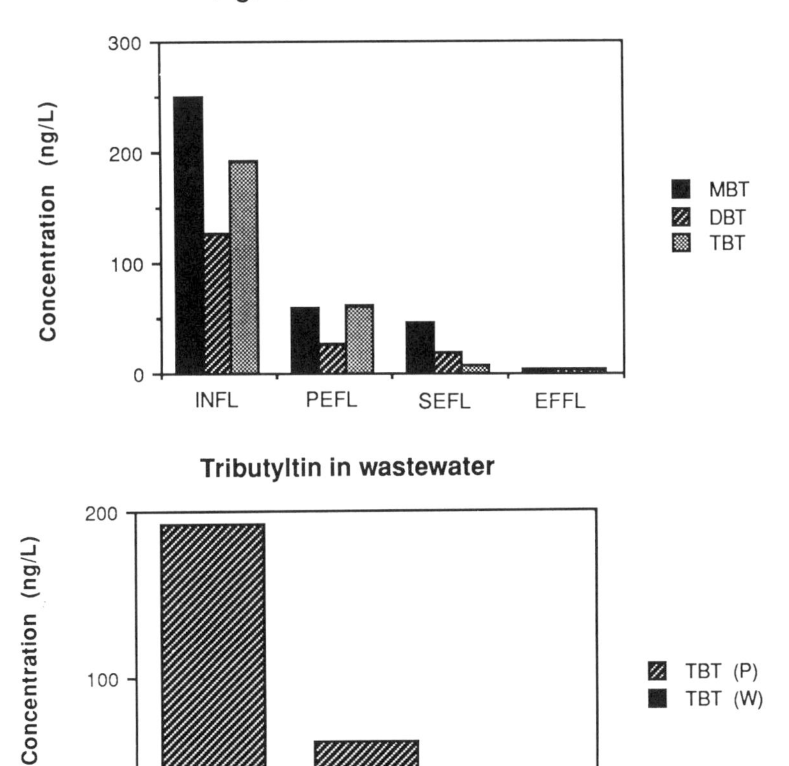

Figure 28.6 (a) Concentrations and fate of butyltin species during the treatment process on 22 February 1988. Concentrations were given in the influent (INFL), in the primary effluent (PEFL), in the secondary effluent (SEFL), and in the effluent of the plant (EFFL). (b) Partitioning of tributyltin (TBT) in wastewater during treatment on that day. Concentrations of dissolved TBT (W) and particulate associated TBT (P) are given in the influent (INFL), in the primary effluent (PEFL), and in the secondary effluent (SEFL).

water (e.g. DOM, lipids content, and salinity) are important as well. The fraction of butyltins that is in association with particles is highest in untreated wastewater, where the concentration of suspended solids, DOM, and lipids is highest as compared with other sites in the plant. In the secondary effluent, the higher proportion of dissolved butyltins was correlated with a decreased concentration of suspended solids, lipids content, and DOM.

Partition coefficients are important because of their relationship to physical adsorption on solids, biomagnification, and lipophilic storage. Distribution ratios of organotins between dissolved and particulate matter could be defined as the concentration of the butyltin species in the solid phase divided by the concentration in the dissolved phase. In the raw wastewater samples the butyltin distribution ratios were estimated for the six sampling days. The TBT distribution ratios varied between 1.7×10^4 and 4.9×10^4 (average value, 3.0×10^4; $\mu g\ kg^{-1}$ TBT in particles per $\mu g\ l^{-1}$ TBT in water); DBT distribution ratios varied between 2.4×10^4 and 6.3×10^4 (average value, 4.4×10^4; μg kg^{-1} DBT in particles per μg l^{-1} DBT in water); and MBT distribution ratios varied between 5.3×10^3 and 2.8×10^4 (average value, 1.7×10^4; μg kg^{-1} MBT in particles per μg l^{-1} MBT in water). This shows that DBT and TBT adsorb more strongly to suspended solids than does MBT. The distribution ratios determined in wastewater are consistent with reported partition coefficients derived from partitioning studies in the water column of marinas or from sediment–water partitioning studies. Published values for TBT sediment–water and particulate–water partition coefficients range from 3.4×10^2 to 1.9×10^6 (Maguire and Tkacz, 1985; Randall and Weber, 1986; Valkirs *et al.*, 1986; Unger *et al.*, 1988; Seligman *et al.*, 1989) with the majority of the values between 1.0×10^3 and 3.0×10^3 (μg kg^{-1} TBT in sediment per μg l^{-1} TBT in water). Data from seawater (Valkirs *et al.*, 1986; Seligman *et al.*, 1989) and from fresh water show that generally <5% of the total TBT was found in the particulate fraction. Hence, due to the very low particulate fraction in the water column of sea- and fresh water, TBT is found primarily in the dissolved form. This is just the opposite to the situation in raw wastewater and in water leached from a landfill as shown in section 28.3.4, Organotins in landfill leachate.

28.3.2. ORGANOTINS IN SEWAGE SLUDGE

As organotins in untreated wastewater are primarily in association with suspended solids, they are transferred into raw sewage sludge, where substantial mean concentrations of MBT, DBT, and TBT in the range of 0.40–0.47 mg kg^{-1} (dry wt) have been determined (Fent *et al.*, 1989, Fent and Müller, 1991). Anaerobically stabilized sludge, which is produced from raw sludge after anaerobic digestion, is used as a soil amendment in agriculture or dumped in landfills or at sea. The chromatogram in Fig. 28.7 shows that, even after 35 d of anaerobic degradation, MBT, DBT, and TBT are present in significant concentrations. This has been observed in all of the 35 sludge samples from 30 treatment plants analyzed in 1988–1990 and 1995. In the treatment plant of Zurich, concentrations of MBT ranged from 0.62 to 0.97 mg kg^{-1} (dry wt), DBT concentrations from 0.64 to 1.24 mg kg^{-1}, and TBT concentrations from 0.61 to 1.51 mg kg (total of organotins, 2.50–3.38 mg kg^{-1}; average dry weight content of the sludges is 6.4%) and average values of MBT, DBT, and TBT were 0.78, 0.98, and 0.99 mg kg^{-1} (total 2.75 mg kg^{-1}, dry wt), respectively. In one sample, mono-, di-, and triphenyltin were also detected. These were also found in 28% of sludge samples surveyed in 1995 (n = 27). Phenyltin levels were between 0.1 and 0.5 mg kg^{-1} (dry wt). Average butyltin levels were also found to be similar to those in 1988–1990 (MBT, DBT, and TBT were 0.5, 1.5, and 1.1 mg kg^{-1}, respectively).

28.3.3. MASS FLUXES AND EFFICIENCY OF ELIMINATION IN THE TREATMENT PROCESS

For the assessment of the environmental impact it is important to know the range of daily and annual inputs of organotins into wastewater and digested sludge. The calculated average daily load of total butyltins for 6 d was 122 g in raw wastewater and 59 g in digested sludge. Taking these loads as representative, it can be calculated that the total annual load of organotins in untreated municipal wastewater in the city of Zurich, Switzerland, is in the range of ~51 kg, of which 17% is TBT. The percentage of MBT, DBT, and TBT is consistent with the percentage of the different organotins consumed worldwide (Fig. 28.1). Based on our measurements, the calculated total annual load of organotins in digested sludge is ~22 kg, of which ~8 kg is TBT. These are, of course, estimates of the annual organotin loading from this city.

Mass fluxes of butyltins in the sewage treatment plant are dominated by the fraction

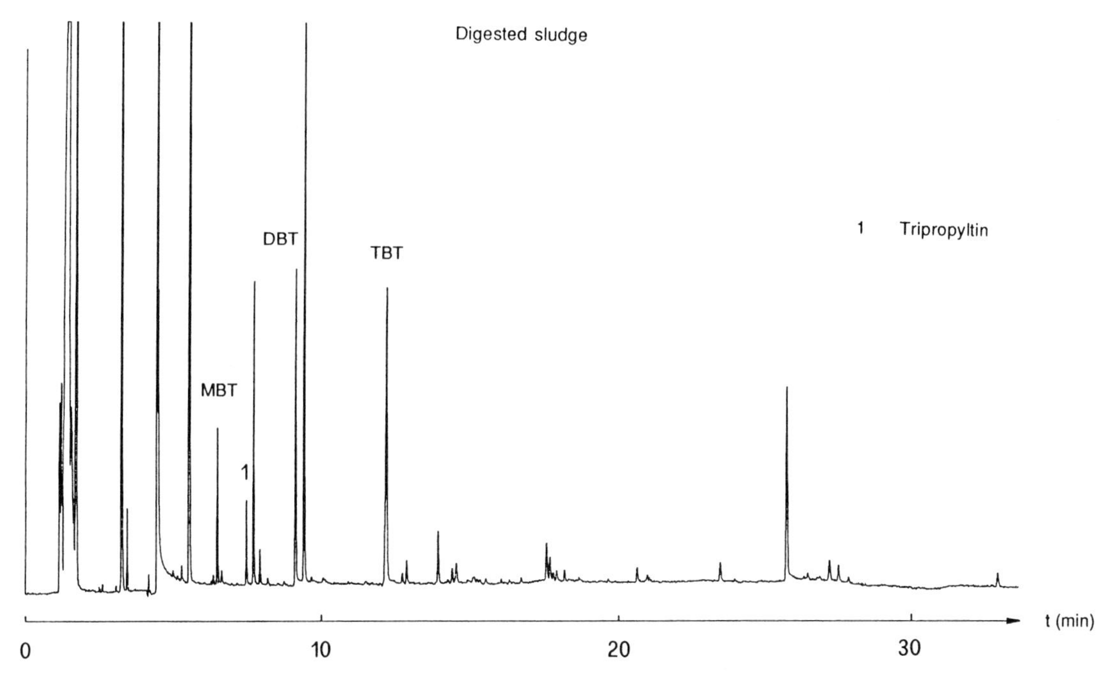

Figure 28.7 Gas chromatogram of a digested sludge sample. Tripropyltin was added as an internal standard prior to extraction. Monobutyltin (MBT), dibutyltin (DBT), and tributyltin (TBT), but no phenyltins, tricyclohexyltin, or dioctyltins were detected.

that is adsorbed on suspended solids. Figure 28.8 shows average loads of total butyltins (percentages of influent load) in the dissolved and particulate fraction through the treatment plant over the sampling period (6 d). Mass fluxes of every compound, MBT, DBT, and TBT, were similar. Seventy-three percent of the total butyltin load was eliminated from raw wastewater by settlement in the primary clarifier. Up to 45% of the incoming butyltin load has been found in raw sludge. Thus, the most important process for organotin elimination in sewage treatment is the incorporation into sludge. Biodegradation by aerobic sludge treatment is of minor importance. This can be deduced from the mass balance, which shows that the net reduction of total butyltins by aerobic digestion was only minimal (7.5% of the incoming load on average). Due to microbial and algal degradation, TBT has a relatively short half-life in the water column of seawater. Half-lives of 13–19 d were determined (Seligman *et al.*, 1986). However, contrary to the situation in the water column, only a minor fraction of the organotins is dissolved in wastewater (Fig. 28.8), and thus bioavailable to the microorganisms. After secondary clarification, only 10% of the influent load was present in wastewater. Excess sludge, recycled into the primary clarifier, contained 9.5% organotin. Thus, primary and secondary settlement and aerobic treatment together resulted in a 90% reduction in total butyltin. As these treatment steps are usual in plants, this may represent the elimination efficiency in plants with very high performance. Filtration led to a further reduction by removal of suspended particulate material, which contained butyltins, giving an elimination of 98% of the incoming total butyltin. Contrary to the plant in Zurich, a much lower elimination of 40% was estimated in a Canadian plant (Chau *et al.*, 1992), or was very minor in Bordeaux, France (Donard *et al.*, 1993).

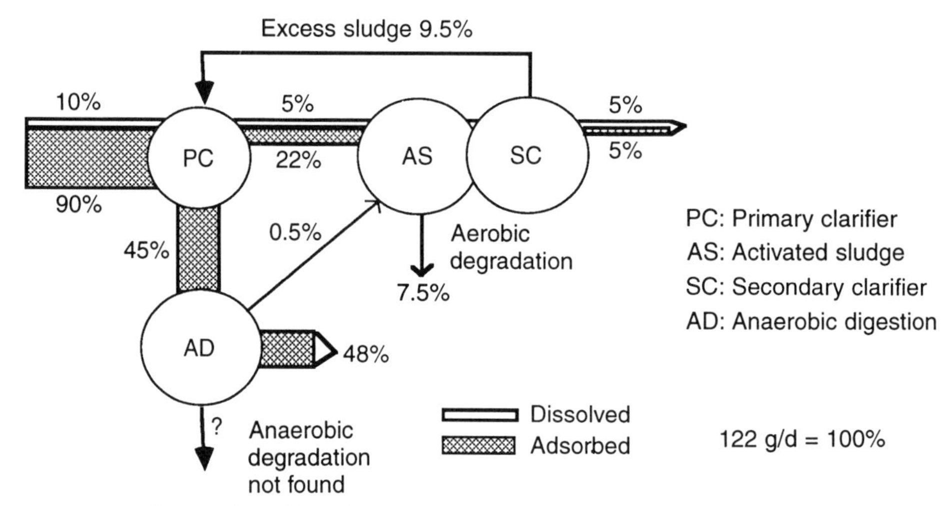

Figure 28.8 Mass fluxes of total butyltins on six sampling days. Estimates of the average loads in the dissolved and in the particulate fraction of wastewater, and average loads in sludge, are given as percentages of the daily incoming organotin load (122 g, representing 100%) [after Fent and Müller (1991)].

In raw sludge, 45% of the incoming load was determined (Fig. 28.8). It should be noted that inputs and outputs in the primary clarifier do not completely balance, because in this field study grab samples rather than 24-h composite samples were taken from raw sludge. In digested sludge produced after 35 d of anaerobic digestion, 48% of the incoming load was determined, and in the digestion supernatant it was 0.5%. Thus, the estimate of the mass balance shows that over the 6-d sampling period, the average load of total butyltins in the digested sludge was not reduced as compared with raw sludge. Even though digested sludge samples and raw sludge samples cannot directly be compared – samples of digested sludge had a much longer mixing period, and both of them were grab samples – it can be concluded from mass balances that anaerobic digestion during sludge treatment does not lead to a significant elimination of butyltin. Our study with laboratory reactors further indicated that degradation of TBT during 12–20 d of sludge treatment is only minimal (up to 30%) (Fent *et al.*, 1991). In summary, the following conclusions can be drawn:

1. Organotins are predominantly eliminated from untreated wastewater in the primary clarifier by incorporation into sewage sludge.
2. Aerobic sludge treatment is a transfer path that is of minor importance, leading to only a minimal reduction of the butyltin load.
3. Anaerobic sludge treatment is not very effective in organotin elimination.
4. Digested sludge has considerable organotin residues.

28.3.4. ORGANOTINS IN LANDFILL LEACHATE

Little is known about the environmental fate of the organotins from the disposal of PVC waste by land burial. It has been assumed that disposal of organotin-containing materials will immobilize these compounds. Studies using clay, topsoil, and sand show that vertical migration was only observed for the sand. In clay soils 95–99% of the trialkylated organotins remained in the upper layer (Lawson, 1986). The data presented here show that organotins are present in the leachate of a Swiss landfill.

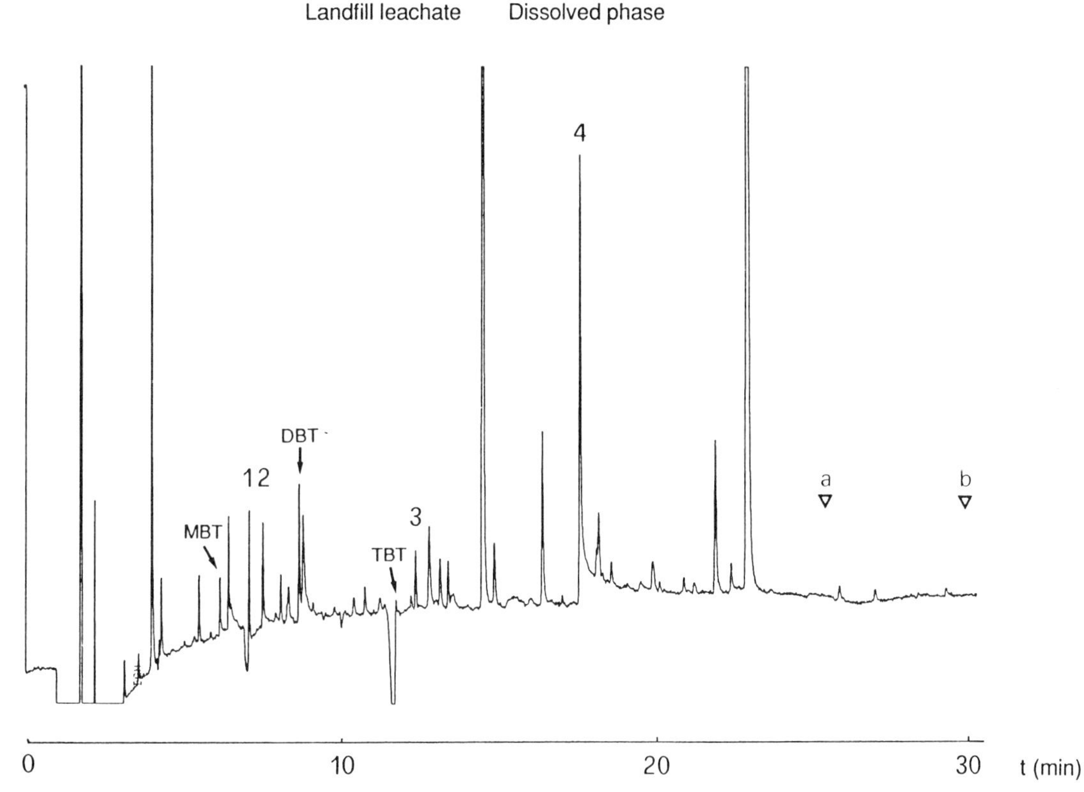

Figure 28.9 HRGC–FPD chromatogram of the dissolved phase of a water sample from a landfill leachate. Four internal standards were added to the samples prior to extraction: 1, tripropyltin; 2, monopentyltin; 3, dipentyltin; and 4, tripentyltin. Monobutyltin (MBT), dibutyltin (DBT), and tributyltin (TBT) were detected. All tin compounds were ethylated. As indicated in the chromatogram, neither dioctyltin (a), triphenyltin (b), mono- or diphenyltin, nor tricyclohexyltin were detected.

Water samples were collected from a landfill in Switzerland. Prior to analysis, samples were filtered through glassfiber filters (pore size 1.2 μm). Residues of MBT, DBT, and TBT were determined both in the dissolved (Fig. 28.9) and in the particulate fraction. Phenyltin compounds, tricyclohexyltin, and dioctyltin were not found. As in wastewater, butyltin species were predominantly associated with suspended solids. Whereas in raw wastewater ~90% of total butyltin was determined in the particulate fraction, in the landfill leachate only 70% was found in the particulate fraction. Concentrations in the dissolved fraction were 40 ng l^{-1} MBT, 66 ng l^{-1} DBT, and 5 ng l^{-1} TBT. In the particulate fraction, levels were 72 ng l^{-1} MBT, 177 ng l^{-1} DBT, and 13 ng l^{-1} TBT (total butyltin, 373 ng l^{-1}). These data indicate that organotins were mobilized from landfills. However, it is unlikely that the organotins present in landfill leachates contribute significantly to the levels found in the environment; and, in this respect, landfill leachates may not be as important as untreated wastewater and sewage sludge.

28.4. DISCUSSION

28.4.1. SOURCES

The use of organotins is so diverse that determination of the actual sources of wastewater pollution by these compounds may be difficult; however, there are a great number of organotin applications that, in turn, could account for input sources into wastewater. As the wastewater and sludge samples were collected in winter to early spring, applications of organotins in agriculture as well as runoff from agricultural fields likely did not contribute to the organotin load. Hence sources of organotins must be found in industrial applications and in industrial and consumer products. The principal commercial use of organotin is as a PVC stabilizer, and the possible route by which organotin compounds may enter the wastewater is leaching and weathering. In addition, wastewater from the PVC processing industry may be important. In my opinion, these are the main sources of MBT and DBT in wastewater. PVC polymer contains ~0.5–10% MBT and DBT by weight (Meinema *et al.*, 1982; Evans and Karpel, 1985; Blunden and Chapman, 1986). MBT and DBT may enter the wastewater not only during the production process, but also by leaching from PVC pipes used for potable water and wastewater and from other PVC materials. It has been reported that leaching of organotins from PVC pipes actually occurs (Komatsu *et al.*, 1970; Boettner *et al.*, 1981). Leaching from 46-m PVC pipes leads to a concentration of 35 mg m^{-3} tin in the water after first use, and to a subsequent constant release of 1 mg m^{-3} (Wu *et al.*, 1989). Leaching of organotin stabilizers from PVC has also been studied in view of the use of PVC in food-contact applications (Koch and Figge, 1971). In addition, organotin-stabilized PVC is not suitable for certain medical applications, due to localized adverse reactions to the organotins (Blunden and Chapman, 1986). The use of mono- and dialkylated organotin as homogeneous catalysts (esterification) results in the incorporation of organotin into the product at low levels, typically in the range of ~0.01–4 g kg^{-1} (Lawson, 1986). Weathering of these products (e.g. urethane, silicone elastomers, and polyester-based resin) may also lead to organotin release into wastewater.

One source of tributyltin input to the environment is suggested to originate from its use as a biocide in the preservation of wood and in the protection of a variety of materials against bacteria, fungi, and insects. This compound is also added to dispersion paints, which may lead to considerable input into wastewater. Its use in textiles as an insecticide (moth control in wool) or microcide or as a fungicide may be of importance, because these biocides could be released to the environment in the washing process. Plastics (e.g. PVC, polyethylene, polyurethane, and silicone) can also be protected from microbial attack by incorporation of TBT in the polymer (Evans and Karpel, 1985). Tributyltin is also added to disinfectants used in hospitals and nurseries, and as a slimicide in paper manufacture. Very high contamination of cooling water from a thermoelectric power plant was reported in Italy (Bacci and Gaggi, 1989). Inputs to the study area from antifouling paints are likely not contributing to the TBT load determined in our studies, because samples were taken in winter to early spring; however, during the summer, and especially in spring and autumn, when pleasure boats are cleaned on land and in docks, TBT from antifouling paints may enter into the wastewater. Triphenyltin, which was detected in a number of sludge samples, could originate from its use as a wood preservative or as a pesticide.

28.4.2. ECOTOXICOLOGICAL CONSEQUENCES

28.4.2a. Wastewater

As the values reported here are derived from Zurich, an industrialized city that has no known exceptionally high input sources, it is believed that this may represent the current situation in industrialized areas. Assuming that these values are representative for

Switzerland, the annual butyltin input into municipal wastewater (2×10^{12} l y^{-1}) will be in the range of >1900 kg y^{-1}. However, it can be assumed that organotin concentrations are even higher in cities that are heavily industrialized or even lower in sparsely industrialized areas. In any case, untreated municipal wastewater gives rise to a significant organotin input into aquatic systems.

The toxicity of wastewater is mainly based on its TBT concentration, since the toxicity of DBT has been suggested to be ~10–100 times lower than that of TBT, and MBT 10–100 times lower than DBT (Hall and Pinkney, 1985). However, it should be noted that the toxicity of MBT and DBT to aquatic organisms is not as well understood. Toxic effects toward the immune system of fish (thymus atrophy) have been shown at 320 μg l^{-1} DBT (Wester and Canton, 1987). Fish early life stages were found to be affected at 0.8 μg l^{-1} TBT (Fent and Meier, 1992). Effects of DBT occurred at concentrations that were three orders of magnitude higher than those of TBT. As with TBT, diorganotin compounds (e.g. DBT) affect mitochondria (causing uncoupling of oxidative phosphorylation and the inhibition of oxygen uptake), and hence affect metabolism (Aldridge and Street, 1964). Dibutyltin and TBT induce thymus atrophy in rats. The effects caused by TBT were less pronounced compared with those caused by DBT, which could indicate that thymus atrophy in rats is caused by the metabolite DBT rather than by TBT (Snoeij *et al.*, 1988). The toxicity of mixtures of MBT, DBT, and TBT at levels found in wastewater is not known; hence, only the toxic effects of TBT will be discussed. As levels of TBT determined in untreated wastewater (64– 217 ng l^{-1} TBT) are below the concentrations that have a toxic effect on sewage bacteria (Argaman *et al.*, 1984), the treatment process is not negatively affected. Untreated municipal wastewater discharges are considered to be acutely toxic to the most susceptible molluscs and gastropods (Beaumont and Budd, 1984; Alzieu, 1986; Bryan *et al.*, 1986; Thain, 1986), zooplankton (U'ren 1983; Bushong *et al.*, 1988), and algae (Walsh *et al.*, 1985; Beaumont and Newman, 1986), dilution in receiving systems likely reduces the concentration to values that are less harmful. However, chronic toxicity to the most sensitive organisms cannot be ruled out. As organotins in raw sewage are predominantly adsorbed onto particles, filter feeding and benthic organisms may be more exposed than others. Trace levels of TBT in the range of a few nanograms per liter found in treated wastewaters elicit chronic toxic effects, predominantly on molluscs and gastropods. Since most data on chronic toxicity, however, are derived from studies with marine organisms, it is not known what effects trace levels of TBT will have on freshwater organisms. Since average treatment plants are able to reduce incoming organotins efficiently, TBT concentrations in the outlet may result in chronic toxicity only to the most susceptible organisms such as oysters, dogwhelks and other gastropods. After dilution in receiving waters, toxicity is further reduced. However, since in sewage effluents ~50% of organotin is in the dissolved form (Fent and Muller, 1991), organotins are bioavailable to most aquatic organisms. It is important to note that treated and untreated wastewaters represent a significant organotin source in surface waters (Fent and Hunn, 1991, 1995; Schebek *et al.*, 1991).

28.4.2b. Sewage sludge

In treatment plants, organotins are transferred by settlement from untreated wastewater into sewage sludge. As anaerobic sludge treatment does not result in a substantial reduction of the organotin load, high concentrations of ~3 mg kg^{-1} (dry wt) of total organotin were found in digested sludge, of which about one-third was TBT. Much higher values up to 30 mg kg^{-1} total organotin (6 mg kg^{-1} TBT) have been reported in other samples (Müller, 1987), and the highest TBT

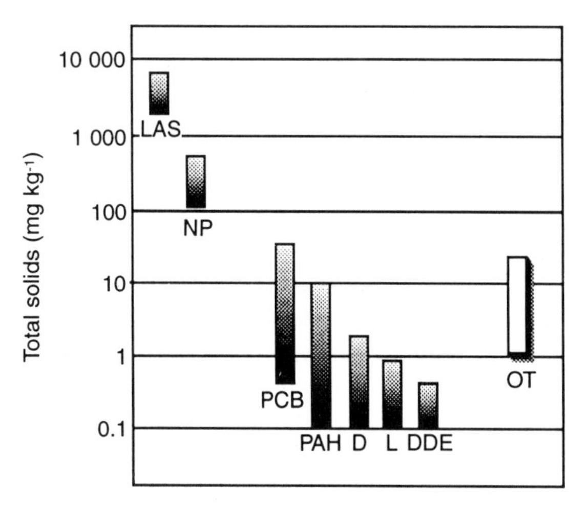

Figure 28.10 Concentration ranges of known organic pollutants in sewage sludges (modified after Giger *et al.*, 1989; and Siegrist *et al.*, 1989). Laundry detergents, linear alkylbenzene sulfonates (LAS), and nonylphenol (NP) are most prominent. Among ubiquitous organic pollutants, polychlorinated biphenyls (PCB), polycyclic aromatic hydrocarbons (PAH), pesticides such as dieldrin (D), lindane (L), and a metabolite of DDT (DDE) are compared with levels of total organotins (OT) found in our studies (Fent and Müller, 1991; Fent, 1995).

residue found in 1995 was 10.3 mg kg^{-1} (Fent, 1995). However, the residues were in the same range as in 1989–1990. The residues of organotin compounds in sewage sludge are in the range of other known ubiquitous organic contaminants, as illustrated in Fig. 28.10. In Switzerland, the annual production of sludge was estimated to be 2.7×10^5 t (dry wt). If this sludge is assumed to be representative for Switzerland, the annual total organotin load in the country would lie in the range of ~800 kg y^{-1}. In Western Europe ~5.5×10^6 t (dry wt) of sewage sludge is disposed of each year, and this level is expected to rise by ~25–50% in the next 10 y (Dean and Suess, 1985); therefore, TBT is of concern in sewage sludge because of its toxicity to aquatic organisms.

Sewage sludge is disposed of in landfills, is dumped at sea, and is used in agriculture as a soil amendment. Dumping and transfer to soil are of ecotoxicological relevance, since these transfer paths give rise to organotin pollution of aquatic and terrestrial systems. Little is known about the environmental fate of the organotins disposed by land burial. I have shown that leaching from landfills that contained a variety of waste material can occur, but little is known about adsoption onto soil and degradation. However, it is unlikely that this source contributes significantly to the levels found in the environment.

In many countries sludge is dumped at sea in significant amounts. In the sea, sludge disperses widely after disposal. Resuspension and remobilization of organotins from sludge is highly probable, thus giving rise to an input source and transfer path at sea not recognized thus far. Desorption of TBT can act as a source for dissolved and particulate-associated forms, which are toxic to organisms in the overlying water. Degradation of dissolved TBT in marine waters (Seligman *et al.*, 1986) would, however, reduce the potential for toxicity and accumulation. Mussels that were placed in cages along known dispersion routes of dumped sludge in the Thames estuary (United Kingdom) have been reported to be under sublethal stress (Whitelaw and Andrews, 1988).

Organotins will be transferred to soils if sewage sludge is used as fertilizer in agriculture. Generally, very little is known about the ecotoxicological consequences of organic pollutants in sewage sludge on soil biota and soil fertility, especially with respect to long-term impacts. A study with detergent-derived organic pollutants and polychlorinated biphenyls (PCB) in sludge-amended soil indicates that these pollutants will accumulate in soils after frequent disposal of sludge, although biodegradation of detergents (but not PCB) occurs (Marcomini *et al.*, 1988). In Switzerland ~40–50% of total sludge produced is used in agriculture. If sludge determined in Zurich is used according to present guidelines (0.25 kg dry solids m^{-2} y^{-1}), the annual organotin input to soil will be ~1 mg m^{-2}, of which 20–30% is TBT. The ecotoxicological consequences of this pollution to soil organisms and to the terrestrial system are largely unknown. Preliminary studies on the microbial activity relevant to fertility of soils have shown that bis-tributyltin oxide (TBTO) at such concentrations had no significant negative impact on soil respiration and total of bacteria (Bollen and Tu, 1972). However, total mold counts and fungi were significantly reduced in soils contaminated with 10 μg l^{-1} TBTO. In addition a number of organisms besides fungi (e.g. algae and invertebrates), which are essential to soil fertility, and which are known to be susceptible to TBT, may be affected. In a study of a terrestrial microcosm, it was shown that 5% of TBTO that was applied on wood blocks as a preservative was released in the upper soil layer and distributed through biota (Gile *et al.*, 1982). In addition, there is a potential for bioaccumulation, since, at least in aquatic organisms, rather high bioaccumulation factors have been determined (Laughlin *et al.*, 1988; Bryan *et al.*, 1989; Fent and Hunn, 1991, 1995). Tributyltin and triphenyltin are degraded slowly and were shown to be adsorbed to soil. Studies of the rates of degradation in soil showed that 50% of both TBT and triphenyltin disappeared from silt and sand in ~20 weeks (Barug and Vonk, 1980). The degradation products of TBT appeared to be DBT derivatives and carbon dioxide.

28.5. CONCLUSIONS

The data presented show that untreated wastewater and anaerobically stabilized sludge are significant sources of organotins entering the aquatic environment via discharge or via dumping. Thus far, little attention has been paid to these sources or to leachates from landfills. There is little information on the total contribution of organotins from these sources to the organotin contamination within the environment. It is therefore necessary to quantify inputs, and to understand the fate and behavior of organotins in different environmental compartments, in order to assess the impact of the growing use of organotins, especially concerning TBT.

The organotins in untreated wastewater, sludge, and landfill leachates are primarily associated with suspended solids. In the treatment plant, organotins are conserved and transferred from wastewater to sewage sludge. Considerable organotin residues occur in digested sewage sludge, because sludge treatment was not very effective in removing the organotin load. Attention should therefore be given to the ecotoxicological implication of organotin-polluted sewage sludge, because this gives rise to additional transfer paths into the aquatic and terrestrial systems.

Regulators have generally considered that the major source of TBT to the marine and freshwater environment is that leached from antifouling paints employed on vessels. In many countries this source has been regulated and controlled. As the measurements in wastewater presented here show, there are additional significant input sources from other organotin applications. Further research and monitoring should show whether organotin loadings from these sources must be reduced by additional regulatory actions.

ACKNOWLEDGMENTS

I thank H. Siegrist, Swiss Federal Institute for Environmental Science and Technology (EAWAG), for providing the wastewater and sludge samples, J. Zeyer (ETH) for the landfill leachate samples, and M.D. Müller, Swiss Federal Research Station for providing the pentyltin standard.

REFERENCES

Aldridge, W.N. and B.W. Street. 1964. Oxidative phosphorylation: biochemical effects and properties of trialkyltins. *Biochem. J.*, **91**, 287–297.

Alzieu, C. 1986. TBT detrimental effects on oyster culture in France – evolution since antifouling paint regulation. In: Proceedings of the Oceans '86 Organotin Symposium, Vol. 4. Marine Technology Society, Washington, DC, pp. 1130–1134.

Alzieu, C. 1991. Environmental problems caused by TBT in France: assessment regulations, prospects. *Mar. Environ. Res.*, **32**, 7–17.

Alzieu, C., J. Sanjuan, P. Michel, M. Borel and J. P. Dreno. 1989. Monitoring and assessment of butyltins in Atlantic coastal waters. *Mar. Pollut. Bull.*, **20**, 22–26.

Alzieu, C., P. Michel, I. Tolosa, E. Bacci, L.D. Mee and J.W. Readman. 1991. Organotin compounds in the Mediterranean: a continuing cause for concern. *Mar. Environ. Res.*, **32**, 261–270.

Argaman, Y., C.E. Hucks and S.E. Shelby. 1984. The effects of organotin on the activated sludge process. *Water Res.*, **18**, 535–542.

Bacci, E. and C. Gaggi. 1989. Organotin compounds in harbour and marina waters from the Northern Tyrrhenian Sea. *Mar. Pollut. Bull*, **20**, 290–292.

Barug, D. and J.W. Vonk. 1980. Studies on the degradation of bis(tributyltin) oxide in soil. *Pestic. Sci.*, **11**, 77–82.

Beaumont, A.R. and M.D. Budd. 1984. High mortality of the larvae of the common mussel at low concentrations of tributyltin. *Mar. Pollut. Bull.*, **15**, 402–405.

Beaumont, A.R. and P.B. Newman. 1986. Low levels of tributyltin reduce growth of marine micro-algae. *Mar. Pollut. Bull.*, **17**, 457–461.

Becker-van Slooten, K., L. Merlini, A.M. Stegmueller, L.F. de Alencastro and J. Tarradellas. 1994. Contamination des boues des stations d'épuration suisses par les organoétains. *Gas-Wasser-Abwasser*, **74**, 104–110.

Blunden, S.J. 1983. The ultraviolet degradation of the methyltin chlorides in carbon tetrachloride and water. *J. Organometal. Chem.*, **248**, 149–160.

Blunden, S.J. and A. Chapman. 1986. Organotin compounds in the environment. In: *Organometallic Compounds in the Environment. Principles and Reactions*, P.J. Craig (Ed.). Longman Group, Harlow, pp. 111–159.

Boettner, E.A., G.L. Ball, Z. Hollingsworth and R. Aquino. 1981. Organic and Organotin Compounds Leached from PVC and CPVC Pipe. Environmental Protection Agency (EPA) Report EPA-600-/1-81-062. EPA Health Effects Research Laboratory, Cincinnati, Ohio.

Bollen, W.B. and C.M. Tu. 1972. Effect of an organotin on microbial activities in soil. *Tin, Its Uses*, **94**, 13–15.

Bryan, G.W., P.E. Gibbs, L.G. Hummerstone and G.R. Burt. 1986. The decline of the gastropod *Nucella lapillus* around south-west England: evidence for the effect of tributyltin from antifouling paints. *J. Mar. Biol. Assoc. UK*, **66**, 611–40.

Bryan, G.W., P.E. Gibbs, L.G. Hummerstone and G.R. Burt. 1989. Uptake and transformation of ^{14}C labelled tributyltin chloride by the dog-whelk, *Nucella lapillus*: importance of absorption from the diet. *Mar. Environ. Res.*, **28**, 241–245.

Bushong, S.J., L.W. Hall Jr, W.S. Hall, W.E. Johnson and R.L. Herman. 1988. Acute toxicity of tributyltin to selected Chesapeake Bay fish and invertebrates. *Water Res.*, **22**, 1027–1032.

Champ, M.A. and D.F. Bleil. 1988. Research Needs Concerning Organotin Compounds Used in Antifouling Paints in Coastal Environments. Science Applications International Corporation. US Department of Commerce, National Oceanic and Atmospheric Administration. Rockville, Maryland (unpaginated).

Chau, Y.K., S. Zhang and R.J. Maguire. 1992. Occurrence of butyltin species in sewage and sludge in Canada. *Sci. Total Environ.*, **121**, 271–281.

Dean, R.B. and M.J. Suess. 1985. The risk to health of chemicals in sewage sludge applied to land. *Waste Manage. Res.*, **3**, 251–278.

Donard, O.F.X., P. Quevauviller and A. Bruchet. 1993. Tin and organotin speciation during wastewater and sludge treatment processes. *Water Res.*, **27**, 1085–1089.

Evans, C.J. and S. Karpel. 1985. Organotin compounds in modern technology. *J. Organometal. Chem.*, **16**, 1–279.

Fent, K. 1989. Organotin speciation in municipal wastewater and sewage sludge: ecotoxicological consequences. *Mar. Environ. Res.*, **28**, 477–483.

Fent, K. 1992. Embryotoxic effects of tributyltin on the minnow *Phoxinus phoxinus*. *Environ. Pollut.*, **76**, 187–194.

Fent, K. 1995. Organotin compounds in municipal wastewater and sewage sludge: contamination, fate in treatment process and ecotoxicological consequences. *Sci. Total Environ.*, in press.

Fent, K. and J. Hunn. 1991. Phenyltins in water, sediment and biota of freshwater marinas. *Environ. Sci. Technol.*, **25**, 956–963.

Fent, K. and J. Hunn. 1995. Organotins in freshwater harbors and rivers: temporal distribution, annual trend and fate. *Environ. Toxicol. Chem.*, **14**, 1123–1132.

Fent, K. and W. Meier. 1992. Tributyltin-induced effects on early life stages of minnows *Phoxinus phoxinus*. *Arch. Environ. Contamin. Toxicol.*, **22**, 428–438.

Fent, K. and M.D. Müller. 1991. Occurrence of organotins in municipal wastewater and sewage sludge and behavior in a treatment plant. *Environ. Sci. Technol.*, **25**, 489–493.

Fent, K., R. Fassbind and H. Siegrist. 1989. Organotins in a municipal wastewater treatment plant. In: Proceedings of the 1st European Conference on Ecotoxicology, 17–19 October 1988, H. Løkke, H. Tyle and F. Bro-Rasmussen (Eds). The Technical University of Denmark, Lyngby, Denmark, pp. 72–80.

Fent, K., J. Hunn, D. Renggli and H.S. Siegrist. 1991. Fate of tributyltin in sewage sludge treatment. *Mar. Environ. Res.*, **32**, 223–231.

Giger, W., A.C. Alder, P.H. Brunner, A. Marcomini and H. Siegrist. 1989. Behavior of LAS in sewage and sludge treatment and in sludge-treated soil. *Tenside Detergents*, **26**, 95–100.

Gile, G.D., J.C. Collins and J.W. Gillet. 1982. Fate and impact of wood preservatives in a terrestrial microcosm. *J. Agr. Food Chem.*, **30**, 295–301.

Guard, H.E., A.B. Cobet and W.M. Coleman, III. 1981. *Science*, **213**, 770–771.

Hall, L. W., Jr and E.A. Pinkney. 1985. Acute and sublethal effects of organotin compounds on aquatic biota: an interpretative literature evaluation. *CRC Crit. Rev. Toxicol.*, **14**, 159–209.

Koch, J. and K. Figge. 1971. Migration of tin-stabilizers from PVC-bottles into beer. *Z. Lebensm. Unters. For.*, **147**, 8–10.

Komatsu, Y., T. Namiki and F. Mori. 1970. Extractability of dialkyltin compounds from rigid poly (vinyl chloride) containers. *Shokuhin Eiseigaku Zasshi*, **11**, 17–22.

Laughlin, R.B., Jr and W. French. 1988. Concentration dependence of bis(tributyl)tin oxide accumulation in the mussel, *Mytilus edulis*. *Environ. Toxicol. Chem.*, **7**, 1021–1026.

Laughlin, R.B., Jr and O. Linden. 1985. Fate and effects of organotin compounds. *Ambio*, **14**, 88–95.

Laughlin, R.B., Jr, H.E. Guard and W.M. Coleman, III. 1986. Tributyltin in seawater speciation and octanol–water partition coefficient. *Environ. Sci. Technol.*, **20**, 201–204.

Lawler, I.F. and J.C. Aldrich. 1987. Sublethal effects of bis(tri-n-butyltin) oxide on *Crassostrea gigas* spat. *Mar. Pollut. Bull.*, **18**, 274–278.

Lawson, G. 1986. Organometallic compounds in polymers – their interactions with the environment. In: *Organometallic Compounds in the Environment. Principles and Reactions*, P.J. Craig (Ed.). Longman, Harlow, pp. 308–344.

Maguire, R.J. 1987. Review: environmental aspects of tributyltin. *Appl. Organometal. Chem.*, **1**, 475–498.

Maguire, R.J. and R.J. Tkacz. 1985. Degradation of the tri-n-butyltin species in water and sediment from Toronto Harbor. *J. Agr. Food Chem.*, **33**, 947–953.

Marcomini, A., P.D. Capel, W. Giger and H. Häni. 1988. Residues of detergent-derived organic pollutants and polychlorinated biphenyls in sludge-amended soil. *Naturwissenschaften*, **75**, 460–462.

Meinema, H.A., T.G. Dam-Meerbeek and J.V. Vonk. 1982. *Evaluation of the Impact of Organotin Compounds on the Aquatic Environment*. Institute of Applied Chemistry TNO, Utrecht, The Netherlands, 126 pp.

Müller, M.D. 1987. Comprehensive trace level determination of organotin compounds in environmental samples using high-resolution gas chromatography with flame photometric detection. *Anal. Chem.*, **59**, 617–623.

Randall, L. and J.H. Weber. 1986. Adsorptive behavior of butyltin compounds under simulated estuarine conditions. *Sci. Total Environ.*, **57**, 191–203.

Ritsema, R., R.W.P.M. Laane and O.F.X. Donard. 1991. Butyltins in marine waters of the Netherlands in 1988 and 1989: concentrations and effects. *Mar. Environ. Res.*, **32**, 243–260.

Schebek, L., M.O. Andreae and H.J. Tobschall. 1991. Methyl- and butyltin compounds in water and sediments of the Rhine river. *Environ. Sci. Technol.*, **25**, 871–878.

Seligman, P.F., A.O. Valkirs and R.F. Lee. 1986. Degradation of tributyltin in San Diego Bay, California, waters. *Environ. Sci. Technol.*, **20**, 1229–1235.

Seligman, P.F., J.G. Grovhoug, A.O. Valkirs, P.M. Stang, R. Fransham, M.O. Stallard, B. Davidson and R.F. Lee. 1989. Distribution and fate of tributyltin in the United States marine environment. *Appl. Organometal. Chem.*, **3**, 31–47.

Snoeij, N.J., A.H. Penninks and W. Seinen. 1988. Dibutyltin and tributyltin compounds induce thymus atrophy in rats due to a selective action on thymic lymphoblasts. *Int. J. Immunopharmacol.*, **10**, 891–899.

Siegrist, H., A.C. Adler, P.H. Brunner and N.W. Giger. 1989. Pathway analysis of selected organic chemicals from sewage to agricultural soil. In: *Sludge Treatment and Use: New Developments, Technological Aspects and Environmental Effects*, A.H. Dirkzwager and P.L. L'Hermite (Eds). Elsevier Applied Science, London, pp. 133–144.

Thain, J.E. 1986. Toxicity of TBT to bivalves: effects of reproduction, growth and survival. In: Proceedings of the Oceans '86 Organotin Symposium, Vol. 4, Marine Technology Society, Washington, DC, pp. 1306–1313.

Thompson, J.A.J., M.G. Sheffer, R.C. Pierce, Y.K. Chau, J.J. Cooney, W.R. Cullen and R.J. Maguire. 1985. Organotin Compounds in the Aquatic Environment: Scientific Criteria for Assessing Their Effects on Environmental Quality. National Research Council of Canada, NRCC Associate Committee on Scientific Criteria for Environmental Quality, Ottawa, 284 pp.

Unger, M.A., W.G. MacIntyre and R.J. Huggett. 1988. Sorption behavior of tributyltin on estuarine and freshwater sediments. *Environ. Toxicol. Chem.*, **7**, 907–915.

U'ren, S.C. 1983. Acute toxicity of bis(tributyltin) oxide to a marine copepod. *Mar. Pollut. Bull.*, **14**, 303–306.

Valkirs, A.O., P.F. Seligman and R.F. Lee. 1986. Butyltin partitioning in marine waters and sediments. In: Proceedings of the Oceans '86 Organotin Symposium, Vol. 4, Marine Technology Society, Washington, DC, pp. 1165–1170.

Valkirs, A.O., P.F. Seligman, P.M. Stang, V. Homer, S.H. Lieberman, G. Vafa and C.A. Dooley. 1986. Measurement of butyltin compounds in San Diego Bay. *Mar. Pollut. Bull.*, **17**, 319–324.

Walsh, G.E., L.L. McLaughlan, E.M. Lores, M.K. Louie and C.H. Deans. 1985. Effects of organotins on growth and survival of two marine diatoms, *Skeletonema costatum* and *Thalassiosira pseudonata*. *Chemosphere*, **14**, 383–392.

Wester, P.W. and J.H. Canton. 1987. Histopathological study of *Poecelia reticulata* (guppy) after long-term exposure to bis(tri-n-butyltin)oxide (TBTO) and di-n-butyltindichloride (DBTC). *Aquat. Toxicol.*, **10**, 143–165.

Whitelaw, K. and M.J. Andrews. 1988. The effects of sewage sludge disposal to sea – the outer Thames Estuary, UK. *Water Sci. Technol.*, **20**, 183–191.

World Health Organization. 1980. *Tin and Organotin Compounds – A Preliminary Review*. Environmental Health Criteria 15. Geneva. 109 pp.

World Health Organization. 1990. International programme on chemical safety. Environmental Health Criteria, 116. In: *Tributyltin Compounds*. Geneva, Switzerland, 273 pp.

Wu, W., R.S. Roberts, Y.C. Chung, W.R. Ernst and S.C. Havlicek. 1989. The extraction of organotin compounds from polyvinyl chloride pipe. *Arch. Environ. Contam. Toxicol.*, **18**, 839–843.

Zuckerman, J.J., R.P. Reisdorf, H.V. Ellis and R.R. Wilkinson. 1978. In: *Organometals and Organometalloids: Occurrence and Fate in the Environment*, F.E. Brinckman and J.M. Bellama (Eds). American Chemical Society, Washington, DC, pp. 388–424.

RESEARCH INFORMATION REQUIREMENTS ASSOCIATED WITH THE ENVIRONMENTAL FATE AND EFFECTS OF ORGANOTIN COMPOUNDS

29

Michael A. Champ[1]* *and Peter F. Seligman*[2]

[1]*Texas Engineering Experiment Station, Washington, DC Office, Texas A&M University System, 4601 North Fairfax Drive, Suite 1130, Arlington, Virginia 22203, USA*

[2]*Naval Command, Control, and Surveillance Center, RDT&E Division, Environmental Sciences, US Navy, San Diego, California 92152-6335, USA*

ABSTRACT

This chapter identifies, delineates, and prioritizes research information requirements relative to the environmental ramifications associated with organotin compounds. The lack of appropriate priorities for strategic planning and managing research has been a

*Company address: Environmental Systems Development Co., PO Box 2439, 7000 Vagabond Drive, Falls Church, Virginia 22042-3934, USA

Organotin. Edited by M.A. Champ and P.F. Seligman. Published in 1996 by Chapman & Hall, London. ISBN 0 412 58240 6

serious problem for the evaluation of highly toxic organotin compounds. Degrees of scientific uncertainty have always reduced or eroded the confidence of policy and decision makers. This uncertainty points to the necessity for improved and standardized risk characterization protocols, which need to be developed to improve ecological risk assessment and management. The focus of future requirements for long-term research information for organotin compounds should be on (1) quantification of the sources and loading levels of TBT in the environment; (2) delineation of the dispersion rates, processes, and transport mechanism; (3) characterization of the exposure pathways to selective target organisms of high risk; (4) assessment of the factors influencing bioavailability and biological uptake mechanisms; (5) identification and characterization of degradation rates and metabolic pathways; (6) quantification of relationships between laboratory studies of cause and effects, and effects found in the field; (7) development of rapid and inexpensive low-level advanced analytical protocols; (8) numerical modeling and prediction of ecosystem concentrations; (9) development of standard reference materials; and (10) risk assessment protocols.

29.1. INTRODUCTION

A number of meetings, workshops, and symposia have been held over the past 15 y to bring together researchers and policy and decision makers so that they could identify, review, and discuss information and research required for understanding and regulating the use of organotin compounds in the environment [Proceedings of the Organotin Symposia of the Oceans '86, '87, '88, '89 Conferences (1986–1989); Proceedings of the Third International Organotin Symposium (Mee and Flower, 1991), and references therein]. The UK Marine Pollution Library at Plymouth has identified >700 papers, reports, or articles related to some aspect of organotin compounds in the environment [United Nations Environment Programme (UNEP), 1989]. Most of these scientific papers, however, do not provide resource managers, policy and decision makers, or research program managers with the kind of information that they need for regulatory actions. The research information is required to support the assessment of human health and environmental risks, to support regulatory activities, and to form the basis for the strategic planning necessary for research and development (Huggett *et al.*, 1992; Abel, Chapter 2).

Without the information they need to set appropriate priorities, research and development (R&D) program managers have often been sidetracked by real or perceived public pressures. Often requests for funding focus on the special interests and capabilities of principal investigators rather than on the development of responsive research projects. When provided with adequate information, program managers would be able to set appropriate priorities that would funnel research funds to projects in critical research areas.

The lack of appropriate priorities for managing research is a serious problem in the evaluation of highly toxic compounds (Champ, 1986; Champ and Lowenstein, 1987). Policy and decision makers cannot make the best use of valid scientific information in the regulatory process as they assess risks to society and the environment and weigh this information against the economic benefits to society from the use of these new high-technology chemicals. The economic benefits of tributyltin (TBT) antifoulants to the economy of the maritime industry over the past few decades have been considerable. In 1990 the International Maritime Organization estimated savings worldwide of ~2.5×10^9 US dollars annually by the use of tributyltin compounds, principally through reductions in fuel consumption of some 7.2×10^6 metric tonnes per year (Milne, 1990).

Degrees of scientific uncertainty have always reduced or eroded the confidence of policy and decision makers. In the absence of irrefutable scientific data, they have added large safety factors in making regulatory or policy decisions. This uncertainty points to the necessity for improved and standardized risk characterization protocols that need to be developed to improve ecological risk assessment and management.

In providing information for policy and regulatory decision making, two types or classes (as related to uses) of research data can be identified.

1. Short-term data are collected to support immediate regulatory or policy- and decision-making actions (i.e. acute or chronic toxicity and background levels are relative to making a decision to restrict or ban the use of a compound).
2. Long-term data are collected to provide a basic scientific understanding of the biological or environmental processes being regulated (i.e. uptake or transport processes, fate, and behavior of a toxic compound in the environment).

Studies used to collect short-term data, such as laboratory toxicity studies, chemical characterization analyses, baseline and monitoring efforts, and field assessment studies, are usually standardized for conformity and comparability of results. Long-term data collection projects are designed to test an hypothesis and usually develop causality or mode-of-action relationships. In this volume, Chapters 2 (Abel) and 3 (Champ and Wade) on policy and regulatory strategies delineate and discuss the short-term data collection requirements of European and US regulatory agencies. The US Environmental Protection Agency (EPA) utilizes as a process its Data Call-in Notices (DCI) to solicit specific information from paint manufacturers for the registration of antifouling paints. For example, the EPA in its TBT DCIs requires a standard suite of basic information, such as standardized dose–response measurements, release rates, toxicity tests, and monitoring data from regions of potential high use.

In the search to measure the effects of organotin compounds, the analytical chemistry of organotin compounds has evolved to levels of detection, sensitivity, and precision at the nanogram-per-liter level (parts-per-trillion level), greatly exceeding the level of detection for most other compounds (see Weber *et al.*, Chapter 4; Matthias, Chapter 5; Unger *et al.*, Chapter 6; Huggett *et al.*, Chapter 7; Siu and Berman, Chapter 8). Organotin compounds are the first group of synthetic high-technology chemicals to be developed and widely used as biocides in the environment that have toxic levels of expression at the parts-per-trillion (ng l^{-1}) level for selected organisms. To appreciate the effective levels of toxicity and environmental concentrations, the measurement of one nanogram per liter of TBT is comparable to measuring 1 second in 31 000 years.

This chapter focuses on the long-term data collection and research information requirements for organotin compounds in coastal waters as a synthesis from several sources: review of the published and gray literature and discussions with researchers, research program managers, and policy and decision makers. In particular, this chapter identifies information requirements for a better understanding of the processes and mechanisms of cause-and-effect relationships and the fate and behavior of organotin compounds in the marine environment.

29.2. RESEARCH INFORMATION REQUIREMENTS

To predict accurately the in-situ effects of the exposure of marine organisms to organotin compounds, particularly tributyltin (TBT) in coastal waters, it is necessary to understand the distribution, transport, fate, and behavior (processes, mechanisms, and rates) of these compounds in the environment and in biota.

It will also be important to understand the factors that influence the partitioning between the biotic and abiotic environments and the biological and chemical processes that degrade TBT to its metabolites (see Hinga *et al.*, 1987).

In 1986, an Interagency Workshop on Aquatic Monitoring and Analysis for Organotin Compounds was held at the US Naval Academy (Landy *et al.*, 1986). The workshop brought together leaders in the detection, quantification, and monitoring of organotin compounds in the aquatic environment. The workshop was co-sponsored by the National Marine Pollution Program Office of the National Oceanic and Atmospheric Administration's (NOAA) National Ocean Service, the US EPA, and the US Navy. The workshop resulted in a publication on the state of the art in analytical methodologies and field sampling strategies (Young *et al.*, 1986). The report also summarized research and monitoring needs as identified by the workshop participants (Young and Champ, 1986). This meeting was followed by the first International Organotin Symposium in Washington, DC, in 1986.

Between the second International Organotin Symposium in Halifax, Nova Scotia, and the third one in Monaco, a consensus developed among researchers for the need to redirect research efforts away from delineating the degree of 'what' the effect of exposure to TBT was (finding deformed oysters) to prediction of the concentrations in given water bodies at which effects would occur. By the Halifax meeting (held in 1987) most participants felt that the hazard of exposure to organotin compounds had been fully demonstrated, particularly in boat harbors and marinas.

The assessment and understanding of biological effects developed very rapidly (see Hall and Bushong, Chapter 9; Laughlin *et al.*, Chapter 10; Waldock *et al.*, Chapter 11; His, Chapter 12; Gibbs and Bryan, Chapter 13; Henderson and Salazar, Chapter 14; Salazar and Salazar, Chapter 15; and Laughlin, Chapter 16).

The participants also recognized that the areas of focus for future studies should be (1) quantification of the sources and loading levels of TBT in the environment; (2) delineation of dispersion and transport mechanisms and pathways; and (3) prediction of the fate, behavior, and subsequent toxicity of organotin compounds that have been introduced into the environment (Champ and Pugh, 1987; Champ and Bleil, 1988).

This chapter has been prepared to identify, delineate, and prioritize research information requirements relative to the environmental ramifications associated with organotin compounds.

29.2.1. SOURCES

The input of organotin compounds from multiple sources to the water column and biota may continue for a long time after the enactment of regulatory actions if bottom sediments serve as temporary reservoirs and release continues from vessels >25 m in length (see Langston and Burt, 1991; Dowson *et al.*, 1993; Langston *et al.*, 1994; Errecalde *et al.*, 1995; Langston and Pope, 1995). To evaluate the probable consequences of a proposed regulatory action, the scientific community needs to be able to understand fully the fate and behavior of organotin compounds following their introduction in the environment. Some questions that must be answered involve where and how much organotin partitions, accumulates, and biodegrades; and how fast and to what organotin compounds are degraded. At least two degradation pathways have been identified, each with a different sequence of intermediate chemical species; how many other degradation pathways are there for organotin compounds within coastal waters?

Organotin compounds enter the aquatic and marine environment and become available to organisms in a variety of ways

(Maguire, 1991; Laughlin, Chapter 16; Roberts *et al.*, Chapter 17; Lee, Chapter 18; Schatzberg, Chapter 19; Seligman *et al.*, Chapters 20 and 21; Harris *et al.*, Chapter 22; Unger *et al.*, Chapter 23). Some of these pathways are not well understood or sufficiently quantified (Cleary *et al.*, 1993; Cooney, 1995; Salazar and Salazar, Chapter 15). There is very little information, for example, on the contribution of nonpoint land and aquatic sources, atmospheric inputs, wastewater treatment discharges, and the quantities that become available to organisms from temporary reservoirs (i.e. resuspension and resolubilization from sediments; see Roberts, Chapter 17). Past studies have clearly shown that the major source of organotin compounds to the marine environment has been antifouling paint leachate from vessels and small pleasure craft (see Huggett *et al.*, Chapter 24; Grovhoug *et al.*, Chapter 25). For this reason, current and planned regulatory strategies have targeted this source.

Most legislation or regulatory strategies that have been proposed or enacted do not totally eliminate the use of TBT-based antifouling paints (boats >25 m in length can use it); therefore, it is quite possible for organotin compounds to continue to enter aquatic and marine environments. Because of the serious concern about the toxicity of TBT to certain sensitive marine organisms, regional databases are needed to quantify total loadings and to identify geographical areas of concern as well as to provide direction for management decisions.

Currently, in the United States, ambient concentrations of organotin compounds are only being monitored in a few harbors where there are US Navy test ships with organotin coatings and at selected estuarine sampling stations of the NOAA National Status and Trends Program (NS&T) and the EPA/EMAP (Environmental Monitoring and Assessment Program). Additional monitoring for TBT in the United States is planned under the US EPA's Data Call-in procedure. Data collected after the first regulatory action have begun to demonstrate a decline in background levels (Hall *et al.*, 1987; Maguire, 1991, and Chapter 26; Champ and Wade, Chapter 3; Waite *et al.*, Chapter 27). The measurement of at least TBT should be included for selected stations in routine coastal water quality monitoring programs. Only when such short-term data are provided can resource managers evaluate the effectiveness of regulatory programs.

Another area that is not well quantified is the contribution of organotin compounds to receiving waters due to episodic events from nonpoint sources such as heavy rainfall and flooding and point sources such as discharges from waste treatment facilities (Fent, Chapter 28). All of these point and nonpoint sources may deliver large quantities of organotin compounds over very short periods of time. Frequently these episodic events occur in the spring months and coincide with the painting of the bottoms of pleasure craft. Unfortunately, the events can coincide with spawning events in rivers and estuaries. The result is an intermittent exposure of sensitive life history stages (i.e. eggs and larvae) to toxic levels of TBT. Studies are needed to provide information on

1. the quantities of atmospheric (if any) and other nonpoint sources of organotin compounds entering the marine environment;
2. the quantities of nonpoint sources of organotin compounds delivered to receiving water during episodic or climatic events;
3. the effect of seasonal variations in temperature and salinity on leaching rates of organotin compounds in antifouling paints;
4. the rates of bioaccumulation and degradation of organotin compounds and their degradation products in saltwater and aerobic or anaerobic sediments from a seasonal perspective;
5. refinement of techniques for measuring release rates to estimate realistic loading factors.

29.2.2. TRANSPORT

To be able to predict downstream concentrations of organotin compounds in receiving waters, it is necessary to understand the factors that influence the partitioning of these compounds, which occur under a wide range of environmental conditions (Roberts *et al.*, Chapter 17). Many factors influence the transport of organotin compounds that are in solution or attached to suspended particulate material or associated with the surface microlayer (Unger *et al.*, Chapter 23).

Within the bottom sediments, organotin species can be partitioned among the nephloid layer, the interstitial pore water, and the subsurface layers. Of major concern is the behavior of organotin species under different physical and environmental conditions and the subsequent variance of partitioning coefficients (Harris *et al.*, Chapter 22). Studies are needed to provide information on

1. the fate of organotin compounds that are introduced into the environment from antifouling paints, which partition in solution in the water column, in the surface microlayer, onto suspended particulate material, and in bottom sediment;
2. the extent to which organotin compounds are distributed with movement of bottom and suspended sediments, either by dredge spoil disposal or natural resuspension of sediments;
3. the role and significance of biological transport as a distribution process, such as perhaps either a tissue burden or absorption on external surfaces of planktonic organisms, with subsequent incorporation in food webs.

29.2.3. EXPOSURE PATHWAYS

It is probable that the critical exposure pathway (to selective target organisms) is the most critical factor in influencing the toxicity of extremely low concentrations of organotin compounds to sensitive marine organisms, as evidenced by water-column concentrations of only a few nanograms per liter, with the surface microlayer or suspended sediment concentrations being two or three orders of magnitude higher. Therefore the actual dose level to organisms in the intertidal zone is not directly related to the water column concentrations but perhaps rather to microlayer and sediment concentrations to which the organism is exposed in a critical pathway. Studies are needed to provide more information on

1. quantification of specific critical exposure pathways such as the surface microlayer, suspended particulate materials, bottom sediments, pore waters, and food webs that affect specific organisms of concern;
2. bacterial products of metabolic debutylation of organotin compounds, which include methyl-dibutyltin and methyl-monobutyltin, and the subsequent effects of exposure to methylated butyltin vs butyltin in the water and the sediment.

29.2.4. BIOAVAILABILITY

Because of the high sensitivity of molluscs to organotin compounds, much of the current research has focused on filter-feeding molluscs and carnivorous gastropods (Spooner *et al.*, 1991; Ten Hallers-Tjabbes, 1994; Ten Hallers-Tjabbes *et al.*, 1994; Guolan and Yong, 1995; Ten Hallers-Tjabbes and Boon, 1995; Waldock *et al.*, Chapter 11; His, Chapter 12; Gibbs and Bryan, Chapter 13) Deposit feeders may also be a significant pathway for the transfer of butyltin from sediment into the food chain (Langston and Burt, 1991), because deposit-feeding molluscs and polychaete and oligochaete worms are important components in the diet of many organisms. These worms feed on benthic sediment and organic materials that have the strongest affinity for butyltins. An additional component in the consideration of bioavailability is the role played by bacteria in sequestering, concentrating, and metabolizing organotin compounds. Bacteria are a food source for deposit-feeding organisms. Studies are needed to provide more information on

1. the roles of bacteria and other micro-organisms in influencing the availability of organotin compounds to marine organisms;
2. the availability of organotin compounds in the surface microlayer of the water column to biota both in the water column and in tidal or wave-flushed areas;
3. the availability of organotin compounds that are adsorbed on suspended or bottom sediments to the biological community;
4. the availability of organotin compounds in planktonic cells or absorbed on surface membranes of consumed organisms;
5. the impact and risk of exposure of coastal birds feeding on molluscs or other organisms exposed to organotin compounds, and subsequent role of seasonal migration as a pathway for transport;
6. the relative bioavailability of butyltin compounds and metabolites in sediments from various sources such as spilled liquid paint, paint sanding dust, and paint chips and scrapings compared with those leaching directly from the immersed cured paint film.

29.2.5. BIOACCUMULATION AND BIOMAGNIFICATION

Studies are needed to assess organotin levels in birds such as oyster catchers or gulls, and whales (Suzuki *et al.*, 1992; Iwata *et al.*, 1994) which are known to feed on invertebrates that have been found to accumulate high levels of organotin compounds, and the subsequent degradation products, in their tissues. Some data have been collected in the United Kingdom that suggest that this is not a problem (R. Abel, UK Department of the Environment, personal communication); however, studies are needed to provide more information on

1. the bioconcentration rates and feeding pathways from the lower trophic level organisms through sea birds and mammals;
2. the key organisms in the bioaccumulation of organotin compounds from water and sediments;
3. the extent of biomagnification of organotin compounds in estuarine and coastal food chains.

29.2.6. DEGRADATION AND METABOLISM

Although work in this area has begun to define degradation rates and metabolic pathways (see Brinckman *et al.*, 1987; Lee, 1991, and Chapter 18; Fukagawa and Suzuki, 1993; Uchida, 1994; Seligman *et al.*, Chapter 21), identification of organisms and mechanisms responsible for degradation of organotin compounds (e.g. production of methylated butyltins vs dibutyl- and monobutyltins) is critical. Studies should also focus on the relative importance of phytoplankton and bacteria in the degradation process. Studies are needed to provide more information on

1. the effects of salinity or temperature changes on the biological and chemical degradation rates of organotin compounds in marine or estuarine waters;
2. the relative importance of ultraviolet light or phytoplankton in the degradation of organotin compounds in the surface microlayer and surface water;
3. the importance of photosynthetic organisms in degrading organotin compounds in the water column and on tidal sediments, and on the environmental conditions under which these organisms function best;
4. the role of sediments in both biological and non-biological degradation.

29.2.7. EFFECTS – TOXICITY STUDIES

The correlation of laboratory toxicity studies to effects found in the field will allow for better dose–response estimates and better definition of cause-and-effect relationships. TBT is toxic to certain organisms, particularly molluscs, in trace quantities because of the characteristics of the physiology of these organisms and critical pathways of exposure as previously discussed in section 29.2.3, Exposure pathways). The relative lack of a monooxygenase (MFO) system in the bivalve molluscs does not permit them to metabolize organotin compounds as readily as other organisms, such as crabs and fish. The mode of action of organotin compounds is strongly membrane oriented (see Eng *et al.*, 1988; Roberts *et al.*, Chapter 17; Lee, Chapter 18).

To determine nonlethal but detrimental effects, toxicity testing in the field and laboratory will have to focus on physiological processes, in particular the reproductive processes. The case in point is the gastropod molluscs, such as the dogwhelk (*Nucella lapillus*). Dogwhelk populations are in jeopardy due to reproductive failure even though the ambient concentrations of TBT are well below the lethal limit for individuals (Gibbs *et al.*, 1991; Spooner *et al.*, 1991; Gibbs and Bryan, Chapter 13).

The two issues provoking most discussion regarding laboratory toxicity testing are the requirement to know the concentration of the actual butyltin species present in exposure solutions during toxicity testing and the need to understand better the interactions of butyltins with sediment and the sediment-dwelling biological community (see Fukagawa and Suzuki, 1993; Kuballa *et al.*, 1995; Roberts, Chapter 17).

A significant body of literature describing acute and chronic toxic effects, lesions, and tissue abnormalities as well as behavioral aberrations resulting from exposure to TBT in solution has been published in the last decade (see Hall and Bushong, Chapter 9; Laughlin *et al.*, Chapter 10). As the ability to determine the actual flowthrough concentrations in the nanogram-per-liter range has improved, toxicity testing should be able to define the environmentally relevant no-effects level (NOEL) for TBT for a wide range of organisms. Studies are also needed that focus on low-level physiological endpoints determined by chronic exposure testing over the full life cycle of the organism. Additional testing for the effect of exposure to TBT on reproductive success, gametogenesis, imposex occurrence, calcification mechanism, and suppression of the immune system will provide needed information on ecosystem risks. Studies are needed to provide more information on

1. the concentrations of organotin compounds that elicit specific responses that range from successfully counteracting the toxin to the complete overwhelming of homeostatic mechanisms;
2. the low-level physiological effects on molluscs of chronic exposure to organotin compounds extending over the full life cycle;
3. the effects on gametogenesis and other aspects of reproductive success in molluscs during chronic exposure to organotin compounds extending over the full life cycle of the mollusc;
4. the effects on the calcification mechanism of molluscs during chronic exposure to organotin compounds extending over the full life cycle of the mollusc;
5. the mechanisms inducing imposex associated with chronic exposure to organotin compounds extending over the full life cycle of the mollusc;
6. the mechanisms inducing suppression of the immune response system due to chronic exposure to organotin compounds over the full life cycle of selected sensitive organisms (in particular, molluscs);
7. the true no-effects level (NOEL) of organotin compounds in coastal waters with respect to molluscs, fish, and crustaceans;
8. the assimilative capacity of coastal ecosystems for TBT using micro- or mesocosms as a model system.

For the accurate assessment of TBT or any xenobiotic compound, it is necessary to determine whether laboratory tests on a few species can adequately represent the response of the coastal ecosystem to that compound.

Over the last decade considerable effort has been expended within the Chesapeake Bay to try to identify the cause or causes of the seriously depressed recruitment of the oyster *Crassostrea virginica* during this period. However, because of its insensitivity to organotin compounds, it is unlikely that organotin compounds are a factor in the reduction of *C. virginica* in the Chesapeake Bay.

29.2.8. ANALYTICAL METHODS

More attention needs to be paid to developing rapid, sensitive, and precise laboratory techniques involving measurement of organotin compounds at nanogram-per-liter concentrations. Without the ability to measure accurately and speciate low levels of organotin compounds, the research community will not be able to delineate exact cause-and-effect relationships in the laboratory or the field (Readman and Mee, 1991).

It is critical that advanced analytical methodologies for the determination of ultratrace butyl- and mixed butylmethyltin species at the nanogram-per-liter level be developed or optimized for marine and estuarine waters and in tissues and sediments. To achieve this requires the following:

1. rapid and inexpensive protocols and analytical methodologies for analysis of organotin compounds;
2. advanced analytical methodologies or protocols for determining butyl- and mixed butylmethyltin species at the nanogram levels in sediment and tissues.

29.2.9. STANDARD REFERENCE MATERIALS (SRMs)

A very high priority should be given to the development and production of certified organotin reference standards or materials for calibration and intercalibration of analytical methods for water, sediment, and tissue, as a mixed reference material for organotin compounds and metabolites. Studies are needed to provide more information on

1. standard reference materials necessary for calibration of laboratory instruments and methods to measure organotin compounds at nanogram-per-liter concentrations;
2. standard reference materials for water, sediment, and tissue necessary to monitor organotin compounds adequately in the environment.

29.2.10. NUMERICAL MODELING

The Dynamic Estuary Model, a pseudo two-dimensional link node model developed by the US EPA, was used to simulate mathematically the hydrodynamics of San Diego Bay; and, in portions of the Chesapeake Bay, to predict the concentrations of TBT that would be expected if the US Navy implemented the plan to paint the entire 600-ship fleet with organotin antifouling paint (Walton *et al.*, 1986; Seligman *et al.*, 1987).

The US EPA has also employed the Dynamic Estuarine Model in the Elizabeth River and Hampton Roads portion of Chesapeake Bay (EPA, 1987a). The impacts of recreational boating activity, commercial shipping, and military ship traffic are being simulated by assumptions regarding several loading levels caused by boating and shipping activity in the area. These estimated concentrations are expected to help EPA make its final regulatory decision regarding continued registration of TBT biocides in antifouling paint (EPA, 1987b).

In vertically well-mixed harbors, this model works reasonably well; however, in complex estuaries with strong temperature or salinity stratification, the model may not work as well. A numerical simulation model of estuarine dispersal of TBT has been developed in the United Kingdom and used in conjunction with available observations to assess the effects of reducing TBT inputs to impacted estuaries (Harris *et al.*, 1991).

Three-dimensional hydrodynamic models are needed to improve dispersion prediction capabilities. In addition, improved fate models are needed that will predict degradation, uptake, partitioning, and sediment loading. Used in conjunction with toxicity and persistence data, numerical models can be effective tools for predictive risk assessments.

29.3. ALTERNATIVES TO ORGANOTIN ANTIFOULING PAINTS

While there is not a product on the market that matches the performance of organotin compounds (TBT) as an antifoulant (Schatzberg, 1987), there are other antifouling chemicals. The primary alternative is copper, usually cuprous oxide. Research is being conducted on alternatives to organotin antifouling paints; however, useful results have not been clearly demonstrated. The three main areas being investigated are (1) alternative toxins, (2) non-toxic compounds, and (3) disruption of the successionary stages of the development of the fouling communities. Specific, or selective, biocides are being investigated and are likely to be employed as an addition to, rather than a replacement for, either organotin or copper. Neither organotin nor copper controls all potential fouling organisms. Copper works well against both algae and calcareous fouling organisms such as barnacles but not well against some bryozoans or bacteria. Organotin is toxic to a wider range of species but does not work as well against algae (Callow *et al.*, 1978).

The ideal environmentally compatible antifouling additive or material must be able to prevent the attachment of any members of the phylogenetic spectrum from bacteria to chordates (tunicates) without being toxic to any of them. The ideal antifouling material should adhere to all anthropogenic surfaces, be highly resistant to abrasion, be non-soluble in seawater, not be inactivated by contact with chemicals in aqueous solution, remain effective indefinitely, be inexpensive, and be easy to apply. In the absence of the ideal substance, research is progressing on the development of alternative toxins and non-toxic materials.

One of the most promising avenues in non-polluting fouling-resistant coatings is the low-surface-energy film. Coatings employing silicone oils have been produced commercially and show promise. Other teflon-like coatings, including perfluorinated alkyl compounds, are also being developed.

Organic molecules that do not include a metal atom are also being tested for antifouling properties. The rubber-based compounds offer abrasion resistance as well as good sound attenuation and are galvanically inert. A coating for aluminum hulls that claims both antifouling and anticorrosive properties has been developed from tar, chlorinated rubber, plasticizer, and hydrocarbon resin.

An undisturbed bacterial film soon embeds the bacterial cells in a matrix of polysaccharides secreted by the bacteria themselves. Current investigations by National Institute of Standards and Technologies (NITS) and several researchers at US universities are focusing on the roll of this polysaccharide as an attractant to further settlement by other organisms. The next organisms to settle in this bacterial film are the single-celled algae. Diatoms, such as the chain-forming diatom *Skelotonema costatum*, live on and in the polysaccharide matrix. These algae have been shown to metabolize and degrade TBT into less toxic chemical species such as dibutyl- and monobutyltin.

This active component of the biofilm on the painted surface may be essential to the subsequent settlement of bryozoans (Stebbing, 1985; William Banta, The American University, personal communication) and barnacles by reducing the concentration of the antifouling agent at the surface of the biofilm. One source of the difference in release rates measured by the US Navy's in-situ measurements of release rate and the EPA – American Society for Testing and Materials release rate tests in the laboratory is the presence of the bacterial–algal biofilm on the TBT-treated Navy ships and its absence in the laboratory tests.

The next stage of succession is the attraction and attachment of the larval stages of multicellular plants (macrophytes) and animals producing hard or calcareous exteriors. This stage exhibits a marked increase in the surface roughness and drag on the painted hull. If the surface could be kept free of the bacterial–algal film, the subsequent settlement might be prevented by failure of the surface to attract the free-swimming larvae of the next biofouling stages.

29.4. RECOMMENDATIONS FOR FUTURE RESEARCH

In order to develop sound policy and regulatory actions, policy and decision makers need clear statements of facts and problems. The scientific community cannot at the present time predict what the consequences to the environment would be for any given regulatory strategy. The experience of regulatory actions in the United Kingdom is illustrative. The initial attempt was to reduce organotin loadings in the environment by a combination of public education and a limitation of the tin content of paints on a weight/percent basis. The result was a dramatic 50% reduction in the concentrations of organotin compounds in the water column; however, that reduction was insufficient to bring the TBT concentrations in impacted estuaries below the levels known to be harmful to the most sensitive species (Abel, Chapter 2; Waite *et al.*, Chapter 27).

A second, more stringent round of legislation was therefore required to protect the environment. The use of TBT in any paint applied to boats <25 m in length was prohibited (with the exception of aluminum-hulled vessels). The second restriction on TBT paint use appears to have further reduced organotin concentrations (Waite *et al.*, 1991, and Chapter 1991; Champ and Wade, Chapter 3), but final results related to subsequent biological effects have not yet been determined.

At present, in the United States many of the coastal states have adopted regulations that should reduce inputs of organotin compounds to the marine environment; however, not a single coastal state (except Michigan) has passed legislation to ban the use of TBT as paint additives in antifouling paints. Owing to the serious public concern about the problem, organotin contamination in estuaries will continue to be a management and research issue, even after legislation is passed and strategies have been implemented to reduce significantly or eliminate certain organotin sources. Various aspects relating to the sources, fates, and effects of organotin compounds in the marine environment are still not well understood or well documented. Therefore, research and monitoring activities to address these information gaps are necessary and should continue in the future.

A summary of the ranked highest priority research areas is given below.

1. Identify and quantify specific and critical exposure pathways and mechanisms of toxicity of organotin compounds to organisms at risk, in particular the role of the surface microlayer, suspended and bottom sediments, pore water, and food chains.
2. Identify and quantify the processes, rates, and mechanisms by which organisms concentrate and degrade organotin compounds.
3. Identify and quantify the key environmental and biological factors and mechanisms that control the degradation of organotin compounds, particularly sediments.

4. Identify the mechanisms of toxicity for sensitive organisms.
5. Develop necessary standard reference materials to allow intercalibration among laboratories and methods.

One of the most important research areas that has been raised by the scientific and regulatory debate over TBT is the need for improved risk assessment methodologies. The complexity of marine and estuarine ecosystems requires an integrated approach to assessing the risks of xenobiotic compounds and of TBT. Assessment methods should include improved measurements of degradation, metabolism, and transport to develop predictive models of exposure. Better methods of field verification and improved testing procedures for chronic toxicity are needed that can predict effects at the population and community levels. Some degree of standardization in ecological risk assessment procedures is needed to provide some consistency to the agencies that must carry out and interpret these risk assessments. Consistent approaches to risk assessment would support improved risk management.

Fundamental needs for long-term data collection related to the process and mechanisms of fouling include the need to

- characterize the nature of the bacterial attachment and the mechanism(s) involved in its reversibility;
- delineate the role(s) of the polysaccharide slime as an attractant to subsequent settlement of other organisms, and determine if the presence of the film is facultative or obligatory;
- determine what part of tributyltin's success as an antifouling agent is due to its effectiveness in suppressing the initial biofilm and what part is due to direct toxicity to higher phylogenetic taxa and determine if the attachment of the bacteria could be prevented, and if that would that be sufficient to prevent subsequent settlement of higher taxa;
- delineate the concentration gradient of organotin compounds across the biofilm.

The following general research areas for antifouling alternatives should be considered:

1. Develop non-toxic, low-surface-energy coatings such as perfluorinated alkyl films.
2. Develop toxic coatings with leachates that are not persistent in the environment. Unsaturated triorganotin coatings based on butyltin compounds are one possible alternative.
3. Develop enzymatic coatings that inhibit fouling utilizing enzymes from organisms that naturally reduce antifouling, such as kelp.

ACKNOWLEDGMENTS

We thank Joyce G. Nuttall for her assistance in the editing of this manuscript. We also thank William A. Bailey, William C. Banta, William R. Blair, Frederick E. Brinckman, Robert J. Huggett, Edward Goldberg, Madelyn Fletcher, and Michael J. Waldock, for their assistance in identifying research information requirements.

REFERENCES

Brinckman, F.E., G.J. Olson, W.R. Blair and E.J. Parks. 1987. Implications of molecular speciation and topology of environmental metals: uptake mechanisms and toxicity of organotins. In: *Aquatic Toxicity and Hazard Assessment*, W.J. Adams, G.A. Chapman and W.G. Landis (Eds). ASTM STP 971. American Society for Testing and Materials, Philadelphia, pp. 219–232.

Callow, M.E., P.A. Millner and L.V. Evans. 1978. Organotin resistance in green seaweeds. In: Ninth International Seaweed Symposium, Santa Barbara, California, 20 August 1977. Science Press, Princeton, New Jersey.

Champ, M.A. 1986. Introduction and overview. In: Proceedings of the Oceans '86 Organotin Symposium, Vol. 4, Marine Technology Society, Washington, DC, pp. i–viii.

Champ, M.A. and D.F. Bleil. 1988. Research Needs Concerning Organotin Compounds Used in Antifouling Paints in Coastal Environments.

NOAA Technical Report Published by the Office of the Chief Scientist, National Ocean Pollution Office. 5 parts plus appendix, paginated separately.

Champ, M.A. and F.L. Lowenstein. 1987. TBT: The dilemma of hi-technology antifouling paints. *Oceanus*, **30**(3), 69–77.

Champ, M.A. and L.W. Pugh. 1987. Tributyltin antifouling paints: introduction and overview. In: Proceedings of the Oceans '87 International Organotin Symposium, Vol. 4, Marine Technology Society, Washington, DC, pp. 1296–1308.

Cleary, J.J., I.R.B. McFadzen and L.D. Peters. 1993. Surface microlayer contamination and toxicity in the North Sea and Plymouth nearshore waters. ICES C.M. Paper (E28), 14 pp.

Cooney, J.J. 1995. Organotin compounds and aquatic bacteria: a review. *Helgolander Meeresuntersuchungen*, **49**(1–4), 663–677.

Dowson, P.H., J.M. Bubb and J.N. Lester. 1993. Depositional profiles and relationships between organotin compounds in freshwater and estuarine sediment cores. *Environ. Monitoring and Assessment*, **28**(2), 145–160.

Eng, G., E.J. Tierney, J.M. Bellama and F.E. Brinckman. 1988. Correlation of molecular total surface area with organotin toxicity for biological and physicochemical applications. *Appl. Organometal. Chem.*, **2**, 171–175.

Errecalde, O., M. Astruc, G. Maury and R. Pinel. 1995. Biotransformation of butyltin compounds using pure strains of microorganisms. *Appl. Organometal. Chem.*, **9**(1), 23–28.

Fukagawa, T. and S. Suzuki. 1993. Cloning of gene responsible for tributyltin chloride (TBTCI) resistance in TBTCI-resistant marine bacterium *Alteromonas* sp. M-1. *Biochem. Biophy. Res. Commun.*, **194**(2), 733–740.

GESAMP – Joint Group of Experts on Scientific Aspects of Marine Pollution (IMO/FAO/UNESCO/WMO/WHO/IAEA/UN/UNEP). 1983. Review of Potentially Harmful Substances – Cadmium, Lead, and Tin. World Health Organization. Reports and Studies No. 22. Geneva, 114 pp.

Gibbs, P.E., G.W. Bryan and P.L. Pascoe. 1991. TBT-induced imposex in the dogwhelk, *Nucella lapillus*: geographical uniformity of the response and effects. *Mar. Environ. Res.*, **32**, 1–5.

Guolan, H. and W. Yong. 1995. Effects of tributyltin chloride on marine bivalve mussels. *Water Res.*, **29**(8), 1877–1884.

Hall, L.W., Jr, M.J. Lenkevich, W.S. Hall, A.E. Pinkney and S.J. Bushong. 1987. Evaluation of butyltin compounds in Maryland waters of Chesapeake Bay. *Mar. Pollut. Bull.*, **18**, 78–83.

Harris, J.R.W., C.C. Hamlin and A.R.D. Stebbing. 1991. A simulation study of the effectiveness of legislation and improved dockyard practice in reducing TBT concentrations in the Tamar Estuary. *Mar. Environ. Res.*, **32**, 279–292.

Hinga, K.R., D. Adelman and M.E.Q. Pilson. 1987. Radiolabeled butyltin studies in the MERL enclosed ecosystem. In: Proceedings of the Oceans '87 International Organotin Symposium, Vol. 4, Marine Technology Society, Washington, DC, pp. 1416–1419.

Huggett, R.J., M.A. Unger, P.F. Seligman and A.O. Valkirs. 1992. The marine biocide tributyltin. *Environ. Sci. Technol.*, **26**, 232–237.

Iwata, H., S. Tanabe, N. Miyazaki and R. Tatsukawa. 1994. Detection of butyltin compound residues in the blubber of marine mammals. *Mar. Pollut. Bull.*, **28**(10), 607–612.

Kuballa, J., R.-D. Wilken, E. Jantzen, K.K. Kwan and Y.K. Chau. 1995. Speciation and genotoxicity of butyltin compounds. *Analyst*, **120**(3), 667–673.

Landy, R.B., S.E. Holm and W.E. Conner (Eds). (1986). Proceedings of the Interagency Workshop on Aquatic Monitoring and Analysis for Organic Compounds. NOAA/National Marine Pollution Office, Rockville, Maryland, 51 pp. + appendices.

Langston, W.J. and G.R. Burt. 1991. Bioavailability and effects of sediment-bound TBT in deposit-feeding clams, *Scrobicularia plana*. *Mar. Environ. Res.*, **32**, 61–77.

Langston, W.J. and N.D. Pope. 1995. Determinants of TBT adsorption and desorption in estuarine sediments. *Mar. Pollut. Bull.*, **31**(1–3), 32–43.

Langston, W.J., G.W. Bryan, G.R. Burt and N.D. Pope. 1994. Effects of sediment metals on estuarine benthic organisms. National Rivers Authority, R&D Note 203, Almondsbury, Bristol, xi, 141 pp.

Lee, R.F. 1991. Metabolism of tributyltin by marine animals and possible linkages to effects. *Mar. Environ. Res.*, **32**, 29–35.

Maguire, R.J. 1991. Aquatic environmental aspects of non-pesticidal organotin compounds. *Water Pollut. Res. J. Can.*, **26**(3), 243–360.

Mee, L.D. and S.W. Fowler (Eds). 1991. Special Issue on Organotin. In: Proceedings of the Third International Organotin Symposium, 17–20 April 1990, Monaco. *Mar. Environ. Res.*, **32**, 292 pp.

Milne, A. 1990. Roughness and drag from the marine chemist's viewpoint. In: Proceedings of the International Workshop on Marine Roughness and Drag. Royal Institution of Naval Architects, London.

Proceedings of the Organotin Symposium of the Oceans '86 Conference. 1986. Washington, DC, 23–25 September 1986 Vol. 4, Marine Technology Society, Washington, DC, and the IEEE, Service Center, 445 Hoes Lane, Piscataway, New Jersey, pp. 1101–1330.

Proceedings of the Organotin Symposium of the Oceans '87 Conference. 1987. Halifax, Nova Scotia, 28 September–1 October, Vol. 4, Marine Technology Society, Washington, DC, and the IEEE Service Center, 445 Hoes Lane, Piscataway, New Jersey, pp. 1296–1454.

Proceedings of the Organotin Symposium of the Oceans '88 Conference. 1988. Baltimore, Maryland, 31 October–2 November, Vol. 4, Marine Technology Society, Washington, DC, and the IEEE Service Center, 445 Hoes Lane, Piscataway, New Jersey.

Proceedings of the Organotin Symposium of the Oceans '89 Conference. 1989. Seattle, Washington, 18–21 September, Vol. 3, Marine Technology Society, Washington, DC, and the IEEE Service Center, Piscataway, New Jersey.

Readman, J.W. and L.D. Mee. 1991. The reliability of analytical data for tributyltin (TBT) in sea water and its implications on water quality criteria. *Mar. Environ. Res.*, **32**, 19–28.

Schatzberg, P. (1987). Organotin antifouling hull paints and the US Navy: a historical perspective. In: Proceedings of the Oceans '87 International Organotin Symposium, Vol. 4, Marine Technology Society, Washington, DC, pp. 1324–1333.

Seligman, P.F., C.M. Adema, P.M. Stang, A.O. Valkirs and J.G. Grovhoug. 1987. Monitoring and prediction of tributyltin in the Elizabeth River and Hampton Roads, Virginia. In: Proceedings of the Oceans '87 International Organotin Symposium, Vol. 4, Marine Technology Society, Washington, DC, pp. 1357–1363.

Spooner, N., P.E. Gibbs, G.W. Bryan and L.J. Goad. (1991). The effect of tributyltin upon steroid titres in the female dogwhelk, *Nucella lapillus*, and the development of imposex. *Mar. Environ. Res.*, **32**, 37–49.

Stebbing, A.R.D. 1985. Organotins and water quality – some lessons to be learned. *Mar. Pollut. Bull.*, **16**, 383–390.

Suzuki, T., R. Matsuda and Y. Saito. 1992. Molecular species of tri-n-butyltin compounds in marine products. *J. Agric. Food Chem.*, **40**(8), 1437–1443.

Ten Hallers-Tjabbes, C.C. 1994. TBT in the open sea: a case for a total ban on the use of TBT antifouling paint. *North Sea Monitor*, **12**(3), 12–14.

Ten Hallers-Tjabbes, C.C., J.F. Kemp and J.P. Boon. 1994. Imposex in whelks (*Buccinum undatum*) from the open North Sea: relation to shipping traffic intensities. *Mar. Pollut. Bull.*, **28**(5), 311–313.

Ten Hallers-Tjabbes, C.C. and J.P. Boon. 1995. Whelks (*Buccinum undatum* L.), dogwhelks (*Nucella lapillus* L.) and TBT – a cause for confusion. *Mar. Pollut. Bull.*, **30**(10), 675–676.

Uchida, M. 1994. Tolerance of marine bacteria for organotin compounds (OTCs) in areas with or without OTC contamination. *Fisheries Sci.*, **60**(3), 267–270.

UNEP (United Nations Environment Programme). 1989. Bibliography on Marine Pollution by Organotin Compounds. MAP Technical Reports Series No. 35. UNEP Regional Seas Directories and Bibliographies, Nairobi, 83 pp.

US EPA (US Environmental Protection Agency). 1987a. Survey of Tributyltin and Dibutyltin Concentrations at Selected Harbors in Chesapeake Bay – Final Report, Chesapeake Bay Program. CBP/TRS 14/87. Annapolis, Maryland. 58 pp. + appendix 40 p.

US EPA. 1987b. Water Quality Advisory for Tributyltin. Office of Water Regulations and Standards, Office of Water Criteria and Standards Division, Washington, DC, 73 pp.

Waite, M.E., M.J. Waldock, J.E. Thain, D.J. Smith and S.M. Milton. 1991. Reductions in TBT concentrations in UK estuaries following legislation in 1986 and 1987. *Mar. Environ. Res.*, **32**, 89–112.

Walton, R., C.M. Adema and P.F. Seligman. 1986. Mathematical modeling of the transport and fate of organotin in harbors. In: Proceedings of the Oceans '86 Organotin Symposium, Vol. 4, Marine Technology Society, Washington, DC, pp. 1297–1301.

Young, D.R. and M.A. Champ. 1986. Field methodologies: subgroup report. In: Interagency Workshop on Aquatic Monitoring and Analysis for Organotin. Sponsored by NOAA/NMPPO, Rockville, Maryland, pp. 7–20.

Young, D.R., P. Schatzberg, F. Brinckman, M.A. Champ, S.E. Holm and R.B. Landy. 1986. Summary report – interagency workshop on aquatic sampling and analysis for organotin compounds. In: Proceedings of the Oceans '86 Organotin Symposium, Vol. 4, Marine Technology Society, Washington, DC, pp. 1135–1140.

INDEX

General references may be found under Organotins; page numbers appearing in **bold** refer to figures and page numbers appearing in *italic* refer to tables.